Transforming Learning. Transforming Lives.

CENTERING ON
Value

CENTERING ON
Choice

CENTERING ON
Engagement

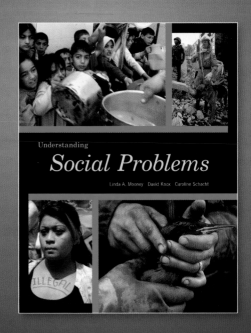

Mooney | Knox | Schacht

Understanding Social Problems, 8th Edition

ISBN 13: 978-1-111-83448-7
Copyright 2013

reflects our commitment to you and your students:

CENTERING ON
Value

ABC® videos, PowerLecture™ for easy lecture preparation, and other teaching and learning supplements are available to enhance your course.

CENTERING ON
Choice

Choose the format that best suits the needs of you and your students:

• Cengage Learning Advantage Edition (loose-leaf format)

• Cengage Learning eBook

• Chapter-by-chapter purchase

• Textbook rental

CENTERING ON
Engagement

Innovative, interactive media resources such as **CourseMate** engage students in the excitement of social problems.

WADSWORTH
CENGAGE Learning

Understanding Social Problems

8th edition

Understanding Social Problems

8th edition

Linda A. Mooney
East Carolina University

David Knox
East Carolina University

Caroline Schacht
East Carolina University

WADSWORTH
CENGAGE Learning

Australia • Brazil • Japan • Korea • Mexico • Singapore • Spain • United Kingdom • United States

***Understanding Social Problems*, Eighth Edition**
Linda A. Mooney, David Knox, and Caroline Schacht

Acquiring Sponsoring Editor: Erin Mitchell

Developmental Editor: Robert Jucha

Assistant Editor: Mallory Ortberg

Editorial Assistant: Mallory Ortberg

Media Editor: John Chell

Marketing Program Manager: Janay Pryor

Content Project Manager: Cheri Palmer

Art Director: Caryl Gorska

Manufacturing Planner: Judy Inouye

Rights Acquisitions Specialist: Dean Dauphinais

Production Service: Jill Traut, MPS North America

Photo Researcher: PreMedia Global

Text Researcher: Pablo D'Stair

Copy Editor: Heather McElwain

Illustrator: MPS Limited, a Macmillan Company

Text Designer: Lisa Buckley

Cover Designer: LF Studio

Cover Image: Upper Left: Nayef Hashlamoun/ X00952/Reuters/Corbis; Upper Right: Owen Franken/Corbis; Lower Left: Tony Savino/Corbis; Lower Right: Gary Braasch/Corbis

Compositor: MPS Limited, a Macmillan Company

For product information and technology assistance, contact us at **Cengage Learning Customer & Sales Support, 1-800-354-9706.**

For permission to use material from this text or product, submit all requests online at **www.cengage.com/permissions.** Further permissions questions can be e-mailed to **permissionrequest@cengage.com.**

Library of Congress Control Number: 2011933357

Student Edition:

ISBN-13: 978-1-111-83448-7

ISBN-10: 1-111-83448-2

Loose-leaf Edition:

ISBN-13: 978-1-111-83773-0

ISBN-10: 1-111-83773-2

Wadsworth
20 Davis Drive
Belmont, CA 94002-3098
USA

Cengage Learning is a leading provider of customized learning solutions with office locations around the globe, including Singapore, the United Kingdom, Australia, Mexico, Brazil, and Japan. Locate your local office at **www.cengage .com/global.**

Cengage Learning products are represented in Canada by Nelson Education, Ltd.

To learn more about Wadsworth, visit **www.cengage.com/wadsworth**

Purchase any of our products at your local college store or at our preferred online store **www.CengageBrain.com.**

Printed in the United States of America
1 2 3 4 5 6 7 16 15 14 13 12

To Margaret Ann Mooney, wife, mother, sister, aunt, grandmother, and great-grandmother. You have enriched all of our lives. Rest in peace Mom.

Brief Contents

Contents

PART 4 Problems of Globalization

Features

Preface

Understanding Social Problems is intended for use in a college-level sociology course. We recognize that many students enrolled in undergraduate sociology classes are not sociology majors. Thus, we have designed our text with the aim of inspiring students—no matter what their academic major or future life path may be—to care about the social problems affecting people throughout the world. In addition to providing a sound theoretical and research basis for sociology majors, *Understanding Social Problems* also speaks to students who are headed for careers in business, psychology, health care, social work, criminal justice, and the nonprofit sector, as well as to those pursuing degrees in education, fine arts, and the humanities or to those who are "undecided." Social problems, after all, affect each and every one of us, directly or indirectly. And everyone—whether a leader in business or politics, a stay-at-home parent, or a student—can become more mindful of how his or her actions (or inactions) perpetuate or alleviate social problems. We hope that *Understanding Social Problems* not only informs but also inspires, planting seeds of social awareness that will grow no matter what academic, occupational, and life path students choose.

New to this Edition

In the eighth edition of *Understanding Social Problems,* Chapter 12, previously titled "Population Growth and Urbanization," has been changed to "Population Growth and Aging." Increased attention to mental health problems in Chapter 2 (previously titled "Problems of Illness and Health Care") is reflected in the new title of the chapter—"Physical and Mental Health and Health Care." Most of the opening vignettes in the eighth edition are new, as are many of the *What Do You Think?* sections, which are designed to engage students in critical thinking. Many of the chapter features (*The Human Side, Social Problems Research Up Close,* and *Self and Society*) have been updated or replaced with new content, and we have also added new photo essays, including one depicting the Deepwater Horizon oil rig explosion and Fukushima nuclear power plant disasters. In this eighth edition, an exciting new feature called *Animals in Society* makes its debut. This boxed feature, found in five chapters, presents issues, problems, policies, and/or programs concerning animals within the context of the social problem discussed in that chapter. This new feature reflects the increasing scholarly attention to the study of the relationship between animals and humans, and the inclusion of *Animals and Society* as a special section of the American Sociological Association. As more than half of U.S. households include companion animals, animal-related topics are likely to appeal to student interest.

The eighth edition has retained pedagogical features that students and professors find useful, including a running glossary, list of key terms, chapter review, and *Test Yourself* sections. Finally, each chapter has new photos and new and updated figures and tables, as well as new and revised material, detailed as follows.

Chapter 1 (Thinking about Social Problems) features new examples of student activism, four new *What Do You Think?* sections, a new opening vignette, and additional "reasons to read this book." An updated *Self and Society* feature presents the newest statistics available on U.S. freshman attitudes toward select social problems, and the *Social Problems Research Up Close* feature includes new data on the sexual behavior of high school students.

Chapter 2 (Physical and Mental Health and Health Care) has a new title to reflect expanded coverage of mental health issues, including new data on the mental well-being of college students. This chapter includes updated information on maternal and child health, HIV/AIDS, and obesity, and new information on antibiotic resistance as a public health threat, health insurance coverage, the Patient Protection and Affordable Care Act, the Centers for Disease Control and Prevention's first report on health disparities and inequalities, new data on life expectancy of U.S. Hispanics, and current research comparing the U.S. health care system with 29 other OECD countries. New *What Do You Think?* sections ask students what they think about (1) how their views about having children would be affected if they (or their partner) had a one in 31 chance of dying from a pregnancy- or childbirth-related cause; (2) proposals to ban the use of tanning beds by minors; and (3) a Texas law that resulted in an HIV-infected man being sentenced to 35 years in prison for spitting on someone because his saliva is considered a "deadly weapon." Finally, a new *Animals and Society* feature looks at "Improving Mental Health through Animal-Assisted Therapy."

Chapter 3 (Alcohol and Other Drugs) features all new data from the most recent reports available on alcohol and drug use including the World Drug Report, the Monitoring the Future survey, the National Survey on Drug Use and Health (NSDUH), the European Monitoring Centre for Drugs and Drug Addiction (EMCDDA), and the 2011 World Health Organization's report on the link between poverty and the global tobacco epidemic.

This revised chapter now includes a discussion of the "Portugal Experiment," a section on drug courts, updated information on the economic costs of alcohol and other drugs, data from the Obama administration's report on the prescription drug crisis in America, and the *Ending the Tobacco Problem: A Blueprint for the Nation*. There are also several new *What Do You Think?* sections on such topics as the legalization of marijuana and Spice (K2), whether or not there should be a health insurance "discount" for non-smokers, and a new drinking game called "icing."

Chapter 4 (Crime and Social Control) includes extensive updated statistics on official crime rates, criminal victimization, prisoners, and the death penalty. This revised chapter has a greater emphasis on white-collar crimes, and covers a number of timely and controversial issues, such as gun control, prison overcrowding, the Tucson shootings, and crime-related issues before the Supreme Court. We have added a new section, "Crime and Victimization," and have expanded the discussions of fear of crime, prevention of juvenile delinquency and adult criminality, gangs, and capital punishment. New or updated *What do You Think?* sections invite students to think about issues related to piracy and American homicides, proposals to make toy guns illegal, the constitutionality of banning violent video games, life sentences for offenders who killed when they were under the age 18, genetic factors that contribute to violent crime, and organ donation by prisoners on death row.

Chapter 5 (Family Problems) includes updated information on changes and trends in U.S. families, including those related to cohabitation, singlehood and age at first marriage, and births to unmarried women. New figures present data on the percentage of U.S. adults who have step-relatives, and the percentage of women and men in five countries who agree that wife beating is acceptable if a wife argues or refuses to have sex with her husband. This revised chapter includes a new section, "Public Attitudes toward Changes in Family Life." We have updated the discussion of parental alienation, and have included a new table on parental alienation behaviors and a new *The Human Side* feature, "Recollections from Adult Children of Parental Alienation." A new *What Do You Think?* section asks why research findings showed that women of color with higher socioeconomic status were more likely to stay in abusive relationships than those with lower socioeconomic status. Finally, this chapter includes new research on three subtypes of male abusers and an *Animals and Society* feature that examines the issue of "Pets and Domestic Violence."

Chapter 6 (Poverty and Economic Inequality) includes recent data on U.S. and global income, wealth inequality, CEO compensation, corporate taxes, homelessness and hate crimes against the homeless, and welfare. This revised chapter presents a new measure of poverty—the Multidimensional Poverty Index (MPI)—as well as the Basic Economic Security Tables for the United States, which provide realistic income levels needed to meet basic living needs. A new *The Human Side* feature describes the experience of "being poor."

Since the last edition of *Understanding Social Problems* was published, the United States and other parts of the world have experienced a prolonged economic downturn. In this revised edition, **Chapter 7** (Work and Unemployment) provides updated coverage of the global economic crisis and the extent of unemployment in the United States and around the world. The *Self and Society* feature, "How Do Your Spending Habits Change in Hard Economic Times?" has been updated with recent Harris Poll data. This revised chapter includes new research on the effects of poor-quality jobs on mental health and an updated section on employment concerns of recent college grads. New Bureau of Labor statistics on fatal and nonfatal workplace injuries and illnesses are presented, and a new photo essay looks at child labor in U.S. agriculture. A new *The Human Side* feature titled "A Worker's Grueling Day" describes a typical day for a worker at a turkey processing plant. New topics include the International Labor Organization's Global Jobs Pact and anti-union legislation in Wisconsin.

Chapter 8 (Problems in Education) includes all new statistics from a variety of sources, including the Organization for Economic Cooperation and Development, the Digest of Education, the Condition of Education, the Children's Defense Fund, the National Assessment of Educational Progress, the 2011 *Quality Counts* report, and the Program for the International Student Assessment. This revised chapter emphasizes the impact of the economy on education, teachers, and college students, and describes *Race to the Top* spending, President Obama's "Blueprint for Education," and the Federal National Educational Technology Plan. There is a new *The Human Side* feature titled, "Dear Mr. President." This chapter also includes a discussion of value-added measurement, as well as recent laws, policies, and proposed legislation in the area of education, including an updated section on common core standards and teacher accountability. The chapter also includes four new *What Do You Think?* sections that ask the following: (1) Is there a need for bilingual education? (2) Are males and females "hardwired" differently? (3) Should teachers' evaluations be made public? and (4) Why are so many college students failing to learn?

Chapter 9 (Race, Ethnicity, and Immigration) includes new data from the 2010 census on the racial and ethnic composition of the United States. This revised chapter also presents new data on the foreign-born U.S. population and on interracial/interethnic marriages in the United States. A new table on racial identification of Hispanics/Latinos in the United States is presented along with an expanded discussion of the U.S. Hispanic/Latino population.

The section, "Myths about Immigration and Immigrants" has been updated and expanded, and the section on immigration policies contains a new discussion of Arizona's immigration law SB 1070. Updated information about hate groups includes a new table on hate group chapters in the United States and two new key terms: *nativist extremist groups* and *Islamophobia*. New or updated *What Do You Think?* sections ask (1) if readers think of Barack Obama mostly as a black person, or mostly as a person of mixed race; (2) if undocumented immigrants who report crimes or commit acts of bravery to save lives should be deported; and (3) if undocumented immigrants should qualify for in-state college tuition.

New to **Chapter 10** (Gender Inequality) are added sections on the complexity of the concept of gender, including a discussion of the "two-spirited person" and a photo essay titled, "The Gender Continuum." *The Human Side* feature is new, and there is a greater emphasis on transgender issues. This revised chapter also has a new section on sexual violence on campus, a new opening vignette, new figures and tables, and updated information from the World Economic Forum, the AAUW, the Global Gender Gap Report, the ACLU, the Global Policy Forum, the United Nations Development Fund for Women, and the State of the World's Children. There are six new *What Do You Think?* sections, facilitating discussions on *bacha posh,* i.e., girls in Afghanistan who dress like boys so they may also enjoy the privileges that accompany being male, liability in sex discrimination cases, support for male versus female athletic programs, and domestic violence as a human rights violation.

In this eighth edition, **Chapter 11** (Sexual Orientation and the Struggle for Equality) features a new *The Human Side,* a new *Social Problems Research Up Close,* and a new *Self and Society.* There are six new *What Do You Think?* sections, three new figures, and two new tables. The revised chapter includes updated discussions of marriage equality, economic discrimination, coming out, the impact of media on LGBT individuals, and a new section, "Prejudice against Lesbians, Gays, and Bisexuals." A number of new concepts and terms are introduced in this updated chapter, including *minority stress theory, oppression, privilege, sexual orientation change efforts (SOCE), gender expression, gender identity, contact hypothesis,* and *LGBTQI.* The chapter also includes recent legal decisions and bills such as the Respect for Marriage Act (RMA), the Every Child Deserves a Family Act, and the Matthew Shepard and James Byrd, Jr. Hate Crimes Prevention Act.

Chapter 12, formerly "Population Growth and Urbanization," has been renamed "Population Growth and Aging." Statistics on population growth projections have been updated and a new *Animals and Society* feature discusses "Pet Overpopulation in the United States." New sections on aging focus on global and U.S. trends in aging, the social and economic challenges of growing elderly populations, and retirement concerns of older Americans and the role of Social Security. The revised chapter examines ageism—prejudice and discrimination against the elderly—and presents the Ageism Survey in the *Self and Society* feature. A new *Social Problems Research Up Close* feature describes the Elder Care Study. New key terms in the chapter include *ageism, ageism by invisibility, baby boomers, Family Support Agreement, sandwich generation,* and *Social Security.*

Chapter 13 (Environmental Problems) includes the following new terms and information about *tar sands* and *tar sands oil, fracking,* and *pinkwashing,* and *Green Revolving Funds* in institutions of higher education. This revised chapter also includes an updated and expanded discussion of nuclear power and a new photo essay titled the "Deepwater Horizon Oil Rig Explosion and Fukushima Nuclear Power Plant Accident." A new *Self and Society* feature assesses attitudes toward government interventions to reduce global warming.

The revised **Chapter 14** (Science and Technology) highlights new information on Web 3.0, telepresence robots, U.S. and world Internet penetration rates, media multitasking, the "quantified self," the Semantic Web, learning health systems, massive security breaches, and a new discussion of the global digital divide. The chapter includes updates on litigation between Google, Microsoft, Novell, and the U.S. government; the Myriad Genetics case; and e-mail privacy case decisions. Five new *What Do You Think?* sections deal with the topics of robots, self-tracking, texting in class, "womb lynching," and Wikileaks. There are two new tables (Attitudes toward Science, and Attitudes about Abortion by Political Affiliation and Religion) and four new figures (Online Activities by Generation, Public Assessment of Scientific Research, People On- and Offline by Country, and the Impact of Media Consumption on 8- to 18-Year Olds). The new *Animals and Society* feature debates the issue of using animals in scientific research.

In the revised **Chapter 15** (Conflict, War, and Terrorism), new topics include the withdrawal from Iraq, the "Arab Spring," the end of civil war in Sri Lanka, the killing of Osama bin Laden and Muammar Gaddafi, repeal of "Don't Ask, Don't Tell," militant groups in Pakistan, and the legality of detaining enemy combatants at Guantanamo. The chapter has a new *Self and Society* feature: "The Nuclear Bomb Quiz." There is also a new *The Human Side* feature called "A Date with Tuesday" and a new *Animals and Society* feature titled "The Unsung Heroes among Us." New *What do You Think?* sections examine issues related to women in combat positions, gays in the military, release of documents through Wikileaks, and targeted killing. This updated chapter presents new statistics on military spending, military suicides, attacks and deaths from terrorism, nuclear talks with Iran, and the killings at Fort Hood. New and updated information for this chapter comes from the latest available data from the United Nations, the Department of Defense, SIPRI, Office of Management and Budget, FAS, Pew Research, the U.S. Department of State, the National Counterterrorism Center, the White House, branches of the military, and NATO.

Features and Pedagogical Aids

We have integrated a number of features and pedagogical aids into the text to help students learn to think about social problems from a sociological perspective. Our mission is to help students think critically about social problems and their implications, and to increase their awareness of how social problems relate to their personal lives.

Exercises and Boxed Features

Animals and Society. New to the eighth edition, a feature called *Animals and Society* examines issues, problems, policies, and/or programs concerning animals within the context of the social problem discussed in that chapter. In this new feature, we examine the following topics: (1) "Improving Mental Health through Animal-Assisted Therapy" (in Chapter 2—Physical and Mental Health and Health Care); (2) "Pets and Domestic Violence" (in Chapter 5—Family Problems); (3) "Pet Overpopulation in the United States" (in Chapter 12—Population Growth and Aging); (4) "The Use of Animals in Scientific Research" (in Chapter 14—Science and Technology); and (5) "The Unsung Heroes among Us" (in Chapter 15—Conflict, War, and Terrorism).

Self and Society. Each chapter includes a social survey designed to help students assess their own attitudes, beliefs, knowledge, or behaviors regarding some aspect of the social problem under discussion. In Chapter 5 (Family Problems), for example, the "Abusive Behavior Inventory" invites students to assess the frequency of various abusive behaviors in their own relationships. The *Self and Society* feature in Chapter 3 (Alcohol and Other Drugs) allows students to measure the consequences of their own drinking behavior and compare it to respondents in a national sample, and students can assess their knowledge of nuclear weapons in Chapter 15 (Conflict, War, and Terrorism).

The Human Side. In addition to the *Self and Society* boxed features, each chapter includes a boxed feature that makes the social problems under discussion more salient by describing personal experiences of individuals who have been affected by them. *The Human Side* feature in Chapter 4 (Crime and Social Control), for example, describes the horrific consequences of being a victim of rape, and *The Human Side* feature in Chapter 9 (Race, Ethnicity, and Immigration) describes the experiences of an immigrant day laborer who was victimized by a violent hate crime. Further, in Chapter 11 (Sexual Orientation and the Struggle for Equality), the founder of the "It Gets Better" campaign expresses his outrage at an email he received.

Social Problems Research Up Close. This feature, found in every chapter, presents examples of social science research and illustrates the sociological enterprise, from theory and data collection to findings and conclusions, thus exposing students to various studies and research methods. The *Social Problems Research Up Close* feature in Chapter 1 discusses social science research, frequently found sections in a research article, and how to read a contingency table. Other *Social Problems Research Up Close* topics include bullying, job loss in midlife, computer hacking, the changing nature of marriage in America, and two-faced racism.

Photo Essay. Chapter 1 features a photo essay on "Students Making a Difference." Chapter 2 (Physical and Mental Health and Health Care) includes a photo essay titled "Modern Animal Food Production: Health and Safety Issues," and "Prison Programs that Work" is in Chapter 4 (Crime and Social Control). In Chapter 6 (Poverty and Economic Inequality), a photo essay covers the topic "Lack of Clean Water and Sanitation among the Poor." Chapter 7 (Work and Unemployment) includes a photo essay on "Child Labor in U.S. Agriculture," and a photo essay in Chapter 10 is titled "The Gender Continuum." Lastly, Chapter 13 (Environmental Problems) depicts the horrors of the "Deepwater Horizon Oil Rig Explosion and Fukushima Nuclear Power Plant Accident."

In-Text Learning Aids

Vignettes. Each chapter begins with a vignette designed to engage students and draw them into the chapter by illustrating the current relevance of the topic under discussion.

For example, Chapter 2 (Physical and Mental Health and Health Care), begins with an account of a man who committed a $1 bank robbery in order to get arrested so he could get the medical care he needed in prison (he had no health insurance and could not afford to pay for medical care). Chapter 9 (Race, Ethnicity, and Immigration) begins with the story of a New Orleans justice of the peace who refused to marry an interracial couple because he doesn't believe in the "mixing of the races," and Chapter 15 (Conflict, War, and Terrorism) opens with a portrayal of a young veteran with PTSD (post-traumatic stress disorder).

Key Terms. Important terms and concepts are highlighted in the text where they first appear. To reemphasize the importance of these words, they are listed at the end of every chapter and are included in the glossary at the end of the text.

Running Glossary. This eighth edition continues the running glossary that highlights the key terms in every chapter by putting the key terms and their definitions in the text margins.

What Do You Think? Sections. Each chapter contains multiple sections called *What Do You Think?* These sections invite students to use critical thinking skills to answer questions about issues related to the chapter content. For example, one *What Do You Think?* feature in Chapter 3 (Alcohol and other Drugs) asks students "Should marijuana be legalized to raise revenues for ailing state economies?" A *What Do You Think?* feature in Chapter 11 (Sexual Orientation and the Struggle for Equality) asks if banning same-sex marriages is based on religious beliefs, and if that is a violation of the second amendment that separates church and state. In Chapter 12 (Population Growth and Aging), a *What Do You Think?* feature asks readers if birthday cards and jokes that make fun of aging are a form of ageism.

Glossary. All key terms are defined in the page margins as well as in the end-of-text glossary.

Understanding [Specific Social Problem] Sections. All too often, students, faced with contradictory theories and study results walk away from social problems courses without any real understanding of their causes and consequences. To address this problem, chapter sections titled "Understanding [specific social problem]" cap the body of each chapter just before the chapter summaries. Unlike the chapter summaries, these sections synthesize the material presented in the chapter, summing up the present state of knowledge and theory on the chapter topic.

Turning to Video is a new feature found at the end of each chapter in which the authors have selected a video from the list of DVDs found in The Wadsworth Sociology Video Library Vol. I and II. The video highlights key concepts discussed in the chapter.

Supplements

The eighth edition of *Understanding Social Problems* comes with a full complement of supplements designed with both faculty and students in mind.

Supplements for the Instructor

Instructor's Resource Manual with Test Bank. This supplement offers instructors learning objectives, key terms, lecture outlines, student projects, classroom activities, Internet and InfoTrac® College Edition exercises, and video suggestions. Test items include multiple-choice and true-false questions with answers and page references, as well as short-answer and essay questions for each chapter. Each multiple-choice item has the question type (factual, applied, or conceptual) indicated. All questions are labeled as new, modified, or pickup, so instructors know if the question is new to this edition of the test bank, modified but picked up from the previous edition of the test bank, or picked up straight from the previous edition of the test bank. Sample syllabi are also included, as well as a resource integration guide, to assist in lesson plan development.

PowerLecture™ and ExamView®. This easy-to-use, one-stop digital library and presentation tool includes the following:

- Preassembled **Microsoft® PowerPoint® lecture slides** with graphics from the text, making it easy for you to assemble, edit, publish, and present custom lectures for your course.
- **ExamView testing software,** which includes all the test items from the printed test bank in electronic format, enabling you to create customized tests of up to 250 items that can be delivered in print or online.

Videos. Adopters of *Understanding Social Problems* have several different video options available with the text.

AIDS in Africa DVD. Southern Africa has been overcome by a pandemic of unparalleled proportions. This documentary series focuses on Namibia, a new democracy, and the many actions that are being taken to control HIV/AIDS there. Included in this series are four documentary films created by the Project Pericles scholars at Elon University.

The Wadsworth Sociology Video Library Vol. I and II. These DVDs drive home the relevance of course topics through short, provocative clips of current and historical events. Perfect for enriching lectures and engaging students in discussion, many of the segments on this volume have been gathered from the BBC Motion Gallery. Ask your Cengage Learning representative for a list of contents.

Supplements for the Student

Study Guide. Each chapter of this critically updated study guide includes a brief chapter outline, learning objectives, key terms, matching exercise, a chapter review fill-in-the-blank exercise, worksheets that students can complete directly in the study guide to help them prepare for exams, Internet activities, InfoTrac College Edition exercises, and a practice test consisting of multiple-choice and true-false questions with answers and page references, as well as short-answer questions and essay questions with page references to enhance and test student understanding of chapter concepts.

Online Resources

Sociology CourseMate. Mooney/Knox/Schacht, *Understanding Social Problems* includes Sociology CourseMate, which helps you make the grade.

Sociology CourseMate includes:

- An interactive eBook, with highlighting, note taking, and search capabilities
- Interactive learning tools including:
  - Quizzes
  - Flash cards
  - Videos
  - And more!

Go to login.cengagebrain.com to access these resources related to your text in Sociology CourseMate.

CourseReader for Sociology. CourseReader for Sociology, first edition, allows you to create a fully customized online reader in minutes. Access a rich collection of thousands of primary and secondary sources, readings, and audio and video selections from multiple disciplines. Each selection includes a descriptive introduction that puts it into context, and every selection is further supported by both critical-thinking and multiple-choice questions designed to reinforce key points. This easy-to-use solution allows you to select exactly the content you need for your courses, and is loaded with convenient pedagogical features like highlighting, printing, note taking, and downloadable MP3 audio files for each reading. You have the freedom to assign and customize individualized content at an affordable price.

WebTutor™ for BlackBoard® and WebCT®. Jump-start your course with customizable, rich, text-specific content within your Course Management System. Whether you want to web-enable your class or put an entire course online, WebTutor delivers. WebTutor offers a wide array of resources including access to the eBook, videos, quizzes, web links, exercises, and more.

Acknowledgments

This text reflects the work of many people. We would like to thank the following for their contributions to the development of this text: Chris Caldeira and Erin Mitchell, Acquisitions Editors; Bob Jucha, Development Editor; Mallory Ortberg, Assistant Editor; John Chell, Associate Media Editor; Cheri Palmer, Content Project Manager; Jill Traut, Project Manager at MPS North America; and Sara Golden, Photo Researcher at PMG. We would also like to acknowledge the support and assistance of Carol L. Jenkins, ks Stanley, John T. Crist, Marieke Van Willigen, and Ronnie Miller. To each we send our heartfelt thanks. Special thanks also to George Glann, whose valuable contributions have assisted in achieving the book's high standard of quality from edition to edition.

Additionally, we are indebted to those who read the manuscript in its various drafts and provided valuable insights and suggestions, many of which have been incorporated into the final manuscript:

Maria D. Cuevas
Yakima Valley Community College

Kim Gilbert
Iowa Lakes Community College

Heather S. Kindell
Morehead State University

John J. Leiker
Utah State University

Sharon A. Nazarchuk
Lackawanna College

Jewrell Rivers
Abraham Baldwin Agricultural College

Michelle Willms
Itawamba Community College

Sally Vyain
Ivy Tech Community College of Indiana

We are also grateful to the reviewers of the previous editions: David Allen, *University of New Orleans;* Patricia Atchison, *Colorado State University;* Wendy Beck, *Eastern Washington University;* Walter Carroll, *Bridgewater State College;* Deanna Chang, *Indiana University of Pennsylvania;* Roland Chilton, *University of Massachusetts;* Verghese Chirayath, *John Carroll University;* Margaret Chok, *Pellissippi State Technical Community College;* Kimberly Clark, *DeKalb College–Central Campus;* Anna M. Cognetto, *Dutchess Community College;* Robert R. Cordell, *West Virginia University at Parkersburg;* Barbara Costello, *Mississippi State University;* William Cross, *Illinois College;* Kim Davies, *Augusta State University;* Doug Degher, *Northern Arizona University;* Katherine Dietrich, *Blinn College;* Jane Ely, *State University of New York–Stony Brook;* William Feigelman, *Nassau Community College;* Joan Ferrante, *Northern Kentucky University;* Robert Gliner, *San Jose State University;* Roberta Goldberg, *Trinity College;* Roger Guy, *Texas Lutheran University;* Julia Hall, *Drexel University;* Millie Harmon, *Chemeketa Community College;* Madonna Harrington-Meyer, *University of Illinois;* Sylvia Jones, *Jefferson Community College;* Nancy Kleniewski, *University of Massachusetts, Lowell;* Daniel Klenow, *North Dakota State University;* Sandra Krell-Andre, *Southeastern Community College;* Pui-Yan Lam, *Eastern Washington University;* Mary Ann Lamanna, *University of Nebraska;*

Phyllis Langton, *George Washington University;* Cooper Lansing, *Erie Community College;* Linda Kaye Larrabee, *Texas Tech University;* Tunga Lergo, *Santa Fe Community College, Main Campus;* Dale Lund, *University of Utah;* Lionel Maldonado, *California State University, San Marcos;* J. Meredith Martin, *University of New Mexico;* Judith Mayo, *Arizona State University;* Peter Meiksins, *Cleveland State University;* JoAnn Miller, *Purdue University;* Clifford Mottaz, *University of Wisconsin–River Falls;* Lynda D. Nyce, *Bluffton College;* Frank J. Page, *University of Utah;* James Peacock, *University of North Carolina;* Barbara Perry, *Northern Arizona University;* Ed Ponczek, *William Rainey Harper College;* Donna Provenza, *California State University at Sacramento;* Cynthia Reynaud, *Louisiana State University;* Carl Marie Rider, *Longwood University;* Jeffrey W. Riemer, *Tennessee Technological University;* Cherylon Robinson, *University of Texas at San Antonio;* Rita Sakitt, *Suffolk County Community College;* Mareleyn Schneider, *Yeshiva University;* Paula Snyder, *Columbus State Community College;* Lawrence Stern, *Collin County Community College;* John Stratton, *University of Iowa;* D. Paul Sullins, *The Catholic University of America;* Vickie Holland Taylor, *Danville Community College;* Joseph Trumino, *St. Vincent's College of St. John's University;* Robert Turley, *Crafton Hills College;* Alice Van Ommeren, *San Joaquin Delta College;* Joseph Vielbig, *Arizona Western University;* Harry L. Vogel, *Kansas State University;* Jay Watterworth, *University of Colorado at Boulder;* Robert Weaver, *Youngstown State University;* Jason Wenzel, *Valencia Community College;* Rose Weitz, *Arizona State University;* Bob Weyer, *County College of Morris;* Oscar Williams, *Diablo Valley College;* Mark Winton, *University of Central Florida;* Diane Zablotsky, *University of North Carolina;* Joan Brehm, *Illinois State University;* Doug Degher, *Northern Arizona University;* Heather Griffiths, *Fayetteville State University;* Amy Holzgang, *Cerritos College;* Janét Hund, *Long Beach City College;* Kathrin Parks, *University of New Mexico;* Craig Robertson, *University of North Alabama;* Matthew Sanderson, *University of Utah;* Jacqueline Steingold, *Wayne State University;* William J. Tinney, Jr., *Black Hills State University.*

Finally, we are interested in ways to improve the text and invite your feedback and suggestions for new ideas and material to be included in subsequent editions.

You can contact us at mooneyl@ecu.edu, knoxd@ecu.edu, or schachtc@ecu.edu.

About the Authors

Linda A. Mooney, PhD, is an associate professor of sociology at East Carolina University in Greenville, North Carolina. In addition to social problems, her specialties include law, criminology, gender, and issues in sexuality. She has published more than 30 professional articles in such journals as *Social Forces, Sociological Inquiry, Sex Roles, Sociological Quarterly,* and *Teaching Sociology.* She has won numerous teaching awards, including the University of North Carolina Board of Governors' Distinguished Professor for Teaching Award.

David Knox, PhD, is professor of sociology at East Carolina University. He has taught Social Problems, Introduction to Sociology, and Sociology of Marriage Problems. He is the author or co-author of 10 books and more than 80 professional articles. His research interests include marriage, family, intimate relationships, and sexual values and behavior.

Caroline Schacht, MA, is a teaching instructor of sociology at East Carolina University. She has taught Introduction to Sociology, Deviant Behavior, Sociology of Food, Sociology of Education, Individuals in Society, and Courtship and Marriage. She has co-authored several textbooks in the areas of social problems, introductory sociology, courtship and marriage, and human sexuality.

Andrew Rich/Getty Images

Thinking about Social Problems

"Unless someone like you cares a whole awful lot, nothing is going to get better. It's not."

—Dr. Seuss, The Lorax

Rubberball/Fotosearch

After the economic turndown of 2008, the U.S. Congress passed and President Obama signed into law the $787 billion American Recovery and Reinvestment Act of 2009. The stimulus package was designed to help failing industries, create jobs, promote consumer spending, rescue the failed housing market, and encourage energy-related investments. As of February 2011, "92.0 percent of the funds, excluding tax benefits, have been made available" (Recovery.gov 2011, p. 1).

IN A FEBRUARY 2011 Gallup Poll, a random sample of Americans were asked, "What do you think is the most important problem facing this country today?" Leading problems included the economy, war and terrorism, health care, immigration, unemployment, government corruption, family decline, poverty, and crime and violence (Gallup 2011). Moreover, survey results indicate that, overall, just 19 percent of Americans were satisfied "with the way things are going in the United States at this time" (Mendes 2011). Compared with previous years, this number is quite low, the second lowest-ever recorded satisfaction rate, the lowest record being in 1992. The most commonly selected problem was unemployment and the lack of jobs, followed by the economy in general, and deficit spending (Gallup 2011). President Obama is keenly aware of the problems the nation has to face. In his 2011 State of the Nation speech, he stated that:

> We should have no illusions about the work ahead of us. Reforming our schools; changing the way we use energy; reducing our deficit—none of this is easy. All of it will take time. And it will be harder because we will argue about everything. The cost. The details. The letter of every law. (Obama 2011)

Problems related to poverty and malnutrition, inadequate education, acquired immunodeficiency syndrome (AIDS), inadequate health care, crime, conflict, oppression of minorities, environmental destruction, and other social issues are both national and international concerns. Such problems present both a threat and a challenge to our national and global society. The primary goal of this textbook is to facilitate increased awareness and understanding of problematic social conditions in U.S. society and throughout the world.

Although the topics covered in this book vary widely, all chapters share common objectives: to explain how social problems are created and maintained; to indicate how they affect individuals, social groups, and societies as a whole; and to examine programs and policies for change. We begin by looking at the nature of social problems.

What Is a Social Problem?

There is no universal, constant, or absolute definition of what constitutes a social problem. Rather, social problems are defined by a combination of objective and subjective criteria that vary across societies, among individuals and groups within a society, and across historical time periods.

Objective and Subjective Elements of Social Problems

objective element of a social problem Awareness of social conditions through one's own life experiences and through reports in the media.

Although social problems take many forms, they all share two important elements: an objective social condition and a subjective interpretation of that social condition. The **objective element of a social problem** refers to the existence of a social condition. We become aware of social conditions through our own life experience, through the media, and through education. We see the homeless, hear gunfire in the streets, and see battered women in hospital emergency rooms. We read about employees losing their jobs as businesses downsize and factories close. In television news reports, we see the anguished faces of parents whose children have been killed by violent youths.

The **subjective element of a social problem** refers to the belief that a particular social condition is harmful to society or to a segment of society and that it should and can be changed. We know that crime, drug addiction, poverty, racism, violence, and pollution exist. These social conditions are not considered social problems, however, unless at least a segment of society believes that these conditions diminish the quality of human life.

By combining these objective and subjective elements, we arrive at the following definition: A **social problem** is a social condition that a segment of society views as harmful to members of society and in need of remedy.

Variability in Definitions of Social Problems

Individuals and groups frequently disagree about what constitutes a social problem. For example, some Americans view the availability of abortion as a social problem, whereas others view restrictions on abortion as a social problem. Similarly, some Americans view homosexuality as a social problem, whereas others view prejudice and discrimination against homosexuals as a social problem. Such variations in what is considered a social problem are due to differences in values, beliefs, and life experiences.

Definitions of social problems vary not only within societies but also across societies and historical time periods. For example, before the 19th century, a husband's legal right and marital obligation was to discipline and control his wife through the use of physical force. Today, the use of physical force is regarded as a social problem rather than a marital right.

Tea drinking is another example of how what is considered a social problem can change over time. In 17th- and 18th-century England, tea drinking was regarded as a "base Indian practice" that was "pernicious to health, obscuring industry, and impoverishing the nation" (Ukers 1935, cited by Troyer & Markle 1984). Today, the English are known for their tradition of drinking tea in the afternoon.

Because social problems can be highly complex, it is helpful to have a framework within which to view them. Sociology provides such a framework. Using a sociological perspective to examine social problems requires knowledge of the basic concepts and tools of sociology. In the remainder of this chapter, we discuss some of these concepts and tools: social structure, culture, the "sociological imagination," major theoretical perspectives, and types of research methods.

> Individuals and groups frequently disagree about what constitutes a social problem. For example, some Americans view the availability of abortion as a social problem, whereas others view restrictions on abortion as a social problem.

What Do You Think? People increasingly are using information technologies (e.g., blogs, web portals, online news feeds) to get their daily news with "traditional media outlets . . . struggling to hold their market share" (Saad 2007, p. 1). For example, in 2009, a number of major dailies closed their doors, including Denver's *Rocky Mountain News*, the *Seattle Post-Intelligencer*, and the *San Francisco Chronicle* (Shaw 2009). If your local print and/or online newspaper folded, where would you go for news? What role do the various media play in our awareness of social problems? Will definitions of social problems change as sources of information change and, if so, in what way?

subjective element of a social problem The belief that a particular social condition is harmful to society, or to a segment of society, and that it should and can be changed.

social problem A social condition that a segment of society views as harmful to members of society and in need of remedy.

Whereas some individuals view homosexual behavior as a social problem, others view homophobia as a social problem. Here, supporters of the repeal of Proposition 8, which banned gay marriages in California, await the appellate court's decision. In August 2010, California U.S. District Judge Vaughn Walker found the new law unconstitutional and overturned Proposition 8 (see Chapter 11). Judge Walker's decision is being appealed.

Elements of Social Structure and Culture

Although society surrounds us and permeates our lives, it is difficult to "see" society. By thinking of society in terms of a picture or image, however, we can visualize society and therefore better understand it. Imagine that society is a coin with two sides: On one side is the structure of society and on the other is the culture of society. Although each side is distinct, both are inseparable from the whole. By looking at the various elements of social structure and culture, we can better understand the root causes of social problems.

Elements of Social Structure

The **structure** of a society refers to the way society is organized. Society is organized into different parts: institutions, social groups, statuses, and roles.

Institutions An **institution** is an established and enduring pattern of social relationships. The five traditional institutions are family, religion, politics, economics, and education, but some sociologists argue that other social institutions, such as science and technology, mass media, medicine, sports, and the military, also play important roles in modern society. Many social problems are generated by inadequacies in various institutions. For example, unemployment may be influenced by the educational institution's failure to prepare individuals for the job market and by alterations in the structure of the economic institution.

Social Groups Institutions are made up of social groups. A **social group** is defined as two or more people who have a common identity, interact, and form a social relationship. For example, the family in which you were reared is a social group that is part of the family institution. The religious association to which you may belong is a social group that is part of the religious institution.

Social groups can be categorized as primary or secondary. **Primary groups,** which tend to involve small numbers of individuals, are characterized by intimate and informal interaction. Families and friends are examples of primary groups. **Secondary groups,** which may involve small or large numbers of individuals, are task-oriented and are characterized by impersonal and formal interaction. Examples of secondary groups include employers and their employees and clerks and their customers.

Statuses Just as institutions consist of social groups, social groups consist of statuses. A **status** is a position that a person occupies within a social group. The statuses we occupy largely define our social identity. The statuses in a family may consist of mother, father, stepmother, stepfather, wife, husband, child, and so on. Statuses can be either ascribed or achieved. An **ascribed status** is one that society assigns to an individual on the basis of factors over which the individual has no control. For example, we have no control over the sex, race, ethnic background, and socioeconomic status into which we are born. Similarly, we are assigned the status of child, teenager, adult, or senior citizen on the basis of our age—something we do not choose or control.

An **achieved status** is assigned on the basis of some characteristic or behavior over which the individual has some control. Whether you achieve the status of college

structure The way society is organized including institutions, social groups, statuses, and roles.

institution An established and enduring pattern of social relationships.

social group Two or more people who have a common identity, interact, and form a social relationship.

primary groups Usually small numbers of individuals characterized by intimate and informal interaction.

secondary groups Involving small or large numbers of individuals, groups that are task-oriented and are characterized by impersonal and formal interaction.

status A position that a person occupies within a social group.

ascribed status A status that society assigns to an individual on the basis of factors over which the individual has no control.

graduate, spouse, parent, bank president, or prison inmate depends largely on your own efforts, behavior, and choices. One's ascribed statuses may affect the likelihood of achieving other statuses, however. For example, if you are born into a poor socioeconomic status, you may find it more difficult to achieve the status of college graduate because of the high cost of a college education.

Every individual has numerous statuses simultaneously. You may be a student, parent, tutor, volunteer fund-raiser, female, and Hispanic. A person's *master status* is the status that is considered the most significant in a person's social identity. In the United States, a person's occupational status is typically regarded as a master status. If you are a full-time student, your master status is likely to be student.

Roles Every status is associated with many **roles,** or the set of rights, obligations, and expectations associated with a status. Roles guide our behavior and allow us to predict the behavior of others. As a student, you are expected to attend class, listen and take notes, study for tests, and complete assignments. Because you know what the role of teacher involves, you can predict that your teacher will lecture, give exams, and assign grades based on your performance on tests.

A single status involves more than one role. For example, the status of prison inmate includes one role for interacting with prison guards and another role for interacting with other prison inmates. Similarly, the status of nurse involves different roles for interacting with physicians and with patients.

Elements of Culture

Whereas the social structure refers to the organization of society, the **culture** refers to the meanings and ways of life that characterize a society. The elements of culture include beliefs, values, norms, sanctions, and symbols.

Beliefs **Beliefs** refer to definitions and explanations about what is assumed to be true. The beliefs of an individual or group influence whether that individual or group views a particular social condition as a social problem. Does secondhand smoke harm nonsmokers? Are nuclear power plants safe? Does violence in movies and on television lead to increased aggression in children? Our beliefs regarding these issues influence whether we view the issues as social problems. Beliefs influence not only how a social condition is interpreted but also the existence of the condition itself. For example, police officers' beliefs about their supervisors' priorities affected officers' problem-solving behavior and the time devoted to it (Engel & Worden 2003). The *Self and Society* feature in this chapter allows you to assess your own beliefs about various social issues and to compare your beliefs with a national sample of first-year college students.

Values **Values** are social agreements about what is considered good and bad, right and wrong, desirable and undesirable. Frequently, social conditions are viewed as social problems when the conditions are incompatible with or contradict closely held values. For example, poverty and homelessness violate the value of human welfare; crime contradicts the values of honesty, private property, and nonviolence; racism, sexism, and heterosexism violate the values of equality and fairness.

Values play an important role not only in the interpretation of a condition as a social problem but also in the development of the social condition itself. Sylvia Ann Hewlett (1992) explains how the American values of freedom and individualism are at the root of many of our social problems:

> There are two sides to the coin of freedom. On the one hand, there is enormous potential for prosperity and personal fulfillment; on the other are all the hazards of untrammeled opportunity and unfettered choice. Free markets can produce grinding poverty as well as spectacular wealth; unregulated industry can create dangerous

> Whereas the social structure refers to the organization of society, the culture refers to the meanings and ways of life that characterize a society.

achieved status A status that society assigns to an individual on the basis of factors over which the individual has some control.

roles The set of rights, obligations, and expectations associated with a status.

culture The meanings and ways of life that characterize a society, including beliefs, values, norms, sanctions, and symbols.

beliefs Definitions and explanations about what is assumed to be true.

values Social agreements about what is considered good and bad, right and wrong, desirable and undesirable.

Indicate whether you agree or disagree with each of the following statements:

Statement	Agree	Disagree
1. Wealthy people should pay a larger share of the taxes than they do now.	_____	_____
2. The federal government should raise taxes to reduce the deficit.	_____	_____
3. Gays and lesbians should have the legal right to adopt a child.	_____	_____
4. Affirmative action in college admissions should be abolished.	_____	_____
5. The federal government is not doing enough to control environmental pollution.	_____	_____
6. A national health care plan is needed to cover everybody's medical costs.	_____	_____
7. The federal government should do more to control the sale of handguns.	_____	_____
8. The chief benefit of a college education is that it increases one's earning power.	_____	_____
9. Addressing global warming should be a federal priority.	_____	_____

Percentage of First-Year College Students Agreeing with Belief Statements*

STATEMENT NUMBER	PERCENTAGE AGREEING IN 2010		
	TOTAL	WOMEN	MEN
1. Wealthy people should pay a larger share of the taxes than they do now.	65.1	66.1	63.9
2. The federal government should raise taxes to reduce the deficit.	31.2	28.7	34.3
3. Gays and lesbians should have the legal right to adopt a child.	74.2	80.0	66.8
4. Affirmative action in college admissions should be abolished.	47.5	43.8	51.9
5. The federal government is not doing enough to control environmental pollution.	78.1	80.8	74.6
6. A national health care plan is needed to cover everybody's medical costs.	63.3	66.4	59.4
7. The federal government should do more to control the sale of handguns.	68.1	75.2	59.3
8. The chief benefit of a college education is that it increases one's earning power.	73.7	72.2	75.6
9. Addressing global warming should be a federal priority.	62.8	66.4	58.4

*Percentages are rounded.
Source: Pryor et al. 2010.

levels of pollution as well as rapid rates of growth; and an unfettered drive for personal fulfillment can have disastrous effects on families and children. Rampant individualism does not bring with it sweet freedom; rather, it explodes in our faces and limits life's potential. (pp. 350–51)

Absent or weak values may contribute to some social problems. For example, many industries do not value protection of the environment and thus contribute to environmental pollution.

Norms and Sanctions **Norms** are socially defined rules of behavior. Norms serve as guidelines for our behavior and for our expectations of the behavior of others.

There are three types of norms: folkways, laws, and mores. *Folkways* refer to the customs and manners of society. In many segments of our society, it is customary to shake hands when being introduced to a new acquaintance, to say "excuse me" after sneezing, and to give presents to family and friends on their birthdays. Although no laws require us to do these things, we are expected to do them because they are part of the cultural tradition, or folkways, of the society in which we live.

Laws are norms that are formalized and backed by political authority. It is normative for a Muslim woman to wear a veil. However, in the United States, failure to remove the

norms Socially defined rules of behavior, including folkways, laws, and mores.

veil for a driver's license photo is grounds for revoking the permit. Such is the case of a Florida woman who brought suit against the state, claiming that her religious rights were being violated because she was required to remove her veil for the driver's license photo (Canedy 2002). She appealed the decision to Florida's District Court of Appeal and lost. The Court recognized, however, "the tension created as a result of choosing between following the dictates of one's religion and the mandates of secular law" (Associated Press 2006).

Mores are norms with a moral basis. Violations of mores may produce shock, horror, and moral indignation. Both littering and child sexual abuse are violations of law, but child sexual abuse is also a violation of our mores because we view such behavior as immoral.

All norms are associated with **sanctions,** or social consequences for conforming to or violating norms. When we conform to a social norm, we may be rewarded by a positive sanction. These may range from an approving smile to a public ceremony in our honor. When we violate a social norm, we may be punished by a negative sanction, which may range from a disapproving look to the death penalty or life in prison. Most sanctions are spontaneous expressions of approval or disapproval by groups or individuals—these are referred to as informal sanctions. Sanctions that are carried out according to some recognized or formal procedure are referred to as formal sanctions. Types of sanctions, then, include positive informal sanctions, positive formal sanctions, negative informal sanctions, and negative formal sanctions (see Table 1.1).

TABLE 1.1 Types and Examples of Sanctions		
	POSITIVE	**NEGATIVE**
Informal	Being praised by one's neighbors for organizing a neighborhood recycling program	Being criticized by one's neighbors for refusing to participate in the neighborhood recycling program
Formal	Being granted a citizen's award for organizing a neighborhood recycling program	Being fined by the city for failing to dispose of trash properly

© Cengage Learning 2013

Symbols A **symbol** is something that represents something else. Without symbols, we could not communicate with one another or live as social beings.

The symbols of a culture include language, gestures, and objects whose meanings the members of a society commonly understand. In our society, a red ribbon tied around a car antenna symbolizes Mothers Against Drunk Driving; a peace sign symbolizes the value of nonviolence; and a white-hooded robe symbolizes the Ku Klux Klan. Sometimes people attach different meanings to the same symbol. The Confederate flag is a symbol of southern pride to some and a symbol of racial bigotry to others.

The elements of the social structure and culture just discussed play a central role in the creation, maintenance, and social response to various social problems. One of the goals of taking a course in social problems is to develop an awareness of how the elements of social structure and culture contribute to social problems. Sociologists refer to this awareness as the "sociological imagination."

The Sociological Imagination

The **sociological imagination,** a term C. Wright Mills (1959) developed, refers to the ability to see the connections between our personal lives and the social world in which we live. When we use our sociological imagination, we are able to distinguish between "private troubles" and "public issues" and to see connections between the events and conditions of our lives and the social and historical context in which we live.

For example, that one person is unemployed constitutes a private trouble. That millions of people are unemployed in the United States constitutes a public issue. Once we understand that other segments of society share personal troubles such as human immunodeficiency virus (HIV) infection, criminal victimization, and poverty, we can look for the elements of social structure and culture that contribute to these public issues and private troubles. If the various elements of social structure and culture contribute to private troubles and public issues, then society's social structure and culture must be changed if these concerns are to be resolved.

sanctions Social consequences for conforming to or violating norms.

symbol Something that represents something else.

sociological imagination The ability to see the connections between our personal lives and the social world in which we live.

> When we use our sociological imagination, we are able to distinguish between "private troubles" and "public issues" and to see connections between the events and conditions of our lives and the social and historical context in which we live.

Rather than viewing the private trouble of being unemployed as a result of an individual's faulty character or lack of job skills, we may understand unemployment as a public issue that results from the failure of the economic and political institutions of society to provide job opportunities to all citizens, as exemplified by the 2009 U.S. recession. Technological innovations emerging from the Industrial Revolution led to machines replacing individual workers. During the economic recession of the 1980s, employers fired employees so the firms could stay in business. Thus, in both these cases, social forces rather than individual skills largely determined whether a person was employed.

Theoretical Perspectives

Theories in sociology provide us with different perspectives with which to view our social world. A perspective is simply a way of looking at the world. A theory is a set of interrelated propositions or principles designed to answer a question or explain a particular phenomenon; it provides us with a perspective. Sociological theories help us to explain and predict the social world in which we live.

Sociology includes three major theoretical perspectives: the structural-functionalist perspective, the conflict perspective, and the symbolic interactionist perspective. Each perspective offers a variety of explanations about the causes of and possible solutions to social problems.

Structural-Functionalist Perspective

The structural-functionalist perspective is based largely on the works of Herbert Spencer, Emile Durkheim, Talcott Parsons, and Robert Merton. According to structural functionalism, society is a system of interconnected parts that work together in harmony to maintain a state of balance and social equilibrium for the whole. For example, each of the social institutions contributes important functions for society: Family provides a context for reproducing, nurturing, and socializing children; education offers a way to transmit a society's skills, knowledge, and culture to its youth; politics provides a means of governing members of society; economics provides for the production, distribution, and consumption of goods and services; and religion provides moral guidance and an outlet for worship of a higher power.

> The structural-functionalist perspective emphasizes the interconnectedness of society by focusing on how each part influences and is influenced by other parts.

The structural-functionalist perspective emphasizes the interconnectedness of society by focusing on how each part influences and is influenced by other parts. For example, the increase in single-parent and dual-earner families has contributed to the number of children who are failing in school because parents have become less available to supervise their children's homework. As a result of changes in technology, colleges are offering more technical programs, and many adults are returning to school to learn new skills that are required in the workplace. The increasing number of women in the workforce has contributed to the formulation of policies against sexual harassment and job discrimination.

Structural functionalists use the terms *functional* and *dysfunctional* to describe the effects of social elements on society. Elements of society are functional if they contribute to social stability and dysfunctional if they disrupt social stability. Some aspects of society can be both functional and dysfunctional. For example, crime is dysfunctional in that it is associated with physical violence, loss of property, and fear. But according to Durkheim and other functionalists, crime is also functional for society because it leads to heightened awareness of shared moral bonds and increased social cohesion.

Sociologists have identified two types of functions: manifest and latent (Merton 1968). **Manifest functions** are consequences that are intended and commonly recognized. **Latent functions** are consequences that are unintended and often hidden. For example, the manifest function of education is to transmit knowledge and skills to society's youth. But public elementary schools also serve as babysitters for employed parents, and colleges offer a place for young adults to meet potential mates. The babysitting and mate-selection functions are not the intended or commonly recognized functions of education; hence, they are latent functions.

What Do You Think? In viewing society as a set of interrelated parts, structural functionalists argue that proposed solutions to social problems may lead to other social problems. For example, urban renewal projects displace residents and break up community cohesion. Racial imbalance in schools led to forced integration, which in turn generated violence and increased hostility between the races. What are some other "solutions" that lead to social problems? Do all solutions come with a price to pay? Can you think of a solution to a social problem that has no negative consequences?

Structural-Functionalist Theories of Social Problems

Two dominant theories of social problems grew out of the structural-functionalist perspective: social pathology and social disorganization.

Social Pathology According to the social pathology model, social problems result from some "sickness" in society. Just as the human body becomes ill when our systems, organs, and cells do not function normally, society becomes "ill" when its parts (i.e., elements of the structure and culture) no longer perform properly. For example, problems such as crime, violence, poverty, and juvenile delinquency are often attributed to the breakdown of the family institution; the decline of the religious institution; and inadequacies in our economic, educational, and political institutions.

Social "illness" also results when members of a society are not adequately socialized to adopt its norms and values. People who do not value honesty, for example, are prone to dishonesties of all sorts. Early theorists attributed the failure in socialization to "sick" people who could not be socialized. Later theorists recognized that failure in the socialization process stemmed from "sick" social conditions, not "sick" people. To prevent or solve social problems, members of society must receive proper socialization and moral education, which may be accomplished in the family, schools, churches, or workplace and/or through the media.

Social Disorganization According to the social disorganization view of social problems, rapid social change (e.g., the cultural revolution of the 1960s) disrupts the norms in a society. When norms become weak or are in conflict with each other, society is in a state of **anomie**, or *normlessness*. Hence, people may steal, physically abuse their spouses or children, abuse drugs, commit rape, or engage in other deviant behavior because the norms regarding these behaviors are weak or conflicting. According to this view, the solution to social problems lies in slowing the pace of social change and strengthening social norms. For example, although the use of alcohol by teenagers is considered a violation of a social norm in our society, this norm is weak. The media portray young people drinking alcohol, teenagers teach each other to drink alcohol and buy fake identification cards (IDs) to purchase alcohol, and parents model drinking behavior by having a few drinks after work or at a social event. Solutions to teenage drinking may involve strengthening norms against it through public education, restricting media depictions of youth and alcohol, imposing stronger sanctions against the use of fake IDs to purchase alcohol, and educating parents to model moderate and responsible drinking behavior.

manifest functions Consequences that are intended and commonly recognized.

latent functions Consequences that are unintended and often hidden.

anomie A state of normlessness in which norms and values are weak or unclear.

Conflict Perspective

Contrary to the structural-functionalism perspective, the conflict perspective views society as composed of different groups and interests competing for power and resources. The conflict perspective explains various aspects of our social world by looking at which groups have power and benefit from a particular social arrangement. For example, feminist theory argues that we live in a patriarchal society—a hierarchical system of organization controlled by men. Although there are many varieties of feminist theory, most would hold that feminism "demands that existing economic, political, and social structures be changed" (Weir and Faulkner 2004, p. xii).

The origins of the conflict perspective can be traced to the classic works of Karl Marx. Marx suggested that all societies go through stages of economic development. As societies evolve from agricultural to industrial, concern over meeting survival needs is replaced by concern over making a profit, the hallmark of a capitalist system. Industrialization leads to the development of two classes of people: the bourgeoisie, or the owners of the means of production (e.g., factories, farms, businesses), and the proletariat, or the workers who earn wages.

> Industrialization leads to the development of two classes of people: the bourgeoisie, or the owners of the means of production (e.g., factories, farms, businesses), and the proletariat, or the workers who earn wages.

The division of society into two broad classes of people—the "haves" and the "have-nots"—is beneficial to the owners of the means of production. The workers, who may earn only subsistence wages, are denied access to the many resources available to the wealthy owners. According to Marx, the bourgeoisie use their power to control the institutions of society to their advantage. For example, Marx suggested that religion serves as an "opiate of the masses" in that it soothes the distress and suffering associated with the working-class lifestyle and focuses the workers' attention on spirituality, God, and the afterlife rather than on worldly concerns such as living conditions. In essence, religion diverts the workers so that they concentrate on being rewarded in heaven for living a moral life rather than on questioning their exploitation.

Conflict Theories of Social Problems

There are two general types of conflict theories of social problems: Marxist and non-Marxist. Marxist theories focus on social conflict that results from economic inequalities; non-Marxist theories focus on social conflict that results from competing values and interests among social groups.

Marxist Conflict Theories According to contemporary Marxist theorists, social problems result from class inequality inherent in a capitalistic system. A system of haves and have-nots may be beneficial to the haves but often translates into poverty for the have-nots. As we will explore later in this textbook, many social problems, including physical and mental illness, low educational achievement, and crime, are linked to poverty.

In addition to creating an impoverished class of people, capitalism also encourages "corporate violence." *Corporate violence* can be defined as actual harm and/or risk of harm inflicted on consumers, workers, and the general public as a result of decisions by corporate executives or managers. Corporate violence can also result from corporate negligence; the quest for profits at any cost; and willful violations of health, safety, and environmental laws (Reiman and Leighton 2010). Our profit-motivated economy encourages individuals who are otherwise good, kind, and law-abiding to knowingly participate in the manufacturing and marketing of defective products, such as brakes on American jets, fuel tanks on automobiles, and contraceptive devices (e.g. intrauterine devices [IUDs]). In 2010, a British Petroleum (BP) oil well off the coast of Louisiana ruptured, killing 11 people and spewing millions of gallons of oil into the Gulf of Mexico (see Chapter 13). Evidence suggests that BP officials knew of the unstable cement seals on the rigs long before what is now being called the worst offshore disaster in U.S. history (Pope 2011). The profit motive has also caused individuals to sell defective medical devices, toxic pesticides, and contaminated foods in the United States and abroad.

Marxist conflict theories also focus on the problem of **alienation,** or powerlessness and meaninglessness in people's lives. In industrialized societies, workers often have little power or control over their jobs, a condition that fosters in them a sense of powerlessness in their lives. The specialized nature of work requires workers to perform limited and repetitive tasks; as a result, workers may come to feel that their lives are meaningless.

Alienation is bred not only in the workplace but also in the classroom. Students have little power over their education and often find that the curriculum is not meaningful to their lives. Like poverty, alienation is linked to other social problems, such as low educational achievement, violence, and suicide.

Marxist explanations of social problems imply that the solution lies in eliminating inequality among classes of people by creating a classless society. The nature of work must also change to avoid alienation. Finally, stronger controls must be applied to corporations to ensure that corporate decisions and practices are based on safety rather than on profit considerations.

Mark Wilson/Getty Images

Preschooler Jacob Hurley, who became seriously ill after eating peanut butter manufactured by the Peanut Corporation of America, is shown sitting with his father Peter Hurley, who is testifying before a House Energy and Commerce Committee hearing on Capitol Hill in Washington, DC, January 2009.

Non-Marxist Conflict Theories Non-Marxist conflict theorists, such as Ralf Dahrendorf, are concerned with conflict that arises when groups have opposing values and interests. For example, anti-abortion activists value the life of unborn embryos and fetuses; pro-choice activists value the right of women to control their own bodies and reproductive decisions. These different value positions reflect different subjective interpretations of what constitutes a social problem. For anti-abortionists, the availability of abortion is the social problem; for pro-choice advocates, the restrictions on abortion are the social problem. Sometimes the social problem is not the conflict itself but rather the way that conflict is expressed. Even most pro-life advocates agree that shooting doctors who perform abortions and blowing up abortion clinics constitute unnecessary violence and lack of respect for life. Value conflicts may occur between diverse categories of people, including nonwhites versus whites, heterosexuals versus homosexuals, young versus old, Democrats versus Republicans, and environmentalists versus industrialists.

Solving the problems that are generated by competing values may involve ensuring that conflicting groups understand each other's views, resolving differences through negotiation or mediation, or agreeing to disagree. Ideally, solutions should be win-win, with both conflicting groups satisfied with the solution. However, outcomes of value conflicts are often influenced by power; the group with the most power may use its position to influence the outcome of value conflicts. For example, when Congress could not get all states to voluntarily increase the legal drinking age to 21, it threatened to withdraw federal highway funds from those that would not comply.

Symbolic Interactionist Perspective

Both the structural-functionalist and the conflict perspectives are concerned with how broad aspects of society, such as institutions and large social groups, influence the social world. This level of sociological analysis is called *macrosociology*: It looks at the big picture of society and suggests how social problems are affected at the institutional level.

alienation A sense of powerlessness and meaninglessness in people's lives.

Microsociology, another level of sociological analysis, is concerned with the social-psychological dynamics of individuals interacting in small groups. Symbolic interactionism reflects the microsociological perspective and was largely influenced by the work of early sociologists and philosophers such as Max Weber, Georg Simmel, Charles Horton Cooley, G. H. Mead, W. I. Thomas, Erving Goffman, and Howard Becker. Symbolic interactionism emphasizes that human behavior is influenced by definitions and meanings that are created and maintained through symbolic interaction with others.

We develop our self-concept by observing how others interact with us and label us. By observing how others view us, we see a reflection of ourselves that Cooley calls the "looking-glass self."

Sociologist W. I. Thomas (1931/1966) emphasized the importance of definitions and meanings in social behavior and its consequences. He suggested that humans respond to their definition of a situation rather than to the objective situation itself. Hence, Thomas noted that situations that we define as real become real in their consequences.

Symbolic interactionism also suggests that social interaction shapes our identity or sense of self. We develop our self-concept by observing how others interact with us and label us. By observing how others view us, we see a reflection of ourselves that Cooley calls the "looking-glass self."

Last, the symbolic interactionist perspective has important implications for how social scientists conduct research. German sociologist Max Weber argued that, to understand individual and group behavior, social scientists must see the world through the eyes of that individual or group. Weber called this approach *verstehen,* which in German means "to understand." *Verstehen* implies that, in conducting research, social scientists must try to understand others' views of reality and the subjective aspects of their experiences, including their symbols, values, attitudes, and beliefs.

Symbolic Interactionist Theories of Social Problems

A basic premise of symbolic interactionist theories of social problems is that a condition must be *defined or recognized* as a social problem for it to *be* a social problem. Three symbolic interactionist theories of social problems are based on this general premise.

Blumer's Stages of a Social Problem Herbert Blumer (1971) suggested that social problems develop in stages. First, social problems pass through the stage of *societal recognition*—the process by which a social problem (for example, drunk driving) is "born." Second, *social legitimation* takes place when the social problem achieves recognition by the larger community, including the media, schools, and churches. As the visibility of traffic fatalities associated with alcohol increased, so did the legitimation of drunk driving as a social problem. The next stage in the development of a social problem involves *mobilization for action,* which occurs when individuals and groups, such as Mothers Against Drunk Driving, become concerned about how to respond to the social condition. This mobilization leads to the *development and implementation of an official plan* for dealing with the problem, involving, for example, highway checkpoints, lower legal blood-alcohol levels, and tougher regulations for driving drunk.

Blumer's stage-development view of social problems is helpful in tracing the development of social problems. For example, although sexual harassment and date rape occurred throughout the 20th century, these issues did not begin to receive recognition as social problems until the 1970s. Social legitimation of these problems was achieved when high schools, colleges, churches, employers, and the media recognized their existence. Organized social groups mobilized to develop and implement plans to deal with these problems. Groups successfully lobbied for the enactment of laws against sexual harassment and the enforcement of sanctions against violators of these laws. Groups also mobilized to provide educational seminars on date rape for high school and college students and to offer support services to victims of date rape.

Some disagree with the symbolic interactionist view that social problems exist only if they are recognized. According to this view, individuals who were victims of date rape in the 1960s may be considered victims of a problem, even though date rape was not recognized at that time as a social problem.

Labeling Theory Labeling theory, a major symbolic interactionist theory of social problems, suggests that a social condition or group is viewed as problematic if it is labeled as such. According to labeling theory, resolving social problems sometimes involves changing the meanings and definitions that are attributed to people and situations. For example, so long as teenagers define drinking alcohol as "cool" and "fun," they will continue to abuse alcohol. So long as our society defines providing sex education and contraceptives to teenagers as inappropriate or immoral, the teenage pregnancy rate in the United States will continue to be higher than that in other industrialized nations.

Social Constructionism Social constructionism is another symbolic interactionist theory of social problems. Similar to labeling theorists and symbolic interactionism in general, social constructionists argue that individuals who interpret the social world around them socially construct reality. Society, therefore, is a social creation rather than an objective given. As such, social constructionists often question the origin and evolution of social problems. For example, most Americans define "drug abuse" as a social problem in the United States but rarely include alcohol or cigarettes in their discussion. A social constructionist would point to the historical roots of alcohol and tobacco use as a means of understanding their legal status. Central to this idea of the social construction of social problems are the media, universities, research institutes, and government agencies, which are often responsible for the public's initial "take" on the problem under discussion.

Table 1.2 summarizes and compares the major theoretical perspectives, their criticisms, and social policy recommendations as they relate to social problems. The study of

TABLE 1.2 Comparison of Theoretical Perspectives

	STRUCTURAL FUNCTIONALISM	CONFLICT THEORY	SYMBOLIC INTERACTIONISM
Representative theorists	Emile Durkheim Talcott Parsons Robert Merton	Karl Marx Ralf Dahrendorf	George H. Mead Charles Cooley Erving Goffman
Society	Society is a set of interrelated parts; cultural consensus exists and leads to social order; natural state of society: balance and harmony.	Society is marked by power struggles over scarce resources; inequities result in conflict; social change is inevitable; natural state of society: imbalance.	Society is a network of interlocking roles; social order is constructed through interaction as individuals, through shared meaning, making sense out of their social world.
Individuals	Individuals are socialized by society's institutions; socialization is the process by which social control is exerted; people need society and its institutions.	People are inherently good but are corrupted by society and its economic structure; institutions are controlled by groups with power; "order" is part of the illusion.	Humans are interpretive and interactive; they are constantly changing as their "social beings" emerge and are molded by changing circumstances.
Cause of social problems?	Rapid social change; social disorganization that disrupts the harmony and balance; inadequate socialization and/or weak institutions.	Inequality; the dominance of groups of people over other groups of people; oppression and exploitation; competition between groups.	Different interpretations of roles; labeling of individuals, groups, or behaviors as deviant; definition of an objective condition as a social problem.
Social policy/ solutions	Repair weak institutions; assure proper socialization; cultivate a strong collective sense of right and wrong.	Minimize competition; create an equitable system for the distribution of resources.	Reduce impact of labeling and associated stigmatization; alter definitions of what is defined as a social problem.
Criticisms	Called "sunshine sociology"; supports the maintenance of the status quo; needs to ask "functional for whom?"; does not deal with issues of power and conflict; incorrectly assumes a consensus.	Utopian model; Marxist states have failed; denies existence of cooperation and equitable exchange; cannot explain cohesion and harmony.	Concentrates on micro issues only; fails to link micro issues to macro-level concerns; too psychological in its approach; assumes label amplified problem.

© Cengage Learning 2013

Each chapter in this book contains a *Social Problems Research Up Close* box that describes a research report or journal article that examines some sociologically significant topic. Some examples of the more prestigious journals in sociology include the *American Sociological Review*, the *American Journal of Sociology*, and *Social Forces*. Journal articles are the primary means by which sociologists, as well as other scientists, exchange ideas and information. Most journal articles begin with *an introduction and review of the literature*. Here, the investigator examines previous research on the topic, identifies specific research areas, and otherwise "sets the stage" for the reader. Often in this section, research hypotheses are set forth, if applicable. A researcher, for example, might hypothesize that the sexual behavior of adolescents has changed over the years as a consequence of increased fear of sexually transmitted diseases and that such changes vary on the basis of sex.

The next major section of a journal article is *sample and methods*. In this section, an investigator describes the characteristics of the sample, if any, and the details of the type of research conducted. The type of data analysis used is also presented in this section (see Appendix). Using the sample research question, a sociologist might obtain data from the Youth Risk Behavior Surveillance Survey collected by the Centers for Disease Control and Prevention.

This self-administered questionnaire is distributed biennially to more than 10,000 high school students across the United States.

The final section of a journal article includes the *findings and conclusions*. The findings of a study describe the results, that is, what the researcher found as a result of the investigation. Findings are then discussed within the context of the hypotheses and the conclusions that can be drawn. Often, research results are presented in tabular form. Reading tables carefully is an important part of drawing accurate conclusions about the research hypotheses. In reading a table, you should follow the steps listed here (see the table within this box):

1. *Read the title of the table and make sure that you understand what the table contains.* The title of the table indicates the unit of analysis (high school students), the dependent variable (sexual risk behaviors), the independent variables (sex and year), and what the numbers represent (percentages).
2. *Read the information contained at the bottom of the table, including the source and any other explanatory information.* For example, the information at the bottom of this table indicates that the data are from the Centers for Disease Control and Prevention, that "sexually active" was defined as having intercourse in the last three months, and that data

on condom use were only from those students who were defined as being currently sexually active.

3. *Examine the row and column headings.* This table looks at the percentage of males and females, over four years, who reported ever having sexual intercourse, having four or more sex partners in a lifetime, being currently sexually active, and using condoms during the last sexual intercourse.
4. *Thoroughly examine the data in the table carefully, looking for patterns between variables.* As indicated in the table, the percentage of males engaging in "risky" sexual behavior has gone down between 2003 and 2009 for three of the four categories and are the lowest recorded over the time period for (1) ever having had sexual intercourse, (2) having four or more partners during lifetime, and (3) being currently sexually active. However, the percentage of males having sex without protection is the highest over the recorded years. The same pattern is not detected for females, where much less variation is recorded over the time period studied for all categories. The difference between male and female sexually risky behaviors should be noted. Contrary to "commonsense" beliefs, males are less likely to be sexually active and more likely to have used a condom during last intercourse.

---

social problems is based on research as well as on theory, however. Indeed, research and theory are intricately related. As Wilson (1983) stated:

> Most of us think of theorizing as quite divorced from the business of gathering facts. It seems to require an abstractness of thought remote from the practical activity of empirical research. But theory building is not a separate activity within sociology. Without theory, the empirical researcher would find it impossible to decide what to observe, how to observe it, or what to make of the observations. (p. 1)

Social Problems Research

Most students taking a course in social problems will not become researchers or conduct research on social problems. Nevertheless, we are all consumers of research that is reported in the media. Politicians, social activist groups, and organizations attempt to justify their decisions, actions, and positions by citing research results. As consumers of research, we need to understand that our personal experiences and casual observations are less reliable than generalizations based on systematic research. One strength of

5. *Use the information you have gathered in Step 4 to address the hypotheses.* Clearly, sexual practices, as hypothesized, have changed over time. For example, both males and females, when comparing data from 2003 to 2009, have a general increase in condom use during sexual intercourse. Further, the percentage of males and females reporting four or more sex partners has decreased in the same time period. Look at the table and see what patterns you detect, and how these patterns address the hypothesis.

6. *Draw conclusions consistent with the information presented.* From the table, can we conclude that sexual practices have changed over time? The answer is probably yes, although the limitations of the survey, the sample, and the measurement techniques used always should be considered. Can we conclude that the observed changes are a consequence of the fear of sexually transmitted diseases? The answer is *no*. Having no measure of fear of sexually transmitted diseases over the time period studied, we are unable to come to such a conclusion. More information, from a variety of sources, is needed. The use of multiple methods and approaches to study a social phenomenon is called *triangulation*.

Percentage of High School Students Reporting Sexually Risky Behaviors, by Sex and Survey Year

SURVEY YEAR	EVER HAD SEXUAL INTERCOURSE	FOUR OR MORE SEX PARTNERS DURING LIFETIME	CURRENTLY SEXUALLY ACTIVE*	CONDOM USED DURING LAST INTERCOURSE†
Male				
2003	48.0	17.5	33.8	68.8
2005	47.9	16.5	33.3	70.0
2007	49.8	17.9	34.3	68.5
2009	46.1	16.2	32.6	70.4
Females				
2003	45.3	11.2	34.6	57.4
2005	45.7	12.0	34.6	55.9
2007	45.9	11.8	35.6	54.9
2009	45.7	11.2	35.7	57.0

*Sexual intercourse during the three months preceding the survey
†Among currently sexually active students
Source: Centers for Disease Control and Prevention 2008; 2010.

scientific research is that it is subjected to critical examination by other researchers (see this chapter's *Social Problems Research Up Close* feature). The more you understand how research is done, the better able you will be to critically examine and question research rather than to passively consume research findings. In the remainder of this section, we discuss the stages of conducting a research study and the various methods of research that sociologists use.

> The more you understand how research is done, the better able you will be to critically examine and question research rather than to passively consume research findings.

Stages of Conducting a Research Study

Sociologists progress through various stages in conducting research on a social problem. In this section, we describe the first four stages: (1) formulating a research question, (2) reviewing the literature, (3) defining variables, and (4) formulating a hypothesis.

Formulating a Research Question A research study usually begins with a research question. Where do research questions originate? How does a particular researcher come to ask a particular research question? In some cases, researchers have a personal interest in a specific topic because of their own life experiences. For example, a researcher who has experienced spouse abuse may wish to do research on such questions as "What factors are associated with domestic violence?" and "How helpful are battered women's shelters in helping abused women break the cycle of abuse in their lives?" Other researchers may ask a particular research question because of their personal values—their concern for humanity and the desire to improve human life. Researchers who are concerned about the spread of HIV infection and AIDS may conduct research on questions such as "How does the use of alcohol influence condom use?" and "What educational strategies are effective for increasing safer sex behavior?" Researchers may also want to test a particular sociological theory, or some aspect of it, to establish its validity or conduct studies to evaluate the effect of a social policy or program. Research questions may also be formulated by the concerns of community groups and social activist organizations in collaboration with academic researchers. Government and industry also hire researchers to answer questions such as "How many children are victimized by episodes of violence at school?" and "What types of computer technologies can protect children against being exposed to pornography on the Internet?"

What Do You Think? In a free society, there must be freedom of information. That is why the U.S. Constitution and, more specifically, the First Amendment protect journalists' sources. If journalists are compelled to reveal their sources, their sources may be unwilling to share information, and that would jeopardize the public's right to know. A journalist cannot reveal information given in confidence without permission from the source or a court order. Do you think sociologists should be granted the same protections as journalists? If a reporter at your school newspaper uncovered a scandal at your university, should he or she be protected by the First Amendment?

Reviewing the Literature After a research question is formulated, researchers review the published material on the topic to find out what is already known about it. Reviewing the literature also provides researchers with ideas about how to conduct their research and helps them formulate new research questions. A literature review serves as an evaluation tool, allowing a comparison of research findings and other sources of information, such as expert opinions, political claims, and journalistic reports.

Defining Variables A **variable** is any measurable event, characteristic, or property that varies or is subject to change. Researchers must operationally define the variables they study. An *operational definition* specifies how a variable is to be measured. For example, an operational definition of the variable "religiosity" might be the number of times the respondent reports going to church or synagogue. Another operational definition of "religiosity" might be the respondent's answer to the question "How important is religion in your life?" (for example, 1 is not important; 2 is somewhat important; 3 is very important).

Operational definitions are particularly important for defining variables that cannot be directly observed. For example, researchers cannot directly observe concepts such as "mental illness," "sexual harassment," "child neglect," "job satisfaction," and "drug abuse." Nor can researchers directly observe perceptions, values, and attitudes.

Formulating a Hypothesis After defining the research variables, researchers may formulate a **hypothesis,** which is a prediction or educated guess about how one variable is related to another variable. The **dependent variable** is the variable that researchers want to explain; that is, it is the variable of interest. The **independent variable** is the variable that is expected to explain change in the dependent variable. In formulating

variable Any measurable event, characteristic, or property that varies or is subject to change.

hypothesis A prediction or educated guess about how one variable is related to another variable.

dependent variable The variable that the researcher wants to explain; the variable of interest.

independent variable The variable that is expected to explain change in the dependent variable.

a hypothesis, researchers predict how the independent variable affects the dependent variable. For example, Kmec (2003) investigated the impact of segregated work environments on minority wages, concluding that "minority concentration in different jobs, occupations, and establishments is a considerable social problem because it perpetuates racial wage inequality" (p. 55). In this example, the independent variable is workplace segregation, and the dependent variable is wages.

From the film *Obedience* © 1968 by Stanley Milgram, copyright renewed 1993 by Alexandra Milgram, and distributed by Penn State Media Sales.

In studying social problems, researchers often assess the effects of several independent variables on one or more dependent variables. A recent study examined the impact of divorce (independent variable) on the emotional well-being of children (the dependent variable). After examining 6,647 adults, 695 of whom had parents who divorced before they were 18, researchers Fuller-Thomson and Dalton (2011) conclude that both sons and daughters have higher rates of suicide ideation than children from intact families. However, once parental addiction and abuse were controlled for, the effects of divorce on female suicide ideation disappeared, but males remained twice as likely as sons from nondivorced families to engage in suicide ideation.

Methods of Data Collection

After identifying a research topic, reviewing the literature, defining the variables, and developing hypotheses, researchers decide which method of data collection to use. Alternatives include experiments, surveys, field research, and secondary data.

Experiments **Experiments** involve manipulating the independent variable to determine how it affects the dependent variable. Experiments require one or more experimental groups that are exposed to the experimental treatment(s) and a control group that is not exposed. After a researcher randomly assigns participants to either an experimental group or a control group, the researcher measures the dependent variable. After the experimental groups are exposed to the treatment, the researcher measures the dependent variable again. If participants have been randomly assigned to the different groups, the researcher may conclude that any difference in the dependent variable among the groups is due to the effect of the independent variable.

An example of a "social problems" experiment on poverty would be to provide welfare payments to one group of unemployed single mothers (experimental group) and no such payments to another group of unemployed single mothers (control group). The independent variable would be welfare payments; the dependent variable would be employment. The researcher's hypothesis would be that mothers in the experimental group would be less likely to have a job after 12 months than mothers in the control group.

The major strength of the experimental method is that it provides evidence for causal relationships, that is, how one variable affects another. A primary weakness is that experiments are often conducted on small samples, usually in artificial laboratory settings; thus, the findings may not be generalized to other people in natural settings.

Surveys **Survey research** involves eliciting information from respondents through questions. An important part of survey research is selecting a sample of those to be questioned. A **sample** is a portion of the population, selected to be representative so that the information from the sample can be generalized to a larger population. For example, instead of asking all abused spouses about their experience, the researcher could ask a representative sample of them and assume that those who were not questioned would give similar responses. After selecting a representative sample, survey researchers either interview people, ask them to complete written questionnaires, or elicit responses to research questions through computers.

In one of the most famous experiments in the social sciences, Stanley Milgram found that 65 percent of a sample of ordinary citizens were willing to use harmful electric shocks—up to 450 volts—on an elderly man with a heart condition simply because the experimenter instructed them to do so. It was later revealed that the man was not really receiving the shocks and that he had been part of the experimental manipulation. The experiment, although providing valuable information, raised many questions on the ethics of scientific research.

experiment A research method that involves manipulating the independent variable to determine how it affects the dependent variable.

survey research A research method that involves eliciting information from respondents through questions.

sample A portion of the population, selected to be representative so that the information from the sample can be generalized to a larger population.

Interviews   In interview survey research, trained interviewers ask respondents a series of questions and make written notes about or tape-record the respondents' answers. Interviews may be conducted over the telephone or face-to-face.

One advantage of interview research is that researchers are able to clarify questions for the respondent and follow up on answers to particular questions. Researchers often conduct face-to-face interviews with groups of individuals who might otherwise be inaccessible. For example, some AIDS-related research attempts to assess the degree to which individuals engage in behavior that places them at high risk for transmitting or contracting HIV. Street youth and intravenous drug users, both high-risk groups for HIV infection, may not have a telephone or address because of their transient lifestyle. These groups may be accessible, however, if the researcher locates their hangouts and conducts face-to-face interviews. Research on drug addicts may also require a face-to-face interview survey design (Jacobs 2003).

The most serious disadvantages of interview research are cost and the lack of privacy and anonymity. Respondents may feel embarrassed or threatened when asked questions that relate to personal issues such as drug use, domestic violence, and sexual behavior. As a result, some respondents may choose not to participate in interview research on sensitive topics. Those who do participate may conceal or alter information or give socially desirable answers to the interviewer's questions (e.g., "No, I do not use drugs").

Questionnaires   Instead of conducting personal or phone interviews, researchers may develop questionnaires that they either mail or give to a sample of respondents. Questionnaire research offers the advantages of being less expensive and less time-consuming than face-to-face or telephone surveys. In addition, questionnaire research provides privacy and anonymity to the research participants. This reduces the likelihood that they will feel threatened or embarrassed when asked personal questions and increases the likelihood that they will provide answers that are not intentionally inaccurate or distorted. A study on the relationship between minority composition of the workplace and the likelihood of workplace drug testing is a case in point. Questionnaires were sent to union leaders of the Communication Workers of America (CWA), asking them about drug-testing policies at their local job sites. Analysis indicated that, as the minority composition of the workplace goes up, the likelihood of *pre-employment testing* and *testing with cause* increases, whereas the likelihood of *random* drug testing decreases (Gee et al. 2006).

The major disadvantage of mail questionnaires is that it is difficult to obtain an adequate response rate. Many people do not want to take the time or make the effort to complete and mail a questionnaire. Others may be unable to read and understand the questionnaire.

"Talking" Computers   A new method of conducting survey research is asking respondents to provide answers to a computer that "talks." Newman et al. (2002) found that syringe exchange program participants were more likely to report "stigmatized behavior" using computer-assisted self-interviewing, but less likely to report "psychological distress" when compared to face-to-face interview respondents. Thus, as in research in general, the reliability of data collected may depend on the interaction between the information sought and the method used.

field research Research that involves observing and studying social behavior in settings in which it occurs naturally.

Field Research   **Field research** involves observing and studying social behavior in settings in which it occurs naturally. Two types of field research are participant observation and nonparticipant observation.

In participant observation research, researchers participate in the phenomenon being studied so as to obtain an insider's perspective on the people and/or behavior being observed. Palacios and Fenwick (2003), two criminologists, attended dozens of raves over a 15-month period to investigate the South Florida drug culture. In nonparticipant observation research, researchers observe the phenomenon being studied without actively participating in the group or the activity. For example, Simi and Futrell (2009) studied white power activists by observing and talking to organizational members but did not participate in any of their unconventional activities.

Sometimes sociologists conduct in-depth detailed analyses or case studies of an individual, group, or event. For example, Fleming (2003) conducted a case study of young auto thieves in British Columbia. He found that, unlike professional thieves, the teenagers' behavior was primarily motivated by thrill-seeking—driving fast, the rush of a possible police pursuit, and the prospect of getting caught.

The main advantage of field research on social problems is that it provides detailed information about the values, rituals, norms, behaviors, symbols, beliefs, and emotions of those being studied. A potential problem with field research is that the researcher's observations may be biased (e.g., the researcher becomes too involved in the group to be objective). In addition, because field research is usually based on small samples, the findings may not be generalizable.

Secondary Data Research Sometimes researchers analyze secondary data, which are data that other researchers or government agencies have already collected or that exist in forms such as historical documents, police reports, school records, and official records of marriages, births, and deaths. Caldas and Bankston (1999) used information from Louisiana's 1990 Graduation Exit Examination to assess the relationship between school achievement and television-viewing habits of more than 40,000 tenth graders. The researchers found that, in general, television viewing is inversely related to academic achievement for whites but has little or no effect on school achievement for Black Americans. A major advantage of using secondary data in studying social problems is that the data are readily accessible, so researchers avoid the time and expense of collecting their own data. Secondary data are also often based on large representative samples. The disadvantage of secondary data is that researchers are limited to the data already collected.

Ten Good Reasons to Read This Book

This textbook approaches the study of social problems with several student benefits in mind:

1. *Understanding that the social world is too complex to be explained by just one theory will expand your thinking about how the world operates.* For example, juvenile delinquency doesn't have just one cause—it is linked to (1) an increased number of youths living in inner-city neighborhoods with little or no parental supervision (social disorganization theory); (2) young people having no legitimate means of acquiring material wealth (anomie theory); (3) youths being angry and frustrated at the inequality and racism in our society (conflict theory); and (4) teachers regarding youths as "no good" and treating them accordingly (labeling theory).
2. *Developing a sociological imagination will help you see the link between private troubles and public issues.* In a society that values personal responsibility, there is a tendency to define failure as a consequence of individual free will. The *sociological imagination* enables us to understand how social forces underlie personal misfortunes and failures, and contribute to personal successes and achievements.
3. *Understanding globalization can help you become a safe, successful, and productive world citizen.* Whether the spread of HIV, war, environmental destruction, human trafficking, or overpopulation, social problems in one part of the world affect other parts of the world. Today's problems call for collective action involving world citizens. There is some indication that students are already responding. In 2009, the

Both structural functionalism and conflict theory address the nature of social change, although in different ways. Durkheim, a structural functionalist, argued that social change, if rapid, was disruptive to society and that the needs of society should take precedence over the desires of individuals; that is, social change should be slow and methodical regardless of popular opinion.

To conflict theorists, social change is a result of the struggle for power by different groups. Specifically, Marx argued that social change was a consequence of the struggle between different economic classes as each strove for supremacy. Marx envisioned social change as primarily a revolutionary process ultimately leading to a utopian society.

Social movements are one means by which social change is realized. A **social movement** is an organized group of individuals with a common purpose to either promote or resist social change through collective action. Some people believe that, to promote social change, one must be in a position of political power and/or have large financial resources. However, the most important prerequisite for becoming actively involved in improving levels of social well-being may be genuine concern and dedication to a social "cause." The following vignettes provide a sampler of college students making a difference:

- Neha Patel of the University of Miami was shocked to realize that recycling was not part of the culture of South Beach, Florida. With a $1,000 grant from Starbucks, she initiated a pilot recycling program called "Raise the Bar" by providing local taverns with recycling bins and an incentive to use them—10 cents per bottle or can up to $100 (Liebowitz 2009).
- In 2006, 33 student activists "hit the road for a seven-week tour of 19 religious and military colleges that discriminate against gay and lesbian students" (Ferrara et al. 2006, p. 1). Although the bus was "tagged" with homophobic slogans and activists were arrested on six campuses, leader Jacob Reitan said the trip was "hugely positive." Activists met and talked with over 10,000 people, including ten school presidents.
- In 2011, over 200 Dickinson College students protested by taking over the campus administration building and demanding that school officials "issue red alerts when an assault occurs and take a stronger stance against sexual offenses, with mandatory and irreversible expulsion for offenders" (Snyder 2011, p. 1). As of this writing, the protests continue, but university officials have made some concessions, including assurances that expulsion of the offender will be the only acceptable response in cases of rape.
- At Middlebury College in Vermont, students successfully convinced the administration that global warming is a real problem that needs to be addressed—immediately. Student activists convinced university officials that the college should invest $11 million in a biomass plant—a plant "fueled by wood chips, grass pellets, and a self-sustaining willow forest" (James 2007, p. 1). Further, after five days of protesting by 1,000 vocal activists, one Middlebury group convinced Vermont Senator Bernie Sanders to reintroduce legislation in Congress that would reduce carbon emissions by 80 percent by the year 2050.
- Ever see a sticker that said "Please Use Revolving Door"? After discovering that eight times as much energy is lost when traditional doors are used versus revolving doors, four Massachusetts Institute of Technology students experimented by putting up signs on campus that read "Help M.I.T. save energy. Please use the revolving door." Use of revolving doors increased by 40 percent. The use of revolving doors over swing doors not only saves the university money but also reduces CO_2 emissions. Soon, all revolving doors on campus will sport the new facility-made signs (Catalysts for Change, 2008).
- In response to accusations of human rights violations of union workers in Coca-Cola bottling plants in Colombia (South America), students at Grinnell College in Iowa formed an anti-Coke campaign. Using the official boycotting policy of the college, the student initiative passed a boycott on all Coca-Cola products in November 2004. Because Coca-Cola had an exclusive contract with Grinnell, Coke and Coke products continued to be sold on campus. However, wherever they were sold, signs read, "The Grinnell College student body has voted to boycott Coca-Cola products. This is a Coca-Cola product" (KillerCoke.org 2005).
- Students at hundreds of campuses are members of anti-sweatshop groups such as Worker Rights Consortium (WRC). The WRC is a student-run watchdog organization that inspects factories worldwide, monitoring the monitors, as part of the anti-sweatshop movement. In 2009, as a result of the WRC and other anti-sweatshop groups, over a dozen schools (e.g., Harvard, Cornell, Georgetown) ended their contracts with collegiate apparel manufacturer Russell Athletic for violations of labor standards (Burns 2009).

fastest growing minor at the University of California at Berkeley was "Global Poverty and Practice" (Anwar 2009).

4. *Understanding that social problems are global in nature is not enough.* You must also understand that the social problems that plague our world are highly interrelated, making it difficult to "fix" one problem without creating others. The invention of the automobile solved the need for more efficient transportation in an industrialized society. However, it also contributed to the deterioration of urban areas as the federal highway system led to the emergence of suburbs, which robbed cities of businesses, workers, and retail establishments. Remember—you can't do just one thing.

social movement An organized group of individuals with a common purpose to either promote or resist social change through collective action.

5. *Although this is a social problems book, it may actually make you more, rather than less, optimistic.* Yes, all the problems discussed in the book are real, and they may seem insurmountable, but they aren't. You'll read about positive social change (for example, the number of people who smoke cigarettes in the United States has dramatically dropped as have rates of homophobia, racism, and sexism). Life expectancy has increased, and more people go to college than ever before. Change for the better can and does happen.

6. *Knowledge is empowering.* Social problems can be frightening, in part, because most people know very little about them beyond what they hear on the news or from their friends. Misinformation can make problems seem worse than they are. The more accurate the information you have, the more you will realize that we, as a society, have the power to solve the problems, and the less alienated you will feel.

7. *The Self and Society exercises increase self-awareness and allow you to position yourself within the social landscape.* For example, earlier in this chapter, you had the opportunity to assess your beliefs about a number of social problems and to compare your responses to a national sample of first-year college students.

8. *The Human Side features make you a more empathetic and compassionate human being.* The study of social problems is always about the quality of life of individuals. By conveying the private pain and personal triumphs associated with social problems, we hope to elicit a level of understanding that may not be attained through the academic study of social problems alone (see this chapter's *The Human Side*).

9. *The Social Problems Research Up Close features teach you the basics of scientific inquiry, making you a smarter consumer of "pop" sociology, psychology, anthropology, and the like.* These boxes demonstrate the scientific enterprise, from theory and data collection to findings and conclusions. Examples of research topics covered include college students' health, alcohol consumption norms among Italian youth, bullying and victimization among minority youth, and computer hackers.

10. *Learning about social problems and their structural and cultural origins helps you—individually or collectively—make a difference in the world.* Individuals can make a difference in society through the choices they make. You may choose to vote for one candidate over another, demand the right to reproductive choice or protest government policies that permit it, drive drunk or stop a friend from driving drunk, repeat a homophobic or racist joke or chastise the person who tells it, and practice safe sex or risk the transmission of sexually transmitted diseases. Collective social action is another, often more powerful way to make a difference. This chapter's photo essay visually portrays students acting collectively to change the world.

What Do You Think? The 2009 Serve America Act was designed to increase "the size of the AmeriCorps service program . . . , expand ways for students to earn money for college, and create opportunities for all Americans to serve in their communities" (Hass 2009, p. 1). One way college students can serve their communities is through service learning. In its simplest form, service learning entails students volunteering in the community and receiving academic credit for their efforts. Universities and colleges are increasingly requiring service learning credits as a criterion for graduation. Do you think that all students should be required to engage in service learning? Why or why not?

Understanding Social Problems

At the end of each chapter, we offer a section titled "Understanding" in which we re-emphasize the social origin of the problem being discussed, the consequences, and the alternative social solutions. Our hope is that readers will end each chapter with a

Student activism is not new nor is it unique to the United States. In the 1930s, the American Youth Congress (AYC) protested racial injustice, educational inequality, and the looming involvement of the United States in WWII. Called the "student brain of the New Deal" by some, the political power of the AYC would not be felt again until the student demonstrations of the 1960s (The Eleanor Roosevelt Papers 2008). Today, however, there is a new activism as students all over the world protest perceived injustices (Rifkind 2009). Aided by new technologies, social networking sites such as Facebook and MySpace allow for "virtual activism" as hundreds of thousands of students join online causes such as "Stop Global Warming" and "Save Darfur."

Jeff Widener/AP Photo

This chapter's photo essay highlights some of the most prominent examples of student activism, past and present. Although the faces have changed over time, the passion and dedication with which students voice their concerns has not.

Antiwar Demonstrations

During the Vietnam War era, students across the United States were vocal about their opposition to America's involvement in the war. In 1970, at Kent State University, the Ohio National Guard opened fire on unarmed student demonstrators, resulting in four deaths and nine injuries. The "Kent State Massacre" sparked campus protests across the nation, leading to the only nationwide student strike in U.S. history. A government report on antiwar demonstrations concluded that the shootings of students by the Ohio National Guard were unjustified (The Scranton Report 1971). No criminal charges were ever filed.

Animal Rights

Across the nation, student advocacy groups are speaking out against animal cruelty. Whether protesting the treatment of circus animals, the use of animals for research, dissection in school classrooms, the auction and shipment of horses to be slaughtered for meat, the confined and cramped spaces in which chickens, pigs, cattle, sheep, and other animals are kept before slaughter, or the abandonment and inhumane treatment of companion animals, student groups such as Students for the Ethical Treatment of Animals (SETA) are organizing to be the voice of rights for animals. Through such venues as weblogs, Facebook, rallies, and community outreach, students are effectively speaking up for those who cannot.

Courtesy of Ioana Samartinean

▲ The Animal Welfare Association of Arizona State University hosts "Meatout Day" to promote veganism and awareness that, as one placard reads, "Flesh is Flesh and Meat is Murder."

Political and Economic Oppression

In 1989, thousands of students from universities across China sat peacefully in Tiananmen Square protesting for democratic reforms and social justice. On June 3rd, tanks entered the square and opened fired on the unarmed students, killing or injuring hundreds, perhaps thousands. There is no official tally of the casualties due to the Chinese government's subsequent clampdown on media and the

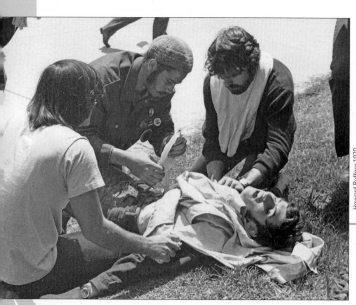

Howard Ruffner 1970

◄ The events of May 4, 1970, led singer/songwriter Neil Young to compose "Ohio" ("Tin soldiers and Nixon coming, we're finally on our own. This summer I hear the drumming, four dead in Ohio . . ."). Here, an injured student lies on the ground as onlookers stare in disbelief.

In 1989, an unknown man brings to a halt the People's Liberation Army as they advanced to disburse peaceful student demonstrations near Tiananmen Square in Peking (Beijing), China. The government's response to the student protests sparked global outrage.

Jack Moebes/CORBIS

reporting of any dissident activities, a policy that continues today. The protests at Tiananmen Square have been described "as the greatest challenge to the communist state in China since the 1949 revolution" (BBC 1989, p. 1).

Marriage Equality

The term marriage equality is fairly new but has become the rallying call of many college and high school students alike. Founded in 1998, the Gay-Straight Alliance Network (GSAN) "is a youth leadership organization that connects school-based Gay-Straight Alliances to each other and to community resources" (GSAN 2009, p. 1). Among other initiatives, GSAN, a member of the grassroots consortium Marriage Equality USA, actively promotes the marriage rights of same-sex couples. Despite some successes (e.g., it is now legal for same-sex partners to marry in several states), the passage of Proposition 8 led to a ban on same-sex marriages in California, leading to protests throughout the country (Garrison 2009; Dolan 2009). The voter initiative remains under appeal.

▲ Four Black American students from what was then called North Carolina Agricultural and Technical College returned to sit at a "whites-only" counter at Woolworth's in Greensboro, North Carolina, on February 2, 1960, setting off one of the most significant protests of the civil rights movement. The counter now sits on display in the Smithsonian Institute in Washington, DC.

Darron R. Silva/Aurora Photos

American Civil Rights Movement

On February 1, 1960, four African American students entered a Greensboro Woolworth's store to buy school supplies (Sykes 1960; Schlosser 2000). If their money was good enough to buy school supplies, why not a cup of coffee, they reasoned? At 4:30 p.m., they sat at the "whites only" lunch counter, intending to place an order. The four young men sat at the counter until closing but were never served. The next day, more students sat at the counter—they too were never served. As news of the "sit-in" spread, students returned to the Greensboro Woolworth's and to other lunch counters across the South. White and Black American students alike from New York to San Francisco began picketing Woolworth's in support of the "Greensboro Four." This one act by four students was the pivotal step in propelling forward what became known as the American civil rights movement (Schlosser 2000).

◀ Students from local high schools and Western Michigan University gather to protest discrimination and abuse against gays.

"sociological imagination" view of the problem and with an idea of how, as a society, we might approach a solution.

Sociologists have been studying social problems since the Industrial Revolution. Industrialization brought about massive social changes: The influence of religion declined, and families became smaller and moved from traditional, rural communities to urban settings. These and other changes have been associated with increases in crime, pollution, divorce, and juvenile delinquency. As these social problems became more widespread, the need to understand their origins and possible solutions became more urgent. The field of sociology developed in response to this urgency. Social problems provided the initial impetus for the development of the field of sociology and continue to be a major focus of sociology.

There is no single agreed-on definition of what constitutes a social problem. Most sociologists agree, however, that all social problems share two important elements: an objective social condition and a subjective interpretation of that condition. Each of the three major theoretical perspectives in sociology—structural-functionalist, conflict, and symbolic interactionist—has its own notion of the causes, consequences, and solutions of social problems.

CHAPTER REVIEW

- **What is a social problem?**
Social problems are defined by a combination of objective and subjective criteria. The objective element of a social problem refers to the existence of a social condition; the subjective element of a social problem refers to the belief that a particular social condition is harmful to society or to a segment of society and that it should and can be changed. By combining these objective and subjective elements, we arrive at the following definition: A social problem is a social condition that a segment of society views as harmful to members of society and in need of remedy.

- **What is meant by the structure of society?**
The structure of a society refers to the way society is organized.

- **What are the components of the structure of society?**
The components are institutions, social groups, statuses, and roles. Institutions are an established and enduring pattern of social relationships and include family, religion, politics, economics, and education. Social groups are defined as two or more people who have a common identity, interact, and form a social relationship. A status is a position that a person occupies within a social group and that can be achieved or ascribed. Every status is associated with many roles, or the set of rights, obligations, and expectations associated with a status.

- **What is meant by the culture of society?**
Whereas social structure refers to the organization of society, culture refers to the meanings and ways of life that characterize a society.

- **What are the components of the culture of society?**
The components are beliefs, values, norms, and symbols. Beliefs refer to definitions and explanations about what is assumed to be true. Values are social agreements about what is considered good and bad, right and wrong, desirable and undesirable. Norms are socially defined rules of behavior. Norms serve as guidelines for our behavior and for our expectations of the behavior of others. Finally, a symbol is something that represents something else.

- **What is the sociological imagination, and why is it important?**
The sociological imagination, a term that C. Wright Mills (1959) developed, refers to the ability to see the connections between our personal lives and the social world in which we live. It is important because, when we use our sociological imagination, we are able to distinguish between "private troubles" and "public issues" and to see connections between the events and conditions of our lives and the social and historical context in which we live.

- **What are the differences between the three sociological perspectives?**
According to structural functionalism, society is a system of interconnected parts that work together in harmony to maintain a state of balance and social equilibrium for the whole. The conflict perspective views society as composed of different groups and interests competing for power and resources. Symbolic interactionism reflects the microsociological perspective and emphasizes that human behavior is influenced by definitions and meanings that are created and maintained through symbolic interaction with others.

- **What are the first four stages of a research study?**
The first four stages of a research study are formulating a research question, reviewing the literature, defining variables, and formulating a hypothesis.

- **How do the various research methods differ from one another?**
Experiments involve manipulating the independent variable to determine how it affects the dependent variable. Survey

research involves eliciting information from respondents through questions. Field research involves observing and studying social behavior in settings in which it occurs naturally. Secondary data are data that other researchers or government agencies have already collected or that exist in forms such as historical documents, police reports, school records, and official records of marriages, births, and deaths.

- • What is a social movement?

Social movements are one means by which social change is realized. A social movement is an organized group of individuals with a common purpose to either promote or resist social change through collective action.

TEST YOURSELF

1. Definitions of social problems are clear and unambiguous.
   a. True
   b. False
2. The social structure of society contains
   a. statuses and roles
   b. institutions and norms
   c. sanctions and social groups
   d. values and beliefs
3. The culture of society refers to its meaning and the ways of life of its members.
   a. True
   b False
4. Alienation
   a. refers to a sense of normlessness
   b. is focused on by symbolic interactionists
   c. can be defined as the powerlessness and meaninglessness in people's lives
   d. is a manifest function of society
5. Blumer's stages of social problems begin with
   a. mobilization for action
   b. societal recognition
   c. social legitimation
   d development and implementation of a plan
6. The independent variable comes first in time; i.e., it precedes the dependent variable.
   a. True
   b. False

7. The third stage in defining a research study is
   a. formulating a hypothesis
   b. reviewing the literature
   c. defining the variables
   d. formulating a research question
8. A sample is a subgroup of the population—the group to whom you actually give the questionnaire.
   a. True
   b. False
9. Studying police behavior by riding along with patrol officers would be an example of
   a. participant observation
   b. nonparticipant observation
   c. field research
   d. both a and c
10. Student benefits of the book include
   a. providing global coverage of social problems
   b. highlighting social problems research
   c encouraging students to take pro-social action
   d. all of the above

Answers: 1. b; 2. a; 3. a; 4. c; 5. b; 6. a; 7. c; 8. a; 9. d; 10. d.

KEY TERMS

MEDIA RESOURCES

Turning to Video

▶ ❚❚ Watch the ABC video *The Tuskegee Experiment* (running time 3:06), available through **CengageBrain.com.** What were the independent and dependent variables in *the Tuskegee Experiment*? Having watched the video, what do you think are some of the ethical issues surrounding the use of experiments on human and other living beings?

Online Study Resources

Log in to **www.cengagebrain.com** to access the resources your instructor has assigned. For this book, you can access:

CourseMate

Access chapter-specific learning tools, including learning objectives, practice quizzes, videos, Internet exercises, flash cards, and glossaries, as well as web links, and more in your Sociology CourseMate.

Gideon Mendel//Documentary/CORBIS

Physical and Mental Health and Health Care

"The defense this nation seeks involves a great deal more than building airplanes, ships, guns, and bombs. We cannot be a strong nation unless we are a healthy nation."

—U.S. President Franklin Roosevelt, 1940

AP Photo/The Gaston Gazette, Ben Goff

James Verone, who had health problems and no medical insurance, committed a one-dollar bank robbery in hopes of receiving free medical care in prison.

IN JUNE 2011, 59-year-old Jerome Verone walked into a bank and handed the teller a note demanding one dollar and warning that he had a gun. The teller handed Verone a single dollar and then called 911. Verone calmly sat on a sofa in the bank and waited for police to arrest him. Why did this man rob a bank for one dollar? The answer is: Verone needed health care, and he was broke and had no health insurance. His plan was to get arrested for robbery and sent to prison, where he could get free health care for his chronic back pain, sore foot that made him limp, and an undiagnosed lump on his chest. Before he robbed the bank, Verone sent a letter to the *Gaston Gazette*, in which he explained, "When you receive this, a bank robbery will have been committed by me" (quoted in Moisse, 2011). Verone wanted people to understand that his motive was medical, not monetary. "The pain was beyond the tolerance I could accept," he explained, "I am of sound mind, but not so much sound body." Verone told reporters he was not afraid of being arrested and sent to prison, explaining, "If you don't have your health, you don't have anything" (Moisse, 2011).

In the United States, access to affordable health care is a pressing concern for millions of Americans, and literally a matter of life or death for some. In this chapter, we address problems of illness throughout the world, and focus on health care in the United States. Taking a sociological look at health issues, we examine why some social groups experience more health problems than others and how social forces affect and are affected by health and illness.

The Global Context: Patterns of Health around the World

developed countries Countries that have relatively high gross national income per capita and have diverse economies made up of many different industries.

developing countries Countries that have relatively low gross national income per capita, with simpler economies that often rely on a few agricultural products.

least developed countries The poorest countries of the world.

morbidity Illnesses, symptoms, and the impairments they produce.

life expectancy The average number of years that individuals born in a given year can expect to live.

Countries are often classified into one of three broad categories according to their economic status: (1) **developed countries** (also known as *high-income countries*) have relatively high gross national income per capita and have diverse economies made up of many different industries; (2) **less developed** or **developing countries** (also known as *middle-income countries*) have relatively low gross national income per capita, and their economies are much simpler, often relying on a few agricultural products; and (3) **least developed countries** (known as *low-income countries*) are the poorest countries of the world. The following sections reveal the striking disparities in patterns of health and illness among developed, less developed, and least developed nations.

Morbidity, Life Expectancy, and Mortality

Three measures of the health of populations are morbidity, life expectancy, and mortality. **Morbidity** refers to illnesses, symptoms, and the impairments they produce. In less developed countries, where poverty and chronic malnutrition are widespread, infectious and parasitic diseases, such as HIV disease, tuberculosis, diarrheal diseases (caused by bacteria, viruses, or parasites), measles, and malaria are much more prevalent than in developed countries, where chronic health problems such as cardiovascular disease and cancer are the major health threats (Weitz 2010).

Life expectancy—the average number of years that individuals born in a given year can expect to live—ranges from 47 in Malawi to 83 in Japan (see Figure 2.1). In

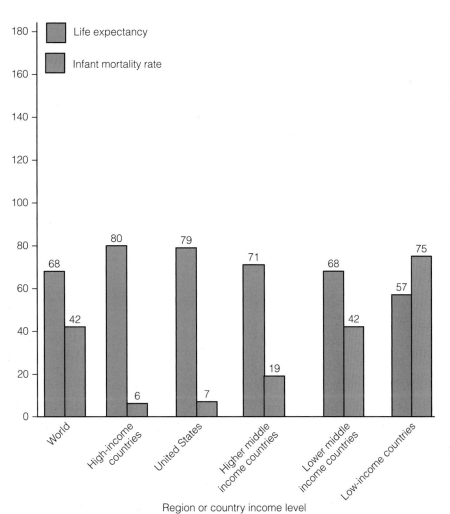

Figure 2.1: Life Expectancy and Infant Mortality Rate by Country Income Level, 2009
Source: World Health Organization 2011b.

12 countries (primarily in Africa), life expectancy is less than 50 years (World Health Organization 2011a).

The leading cause of **mortality,** or death, worldwide is heart disease, followed by stroke and respiratory infections. In low-income countries, the three top causes of death are respiratory infections, diarrheal diseases, and HIV/AIDS (World Health Organization 2011b). The World Health Organization (2009a) has identified high blood pressure as the leading global mortality risk factor (responsible for 13 percent of deaths globally), followed by tobacco use (9 percent), high blood glucose (6 percent), physical inactivity (6 percent), being overweight and obesity (5 percent). These risk factors contribute to chronic diseases such as heart disease, diabetes, and cancers in countries across all income groups: high, middle, and low.

What Do You Think? Data on deaths from international terrorism and tobacco-related deaths in 37 developed and eastern European countries revealed that tobacco-related deaths outnumbered terrorist deaths by about a whopping 5,700 times (Thomson & Wilson 2005). The number of tobacco deaths was equivalent to the impact of a September 11, 2001–type terrorist attack every 14 hours! Given that tobacco-related deaths grossly outnumber terrorism-related deaths, why hasn't the U.S. government waged a "war on tobacco" on a scale similar to its "war on terrorism"?

mortality Death.

TABLE 2.1 Top Three Causes of Death by Selected Age Groups, United States

AGE (YEARS)	LEADING CAUSES OF DEATH		
	FIRST	SECOND	THIRD
1 to 14	Unintentional injuries	Cancer	Congenital/chromosomal abnormalities
15 to 24	Unintentional injuries	Homicide	Suicide
25 to 44	Unintentional injuries	Cancer	Heart disease
45 to 64	Cancer	Heart disease	Unintentional injury
65+	Heart disease	Cancer	Stroke

Source: National Center for Health Statistics 2011.

When Tanzanian mothers are in labor, they often say to their older children, "I'm going to go and fetch the new baby; it is a dangerous journey and I may not return."

TABLE 2.2 Trained Childbirth Assistance and Lifetime Chance of Maternal Mortality by Region

	PERCENTAGE OF BIRTHS ATTENDED BY SKILLED PERSONNEL	ADULT LIFETIME CHANCE OF MATERNAL MORTALITY*
Developed countries	99	1 in 4,300
Developing countries	63	1 in 120
Sub-Saharan Africa	46	1 in 31

*Probability that a 15-year-old female will die eventually from a maternal cause.
Source: UNICEF 2010; World Health Organization 2010b.

infant mortality rate The number of deaths of live-born infants under 1 year of age per 1,000 live births (in any given year).

under-5 mortality rate The rate of deaths of children under age 5.

maternal mortality rate A measure of deaths that result from complications associated with pregnancy, childbirth, and unsafe abortion.

In the United States, the leading cause of death for both women and men is heart disease, followed by cancer and stroke (National Center for Health Statistics 2011). As shown in Table 2.1, U.S. mortality patterns vary by age.

Mortality Rates among Infants and Children The **infant mortality rate,** the number of deaths of live-born infants under 1 year of age per 1,000 live births (in any given year), ranges from an average of 6 in high-income nations to 75 in low-income nations (World Health Organization 2011b). Death rates of children under age 5 (**under-5 mortality rate**) are similarly much higher in low-income countries. One-half of all child deaths occur in sub-Saharan Africa where, in 2008, one in seven children died before their 5th birthday (UNICEF 2010).

One of the major causes of infant and child death worldwide is diarrhea, resulting from poor water quality and sanitation. More than one-third of the world's population—2.6 billion people—does not have access to adequate sanitation facilities (World Health Organization & UNICEF 2010). Another major contributing factor to deaths of infants and children is undernutrition. In the developing world, one in four children under age 5 is underweight (UNICEF 2010).

Maternal Mortality Rates The **maternal mortality rate** is a measure of deaths that result from complications associated with pregnancy and childbirth, such as hemorrhage (severe loss of blood) or infection. Women in the United States and other developed countries generally do not experience pregnancy and childbirth as life threatening. But for women ages 15 to 49 in developing countries, maternal mortality is the leading cause of death and disability. When Tanzanian mothers are in labor, they often say to their older children, "I'm going to go and fetch the new baby; it is a dangerous journey and I may not return" (Grossman 2009). More than a half million women die every year from childbirth or pregnancy-related causes. And for every maternal death, 15 to 30 women suffer from chronic illness and disability caused by pregnancy or childbirth (Paruzzolo et al. 2010).

Rates of maternal mortality show a greater disparity between rich and poor countries than any of the other societal health measures. Nearly all (99 percent) maternal deaths occur in low-income countries (World Health Organization 2010a). High maternal mortality rates in less developed countries are related to poor-quality and inaccessible health care; most women give birth without the assistance of trained personnel (see Table 2.2). High maternal mortality rates are also linked to malnutrition and poor sanitation and to pregnancy and childbearing at early ages. Women in many countries also lack access to family planning services and/or do not have the support of their male partners to use contraceptive methods such as condoms. Consequently, many women resort to abortion to limit their childbearing, even in countries where abortion is illegal and unsafe.

Sociological Theories of Illness and Health Care

The three major sociological theories—structural functionalism, conflict theory, and symbolic interactionism—each contribute to our understanding of illness and health care.

Structural-Functionalist Perspective

According to the structural-functionalist perspective, health care is a social institution that functions to maintain the well-being of societal members and, consequently, of the social system as a whole. Thus, this perspective points to how failures in the health care system affect not only the well-being of individuals, but also the health of other social institutions, such as the economy and the family.

The structural-functionalist perspective examines how changes in society affect health. As societies develop and provide better living conditions, life expectancy increases and birthrates decrease (Weitz 2010). At the same time, the main causes of death and disability shift from infectious disease, and infant, child, and maternal mortality to chronic, noninfectious illness and disease such as cancer, heart disease, Alzheimer's disease, and arthritis.

Just as social change affects health, health concerns may lead to social change. The emergence of HIV and AIDS in the U.S. gay male population helped unite and mobilize gay rights activists. Concern over the effects of exposure to tobacco smoke—the greatest cause of disease and death in the United States and other developed countries—led to legislation banning smoking in public places.

Finally, the structural-functionalist perspective draws attention to latent dysfunctions, or unintended and often unrecognized negative consequences of social patterns or behavior. For example, the use of antibiotics—in prescriptions; soaps, hand wipes, and cleaning agents; and factory farm animal feed—has produced a serious unintended consequence: the emergence of antibiotic-resistant bacteria. Many infectious diseases are becoming increasingly difficult to treat because these infections do not respond to standard antibiotic treatment (Nader 2011). In 2011, the World Health Organization, which claims antibiotic resistance to be one of the three most serious public health threats, dedicated its annual World Health Day to raising awareness about the growing threat of antibiotic-resistant infections.

Ronda Bailey-Wade

After scraping his leg in a scooter accident, 9-year-old Brock Wade developed a life-threatening bacterial infection that was resistant to antibiotics. After more than a month in a hospital, Brock, against all odds, recovered and was well enough to return home.

Conflict Perspective

The conflict perspective focuses on how wealth, status, power, and the profit motive influence illness and health care. Worldwide, the poor experience more health problems and have less access to quality medical care (Feachum 2000). The conflict perspective points to ways in which powerful groups and wealthy corporations influence health-related policies and laws through lobbying and financial contributions to politicians and political candidates. Private health insurance companies have much to lose if the United States adopts a national public health insurance program or even a public insurance option,

Photo Essay

Modern Animal Food Production: Health and Safety Issues

Every year, one out of six Americans (48 million people) becomes ill with food-borne illnesses: 128,000 are hospitalized, and 3,000 Americans die (Katel 2010). Many cases of "food poisoning" stem from modern methods of raising and processing food animals. Increasingly, food animals are not raised in expansive meadows or pastures; rather, they are raised in concentrated animal feeding operations (CAFOs), also known as "factory farms," giant corporate-controlled livestock farms where large numbers (sometimes tens or hundreds of thousands) of animals—typically cows, hogs, turkeys, or chickens—are "produced" in factory-like settings, often indoors, to maximize production and profits.

Daniel Pepper/Getty Images

▲ To prevent disease from spreading among animals living in crowded conditions, factory-farmed animals are fed diets laced with antibiotics.

Kelley McCall/AP Photo

▲ The hogs in this concentrated animal feeding operation (CAFO) will live their entire lives, from birth to slaughter, inside a crowded, controlled indoor environment

Conditions at factory farms are crowded and often unsanitary. In 2010, at least 1,600 Americans were sickened by salmonella-contaminated eggs, which lead to the biggest egg recall in U.S. history. Government inspectors traced the egg contamination to egg processors in Iowa, where they found disturbing health violations:

"Chicken manure located in the manure pits below the egg laying operation was observed to be approximately 4 feet high to 8 feet high. . . . Live and dead flies too numerous to count were observed. . . . In addition, live and dead maggots too numerous to count were observed," as well as "holes appearing to be rodent burrows located

along the second floor baseboards." (reported in Katel 2010, p. 1)

The diet of factory-farmed animals consists largely of corn, which is cheap and efficient in fattening the animals. The digestive system of cows is designed for grass; corn makes cows sick and susceptible to disease. Factory-farmed chickens, turkeys, and hogs, and farm-raised fish are also susceptible to disease because of the crowded and unsanitary living conditions. To prevent the spread of disease in CAFOs or fish farms, and to speed up growth, food animals are fed antibiotics, which contributes to the emergence of superresistant bacterial infections that will not respond to treatment.

About a third of U.S. dairy cows are given recombinant bovine growth hormone (rBGH), manufactured by Monsanto and sold under

the trade name Posilac, to increase milk production. Although the Food and Drug Administration (FDA) approved Posilac in 1993, and has supported Monsanto's claims that milk containing rBGH is safe for consumers, some experts warn that rBGH raises the risk of breast, colon, and prostate cancer (Epstein 2006). Other countries, including all of Europe, Canada, Australia, New Zealand, and Japan, have banned milk containing rBGH.

In modern slaughterhouse and meatpacking production, meat can be contaminated

▼ In 2010, 380 million eggs were recalled after several hundred people became ill with salmonella traced to eggs.

AP Photo/Reed Saxon

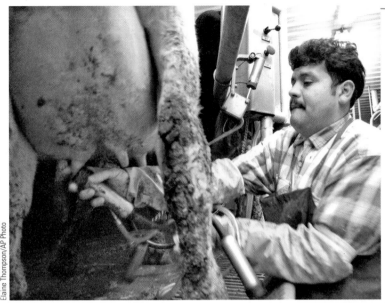

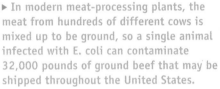

▶ In modern meat-processing plants, the meat from hundreds of different cows is mixed up to be ground, so a single animal infected with E. coli can contaminate 32,000 pounds of ground beef that may be shipped throughout the United States.

◀ Cows are given hormones (through shots or ear implants) to increase milk production.

▲ As workers remove the stomach and intestines of beef cattle, fecal matter can spill out and contaminate the meat.

with fecal matter. In the slaughterhouse, if the animal's hide has not been adequately cleaned, chunks of manure may fall from it onto the meat.

When the cow's stomach and intestines are removed, the fecal matter in the digestive system may spill out and contaminate the meat as well. Fecal matter contains the microbe Escherichia coli O157:H7, which can cause serious illness and death in humans. Because of the mass-production techniques of modern meat processing, a single cow infected with E. coli O157:H7 can contaminate 32,000 pounds of ground beef (Schlosser 2002). The first widespread deadly outbreak of food-borne illness to attract national attention occurred in 1993, when hamburgers sold by the fast-food chain Jack in the Box in Washington, Idaho, California, and Nevada were found to be contaminated with E. coli. Four children died, and nearly 1,000 people were sickened, many requiring hospitalization.

Health problems also result from the massive quantities of animal waste that are produced and stored around factory farms. A single dairy cow produces more than 20 tons of manure annually, and a hog can produce more than two tons (Weeks 2007). Manure is often stored in lagoons, which can leak or break, contaminating groundwater. Livestock factories also pose a threat to air quality. People who live near large livestock farms complain about headaches, runny noses, sore throats, nausea, stomach cramps, diarrhea, burning eyes, coughing, bronchitis, and shortness of breath (Singer & Mason 2006; Weeks 2007).

In an effort to make our food safer, in January 2011, President Obama signed into law the Food Safety Modernization Act, which expands the regulatory powers of the FDA. For example, the FDA now can require food facilities to disclose their food test results, speed up the food recall process, and require grocery stores to post notices about food recalls. However, the FDA oversees the production of all food products except meat, poultry, and dairy, which fall under the jurisdiction of the U.S. Department of Agriculture (USDA).

◀ People who live near animal waste lagoons endure an awful stench and suffer from a number of health problems from the fumes, including nausea and headaches.

and have spent millions of dollars opposing such proposals (Mayer 2009). Corporations also hire public relations (PR) companies to influence public opinion about health care issues. In his book *Deadly Spin* (2010), insurance industry insider Wendell Potter describes how the insurance industry hired a PR firm to manipulate public opinion on health care reform in part by discrediting Michael Moore's 2007 documentary *Sicko* (see also this chapter's *Human Side* feature).

The conflict perspective criticizes the pharmaceutical and health care industry for placing profits above people. "Drugmakers, device makers, and insurers decide which products to develop based not on what patients need, but on what their marketers tell them will sell—and produce the highest profit" (Mahar 2006, p. xviii). For example, not enough drugs are being developed to combat the growing public health threat of antibiotic-resistant infections, in part because pharmaceutical companies do not have a financial incentive. "Antibiotics . . . have a poor return on investment because they are taken for a short period of time and cure their target disease. In contrast, drugs that treat chronic illness, such as high blood pressure, are taken daily for the rest of a patient's life" (Braine 2011).

Many industries place profit above health considerations of workers and consumers. Chapter 7, "Work and Unemployment," discusses how employers often cut costs by neglecting to provide adequate safety measures for their employees. Chapter 13, "Environmental Problems," looks at how corporations often ignore environmental laws and policies, exposing the public to harmful pollution. The food industry is more concerned about profits than about public health. For example, most meat and dairy producers routinely feed antibiotics to animals, which humans then consume. Antibiotics in animal feed have contributed to the development of strains of antibiotic-resistant bacteria in humans (see this chapter's Photo Essay). Efforts to limit the use of antibiotics in animal feed have been blocked by the pharmaceutical and livestock industry lobbies, whose profits would be threatened if antibiotic use in food animals was limited (Katel 2010).

Symbolic Interactionist Perspective

Symbolic interactionists focus on (1) how meanings, definitions, and labels influence health, illness, and health care; and (2) how such meanings are learned through interaction with others and through media messages and portrayals. According to the symbolic interactionist perspective of illness, "there are no illnesses or diseases in nature. There are only conditions that society, or groups within it, has come to define as illness or disease" (Goldstein 1999, p. 31). Psychiatrist Thomas Szasz (1961/1970) argued that what we call "mental illness" is no more than a label conferred on those individuals who are "different," that is, those who do not conform to society's definitions of appropriate behavior.

Defining or labeling behaviors and conditions as medical problems is part of a trend known as **medicalization**. Behaviors and conditions that have undergone medicalization include post-traumatic stress disorder, premenstrual syndrome, menopause, childbirth, attention-deficit/hyperactivity disorder; and even the natural process of dying. Conflict theorists view medicalization as resulting from the medical profession's domination and pursuit of profits. A symbolic interactionist perspective suggests that medicalization results from the efforts of sufferers to "translate their individual experiences of distress into shared experiences of illness" (Barker 2002, p. 295).

According to symbolic interactionism, conceptions of health and illness are socially constructed. It follows, then, that definitions of health and illness vary over time and from society to society. In some countries, being fat is a sign of health and wellness; in others, it is an indication of mental illness or a lack of self-control. Among some cultural groups, perceiving visions or voices of religious figures is considered a normal religious experience, whereas such "hallucinations" would be indicative of mental illness in other cultures. In 18th- and 19th-century America, masturbation was considered an unhealthy act that

medicalization Defining or labeling behaviors and conditions as medical problems.

caused a range of physical and mental health problems (Allen 2000). Today, most health professionals agree that masturbation is a normal, healthy aspect of sexual expression.

Symbolic interactionism draws attention to the effects that meanings and labels have on health and health-risk behaviors. For example, among white Americans, having a "tan" is culturally defined as youthful and attractive, and so many white Americans sunbathe and use tanning beds—behaviors that increase one's risk of developing skin cancer. Meanings and labels also affect health policies. After the International Agency for Research on Cancer issued a 2009 report labeling tanning beds as "carcinogenic to humans," many states proposed and/or enacted legislation to restrict the use of tanning beds among minors by, for example, requiring parental permission (National Conference of State Legislators 2010; Reinberg 2009).

What Do You Think? The risk of developing melanoma (skin cancer) increases 75 percent when individuals use tanning beds before age 30 (Reinberg 2009). Some health advocates are calling for a total ban on the use of tanning beds by minors. Would you support such a ban? Why or why not?

Symbolic interactionists also focus on the stigmatization of individuals who are in poor health or who lack health insurance. A **stigma** refers to a discrediting label that affects an individual's self-concept and disqualifies that person from full social acceptance. (Originally, the word *stigma* referred to a mark burned into the skin of a criminal or slave.) The stigma associated with poor health often results in prejudice and discrimination against individuals with mental illnesses, drug addictions, physical deformities and impairments, missing or decayed teeth, obesity, HIV infection and AIDS, and other health conditions. Further, a study of U.S. adults without insurance found that "uninsured Americans . . . noted the stigma of lacking health insurance, citing medical providers who treat them like 'losers' because they are uninsured" (Sered & Fernandopulle 2005, p. 16).

The stigma associated with health problems and/or lack of health insurance implies that individuals—rather than society—are responsible for their health. In U.S. culture, "sickness increasingly seems to be construed as a personal failure—a failure of ethical virtue, a failure to take care of oneself 'properly' by eating the 'right' foods or getting 'enough' exercise, a failure to get a Pap smear, a failure to control sexual promiscuity, genetic failure, a failure of will, or a failure of commitment—rather than society's failure to provide basic services to all of its citizens" (Sered & Fernandopulle 2005, p. 16). One stigmatized population is those with HIV/AIDS, a topic we discuss in the next section.

HIV/AIDS: A Global Health Concern

HIV, the virus that causes AIDS (acquired immunodeficiency syndrome) continues to threaten the health of populations around the world. More than 33 million people are living with HIV/AIDS, most of whom (97 percent) live in low- or middle-income countries, particularly in sub-Saharan Africa (Kaiser Family Foundation 2010). Most people infected with HIV do not know it.

Most people infected with HIV do not know it.

Worldwide, most cases of HIV infection are transmitted through heterosexual contact, and slightly more than half of all people living with HIV/AIDS worldwide are women (Kaiser Family Foundation 2010). HIV is transmitted through sexual contact (anal or vaginal intercourse; oral sex), sharing intravenous needles, perinatal transmission (from infected mother to fetus or newborn), blood transfusions or blood products, and breast milk (Centers for Disease Control and Prevention 2010a).

stigma A discrediting label that affects an individual's self-concept and disqualifies that person from full social acceptance.

HIV/AIDS in Africa

HIV/AIDS is the leading cause of death in Africa. More than two-thirds of people with HIV live in sub-Saharan Africa, where one in 20 adults (ages 15 to 49) has HIV and the HIV prevalence rate among women is more than double that of men (Kaiser Family Foundation 2010). In nine countries, more than one in ten adults is estimated to be HIV-positive. The highest prevalence rate is in Swaziland, where one in four adults is living with HIV.

The HIV/AIDS epidemic creates an enormous burden for the limited health care resources of poor countries. Economic development is threatened by the HIV epidemic, which diverts national funds to health-related needs and reduces the size of a nation's workforce. There are 16.6 million AIDS orphans (one or both parents have died of AIDS); most live in sub-Saharan Africa (Kaiser Family Foundation 2010).

HIV/AIDS in the United States

Although heterosexual transmission is the predominant mode of HIV transmission worldwide, in the United States, more than half (54 percent) of HIV diagnoses (in 2008) involved transmission through male-to-male sexual contact (Centers for Disease Control and Prevention 2010b). Among men with HIV/AIDS, the primary mode of transmission is through male-to-male sexual contact, followed by heterosexual contact and injection drug use. Among women with HIV/AIDS, the primary mode of transmission is through heterosexual contact, followed by injection drug use. HIV infection rates among African Americans are seven times higher than among whites, and rates for Latinos are three times higher. Nearly half (48 percent) of people living with HIV/AIDS in the United States are black/African American.

Childhood obesity is becoming more common throughout the developed world. At 8 years of age, Connor McCreaddie, shown here with his mother, weighed 218 pounds.

Scott Heppell/AP Photo

The Growing Problem of Obesity

Obesity is a major health problem throughout the industrialized world and is increasing in developing countries. In the United States, between 1988 to 1994 and 2007 to 2008, the prevalence of obesity almost doubled, from 11 percent to 20 percent among children 6 to 11 years of age, increased from 11 percent to 18 percent among adolescents 12 to 19 years of age, and rose from 22 percent to 34 percent among adults 20 years of age and over (National Center for Health Statistics 2011). Two-thirds of U.S. adults are either overweight or obese (based on body mass index, or BMI $\geq$ 30) and nearly one in five youths ages 6 to 19 are obese. Obesity, which can lead to heart disease, stroke, some cancers, diabetes, and other health problems, is the second biggest cause of preventable deaths in the United States (second only to tobacco use) (Stein & Connolly 2004). Researchers predict that obesity will shorten the average U.S. life expectancy by at least two to five years over the next 50 years, reversing the mostly steady increase in life expectancy that has occurred over the past two centuries (Olshansky et al. 2005).

Two-thirds of U.S. adults are either overweight or obese.

Although genetics and certain medical conditions contribute to obesity, two social and lifestyle factors that play a major role in the obesity epidemic are patterns of food consumption and lack of physical activity. In the last few decades, consumption of snacks, sugary drinks, "fast food" (often high in fat and calories), and supersized portions has increased, while levels of physical activity have decreased.

> **What Do You Think?** In 2007, 8-year-old Connor McCreaddie of the United Kingdom weighed 218 pounds. A child protection conference was held to determine whether Connor should be removed from his home and placed into foster care, where his diet would be carefully controlled. This decision involved determining whether Connor's mother was abusing him by providing Connor with excessive high-calorie food. In this case, Connor's mother was allowed to keep custody of her son (*Guardian* 2007). In a similar case in North Carolina, a mother whose 7-year-old son weighed more than 250 pounds reported that the local Division of Social Services threatened to take her child away if he did not lose weight (Associated Press 2007). Do you think that severely obese children should be considered as victims of child abuse and taken from their parents and placed in foster care?

Obesity is also related to socioeconomic status. In less developed countries, poverty is associated with undernutrition and starvation. In the United States, however, being poor increases one's risk of being overweight or obese. High-calorie processed foods tend to be more affordable than fresh vegetables, fruits, and lean meats or fish. Residents of low-income areas often lack access to large grocery stores that sell a variety of foods, and instead rely on neighborhood fast-food chains and convenience stores that sell mostly high-calorie processed food.

> Mental illness is a "hidden epidemic" because the shame and embarrassment associated with mental problems discourage people from acknowledging and talking about them.

Mental Illness: The Hidden Epidemic

What it means to be mentally healthy varies across cultures. In the United States, **mental health** is defined as the successful performance of mental function, resulting in productive activities, fulfilling relationships with other people, and the ability to adapt to change and to cope with adversity (U.S. Department of Health and Human Services 2001). **Mental illness** refers collectively to all mental disorders, which are characterized by sustained patterns of abnormal thinking, mood (emotions), or behaviors that are accompanied by significant distress and/or impairment in daily functioning. The most common mental disorders are anxiety disorders, mood (depressive and bipolar) disorders, and impulse control disorders (Substance Abuse and Mental Health Services Administration 2010a).

Mental illness is a "hidden epidemic" because the shame and embarrassment associated with mental problems discourage people from acknowledging and talking about them. Negative stereotypes of people with mental illness contribute to its stigma. People are twice as likely today than they were in the 1950s to believe that people with mental illness are violent. The reality is that the vast majority of people with mental illness are not violent, though they are 2.5 times more likely to be victims of violence than members of the general population (Dingfelder 2009a).

Extent and Impact of Mental Illness

Prevalence rates of mental illness in the United States are based primarily on results from the National Survey on Drug Use and Health, which does not include people who are homeless, in an institution (i.e., jail or prison, hospital, or long-term care facility), or

mental health The successful performance of mental function, resulting in productive activities, fulfilling relationships with other people, and the ability to adapt to change and to cope with adversity.

mental illness Refers collectively to all mental disorders, which are characterized by sustained patterns of abnormal thinking, mood (emotions), or behaviors that are accompanied by significant distress and/or impairment in daily functioning.

CLASSIFICATION	DESCRIPTION
Anxiety disorders	Disorders characterized by anxiety that is manifest in phobias, panic attacks, or obsessive-compulsive disorder
Dissociative disorders	Problems involving a splitting or dissociation of normal consciousness, such as amnesia and multiple personality
Disorders first evident in infancy, childhood, or adolescence	Disorders including mental retardation, attention-deficit/hyperactivity, and stuttering
Eating or sleeping disorders	Disorders including anorexia, bulimia, and insomnia
Impulse control disorders	Problems involving the inability to control undesirable impulses, such as kleptomania, pyromania, and pathological gambling
Mood disorders	Emotional disorders such as major depression and bipolar (manic-depressive) disorder
Organic mental disorders	Psychological or behavioral disorders associated with dysfunctions of the brain caused by aging, disease, or brain damage (such as Alzheimer's disease)
Personality disorders	Maladaptive personality traits that are generally resistant to treatment, such as paranoid and antisocial personality types
Schizophrenia and other psychotic disorders	Disorders with symptoms such as delusions or hallucinations
Somatoform disorders	Psychological problems that present themselves as symptoms of physical disease, such as hypochondria
Substance-related disorders	Disorders resulting from abuse of alcohol and/or drugs, such as barbiturates, cocaine, or amphetamines

© Cengage Learning 2013

in the military. Among the noninstitutionalized civilian population, in any given year, one in five U.S. adults experiences mental illness; 4.8 percent of U.S. adults experience serious mental illness. In 2009, 12 percent of the U.S. youth population (ages 12 to 17) received treatment or counseling for behavioral or emotional problems. The most common reasons for receiving services among youths was feeling depressed (46.7 percent) (Substance Abuse and Mental Health Services Administration 2010b).

An annual nationwide survey of first-year college students in 2010 found that the percentage of students reporting that their emotional health was in the "highest 10 percent" or "above average" when compared to their peers was at the lowest point since researchers began asking the question in 1985 (Pryor et al. 2010). This survey found that more students reported having attention-deficit/hyperactivity disorder (ADHD) (5.0 percent) or a psychological disorder (3.8 percent) than any other disability or medical condition. A survey of 320 college counseling centers found that one in ten college students sought counseling in the past year; 44 percent of these students had "severe psychological problems" (Gallagher 2010). To assess your attitudes toward seeking professional psychological help, see this chapter's *Self and Society* feature.

Untreated mental illness can lead to poor educational achievement, lost productivity, unsuccessful relationships, significant distress, violence and abuse, incarceration, unemployment, homelessness, and poverty. Half of students identified as having emotional disturbances drop out of high school (Gruttadaro 2005). Individuals with a mental disorder, most commonly a depressive or substance abuse disorder, commit most suicides (more than 90 percent) in the United States (National Institute of Mental Health 2008).

What Do You Think? The age group with the highest rate of "serious psychological distress" is between ages 18 and 25. Why do you think this is so?

For each of the following statements, indicate your level of agreement according to the following scale: agree, partly agree, partly disagree, or disagree.

1. If I believed I was having a mental breakdown, my first inclination would be to get professional attention. (S) _____
2. The idea of talking about problems with a psychologist strikes me as a poor way to get rid of emotional conflicts. (R) _____
3. If I were experiencing a serious emotional crisis at this point in my life, I would be confident that I could find relief in psychotherapy. (S) _____
4. There is something admirable in the attitude of a person who is willing to cope with conflicts and fears without resorting to professional help. (S) _____
5. I would want to get psychological help if I were worried or upset for a long period of time. (S) _____
6. I might want to have psychological counseling in the future. (S) _____
7. A person with an emotional problem is not likely to solve it alone; the person is likely to solve it with professional help. (S) _____
8. Considering the time and expense involved in psychotherapy, it would have doubtful value for a person like me. (R) _____
9. A person should work out his or her own problems; getting psychological counseling would be a last resort. (R) _____
10. Personal and emotional troubles, like many things, tend to work out by themselves. (R) _____

Scoring: Straight items (S) are scored 3, 2, 1, 0, and reversal items (R) are scored 0, 1, 2, 3, respectively, for the response alternatives *agree, partly agree, partly disagree,* and *disagree.* Add the responses for each item to find your total score, which will range from 0 to 30. The higher your score, the more positive your attitudes are toward seeking professional psychological help; the lower your score, the more negative your attitudes are toward seeking professional psychological help.

In a study of 389 undergraduate students (primarily 18-year-old freshmen), the average score on this scale was 17.45 (Fischer & Farina 1995). Women were more likely to have higher scores, reflecting more positive attitudes. For women in the study, the average score was 19.08, compared with 15.46 for men.

Source: Reprinted by permission of the American College Personnel Association.

Causes of Mental Illness

Stigma surrounding mental illness is partly due to misconceptions about their causes, such as the misconception that mental illness is caused by personal weakness, or results from engaging in immoral behavior. In some cultures, people with mental illness are viewed as being possessed by evil spirits or supernatural forces. In some parts of Somalia, people with severe mental problems are subjected to the "hyena cure" that involves dropping the mentally ill person into a pit with one or more hyenas that have been starved of food. It is believed the hyenas will scare away the evil spirits that inhabit the person (World Health Organization 2010b).

Scientific and medical explanations of mental illness focus on genetic, neurological conditions. Social and environmental influences, such as poverty, relationship abuse, job loss, divorce, the death of a loved one, devastation from a natural disaster such as flood or earthquake, the onset of illness or disabling injury, and the trauma of war, also can trigger mental health problems. Broadly speaking, "mental health is impacted detrimentally when civil, cultural, economic, political, and social rights are infringed" (World Health Organization 2010b, p. xxvi).

Social Factors and Lifestyle Behaviors Associated with Health and Illness

Health problems are linked to lifestyle behaviors such as excessive alcohol consumption, cigarette smoking, unprotected sexual intercourse, and inadequate consumption of fruits and vegetables (see this chapter's *Social Problems Research Up Close* feature). However, health and illness are also affected by social factors such as globalization, social class and poverty, education, race/ethnicity, and gender.

The National College Health Assessment is a survey developed by the American College Health Association to assess the health status of college students across the country.

Sample and Methods

A total of 42 postsecondary institutions self-selected to participate in the Fall 2010 National College Health Assessment. Data from institutions that did not survey all students or use random sampling techniques were not used, yielding a final sample of 30,093 students at 39 campuses (American College Health Association 2011). The response rate was 78 percent for schools using paper surveys and 19 percent for schools conducting web-based surveys. The survey contains questions that assess student health status and health problems, risk and protective behaviors, and health impediments to academic performance.

TABLE 1 Top Six Self-Reported Physical Health Problems Students Had Diagnosed or Treated in the Past Year

HEALTH PROBLEM	RANK	PERCENTAGE
Allergies	1	22.0
Sinus infection	2	17.1
Back pain	3	12.0
Strep throat	4	11.2
Asthma	5	9.1
Urinary tract infection	6	9.0

Adapted from: American College Health Association. 2011. American College Health Association National College Health Assessment Fall 2010 reference group data report. Baltimore: American College Health Association.

Selected Findings and Conclusions

- *Most commonly reported health problems.* The most commonly reported physical health problems of college students are allergies and sinus infections (see Table 1 within this box).

- *Alcohol, tobacco, and marijuana use.* The majority of college students (60 percent) reported having consumed alcohol in the past

Globalization and Health

Globalization, broadly defined as the growing economic, political, and social interconnectedness among societies throughout the world, has had both positive and negative effects on health. On the positive side, globalized communications technology is used to monitor and report on outbreaks of disease, disseminate guidelines for controlling and treating disease, and share medical knowledge and research findings (Lee 2003).

On the negative side, aspects of globalization such as increased travel and the expansion of trade and transnational corporations have been linked to a number of health problems. Increased business travel and tourism facilitates the spread of infectious disease. In just the first two months of the swine flu pandemic of 2009, the disease spread to infect nearly 600,000 people in more than 70 countries around the world.

Increased international trade has expanded the range of goods available to consumers, but at a cost to global health. The increased transportation of goods by air, sea, and land contributes to pollution caused by the burning of fossil fuels. In addition, the expansion of international trade of harmful products such as tobacco, alcohol, and sugary drinks and processed or "fast" foods is associated with a worldwide rise in cancer, heart disease, stroke, obesity, and diabetes (Hawkes 2006; World Health Organization 2002).

Poverty and Health

globalization The growing economic, political, and social interconnectedness among societies throughout the world.

Poverty is associated with malnutrition, hazardous housing and working conditions, unsafe water and sanitation, and lack of access to medical care (see also Chapter 6). In low-income countries, for example, people with cancer lack access to medical treatment and

30 days, 15 percent said that they used cigarettes, and 14 percent had used marijuana.

- *Sexual health and condom use.* Two-thirds (66 percent) of college students reported that they had at least one sexual partner (someone with whom they had oral sex or vaginal or anal intercourse) in the past year. Among sexually active students, 54 percent said they used a condom the last time they had vaginal intercourse and 29 percent used a condom during anal intercourse; 16 percent reported using (or their partner used) emergency contraception (the "morning-after pill") within the past year.
- *Nutrition, exercise, and weight.* Only 5 percent of college students reported that they ate the recommended five or more servings of fruits and vegetables daily; 57 percent reported that they exercised moderately for at least 30 minutes one to four days in the last seven days. One-third of college students are either overweight or obese.

TABLE 2 Percentage of Students Experiencing Selected Mental Health Difficulties in the Past 12 Months

MENTAL HEALTH DIFFICULTY	PERCENTAGE
Felt so depressed it was difficult to function	13.7
Felt overwhelming anxiety	17.2
Seriously considered suicide	3.7
Felt things were hopeless	19.8

Adapted from: American College Health Association. 2011. American College Health Association National College Health Assessment Fall 2010 reference group data report. Baltimore: American College Health Association. Note: Some percentages are rounded.

- *Mental health.* Table 2 within this box presents the percentage of students who reported experiencing various mental health difficulties in the past 12 months.

Discussion

The results of the National College Health Assessment reveal the extent and types of health problems and risk behaviors of college students. These data can be used to help colleges and universities design and implement health services that meet the needs of college students.

therefore have lower survival rates compared with cancer patients in high-income countries (Farmer et al., 2010). In the United States, low socioeconomic status is associated with higher incidence and prevalence of health problems, disease, and death. Poverty is associated with higher rates of health-risk behaviors such as smoking, drinking alcohol, being overweight, and being physically inactive. The poor are also exposed to more environmental health hazards, and have unequal access to and use of medical care (Lantz et al. 1998). In the United States, adults living below the poverty level are five times more likely to report their health as being fair or poor compared with adults with family income at least four times the poverty level (National Center for Health Statistics 2011). In addition, members of the lower class are subjected to the most stress and have the fewest resources to cope with it (Cockerham 2007). Stress has been linked to a variety of physical and mental health problems, including high blood pressure, cancer, chronic fatigue, and substance abuse. U.S. adults living below the poverty threshold are more than eight times as likely to report serious psychological distress as adults in families with an income at least four times the poverty level (National Center for Health Statistics 2011).

Just as poverty contributes to health problems, health problems contribute to poverty. Physical and mental health problems can limit one's ability to pursue education or vocational training and to find or keep employment. The high cost of health care not only deepens the poverty of people who are already barely getting by but also can financially devastate middle-class families. Later in this chapter, we look more closely at the high cost of health care and its consequences for individuals and families.

Education and Health

Although economic resources are important influences on health, education seems to be the strongest single predictor of good health. A *New York Times* report on the link between education and lifespan concluded that: "The one social factor that researchers

Education seems to be the strongest single predictor of good health.

agree is consistently linked to longer lives in every country where it has been studied is education" (Kolata 2007, p. A1). Individuals with low levels of education are more likely to engage in health-risk behaviors such as smoking and heavy drinking. Women with less education are less likely to seek prenatal care and are more likely to smoke during pregnancy, which helps explain why low birth weight and infant mortality are more common among children of less educated mothers (Children's Defense Fund 2000).

In some cases, lack of education means that individuals do not know about health risks or how to avoid them. A national survey in India found that only 18 percent of illiterate women had heard of AIDS, compared with 92 percent of women who had completed high school (Ninan 2003).

Gender and Health

In many societies, women and girls are viewed and treated as socially inferior, and are denied equal access to nutrition and health care. Women's health is also affected by traditional family responsibilities. Women do most of the food preparation and, where solid fuels are used for cooking indoors, women are more likely than men to suffer from respiratory problems due to exposure to indoor air pollution. Gender inequality also exposes women to domestic and sexual exploitation, increasing women's risk of physical injury, and of acquiring HIV and other sexually transmissible infections (World Health Organization 2009b). In the United States, at least one in three women has been beaten, coerced into sex, or abused in some way—most often by someone the woman knows (Alan Guttmacher Institute 2004). "Although neither health care workers nor the general public typically thinks of battering as a health problem, it is a major cause of injury, disability, and death among American women, as among women worldwide" (Weitz 2010, p. 55).

In the United States before the 20th century, the life expectancy of U.S. women was shorter than that of men because of the high rate of maternal mortality that resulted from complications of pregnancy and childbirth. But today, life expectancy of U.S. women (80.4 years) is greater than that of U.S. men (75.4 years) (National Center for Health Statistics 2011). Lower life expectancy for U.S. men is due to a number of factors. Men tend to work in more dangerous jobs than women, such as agriculture, construction, and the military. In addition, "beliefs about masculinity and manhood that are deeply rooted in culture . . . play a role in shaping the behavioral patterns of men in ways that have consequences for their health" (Williams 2003, p. 726). Men are socialized to be strong, independent, competitive, and aggressive and to avoid expressions of emotion or vulnerability that could be construed as weakness. These male gender expectations can lead men to take actions that harm themselves or to refrain from engaging in health-protective behaviors. For example, socialization to be aggressive and competitive leads to risky behaviors (such as dangerous sports, fast driving, and violence) that contribute to men's higher risk of injuries and accidents. Men are more likely than women to smoke cigarettes and to abuse alcohol and drugs but are less likely than women to visit a doctor and to adhere to medical regimens (Williams 2003).

Regarding mental health, a national survey found that U.S. females were more likely than males to experience severe psychological distress (Substance Abuse and Mental Health Services Administration 2010b). However, rates of mental illness are similar for women and men, although they differ in the types of mental illness they experience; women have higher rates of mood and anxiety disorders, and men have higher rates of personality and substance-related disorders. Although women are more likely to attempt suicide, men are more likely to succeed at it because they use deadlier methods. Biological differences can account for some gender differences in mental health. For

Robin Nelson/PhotoEdit

Physical abuse is a major cause of injury, disability, and death among women.

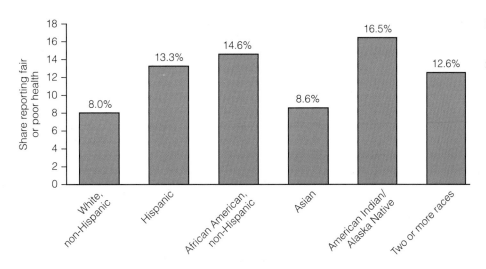

Figure 2.2: **Fair or Poor Health Status by Race/Ethnicity**
Source: James et al. 2007

example, hormonal changes after childbirth can result in some women suffering from postpartum depression. Gender differences in mental health can also be attributed to gender roles. For example, the unequal status of women and the strain of doing the majority of housework and child care may predispose women to experience greater psychological distress.

Race, Ethnicity, and Health

U.S. racial and ethnic minorities are more likely than non-Hispanic whites to rate their health as fair or poor (see Figure 2.2). Non-Hispanic black men and women have higher rates of hypertension, heart disease, and stroke and lower life expectancy compared with their white counterparts (see Table 2.4). But a surprising report that is the first to calculate U.S. Hispanic life expectancy found that U.S. Hispanics outlive whites by more than two years and blacks by more than seven (Arias 2010). One theory for this "Hispanic paradox"—longevity among a population with a large share of poor, undereducated members—is that Hispanics who immigrate to the United States are among the healthiest from their countries.

The highest rates of obesity are among Black Americans, followed by Hispanics. Infants born to black women are 2.4 times more likely to die than infants born to non-Hispanic white women (Centers for Disease Control and Prevention 2011). Infant mortality rates among some racial/ethnic minorities (U.S. Asians, Mexicans, Central and South Americans, and Cubans) are lower than that of non-Hispanic whites because, even though the minorities have lower socioeconomic status, they have strong networks of family and community support.

Health disparities are largely due to substantial racial/ethnic differences in income, education, housing, and access to health care. Racial and ethnic minorities are less likely than whites to have health insurance and so are less likely to receive

TABLE 2.4 Life Expectancy in the United States by Sex and Race/Hispanic Origin*							
ALL RACES		**NON-HISPANIC BLACK**		**NON-HISPANIC WHITE**		**HISPANIC**	
FEMALE	MALE	FEMALE	MALE	FEMALE	MALE	FEMALE	MALE
80.4	75.4	76.8	70.0	80.8	75.9	83.1	77.9

Source: Arias 2010.

*For individuals born in 2006.

preventive services (such as colon cancer screening), medical treatment for chronic conditions, and prenatal care. The poorer health of minorities is also related to the fact that minorities are more likely than whites to live in environments where they are exposed to hazards such as toxic chemicals and other environmental hazards (see also Chapter 13).

Health disparities are sometimes explained by differences in lifestyle behaviors. Compared with white Americans, Native Americans/Alaskan Natives have the highest death rate from motor vehicle crashes, because they have the highest rates of alcohol-impaired driving as well as seatbelt nonuse (Centers for Disease Control and Prevention 2011). Black Americans have the highest rate of obesity in part because of racial differences in eating behavior. But lifestyle behaviors are often influenced by social factors. Blacks, on average, have lower incomes than whites, and so are more likely to choose foods that are more affordable, and junk foods and fast foods tend to be cheaper than fruits, vegetables, and lean sources of protein. However, racial disparities in obesity persist even after controlling for family income (Centers for Disease Control and Prevention 2011).

Another factor that contributes to racial/ethnic health disparities is stress resulting from prejudice and discrimination. In an attempt to explain why high blood pressure afflicts more African Americans (40 percent) than whites (28 percent), a researcher at the University of Florida studied disparities in blood pressure among adults in Puerto Rico, where people are viewed according to their skin color as being *blanco* (white), *trigueno* (intermediate), or *negro* (black). People who are *blanco* belong to a privileged class; people who are *negro* are the stigmatized minority group. Mulligan found that people who were both classified as *negro* and had high incomes or education had the highest blood pressure. "That seems a little unexpected, because being better off economically is often associated with better health. . . . But these people often spoke about frustrating daily interactions, where others treated them badly or looked down on them as if they were poor" (Mulligan, quoted in Fischman 2010). This frustration led to stress, and stress produces high blood pressure.

Regarding mental health, research finds no significant difference among races in their overall rates of mental illness (Cockerham 2007). Differences that do exist are often associated more with social class than with race or ethnicity. However, some studies suggest that minorities have a higher risk for mental disorders, such as anxiety and depression, in part because of racism and discrimination, which adversely affect physical and mental health (U.S. Department of Health and Human Services 2001). Minorities also have less access to mental health services, are less likely to receive needed mental health services, often receive lower-quality mental health care, and are underrepresented in mental health research (U.S. Department of Health and Human Services 2001).

Problems in U.S. Health Care

The World Health Organization (2000) did an analysis of the world's health systems and found that, although the United States spends a higher portion of its gross domestic product (GDP) on health care than any other country, it ranks 37 out of 191 countries according to its performance. The analysis concluded that France provides the best overall health care among major countries, followed by Italy, Spain, Oman, Austria, and Japan. A more recent comparison of health care in 30 industrialized countries found that health care spending in the United States is significantly higher than in other industrialized countries, both per capita and as a percent of GDP, and yet the nation ranks in the bottom quartile in life expectancy among Organisation for Economic Co-operation and Development (OECD) countries (Anderson & Squires 2010). In another study of six countries—Australia, Canada, Germany, New Zealand, the United Kingdom, and the United States—the United States ranked last on dimensions of access, patient safety, efficiency, and equity (Davis et al. 2007).

After presenting a brief overview of U.S. health care, we address some of the major health care problems in the United States—inadequate health insurance coverage, the high cost of medical care and insurance, and inadequate mental health care.

U.S. Health Care: An Overview

In the United States, there is no one health care system; rather, health care is offered through various private and public means (see Figure 2.3).

In traditional health insurance plans, the insured choose their health care providers, whose fees are reimbursed by the insurance company. Insured individuals typically pay an out-of-pocket "deductible" (usually ranging from a few hundred to a thousand dollars or more per year per person) as well as a percentage of medical expenses (e.g., 20 percent) until a maximum out-of-pocket expense amount is reached (after which insurance will cover 100 percent of medical costs up to a limit). *Health maintenance organizations (HMOs)* are group plans in which people pay a monthly premium for comprehensive health care services, with an emphasis on preventive health care. *Preferred provider organizations (PPOs)* sell group health insurance to employers who agree to send their employees to certain health care providers or hospitals in return for cost discounts. Most insurance companies control costs through **managed care,** which involves monitoring and controlling the decisions of health care providers. The insurance company may, for example, require doctors to receive approval before they can hospitalize a patient, perform surgery, or order an expensive diagnostic test.

Medicare **Medicare** is funded by the federal government and reimburses the elderly and people with certain disabilities for their health care. Individuals contribute payroll taxes to Medicare throughout their working lives and generally become eligible for Medicare when they reach 65, regardless of their income or health status. Medicare consists of four separate programs: Part A is hospital insurance for inpatient care, which is free, but enrollees may pay a deductible and a co-payment. Part B is a supplementary medical insurance program, which helps pay for physician, outpatient, and other services. Part B is voluntary and is not free; enrollees must pay a monthly premium as well as a co-payment for services. Medicare does not cover long-term nursing home care, dental care, eyeglasses, and other types of services, which is why many individuals who receive Medicare also enroll in Part C, which allows beneficiaries to purchase private supplementary insurance that receives payments from Medicare. Part D is an outpatient drug benefit that is voluntary and requires enrollees to pay a monthly premium, meet an annual deductible, and pay coinsurance for their prescriptions.

Medicaid and SCHIP **Medicaid,** which provides health care coverage for the poor, is jointly funded by the federal and state governments. Eligibility rules and benefits vary from state to state, and, in many states, Medicaid provides health care only for those who are well below the federal poverty level. The **State Children's Health Insurance Program (SCHIP)** provides health coverage to children without insurance, many of whom come from families with income too high to qualify for Medicaid but too low to afford private health insurance. Under this initiative, states receive matching federal funds to provide medical insurance to children without insurance.

Workers' Compensation **Workers' compensation** (also known as *workers' comp*) is an insurance program that provides medical and living expenses for people with work-related injuries or illnesses. Employers pay a certain amount into their state's workers' compensation insurance pool, and workers injured on the job can apply to that pool for

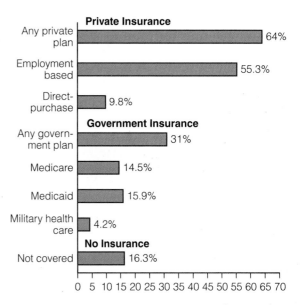

Figure 2.3: Coverage by Type of Health Insurance, 2010
Source: DeNavas-Walt et al. 2011.

managed care Any medical insurance plan that controls costs through monitoring and controlling the decisions of health care providers.

Medicare A federally funded program that provides health insurance benefits to the elderly, disabled, and those with advanced kidney disease.

Medicaid A public health insurance program, jointly funded by the federal and state governments, that provides health insurance coverage for the poor who meet eligibility requirements.

State Children's Health Insurance Program (SCHIP) A public health insurance program, jointly funded by the federal and state governments, that provides health insurance coverage for children whose families meet income eligibility standards.

workers' compensation Also known as workers' comp, an insurance program that provides medical workers' compensation and living expenses for people with work-related injuries or illnesses.

medical expenses and for compensation for work days lost. In exchange for that benefit, workers cannot sue their employers for damages. However, not all employers acquire workers' compensation insurance, even in states where it is legally required. Further, many employees with work-related illness or injuries do not apply for workers' compensation benefits because (1) they fear getting fired for making a claim, (2) they are not aware that they are covered by workers' comp, and/or (3) the employer offers incentives (i.e., bonuses) to employees when no workers' comp claims are filed in a given period of time (Sered & Fernandopulle 2005). Even when employees file a workers' comp claim, the coverage the employee receives rarely covers the cost of the employee's injury or illness. "The typical scenario . . . is one in which the workers' comp insurance company delays accepting and paying the worker's claim. In the meantime, the injured employee racks up medical and other bills and then, in desperation, accepts a lump sum monetary settlement that does not come close to covering medical expenses or replacing lost wages" (Sered & Fernandopulle 2005, p. 94).

Military Health Care Military health care includes Civilian Health and Medical Program of the Uniformed Services (CHAMPUS), Civilian Health and Medical Program of the Department of Veterans Affairs (CHAMPVA), and care provided by the Department of Defense and the Department of Veterans Affairs. A series of 2007 *Washington Post* reports brought attention to the abysmal conditions at military medical facilities and Veterans Administration (VA) hospitals around the country, describing medical facilities infested with mice and mold and characterized by "indifferent, untrained staff; lost paperwork; medical appointments that drop from computers; and long waits for consultations" (Hull & Priest 2007). Other reports of military medical facilities described peeling paint, asbestos, overflowing trash, fruit fly infestations, no nurses, and lack of blankets and linens.

Many veterans have complained that they have not received the benefits they deserve or that they have had long waits to get benefits. Mental health care for military personnel and military veterans is also inadequate. Only about half of veterans with post-traumatic stress disorder seek treatment, either because of the stigma of having mental health problems or because help was not available due to a shortage of military mental health professions (Dingfelder 2009b).

Inadequate Health Insurance Coverage

Many other countries, including 31 European countries and Canada, have national health insurance systems that provide **universal health care**—health care to all citizens. National health insurance is typically administered and paid for by government and funded by taxes or social security contributions. Despite differences in how national health insurance works in various countries, typically, the government (1) directly controls the financing and organization of health services, (2) directly pays providers, (3) owns most of the medical facilities (Canada is an exception), (4) guarantees universal access to health care, (5) allows private care for individuals who are willing to pay for their medical expenses, and (6) allows individuals to supplement their national health care with private insurance as an upgrade to a higher class of service and a larger range of services (Cockerham 2007; Quadagno 2004). Any rationing of health care in countries with national health insurance is done on the basis of medical need, not ability to pay.

Unlike other industrialized countries that provide universal health care coverage, 16.3 percent of Americans (49.9 million people) did not have health insurance coverage in 2010 (DeNavas-Walt et al. 2011). Those who have health insurance have no guarantee that their coverage will continue when they most need it, as health insurance companies have frequently denied or limited treatment, covered only less expensive drugs, and even terminated policies to avoid paying claims (Orr 2009).

Disparities in Health Insurance Coverage Non-Hispanic whites are more likely than racial and ethnic minorities to have health insurance. Hispanics have the highest rate of uninsured, with nearly a third of Hispanics lacking health insurance in 2010 (DeNavas-Walt et al. 2011).

universal health care A system of health care, typically financed by the government, that ensures health care coverage for all citizens.

Of all age groups, young adults aged 19 to 25 are the least likely to have health insurance. In 2010, nearly one in three (29.7 percent) young adults was uninsured (DeNavas-Walt et al. 2011).

Of all age groups, young adults aged 19 to 25 are the least likely to have health insurance.

Employed individuals and individuals with higher incomes are more likely to have health insurance. However, employment is no guarantee of health care coverage; in 2010, 15 percent of full-time workers were uninsured (DeNavas-Walt et al. 2011). Some businesses do not offer health benefits to their employees; some employees are not eligible for health benefits because of waiting periods or part-time status, and some employees who are eligible may not enroll in employer-provided health insurance because they cannot afford their share of the premiums.

Inadequate Insurance for the Poor Many Americans believe that Medicaid and SCHIP—public health insurance programs for the poor—cover all low-income children, adults, and families. But Medicaid eligibility levels are set so low that many low-income adults are not eligible. Because Medicaid makes up the largest portion of state budgets, states struggling with budget deficits have cut back on reimbursement rates, causing some doctors and hospitals to stop accepting Medicaid patients. Other states have cut the types of services Medicaid covers (Vestal 2011).

Consequences of Inadequate Health Insurance An estimated 45,000 deaths per year in the United States are attributable to lack of health insurance (Park 2009).

Individuals who lack health insurance are less likely to receive preventive care, are more likely to be hospitalized for avoidable health problems, and are more likely to have disease diagnosed in the late stages. The uninsured are three times more likely than the insured to be unable to pay for basic necessities because of their medical bills (Kaiser Commission on Medicaid and the Uninsured 2010).

An estimated 45,000 deaths per year in the United States are attributable to lack of health insurance.

Because most health care providers do not accept patients who do not have insurance, many individuals without insurance resort to using the local hospital emergency room (Scal & Town 2007). The federal Emergency Medical Treatment and Active Labor Act requires hospitals to assess all patients who come to their emergency rooms to determine whether an emergency medical condition exists and, if it does, to stabilize patients before transferring them to another facility. Hospital patients without insurance are almost always billed at a much higher cost than the prices negotiated by insurance companies.

Individuals who lack dental insurance commonly have untreated dental problems, which can lead to or exacerbate other health problems:

> Because they affect the ability to chew, untreated dental problems tend to exacerbate conditions such as diabetes or heart disease. . . . Missing and rotten teeth make it painful if not impossible to chew fruits, whole grain foods, salads, or many of the fiber-rich foods recommended by doctors and nutrition experts. (Sered & Fernandopulle 2005, pp. 166–67)

In their book *Uninsured in America*, Sered and Fernandopulle (2005) described one interviewee who "covered her mouth with her hand during our entire interview because she was embarrassed about her rotting teeth" and another interviewee "used his pliers to yank out decayed and aching teeth" (p. 166). The authors note that "almost every time we asked interviewees what their first priority would be if the president established universal health coverage tomorrow, the immediate answer was 'my teeth'" (p. 166).

The High Cost of Health Care

Total U.S. spending on health care in 2009 was $2.5 trillion, which translates into $8,086 per person and more than 17.6 percent of gross domestic product (GDP)—far more than any other industrialized nation (Centers for Medicare and Medicaid Services 2011). Yet virtually every other wealthy nation has better health outcomes, as measured by life expectancy and infant mortality.

Several factors have contributed to escalating medical costs. These include increased longevity; and the high cost of hospital services and medical technology, prescription drugs, and health insurance.

Increased Longevity
Because of improved sanitation and medical advances, people are living longer today than in previous generations. People older than age 65 use medical services more than younger individuals and are also more likely to take prescription medicine on a daily basis.

Cost of Hospital Services and Medical Technology
Compared to other OECD countries, hospitalizations in the United States are less frequent and shorter. Yet, spending per hospital discharge in the United States ($17,126) is 2.5 times higher than the OECD median ($6,867) (Anderson & Squires 2010). The United States has high rates of performing costly medical procedures involving advanced medical technology, such as coronary procedures (angioplasty, stenting, and cardiac catheterizations), knee replacements, and dialysis (Anderson & Squires 2010). Consider the advancements in the treatment of preterm babies, for which very little could be done in 1950. By 1990, special ventilators and neonatal intensive care became standard treatment for preterm babies in the United States (Kaiser Family Foundation 2007).

Cost of Prescription Drugs
Per capita spending on pharmaceuticals is higher in the United States than in any other OECD country (Anderson & Squires 2010). The United States pays 81 percent more for patented brand-name prescription drugs than Canada and six western European nations (Sager & Socolar 2004). The pharmaceutical industry, which is among the most profitable industries in the United States, argues that U.S. drug prices are high because of the high cost of researching and developing (R & D) new drugs. But a critical analysis reveals that the industry purposely overestimates R & D costs to justify their high drug prices (Light & Warburton 2011). Further, most large drug companies pay substantially more for marketing, advertising, and administration than for research and development (Families USA 2007).

Cost of Health Insurance
In 2010, the average annual premiums for employer-sponsored coverage were $5,049 for an individual (worker's share = $899) and $13,770

Expensive medical technology contributes to high health care spending.

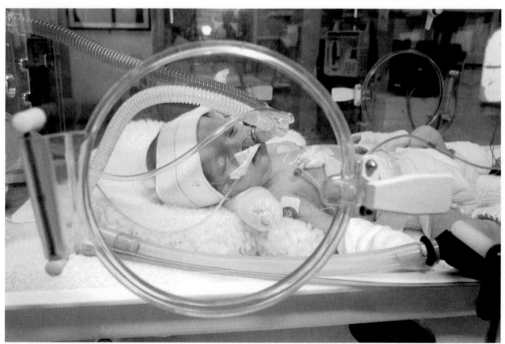

for a family (worker's share = $3,997). From 2000 to 2010, average premiums for family coverage rose 114 percent (Kaiser Family Foundation/HRET 2010). U.S. health insurance is costly largely because of high administrative expenses, which per capita are six times higher than in western European nations (The National Coalition on Health Care 2009). Harrison (2008) explains the reason:

> The United States has the most bureaucratic health care system in the world, including over 1,500 different companies, each offering multiple plans, each with its own marketing program and enrollment procedures, its own paperwork and policies, its CEO salaries, sales commissions, and other nonclinical costs—and, of course, if it is a for-profit company, its profits.

Consequences of the High Cost of Health Care for Individuals and Families When an uninsured driver hit 12-year-old Candice Jackson while she was getting off a school bus, she spent four months in the hospital and incurred about $90,000 in uncovered medical bills. Her mother needed two knee replacements, adding another $20,000 to the family's medical bills. A few years later, at age 16, Candice swerved off the road to avoid hitting a deer and sustained a head injury that required brain surgery. In anticipation of the medical bills from Candice's latest mishap, Candice's dad, Lanny Jackson, who works in the service center of a car dealership, felt forced to file for bankruptcy (Springen 2006).

Lanny Jackson is not alone. One study found that medical bills, as well as income lost due to illness, contributed to two-thirds of all bankruptcies in 2007 (Himmelstein et al. 2009). Most medical debtors were well-educated homeowners with middle-class jobs, and three-fourths had health insurance. Having insurance does not guarantee that one is protected against financial devastation resulting from illness or injury, because even the insured typically must pay co-payments, deductibles, and exclusions. In addition, the link between coverage and employment means that insurance is often lost when it is needed the most—when workers lose their jobs because of medical problems. Although the Consolidated Omnibus Budget Reconciliation Act (COBRA) allows people to continue their insurance coverage when they lose a job, the premiums for a family typically exceed $1,000, whereas the average unemployment insurance payment is $1,425 (Orr 2009).

> Having insurance does not guarantee that one is protected against financial devastation resulting from illness or injury, because even the insured typically must pay co-payments, deductibles, and exclusions.

Many individuals forgo needed medicine and/or medical care when they cannot afford to pay for it (see Table 2.5). Forgoing medicine or medical care often exacerbates a medical condition, leading to even higher medical costs, or tragically, leading to death.

TABLE 2.5 Cutting Back on Medical Care Due to Cost

In the past 12 months, because of cost, have you or another family member living in your household . . . ?

	PERCENT SAYING "YES"
Relied on home remedies or over-the-counter (OTC) drugs instead of seeing a doctor	37%
Skipped dental care or checkups	35%
Put off or postponed getting health care you needed	31%
Skipped recommended medical test or treatment	27%
Not filled a prescription for medicine	26%
Cut pills in half or skipped doses of medicine	19%
Had problems getting mental health care	8%
Did ANY of the above	55%

Source: Adapted from Kaiser Family Foundation 2009a.

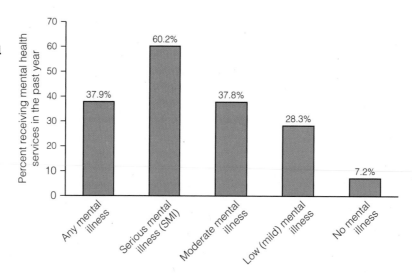

Figure 2.4: **Receipt of Mental Health Services among U.S. Adults, by Level of Mental Illness, 2009**
Source: Substance Abuse and Mental Health Services Administration 2010b.

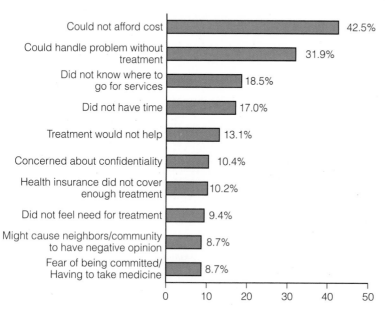

Figure 2.5: **Reasons for Not Receiving Mental Health Services in the Past Year among Adults with an Unmet Need for Mental Health Care, 2009**
Source: Substance Abuse and Mental Health Services Administration 2010b.

deinstitutionalization The removal of individuals with psychiatric disorders from mental hospitals and large residential institutions to outpatient community mental health centers.

Inadequate Mental Health Care

Since the 1960s, U.S. mental health policy has focused on reducing costly and often neglectful institutional care and on providing more services in the community. This movement, known as **deinstitutionalization,** has resulted in a significant decrease in the number of mental health facilities with 24-hour or residential treatment, the number of psychiatric treatment beds available, and in admissions to state and county mental hospitals in the last few decades. The most frequently used source of care for mental health problems has become primary care and general doctors and nurses. Other "nonspecialty" care providers include community health centers, schools, nursing homes, correctional institutions, and emergency rooms. In 2005, 8.1 percent of all visits to hospital emergency rooms involved a primary or secondary diagnosis of a mental health disorder (Substance Abuse and Mental Health Services Administration 2010a). This fragmented system of mental health care leaves many people with mental health problems to fall through the cracks. Among the 11 million U.S. adults with "serious mental illness" in 2009, 40 percent received no mental health services in the past year (Substance Abuse and Mental Health Services Administration 2010b) (see Figure 2.4). The most common reason why adults with an unmet need for mental health treatment did not receive it is that they "could not afford the cost" (see Figure 2.5).

Mental health services are often inaccessible, especially in rural areas. In most states, services are available from "9 to 5"; the system is "closed" in the evenings and on weekends when many people with mental illness experience the greatest need. Across the nation, people with severe mental illness end up in jails and prisons, homeless shelters, and hospital emergency rooms. Many children with untreated mental disorders drop out of school or end up in foster care or the juvenile justice system. Given the increasing growth of minority populations, another deficit in the mental health system is the inadequate number of mental health clinicians who speak the client's language and who are aware of cultural norms and values of minority populations (U.S. Department of Health and Human Services 2001).

Strategies for Action: Improving Health and Health Care

Two broad approaches to improving the health of populations are selective primary health care and comprehensive primary health care (Sanders & Chopra 2003). **Selective primary health care** focuses on interventions that target specific health problems, such as promoting condom use to prevent HIV infections and providing immunizations against childhood diseases to promote child survival. **Comprehensive primary health care** focuses on the broader social determinants of health, such as poverty and economic inequality, gender inequality, racial/ethnic discrimination, and environmental pollution. Many strategies to alleviate social problems in subsequent chapters of this textbook are also important elements to a comprehensive primary health care approach.

Improving Maternal and Child Health

Between 1990 and 2008, 147 countries had a decline in maternal mortality rates; 23 countries had an increase (World Health Organization 2010b). Over the same time period, the global under-5 mortality rate dropped from 90 deaths per 1,000 live births to 65 (UNICEF 2010).

Declines in maternal and child mortality are due to several factors. Between 1990 and 2008, the percentage of deliveries in developing countries that were attended by skilled health workers increased from 53 percent to 63 percent. Yet, in Africa and Southeast Asia, less than half of women receive skilled care during childbirth (World Health Organization 2010b). Contraceptive use among women in developing countries also increased from 52 percent in 1990 to 62 percent in 2008. But there is still an unmet need for family planning services: Nearly a quarter of women in Africa wanting to delay or stop childbearing are not using contraceptives (World Health Organization 2011b). Family planning reduces maternal mortality by reducing the number of unintended pregnancies and by enabling women to space births two to three years apart, which decreases infant mortality significantly (Murphy 2003).

Improvements in maternal and child health are also due to the decline in child marriages, as early childbirth is associated with higher health risks to women and infants. Over the last few decades, the proportion of women in the developing world who were married before age 18 declined from nearly half to about a third (UNICEF 2010). Yet, more than half of the girls in nine countries (most in Africa) are married before age 18 (International Center for Research on Women 2010). Some women's health advocates are fighting to pass legislation aimed at preventing "child marriage."

Another strategy to improve the health of women and children is to provide women with education and income-producing opportunities. Promoting women's education increases the status and power of women to control their reproductive lives, exposes women to information about health issues, and also delays marriage and childbearing. In many developing countries, women's lack of power and status means that they have little control over health-related decisions. Men make the decisions about whether or when their wives (or partners) will have sexual relations, use contraception, or use health services.

Improvements in child health have been made through immunization programs, which reduced the number of measles deaths from 733,000 in 1990 to 164,000 in 2008 (UNICEF 2010). Other strategies beneficial to women and children include providing mosquito nets to prevent malaria and providing HIV-infected pregnant women with antiretroviral medication (which reduces risk of mother-child transmission of HIV). Ensuring access to adequate nutrition, clean water, and sanitation are also important (see Chapter 6). But still, millions of women and children lack immunizations and other basic health services. Providing these services requires funding, but the cost of providing basic health services

Keith Levit/PhotoLibrary

Childbearing at an early age involves higher health risks for women and infants.

selective primary health care An approach to health care that focuses on using specific interventions to target specific health problems.

comprehensive primary health care An approach to health care that focuses on the broader social determinants of health, such as poverty and economic inequality, gender inequality, environment, and community development.

for mothers and infants in low-income countries is only $3 per person (Oxfam GB 2004). The question is, do the rich countries of the world have the political will to support efforts to protect the health and lives of women and infants in the developing world?

HIV/AIDS Prevention and Alleviation Strategies

As of this writing, there is no vaccine to prevent HIV infection. As researchers continue to work on developing such a vaccine, a number of other strategies are available to help prevent and treat HIV/AIDS.

HIV/AIDS Education and Access to Condoms HIV/AIDS prevention efforts include educating populations about how HIV is transmitted and how to protect against HIV transmission, and providing access to condoms as a means of preventing HIV transmission. Many Americans—especially adolescents and young adults—engage in high-risk behavior such as not using a condom during vaginal or anal intercourse (American College Health Association 2011).

A national survey found that many U.S. adults either do not know how HIV is transmitted or mistakenly believe that HIV can be transmitted through sharing a drinking glass, touching a toilet seat, or swimming in a pool with someone who is HIV positive (Kaiser Family Foundation 2009b). Nearly one in five is unaware that there is no cure for AIDS. More than half do not know that a pregnant woman with HIV infection can take medication to reduce the risk of her baby being born with HIV infection.

What Do You Think? In 2008, 42-year old Willie Campbell was sentenced to 35 years in prison for harassing a public servant with what Texas law views as a "deadly weapon": his saliva (Kovach 2008). Campbell, who is HIV-positive, had spat in the face of a police officer who was arresting him for public intoxication. The officer did not become infected with HIV, which is not surprising given that saliva has never been shown to transmit HIV. What do you think of a law that labels the saliva of an HIV-infected person as a "deadly weapon?"

HIV/AIDS education occurs through media and public service announcements, faith-based groups, health care providers, schools, and the workplace. With the HIV infection rate growing among the older-than-50 population, HIV/AIDS education is also taking place in some senior centers (Goldberg 2005). Some HIV/AIDS education is based on the ABC approach—a prevention strategy that involves three elements: A = Abstain; B = Be faithful; C = Use condoms (Halperin et al. 2004).

Providing education that advocates condom use and providing youth with access to condoms are controversial topics. Many conservatives believe that promoting use of condoms sends the "wrong message" that sex outside marriage is OK. Another controversy involves the question of whether to provide condoms to prison inmates. Vermont and Mississippi allow condom distribution in prisons, as do Canada, most of western Europe, and parts of Latin America. One deputy at the Los Angeles Sheriff's Department, which allows only homosexual inmates to receive condoms provided by a local nonprofit organization, said, "We're not promoting sex; we're promoting health" (Sanders 2005).

HIV Testing and Treatment About one in five HIV-infected people in the United States have not been diagnosed and do not know they are infected (MMWR 2011). Less than half (40 percent) of U.S. adults have ever been tested for HIV (Pleis, Ward, & Lucas 2010). Among the young adults ages 18 to 29 who have not been tested for HIV, the most common reason cited is that they do not think they are at risk (see Figure 2.6). HIV testing is important because, when antiretroviral treatment is begun early, it lowers the amount of virus in the blood and genital secretions, which slows the progression to AIDS and reduces the risk for transmission to uninfected individuals. Although individuals who

Public health messages encourage people to get tested for HIV.

know they are HIV-infected sometimes choose to continue to engage in risky behavior that exposes others to the virus, many people who know they are HIV-infected take precautions to avoid transmitting the virus to others.

The Fight against HIV/AIDS Stigma and Discrimination The HIV/AIDS-related stigma stems from societal views that people with HIV/AIDS are immoral and shameful, and results in discrimination in employment, housing, social relationships, and medical care. A survey of 1,000 physicians and nurses in Nigeria found that 1 in 10 admitted to refusing care for an HIV/AIDS patient or had denied HIV/AIDS patients admission to a hospital, and 20 percent believed that people living with HIV/AIDS have behaved immorally and deserved their fate (AVERT 2004). The stigma surrounding HIV/AIDS has also led to acts of violence against people perceived to be infected with HIV.

HIV/AIDS stigma and discrimination can deter people from getting tested for the disease, can make them less likely to acknowledge their risk of infection, and can discourage those who are HIV-positive from discussing their HIV status with their sexual and needle-sharing partners. Combating the stigma and discrimination against people who are affected by HIV/AIDS is crucial to improving care, quality of life, and emotional health for people living with HIV and AIDS and to reducing the number of new HIV infections.

Fighting antigay prejudice and discrimination is also important in efforts to support the well-being of individuals diagnosed with HIV/AIDS. In Africa, where about one-half of the nations have laws that criminalize same-sex sexual behavior, "fear of arrest prevents people from attending meetings or socializing in locations where their sexual identities become suspect. These are precisely the locations, however, where HIV prevention

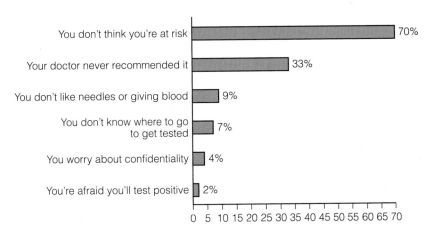

Figure 2.6 Reported Reasons for Not Being HIV-Tested among Young U.S. Adults
Source: Kaiser Family Foundation 2009a.

Note: Percent is among the 45% of those ages 18-29 who have never been tested.

Per-Anders Pettersson/Getty Images

Cynthia Leshomo is the 2005 winner of the Miss HIV Stigma Free Pageant for HIV-positive women. First held in 2002, the Miss HIV pageant is a way of showing that HIV-positive individuals need not be ashamed and that, with treatment, they can look good and lead productive lives.

training, counseling, and materials (informational brochures, condoms . . . etc.) are available" (Johnson 2007, pp. 46–47).

Needle Exchange Programs To reduce transmission of HIV among injection drug users, their sex partners, and their children, some countries and U.S. communities have established **needle exchange programs** (also known as syringe exchange programs), which provide new, sterile syringes in exchange for used, contaminated syringes. Many needle exchange programs also provide other social and health services, such as referrals to drug counseling and treatment, HIV testing and screening for other sexually transmissible diseases, hepatitis vaccinations, and condoms. The American Medical Association, the American Public Health Association, and the World Health Organization have endorsed needle exchange as an effective means of HIV prevention. Needle exchange programs also protect public health by providing safe disposal of potentially infectious syringes.

In Canada, sterile injection equipment is available to drug users in pharmacies and through numerous needle exchange programs. In contrast, most U.S. states prohibit the sale or possession of sterile needles or syringes without a medical prescription.

In 2009, 184 needle exchange programs were operating in the United States (Centers for Disease Control and Prevention 2010c). Although state and local public funding for needle exchange programs has increased in recent years, the United States is the only country in the world to explicitly ban the use of federal funds for needle exchange (Human Rights Watch 2005).

Financial and Medical Aid to Developing Countries Life-extending treatment for individuals infected with HIV is not affordable for many people in the developing world. Developing countries—those hardest hit by HIV/AIDS—depend on aid from wealthier countries to help provide medications, HIV/AIDS education programs, and condoms. In 2002, the United Nations helped to create the Global Fund to Fight AIDS, Tuberculosis, and Malaria to help poor countries fight these diseases. Although total global spending on HIV prevention and treatment rose from 300 million in 1996, to 15.9 billion in 2009, more than 60 percent of people needing antiretroviral therapy (medication to attack the HIV virus) have not received it (Kaiser Family Foundation 2010).

Fighting the Growing Problem of Obesity

In general, reducing and preventing obesity requires encouraging people to (1) eat a diet with sensible portions, with lots of high-fiber fruits and vegetables, and with minimal sugar and fat, and (2) engage in regular physical activity. Some of the strategies to achieve these goals include the following:

Restrictions on Advertisements In response to concerns about childhood obesity, Ireland has banned advertising of candy and fast food on television, Great Britain has banned advertising for junk food on children's television programming, and Sweden and Norway prohibit advertising that targets children. In the United States, the 2006 Children's Food and Beverage Advertising Initiative (CFBAI) involves fast food companies voluntarily agreeing to devote at least half of their advertising directed toward children under 12 to promote healthier dietary choices. According to SourceWatch (2010), "corporations and industries typically enact such voluntary 'initiatives,' or codes of conduct when they are faced with public efforts to tax or regulate some harmful aspect of their business. Such voluntary codes create the illusion of responsible behavior by the industry, while minimally hampering its advertising, sales or marketing practices while at the same time tamping down public discontent and staving off government legislation to regulate their behavior" (n.p.).

needle exchange programs Programs designed to reduce transmission of HIV by providing intravenous drug users with new, sterile syringes in exchange for used, contaminated syringes.

Public Education A variety of public education strategies are being used to inform the public about the importance of exercise and diet and their effects on health. France requires that advertisements promoting processed, sweetened, or salted food and drinks on television, radio, billboards, and the Internet must include one of four health messages: "For your health, eat at least five fruits and vegetables a day," "For your health, undertake regular physical activity," "For your health, avoid eating too much fat, too much sugar, too much salt," or "For your health, avoid snacking between meals" (Combes 2007).

Because many Americans do not know how many calories, fat grams, sodium, and so on, are in the foods they eat, consumers are encouraged to read the nutritional labels on packaged foods. Menu labeling—the posting of nutrition information on menus and menu boards—is required in Seattle, New York City, and San Francisco, and several other localities and states have considered similar legislation (Trust for America's Health 2008). Proposed federal legislation called the Menu Education and Labeling (MEAL) Act, if passed, would require all chain restaurants to list nutritional information for all meals on the menu.

School Nutrition and Physical Activity Programs A number of states and school districts have implemented or considered bans or restrictions on vending machine sales of sugary drinks and "junk food" snacks. In an effort to increase consumption of produce, some schools have "farm-to-school" programs that consist of school gardens and the opportunity to purchase fresh, locally grown produce to use in school breakfast and lunch programs. And many states have recess and/or physical activity requirements, although these requirements are often inadequate or not enforced. The Healthy School Meals Act of 2010, if passed, will encourage the inclusion of healthful plant-based options in the National School Lunch and Breakfast Programs.

Interventions to Treat Obesity Interventions to treat obesity include weight-loss or fitness clubs, nutrition and weight-loss counseling, weight-loss medications, and surgical procedures. In 2004, the U.S. government classified obesity as an illness, so Medicare could cover obesity treatments. The proposed Medicaid Obesity Treatment Act would require Medicaid coverage of prescription drugs to treat obesity. Some private insurers also cover treatment for obesity; at least seven states require it (Cotton et al. 2007).

Strategies to Improve Mental Health Care

In 2011, 22-year-old Jared Loughner shot and killed six people and injured several more in Tucson, Arizona. Loughner had shown signs of mental instability prior to the shootings, but had not been diagnosed or treated for mental illness. The Tucson shootings drew national attention to the importance of mental health treatment, at a time when many states were cutting mental health funding and services. Two areas for improving mental health care in the United States are eliminating the stigma associated with mental illness and improving access to mental health services.

Eliminating the Stigma of Mental Illness The first White House conference on mental health called for a national campaign to eliminate the stigma associated with mental illness. Fearing the negative label of "mental illness" and the social rejection and stigmatization associated with mental illness, individuals are reluctant to seek psychological services (Komiya et al. 2000). In a study of eighth graders, boys were more likely than girls to agree with the statement, "Seeing a counselor for emotional problems makes people think you are weird or different" (Chandra & Minkovitz 2006). In the same study, more boys than girls (38 percent versus 23 percent) reported that they were not willing to use mental health services. The most frequently cited reason was "Too embarrassed by what other kids would say." This chapter's *Animals and Society* feature suggests that the use of animals in mental health services can increase motivation to participate in mental health counseling or therapy.

A growing trend in mental health services involves the use of therapy animals, primarily dogs, cats, and horses, to improve the functioning of individuals with a variety of mental health conditions, including anxiety, depression, family problems, autism, eating disorders, post-traumatic stress, attention-deficit disorders, and many others (Fine 2010; Peters 2011). Animal-assisted therapy (AAT) is used in a variety of mental health care settings, including private therapy offices, mental health clinics, hospitals, and long-term care and residential facilities. Some school systems use AAT to assist children with behavioral disorders (e.g., attention-deficit/hyperactivity disorder, autism spectrum disorders). In the United States, AAT was used as early as 1919, when dogs were used with psychiatric patients at St. Elizabeth's Hospital in Washington, DC. The use of animals in mental health treatment gained increased attention with the 1969 publication of *Pet-Oriented Child Psychotherapy* by Dr. Boris Levinson, who had discovered the therapeutic value of animals quite by accident when a young therapy patient arrived early for his appointment and met and embraced Levinson's dog, Jingles, who was in the office that day. Levinson observed the powerful impact the dog had on the boy and how the dog helped Levinson develop a rapport with his client (Urichuk with Anderson 2003).

Animals can contribute to therapy in a variety of ways, including (1) reducing anxiety, (2) helping the trust- and rapport-building

AP Photo/Tim Roske

Therapy dogs are used in a variety of settings, including clinics, private medical offices, nursing homes, schools, and hospitals.

process between the client and the therapist, (3) increasing motivation to attend and participate in therapy because of the desire to spend time with the therapy animal, and (4) stimulating conversation about difficult topics. One clinician reported an experience working with a child who had been traumatized by sexual abuse:

Fearing the negative label of "mental illness" and the social rejection and stigmatization associated with mental illness, individuals are reluctant to seek psychological services.

Reducing the stigma associated with mental illness might be achieved through encouraging individuals to seek treatment and making treatment accessible and affordable. This is because, "Effective interventions help people to understand that mental disorders are not character flaws but are legitimate illnesses that respond to specific treatments, just as other health conditions respond to medical interventions" (U.S. Department of Health and Human Services 1999, p. viii).

The National Alliance on Mental Illness (NAMI) has a StigmaBusters campaign, whereby the public submits instances of media content that stigmatize individuals with mental illness to StigmaBusters, which then conveys their concerns to media organizations and corporations, urging them to avoid stigmatizing portrayals of mental illness.

I told one child that Buster [a dog] had a nightmare. I then asked the child, "What do you think Buster's nightmare was about?" The child said, "The nightmare was about being afraid of getting hurt again by someone mean." (Cited in Kruger & Serpell 2010, p. 39)

AAT is also used with individuals with severe mental disorders. Marsha was a 23-year-old woman who was diagnosed with catatonic schizophrenia. She was treated with medication and electroshock therapy, without improvement. She was withdrawn, frozen, and nearly mute. A therapy dog was introduced into her treatment:

At first there was no improvement in Marsha's behavior. . . . She remained very withdrawn and the only signs of communication were when she was with the dog. When the dog was taken away, she would get off her chair and go after it. She began to walk the dog a little and . . . was given a written schedule of the hours when the dog would come and visit her; she began to look forward to the visits and to talk about the dog with the other patients. Six days after the introduction of the dog Marsha suddenly showed marked improvement and shortly thereafter she was discharged. (Cited in Urichuk with Anderson 2003, pp. 107–108)

The most common use of AAT in mental health services involves therapists working in partnership with their own pet that has been evaluated and certified as appropriate for therapy work (Chandler 2005). Dogs used as therapy animals must meet rigorous requirements through organizations such as Therapy Dogs Inc., Delta Society®, and Therapy Dogs International. For the dogs, this includes a temperament test, obedience class training, and additional AAT training in which the dogs learn things like not reacting to loud noises, how to ride on elevators, and being comfortable around patients who use wheelchairs or walkers. In addition, the dogs must maintain good health and remain current on vaccinations.

Not all AAT involves using specially trained or certified animals. Some individuals achieve improvements in mental health functioning and well-being as a result of interacting with and taking care of farm animals (Arehart-Treichel 2008). One case study describes how Mark, a young teenager with autism, benefited through his interactions with donkeys:

At first Mark could not get near the donkeys. Naturally wary of people, the donkeys ran away from Mark as he marched after them. . . . But Mark was motivated, and with guidance and patience has learnt to approach the donkeys slowly and gently. He has become aware of the donkeys' feelings and of how his actions impact them; he has built a relationship with the donkeys based upon mutual trust and respect. He is now rewarded each week by Ceilidh running up to him to have her face rubbed. This is a new experience for Mark that we are working on transferring to his human relationships. (Cited in Urichuk with Anderson 2003, p. 80)

Small animals such as rabbits, guinea pigs, and birds can also be used as therapy animals:

Marta was an 8-year-old diagnosed as an emotionally disturbed child of a strict and abusive mother. She was aggressive and hyperactive, sexually precocious, and had temper tantrums. In her first few months at the residential school, no one could get her to talk about her relationship with her mother. In the first session with a small furry rabbit, she held him in her lap and stroked him, telling the therapist that the rabbit's ears had been chewed by the mother rabbit. (The rabbit's ears were normal). The therapist asked her why this was so. Marta responded, "The mother rabbit chewed the baby rabbit's ears all up. She wanted the baby to leave home." The therapist then asked, "How did the baby rabbit feel?" In answering, Marta said "Sad. The baby rabbit loves the mother rabbit but the mother rabbit no longer loves the baby." This dialogue about the rabbit was an opener for Marta to then talk of her own feelings about the mother who badly beat her. (Cited in Urichuk with Anderson 2003, pp. 64–65)

It is important to note that AAT is not appropriate for individuals who are fearful of or allergic to animals. Even with careful selection and training of therapy animals, there is some risk that a therapy animal could injure a client, and also risk that a client could hurt the animal. But with close supervision and careful management, animal-assisted therapy provides opportunities for improving the well-being and functioning of children and adults with a variety of mental health problems.

"Real Men, Real Depression" is an anti-stigma public education campaign that includes print, television, and radio public service announcements. "Breaking the Silence," a curriculum for elementary, middle, and high schools available through NAMI, uses true stories, activities, a board game, and posters to debunk myths about mental illness and sensitize students to the pain that words such as *psycho* and *schizo* as well as frightening or comic media images of mentally ill people can cause (Harrison 2002). The Department of Defense has launched an anti-stigma campaign called "Real Warriors, Real Battles, Real Strength" designed to assure military personnel that seeking mental health treatment will not harm their career and to publicize stories of military personnel who have been successfully treated for mental health problems (Dingfelder 2009b). Effective anti-stigma campaigns not only focus on eradicating negative stereotypes of people with mental illness, but also emphasize the positive accomplishments and contributions of people with mental illness.

Real Men. Real Depression.

It takes courage to ask for help. These men did.

NIMH

This public education brochure on men and depression is available from the National Institute of Mental Health (NIMH), www.nimh.nih.gov.

Improving Access to Mental Health Care The most frequently used source of care for mental health problems among U.S. adults is primary care and general medical care providers (who are not specialists in mental health care) (Substance Abuse and Mental Health Services Administration 2010a). Primary care physicians who try to obtain outpatient mental health services for their patients are often unsuccessful because of shortages in mental health professionals (e.g., psychiatrists, psychologists, psychiatric nurses, social workers) and the lack of adequate health insurance. Thus, improving access to mental health services involves (1) recruiting more mental health professionals, especially those willing to serve in rural and impoverished communities and who have cultural competency to work with clients from diverse cultural backgrounds, and (2) improving insurance coverage for mental health issues.

Until the mid-1990s, private insurance plans had co-pays or deductibles and more restrictions for mental health treatment. Since then, a number of state and federal laws have attempted to rectify this inequality by requiring health insurance plans to treat mental illness and physical illness equally—a concept known as **parity**. The 2008 Mental Health Parity and Addiction Equity Act requires equality for deductibles, co-payments, coinsurance, and covered hospital days or outpatient visits. This parity law has significant limitations: It does not require insurance companies to offer mental health treatment coverage, or to cover every condition listed in the American Psychiatric Association's *Diagnostic and Statistical Manual of Mental Disorders* (DSM-IV), and it does not apply to employers with fewer than 50 employees (Jenkins 2008). Most states have enacted mental health parity laws, but many of these laws do not address substance abuse, are limited to the more serious mental illnesses, apply only to government employees, or exclude small businesses.

U.S. Federal Health Care Reform

Since 1912, when Theodore Roosevelt first proposed a national health insurance plan, the Truman, Nixon, Carter, Clinton, and Obama administrations have advocated the idea of health care for all Americans. Health care reform efforts in the United States generally fall into one of three categories: (1) creation of a public single-payer health program; (2) expansion of existing government health insurance programs; and (3) making private insurance more affordable through tax credits or deductions and other means. Following much heated debate, in March 2010, the Patient Protection and Affordable Care Act (PPACA), commonly referred to as the Affordable Care Act or "Obamacare," was passed by Congress and signed into law by President Obama. Just a few of the many provisions of the recent health care reform legislation include:

- Establishing an "individual mandate" that requires U.S. citizens and legal residents to have health insurance or face a tax penalty (exemptions granted for financial hardship).
- Creating health insurance exchanges: online marketplaces where consumers can shop for, compare, and enroll in insurance plans.
- Providing tax credits to businesses that provide insurance to their employees.

parity In health care, a concept requiring equality between mental health care insurance coverage and other health care coverage.

- Requiring health insurance plans to provide dependent coverage for children up to age 26.
- Prohibiting health insurance plans from placing lifetime limits on the dollar value of coverage; restricting annual limits on coverage; prohibiting insurers from canceling coverage except in cases of fraud; and prohibiting denial of insurance due to preexisting conditions.
- Requiring insurance companies to use a certain percentage of the premiums they collect on medical care, as opposed to administrative expenses and profits.
- Expanding Medicaid to cover more low-income individuals/families.
- Providing discounts on brand-name prescription drugs and free preventive services and annual wellness exams for Medicare enrollees, and raising Medicare premiums for some higher-income seniors.

The Congressional Budget Office (CBO) estimates the new health reform law will provide coverage to an additional 32 million when fully implemented in 2019. But the new law is hotly debated, and public views on the law are divided. One poll found that half of Americans had an unfavorable opinion of the health reform law; 41 percent had a favorable opinion of it (Kaiser Family Foundation/Harvard School of Public Health 2011). Although some who are opposed to the PPACA believe the law goes too far, others believe it does not go far enough. Regarding American's views about what lawmakers should do about health reform, 28 percent want the law expanded; 23 percent want the law repealed and replaced with a Republican-sponsored alternative; 20 percent want the law repealed and not replaced; and 19 percent say leave the law as is (Kaiser Family Foundation/Harvard School of Public Health 2011).

Opponents of the law claim that "Obamacare" will increase the budget deficit and will "kill jobs," but the Congressional Budget Office finds that the law will *reduce* deficits and will not hurt jobs (Van de Water 2011). At this writing, the future of the PPACA is uncertain: The Republican-dominated House has threatened to defund health care reform, and two federal judges have ruled against the PPACA, claiming that it is unconstitutional for Congress to require people to have health insurance; three federal judges have upheld the law. Just as it seemed inevitable that the fight over health care reform would reach the Supreme Court, Obama announced in March 2011 that he would agree to amend the PPACA (with the Empowering States to be Innovative Amendment) to allow states to opt out of the PPACA and set up their own health care system as long as it achieves the same goals of expanding care without increasing costs.

Meanwhile, other health care reform proposals aim to create a **single-payer health care** system in which a single tax-financed public insurance program replaces private insurance companies. Advocates of single-payer health care financing argue that administration costs of private insurance consume nearly a third of Americans' health dollars. These costs include insurance company overhead, underwriting, billing, sales and marketing departments, exorbitant executive pay, and profits. In addition, hospitals and doctors must pay administrative staff to deal with the various billing policies and procedures of different insurers. Supporters argue that single-payer health care would save more than $400 billion per year—enough to provide health coverage to everyone without any additional expense.

The United States National Health Care Act, also known as the Expanded and Improved Medicare for All Act, introduced by Representative John Conyers (D-Michigan), would expand Medicare to every U.S. resident. In May 2011, Vermont became the first state to pass legislation to establish state-level single-payer health care.

Opponents argue that a single-payer national health insurance program would amount to a "government takeover" of health care and would result in higher costs, less choice, rationing, and excessive bureaucracy—the very outcomes that have resulted from corporatized medicine (Nader 2009). The recent rise of the grassroots Tea Party political movement has fueled opposition to "big government," viewing government "takeover" of health care as an intrusion into individual freedoms. Skepticism over health care reform

single-payer health care A health care system in which a single tax-financed public insurance program replaces private insurance companies.

Courtesy of Emily Potter

Wendell Potter is the former vice president of corporate communications at CIGNA, one of the largest health insurance companies in the United States. In the following account, Mr. Potter describes how the insurance industry influences the health reform debate and explains why he decided to speak out against the insurance industry.

I'm the former insurance industry insider now speaking out about how big for-profit insurers have hijacked our health care system and turned it into a giant ATM for Wall Street investors, and how the industry is using its massive wealth and influence to determine what is (and is not) included in the health care reform legislation. . . .

Although . . . I had a great career in the insurance industry (four years at Humana and nearly 15 at CIGNA), in recent years I had grown increasingly uncomfortable serving as one of the industry's top PR executives. My responsibilities at CIGNA . . . included serving as the company's chief spokesman to the media on all corporate and financial matters. I also served on a lot of trade association committees and industry-financed coalitions, many of which were essentially front groups for insurers. So I was in a unique position to see not only how Wall Street analysts and investors influence decisions insurance company executives make but also how the industry has carried out behind-the-scenes PR and lobbying campaigns to kill or weaken any health care reform efforts that threatened insurers' profitability.

I also have seen how the industry's practices . . . have contributed to the tragedy of nearly 50 million people being uninsured as well as to the growing number of Americans who, because insurers now require them to pay thousands of dollars out of their own pockets before their coverage kicks in—are underinsured. An estimated 25 million of us now fall into that category.

What I saw happening over the past few years was a steady movement away from the concept of insurance and toward "individual responsibility," a term used a lot by insurers. . . . This is playing out as a continuous shifting of the financial burden of health care costs away from insurers and employers and onto the backs of individuals. As a result, more and more sick people are not going to the doctor or picking up their prescriptions because of costs. If they are unfortunate enough to become seriously ill or injured, many people enrolled in these plans find themselves on the hook for such high medical bills that they are losing their homes to foreclosure or being forced into bankruptcy. As an industry spokesman, I was expected to put a positive spin on this trend that the industry created and euphemistically refers to as "consumerism" and to promote so-called "consumer-driven" health plans. . . . I thought I could live with being a well-paid huckster and hang in

also stems from questions about how much reform will cost and who will pay for it. Although the answers to the questions of cost are being debated, what is certain is that we are already paying a high price for the system we have and, in the words of Obama, "If we don't change, we can't expect a different result."

Dr. Marcia Angell (2003), lecturer at Harvard Medical School and former editor of the *New England Journal of Medicine,* makes the following plea for a national health program:

> We live in a country that tolerates enormous disparities in income, material possessions, and social privilege. That may be an inevitable consequence of a free-market economy. But those disparities should not extend to denying some of our citizens certain essential services because of their income or social status. One of those services is health care. Others are education, clean water and air, equal justice, and protection from crime, all of which we already acknowledge are public responsibilities. We need to acknowledge the same thing for health care.

There is considerable support for single-payer health care among doctors, nurses, and the general public. The insurance industry opposes the adoption of such a system because the private health insurance industry would be virtually eliminated, and so spends a great deal of money on lobbying, political contributions, and public relations to influence the health reform debate (see this chapter's *The Human Side* feature).

there a few more years until I could retire. I probably would have if I hadn't made a . . . decision a couple of years ago that changed the direction of my life. While visiting my folks in northeast Tennessee. . . I read in the local paper about a health "expedition" being held that weekend a few miles up U.S. 23 in Wise, Virginia. Doctors, nurses, and other medical professionals were volunteering their time to provide free medical care to people who lived in the area. . . . Remote Area Medical, a nonprofit group whose original mission was to provide free care to people in remote villages in South America, was organizing the expedition. I decided to check it out. . . .

Nothing could have prepared me for what I saw when I reached the Wise County Fairgrounds, where the expedition was being held. Hundreds of people had camped out all night in the parking lot to be assured of seeing a doctor or dentist when the gates opened. By the time I got there, long lines of people stretched from every animal stall and tent where the volunteers were treating patients.

That scene was so visually and emotionally stunning, it was all I could do to hold back tears. How could it be that citizens of the richest nation in the world were being treated this way?

A couple of weeks later I was boarding a corporate jet. . . . When the flight attendant served my lunch on gold-rimmed china and gave me a gold-plated knife and fork to eat it with, I realized for the first time that someone's insurance premiums were paying for me to travel in such luxury. I also realized that one of the reasons those people in Wise County had to wait in long lines to be treated in animal stalls was because our Wall Street–driven health care system has created one of the most inequitable health care systems on the planet.

Although I quit my job last year, I did not make a final decision to speak out as a former insider until recently when it became clear to me that the insurance industry and its allies (often including drug and medical device makers, business groups, and even the American Medical Association) were succeeding in shaping the current debate on health care reform. . . . Remember this: Whenever you hear a politician or pundit use the term *government-run health care* and warn that the creation of a public health insurance option that would compete with private insurers (or heaven forbid, a single-payer system like the one Canada has) will "lead us down the path to socialism," know that the original source of the sound bite most likely was some flack like I used to be.

Bottom line: I ultimately decided the stakes are too high for me to just sit on the sidelines and let the special interests win again. So I have joined forces with thousands of other Americans who are trying to persuade our lawmakers to listen to us for a change, not just to the insurance and drug company executives who are spending millions to shape reform to benefit them and the Wall Street hedge fund managers they are beholden to. . . .

The people of Wise County and every county deserve much better than to be left behind to suffer or die ahead of their time due to Wall Street's efforts to keep our government from ensuring that all Americans have real access to first-class health care.

Source: Potter, Wendell, "The Health Care Industry vs. Health Reform." Copyright © 2009 Center for Media and Democracy. Reprinted by permission. Available at http://www.prwatch.org

What Do You Think? In 2005, Representative Pete Stark (D-California) proposed an amendment to the U.S. Constitution to guarantee health care as a right for every American. Stark argued that "the health of every American is vital to their unalienable rights of 'life, liberty, and the pursuit of happiness.' . . . To ensure these rights are fully enjoyed, we must be certain that every American can access quality health care—regardless of their income, race, education or job status" ("Stark Introduces Constitutional Amendment" 2005). Do you favor or oppose amending the Constitution to guarantee health care as a right for every American?

Understanding Problems of Illness and Health Care

Although human health has probably improved more over the past half-century than over the previous three millennia, the gap in health between rich and poor remains wide, and the very poor suffer appallingly (Feachum 2000). Poor countries need economic and material assistance to alleviate problems such as HIV/AIDS, high maternal and infant mortality rates, and malaria. Cancer, once viewed as a disease that affects primarily wealthy countries, has now become prevalent in low-income countries where treatment is either not available or is not affordable (Farmer et al. 2010). The wealthy countries of the world do have resources to make a difference in the health of the world. When a tsunami took the lives of thousands of people in 2004, media attention to the tragedy elicited an outpouring of aid throughout the world to help affected regions. Meanwhile, malaria, which has been referred to as "a silent tsunami," takes the lives of more than 150,000 African children each month, which is about the same as the death toll of the southern Asia tsunami disaster (Sachs 2005). Yet malaria continues to receive comparatively scant public attention, even though it is largely preventable with $5 mosquito bed nets, and is treatable with medicines at roughly $1 per dose.

Although poverty may be the most powerful social factor affecting health, other social factors that affect health include globalization, increased longevity, family structure, gender, education, and race or ethnicity. Although individuals make choices that affect their health—choices such as whether to smoke, exercise, eat a healthy diet, engage in risky sexual activity, wear a seat belt, and so on—those choices are also influenced by social, economic, and political forces that must be addressed if the goal is to improve the health not only of individuals but also of entire populations. By focusing on individual behaviors that affect health and illness, we often overlook social causes of health problems (Link & Phelan 2001). Despite the significance of recent health care reform efforts to improve Americans' health insurance coverage, access to health care is only one piece of the puzzle: "Health and longevity are also profoundly influenced by where and how Americans live, learn, work, and play" (Williams et al. 2010, p. 1,481).

A sociological view of illness and health care looks not only at the social causes, but also at the social *consequences* of health problems—consequences that potentially affect us all. In *Uninsured in America* (2005), Sered and Fernandopulle explain:

> If millions of American children do not have reliable, basic health care, all children who attend American schools are at risk through daily exposure to untreated disease. If millions of restaurant and food industry workers do not have health insurance, people preparing food and waiting tables are sharing their health problems with everyone they serve. . . . If tens of millions of Americans go without basic and preventive care, we all pay the bill when their health problems turn into complex medical emergencies necessitating expensive . . . treatment. (p. 20)

Although certain changes in medical practices and policies may help to improve world health, "the health sector should be seen as an important, but not the sole, force in the movement toward global health" (Lerer et al. 1998, p. 18). A comprehensive approach to improving the health of a society requires addressing diverse issues such as poverty and economic inequality, gender inequality, population growth, environmental issues, education, housing, energy, water and sanitation, agriculture, and workplace safety.

Improving the health of the world also means seeking nonmilitary solutions to international conflicts. In addition to the deaths, injuries, and illnesses that result from combat, war diverts economic resources from health programs, leads to hunger and disease caused by the destruction of infrastructure, causes psychological trauma, and contributes to environmental pollution (Sidel & Levy 2002). Thus, "the prevention of war . . . is surely one of the most critical steps mankind can make to protect public health" (White 2003, p. 228).

The World Health Organization (1946) defined **health** as "a state of complete physical, mental, and social well-being" (p. 3). Based on this definition, we conclude this chapter with the suggestion that the study of social problems is, essentially, the study of health problems, as each social problem is concerned with the physical, mental, and social well-being of humans and the social groups of which they are a part. As you read other chapters in this book, consider how the problems in each chapter affect the health of individuals, families, populations, and nations.

health According to the World Health Organization, "a state of complete physical, mental, and social well-being."

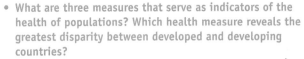
- **What are three measures that serve as indicators of the health of populations? Which health measure reveals the greatest disparity between developed and developing countries?**

Measures of health that serve as indicators of the health of populations include morbidity, life expectancy, and mortality rates (including infant and under-5 childhood mortality rates and maternal mortality rates). Maternal mortality rates reveal the greatest disparity between developed and developing countries.

- **Which theoretical perspective criticizes the pharmaceutical and health care industry for placing profits above people?**

The conflict perspective criticizes the pharmaceutical and health care industry for placing profits above people. For example, pharmaceutical companies' research and development budgets are spent not according to public health needs but rather according to calculations about maximizing profits. Because the masses of people in developing countries lack the resources to pay high prices for medication, pharmaceutical companies do not see the development of drugs for diseases of poor countries as a profitable investment.

- **Where is HIV/AIDS most prevalent in the world?**

HIV/AIDS is most prevalent in Africa, particularly sub-Saharan Africa, where one in 20 adults is infected with HIV. Two-thirds of people with HIV live in sub-Saharan Africa.

- **What is the second biggest cause of preventable deaths in the United States (second only to tobacco)?**

Obesity, which can lead to heart disease, hypertension, diabetes, and other health problems, is the second biggest cause of preventable deaths in the United States.

- **Why is mental illness referred to as a "hidden epidemic"?**

Mental illness is a "hidden epidemic" because the shame and embarrassment associated with mental problems discourage people from acknowledging and talking about mental illness. Because male gender expectations associate masculinity with emotional strength, men are particularly prone to deny or ignore mental problems.

- **What features of globalization have contributed to health problems?**

Features of globalization that have been linked to problems in health are increased travel and the expansion of trade and transnational corporations.

- **According to a World Health Organization analysis of the world's health systems, which country provides the best overall health care?**

The World Health Organization found that France provides the best overall health care among major countries, followed by Italy, Spain, Oman, Austria, and Japan. The United States ranked 37 out of 191 countries, despite the fact that the United States spends a higher portion of its gross domestic product on health care than any other country.

- **What is the difference between selective primary health care and comprehensive primary health care?**

Selective primary health care focuses on using specific interventions to target specific health problems, such as promoting condom use to prevent HIV infections and providing immunizations against childhood diseases to promote child survival. In contrast, comprehensive primary health care focuses on the broader social determinants of health, such as poverty and economic inequality, gender inequality, environment, and community development.

- **What are the categories of health care reform efforts in the United States?**

The goals of health care reform efforts in the United States generally fall into one of three categories: (1) creation of a universal health program; (2) expansion of existing government health insurance programs; and (3) implementing tax incentives and other strategies to make private insurance more affordable.

- **How does the World Health Organization define health?**

Health, according to the World Health Organization, is "a state of complete physical, mental, and social well-being." Based on this definition, we suggest that the study of social problems is, essentially, the study of health problems, because each social problem is concerned with the physical, mental, and social well-being of humans and the social groups of which they are a part.

TEST YOURSELF

1. Worldwide, the leading cause of death is
   a. HIV/AIDS
   b. traffic accidents
   c. heart disease
   d. cancer
2. The United States has the lowest infant mortality rate in the world.
   a. True
   b. False
3. Worldwide, the predominant mode of HIV transmission is through
   a. heterosexual contact
   b. needle-sharing
   c. same-sex sexual contact
   d. blood transfusions

4. In the United States, ___ of adults are either overweight or obese.
    a. 10 percent
    b. One-third
    c. Two-thirds
    d. 20 percent
5. What age group has the highest rate of "serious psychological stress?"
    a. 12 to 17
    b. 18 to 25
    c. 26 to 64
    d. 65 and over
6. The United States spends far more per person on health care than does any other industrialized nation.
    a. True
    b. False
7. All U.S. children living below the poverty line are covered by publicly funded health insurance.
    a. True
    b. False
8. Of all age groups in the United States, young adults ages 18 to 24 are the least likely to have health insurance.
    a. True
    b. False
9. One study found that medical bills, as well as income lost due to illness, contributed to two-thirds of all bankruptcies in 2007. Most medical debtors
    a. were educated homeowners with middle-class jobs
    b. had health insurance
    c. both a and b
    d. neither a nor b
10. A health care system in which a single tax-financed public insurance program replaces private insurance companies is called which of the following?
    a. Constitutional care
    b. Managed care
    c. Obamacare
    d. Single-payer health care

Answers: 1. c; 2. b; 3. a; 4. c; 5. b; 6. a; 7. b; 8. a; 9. c; 10. d.

KEY TERMS

comprehensive primary health care   51
deinstitutionalization   50
developed countries   28
developing countries   28
globalization   40
health   62
infant mortality rate   30
least developed countries   28
life expectancy   28
managed care   45

maternal mortality rate   30
Medicaid   45
medicalization   34
Medicare   45
mental health   37
mental illness   37
morbidity   28
mortality   29
needle exchange programs   54
parity   58

selective primary health care   51
single-payer health care   59
State Children's Health Insurance Program (SCHIP)   45
stigma   35
under-5 mortality rate   30
universal health care   46
workers' compensation   45

MEDIA RESOURCES

Turning to Video

 Watch the BBC video *Young and Uninsured* (running time 2:22), available through **CengageBrain.com.**

This video describes the need for young adults to have health insurance. How many young college-age adults do you know who have had a serious injury or medical problem?

Online Study Resources

Log in to **www.cengagebrain.com** to access the resources your instructor has assigned. For this book, you can access:

CourseMate

Access chapter-specific learning tools, including learning objectives, practice quizzes, videos, Internet exercises, flash cards, and glossaries, web links, and more in your Sociology CourseMate.

Westend61/Getty Images

3

Alcohol and Other Drugs

"Substance abuse, the nation's number one preventable health problem, places an enormous burden on American society, harming health, family life, the economy, and public safety, and threatening many other aspects of life."

—The Robert Wood Johnson Foundation, Institute for Health Policy, Brandeis University

Andre Jenny/Alamy

Elena Shapiro was just one of many people killed on the road by drunk drivers in 2009, and despite Raymond Cook's relatively light sentence, at least he was caught and prosecuted. The average drunk driver has driven over 80 times before being arrested for the first time.

ALL THAT MARKED the death of 20-year-old ballerina Elena Shapiro was a teddy bear, a dimly lit candle, flower petals, and a ballet slipper. This was the intersection where it happened—where Raymond Cook, a prominent plastic surgeon, drove his black Mercedes into Elena Shapiro's Elantra at close to 80 mph. Her car crumpled like a piece of paper, and Elena, pinned inside, was hit with such force that strands of her long blond hair were found in the back seat. The trunk had been pushed into the cabin of her car; a flip-flop remained on the console; the keys dangled from the ignition; and blood stains covered the grey upholstery.

Raymond Cook voluntarily surrendered his medical license and went into an alcohol rehabilitation facility. He was charged with second-degree murder. The jury never heard that this was not the first time he had been charged with drunk driving or the first time he had caused an accident. He also had a history of speeding and, on that awful night, a radar detector was found in his car.

Throughout the trial he remained emotionless as witness after witness testified that they had tried to stop him from drinking and driving and that he was visibly drunk at the country club that afternoon and, later that evening, at a bar and restaurant.

In the end, he was convicted of involuntary manslaughter and received a three- to four-year maximum sentence. In three to four years, or less, Raymond Cook will start his life over. What would have happened over those same three to four years for Elena? A new love? Her firstborn child? A starring role in Tchaikovsky's *Sleeping Beauty*? We'll never know.

Drug-related deaths are just one of the many negative consequences that can result from alcohol and drug abuse. The abuse of alcohol and other drugs is a social problem when it interferes with the well-being of individuals and/or the societies in which they live—when it jeopardizes health, safety, work and academic success, family, and friends. But managing the drug problem is a difficult undertaking. In dealing with drugs, a society must balance individual rights and civil liberties against the personal and social harm that drugs promote—crack babies, suicide, drunk driving, industrial accidents, mental illness, unemployment, and teenage addiction. When to regulate, what to regulate, and who should regulate are complex social issues. Our discussion begins by looking at how drugs are used and regulated in other societies.

The Global Context: Drug Use and Abuse

Pharmacologically, a **drug** is any substance other than food that alters the structure or functioning of a living organism when it enters the bloodstream. Using this definition, everything from vitamins to aspirin is a drug. Sociologically, the term *drug* refers to any chemical substance that (1) has a direct effect on the user's physical, psychological, and/or intellectual functioning; (2) has the potential to be abused; and (3) has adverse consequences for the individual and/or society. Societies vary in how they define and respond to drug use. Thus, drug use is influenced by the social context of the particular society in which it occurs.

drug Any substance other than food that alters the structure or functioning of a living organism when it enters the bloodstream.

Drug Use and Abuse Around the World

Globally, 4 to 6 percent of the world's population between the ages of 15 and 64—over 200 million people—reported using at least one illicit drug in the previous year (WDR 2010). According to the most recent report, cannabis (i.e., marijuana and hashish) remains by far the most widely used illegal drug, followed by amphetamine-type stimulants (ATS), cocaine, and opiate-based drugs.

Worldwide, 55 percent of adults have consumed alcohol at some time in their life, and 2.4 million people die from alcohol-related use every year (WHO 2011a). Alcohol use is much higher in North and South American countries, European countries, New Zealand, and Japan than in Middle Eastern or African countries, or China (Degenhardt et al. 2008). There are approximately one billion smokers in the world and 80 percent of them come from low- and middle-income countries (WHO 2011b).

> There are approximately one billion smokers in the world and 80 percent of them come from low- and middle-income countries.

Illicit drug use varies by location as well. For example, an estimated 6.8 percent of adult Europeans have used cannabis in the previous year, 2.2 percent in Central America, and 1.2 percent in Asia (WDR 2010; EMCDDA 2010). Lifetime use of amphetamines among the 15 to 64 population is as low as 0.1 percent of the population in Greece and as high as 12.3 percent of the population in the United Kingdom (EMCDDA 2010). Moreover, lifetime use of *any* illicit drug by 15- and 16-year-olds ranges from 46 percent in the Czech Republic, to 5 percent in Armenia, with the U.S. rates at about 39 percent (ESPAD 2009; MTF 2011).

Finally, drug use varies over time. In 1974, 38 percent of high school seniors reported "bingeing" in the 30 days prior to the survey; by 1992, that number had dropped to a low of 28 percent. Similarly, 30-day prevalence of smoking by high school seniors in 1976 was 39 percent. By 1992, the 30-day prevalence of smoking for 12th graders had dropped to 28 percent but increased by nearly a third over the next several years, peaking in 1997, before dropping again. Today, about 21 percent of 12th graders report smoking in the previous month (MTF 2011).

Some have argued that differences in drug use can be attributed to variations in drug policies. The Netherlands, for example, has had an official government policy of treating the use of drugs such as marijuana and hashish as a public health issue rather than a criminal justice issue since the mid-1970s. Treatment of the drug user and prevention of future drug use are prioritized over the more punitive response of imprisonment found in many other countries.

In the first decade of the policy, drug use did not appear to increase. However, increases in marijuana use were reported in the early 1990s, with the advent of "cannabis cafés." These coffee shops sell small amounts of marijuana for personal use and, presumably, prevent casual marijuana users from coming into contact with "hard drug" dealers (MacCoun & Reuter 2001; Drug Policy Alliance 2003). Some evidence suggests that marijuana use among Dutch youth is decreasing, rebutting those who would argue that liberal drug policies result in increased drug abuse (Sheldon 2000; Burke 2006). Nonetheless, concerns over the "tolerance" policy and the millions of drug tourists who visit the Netherlands annually have led the incoming conservative government to make plans to require that coffee shops be "member-only" establishments (Pignal 2010).

Historically, Great Britain has also adopted a "medical model," particularly in regard to heroin and cocaine. As early as the 1960s, English doctors prescribed opiates and cocaine for their drug-addicted patients who were unlikely to quit using drugs on their

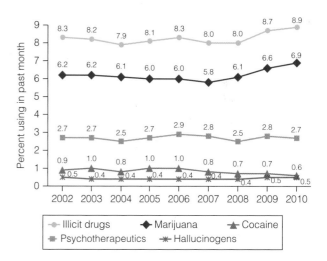

Figure 3.1: **Past Month Use of Selected Illicit Drugs among People Aged 12 or Older, 2002–2010**
Source: NSDUH 2011.

own and for the treatment of withdrawal symptoms. By the 1970s, however, British laws had become more restrictive, making it difficult for either physicians or users to obtain drugs legally. Today, British government policy provides for limited distribution of drugs by licensed drug treatment specialists to addicts who might otherwise resort to crime to support their habits (Abadinsky 2008).

What Do You Think? Proposition 19, if passed, would have made California the first state in the Union to legalize the use and possession of marijuana. The initiative "would have allowed adults age 21 and older to possess and grow small amounts of marijuana . . ." (*Leff & Wohlsen 2010, p. 1*). The measure, however, failed—54 percent to 46 percent. Some observers note that the measure failed because older, more conservative voters came out in droves to vote whereas many younger, pro-legalization voters remained at home. Do you think marijuana should be legalized? If so, what age limits, if any, would you impose?

In stark contrast to such health-based policies, many other countries execute drug users and/or dealers or subject them to corporal punishment, which may include whipping, stoning, beating, and torture. Such policies are found primarily in less developed nations, where religious and cultural prohibitions condemn the use of any type of drug, including alcohol and tobacco. For example, convicted drug offenders in Bangladesh, China, India, Laos, and Pakistan are subject to severe punishments including the threat of death (CNN 2009).

Drug Use and Abuse in the United States

In the United States, cultural definitions of drug use are contradictory—condemning it on the one hand (e.g., heroin), yet encouraging and tolerating it on the other (e.g., alcohol). At various times in U.S. history, many drugs that are illegal today were legal and readily available. In the 1800s and the early 1900s, opium was routinely used in medicines as a pain reliever, and morphine was taken as a treatment for dysentery and fatigue. Amphetamine-based inhalers were legally available until 1949, and cocaine was an active ingredient in Coca-Cola until 1906, when it was replaced with another drug—caffeine (Witters et al. 1992; Abadinsky 2008). Not surprisingly, Americans' concerns with drugs have varied over the years. In the 1970s, when drug use was at its highest, concern over drugs was relatively low. However, in 2010, nearly two-thirds of Americans believe that drug abuse is a significant problem in America (Reid 2010).

Use of illicit drugs in the United States is a fairly common phenomenon. For example, according to the National Survey on Drug Use and Health (NSDUH 2011), in 2010, 15.7 percent of the 12 and over population reported using marijuana in the previous year on almost a daily basis, i.e., 300 or more days in the year prior to the survey. Further, in 2010, 51.8 percent of Americans aged 12 or older reported consuming alcohol in the past month, and nearly 69.6 million Americans—27.4 percent of the 12 and older population—reported past-month use of tobacco products (NSDUH 2011).

Sociological Theories of Drug Use and Abuse

Drug abuse occurs when acceptable social standards of drug use are violated, resulting in adverse physiological, psychological, and/or social consequences. For example, when an individual's drug use leads to hospitalization, arrest, or divorce, such use is usually considered abusive. Drug abuse, however, does not always entail drug addiction. Drug addiction, or **chemical dependency,** refers to a condition in which drug use is compulsive—users are

drug abuse The violation of social standards of acceptable drug use, resulting in adverse physiological, psychological, and/or social consequences.

chemical dependency A condition in which drug use is compulsive and users are unable to stop because of physical and/or psychological dependency.

unable to stop because of their dependency. The dependency may be psychological (the individual needs the drug to achieve a feeling of well-being) and/or physical (withdrawal symptoms occur when the individual stops taking the drug). For example, withdrawal from marijuana includes depression, anger, decreased appetite, and restlessness (Zickler 2003). In 2010, more than 22.1 million Americans, 8.7 percent of the population 12 or older, were defined as being dependent on or abusers of alcohol and/or other drugs. Of that number, 15 million were dependent on or abused alcohol only, 4.2 million were dependent on or abused illicit drugs but not alcohol, and 2.9 million were dependent on or abused both illicit drugs and alcohol (NSDUH 2011). Individuals who are dependent on or abuse illicit drugs and/or alcohol are disproportionately male, American Indians, or Alaska natives, unemployed, and between the ages of 18 and 25 (NSDUH 2011).

Various theories provide explanations for why some people use and abuse drugs. Drug use is not simply a matter of individual choice. Theories of drug use explain how structural and cultural forces as well as biological and psychological factors influence drug use and society's responses to it.

Structural-Functionalist Perspective

Structural functionalists argue that drug abuse is a response to weakening societal norms. As society becomes more complex and as rapid social change occurs, norms and values become unclear and ambiguous, resulting in anomie—a state of normlessness. Anomie may exist at the societal level, resulting in social strains and inconsistencies that lead to drug use. For example, research indicates that increased alcohol consumption in the 1830s and the 1960s was a response to rapid social change and the resulting stress (Rorabaugh 1979). Anomie produces inconsistencies in cultural norms regarding drug use. For example, although public health officials and health care professionals warn of the dangers of alcohol and tobacco use, advertisers glorify the use of alcohol and tobacco, and the U.S. government subsidizes the alcohol and tobacco industries. Furthermore, cultural traditions, such as giving away cigars to celebrate the birth of a child and toasting a bride and groom with champagne, persist.

> Anomie produces inconsistencies in cultural norms regarding drug use. . . . [P]ublic health officials and health care professionals warn of the dangers of alcohol and tobacco use, advertisers glorify the use of alcohol and tobacco, and the U.S. government subsidizes the alcohol and tobacco industries.

Anomie may also exist at the individual level, as when a person suffers feelings of estrangement, isolation, and turmoil over appropriate and inappropriate behavior. An adolescent whose parents are experiencing a divorce, who is separated from friends and family as a consequence of moving, or who lacks parental supervision and discipline may be more vulnerable to drug use because of such conditions. Thus, from a structural-functionalist perspective, drug use is a response to the absence of a perceived bond between the individual and society and to the weakening of a consensus regarding what is considered acceptable.

Consistent with this perspective, in a national poll of Americans 18 years or older, peer pressure and lack of parental supervision were the two most common responses given for why teenagers take drugs (Pew Research Center 2002). Similarly, the importance of the family in deterring drug use is highlighted in the national youth media campaign—"Parents. The Anti-Drug" (ONDCP 2009).

Conflict Perspective

The conflict perspective emphasizes the importance of power differentials in influencing drug use behavior and societal values concerning drug use. From a conflict perspective, drug use occurs as a response to the inequality perpetuated by a capitalist system. Societal members, alienated from work, friends, and family as well as from society and

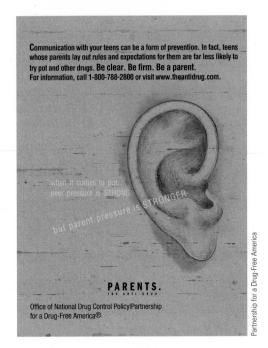

This poster from the Office of National Drug Control Policy's National Youth Anti-Drug Media Campaign emphasizes the importance of a close relationship between parent and child in the fight against drug use by youths.

its institutions, turn to drugs as a means of escaping the oppression and frustration caused by the inequality they experience. Furthermore, conflict theorists emphasize that the most powerful members of society influence the definitions of which drugs are illegal and the penalties associated with illegal drug production, sales, and use.

For example, alcohol is legal because it is often consumed by those who have the power and influence to define its acceptability—white males (NSDUH 2010). This group also disproportionately profits from the sale and distribution of alcohol and can afford powerful lobbying groups in Washington D.C., to guard the alcohol industry's interests. Because this group also commonly uses tobacco and caffeine, societal definitions of these substances are also relatively accepting. Conversely, minority group members disproportionately use crack cocaine rather than powder cocaine (Mauer 2009). Although the pharmacological properties of the two drugs are the same, possession of 5 grams of crack cocaine carries the same penalty under federal law as possession of 500 grams of powdered cocaine (Taifia 2006). In 2010, Congress voted to change the 1986 law that established the 100 to 1 ratio sentencing disparity but also eliminates the five-year mandatory minimum for first time possession of crack cocaine. Of people convicted of crack offenses, 80 percent are African American (Abrams 2010).

The use of opium by Chinese immigrants in the 1800s provides a historical example. The Chinese, who had been brought to the United States to work on the railroads, regularly smoked opium as part of their cultural tradition. As unemployment among white workers increased, however, so did resentment of Chinese laborers. Attacking the use of opium became a convenient means of attacking the Chinese, and in 1877, Nevada became the first of many states to prohibit opium use. As Morgan (1978) observed:

> The first opium laws in California were not the result of a moral crusade against the drug itself. Instead, it represented a coercive action directed against a vice that was merely an appendage of the real menace—the Chinese—and not the Chinese per se, but the laboring "Chinamen" who threatened the economic security of the white working class. (p. 59)

The criminalization of other drugs, including cocaine, heroin, and marijuana, follows similar patterns of social control of the powerless, political opponents, and/or minorities. In the 1940s, marijuana was used primarily by minority group members, and users faced severe criminal penalties. However, after white, middle-class college students began to use marijuana in the 1970s, the government reduced the penalties associated with its use. Although the nature and pharmacological properties of the drug had not changed, the population of users was now connected to power and influence. Thus, conflict theorists regard the regulation of certain drugs, as well as drug use itself, as a reflection of differences in the political, economic, and social power of various interest groups.

Symbolic Interactionist Perspective

Symbolic interactionism, which emphasizes the importance of definitions and labeling, concentrates on the social meanings associated with drug use. If the initial drug use experience is defined as pleasurable, it is likely to recur, and the individual may earn the label of "drug user" over time. If this definition is internalized so that the individual assumes an identity of a drug user, the behavior will probably continue and may even escalate. Conversely, Copes et al. (2008) observed that respondents who self-identified as "hustlers" rather than "crackheads" were less likely to fall prey to the debilitating effects of the drug for ". . . [s]lipping into uncontrollable addiction is antithetical to the hustler identity . . ." (p. 256).

Drug use is also learned through symbolic interaction in small groups. In a study of binge drinking, researchers found that students who believed that their friends were binge drinking were more likely to also engage in binge drinking themselves

(Weitzman et al. 2003). First-time users learn not only the motivations for drug use and its techniques but also what to experience. Becker (1966) explained how marijuana users learn to ingest the drug. A novice being coached by a regular user reported the experience:

> I was smoking like I did an ordinary cigarette. He said, "No, don't do it like that." He said, "Suck it, you know, draw in and hold it in your lungs . . . for a period of time." I said, "Is there any limit of time to hold it?" He said, "No, just till you feel that you want to let it out, let it out." So I did that three or four times. (p. 47)

Marijuana users not only learn the way to ingest the smoke but also to label the experience positively. When peers define certain drugs, behaviors, and experiences as not only acceptable but also pleasurable, drug use is likely to continue.

Interactionists also emphasize that symbols can be manipulated and used for political and economic agendas. The popular DARE (Drug Abuse Resistance Education) program, with its antidrug emphasis fostered by local schools and police, carries a powerful symbolic value with which politicians want the public to identify. "Thus, ameliorative programs which are imbued with these potent symbolic qualities (like DARE's links to schools and police) are virtually assured widespread public acceptance (regardless of actual effectiveness) which in turn advances the interests of political leaders who benefit from being associated with highly visible, popular symbolic programs" (Wysong et al. 1994, p. 461). Ironically, a meta-analysis of the program led West and O'Neal (2004) to conclude that DARE does not significantly prevent drug use among school-aged children.

Biological and Psychological Theories

Drug use and addiction are probably the result of a complex interplay of social, psychological, and biological forces. For example, some researchers suggest that drug use and addiction are caused by a "bio-behavioral disorder," which combines biological and psychological factors (Margolis & Zweben 2011). Biological research has primarily concentrated on the role of genetics in predisposing an individual to drug use. Research indicates that severe, early-onset alcoholism may be genetically predisposed, with some men having 10 times the risk of addiction as those without a genetic predisposition. Interestingly, other problems such as depression, chronic anxiety, and attention deficit disorder are also linked to the likelihood of addiction. Nonetheless, researchers warn, "Nobody is predestined to be an alcoholic" (Firshein 2003).

Biological theories of drug use also hypothesize that some individuals are physiologically predisposed to experience more pleasure from drugs than others and, consequently, are more likely to be drug users. According to these theories, the central nervous system, which is composed primarily of the brain and spinal cord, processes drugs through neurotransmitters in a way that produces an unusually euphoric experience. Individuals not so physiologically inclined reported less pleasant experiences and are less likely to continue use (Jarvik 1990; National Institute on Alcohol Abuse and Alcoholism 2000).

Psychological explanations focus on the tendency of certain personality types to be more susceptible to drug use. Individuals who are particularly prone to anxiety may be more likely to use drugs as a way to relax, gain self-confidence, or ease tension. For example, research indicates that child maltreatment, particularly among females, contributes to alcohol and drug abuse that extends into adulthood (Gilbert et al. 2009).

Psychological theories of drug abuse also emphasize that drug use may be maintained by positive or negative reinforcement. Thus, for example, cocaine use may be maintained as a result of the rewarding "high" it produces—a positive reinforcement. Alternatively, heroin use, often associated with severe withdrawal symptoms, may continue as a result of a negative reinforcement, that is, the distress the user feels when faced with withdrawal (Abadinsky 2008). Reinforcement may come from a variety of sources including the media. In a study of the portrayal of alcohol use in 601 contemporary movies, researchers found that exposure to alcohol use in the movies was positively associated with early-onset drinking (Sargant et al. 2006).

Indicate whether you have or have not experienced any of the following in the last 12 months as a consequence of your own drinking. When finished, compare your responses to those of a national sample of college students.

Consequence	Yes	No
1. Did something you later regretted	_____	_____
2. Forgot where you were or what you did	_____	_____
3. Got in trouble with the police	_____	_____
4. Had sex with someone without giving your consent	_____	_____
5. Had sex with someone without getting their consent	_____	_____
6. Had unprotected sex	_____	_____
7. Physically injured yourself	_____	_____
8. Physically injured another person	_____	_____
9. Seriously considered suicide	_____	_____
10. Reported one or more of the above	_____	_____

These survey items are from the American College Health Association's National College Health Assessment II (2011). The following data are for 2010. Only "those institutions that surveyed all students, or used a random sampling technique, are included in the analysis, yielding a final data set consisting of 30,093 students and 39 schools"

(American College Health Association 2011, p. 2). All students are from public or private, two- or four-year colleges and universities.

Consequence	Percentage Reporting Consequence		
	Females	Males	Total
1. Did something you later regretted	32.1	32.7	32.4
2. Forgot where you were or what you did	27.9	32.7	29.7
3. Got in trouble with the police	2.9	6.7	4.3
4. Had sex with someone without giving your consent	2.2	1.8	2.1
5. Had sex with someone without getting their consent	0.3	1.0	0.6
6. Had unprotected sex	15.1	17.6	16.1
7. Physically injured yourself	14.2	16.6	15.1
8. Physically injured another person	1.6	4.3	2.7
9. Seriously considered suicide	1.6	2.0	1.8
10. Reported one or more of the above	47.4	52.2	49.2

Source: American College Health Association. 2011. *National College Health Assessment II*. Reference Group Executive Summary, Fall 2010. Baltimore, Maryland: American College Health Association. http://www.acha-ncha.org/docs/ACHA-NCHA-II_ReferenceGroup_ExecutiveSummary_Spring2010.pdf

Frequently Used Legal Drugs

Social definitions regarding which drugs are legal or illegal vary over time, circumstance, and societal forces. In the United States, two of the most dangerous and widely abused drugs, alcohol and tobacco, are legal.

Alcohol: The Drug of Choice

Americans' attitudes toward alcohol have a long and varied history. Although alcohol was a common beverage in early America, by 1920, the federal government had prohibited its manufacture, sale, and distribution through the passage of the Eighteenth Amendment to the U.S. Constitution. Many have argued that Prohibition, like the opium regulations of the late 1800s, was in fact a "moral crusade" (Gusfield 1963) against immigrant groups who were more likely to use alcohol. The amendment had little popular support and was repealed in 1933. Today, the U.S. population is experiencing a resurgence of concern about alcohol. What has been called a "new temperance" has manifested itself in federally mandated 21-year-old drinking age laws, warning labels on alcohol bottles, increased concern over fetal alcohol syndrome and underage drinking, stricter enforcement of drinking and driving regulations (e.g., checkpoint traffic stops), and zero-tolerance policies. Such practices may have had an effect on drinking norms. Between 2002 and 2010, the rate of heavy alcohol use for 12 to 20-year-olds decreased by about 1.1 percent (NSDUH 2011).

Despite such restrictive policies, alcohol remains the most widely used and abused drug in America. According to a recent poll, 67 percent of U.S. adults drink alcohol, the highest number recorded since 1985 (Newport 2011). Although most people who drink alcohol do so moderately and experience few negative effects (see this chapter's *Self and Society* feature), people with alcoholism are psychologically and physically addicted to alcohol and suffer various degrees of physical, economic, psychological, and personal harm.

The National Survey on Drug Use and Health, conducted by the U.S. Department of Health and Human Services, reported that 131.3 million Americans age 12 and older consumed alcohol at least once in the month preceding the survey; that is, they were *current users* (NSDUH 2011). Of this number, 6.7 percent reported **heavy drinking**, and 23.1 percent—58.6 million people—reported **binge drinking.**

Even more troubling were the 10.0 million current users of alcohol who were 12 to 20 years old—underage drinkers—many of whom got their alcohol for free from adults—often parents. Nearly 17 percent were binge drinkers and 5.1 percent were heavy drinkers (NSDUH 2011). Although teen drinking has decreased in recent years, binge drinking in college continues to attract the public's attention. The likelihood of a college student binge drinking is impacted by environmental variables including place of residence (e.g., on campus versus off campus); cost and availability of alcohol; campus, local, and state alcohol policies; age, gender, and ethnic and racial makeup of the student population; prevention strategies; and the college drinking culture (Wechsler & Nelson 2008).

What Do You Think? There are many different drinking games and, up until recently, beer pong was one of the most popular. But there's a new game in town—"icing." The rules are easy (Goodman 2010): You hand a friend a Smirnoff Ice Malt and they have to drop on one knee and drink it down. The only exception is if the intended "icing" victim already has a Smirnoff Ice in hand (called an "ice block"), in which case the "attacker" must drink both bottles of the sugary drink or, in some versions, the two players each drink one of the brews down. There are even several instructional videos on YouTube. Do you think that "icing" encourages binge drinking and, if so, what recommendations would you make to limit the severity of the consequences?

Many binge drinkers began drinking in high school, with almost one-third having their first drink before age 13. Research indicates that the younger the age of onset, the higher the probability that an individual will develop a drinking disorder at some time in his or her life (Hingson et al. 2006; Behrendt et al. 2009). For example, an individual's chance of becoming dependent on alcohol is 40 percent if the person's drinking began before the age of 13. Additional results from the National Survey on Drug Use and Health (2011) include the following:

> Research indicates that the younger the age of onset, the higher the probability that an individual will develop a drinking disorder at some time in his or her life.

- The highest levels of both heavy and binge drinking are among 21- to 25-year-olds; people 65 or older had the lowest rates of binge drinking.
- Rates of alcohol use are higher among the full-time employed adults than among the unemployed; however, patterns of heavy or binge drinking are highest among the unemployed.
- Binge drinking rates for full-time college students decreased between 2009 and 2010; however, the rate of heavy drinking among those enrolled full-time in college is higher than for those not enrolled in college full-time.
- Asians are least likely to report binge drinking, and Hispanics or Latinos are most likely to report it.
- Underage current drinkers, i.e., those between the ages of 12 and 20, were most likely from the Northeast, followed by the Midwest, West, and South.
- More males than females age 12 to 20 reported binge drinking, heavy drinking, and current alcohol use.

Despite the fact that males are more likely than females to abuse alcohol, some evidence suggests that female drinking is on the rise (Armstrong & McCarroll 2004; Zailckas 2005).

heavy drinking As defined by the U.S. Department of Health and Human Services, five or more drinks on the same occasion on each of five or more days in the past 30 days prior to the National Survey on Drug Use and Health.

binge drinking As defined by the U.S. Department of Health and Human Services, drinking five or more drinks on the same occasion on at least one day in the past 30 days prior to the National Survey on Drug Use and Health.

Andy Kropa/Redux

For many students, tailgating at football games is an essential part of college life. But tailgating, going "downtown," the fraternity or sorority party, happy hour, Greek initiation rituals, and all the other events deemed suitable for drinking alcohol, have led to high frequency and consumption rates for college students.

The Tobacco Epidemic

Native Americans first cultivated tobacco and introduced it to the European settlers in the 1500s. The Europeans believed that tobacco had medicinal properties, and its use spread throughout Europe, ensuring the economic success of the colonies in the New World. Tobacco was initially used primarily through chewing and snuffing, but smoking became more popular in time, even though scientific evidence that linked tobacco smoking to lung cancer existed as early as 1859 (Feagin & Feagin 1994). However, the U.S. Surgeon General did not conclude that tobacco products are addictive and that nicotine causes dependency until 1989.

Globally, over 80 percent of the 1.1 billion smokers in the world live in low- or middle-income countries. Researchers using what is called **meta-analysis** examined 125 scientific papers that included 31,146,096 respondents worldwide (WHO 2011c). The results are counterintuitive. Despite the obvious additional cost of using tobacco products, the analysis found strong evidence of an inverse relationship between income and smoking, i.e., as income goes down, the prevalence of smoking goes up—a relationship that is stronger for younger age groups than older. The authors suggest a four-stage explanatory model:

> In earlier stages, smoking disseminates among higher-income groups who are more open to innovation. During the intermediate stages, smoking diffuses to the rest of the population. Later, smoking declines among the high-income level strata, as they are concerned with health, fitness, and the harm of smoking. Only after a long history of cigarette consumption, when all SES [socio-economic status] groups have been similarly exposed to smoking, does the inverse social status gradient emerge. (p. 27)

Tobacco is one of the most widely used drugs in the United States. According to a U.S. Department of Health and Human Services survey, 69.6 million Americans—27.4 percent of those 12 and older—are current tobacco users (NSDUH 2011). Use of all tobacco products, including smokeless tobacco (8.9 million users), cigars (13.2 million users), pipe tobacco (2.2 million users), and cigarettes (58.3 million), is higher for high school graduates than for college graduates, males, and Americans Indians and Alaska natives (NSDUH 2011).

In 2010, 10.7 percent of the 12 and 17 population reported use of a tobacco product in the past month (NSDUH 2011). Research evidence suggests that youth develop attitudes and beliefs about tobacco products at an early age (Freeman et al. 2005). Advertising of tobacco products continues to have an influence on youth, despite the 2009 Family Smoking Prevention and Tobacco Control Act, which, among other provisions, outlawed flavored cigarettes most often marketed to children. Tobacco company executives, however, have argued that the 2009 act only covers "cigarettes, cigarette tobacco, roll-your-own tobacco, and smokeless tobacco" and that cigars are excluded from the control of the Food and Drug Administration (Myers 2011, p.1). Several tobacco companies have begun selling small, flavored cigars, heightening fears that marketing tobacco products to children is likely to continue.

There is also considerable evidence that cigarette advertisers target minorities. For example, Primack et al. (2007) found that tobacco advertisements in African American communities were 2.6 times higher per person than in white communities. Further, the likelihood of tobacco-related billboards was 70 percent higher in African American than in white communities. Also, the Food and Drug Administration's (FDA 2011a)

meta-analysis Meta-analysis combines the results of several studies addressing a research question; i.e., it is the analysis of analyses.

Tobacco Products Scientific Advisory Committee reports that "menthol cigarettes are marketed disproportionately to younger smokers" *and* "disproportionately marketed per capita to African Americans" (p. 40–41). Finally, the tobacco industry has developed advertising campaigns targeting women, including Philip Morris's "purse packs" modeled after cosmetic cases and half the size of a regular package of cigarettes (Tobacco Free Kids 2009). Such initiatives are also taking place in developing countries where:

> Because most women currently do not use tobacco, the tobacco industry aggressively markets to them to tap this potential new market. Advertising, promotion, and sponsorship, including charitable donations to women's causes, weaken

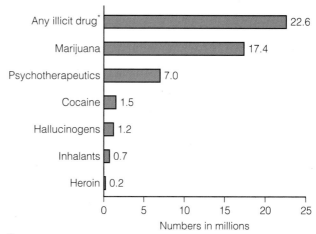

* Illicit drugs include marijuana/hashish, cocaine (including crack), heroin, hallucinogens, inhalants, or prescription-type psychotherapeutics used nonmedically.

Figure 3.2: Past-Month Illicit Drug Use among People Aged 12 or Older, 2010
Source: NSDUH 2011.

cultural opposition to women using tobacco. Product design and marketing, including the use of attractive models in advertising and brands marketed specifically to women, are explicitly crafted to encourage women to smoke (WHO 2008, p. 16–17).

Frequently Used Illegal Drugs

More than 22.6 million people in the United States are current illicit drug users, representing 8.9 percent of the population age 12 and older. Users of illegal drugs, although varying by type of drug used, are more likely to be male, to be young, and to be a member of a minority group (NSDUH 2011).

Marijuana Madness

Marijuana is the most commonly used and most heavily trafficked illicit drug in the world (see Table 3.1 for a list of commonly abused drugs, their commercial and street names, and their intoxication and health effects). Globally, there are between 129 and 191 million marijuana users representing 2.9 to 4.3 percent of the world's 15- to 64-year-old population. Regionally, marijuana is also the most dominant illicit drug, and its cultivation and consumption are particularly high in the Americas. For example, the largest producers of marijuana in the world are Mexico, followed by Paraguay and the United States (WDR 2010).

Marijuana's active ingredient is THC (Δ^9-tetrahydrocannabinol), which in varying amounts can act as a sedative or a hallucinogen. When just the top of the marijuana plant is sold, it is called hashish. Hashish is much more potent than marijuana, which comes from the entire plant. Marijuana use dates back to 2737 b.c. in China, and marijuana has a long tradition of use in India, the Middle East, and Europe. In North America, hemp, as it was then called, was used to make rope and as a treatment for various ailments. Nevertheless, in 1937, Congress passed the Marijuana Tax Act, which restricted the use of marijuana; the law was passed as a result of a media campaign that portrayed marijuana users as "dope fiends" and, as conflict theorists note, was enacted at a time of growing sentiment against Mexican immigrants (Witters et al. 1992).

There are more than 17.4 million current marijuana users, representing 6.9 percent of the U.S. population age 12 and older (NSDUH 2011). According to the most recent Monitoring the Future (MTF) survey, *daily marijuana use* increased dramatically in 8th, 10th, and 12th grades, with daily use estimated to be 1.2 percent, 3.3 percent, and 6.1 percent, respectively. As the study notes, nearly one in 16 high school seniors is a current daily, or near-daily, marijuana user (MTF 2011). Further, of 12 and older first-time

TABLE 3.1 Commonly Abused Drugs

SUBSTANCES: CATEGORY AND NAME	EXAMPLES OF COMMERCIAL AND STREET NAMES	DEA SCHEDULE*/ HOW ADMINISTERED**	ACUTE EFFECTS/HEALTH RISKS
Tobacco			
Nicotine	Found in cigarettes, cigars, bidis, and smokeless tobacco (snuff, spit tobacco, chew)	Not scheduled/smoked, snorted, chewed	Increased blood pressure, and heart rate/chronic lung disease; cardiovascular disease; stroke; cancers of the mouth, pharynx, larynx, esophagus, stomach, pancreas, cervix, kidney, bladder, and acute myeloid leukemia; adverse pregnancy outcomes; addiction
Alcohol			
Alcohol (ethyl alcohol)	Found in liquor, beer, and wine	Not scheduled/swallowed	In low doses, euphoria, mild stimulation, relaxation, lowered inhibitions; in higher doses, drowsiness, slurred speech, nausea, emotional volatility, loss of coordination, visual distortions, impaired memory, sexual dysfunction, loss of consciousness/increased risk of injuries, violence, fetal damage (in pregnant women); depression; neurologic deficits; hypertension; liver and heart disease; addiction; fatal overdose
Cannabinoids			
marijuana	Blunt, dope, ganja, grass, herb, joint, bud, Mary Jane, pot, reefer, green, trees, smoke, sinsemilla, skunk, weed	I/smoked, swallowed	Euphoria; relaxation; slowed reaction time; distorted sensory perception; impaired balance and coordination; increased heart rate and appetite; impaired learning, memory; anxiety; panic attacks; psychosis/cough, frequent respiratory infections; possible mental health decline; addiction
hashish	Boom, gangster, hash, hash oil, hemp	I/smoked, swallowed	
Opioids			
Heroin	Diacetylmorphine: smack, horse, brown sugar, dope, H, junk, skag, skunk, white horse, China white; cheese (with OTC cold medicine and antihistamine)	I/injected, smoked, snorted	Euphoria; drowsiness; impaired coordination; dizziness; confusion; nausea; sedation; feeling of heaviness in the body; slowed or arrested breathing/constipation; endocarditis; hepatitis; HIV; addiction; fatal overdose
Opium	Laudanum, paregoric: big O, black stuff, block, gum, hop	II, III, V/swallowed, smoked	
Stimulants			
Cocaine	Cocaine hydrochloride: blow, bump, C, candy, Charlie, coke, crack, flake, rock, snow, toot	II/snorted, smoked, injected	Increased heart rate, blood pressure, body temperature, metabolism; feelings of exhilaration; increased energy, mental alertness; tremors; reduced appetite; irritability; anxiety; panic; paranoia; violent behavior; psychosis/weight loss, insomnia; cardiac or cardiovascular complications; stroke; seizures; addiction
Amphetamine	Biphetamine, Dexedrine: bennies, black beauties, crosses, hearts, LA turnaround, speed, truck drivers, uppers	II/swallowed, snorted, smoked, injected	**Also, for cocaine**—nasal damage from snorting
Methamphetamine	Desoxyn: meth, ice, crank, chalk, crystal, fire, glass, go fast, speed	II/swallowed, snorted, smoked, injected	**Also, for methamphetamine**—severe dental problems
Club Drugs			
MDMA (methylenedioxy-methamphetamine)	Ecstasy, Adam, clarity, Eve, lover's speed, peace, uppers	I/swallowed, snorted, injected	**MDMA**—mild hallucinogenic effects; increased tactile sensitivity; empathic feelings; lowered inhibition; anxiety; chills; sweating; teeth clenching; muscle cramping/sleep disturbances; depression; impaired memory; hyperthermia; addiction

TABLE 3.1 *(Continued)*

SUBSTANCES: CATEGORY AND NAME	EXAMPLES OF COMMERCIAL AND STREET NAMES	DEA SCHEDULE*/ HOW ADMINISTERED**	ACUTE EFFECTS/HEALTH RISKS
Flunitrazepam***	Rohypnol: forget-me pill, Mexican Valium, R2, roach, Roche, roofies, roofinol, rope, rophies	IV/swallowed, snorted	**Flunitrazepam**—sedation; muscle relaxation; confusion; memory loss; dizziness; impaired coordination/addiction
GHB***	Gamma-hydroxybutyrate: G, Georgia home boy, grievous bodily harm, liquid Ecstasy, soap, scoop, goop, liquid X	I/swallowed	**GHB**—drowsiness; nausea; headache; disorientation; loss of coordination; memory loss/unconsciousness; seizures; coma
Dissociative Drugs			
Ketamine	Ketalar SV: cat Valium, K, Special K, vitamin K	III/injected, snorted, smoked	Feelings of being separate from one's body and environment; impaired motor function/anxiety; tremors; numbness; memory loss; nausea
PCP and analogs	Phencyclidine: angel dust, boat, hog, love boat, peace pill	I, II/swallowed, smoked, injected	**Also, for ketamine**—analgesia; impaired memory; delirium; respiratory depression and arrest; death
Salvia divinorum	Salvia, Shepherdess's Herb, Maria Pastora, magic mint, Sally-D	Not scheduled/chewed, swallowed, smoked	**Also, for PCP and analogs**—analgesia; psychosis; aggression; violence; slurred speech; loss of coordination; hallucinations
Dextromethorphan (DXM)	Found in some cough and cold medications: Robotripping, Robo, Triple C	Not scheduled/swallowed	**Also, for DXM**—euphoria; slurred speech; confusion; dizziness; distorted visual perceptions
Hallucinogens			
LSD	Lysergic acid diethylamide: acid, blotter, cubes, microdot yellow sunshine, blue heaven	I/swallowed, absorbed through mouth tissues	Altered states of perception and feeling; hallucinations; nausea
Mescaline	Buttons, cactus, mesc, peyote	I/swallowed, smoked	**Also, LSD and mescaline**—increased body temperature, heart rate, blood pressure; loss of appetite; sweating; sleeplessness; numbness, dizziness, weakness, tremors; impulsive behavior; rapid shifts in emotion
Psilocybin	Magic mushrooms, purple passion, shrooms, little smoke	I/swallowed	**Also, for LSD**—flashbacks, hallucinogen persisting perception disorder **Also for psilocybin**—nervousness; paranoia; panic
Other Compounds			
Anabolic steroids	Anadrol, Oxandrin, Durabolin, Depo-Testosterone, Equipoise: roids, juice, gym candy, pumpers	III/injected, swallowed, applied to skin	**Steroids**—no intoxication effects/hypertension; blood clotting and cholesterol changes; liver cysts; hostility and aggression; acne; in adolescents—premature stoppage of growth; in males—prostate cancer, reduced sperm production, shrunken testicles, breast enlargement; in females—menstrual irregularities, development of beard and other masculine characteristics
Inhalants	Solvents (paint thinners, gasoline, glues); gases (butane, propane, aerosol propellants, nitrous oxide); nitrites (isoamyl, isobutyl, cyclohexyl): laughing gas, poppers, snappers, whippets	Not scheduled/inhaled through nose or mouth	**Inhalants** (varies by chemical)—stimulation; loss of inhibition; headache; nausea or vomiting; slurred speech; loss of motor coordination; wheezing/cramps; muscle weakness; depression; memory impairment; damage to cardiovascular and nervous systems; unconsciousness; sudden death

* Schedule I and II drugs have a high potential for abuse. They require greater storage security and have a quota on manufacturing, among other restrictions. Schedule I drugs are available for research only and have no approved medical use; Schedule II drugs are available only by prescription (unrefillable) and require a form for ordering. Schedule III and IV drugs are available by prescription, may have five refills in six months, and may be ordered orally. Some Schedule V drugs are available over the counter.

** Some of the health risks are directly related to the route of drug administration. For example, injection drug use can increase the risk of infection through needle contamination with staphylococci, HIV, hepatitis, and other organisms.

*** Associated with sexual assaults.

Source: NIDA (2010).

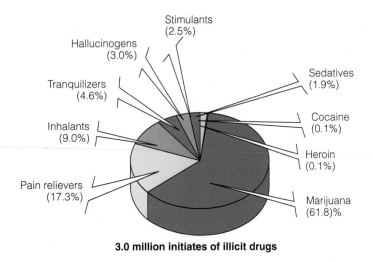

3.0 million initiates of illicit drugs

Note: The percentages do not add to 100 percent due to rounding or because a small number of respondents initiated multiple drugs on the same day. The first specific drug refers to the one that was used on the occasion of first-time use of any illicit drug.

Figure 3.3: First Illicit Drug Used among Those Who Initiated Illicit Drug Use in the Past year, Aged 12 and Older, 2010
Source: NSDUH 2011.

users of an illicit drug, nearly 61.8 percent initiate with marijuana (NSDUH 2011) (see Figure 3.3).

In the present debate over the legalization of marijuana, many express fears that it is a **gateway drug,** the use of which causes progression to other drugs. "The gateway hypothesis holds that consumption of abusable drugs progresses in orderly fashion through several discrete stages. The entire sequence, which is exhibited by only a small minority of drug users, begins with beer or wine and moves progressively through hard liquor or tobacco, marijuana, and finally hard drugs" (Tarter et al. 2006, p. 2,134). Most research suggests, however, that people who experiment with one drug are more likely to experiment with another. Indeed, most drug users use several drugs at the same time. As Lee and Abdel-Ghany (2004) note, there is a strong "contemporaneous relationship between smoking cigarettes, drinking alcohol, smoking marijuana, and using cocaine" (p. 454).

> **What Do You Think?** K2, sometimes called Spice, is an herbal concoction that contains synthetic chemicals that, when smoked, act on the cannabinoid receptors in the brain and produces a "high" similar to marijuana (Fay 2010). Often sold as incense, according to one convenience store owner, K2 is " . . . selling like crazy. . . . more than cigarettes, more than pipe tobacco, posters or T-shirts." (McDonald 2010, p. 1). However, in 2011, the U.S. Drug Enforcement Administration (DEA) exercised its authority to temporarily ban the synthetic chemicals used in making K2 (DEA 2011). Although there are health concerns (see Physical and Mental Health Costs later in this chapter), it could be argued that the risks of K2 are no greater than that of alcohol or tobacco use and perhaps less. Do you think that K2 should be banned and, if so, why?

Cocaine: From Coca-Cola to Crack

Cocaine is classified as a stimulant and, as such, produces feelings of excitation, alertness, and euphoria. Although prescription stimulants such as methamphetamine and dextroamphetamine are commonly abused, societal concern over drug abuse has focused on cocaine over the last 20 years. Its increased use, addictive qualities, physiological effects, and worldwide distribution have fueled such concerns. More than any other single substance, cocaine led to the early phases of the war on drugs.

Cocaine, which is made from the coca plant, has been used for thousands of years. Coca leaves were used in the original formula for Coca-Cola but, in the early 1900s, anti-cocaine sentiment emerged as a response to the heavy use of cocaine among urban blacks, poor whites, and criminals (Witters et al. 1992; Thio 2007; Friedman-Rudovsky 2009). Cocaine was outlawed in 1914, by the Harrison Narcotics Tax Act, but its use and effects continued to be misunderstood. For example, a 1982 *Scientific American* article suggested that cocaine was no more habit-forming than potato chips (Van Dyck & Byck

gateway drug A drug (e.g., marijuana) that is believed to lead to the use of other drugs (e.g., cocaine).

1982). As demand and then supply increased, prices fell from $100 a dose to $10 a dose, and "from 1978 to 1987, the U.S. experienced the largest cocaine epidemic in history" (Witters et al. 1992, p. 256).

Globally, cocaine use has decreased, particularly in North America where the largest number of cocaine users is found (WDR 2010). This general decrease varies across regions, with cocaine use continuing to decline in developed countries but increasing in developing counties. Europe is the exception. In Europe, cocaine use has increased over the last decade and is now the second most commonly used illicit drug after cannabis (EMCDDA 2010).

According to the National Survey on Drug Use and Health, 1.5 million Americans 12 years and older are current cocaine users, representing 0.6 percent of that population (NSDUH 2011). In 2009, the current prevalence of cocaine use by 8th, 10th, and 12th graders was 0.9 percent. The percentage of 12th graders indicating that getting cocaine is "fairly easy" or "very easy" decreased between 2007 (47.1 percent) and 2010 (36.0 percent), with 58.1 percent of 12th graders responding that trying cocaine once or twice is a "great risk" (MTF 2011).

Crack is a crystallized product made by boiling a mixture of baking soda, water, and cocaine. The result, also called rock, base, and gravel, is relatively inexpensive and was not popular until the mid-1980s. Crack is one of the most dangerous drugs to surface in decades. Crack dealers often give drug users their first few "hits" free, knowing the drug's intense high and addictive qualities are likely to produce returning customers. An addiction to crack can take six to ten weeks; an addiction to pure cocaine can take three to four years (Thio 2007).

According to the National Survey on Drug Use and Health, the number of people 12 and older who used crack cocaine for the first time decreased from 337,000 to 83,000 between 2002 and 2010 (NSDUH 2011). Results from the Monitoring the Future survey indicate that this general decline in crack use is reflected in the 8th, 10th, and 12th grade population (MTF 2011). When high school seniors were asked, *"How much do you think people risk harming themselves (physically or in other ways) if they try crack once or twice?*—nearly 50 percent responded at "great risk."

Methamphetamine: The Meth Epidemic

Methamphetamine is a central nervous system stimulant that is highly addictive. Although the drug has only recently become popular, it is not new.

During the Second World War, soldiers on both sides used it to reduce fatigue and enhance performance. Hitler was widely believed to be a meth addict. Later, in the 1960s, President John Kennedy also used the drug and soon it caught on among so-called "speed freaks." But, because it was extremely expensive as well as difficult to obtain, meth was never close to being as widely used as cocaine (Thio 2007, p. 276).

Today, methamphetamine is relatively inexpensive and easily obtained, with more than 36.8 percent of high school seniors reporting that "crystal meth," i.e., methamphetamine in its crystalline form, is "fairly easy" or "very easy" to get (MTF 2011). Ease of obtaining the drug is due, in part, to the large amounts of methamphetamine smuggled into the United States from Mexico, and the ease of production. Because methamphetamine can be made from cold medications such as Sudafed, the U.S. Congress passed the Comprehensive Methamphetamine Control Act of 1996 that made obtaining the chemicals needed to make methamphetamine more difficult (ONDCP 2006; Thio 2007). In 2006, the Combat Methamphetamine Epidemic Act, which further articulated standards for selling over-the-counter medications used in methamphetamine production, went into effect.

Gone are the days, however, of "filthy containers simmering over open flames, cans of flammable liquids and hundreds of pills, . . . and foul odors, sometimes spark[ing] explosions . . . " (AP 2010, p. 1). Producers realized that by making smaller amounts at a time, they could use less pseudoephedrine, thereby "flying under the radar" of the 1996 and 2006 anti-meth laws. The new "shake-and-bake" process is so easy that drug users can make crystal meth while driving—add a couple of handfuls of cold medications to a

crack A crystallized illegal drug product produced by boiling a mixture of baking soda, water, and cocaine.

For those who use methamphetamine, the physical transformation is remarkable. The time lapse between the before (left) and after (right) pictures of this methamphetamine user is only three years, five months.

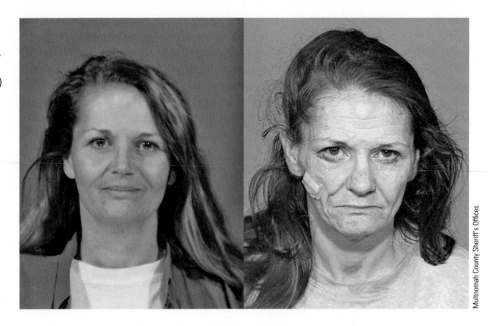

Multnomah County Sheriff's Offices

two-liter bottle, pour in some chemicals, and shake. Not surprisingly, after years of declining methamphetamine-related arrests, the number of do-it-yourself users has contributed to a recent increase in seizures and arrests (AP 2010, p.1).

The new "shake-and-bake" process is so easy that drug users can make crystal meth while driving—add a couple of handfuls of cold medication to a two-liter bottle, pour in some chemicals, and shake.

Globally, in terms of consumption, with the exception of Czech and Slovak countries, use of methamphetamine in Europe is fairly low (EMCDDA 2009). Further, the prevalence of monthly methamphetamine use in the United States for the 12 and older population has remained stable in recent years, although decreasing by 0.1 percent between 2009 (502,000 people) and 2010 (353,000 people). (NSDUH 2011). In 2010, 40.9 percent of past-year methamphetamine users reported that they obtained the methamphetamine for free from a friend or relative. Lifetime prevalence of methamphetamine for 8th, 10th, and 12th graders is 2.2 percent (MTF 2011).

Other Illegal Drugs

Other drugs abused throughout the world include **club drugs** (e.g., LSD and Ecstasy), heroin, "bath salts," prescription drugs (e.g., tranquilizers and amphetamines), and inhalants (e.g., glue).

Club Drugs Worldwide, over 10.5 million people between the ages of 15 and 64 use Ecstasy, the most common name for MDMA, at least once a year. Of those 10.5 million users, the highest use of this drug is in Oceania and the lowest in Africa (WDR 2010). In the United States in 2010, 695,000 people 12 and older (0.3 percent) reported past-month Ecstasy use (NSDUH 2011). Lifetime prevalence, i.e., whether or not they have ever used Ecstasy, by 8th, 10th, and 12th graders is 5.5 percent (MTF 2011).

Ecstasy produces a visual albeit mild effect when ingested; the hallucinatory effects of ketamine and LSD are notable. The use of ketamine, an animal tranquilizer, is growing globally with, for example, rates for the under-21 population in Hong Kong reaching 73 percent in 2006 (Brownell 2009). In the United States, ketamine use by 8th, 10th, and 12th graders has decreased in recent years to annual prevalence rates of 1.0 percent, 1.1 percent, and 1.6 percent, respectively (MTF 2011).

LSD is a synthetic hallucinogen, although many other hallucinogens are produced naturally (e.g., salvia). In 2010, among people aged 12 or older, 377,000 reported past-year

club drugs A general term for illicit, often synthetic, drugs commonly used at nightclubs or all-night dances called "raves."

initiation of LSD. The average age of first use was 19.1 years (NSDUH 2011). Between 2002 and 2010, youth between the ages of 12 and 17 reported lower rates of perceived risk of LSD use. According to the National Survey on Drug Use and Health (2011), 14.8 percent of 12 to 17 year olds believe that LSD would be fairly or very easily available.

Salvia is an herb grown in Mexico and Central and South America and imported into the United States (DEA 2011). The plant as well as seeds, cuttings, and leaves, are sold over the Internet and in tobacco shops. The herb is legal in most of the United States, although health concerns have led to legislation to control the substance in 16 states. When 12th graders were asked about the harmfulness of the drug, 39.8 percent responded that saliva is a "great risk" either physically or in other ways if used once or twice (MTF 2011).

GHB and Rohypnol are often called **date-rape drugs** because of their use in rendering victims incapable of resisting sexual assaults. In 1990, the Food and Drug Administration banned GHB, a central nervous system depressant, although kits containing all the necessary ingredients to manufacture the drug continued to be available on the Internet. On February 18, 2000, President Clinton signed a bill that made GHB a controlled substance and thus illegal to manufacture, possess, or sell. Nonetheless, in 2010, 1.4 percent of 12th graders, 0.6 percent of 10th graders, and 0.6 percent of 8th graders reported past-year use of GHB (MTF 2011).

Rohypnol, presently illegal in the United States, is lawfully sold in Europe and Mexico for the short-term treatment of insomnia (DEA 2011). It belongs to a class of drugs known as benzodiazepines, which also includes common prescription drugs such as Valium, Halcion, and Xanax. Rohypnol is a tasteless and odorless depressant. The effects of Rohypnol begin within 20 minutes and can incapacitate a victim for up to 12 hours (Abadinsky 2008). In 2010, 0.8 percent of 8th, 10th, and 12th graders reported past-year use of Rohypnol (MTF 2011).

Heroin Heroin is an analgesic—that is, a painkiller—and is the most commonly abused class of drugs called opiates. Most heroin comes from the poppy fields of Afghanistan, which controls 90 percent of the global opium market and provides the Taliban with an estimated $300 million a year (Nordland 2010; Filkins 2009). Highly addictive, heroin can be injected, snorted, or smoked. If intravenous injection is used, the onset of the euphoric effects is felt within 7 to 8 seconds; if heroin is snorted or smoked, peak effects are felt within 10 to 15 minutes (NIDA 2005).

Over the last 10 years in Europe and Asia, opiates such as heroin have been the main problem drugs as indicated by the demand for treatment. In Europe, 55 percent of those who request drug treatment abuse opiates and, in Asia, 60 percent of those who request treatment abuse opiates (WDR 2011). In the United States, between 2009 and 2010, the number of people who were dependent on or abused heroin decreased from 399,000 to 359,000 (NSDUH 2011). Despite the decrease, the relatively high numbers are a result of heroin use by suburban, middle class white students.

> The explosion of heroin in suburban America isn't by accident. Rather, it is the plan of drug lords from Mexico and Columbia, who strategically market the drug to Middle America with new, sophisticated techniques. Packets of heroin are now stamped with popular brand names like Chevrolet or Prada, or marketed using blockbuster movie names aimed at young people, like the *Twilight* series. Dealers even give it away for free in the suburbs at first. Once the kids are hooked, they sell it to them, dirt cheap. In fact, kids can buy a small bag of heroin for as little as $5. It's cheaper than movie tickets or a six-pack of beer. (Alfonsi and Siegel 2010, p. 1)

"Bath Salts" Often sold under such innocuous-sounding names as Ivory Wave, Vanilla Sky, Bliss, White Lightening, Hurricane Charlie, and Red Dove, bath salts as they are called are one of the newest designer drugs. Bath salts often contain dangerous chemicals that act as stimulants that result in paranoia, rapid heart rate and chest pains, hallucinations, and suicidal thoughts, all of which may lead to a visit to the local emergency

date-rape drugs Drugs that are used to render victims incapable of resisting sexual assaults.

Sakchai Lalit/AP Photo

room or a call to the poison control center (McMillen 2011). Calls to poison control centers about bath salts increased from 303 in all of 2010 to 3,470 in the first six months of 2011. Although presently banned in 28 states, bath salts remain legal in others and are sold in convenience stores, tobacco shops, and over the Internet. Bath salts sell for between $25 and $50 per 50-milligram packet (Goodnough & Zezima 2011).

Psychotherapeutic Drugs Estimates of lifetime use of psychotherapeutic drugs—that is, nonmedical use of any prescription pain reliever, stimulant, sedative, or tranquilizer—decreased between 2008 and 2009 (NSDUH 2011). Approximately 7 million people, 2.7 percent of the U.S. population 12 and older, reported current use of a psychotherapeutic drug for nonmedical reasons in 2010. Of these users, 5.1 million used pain relievers, 2.1 million used tranquilizers, 1.1 million used stimulants, and 374,000 used sedatives. Over half of those who reported prescription drug use for nonmedical reasons reported that they received the drugs from friends and relatives for free (NSDUH 2011). In 2011, President Obama introduced a four-pronged national plan aimed at fighting "America's Prescription Drug Abuse Crisis" through increased awareness of the problem through education, prescription drug monitoring programs, the development of proper drug-disposal facilities, and increased law enforcement focused on shutting down "pill mills" and "doctor shoppers" (Obama 2011).

Recently, there has been an increase in the use of pain-relieving synthetic opioids such as codeine and OxyContin (WDR 2008). In 2009, there were 584,000 new users of OxyContin aged 12 and older. The average age of first-time OxyContin users between the ages of 12 and 49 was 22.3 years, and of those between 18 and 22 years of age, current OxyContin use increased from 0.2 percent in 2008 to 0.6 in 2009. Figure 3.4 indicates that of the 14 most commonly abused drugs by high school seniors, eight of them are prescriptions or bought over the counter (MTF 2011).

Throughout much of the developing world, homeless children "huff" glue to escape hunger pains and the horrors of living on the street. Globally, UNICEF estimates that there are more than 200 million homeless children—the equivalent of two-thirds the population of the United States.

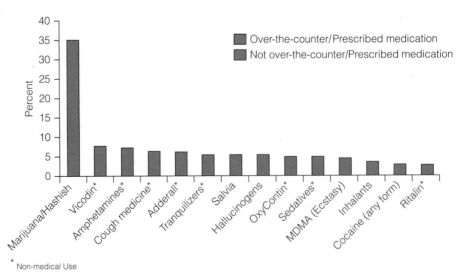

Figure 3.4: Prevalence of Past-Year Drug Abuse among 12th Graders, 2009 In 2009, eight of the top 14 drugs that high school seniors most commonly abused were either prescribed or over-the-counter medications (red bars).
Source: MTF 2010.

* Non-medical Use

Note: Categories are not mutually exclusive.

Inhalants Inhalants act on the central nervous system with users reporting a psychoactive, mind-altering effect (Abadinsky 2008). Common inhalants include adhesives (e.g., rubber cement), food products (e.g., vegetable cooking spray), aerosols (e.g., hair spray and air fresheners), anesthetics (ether), and cleaning agents (e.g., spot remover). In total, over 1,000 household products are currently abused.

Approximately 8.5 percent of the population 12 and older has reported trying inhalants at least once in their lifetime (NSDUH 2011).

Youth, however, are particularly prone to inhalant use, erroneously believing it is harmless or that prolonged use is necessary for any harm to result. For example, in 2010, among people between the ages of 12 and 49, the average age of first-time use of an inhalant was 16.3, compared to 21.2 for cocaine, 21 for pain relievers, and 24.6 for tranquilizers (NSDUH 2011). In 2009, 12.1 percent of 8th, 10th, and 12th grade students reported lifetime use of an inhalant, nearly twice the rate of European 15- and 16-year-olds (MTF 2011; ESPAD 2009).

Societal Consequences of Drug Use and Abuse

Drugs are a social problem not only because of their adverse effects on individuals but also because of the negative consequences their use has for society as a whole. Everyone is a victim of drug abuse. Drugs contribute to problems within the family (although families can also lower the likelihood of abusive drug use—see this chapter's *Social Problems Research Up Close*) and to escalating crime rates, are tremendously costly, and place a heavy strain on the environment. Drug abuse also has serious consequences for health at both the individual and the societal level.

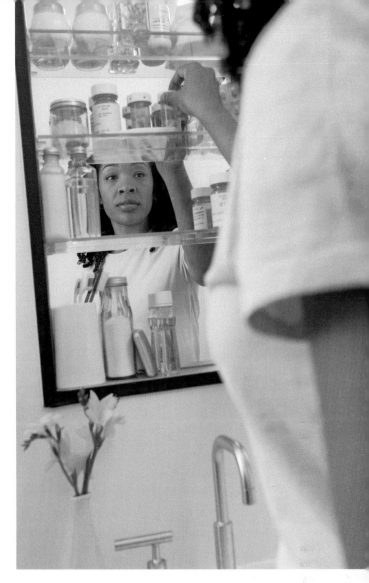

In an effort to reduce the amount of prescription drugs available, law enforcement and other volunteers have taken to going house to house to collect expired medicines which are then disposed off. Here a woman checks the expiration dates on her prescription drug containers.

The Cost to the Family

The cost of drug abuse to families is incalculable. It is estimated that, in the United States, one in 10 children under the age of 18 lives with at least one parent in need of treatment for drug or alcohol dependency (SAMHSA 2009). Children raised in such homes are more likely to (1) live in an environment riddled with conflict, (2) have a higher probability of physical illness including injuries or death from an automobile accident, and (3) are more likely to be victims of child abuse and neglect (SAMHSA 2007; 2009). Children of alcoholics, if using alcohol or other drugs, are more likely to have alcohol or drug problems than children of nonalcoholics (SAMHSA 2007). Further, children of alcoholics, and particularly female children of alcoholics, suffer from "significant mental health consequences . . . that persist far into adulthood" (Balsa et al. 2009, p. 55).

Parents who report abusing alcohol in the past year are also more likely to report cigarette and illicit drug use than parents who did not report alcohol abuse in the previous year. They were also more likely to report "household turbulence," including yelling, serious arguments, and violence (NSDUH 2004). Moreover, alcohol abuse is the single most common trait associated with wife abuse (Flanzer 2005). For example, a study of 634 couples who were assessed at the time of their marriage and again at the time of their first, second, and fourth anniversaries, indicates that both husbands and wives were

There's some debate over the relationship between onset of drinking, parents' alcohol consumption patterns, and the context in which drinking first occurs. For example, some research suggests that the more alcohol consumed in the home, the higher the likelihood that children will grow up and engage in excessive drinking behavior. Another body of research suggests that when children use alcohol in the home, they're less likely to abuse it in the future. The following ethnography looks at the relationship between onset of drinking, parental drinking behavior, type of alcohol consumed, and the social context in which drinking occurs.

Sample and Methods

The researchers used "anthropological ethnographic interviewing" to "understand the world as seen by the respondent within the context of the respondents everyday life" (Strunin et al. 2011, p. 346). The sample was composed of 160 Italian youth and young adults—half female and half male—divided into two age groups, 16- to 18-year-olds and 25- to 30-year-olds, individuals in each age group having a variety of consumption rates.

To identify prospective subjects, an alcohol screening test was administered in two high school locations for current and former students. Further, to assure the collection of comparable data, an interview guide was developed to make sure that all topics were covered and that the questions were asked in a specific order. A quantity–frequency index

was used to measure self-reported drinking behavior. Lastly, students were asked about their family histories, extended family members, as well as questions about their beliefs, norms, and behaviors related to alcohol use. The interviews were audiotaped, transcribed, and translated into English.

Findings and Conclusions

Youth who described drinking as part of their Italian heritage also described it in terms of being "culturally normative." For example, a young female adult described her parents' attitude toward alcohol in the following way: ". . . a simple attitude. Alcohol is a pleasure, and has to be lived as something positive, this is the type of values we have at home. A wine culture, wine at meals for the pleasure of it, no abuse because that can cause health problems" (p. 349).

This research points to the importance of the social context in which drinking takes place. For example, even though parents who allowed their children to drink with meals may be seen as permissive, they were also more likely to talk to their children about the dangers of alcohol, including such health issues as "addiction, liver, and heart diseases . . . and the dangers of drinking and driving" (p. 352). Further, youth who were allowed to drink with meals in their homes were more likely to have their first drink within a family setting and less likely to mix types of alcohol. Alternatively, the first drink experiences of those not allowed to drink at home

included drinking large amounts of more than one type of alcohol and getting drunk.

There are also differences in the likelihood of binge drinking, defined as drinking five or more drinks on one occasion, and age of getting drunk for the first time. Those who consumed alcohol with meals in the home delayed binge drinking and getting drunk when compared to respondents who were not allowed to drink in the home. Further, only those who did not drink alcohol in the home got drunk on their first occasion of drinking.

The authors conclude that, "the introduction of alcohol, typically wine, in a family setting may protect against harmful drinking including binge drinking and drunkenness" (Strunin et al. 2011, p. 354). The researchers also suggest that many of their findings can be generalized to other populations. For example, the United States does not have a tradition of permitting children to consume alcohol with meals and, perhaps consequently, we have one of the highest adolescent consumption rates in the world. Moderate drinking at home within the context of the family setting could lower excessive drinking rates in the United States and teach healthier drinking patterns.

Source: Strunin, Lee, Kirstin Lindeman, Enrico Tempesta, Pierluigi Ascani, Simona Avan, and Luca Parisi. 2010. "Familial Drinking in Italy: Harmful or Protective Factors?" *Addiction Research and Theory* 18(3):344–358.

impacted by the alcoholism of their opposite sex parent (Kearns-Bodkin and Leonard 2008):

> For husbands, alcoholism in the mother was associated with lower marital satisfaction. For wives, alcoholism in the father was related to lower marital intimacy. Husbands' physical aggression was influenced by his mother's and father's alcoholism; high levels of physical aggression were present among men with alcoholic mothers and nonalcoholic fathers. Interestingly, wives' experience of her husband's aggression was also highest among women with alcoholic mothers and nonalcoholic fathers.

Crime and Violence

The drug behavior of individuals arrested, incarcerated, and in drug treatment programs also provides evidence of a link between drugs and crime. For example, surveys indicate that about 27 percent of victims of violent crime report that the offender was involved with alcohol or drugs. Further, parents on probation or parole (see Chapter 4) not only use tobacco, alcohol, and illicit drugs at higher rates than parents not on probation or parole, they're also more likely to abuse or be addicted to alcohol or illicit drugs (SAMHSA 2010). Similarly, juvenile delinquency is associated with drug use.

In an interdisciplinary study of 1,354 male and female serious juvenile offenders between the ages of 14 and 18 seven years after conviction, Mulvey (2011) concludes that although substance abuse is a strong predictor of both self-reported delinquent behavior and number of arrests, substance abuse treatment reduces the incidences of drug use and nondrug-related crime.

The relationship between crime and drug use, however, is complex. Sociologists disagree as to whether drugs actually "cause" crime or whether, instead, criminal activity leads to drug involvement. Alternatively, as Siegel (2006) noted, criminal involvement and drug use can occur at the same time; that is, someone can take drugs and commit crimes out of the desire to engage in risk-taking behaviors. Furthermore, because both crime and drug use are associated with low socioeconomic status, poverty may actually be the more powerful explanatory variable.

In addition to the hypothesized crime–drug use link, some criminal offenses are defined by use of drugs: possession, cultivation, production, and sale of controlled substances; public intoxication; drunk and disorderly conduct; and driving while intoxicated. Driving while intoxicated is one of the most common drug-related crimes. According to the National Highway Traffic Safety Administration, in 2009, nearly 11,000 people died in alcohol-related crashes—one-third of all traffic-related deaths. Alcohol is not the only drug that impairs driving. In 2010, 10.6 million people 12 years of age and older reported driving while under the influence of an illicit drug. As with alcohol-impaired drivers, the rate is highest for 18- to 25-year-olds (NSDUH 2011).

The High Price of Alcohol and Other Drugs

A report by the National Center on Addiction and Substance Abuse at Columbia University (CASA 2009) set the total annual cost of substance abuse and addiction in the United States at $467.7 billion. More importantly, the report contends that, for every dollar spent on drug abuse by federal and state governments

> . . . 95.6 cents went to shoveling up the wreckage and only 1.9 cents on prevention and treatment, 0.4 cents on research, 1.4 cents on taxation or regulation and 0.7 cents on interdiction. Under any circumstances spending more than 95 percent of taxpayer dollars on the consequences of tobacco, alcohol, and other drug abuse and addiction and less than 2 percent to relieve individuals and taxpayers of this burden would be considered a reckless misallocation of public funds. In these economic times, such upside-down-cake public policy is unconscionable. (CASA 2009, p. i)

At the federal level, the cost of "shoveling up the wreckage" includes the cost of (1) health care due to substance abuse and addiction (the highest proportion of wreckage spending), (2) adult and juvenile crime (e.g., corrections), (3) child and family assistance programs (e.g., welfare), (4) education (e.g., Safe School Initiatives), (5) public safety (e.g., drug enforcement), (6) mental health and developmental disabilities (e.g., treatment of addiction), and (7) the federal workforce (e.g., loss of productivity). The report concludes that prevention programs must become a priority to reduce the economic costs of drug abuse.

What Do You Think? Most people assume that smokers cost tax payers millions of dollars in associated health risks—lung cancer, chronic obstructive pulmonary disease, heart attacks, hypertension, just to name a few (see Figure 3.3). In fact, it is estimated that smokers cost the "country $96 billion a year in direct health care costs and another $97 billion a year in loss of productivity" (Werner 2009, p. 1). Economists, however, are quick to point out that smokers may in fact not be an economic burden. For example, a 2008 Dutch study found that health care for smokers from age 20 on costs $326,000; for nonsmokers, $417,000. Why? Nonsmokers live longer. Given the results of this study, should smokers have lower health insurance premiums than nonsmokers?

Physical and Mental Health Costs

Cigarette smoking is the leading preventable cause of disease and death in the world. In the 20th century, 100 million people died worldwide in what is being called the "tobacco epidemic" (WHO 2008). According to the World Health Organization, the global tobacco epidemic kills 5.4 million people a year and, unless significant policy changes occur, by 2030, there will be more than 8 million deaths a year, 80 percent in developing countries. Further, another 8.6 million people annually are living with serious diseases as a result of smoking (CDC 2011a) (see Figure 3.5).

According to the World Health Organization (2008), there are six state initiatives which, if adopted, could potentially reverse the worldwide tobacco crisis by: (1) developing policies that prevent tobacco use, (2) protecting people from tobacco smoke by developing smoke-free laws, (3) providing help to people who want to quit using tobacco products, (4) publicizing the dangers of tobacco products, (5) banning, and enforcing existing bans, which prohibit "tobacco advertising, promotion, and sponsorship" (p. 11), and (6) increasing the cost of tobacco products by raising taxes. It should be noted, however, that the health impact of smoking goes beyond consumption and the effects of secondhand smoke. For example, children who work in tobacco fields are at risk for "green tobacco sickness," a disease caused by the absorption of nicotine through the skin by handling wet tobacco leaves (WHO 2011b).

Annually, alcohol abuse is responsible for over 2.5 million deaths, 4 percent of all deaths worldwide (Nebehay 2011). Alcohol kills more people than AIDS, tuberculosis, or violence, and is responsible for 80,000 deaths annually in the United States (WHO 2011d). Although alcohol abuse is the third leading preventable cause of death in the United States (CDC 2011b), recently, deaths from illicit drug use have exceeded deaths from alcohol abuse (CDC 2011c).

Early and continued abuse of alcohol is related to cancer, cirrhosis of the liver, gastrointestinal diseases, epilepsy, intentional and unintentional injuries, violence, and cardiovascular diseases (WHO 2011d). Maternal prenatal alcohol use is associated with one of the leading preventable causes of birth defects and developmental disabilities in children—**fetal alcohol syndrome**—a syndrome characterized by serious physical and mental handicaps, including low birth weight, facial deformities, mental retardation, and hearing and vision problems (CDC 2006).

Heavy alcohol and drug use are also associated with negative consequences for an individual's mental health. For example, nonsmokers exposed to secondhand smoke are twice as likely to suffer from depression as nonsmokers not exposed to secondhand smoke (Elias 2009). Although causal ordering is difficult to determine, research also indicates that several disorders as identified by the American Psychiatric Association are associated with increased risks of drug abuse including antisocial personality disorder (15.5 percent increase), manic episodes (14.5 percent increase), and schizophrenia (10.1 percent increase) (National Drug Intelligence Center 2004).

The Cost of Drug Use on the Environment

Although not something usually considered, the production of illegal drugs has a tremendous impact on the environment. Much of the impact is a consequence of the cultivation of marijuana, cocaine, and opium. For example, the Colombian government estimates that during the decades of 1988 to 2008, nearly 5.4 million acres of rainforest (an area the size of New Jersey) were destroyed because of illegal drug production. Further, studies have shown that as much as 25 percent of the deforestation that takes

> According to the World Health Organization, the global tobacco epidemic kills 5.4 million people a year and, unless significant policy changes occur, by 2030, there will be more than 8 million deaths a year, 80 percent in developing countries.

fetal alcohol syndrome A syndrome characterized by serious physical and mental handicaps as a result of maternal drinking during pregnancy.

place in Peru is "associated with clear cutting and burning for planting coca bushes" (DEA 2010, p. 1).

In the United States and Mexico, outdoor cannabis cultivation leads to contaminated water, clear-cutting of natural vegetation, the disposable of nonbiodegradable materials, and the diversion of natural waterways often polluted with toxic chemicals, which endanger fish and other wildlife (U.S. Department of Justice 2010). Fish and wildlife aren't the only ones who should worry about water quality. A recent study of the quality of water at a Spanish national park found traces of eight illegal drugs in the waterways including Ecstasy, cocaine, and methamphetamines (Sohn 2010).

Treatment Alternatives

In 2010, 2.6 million people aged 12 or older were treated for some kind of problem associated with the use of alcohol or illicit drugs (NSDUH 2011) (see Figure 3.6). Individuals who are interested in overcoming problem drug use have a number of treatment alternatives from which to choose. Some options include family therapy, counseling, private and state treatment facilities, community care programs, pharmacotherapy (i.e., use of treatment medications), behavior modification, drug maintenance programs, and employee assistance programs. Three commonly used techniques are inpatient or outpatient treatment, peer support groups, and drug courts.

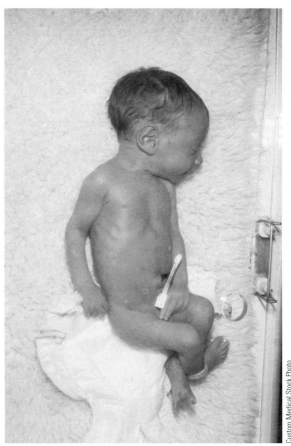

Fetal alcohol syndrome includes a mix of birth defects: mild to severe mental retardation, low birth weight, central nervous system dysfunction, and malformations of the skull and face. This syndrome is a result of women consuming large amounts of alcohol during pregnancy. One in eight women who are pregnant continue to consume alcohol.

Inpatient and Outpatient Treatment

Inpatient treatment refers to treatment of drug dependence in a hospital and, most importantly, includes medical supervision of detoxification. Most inpatient programs last between 30 and 90 days and target individuals whose withdrawal symptoms require close monitoring (e.g., alcoholics, cocaine addicts). Some drug-dependent patients, however, can be safely treated as outpatients. Outpatient treatment allows individuals to remain in their home and work environments and is often less expensive. In outpatient treatment, patients are under the care of a physician who evaluates their progress regularly, prescribes needed medication, and watches for signs of a relapse.

The longer patients stay in treatment, the greater the likelihood of a successful recovery (NIDA 2006). A recent study of the effectiveness of inpatient and outpatient treatment identifies other variables that predict successful drug treatment. Nearly 400 parole violators who opted for a year-long substance abuse program rather than returning to prison participated in a four-stage program. Phase 1 included 90 days of inpatient substance abuse treatment. Phase 2 included 90 days of outpatient treatment in an alcohol treatment facility accompanied by electronic monitoring, curfews, and weekly drug tests. Phase 3 included 90 days of outpatient treatment with individual and group counseling, and the opportunity to seek employment. Phase 4 included 90 days of outpatient treatment with a special emphasis on reintegration into the community (Zanis et al. 2009).

Results indicated that five variables were independent predictors of completion failure. Offenders who were the least likely to complete treatment (1) had a history of significant problems with their mothers, (2) had problems with their sexual partners in the 30 days prior to admission to the program, (3) had longer periods of incarceration, (4) had used heroin in the 30 days prior to admission to the program, and (5) were younger in age than those who successfully completed the program. Sadly, less than 33 percent of the participants completed the 12-month program.

Figure 3.5: Diseases Caused by Smoking and by Secondhand Smoke (Children and Adults)
Source: WHO 2008. From *WHO Report on the Global Tobacco Epidemic, 2008: The MPOWER package* (p. 11). Geneva, World Health Organization, 2008. Reprinted with permission.

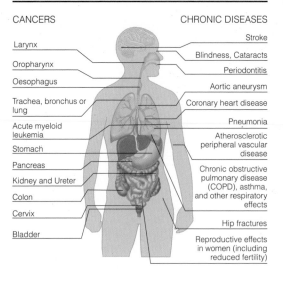

DISEASES CAUSED BY SMOKING

CANCERS — CHRONIC DISEASES

Larynx

Oropharynx

Oesophagus

Trachea, bronchus or lung

Acute myeloid leukemia

Stomach

Pancreas

Kidney and Ureter

Colon

Cervix

Bladder

Stroke

Blindness, Cataracts

Periodontitis

Aortic aneurysm

Coronary heart disease

Pneumonia

Atherosclerotic peripheral vascular disease

Chronic obstructive pulmonary disease (COPD), asthma, and other respiratory effects

Hip fractures

Reproductive effects in women (including reduced fertility)

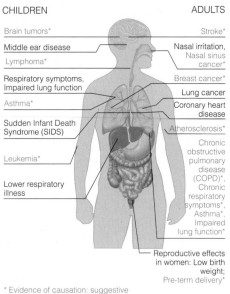

DISEASES CAUSED BY SECOND-HAND SMOKE

CHILDREN — ADULTS

Brain tumors*

Middle ear disease

Lymphoma*

Respiratory symptoms, Impaired lung function

Asthma*

Sudden Infant Death Syndrome (SIDS)

Leukemia*

Lower respiratory illness

Stroke*

Nasal irritation, Nasal sinus cancer*

Breast cancer*

Lung cancer

Coronary heart disease

Atherosclerosis*

Chronic obstructive pulmonary disease (COPD)*, Chronic respiratory symptoms*, Asthma*, Impaired lung function*

Reproductive effects in women: Low birth weight; Pre-term delivery*

* Evidence of causation: suggestive
Evidence of causation: sufficient

Peer Support Groups

Twelve-Step Programs Both Alcoholics Anonymous (AA) and Narcotics Anonymous (NA) are voluntary associations whose only membership requirement is the desire to stop drinking or taking drugs. AA and NA are self-help groups in that nonprofessionals operate them, offer "sponsors" to each new member, and proceed along a continuum of 12 steps to recovery. Members are immediately immersed in a fellowship of caring individuals with whom they meet daily or weekly to affirm their commitment. Some have argued that AA and NA members trade their addiction to drugs for feelings of interpersonal connectedness by bonding with other group members. In a survey of recovering addicts, more than 50 percent reported using a self-help program such as AA in their recovery (Willing 2002). AA boasts over 115,000 groups where over 2.1 million members meet in 150 countries. Women make up about 35 percent of AA membership (Alcoholics Anonymous 2010).

Symbolic interactionists emphasize that AA and NA provide social contexts in which people develop new meanings. Others who offer positive labels, encouragement, and social support for sobriety surround abusers. Sponsors tell the new members that they can be successful in controlling alcohol and/or drugs "one day at a time" and provide regular interpersonal reinforcement for doing so. Some research indicates that mutual support programs work. For example, in a study assessing the effectiveness of such groups, Kelly

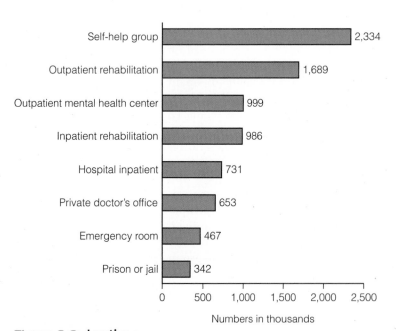

Numbers in thousands

Self-help group — 2,334
Outpatient rehabilitation — 1,689
Outpatient mental health center — 999
Inpatient rehabilitation — 986
Hospital inpatient — 731
Private doctor's office — 653
Emergency room — 467
Prison or jail — 342

Figure 3.6: Location Where Last-Year Substance Use Treatment Was Received among Persons 12 and Older, 2010
Source: NSDUH 2011.

et al. (2006) concluded that involvement in such groups may be very successful—for both males and females—even when participation is limited.

Therapeutic Communities In **therapeutic communities,** which house between 35 and 500 people for up to 15 months, participants abstain from drugs, develop marketable skills, and receive counseling. Synanon, which was established in 1958, was the first therapeutic community for alcoholics and was later expanded to include other drug users. More than 400 residential treatment centers are now in existence, including Daytop Village and Phoenix House, the largest therapeutic communities in the country. Phoenix Houses serve more than 7,000 men, women, and teens a day in over 120 locations in 11 states (Phoenix House 2011). The longer a person stays at such a facility, the greater the chance of overcoming dependency. Living with a partner before entering the program and having a strong self-concept are also predictive of success (Dekel et al. 2004). Symbolic interactionists argue that behavioral changes appear to be a consequence of revised self-definition and the positive expectations of others. An assessment of 33 therapeutic community facilities in Peru indicated that therapeutic communities had a significant positive impact on the over 500 former clients interviewed six months after release (Johnson et al. 2008).

Drug Courts

Concern over the punitive treatment of drug offenders and the failure of the criminal justice system to reduce recidivism rates led to the development of **drug courts.** A recent report by The Sentencing Project (King and Pasquarella 2009), entitled *Drug Courts: A Review of the Evidence,* identifies two types of drug courts—deferred prosecution programs and postadjudication programs.

In a deferred prosecution or diversion setting, defendants who meet certain eligibility requirements are diverted into the drug court system prior to pleading to a charge. Defendants are not required to plead guilty, and those who complete the drug court program are not prosecuted further. Failure to complete the program, however, results in prosecution. Alternatively, in the postadjudication model, defendants must plead guilty to their charges, but their sentences are deferred or suspended while they participate in the drug court program. Successful completion of the program results in a waived sentence and sometimes an expungement of the offense. However, in cases where individuals fail to meet the requirements of the drug court (such as a habitual recurrence of drug use), they will be returned to the criminal court to face sentencing on the guilty plea.

For an offender to be considered successful, they must complete the entire six-month to one-year program and remain drug-free without arrests for a specified period of time. According to *Drug Courts: A Review of the Evidence,* drug court graduates, when compared to their traditional court counterparts, are less likely to reoffend and/or have longer periods of time between arrests. Further, evaluations of the costs of such programs, in general, find that drug courts are less expensive than probation or institutionalization. Although the average annual cost of incarceration is $23,000 per inmate, the average annual cost of an individual participating in a drug court program is $4,300 per person. As of this writing, there are over 1,600 drug courts in all 50 states.

Strategies for Action: America Responds

Drug use is a complex social issue that is exacerbated by the structural and cultural forces of society that contribute to its existence. Although the structure of society perpetuates a system of inequality, creating in some the need to escape, the culture of society, through the media and normative contradictions, sends mixed messages about the acceptability of drug use. Thus, trying to end drug use by developing programs, laws, or initiatives may be

therapeutic communities Organizations in which approximately 35 to 500 individuals reside for up to 15 months to abstain from drugs, develop marketable skills, and receive counseling.

drug courts Special courts that divert drug offenders to treatment programs in lieu of probation or incarceration.

unrealistic. Nevertheless, numerous social policies have been implemented or proposed to help control drug use and its negative consequences with various levels of success.

Alcohol and Tobacco

Although there may be some overlap (e.g., education), strategies to deal with alcohol and tobacco abuse are often different from those initiated to deal with illegal drugs. Prohibition, the largest social policy attempt to control a drug in the United States, was a failure, and criminalizing tobacco is likely to be just as successful. However, research has identified several promising strategies in reducing alcohol and tobacco use including economic incentives, government regulations, legal sanctions, and education and treatment.

Economic Incentives One method of reducing alcohol and tobacco use is to increase the cost of the product. After examining the cost of cigarettes by state, Dinno and Glantz (2009) conclude that increased cigarette prices are associated with reduced prevalence and consumption rates. The benefit of increasing the cost of cigarettes, however, varies by a state's current smoking and tax rate—"A $1.00 increase in South Carolina's $0.07 cigarette tax, which is the lowest in the nation and has not increased since 1977, could increase the state's annual revenue by $180 million . . . and in 5 years would result in 78,200 fewer smokers and prevent more than 15,700 smoking-related deaths" (CASA 2009, p. 55). Similarly, Williams et al. (2005) note that "increasing the price of alcohol which can be achieved by eliminating price specials and promotions, or by raising price excise taxes, would lead to a reduction in both moderate and heavy drinking by college students" (p. 88). Other examples of economic incentives include reimbursement for smoking or alcohol cessation programs, reduced health insurance premiums for nonsmokers, and reduced car insurance premiums for nondrinkers.

What Do You Think? To some, no doubt, the mixture of alcohol and caffeine sounds like a dream come true—but not to the Food and Drug Administration (FDA). Producers of several new beers with names like Joose, Four Loko, and Moonshot '69 were recently told by the FDA to either remove the caffeine from their beers or stop selling them (Zezima 2010). One argument is that the caffeinated, often fruit-flavored beers are not safe. As one emergency room doctor stated, "It's a recipe for disaster because your body's natural defense is to get sleepy and not want to drink, but in this case you're tricking the body with caffeine"(Goodnough 2010, p. 1). With everything from Red Bull to Mountain Dew containing caffeine, should these new beers be singled out and taken off the market? What do you think?

Government Regulation Federal, state, and local governmental regulations have each had some success in reducing tobacco and alcohol use and the problems associated with them. In 1984, states raised the legal drinking age to 21, under threat of losing federal highway funds. According to Mothers Against Drunk Driving (MADD), the minimum legal drinking age prevented over 25,000 drunken driving deaths and decreased alcohol-related car accidents by 16 percent (MADD 2011). It should be noted, however, that some research questions the law's impact (Miron & Tetelbaum 2009). Similarly, clean air laws restrict smoking in the workplace, bars, restaurants, and the like, and reduce consumption rates as well as secondhand smoke exposure.

One of the most important pieces of legislation in recent years is the Family Smoking Prevention and Tobacco Control Act of 2009. The law gives authority to the Food and Drug Administration to regulate the manufacturing (e.g., tobacco companies must now disclose ingredients in their products), marketing (e.g., tobacco names or logos may no longer be used to sponsor sporting or entertainment events), and sale of tobacco products (e.g., terms like *light, mild,* and *low tar* may no longer be used). Additionally, the law requires that warning labels must be strengthened and cover at least the top half of the front and back panels of the packaging by Fall, 2012 (Myers 2009a; Fact Sheet 2009; FDA 2011b). Worldwide, evidence suggests that where tobacco package warnings are used, knowledge concerning the risks of tobacco use is increased (WHO 2011e). Nonetheless, it is estimated that tobacco companies spend $35 million *a day* in cigarette advertising and marketing (Layton 2010).

Legal Action Federal and state governments, as well as smokers, ex-smokers, and the families of victims of smoking have taken legal action against tobacco companies. The 1990s brought billion-dollar judgments against the tobacco industry on behalf of states suing for reimbursement of smoking-related health care costs and families seeking punitive damages for the deaths of loved ones (Timeline 2001). In 1998, tobacco manufacturers reached a settlement with 46 states, agreeing to pay billions of dollars for reimbursement of state smoking-related health costs. The settlement also restricted the marketing, promotion, and advertising of tobacco products directed toward minors (Wilson 1999). More recently, after years of litigation, the District of Columbia U.S. Court of Appeals held that "the tobacco industry had and continues to engage in a massive, decades-long campaign to defraud the American public . . . including falsely denying that nicotine is addictive, falsely representing that 'light,' and 'low tar' cigarettes present fewer health risks, falsely denying that they market to kids, and falsely denying that secondhand smoke causes disease" (Myers 2009b, p. 1).

Suits against retailers, distributors, and manufacturers of alcohol are more recent and often modeled after tobacco litigation. These suits primarily concern accusations of unlawful marketing, sales to underage drinkers, and failure to adequately warn of the risks of alcohol. To date, legal action against alcohol manufacturers have been much less successful than suits against the tobacco industry (Willing 2004; Marin Institute 2006).

So what can be done about the nation's number one cause of preventable deaths? To a large extent, experts' recommendations mirror those made by the World Health Organization: develop and enforce tobacco controls, pass clean air laws, initiate school- and health-based smoking prevention programs for children and adolescents, and use warning labels with particular attention to cigarette packaging. Further, in *Ending the Tobacco Problem, A Blueprint for the Nation*, the researchers note that a fundamental change is needed—"[I]t is time for Congress and other policy makers to change the legal structure of tobacco policy, thereby laying the foundation for a strategic initiative to end the nation's tobacco problem, that is, reducing tobacco use to a level that is insignificant from a public health standpoint" (Bonnie et al. 2007, p. 271).

Prevention Although impossible to discuss all the programs and policies that are successful in reducing alcohol and tobacco consumption, one is worth noting. There is a large body of evidence that exposure to cigarette or alcohol advertisements increases drug use and, conversely, exposure to anti-smoking or anti-drinking advertisements reduce drug use (Terry-McElrath et al. 2007; Aloise-Young et al. 2006; Wakefield et al. 2008; Snyder et al. 2006). Despite the effectiveness of anti-smoking and anti-drinking campaigns worldwide, for example, less than 5 percent of countries have total bans on the marketing, promotion, and sponsorship of tobacco products (WHO 2008). Anti-smoking and anti-drinking campaigns should be directed toward children. According to the National Center on Addiction and Substance Abuse at Columbia University, "a child who reached age 21 without smoking, using illicit drugs, or abusing alcohol is virtually certain never to do so" (CASA 2009, p. 6).

Illegal Drugs

The War on Drugs In the 1980s, the federal government declared a "war on drugs," which was based on the belief that controlling drug availability would limit drug use and, in turn, drug-related problems. In contrast to a **harm reduction** position, which focuses on minimizing the costs of drug use for both user and society (e.g., distributing

harm reduction A recent public health position that advocates reducing the harmful consequences of drug use for the user as well as for society as a whole.

Anti-smoking advertisements such as these have been released by the FDA for public comment and will be required by September 2012. It is hoped that such images will reduce smoking and limit the number of new smokers. Other images picture a toe-tagged corpse, a man dying of cancer, and a graveyard.

clean syringes to decrease the risk of HIV infection), this "zero-tolerance" approach advocates get-tough law enforcement policies, and is responsible for the dramatic increase in the jail and prison population. In 1980, there were an estimated 40,000 drug offenders in jail or prison; in 2009, there were more than half a million. Interestingly, as methamphetamine enforcement steps up and crack cocaine use declines, the number of black compared to white drug offenders in prison has declined (Fears 2009).

The harsher penalties enacted as part of the "war on drugs" required prison sentences for almost all drug offenders—first-time or repeat—and limited judicial discretion in deciding what best served the public's interest. The "Rockefeller drug laws" as they were called resulted in a disproportionate number of Hispanics and African Americans receiving "excessively long and unnecessary prison sentences" for even the most minor drug offenders (Human Rights Watch 2007, p. 1). In response to public outcries and accusations of institutional racism (see Chapter 9), reform of such laws began (Peters 2009).

What Do You Think? To recruit members and intimidate enemies, Mexican drug cartel leaders have posted videos on YouTube (Jervis 2009). Many are so gruesome that they are preceded by a warning that the content may be inappropriate for viewers under the age of 18. Gang members also post crimes by their rival gangs to publicize their misdeeds (Cattan 2010). Although YouTube officials have alerted law enforcement agencies about these videos, it could be argued that allowing them to be broadcast is "aiding and abetting" the enemy in the war against drugs. Do you think drug cartel leaders should be able to post such videos on the YouTube website? Why or why not?

There is also the question as to whether or not the war on drugs is working—is it stopping the flow of illegal drugs into the United States and lowering drug-related problems? R. Gil Kerlikowske, Director of the White House Office of National Drug Control Policy, recently stated that after 40 years and $1 trillion dollars, " . . . the concern about drugs and drug problems is, if anything, magnified, intensified" (cited in Mendoza 2010, p. 1). Further, a report by the National Center on Addiction and Substance Abuse at Columbia University concluded that policies that allocate billions of dollars to "disrupt and deter" the flow of illegal drugs into the United States have little impact on drug use and addiction (CASA 2009, p. 58). For example, 65 percent of all inmates are addicted to alcohol or other drugs, and only 11 percent receive any kind of substance abuse treatment (CASA 2010). The "war on drugs" will do very little, if anything, for these and future drug-involved inmates.

There are also concerns that present policies are not only ineffective but create collateral damage. *In America's Longest War* (1994), Yale law professor Steven Duke and coauthor Albert C. Gross argued that the war on drugs, much like Prohibition, has intensified other social problems: drug-related gang violence and turf wars, the creation of syndicate-controlled black markets, unemployment, the spread of AIDS, overcrowded prisons, corrupt law enforcement officials, and the diversion of police from other serious crimes. Further, the war on drugs has had a tremendous impact on families: Conviction of a drug offense can lead to a parent being in prison, eviction from public housing, deportation, and permanent exclusion from public assistance (Caulkins et al. 2005).

demand reduction One of two strategies in the U.S. war on drugs (the other is supply reduction), demand reduction focuses on reducing the demand for drugs through treatment, prevention, and research.

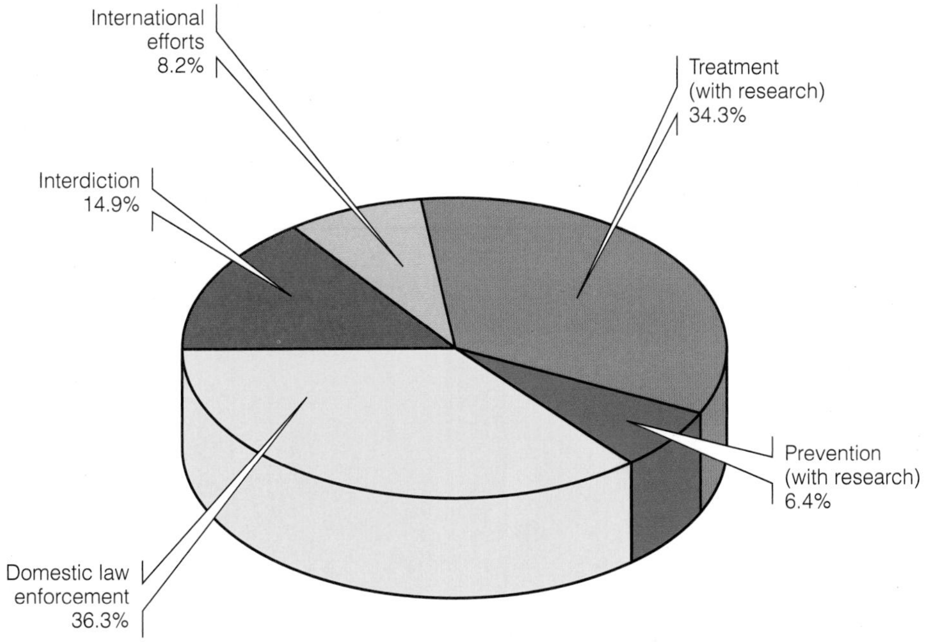

International
efforts
8.2%

Interdiction
14.9%

Treatment
(with research)
34.3%

Figure 3.7: **Federal Drug Control Spending by Function, Fiscal Year 2009**
Source: ONDCP 2008.

Prevention
(with research)
6.4%

Domestic law
enforcement
36.3%

Finally, the war on drugs continues at an astronomical societal cost—a requested $26.2 billion in fiscal year 2012 (ONDCP 2011). The U.S. policy on fighting drugs is two-pronged. First is **demand reduction,** which entails reducing the demand for drugs through treatment and prevention (see Figure 3.7). The second strategy is **supply reduction.** A much more punitive strategy, supply reduction relies on international efforts, interdiction, and domestic law enforcement to reduce the supply of illegal drugs. In 2012, 40.6 percent of drug control spending will be focused on demand reduction and 59.4 percent on supply reduction—a more balanced approach than in previous years (ONDCP 2011).

A 2010 survey of 1,003 U.S. adults echoes politicians' concerns about the effectiveness of the war on drugs. Sixty-three percent of Democrats, 64 percent of Republicans, and 70 percent of Independents believe that the "war on drugs" has been an abject failure (Reid 2010). These and similar survey results indicate that many people around the country (1) believe that the war on drugs—and supply reduction efforts in particular—should be abandoned, and (2) favor deregulation.

Deregulation or Legalization Given the questionable successes of the war on drugs, it is not surprising that many advocate alternative measures to the rather punitive emphasis of the last several decades.

Deregulation is the reduction of government control over certain drugs. For example, although individuals must be 21 years old to purchase alcohol and 18 to purchase cigarettes, both substances are legal and can be purchased freely. In some states, possession of marijuana in small amounts is a misdemeanor rather than a felony, and in 16 states, marijuana is lawfully used for medical purposes. Deregulation is popular in other countries as well. For example, personal possession of any drug, even those considered the most dangerous, is legal in Spain, Italy, the Baltic States, and the Czech Republic (*The Economist* 2009).

Proponents for the **legalization** of drugs affirm the right of adults to make informed choices. They also argue that the tremendous revenues realized from drug taxes could be used to benefit all citizens, that purity and safety controls could be implemented,

[T]he war on drugs, much like Prohibition, has intensified other social problems: drug-related gang violence and turf wars, the creation of syndicate-controlled black markets, unemployment, the spread of AIDS, overcrowded prisons, corrupt law enforcement officials, and the diversion of police from other serious crimes.

supply reduction One of two strategies in the U.S. war on drugs (the other is demand reduction), supply reduction concentrates on reducing the supply of drugs available on the streets through international efforts, interdiction, and domestic law enforcement.

deregulation The reduction of government control over, for example, certain drugs.

legalization Making prohibited behaviors legal; for example, legalizing drug use or prostitution.

According to journalist Scott Kraft, new "fencing and high-tech devices make it difficult for drug traffickers to cross the [Mexican–U.S.] border" (Kraft 2009, p. 1). As a result, smugglers have resorted to carrying their illegal imports on their backs and crossing the desert on foot. This chapter's The Human Side excerpt vividly describes the efforts of border patrol agents who have had to use "tracking skills borrowed a century ago from Native Americans: 'cutting for sign' to detect where someone has crossed the Earth's surface, and 'pushing sign,' tracking that person down."

N.M.—Bill Fraley knelt to examine the brown, pebbled soil, like an art professor studying a familiar drawing.

"See those two fine lines?" he said, passing a finger over two shoe prints, each with washboard rows of ridges. His hand moved to another heel print a few inches away. "And there's a doper lug," the heel imprint of a boot sometimes worn by drug smugglers. . . .

A few steps away, a 5-foot barbed-wire fence cut through the cactus and greasewood, separating the United States from Mexico. The Border Patrol agent stood and tipped back the brim of his Stratton cowboy hat, eyes hidden behind aviator sunglasses. A satisfied expression hung on his chiseled face.

There were at least three of them, he figured. "It rained all day yesterday and these signs are on top of the rain," he said. "So I'd say they crossed yesterday, between 6 and 7.

And it looks like they've got heavy loads of dope on them."

Tuesday, 10 a.m.

To Fraley, 50, sign-cutting is both art and science. He looks for footprints, though he usually finds just fragments. He looks for disturbances: turned-over rocks, broken twigs, bent barbed wire. He looks for chewed gum, a cigarette butt, the residue of a line of cocaine snorted on a rock. He looks for clues to fix the time. . . .

"What I really love," Fraley said, "is when you come across an ant pile that's been stepped in. Ants will rebuild an anthill in an hour, so if you see a footprint in an anthill you'd better look up—because you're likely to be looking right at your adversaries. . . ."

This group, [Border Patrol agent] Parra and Fraley agreed, would probably spend tonight moving through the Animas Mountains, which appeared in the distance, bathed in the blue shadows of puffy clouds. Tracking them through that terrain at night would be impossible. But eventually, the smugglers would have to drop down into the valley and cross westward to the Peloncillo Mountains. That's when the Border Patrol would have its best chance to catch them.

Wednesday, 1 p.m.

Parra's radio crackled with good news. Seismic sensors in the Animas Mountains had recorded movement overnight, and that

morning, agents on horseback had picked up the sign. It was the same group. . . .

About 3 p.m., Parra, 43, pulled to a stop and met up with the horse patrol unit leader, Lawrence "Junior" Helbig. Helbig and his partner had spent much of the day tracking the footprints.

"It's the same group," Helbig said. "The odds of having several fine lines and a doper lug together are just too high."

Fraley had confirmed three sets of footprints; now Helbig had spotted two more. A late-morning shower had mucked up the trail, but the color of the soil in the prints suggested they were only about four hours old. . . .

Two weeks earlier, Helbig had tracked five smugglers into a nearby thicket. The smugglers jumped out and "quailed," running in all directions. Agents caught them and found several hundred pounds of marijuana as well as an AK-47 rifle. One of the smugglers said the weapon was for protection—from other drug smugglers.

"Thank goodness they're not real violent toward us yet," Helbig said. "But there's a reason we carry a sidearm."

Wednesday, 8:15 p.m.

Parra, tired after a 14-hour day, headed home. . . . Jose Portillo, 36, the night supervisor, set a trap.

He assigned two agents to hide on one side of County Road 1. Using thermal imaging binoculars, they would try to pick up the

and that legalization would expand the number of distributors, thereby increasing competition and reducing prices. Drugs would thus be safer, drug-related crimes would be reduced, and production and distribution of previously controlled substances would be taken out of the hands of the underworld.

Those in favor of legalization also suggest that the greater availability of drugs would not increase demand, pointing to countries where some drugs have already been decriminalized. **Decriminalization,** which entails removing state penalties for certain drugs, promotes a medical rather than criminal approach to drug use that encourages users to seek treatment and adopt preventive practices. For example, in 2001, Portugal, "became the first European country to officially abolished all criminal penalties for personal possession of drugs, including marijuana, cocaine, heroin, and methamphetamine" (Szalavitz 2009, p. 1). Despite fears of negative consequences, a report by the Cato Institute concludes that decriminalization of personal possession of drugs was responsible for reducing new cases of HIV infections, decreasing drug use among Portuguese teens, and doubling the number of people seeking treatment for drug addiction (Greenwald 2009).

Opponents of legalization argue that it would be construed as government approval of drug use and, as a consequence, drug experimentation and abuse would increase. Furthermore, although the legalization of drugs would result in substantial revenues for the government, drug trafficking and black markets would still flourish because all drugs would not be decriminalized (e.g., crack). Legalization would also require an extensive

decriminalization Entails removing state penalties for certain drugs, and promotes a medical rather than criminal approach to drug use that encourages users to seek treatment and adopt preventive practices.

smugglers as they descended into the valley. Then they would radio another two-man unit, this one armed with M-4 rifles. If all went according to plan, the smugglers would never make it across County Road 1. . . .

At 9:15 p.m., Portillo checked in with his two teams.

Nothing.

"Tonight is the night to catch them," Portillo said, gazing out his windshield, Orion shimmering in the sky. "It's harder after this."

At 1:46 a.m., Portillo reached for the radio handset.

"Let's pull out," he told his surveillance units.

He didn't try to hide his disappointment. "They've crossed by now," he said. "They must have taken a different route."

Pause.

"Tomorrow, it's do or die."

Thursday, 10 a.m.

Parra read the overnight report: The smugglers had crossed County Road 1 several miles from the stakeout, and made it to the Peloncillo Mountains. Rogelio Villa and his partner, on foot in the Peloncillo range, picked up the sign. "We've got our guys over here," Villa radioed Parra. "They were definitely here late last night or early this morning. Looks like there are four or five of them."

The footprints were different, but Parra wasn't worried. Smugglers often swap out

their boots. "It's likely these are the same guys," he said.

Traffickers know that footprints can give them away. So they walk on rocks, where they don't leave prints. They walk backward. They wear boots like those worn by Border Patrol agents. They tie strips of carpet to their soles to avoid leaving clear prints on dirt roads. ("I've even seen them take the hoofs from cattle and glue them to their shoes," one agent said.)

Two agents jumped ahead to see how far the smugglers had gotten. At a cattle watering tank, they came upon a rancher's motion-activated game camera. An agent took the memory card out of the camera and put it in his own.

The photo that popped up was clear: a muscular, dark-haired man with a short beard, wearing black jeans and a sweatshirt under his striped shirt. A water bottle in his hand had been shrouded in black cloth, to avoid a reflection that might give his position away. On his back was a parcel, about 3 feet square. Marijuana. . . .

Friday, noon

Two infrared scopes aimed at Weatherby Canyon had detected no movement the night before. Parra returned to look for signs that the smugglers had come down the mountain. Two younger agents searched for an hour and found nothing.

Within 10 minutes of arriving, though, Parra picked up the footprints. . . .

"I'm leaning toward thinking they're still up on top of those hills," he said, lighting a Marlboro. "We were keeping this area very hot last night, and they could have stayed up."

Later, Parra reconsidered when a nearby rancher reported that his dogs had barked loudly at 3 a.m. "I don't see how we would have missed them," he said. "But the dopers must have been moving."

Saturday morning

Tim Lowe, the day supervisor, dispatched two agents on ATVs to Weatherby.

A scope had been deployed there briefly the night before, "but they 10-3ed it," Lowe said, using the code for terminating an operation. No one had been able to pick up the sign again.

Sunday afternoon

The exit for Steins Ghost Town on Interstate 10 leads to a cemetery of weathered crosses. Next to the cemetery, pieces of cloth and shoulder straps made of old blankets lay on the ground.

The drugs were gone, likely bound for Tucson and points west.

So were the smugglers, headed back to Mexico to collect their paychecks and pick up another load.

Back on the border, the day shift was out—cutting for new sign.

Source: Kraft, Scott. 2009. "Pursuing Smugglers, Border Agents Become Trackers." *Los Angeles Times*, May 12. Copyright © 2009 by the *Los Angeles Times*. Reprinted by permission.

and costly bureaucracy to regulate the manufacture, sale, and distribution of drugs. Finally, the position that drug use is an individual's right cannot guarantee that others will not be harmed. It is illogical to assume that a greater availability of drugs will translate into a safer society.

State Initiatives

Several initiatives have resulted in statewide referendums concerning the cost-effectiveness of government policies. For example, as a result of the Substance Abuse and Crime Prevention Act of 2000 (Proposition 36), California (as well as many other states) now requires that nonviolent first- and second-time minor drug offenders receive treatment, including job training, therapy, literacy education, and family counseling rather than jail time. The act, passed by 61 percent of California voters, permanently changed state law. Every year, over 30,000 drug offenders enter treatment, about half of them for the first times (Wheeler 2008).

Lastly, over the past decade, voters and state governments have enacted significant drug policy reforms. For example, Connecticut passed significant overdose prevention legislation, following in the footsteps of New Mexico, Texas and Kansas passed legislation providing for treatment instead of incarceration for first-time drug offenders, and Illinois passed legislation allowing for the sale of sterile syringes without a prescription (Drug Policy Alliance 2007). Further, many states have active drug policy reform organizations. A Better Way Foundation in Connecticut, for example, is a non-profit organization "dedicated to shifting current drug policy from a paradigm that prioritizes incarceration to one that prioritizes public health, treatment, and public safety" (Better Way Foundation 2009, p. 1).

Understanding Alcohol and Other Drug Use

In summary, substance abuse—that is, drugs and their use—is socially defined. As the structure of society changes, the acceptability of one drug or another changes as well. As conflict theorists assert, the status of a drug as legal or illegal is intricately linked to those who have the power to define acceptable and unacceptable drug use. There is also little doubt that rapid social change, anomie, alienation, and inequality further drug use and abuse. Symbolic interactionism also plays a significant role in the process: If people are labeled "drug users" and are expected to behave accordingly, then drug use is likely to continue. If people experience positive reinforcement of such behaviors and/or have a biological predisposition to use drugs, the probability of their drug involvement is even higher. Thus, the theories of drug use complement rather than contradict one another.

There are two issues that need to be addressed in understanding drug use. The first is at the micro level—why does a given individual use alcohol or other drugs? Many individuals at high risk for drug use have been "failed by society"—they are living in poverty, unemployed, victims of abuse, dependents of addicted and neglectful parents, and the like. Despite the social origins of drug use, many treatment alternatives, emanating from a clinical model of drug use, assume that the origin of the problem lies within the individual rather than in the structure and culture of society. Although admittedly the problem may lie within the individual when treatment occurs, policies that address the social causes of drug abuse must be a priority in dealing with the drug problem in the United States.

The second question, related to the first, asks why drug use varies so dramatically across societies, often independent of a country's drug policies. The United States metes out some of the most severe penalties for drug violations in the world, but has one of the highest rates of marijuana and cocaine use. On the other hand, as mentioned earlier, Portugal decriminalized personal possession of all drugs, and youth drug use is down (Szalavitz 2009). Most compellingly, a 2008 World Health Organization survey of 17 countries concluded that there is no link between the harshness of drug policies and the consumption rates of its citizenry (Degenhardt et al. 2008).

That said, what is needed is a more balanced approach—one that acknowledges that not all drugs have the same impact on society or on the individuals that use them. The present administration appears to be leaning this way. For example, in a break from the previous administration, President Obama supported federally funded needle exchange programs—a staple of harm reduction advocates—and has stated that it is "entirely appropriate" to use marijuana for the same purposes and under the same controls as other drugs prescribed by a physician (Heinrich 2009; Dinan and Conery 2009). On the other hand, there is little doubt that the present administration is serious about reducing the supply of illegal drugs entering the United States. Nearly 60 percent of the proposed 2012 drug budget is allocated to domestic law enforcement, interdiction, and international efforts in the fight against drugs. It should again, however, be noted that the proposed distribution of funding for harm reduction versus supply reduction is the most balanced in U.S. history. Only time will tell if this new approach to drug control, one that reflects both a public health and criminal justice position, will be more successful than the policies of previous administrations.

[A] 2008 World Health Organization survey of 17 countries concluded that there is no link between the harshness of drug policies and the consumption rates of its citizenry.

- **What is a drug, and what is meant by drug abuse?**
Sociologically, the term *drug* refers to any chemical substance that (1) has a direct effect on the user's physical, psychological, and/or intellectual functioning; (2) has the potential to be abused; and (3) has adverse consequences for the individual and/or society. Drug abuse occurs when acceptable social standards of drug use are violated, resulting in adverse physiological, psychological, and/or social consequences.

- **How do the three sociological theories of society explain drug use?**
Structural functionalists argue that drug abuse is a response to the weakening of norms in society, leading to a condition known as anomie or normlessness. From a conflict perspective, drug use occurs as a response to the inequality perpetuated by a capitalist system as societal members respond to alienation from their work, family, and friends. Symbolic interactionism concentrates on the social meanings associated with drug use. If the initial drug use experience is defined as pleasurable, it is likely to recur, and over time, the individual may earn the label of "drug user."

- **What are the most frequently used legal and illegal drugs?**
Alcohol is the most commonly used and abused legal drug in America. The use of tobacco products is also very high, with 23 percent of Americans reporting that they currently smoke cigarettes. Marijuana is the most commonly used illicit drug, with 166 million marijuana users, representing 3.9 percent of the world's population between the ages of 15 and 64.

- **What are the consequences of drug use?**
The consequences of drug use are fivefold. First is the cost to the family, often manifesting itself in higher rates of divorce, spouse abuse, child abuse, and child neglect. Second is the relationship between drugs and crime. Those arrested have disproportionately higher rates of drug use. Although drug users commit more crimes, sociologists disagree as to whether drugs actually "cause" crime or whether, instead, criminal activity leads to drug involvement. Third are the economic costs (e.g., loss of productivity), which are in the billions. Then there are the health costs of abusing drugs, including shortened life expectancy; higher morbidity (e.g., cirrhosis of the liver and lung cancer); exposure to HIV infection, hepatitis, and other diseases through shared needles; a weakened immune system; birth defects such as fetal alcohol syndrome; drug addiction in children; and higher death rates. Finally, illegal drug production takes its toll on the environment, which impacts all Americans.

- **What treatment alternatives are available for drug users?**
Although there are many ways to treat drug abuse, two methods stand out. The inpatient–outpatient model entails medical supervision of detoxification and may or may not include hospitalization. Twelve-step programs such as Alcoholics Anonymous (AA) and Narcotics Anonymous (NA) are particularly popular, as are therapeutic communities. Therapeutic communities are residential facilities where drug users learn to redefine themselves and their behavior as a response to the expectations of others and self-definition. Finally, drug courts are used as an alternative to the traditional punitive methods of probation and incarceration.

- **What can be done about the drug problem?**
First, there are government regulations limiting the use (e.g., the law establishing the 21-year-old drinking age) and distribution (e.g., prohibitions about importing drugs) of legal and illegal drugs. The government also imposes sanctions on those who violate drug regulations and provides treatment facilities for other offenders. Economic incentives (e.g., cost) and prevention programs have also been found to impact consumption rates. Finally, legal action holding companies responsible for the consequences for their product—for example, class-action suits against tobacco producers—have been fairly successful.

TEST YOURSELF

1. "Cannabis cafés" are commonplace throughout England.
   a. True
   b. False
2. The most used illicit drug in the world is
   a. heroin
   b. marijuana
   c. cocaine
   d. methamphetamine
3. What theory would argue that the continued legality of alcohol is a consequence of corporate greed?
   a. Structural functionalism
   b. Symbolic interactionism
   c. Reinforcement theory
   d. Conflict theory
4. Cigarettes smoking is
   a. the third leading cause of preventable death in the United States
   b. not addictive
   c. the most common use of tobacco products
   d. increasing in the United States
5. In the United States, drinking is highest among young, nonwhite males.
   a. True
   b. False

6. Which European country decriminalized personal possession of all drugs, including marijuana, cocaine, heroin, and methamphetamine?
   a. Spain
   b. France
   c. Portugal
   d. The Netherlands
7. The active ingredient in marijuana, THC, can act as a sedative or a hallucinogen.
   a. True
   b. False
8. According to the proposed 2012 budget, most federal drug control dollars are allocated to
   a. international efforts
   b. domestic law enforcement
   c. prevention and research
   d. treatment and research
9. Decriminalization refers to the removal of penalties for certain drugs.
   a. True
   b. False
10. The two-pronged drug control strategy of the U.S. government entails supply reduction and harm reduction.
   a. True
   b. False

Answers: 1: b; 2: b; 3: d; 4: c; 5: b; 6: c; 7: a; 8: b; 9: a; 10: b.

KEY TERMS

binge drinking 73
chemical dependency 68
club drugs 80
crack 79
date-rape drugs 81
decriminalization 94
demand reduction 93

deregulation 93
drug 66
drug abuse 68
drug courts 89
fetal alcohol syndrome 86
gateway drug 78
harm reduction 91

heavy drinking 73
legalization 93
meta-analysis 74
supply reduction 93
therapeutic communities 89

MEDIA RESOURCES

Turning to Video

▶❙❙ Watch the video on *Drinking on Campus: Recovery Dorms* (running time 4:12), available through **CengageBrain.com.** This video describes "sober dorms" and the help university's are offering students with drinking problems. What kind of help does your school offer students with alcohol or drug problems?

Online Study Resources

Log in to **www.cengagebrain.com** to access the resources your instructor has assigned. For this book, you can access:

CourseMate

Access chapter-specific learning tools including learning objectives, practice quizzes, videos, Internet exercises, flash cards, and glossaries, as well as web links, and more in your Sociology CourseMate.

4

Crime and
Social Control

"Unjust social arrangements are themselves a kind of extortion, even violence."

—John Rawls, *A Theory of Justice.*

CamEl Creative/Workbook Stock/Getty Images

IT MAY HAVE started like any other workday, but for health care professionals in nine cities throughout the United States, the day ended quite differently. On February 17, 2011, over 700 law enforcement agents, guided by members of a special task force, arrested 111 doctors, nurses, physical therapists, and health care provider executives, charging them with fraudulently billing Medicare for over $225 million (Kennedy 2011; Weaver 2011; Serrano 2011). There's the Detroit podiatrist who charged Medicare $700,000 for clipping patients' toenails billed as "partial toenail removal," and the New York proctologist who billed Medicare some $6.5 million for hemorrhoid removal procedures, most of which were never performed (Kennedy 2011, p. 1). Other accusations included kickbacks for recommending the use of unnecessary medical equipment, billing for procedures and therapies that were never performed, recruiting patients into health care scams, writing fake prescriptions, and money laundering. Said Daniel Levinson of the federal Department of Health and Human Services, "We will not tolerate criminals lining their pockets at the expense of Medicare patients and taxpayers" (quoted in Serrano 2011, p. 1).

The U.S. Department of Justice's Task Force on Medicare Fraud, working with the Health and Human Services Department, was established in 2007, and since that time has charged nearly 1,000 people with "false billing schemes totaling more than $2.3 billion" and resulting in 750 convictions (Serrano 2011, p. 1). The problem, however, is greater than simply arresting violators. Prevention is always better than prosecution, and as one Federal Bureau of Investigation (FBI) agent suggested, Medicare "should conduct comprehensive criminal background checks on providers, stop the practice of paying claims within 14 days, suspend payments for questionable charges, and change the system of using Social Security numbers for beneficiaries to prevent identity theft" (Weaver 2011, p. 1).

The U.S. Department of Justice and the FBI are just two parts of the massive bureaucracy called the criminal justice system, which includes law enforcement, courts, and corrections. In this chapter, we examine the criminal justice system as well as theories, types, and demographic patterns of criminal behavior. The economic, social, and psychological costs of crime are also examined. The chapter concludes with a discussion of social control, including policies and prevention programs designed to reduce crime in the United States.

The Global Context: International Crime and Social Control

Several facts about crime are true throughout the world. First, crime is ubiquitous—there is no country where crime does not exist. Second, most countries have the same components in their criminal justice systems: police, courts, and prisons. Third, adult males make up the largest category of crime suspects worldwide. Fourth, in all countries, theft is the most common crime committed, whereas violent crime is a relatively rare event.

Even so, dramatic differences do exist in international crime rates, although comparisons are made difficult by variations in measurement and crime definitions (Siegel 2006). Because of these difficulties, Winslow and Zhang (2008, pp. 30–32) created a global crime database from data from the United Nations and INTERPOL—the international police agency. In their discussion of the "United States versus the World," the authors reveal some interesting and often counterintuitive findings. First, the United States does not have the highest crime rate in the world. Using their database, the United States ranks 12th among 165 nations, with Sweden, Denmark, Australia, and Great Britain, in rank order, each having a higher crime rate than the United States.

Winslow and Zhang (2008) also examined crime rates by dividing them into types of crime—violent crime or property crime. Violent crime, as discussed later in the chapter, includes murder, rape, robbery, and aggravated assault. When one compares the United States to other countries, the United States once again is not in the top 10. Several developing countries (e.g., Namibia and Swaziland) as well as developed countries (e.g., Australia and Sweden) have higher violent crime rates than the United States. Property crimes show a similar pattern. Based on the global crime database created by Winslow and Zhang (2008), the United States ranks 13th in property crimes (car theft, burglary, and larceny) with Sweden, Denmark, Australia, and Great Britain topping the list.

Violent crime and property crimes represent just two types of crime that take place worldwide. Although we are concerned about these types of crimes and the possibility of victimization, INTERPOL has identified six global priority areas (INTERPOL 2011a): (1) drugs and criminal organizations (e.g., drug trafficking), (2) financial and high-tech crimes (e.g., counterfeiting, fraud, and cybercrime), (3) tracking of fugitives, (4) public safety and countering terrorism (discussed in Chapter 15), (5) trafficking in human beings, and 6) fighting corruption (e.g. enforcing the rule of law). Each of these priority areas contains a relatively new category of crimes—transnational crime. As defined by the U.S. Department of Justice, **transnational crime** is "organized criminal activity across one or more national borders" (U.S. Department of Justice 2003). The significance of transnational crime should not be minimized. As Shelley states (2007):

> Transnational crime will be a defining issue of the 21st century for policy makers— as defining as the Cold War was for the 20th century and colonialism was for the 19th. Terrorists and transnational crime groups will proliferate because these crime groups are major beneficiaries of globalization. They take advantage of increased travel, trade, rapid money movements, telecommunications and computer links, and are well positioned for growth. (p. 1)

For example, the Internet has led to an explosive growth in child pornography. In 2005, a United Nations expert on the subject told the 53-nation U.N. Commission on Human Rights that governments must act now to curb the proliferation of child pornography (Klapper 2005). Of late, several successes have been recorded. In 2010, more than 50 people were arrested in an international child pornography ring that ". . . at its peak had more than 1,000 members trading millions of sexually explicit images" (Wilson 2010, p. 1).

. . . 12.5 million people in the world have been victims of human trafficking . . . [and] of that number, it is estimated that 80.0 percent are women and girls, and 50.0 percent are minors.

Human trafficking is another example of transnational crime. According to the Polaris Project, a U.S.-based organization fighting human trafficking and slavery in all its forms, 12.5 million people in the world have been victims of human trafficking (Polaris Project 2011). Of that number, it is estimated that 80 percent are women and girls, and 50 percent are minors (U.S. Department of State 2008). Although the majority of people are trafficked into commercial sexual exploitation, others are trafficked into forced labor, sexual servitude, and for use as child soldiers.

AP Photo/Pavel Rahman

A young boy, a victim of human trafficking, is forced to work in a balloon factory outside of Dhaka, Bangladesh. Human trafficking is estimated to be multibillion-dollar enterprise, approaching the international value of drug and arms trafficking (INTERPOL 2011c).

transnational crime Criminal activity that occurs across one or more national borders.

What Do You Think? According to the International Maritime Bureau (IMB), hijacking of ships and other vessels reached an all time high in 2010—53 ships hijacked, and 1,181 hostages held—86 percent by Somali pirates (ICC 2011). For example, in 2011, Somali pirates hijacked an American yacht and killed four of its passengers—the first time Americans were killed in a maritime piracy incident (MacAskill & Rice 2011). In federal law, death resulting from aircraft piracy, or attempted hijacking of an aircraft, carries the death penalty. Do you think that deaths that result from maritime hijacking should carry the death penalty? What if the death is not of a hostage but of a fellow hijacker?

Sources of Crime Statistics

The U.S. government spends millions of dollars annually to compile and analyze crime statistics. A **crime** is a violation of a federal, state, or local criminal law. For a violation to be a crime, however, the offender must have acted voluntarily and with intent and have no legally acceptable excuse (e.g., insanity) or justification (e.g., self-defense) for their behavior. The three major types of statistics used to measure crime are official statistics, victimization surveys, and self-report offender surveys.

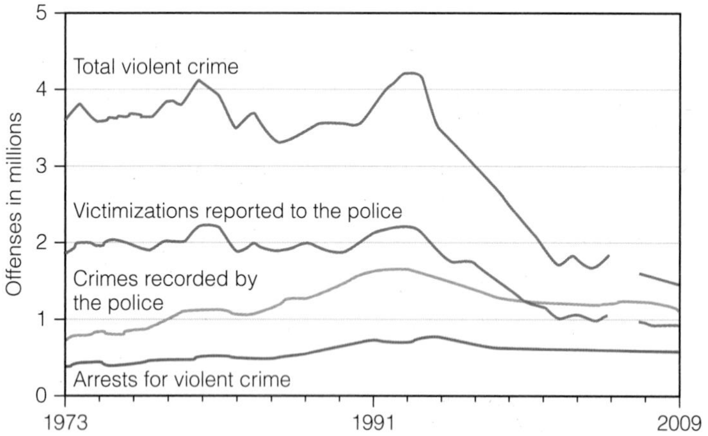

Figure 4.1: Four Measures of Serious Violent Crime
The serious violent crimes included are rape, robbery, aggravated assault, and homicide.
Source: BJS 2010a.

Official Statistics

Local sheriffs' departments and police departments throughout the United States collect information on the number of reported crimes and arrests and voluntarily report them to the FBI. The FBI then compiles these statistics annually and publishes them, in summary form, in the Uniform Crime Reports (UCR). The UCR lists **crime rates** or the number of crimes committed per 100,000 population, the actual number of crimes and the percentage of change over time, and clearance rates. **Clearance rates** measure the percentage of cases in which an arrest and official charge have been made and the case has been turned over to the courts.

These statistics have several shortcomings. For example, many incidents of crime go unreported. In 2009, nearly half of all violent crimes and about 40 percent of all property crimes went unreported (BJS 2010a). Even if a crime is reported, the police may not record it (see Figure 4.1). Alternatively, some rates may be exaggerated. Motivation for such distortions may come from the public (e.g., demanding that something be done), from political officials (e.g., election of a sheriff), and/or from organizational pressures (e.g., budget requests). For example, a police department may "crack down" on drug-related crimes in an election year. The result is an increase in the recorded number of these offenses for that year. Such an increase reflects a change in the behavior of law enforcement personnel, not a change in the number of drug violations. Thus, official crime statistics may be a better indicator of what police are doing rather than of what criminals are doing.

In the 1970s, the "law enforcement community called for a thorough evaluation of the UCR Program to recommend an expanded and enhanced data collection system to meet the needs of law enforcement in the 21st century" (FBI 2009a, p. 1). The result, the National Incident-Based Reporting System (NIBRS), requires that law enforcement agencies provide extensive information on each criminal incident and arrest for 22 offenses (Group A) and arrestee information on 11 lesser offenses (Group B). Thus far, the FBI has certified 36 states for NIBRS participation. The hope is that, once implemented nationally, the NIBRS will provide more reliable and comprehensive crime data.

crime An act, or the omission of an act, that is a violation of a federal, state, or local criminal law for which the state can apply sanctions.

crime rate The number of crimes committed per 100,000 population.

clearance rate The percentage of crimes in which an arrest and official charge have been made and the case has been turned over to the courts.

Read each of the following questions. Since the age of 16, if you have ever engaged in the behavior described, place a "1" in the space provided. If you have not engaged in the behavior, put a "0" in the space provided. After completing the survey, read the section on interpretation to see what your answers mean.

Questions

1. Have you ever been in possession of drug paraphernalia? _____
2. Have you ever lied about your age or about anything else when making application to rent an automobile? _____
3. Have you ever obtained a false ID to gain entry to a bar or event? _____
4. Have you ever tampered with a coin-operated vending machine or parking meter? _____
5. Have you ever shared, given, or shown pornographic material to someone under 18? _____
6. Have you ever begun and/or participated in a basketball, baseball, or football pool? _____
7. Have you ever used "filthy, obscene, annoying, or offensive" language while on the telephone? _____
8. Have you ever given or sold a beer to someone under the age of 21? _____
9. Have you ever been on someone else's property (land, house, boat, structure, and so on) without that person's permission? _____
10. Have you ever forwarded a chain letter with the intent to profit from it? _____
11. Have you ever improperly gained access to someone else's e-mail or other computer account? _____
12. Have you ever written a check for over $150 when you knew it was bad? _____

Interpretation

Each of the activities described in these questions represents criminal behavior that was subject to fines, imprisonment, or both under the laws of Florida in 2008. For each activity, the following table lists the maximum prison sentence and/or fine for a first-time offender. To calculate your "prison time" and/or fines, sum the numbers corresponding to each activity you have engaged in.

Offense	Maximum Prison Sentence	Maximum Fine
1. Possession of drug paraphernalia	1 year	$1,000
2. Fraud	5 years	$5,000
3. Possession of false ID or driver's license	5 years	$5,000
4. Larceny	2 months	$500
5. Protection of minors from obscenity	5 years	$5,000
6. Illegal gambling	2 months	$500
7. Harassing/obscene telecommunications	2 months	$500
8. Illegal distribution of alcohol	2 months	$500
9. Trespassing	1 year	$1,000
10. Illegal gambling	1 year	$1,000
11. Illegal misappropriation of cyber communication	5 years	$5,000
12. Worthless checks	5 years	$5,000

Source: Florida Criminal Code 2009.

Victimization Surveys

Acknowledging "the dark figure of crime," that is, the tendency for so many crimes to go unreported and thus undetected by the UCR, the U.S. Department of Justice conducts the National Crime Victimization Survey (NCVS). Begun in 1973 and conducted annually, the NCVS interviews about 100,000 people about their experiences as victims of crime. Interviewers collect a variety of information, including the victim's background (e.g., age, race and ethnicity, sex, marital status, education, and area of residence), relationship to the offender (stranger or nonstranger), and the extent to which the victim was harmed.

In 2010, the latest year for which victimization data are available, there were 3.8 million violent crimes and 14.8 million property crimes (Truman 2011). All of the major violent and property crimes (see Types of Crime) measured by the Bureau of Justice Statistics dropped between 2001 and 2010. Nonetheless, in 2010, there was one rape or sexual assault, two robberies, three aggravated assaults, and ten simple assaults for every 1,000 people 12 and older in the United States (BJS 2011a). Although victimization surveys provide detailed information about crime victims, they provide less reliable data on offenders.

> . . . [I]n 2010, there was one rape or sexual assault, two robberies, three aggravated assaults, and ten simple assaults for every 1,000 people 12 and older in the United States.

Self-Report Offender Surveys

Self-report surveys ask offenders about their criminal behavior. The sample may consist of a population with known police records, such as a prison population, or it may include respondents from the general population, such as college students.

Self-report data compensate for many of the problems associated with official statistics but are still subject to exaggerations and concealment. The Criminal Activities Survey in this chapter's *Self and Society* feature asks you to indicate whether you have engaged in a variety of illegal activities.

Self-report surveys reveal that virtually every adult has engaged in some type of criminal activity. Why then is only a fraction of the population labeled criminal? Like a funnel, which is large at one end and small at the other, only a small proportion of the total population of law violators are ever convicted of a crime. For individuals to be officially labeled criminals, (1) their behavior must become known to have occurred; (2) the behavior must come to the attention of the police, who then file a report, conduct an investigation, and make an arrest; and finally, (3) the arrestee must go through a preliminary hearing, an arraignment, and a trial and may or may not be convicted. At every stage of the process, offenders may be "funneled" out. As Figure 4.1 indicates, the measures of crime used at various points in time lead to different results.

Sociological Theories of Crime

Some explanations of crime focus on psychological aspects of the offenders, such as psychopathic personalities, unhealthy relationships with parents, and mental illness. Other crime theories focus on the role of biological variables, such as central nervous system malfunctioning, stress hormones, vitamin or mineral deficiencies, chromosomal abnormalities, and a genetic predisposition toward aggression. Sociological theories of crime and violence emphasize the role of social factors in criminal behavior and societal responses to it.

Structural-Functionalist Perspective

According to Durkheim and other structural functionalists, crime is functional for society. One of the functions of crime and other deviant behavior is that it strengthens group cohesion: "The deviant individual violates rules of conduct that the rest of the community holds in high respect; and when these people come together to express their outrage over the offense . . . they develop a tighter bond of solidarity than existed earlier" (Erikson 1966, p. 4).

Crime can also lead to social change. For example, an episode of local violence may "achieve broad improvements in city services . . . be a catalyst for making public agencies more effective and responsive, for strengthening families and social institutions, and for creating public-private partnerships" (National Research Council 1994, pp. 9–10).

Although structural functionalism as a theoretical perspective deals directly with some aspects of crime, it is not a theory of crime per se. Three major theories of crime have developed from structural functionalism, however. The first, called strain theory, was developed by Robert Merton (1957) and uses Durkheim's concept of *anomie,* or normlessness. Merton argued that, when the structure of society limits legitimate means (e.g., a job) of acquiring culturally defined goals (e.g., money), the resulting strain may lead to crime. For example, rapid economic social change in Russia, and specifically high rates of unemployment, has lead to increases in the homicide rates (Pridemore & Kim 2007).

Individuals, then, must adapt to the inconsistency between means and goals in a society that socializes everyone into wanting the same thing but provides opportunities for only some (see Table 4.1). *Conformity* occurs when individuals accept the culturally defined goals and the socially legitimate means of achieving them. Merton suggested that most individuals, even those who do not have easy access to the means and goals, remain conformists. *Innovation* occurs when an individual accepts the goals of society but rejects or lacks the socially legitimate means of achieving them. Innovation, the mode of adaptation most associated with criminal behavior, explains the high rate of crime committed by uneducated and poor individuals who do not have access to legitimate means of achieving the social goals of wealth and power.

Another adaptation is *ritualism,* in which, for example, individuals accept a lifestyle of hard work but reject the cultural goal of monetary rewards. Ritualists go through the motions of getting an education and working hard, yet they are not committed to the goal of accumulating wealth or power. *Retreatism* involves rejecting both the cultural goal of success and the socially legitimate means of achieving it. Retreatists withdraw or retreat from society and may become alcoholics, drug addicts, or vagrants. Finally, *rebellion* occurs when individuals reject both culturally defined goals and means and substitute new goals and means. For example, rebels may use social or political activism to replace the goal of personal wealth with the goal of social justice and equality.

Whereas strain theory explains criminal behavior as a result of blocked opportunities, subcultural theories argue that certain groups or subcultures in society have values and attitudes that are conducive to crime and violence. Members of these groups and subcultures, as well as other individuals who interact with them, may adopt the crime-promoting attitudes and values of the group. For example, Kubrin and Weitzer (2003) found that retaliatory homicide is a response to subcultural norms of violence that exist in some neighborhoods.

However, if blocked opportunities and subcultural values are responsible for crime, why don't all members of the affected groups become criminals? Control theory may answer that question. Hirschi (1969), consistent with Durkheim's emphasis on social solidarity, suggests that a strong social bond between individuals and the social order constrains some individuals from violating social norms. Hirschi identified four elements of the social bond: attachment to significant others, commitment to conventional goals, involvement in conventional activities, and belief in the moral standards of society. Several empirical tests of Hirschi's theory support the notion that the higher the attachment, commitment, involvement, and belief, the higher the social bond and the lower the probability of criminal behavior. For example, Ford (2005), using data from the National Youth Survey, concludes that a strong family bond lowers the probability of adolescent substance use and delinquency, and Bell (2009) reports that weaker attachment to parents is associated with a greater likelihood of gang membership for both males and females.

TABLE 4.1 Merton's Strain Theory

MODE OF ADAPTATION	CULTURALLY DEFINED GOALS	STRUCTURALLY DEFINED MEANS
1. Conformity	+	+
2. Innovation	+	−
3. Ritualism	−	+
4. Retreatism	−	−
5. Rebellion	±	±

+ = acceptance of/access to; − = rejection of/lack of access to; ± = rejection of culturally defined goals and structurally defined means and replacement with new goals and means;

Source: Adapted with permission of The Free Press, of Simon & Schuster Adult Publishing Group, from Robert K. Merton's *Social Theory and Social Structure* (1957). Copyright © 1957 by The Free Press; copyright renewed 1985 by Robert K. Merton. All rights reserved.

Conflict Perspective

Conflict theories of crime suggest that deviance is inevitable whenever two groups have differing degrees of power; in addition, the more inequality there is in a society, the greater the crime rate in that society. Social inequality leads individuals to commit crimes such as larceny and burglary as a means of economic survival. Other individuals, who are angry and frustrated by their low position in the socioeconomic hierarchy, express their rage and frustration through crimes such as drug use, assault, and homicide. In Argentina, for example, the soaring violent crime rate is hypothesized to be "a product of the enormous imbalance in income distribution . . . between the rich and the poor" (Pertossi 2000).

According to the conflict perspective, those in power define what is criminal and what is not, and these definitions reflect the interests of the ruling class. Laws against vagrancy, for example, penalize individuals who do not contribute to the capitalist system of work and consumerism. Furthermore, D'Alessio and Stolzenberg (2002, p. 178) found that "in cities with high unemployment, unemployed defendants have a substantially

Yadid Levy/Alamy

higher probability of pretrial detention" than employed defendants. Rather than viewing law as a mechanism that protects all members of society, conflict theorists focus on how laws are created by those in power to protect the ruling class. Wealthy corporations contribute money to campaigns to influence politicians to enact tax laws that serve corporate interests (Reiman & Leighton 2010).

In addition, conflict theorists argue that law enforcement is applied differentially, penalizing those without power and benefiting those with power. For example, a 2009 report by the National Council on Crime and Delinquency found that ". . . in arrests, court processing and sentencing, new admissions and ongoing populations in prison and jails, probation and parole, capital punishment, and recidivism . . . persons of color, particularly African Americans, are more likely to receive less favorable results than their white counterparts" (Hartney & Vuong 2009, p. 2). Female prostitutes are more likely to be arrested than are the men who seek their services. Rape laws originated to serve the interests of husbands and fathers who wanted to protect their property—wives and unmarried daughters. Finally, unlike street criminals, corporate criminals are most often punished by fines rather than lengthy prison terms.

Societal beliefs also reflect power differentials. For example, "rape myths" are perpetuated by the male-dominated culture to foster the belief that women are to blame for their own victimization, thereby, in the minds of many, exonerating the offenders. Such myths include the notion that when a woman says *no* she means *yes,* that "good girls" don't get raped, that appearance indicates willingness, and that women secretly want to be raped. Not surprisingly, there is less rape in societies where women and men have greater equality.

To Marxists, the cultural definition of women as property contributes to the high rates of female criminality and, specifically, involvement in prostitution, drug abuse, and petty theft. In the Netherlands, prostitution has been legal since 2000. Amsterdam's "redlight district" is famous for its displays of "window prostitutes."

Symbolic Interactionist Perspective

Two important theories of crime emanate from the symbolic interactionist perspective. The first, labeling theory, focuses on two questions: How do crime and deviance come to be defined as such, and what are the effects of being labeled criminal or deviant? According to Howard Becker (1963):

> Social groups create deviance by making rules whose infractions constitute deviance, and by applying those rules to particular people and labeling them as outsiders. From this point of view, deviance is not a quality of the act a person commits, but rather a consequence of the application by others of rules and sanctions to an "offender." The deviant is one to whom the label has successfully been applied; deviant behavior is behavior that people so label. (p. 238)

Labeling theorists make a distinction between **primary deviance,** which is deviant behavior committed before a person is caught and labeled an offender, and **secondary deviance,** which is deviance that results from being caught and labeled. After a person violates the law and is apprehended, that person is stigmatized as a criminal. This deviant label often dominates the social identity of the person to whom it is applied and becomes the person's "master status," that is, the primary basis on which the person is defined by others.

Being labeled as deviant often leads to further deviant behavior because (1) the person who is labeled as deviant is often denied opportunities for engaging in nondeviant behavior, and (2) the labeled person internalizes the deviant label, adopts a deviant self-concept, and acts accordingly. For example, a teenager who is caught selling drugs at school may be expelled and thus denied opportunities to participate in nondeviant school activities (e.g., sports and clubs) and to associate with nondeviant peer groups. The labeled and stigmatized teenager may also adopt the self-concept of

primary deviance Deviant behavior committed before a person is caught and labeled an offender.

secondary deviance Deviant behavior that results from being caught and labeled as an offender.

a "druggie" or "pusher" and continue to pursue drug-related activities and membership in the drug culture.

The assignment of meaning and definitions learned from others is also central to the second symbolic interactionist theory of crime, differential association. Edwin Sutherland (1939) proposed that, through interaction with others, individuals learn the values and attitudes associated with crime as well as the techniques and motivations for criminal behavior. Individuals who are exposed to more definitions favorable to law violation (e.g., "crime pays") than to unfavorable ones (e.g., "do the crime, you'll do the time") are more likely to engage in criminal behavior. Thus, children who see their parents benefit from crime or who live in high-crime neighborhoods where success is associated with illegal behavior are more likely to engage in criminal behavior.

> Individuals who are exposed to more definitions favorable to law violation (e.g., "crime pays") than to unfavorable ones (e.g., "do the crime, you'll do the time") are more likely to engage in criminal behavior.

Unfavorable definitions come from a variety of sources. Of particular concern in recent years is the role of video games in promoting criminal or violent behavior such as the gratuitously violent video game, *Postal 2,* in which players can set harmless bystanders on fire. In response to this and other violent video games, many states now require a video rating system that differentiates between cartoon violence, fantasy violence, intense violence, and sexual violence. In 2010, the U.S. Supreme Court heard arguments concerning the constitutionality of a California law that bans the sale of violent video games to minors.

What Do You Think? In 2011, a Democratic lawmaker in Hawaii's State House introduced a bill that would make it a crime to sell a toy gun to anyone under the age of 18 (Sakahara 2011). The proposed legislation carries a penalty of a $2,000 fine, jail time, or both. Would you support such a law? Do you think that exposure to violent toys, games, movies, and the like leads to aggressiveness? If so, could that explain gender differences in violent crime rates?

Types of Crime

The FBI identifies eight index offenses as the most serious crimes in the United States. The **index offenses**, or street crimes as they are often called, can be against a person (called violent or personal crimes) or against property (see Table 4.2).

Other types of crime include vice crime (such as drug use, gambling, and prostitution), organized crime, white-collar crime, computer crime, and juvenile delinquency. Hate crimes are discussed in Chapter 9.

Street Crime: Violent Offenses

The most recent data available from the FBI's Uniform Crime Reports indicate that the 2010 violent crime rate decreased from the previous year by 6.0 percent. Remember, however, that crime statistics represent only those crimes *reported* to the police: 1.2 million violent crimes in 2010.

Violent crime includes homicide, assault, rape, and robbery. *Homicide* refers to the willful or nonnegligent killing of one human being by another individual or group of individuals. Although homicide is the most serious of the violent crimes, it is also the least common, accounting for 1.2 percent of all violent crimes (FBI 2011). A typical homicide scenario includes a male killing a male with a handgun after a heated argument. The victim and offender are disproportionately young and of minority status. When a woman is

index offenses Crimes identified by the FBI as the most serious, including personal or violent crimes (homicide, assault, rape, and robbery) and property crimes (larceny, motor vehicle theft, burglary, and arson).

TABLE 4.2 Index Crime Rates, Percentage Change, and Clearance Rates, 2010

	RATE PER 100,000, 2010	PERCENTAGE CHANGE IN RATE, 2001–2010	PERCENTAGE CLEARED, 2010
Violent crime			
Murder	4.8	−15.0	64.8
Forcible rape	27.5**	−13.8	40.3
Robbery	119.1	−19.7	28.2
Aggravated assault	252.3	−20.8	56.4
Total	403.6	−20.0	47.2
Property crime			
Burglary	600.6	−5.7	12.4
Larceny/theft	2,003.5	−19.4	21.1
Motor vehicle theft	238.8	−44.5	11.8
Arson	N/A*	N/A	19.0
Total†	2,941.9	−13.0	18.3

Source: FBI 2011.

*Arson rates per 100,000 are calculated independently because population coverage for arson is lower than for the other index offenses.

**per 100,000 female inhabitants

†Property crime totals do not include arson.

murdered and the victim–offender relationship is known, she is most likely to have been killed by her husband or boyfriend (FBI 2011).

Mass murders have more than one victim in a killing event. In 2011, Arizona representative Gabrielle "Gabby" Giffords and 17 others were shot outside of a Tucson supermarket where representative Giffords was meeting with constituents. The alleged shooter, Jared Lee Loughner, has been found incompetent to stand trial and has been committed to a federal prison hospital for treatment and observation.

Unlike mass murder, serial murder is the "unlawful killing of two or more victims by the same offender(s), in separate events" (U.S. Department of Justice 2008, p. 1). The most well-known serial killers, who were responsible for some of the most horrific episodes of homicide, are Ted Bundy, Kenneth Bianchi, and Jeffrey Dahmer. More recently, Dennis Rader, the self-proclaimed "BTK" (bind, torture, kill) killer was captured. Accused of killing 10 people (2 men and 8 women) between 1974 and 1991, Rader was convicted of murder and received 10 consecutive life sentences with no chance of parole for 175 years (Coates 2005; Romano 2005).

Another form of violent crime, *aggravated assault*, involves attacking a person with the intent to cause serious bodily injury. Like homicide, aggravated assault occurs most often between members of the same race and, as with violent crime in general, is more likely to occur in warm weather months. In 2010, the assault rate was over 50 times the murder rate, with assaults making up an estimated 62.5 percent of all violent crimes (FBI 2011).

Rape is also classified as a violent crime and is also intraracial; that is, the victim and offender tend to be from the same racial group. The FBI definition of *rape* contains three elements: sexual penetration, force or the threat of force, and nonconsent of the victim. In 2010, 84,767 forcible rapes were reported in the United States (FBI 2011). Rapes are more likely to occur in warm months, in part because of the greater ease of victimization. People are outside more and later, doors are open, windows are unlocked, and so forth.

Perhaps as much as 80 percent of all rapes are **acquaintance rapes**—rapes committed by someone the victim knows. Although acquaintance rapes are the most likely to occur, they are the least likely to be reported and the most difficult to prosecute. Unless the rape is what Williams (1984) calls a **classic rape**—that is, the rapist was a stranger who used a weapon and the attack resulted in serious bodily injury—women hesitate to report the crime out of fear of not being believed. The increased use of "rape drugs," such as Rohypnol, may lower reporting levels even further (see Chapter 3). This chapter's *The Human Side* poignantly describes the impact of rape on one woman's life.

Robbery, unlike simple theft, also involves force or the threat of force or putting a victim in fear and is thus considered a violent crime. Officially, in 2010, more than 367,832 robberies took place in the United States. Robberies are most often (41.4 percent) committed with the use of a firearm and occur disproportionately in southern states (FBI 2011). Robbers and thus robberies vary dramatically in type, from opportunistic robberies whose victims are easily accessible and that yield only a small amount of money, to professional robberies of commercial establishments, such as banks, jewelry stores, and convenience stores. According to the FBI, in 2010, the average dollar value lost per robbery was $1,239. Of different victim categories, the highest average dollar loss was $4,410 per bank robbery (FBI 2011).

acquaintance rape Rape committed by someone known to the victim.

classic rape Rape committed by a stranger, with the use of a weapon, resulting in serious bodily injury to the victim.

In many jurisdictions, victims are allowed to make statements to the court describing the impact of their victimization on their life and the lives of friends and family. Here, 23-year-old Jessica poignantly describes the often unseen consequences of violent crime, in this case, the impact of a brutal rape.

My name is Jessica. I once knew what that meant. Now all I can tell you is that I am still Jessica; however, this no longer holds any meaning to me. I have lost my identity in the cruelest of ways.

I was a 23-year-old single mother, a sister, a daughter, a girlfriend, partner, friend, and confidant to many people. I was strong, independent, and willing to make an effort. I was attempting to hold down a second job to make a better life for my child and a better person out of me. I was destroyed.

I was raped. Not once, not twice, but so many times and in so many ways that it all becomes a blur. These images haunt my days, my nights, my dreams, and my realities. I am no longer the person I was before. I was once the person that people could rely on. Now I am a shell of my former self, a speck of the brave person that was Jessica. I had my way of life, my self-esteem, my respect, and my dignity stripped from me in the most terrifying of situations.

My trust in people is all but destroyed. I even have trouble enjoying a quiet drink with my partner, family, or friends without feeling anxious and wary. I am constantly looking over my shoulder, fearing there is someone there who wants to hurt me. This is only the beginning. I fluctuate from extreme insomnia to extreme fatigue. My motivation is gone. My joy of motherhood is waning. My ability to love and care for others are disappearing. My trust in the justice system is all but gone. I feel like I tried so hard and was beaten down. I feel weak and vulnerable.

I suffered physical trauma to my shoulder, knees, and feet from being dragged along carpet. I suffered cuts and bruising to my genitals from continuous rapes. I suffered marks to my ankles and wrists from being bound, and I lost chunks of hair from being gagged with tape around my head. I thought I was going to die. I take medication to sleep, to relax, and to stay relatively sane. I live as a corpse with no visible future to aim for. I have cut myself several times to try and take away the pain in my heart. I don't have the strength to end my own life.

What I experienced was like looking into the eyes of Satan himself. I am not a religious person but I know that I have seen the very depths of hell.

I wish that my loved ones didn't know what happened to me. To see the horror in their eyes and feel the pain in their hearts is unbearable. My beautiful daughter doesn't know the beginning of what I suffered but she suffered too. She feels like she was abandoned by her mother and required counseling.

The trauma happened to my partner. I can see it in the way he looks at me. Our communication, love, sex, and friendship is going to take a long time to repair. He has been so deeply affected by my experience but I can't help him and he can't help me. Nobody can.

I was burgled at work on my third day of my second job. I was bound and gagged. I was abducted. I was lied to. I was confused. I was cold and alone. I WAS RAPED. This man, who does not deserve a name, hurt me in the most unimaginable of ways. Yes I survived. Yes I am alive, I just don't live. I am existing. I am empty. . . .

I am a real person who went through torture. I am not a statistic or a nameless face on the street. I am your mother, your sister, your daughter, and your friend. What happened to me was real. At least give me the satisfaction of seeing this man put in jail for the maximum term of 25 years. After all, if I don't qualify as a victim of the most violent, serious, and heinous of crimes, then who the hell does?

I am Jessica.

Source: *The Herald Sun* 2008.

What Do You Think? Forensic psychiatrist, William Bernet of Vanderbilt University, argues that the combination of two factors helps explain abrupt, violent behavior (Hagerty 2010). The first is a variant of the MAO-A gene sometimes called the "warrior gene" because of its association with violence. The second factor is a history of child abuse. Sociologically, the question is this: If it can be scientifically established that violent behavior is, in whole or in part, a consequence of an individual's genetic makeup, what is the implication for criminal justice policy? Should society hold offenders responsible for factors they have no control over? Should parents be required by law to have their children's DNA tested? Should people who have the variant gene be sterilized? What do you think?

Street Crime: Property Offenses

Property crimes are those in which someone's property is damaged, destroyed, or stolen; they include larceny, motor vehicle theft, burglary, and arson. The number of property crimes has gone down since 1998, with an almost 13 percent decrease in the last decade. **Larceny,** or simple theft, accounts for more than two-thirds of all property

larceny Larceny is simple theft; it does not entail force or the use of force, or breaking and entering.

In this picture from the video game *Postal 2*, a man holding a shovel has just killed innocent bystanders, predominantly women. *Postal 2 was* one of the video games considered as part of the deliberation of the U.S. Supreme Court on whether or not video game content is protected under the First Amendment's right to free speech. In 2011, the Court ruled that video games, as books and plays, are "art" and, therefore, protected by the First Amendment.

arrests (FBI 2011), and is the most common index offense. In 2010, the average dollar value lost per larceny incident was $988. Examples of larcenies include purse snatching, theft of a bicycle, pickpocketing, theft from a coin-operated machine, and shoplifting. In 2010, an estimated 6.2 million larcenies were reported in the United States (FBI 2011).

Larcenies involving automobiles and auto accessories are the largest category of thefts. However, because of the cost involved, *motor vehicle theft* is considered a separate index offense. Numbering nearly 740,000 in 2010, the motor vehicle theft rate has decreased 40.1 percent since 2001 (FBI 2011). Because of insurance requirements, vehicle theft is one of the most highly reported index crimes, and, consequently, estimates between the FBI's Uniform Crime Reports and the National Crime Victimization Survey are fairly compatible. Less than 11.8 percent of motor vehicle thefts are cleared.

Burglary, which is the second most common index offense after larceny, entails entering a structure, usually a house, with the intent to commit a crime while inside. Official statistics indicate that, in 2010, more than 2.1 million burglaries occurred, a rate of 699.6 per 100,000 population (FBI 2011). Most burglaries are residential rather than commercial and take place during the day when houses are unoccupied. The most common type of burglary is forcible entry, followed by unlawful entry.

Arson involves the malicious burning of the property of another. Estimating the frequency and nature of arson is difficult given the legal requirement of "maliciousness." Of the reported cases of arson, 45.5 percent involved structures (most of which were residential), and 26 percent involved movable property (e.g., boat or car), with the remainder being miscellaneous property (e.g., crops or timber). In 2010, the average dollar amount of damage as a result of arson was $17,612 (FBI 2011).

Looking across index offense categories, it should be noted that all violent and property crime rates have decreased between 2009 and 2010. Interestingly, despite the economic downturn and the oft-cited relationship between the economy and crime, 2010 crime data indicates the same trend—each of the eight index crime rates decreased between 2009 and 2010 (FBI 2011).

Vice Crime

Vice crimes, often thought of as crimes against morality, are illegal activities that have no complaining participant(s) and are often called **victimless crimes**. Examples of vice crimes include using illegal drugs, engaging in or soliciting prostitution, illegal gambling, and pornography.

Most Americans view drug use as socially disruptive (see Chapter 3). There is less consensus, however, nationally or internationally, that gambling and prostitution are problematic. For example, the Netherlands fully legalized prostitution in 2000, hoping to cut the ties between the sex trade and organized crime—a link that remains. Alternatively, Sweden penalizes clients of prostitutes and treats sex workers as victims, an approach several European countries including England and Wales are now adopting (*The Economist* 2008).

In the United States, prostitution is illegal with the exception of several counties in Nevada. Despite its illegal status, it is a multimillion-dollar industry, with 62,668 arrests for prostitution and commercial vice in 2010 (FBI 2011). Motivated by profit, smugglers traffic thousands of women and children into the United States for purposes of prostitution. Trafficking *within* the United States also occurs. In 2003, the FBI established the Innocence Lost National Initiative to address domestic sex trafficking of children. Federal, state, and local law enforcement efforts have resulted in the recovery of 1,100 child victims. It is estimated that there are "tens of thousands" of child sex trafficking victims in the United States (FBI 2008).

Gambling is legal in many U.S. states including casinos in Nevada, New Jersey, Connecticut, North Carolina, and other states, as well as state lotteries, bingo parlors, horse and dog racing, and jai alai. In addition, although illegal in the United States, online gambling flourishes on offshore gambling sites. However, because of the potential for generating revenues, Congress is presently reevaluating its 2006 ban on online gambling (Chan 2010). Some have argued that there is little difference, other than societal definitions of acceptable and unacceptable behavior, between gambling and other risky ventures such as investing in the stock market. Conflict theorists are quick to note that the difference is who is making the wager.

Pornography, particularly Internet pornography, is a growing international problem. Regulation is made difficult by fears of government censorship and legal wrangling as to what constitutes "obscenity." For many, the concern with pornography is not its consumption per se but the possible effects of viewing or reading pornography—increased sexual aggression. Although the literature on this topic is mixed, Conklin (2007, p. 221) concluded that there is no "consistent evidence that nonviolent pornography causes sex crimes."

Organized crime refers to criminal activity conducted by members of a hierarchically arranged structure devoted primarily to making money through illegal means. Although discussed under victimless crimes because of its association with prostitution, drugs, and gambling, organized crime groups often use coercive tactics. For example, organized crime groups may force legitimate businesses to pay "protection money" by threatening vandalism or violence.

The traditional notion of organized crime is the Mafia, a national band of interlocked Italian families. However, members of many ethnic groups engage in organized crime in the United States:

> Chinese, Vietnamese, Korean, and Japanese gangs have been found on the East and West coasts, active in smuggling drugs and extorting money from businesses in their communities. Scores of other groups can be found in various cities: Israelis dominating insurance fraud in Los Angeles, Cubans running illegal gambling operations in Miami, Canadians engaging in gun smuggling and money laundering in Miami, Russians carrying out extortion and contract murders in New York. (Thio 2007, p. 374)

Organized crime also occurs in other countries. For example, with more than 90,000 members and associates in 3,000 crime groups, the Japanese Yakuza are one of the largest crime organizations in the world. The young men who join the Yakuza tend to be from

victimless crimes Illegal activities that have no complaining participant(s) and are often thought of as crimes against morality, such as prostitution.

organized crime Criminal activity conducted by members of a hierarchically arranged structure devoted primarily to making money through illegal means.

WikiLeaks founder Julian Assange has been called everything from a hero, to a thief, public servant, spy, and traitor. In 2010, Assange dumped over a half million documents, many previously classified, into the public eye via the Internet.

Transnational crime organizations are involved in many types of transnational crime including money laundering, narcotics, arms smuggling, and trafficking in people.

the lower class and must undergo a training period of five years. During this apprenticeship, members learn absolute loyalty to their superiors as well as the other norms and values of the group. The Yakuza are involved in drugs, illegal gambling, pornography, and prostitution, as well as several legitimate businesses. Interestingly, the Yakuza proudly display their name at their "corporate" headquarters, and recruits wear lapel pins identifying themselves as members (Thio 2004; Winslow & Zhang 2008).

Unlike traditional crime organizations that are hierarchically arranged, transnational crime organizations tend to be decentralized and less likely to operate through legitimate businesses. Transnational crime organizations, like transnational crime in general, directly or indirectly involve more than one country. Transnational crime organizations are a growing threat to the United States and to global security. As Wagley (2006) explained:

> The end of the Cold War—along with increasing globalization beginning in the 1990s—has helped criminal organizations expand their activities and gain global reach. Criminal networks are believed to have benefited from the weakening of certain government institutions, more open borders, and the resurgence of ethnic and regional conflicts across the former Soviet Union and many other regions. Transnational criminal organizations have also exploited expanding financial markets and rapid technological developments. (p. 1)

Transnational crime organizations are also less likely than the traditional crime "families" to develop around a family or ethnic structure. Transnational crime organizations are involved in many types of transnational crime including money laundering, narcotics, arms smuggling, and trafficking in people. Further, terrorists are increasingly supporting themselves through transnational organized crime groups. For example, the 2003 bombing of a Madrid commuter train was financed through drug dealing (Wagley 2006).

White-Collar Crime

white-collar crime Includes both *occupational crime*, in which individuals commit crimes in the course of their employment, and *corporate crime*, in which corporations violate the law in the interest of maximizing profit.

White-collar crime includes both *occupational crime*, in which individuals commit crimes in the course of their employment, and *corporate crime*, in which corporations violate the law in the interest of maximizing profit. Occupational crime is motivated by individual gain. Employee theft of merchandise, or pilferage, is one of the most common types of occupational crime. Other examples include embezzlement, forgery and counterfeiting, and insurance fraud. Price fixing, antitrust violations, and security fraud are all examples of corporate crime, that is, crime that benefits the organization.

In recent years, several officers of major corporations, including Enron, WorldCom, Adelphia, and ImClone, have been charged with securities fraud, tax evasion, and insider trading. Joseph Cassano, a former executive at AIG (American International Group, Inc.), was investigated by the FBI, the U.S. Securities and Exchange Commission, and the British Serious Fraud Office for misrepresenting the extent of AIG's losses after the company insured a trillion dollars worth of risky loans held by banks. Despite being called the man who brought down AIG, the U.S. Department of Justice decided not to file criminal charges against Cassano. AIG has received over $180 billion in bailout money from the U.S. government to meet its financial obligations, and is often "credited" with playing a major role in the U.S. economic crisis (Schecter et al. 2009; Salow 2009).

In 2002, President Bush signed the Sarbanes-Oxley Act, which significantly increased penalties for white-collar offenders. Nonetheless, many white-collar criminals go unpunished. First, many companies, not wishing the bad publicity surrounding a scandal, simply dismiss the parties involved rather than press charges. Second, many white-collar crimes, as traditional crimes, go undetected. In a survey of a representative sample of

2,503 U.S. households, the National White Collar Crime Center (NWCCC) found that nearly one in four households (24 percent) had been the victim of some type of white-collar crime in the last year. However, only 11.7 percent of the crimes were brought to the attention of a law enforcement agency (NWCCC 2010).

Third, federal prosecutions of white-collar criminals have generally decreased recently. Few believe the decrease is a result of a lower prevalence of white-collar crime offenses. Two forces appear to be in operation. First, white-collar crimes are becoming increasingly complex, making prosecution a time- and resource-intensive endeavor. Second, the decrease in white-collar crime prosecutions represents a shift in priorities (Marks 2006). After the events of 9/11, nearly one-third of FBI agents were moved from criminal programs to terrorism and intelligence duties, leaving "the bureau seriously exposed in investigating areas such as white-collar crime . . . " (Lichtblau et al. 2008, p. 1).

Today, no doubt triggered by investigations of mortgage fraud on the heels of the subprime mortgage crisis, the Obama administration has increased its attention on financial crimes. For example, in 2011, a federal investigation found that Goldman Sachs had defrauded their customers and misled Congress. The results of the investigation lead to a New York prosecutor subpoenaing the bank and investment firm's records (Hilzenrath 2011). Also in 2011, the Securities and Exchange Commission brought charges against a Food and Drug Administration (FDA) chemist for **insider trading**, that is, using privileged (i.e., nonpublic) information as an employee of the FDA to gain an unfair advantage in buying, selling, and trading securities, in this case, pharmaceutical stocks (Mundy & Kendall 2011).

Corporate violence, a form of corporate crime, refers to the production of unsafe products and the failure of corporations to provide a safe working environment for their employees. Corporate violence is the result of negligence, the pursuit of profit at any cost, and intentional violations of health, safety, and environmental regulations. For example, in 2007, contaminated pet food was responsible for the deaths of hundreds of dogs and cats. In 2008, two Chinese executives and a U.S. company president and CEO were indicted for scheming to import tainted products used in the manufacturing of pet food (FDA 2008). Similarly, in 2009, a Chinese court sentenced two men to death and one to life in prison for endangering public safety by producing and selling contaminated dairy products. Three hundred thousand children became ill as a result of melamine-laced milk, and at least six died (Barboza 2009).

Two more recent events, both having international repercussions, have likely colored the public's view of the need for increased corporate responsibility. The first entails the accusation of sudden acceleration problems in Toyota vehicles and, the second, the Gulf Oil spill. Although recent reports suggest there is little scientific evidence tying car accidents to sudden acceleration issues in Toyota vehicles, documents have surfaced that do detail Toyota Motor Corporation's delay in recalling millions of vehicles thought to be defective (Frank 2011; Maynard 2010).

Finally, the British Petroleum (BP) oil spill off the coast of Louisiana epitomizes, perhaps more than any other case in history, the public's increasing concern with corporate violence and the need for corporate responsibility. A recent Gallup Poll indicates that the majority of Americans believe that BP should ". . . pay for all financial losses resulting from the Gulf Coast oil spill, including wages of workers put out of work, even if those payments ultimately drive the company out of business," and that President Obama ". . . has not been tough enough in his dealings with BP" (Newport 2010) (see Chapter 13). Table 4.3 summarizes some of the major categories of white-collar crime.

TABLE 4.3 Types of White-Collar Crime	
CRIMES AGAINST CONSUMERS	**CRIMES AGAINST EMPLOYEES**
Deceptive advertising	Health and safety violations
Antitrust violations	Wage and hour violations
Dangerous products	Discriminatory hiring practices
Manufacturer kickbacks	Illegal labor practices
Physician insurance fraud	Unlawful surveillance practices
CRIMES AGAINST THE PUBLIC	**CRIMES AGAINST EMPLOYERS**
Toxic waste disposal	Embezzlement
Pollution violations	Pilferage
Tax fraud	Misappropriation of government funds
Security violations	Counterfeit production of goods
Police brutality	Business credit fraud

© Cengage Learning 2013

insider trading The use of privileged (i.e., nonpublic) information by an employee of an organization that gives that employee an unfair advantage in buying, selling, and trading stocks or other securities.

corporate violence The production of unsafe products and the failure of corporations to provide a safe working environment for their employees.

Computer Crime

Computer crime refers to any violation of the law in which a computer is the target or means of criminal activity. Sometimes called cybercrime, computer crime is one of the fastest-growing types of crime in the United States. Hacking, or unauthorized computer intrusion, is one type of computer crime. In 2011, the Sony server was attacked by hackers who stole information from up to as many as 100 million customer accounts located on the Sony PlayStation network (Pham 2011).

In 2010, the Consumer Sentinel Network (CSN) logged 1.3 million consumer complaints, with identity theft being the most common complaint category (FTC 2011). **Identity theft** is the use of someone else's identification (e.g., Social Security number or birth date) to obtain credit or other economic rewards. Although mail theft is one of the most common modes of obtaining the needed information, new technologies have contributed to the increased rate of identity theft. For example, in 2006, a Department of Veterans Affairs official downloaded the personnel records of more than 26 million veterans to his laptop computer, which was then stolen, "exposing all the information necessary to swipe the identity of virtually every person released from military service since 1975" (Levy 2006).

The AMBER Alert Logo is a Registered Trademark of the U.S. Department of Justice

The AMBER Alert program is named after 9-year-old Amber Hagerman who was murdered after being abducted while riding her bike near her grandparent's home in Arlington, Texas. The program is a voluntary alliance between law enforcement, the news media, and transportation agencies, which alerts the public about the most serious child abduction cases.

Identity theft is just one category of computer crime. Another category is Internet fraud. According to the Internet Crime Complaint Center (ICCC 2011), the most common type of Internet fraud is nondelivery of merchandise or payment, followed by FBI-related scams (e.g. impersonating a FBI agent), which comprises about 13 percent of all complaints. Other Internet fraud categories include check fraud, confidence fraud, Nigerian letter fraud (i.e., a letter offering the recipient the "opportunity" to share in millions of dollars being illegally transferred to the United States—just send us your bank account numbers!), computer fraud, and credit/debit card fraud.

Another type of computer crime is online child sexual exploitation. According to the National Center for Missing and Exploited Children (NCMEC), more than 30 million children younger than age 18 are on the "net" and one in seven "receives a sexual solicitation online which includes a request to engage in sexual activity, a request to engage in sexual talk, or a request to give out personal sexual information" (NCMEC 2007, p. 1). In 1998, the U.S. Department of Justice created the Internet Crimes Against Children Task Force Program. Since its inception, the Task Force has investigated over 236,000 child sexual exploitation complaints, resulting in the arrest of more than 24,000 offenders (U.S. Department of Justice 2011).

> **What Do You Think?** Over 10 years ago, two brothers were roughhousing when suddenly it ". . . escalated from goofing around with a blowgun to an angry threat with a bow and arrow to the fatal thrust of a hunting knife" (Liptak & Petak 2011, p. 1). The surviving brother, who was 14 years old at the time and in seventh grade, is now in a maximum-security prison serving a life sentence without the possibility of parole. In 2009, the U.S. Supreme Court held that life sentences for juveniles without the possibility of parole were in violation of the Eighth Amendment's prohibition of cruel and unusual punishment—except in cases that involved a killing. Do you think juveniles who kill should be able to be sentenced to life imprisonment without the possibility of parole? At what age should full legal responsibility begin?

computer crime Any violation of the law in which a computer is the target or means of criminal activity.

identity theft The use of someone else's identification (e.g., Social Security number, birth date) to obtain credit or other economic rewards.

Juvenile Delinquency

In general, children younger than age 18 are handled by the juvenile courts, either as status offenders or as delinquent offenders. A *status offense* is a violation that can be committed only by a juvenile, such as running away from home, truancy, and underage drinking. A *delinquent offense* is an offense that would be a crime if committed by an adult, such as the eight index offenses. The most common status offenses handled in juvenile court are underage drinking, truancy, and running away. In 2010, 12.6 percent of all arrests (excluding traffic violations) were of offenders younger than age 18 (FBI 2011). As is the case with adults, juveniles commit more property crimes than violent crimes.

Although the number of juveniles arrested for violent crimes decreased by 4.3 percent between 2009 and 2010 (FBI 2011), Americans remain concerned about juvenile violence and, specifically, the high rate of gang-related violence. The growth of gangs is, in part, a function of two interrelated social forces: the increased availability of guns in the 1980s, and the lucrative and expanding drug trade. In 2009, there were over 28,000 gangs and three-quarter of a million gang members throughout the United States (Egley & Howell 2011). Gang activity is more likely to be found in large urban areas. For example, between 2005 and 2009, 86.3 percent of law enforcement in large cities reported the presence of gang activity compared to 51.8 percent in suburban counties, 32.9 percent in small cities, and 17 percent in rural counties. Gangs are estimated to be responsible for as much as 80 percent of all crime in some communities, and are becoming more violent (National Gang Intelligence Center 2009). When law enforcement officials were asked about factors influencing gang violence, the most common reason given was drug-related factors (Egley & Howell 2011) (see Figure 4.2).

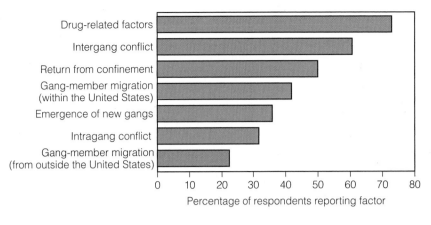

Figure 4.2: Factors Influencing Local Gang Violence, 2009
Source: Office of Juvenile Justice and Delinquency Prevention 2011.

Demographic Patterns of Crime

Although virtually everyone violates a law at some time, individuals with certain demographic characteristics are disproportionately represented in the crime statistics. Victims, for example, are disproportionately young, lower-class, minority males from urban areas. Similarly, the probability of being an offender varies by gender, age, race, social class, and region (see Figure 4.3). This section ends with a discussion of crime and victimization.

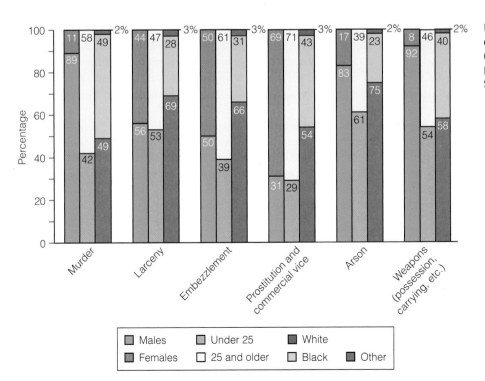

Figure 4.3: Percentage of Arrests for Selected Crimes by Sex, Age, and Race, 2010
Source: FBI 2011.

Gender and Crime

It is a universal truth that women everywhere are less likely to commit crime than men. In the United States, both official statistics and self-report data indicate that females commit fewer violent crimes than males. In 2010, males accounted for 74.5 percent of all arrests, 80.5 percent of all arrests for violent crime, and 62.4 percent of all arrests for property crimes (FBI 2011). Not only are females less likely than males to commit serious crimes, but also the monetary value of female involvement in theft, property damage, and illegal drugs is typically far less than that for similar offenses committed by males.

Nonetheless, rates of female criminality have increased dramatically over the last decade. For example, between 2001 and 2010, arrest rates for women increased for robbery (29.1 percent), burglary (19.4 percent), larceny (31.7 percent), and drug abuse violations (13.7 percent) (FBI 2011). Heimer et al. (2005) argue that such increases are a function of the *economic marginalization* of women relative to men, that is, female criminality goes up when women's economic circumstances in relation to men's decline.

The recent increase in crimes committed by females has led to growth of feminist criminology. **Feminist criminology** focuses on how the subordinate position of women in society affects their criminal behavior and victimization. For example, Chesney-Lind and Shelden (2004) reported that arrest rates for runaway juvenile females are higher than those for males not only because girls are more likely to run away as a consequence of sexual abuse in the home but also because police with paternalistic attitudes are more likely to arrest female runaways than male runaways. Feminist criminology, concentrating on gender inequality in society, thus adds insights into understanding crime and violence that are often neglected by traditional theories of crime. Feminist criminology has also had an impact on public policy. Mandatory arrest for domestic violence offenders, the development of rape shield laws, public support for battered women's shelters, laws against sexual harassment, and the repeal of the spousal exception in rape cases are all, according to Winslow and Zhang (2008), outcomes of feminist criminology.

> . . . the development of rape shield laws, public support for battered women's shelters, laws against sexual harassment, and the repeal of the spousal exception in rape cases are all . . . outcomes of feminist criminology.

Age and Crime

In general, criminal activity is more prevalent among younger people than among older people. In 2010, 42.1 percent of all arrests in the United States were of people younger than age 25 (FBI 2011). Although those younger than age 25 made up over half of all arrests in the United States for crimes such as robbery, burglary, motor vehicle theft, and arson, those younger than age 25 were significantly less likely to be arrested for crimes such as embezzlement, fraud, and forgery and counterfeiting. Those 65 and older made up less than 1 percent of total arrests for the same year (FBI 2011).

Why is criminal activity more prevalent among individuals in their late teens and early twenties? One reason is that juveniles are insulated from many of the legal penalties for criminal behavior. Younger individuals are also more likely to be unemployed or employed in low-wage jobs. Thus, as strain theorists argue, they have less access to legitimate means for acquiring material goods.

Some research suggests, however, that high school students who have jobs become more, rather than less, involved in crime (Felson 2002). In earlier generations, teenagers who worked did so to support themselves and/or their families. Today, teenagers who work typically spend their earnings on recreation and "extras," including car payments and gasoline. The increased mobility associated with having a vehicle also increases the opportunity for criminal behavior and reduces parental control.

Other hypothesized reasons for the age–crime relationship are also linked to specific theories of criminal behaviors. For example, conflict theorists would argue that teenagers and young adults have less power in society than their middle-aged and elderly

feminist criminology An approach that focuses on how the subordinate position of women in society affects their criminal behavior and victimization.

counterparts. One manifestation of this lack of power is that the police, using a mental map of who is a "typical offender," are more likely to have teenagers and young adults in their suspect pool. With increased surveillance of teenagers and young adults comes increased detection of criminal involvement—a self-fulfilling prophecy.

What Do You Think? Crime statistics are sensitive to demographic changes. For example, crime rates in the United States began to rise in the 1960s, as baby boom teenagers entered high school. That said, FBI arrest data indicate that offenders are disproportionately young males (FBI 2011). Do you think that crime rates are a function of the number of young males in the general population? If so, do you think their elevated numbers reflect their higher rates of criminal involvement, or institutional bias in the criminal justice system?

Race, Social Class, and Crime

Race is a factor in who gets arrested. Minorities are disproportionately represented in official statistics. For example, African Americans represent about 13 percent of the population but account for 38.1 percent of all arrests for violent index offenses, and 28.9 percent of all arrests for property index offenses (FBI 2011). They have 3.5 times the arrest rate of whites for drug offenses, are six times more likely to be admitted to prison and, if admitted to prison for a violent crime, receive longer sentences than their white counterparts (Hartney & Vuong 2009).

Nevertheless, it is inaccurate to conclude that race and crime are causally related. First, official statistics reflect the behaviors and policies of criminal justice actors. Thus, the high rate of arrests, conviction, and incarceration of minorities may be a consequence of individual and institutional bias against minorities. For example, **racial profiling—** the practice of targeting suspects on the basis of race—may be responsible for their higher arrest rates. Proponents of the practice argue that because race, like gender, is a significant predictor of who commits crime, the practice should be allowed. Opponents hold that racial profiling is little more than discrimination and should, therefore, be abolished. Presently, more than 20 states have laws that prohibit racial profiling and/or require that state jurisdictions collect data on police stops and searches (IRJ 2011).

Second, race and social class are closely related in that nonwhites are overrepresented in the lower classes. Because lower-class members lack legitimate means to acquire material goods, they may turn to instrumental, or economically motivated, crimes. In addition, although the "haves" typically earn social respect through their socioeconomic status, educational achievement, and occupational role, the "have-nots" more often live in communities where respect is based on physical strength and violence, as subcultural theorists argue. For example, Kubrin (2005) examined the "street code" of inner-city black neighborhoods by analyzing rap music lyrics. Her results indicate that "lyrics instruct listeners that toughness and the willingness to use violence are central

Mark Richards/Photo Edit

Most street gangs are local or neighborhood gangs that operate from a single location, ranging in size from just a few to several hundred members. Here, members of the Death Squad gang flash signs while at a city park in Los Angeles.

racial profiling The law enforcement practice of targeting suspects on the basis of race.

> [African Americans] . . . have 3.5 times the arrest rate of whites for drug offenses, are six times more likely to be admitted to prison, and, if admitted to prison for a violent crime, receive longer sentences than their white counterparts.

to establishing viable masculine identity, gaining respect, and building a reputation" (p. 375). This chapter's *Social Problems Research Up Close* feature examines violence and "being tough" in a sample of minority youth.

A third hypothesis is that criminal justice system contact, which is higher for nonwhites, may actually act as the independent variable; that is, it may lead to a lower position in the stratification system. Kerley and colleagues (2004) found that "contact with the criminal justice system, especially when it occurs early in life, is a major life event that has a deleterious effect on individuals' subsequent income level" (p. 549).

Some research indicates, however, that even when social class backgrounds of blacks and whites are comparable, blacks have higher rates of criminality. Sampson et al. (2005), using data from nearly 3,000 respondents aged 18 to 25, found that the likelihood of self-reported violence by blacks was 85 percent higher than for whites. Interestingly, the likelihood of Latino self-reported violence was 10 percent less than that reported by whites.

Region and Crime

In general, crime rates and, in particular, violent crime rates, are higher in metropolitan areas than in non-metropolitan areas. In 2010, the violent crime rate in metropolitan counties was 140.9 per 100,000 population; in non-metropolitan counties, it was 107.7 per 100,000 population (FBI 2011).

Higher crime rates in urban areas result from several factors. First, social control is a function of small intimate groups that socialize their members to engage in law-abiding behavior, expressing approval for their doing so and disapproval for their noncompliance. In large urban areas, people are less likely to know one another and thus are not influenced by the approval or disapproval of strangers. Demographic factors also explain why crime rates are higher in urban areas: Large cities have large concentrations of poor, unemployed, and minority individuals. Finally, some of the nation's most violent cities, including the 10 most dangerous cities in the United States, have been identified by the U.S. Department of Justice as transit points for Mexican drug cartels (Greenburg 2009).

Crime rates also vary by region of the country. In 2010, both violent and property crimes were highest in southern states, followed by western, midwestern, and northeastern states. Violent crime is particularly high in the South, with 43.8 percent of all murders and 43.6 percent of all aggravated assaults recorded in southern states (FBI 2011). The high rate of southern lethal violence has been linked to high rates of poverty and minority populations in the South, a southern "subculture of violence," higher rates of gun ownership, and a warmer climate that facilitates victimization by increasing the frequency of social interaction.

Crime and Victimization

According to the National Crime Victimization Survey (NCVS), 40 percent of male victims of violent crime knew their offenders compared to 64 percent of female victims of violent crime (Truman 2011). Victims who knew their offenders were most likely to classify them as "friends or acquaintances." These results are, in part, a function of the tendency for females to be much more likely to report victimization by an intimate partner than males.

Victims of violent crimes are most often under the age of 25, and least often over the age of 65. According to the NCVS, weapons were used in 22 percent of all violent crimes. Blacks were more often victims of violent crime, including robbery, rape, and aggravated assault, than whites. Further, Blacks were more likely to be victims of violent crime in general, and to be victims of robbery, rape, simple assault, and aggravated

A large body of research documents that fighting in adolescence is a fairly common event, although frequencies vary significantly by race, age, and sex. For example, black youth are more likely to report being involved in at least one physical fight in the previous year when compared to their white counterparts (CDCP 2008). Thus, the present study is an important one for it examines the relationship between select explanatory variables (both risks and assets) and the likelihood of fighting among a sample of at-risk minority youth (Wright & Fitzpatrick 2006).

Sample and Methods

All respondents were in grades 5 through 12 and were enrolled in a central Alabama school system (Wright & Fitzpatrick 2006). The school district sent home a letter detailing the purpose of the study, and parents were asked to give permission for their child's inclusion in the study and referred to a sample questionnaire that was on file and available for review. The final sample consisted of 1,642 African American youth (51 percent female) with a median age of 14 years. Participation in the survey was voluntary, and the response rate was 65 percent.

The dependent variable is fighting and was measured by asking respondents the frequency of their fighting in the last 30 days. Socio-demographic variables include sex, age, mother's and father's education, and mother's and father's occupational status. Risk factors, that is, factors associated with an increase in the likelihood of fighting, include academic performance (poor grades), family intactness (zero or one parent in the home),

parental violence (physical assault from parent or other adult guardian), and a composite measure of gang affiliation (being a member, being asked to be a member, or having friends who are members).

Asset variables are variables that are predicted to decrease incidents of fighting. The eight asset variables were divided into three categories. The first category is self-esteem measured by a respondent's sense of satisfaction, pride, worth, and respect. Parental involvement was measured by a respondent's (1) parents monitoring of where their child goes with their friends, (2) frequency of talking to parents about problems, and (3) frequency of eating dinner with the family. School involvement includes teacher attention, respondent's involvement in school activities and clubs, and self-reported happiness with school.

Results and Conclusions

Frequency of fighting was higher for elementary and middle school students than for high school students, with the highest fighting frequency occurring in middle schools. Across school type, 15 percent or more of the students reported the highest response category of fighting—six or more times over the last 30 days.

Data analysis also indicates that fighting is negatively associated with family intactness and self-esteem, with two-parent homes and high self-esteem leading to lower probabilities of fighting. Alternatively, parental violence and gang affiliation are associated with increased probabilities of fighting. Talking to parents about problems and having

parents who monitor activities with friends are significantly associated with decreased rates of fighting. Additional asset variables associated with decreased levels of fighting include being happy with school, attention from teachers, and involvement with school clubs. As the authors note, interestingly, higher involvement in sports was associated with higher rates of fighting.

When multiple variables were analyzed at the same time, three of the four risk factors were significantly associated with higher rates of fighting—lower grades, exposure to violence in the family, and gang affiliation. Of the asset variables, a lack of parental monitoring and being unhappy at school were predictive of increased fighting behavior. Note that low self-esteem, when in the presence of other variables, is not associated with youthful fighting.

The authors concluded that the "risk and asset" model they present has practical implications in terms of organizing intervention techniques. Risk factors need to be "suppressed or eliminated" and asset factors need to be "encouraged or facilitated" (Wright & Fitzpatrick 2006, p. 260). For example, results from the present study show that "parental monitoring and being happy at school were associated with lower frequency of fighting, suggesting the importance of continued support for outreach to parents and further efforts to reduce or eliminate the community factors that promote proliferation of gangs" (Wright & Fitzpatrick 2006, p. 251).

Source: Wright and Fitzpatrick 2006.

assault when compared to Hispanics (Truman 2011). Interestingly, American Indians and Alaskan Natives had the highest victimization rate, followed by persons of two or more races (Truman 2011).

Households with incomes of less than $7,500 per year were over 2.5 times more likely to experience burglary than households with annual earnings of $75,000 or more. Households with incomes below $7,500 were also more likely to be the victims of property theft. Further, larger households were more likely to report being the victim of a property crime than smaller households. Consistent with the FBI's Uniform Crime Report, according to the NCVS, both nonlethal violent crime and property crime decreased between 2009 and 2010 (Truman 2011).

The Costs of Crime and Social Control

The costs of crime and violence are difficult to quantify but minimally include physical injury and loss of life, economic losses, and social and psychological costs.

Physical Injury and Loss of Life

Crime often results in physical injury and loss of life. For example, homicide is the second most common cause of death among 15- to 25-year-olds, exceeded only by accidental death (U.S. Census Bureau 2011). In 2010, 14,748 people were the victims of known homicides in the United States (FBI 2011). That number is dwarfed, however, by the deaths that take place as a consequence of white-collar crime. Criminologist Steven Barkan (2006), who collected data from a variety of sources, reported that annually there are (1) 56,425 workplace-related deaths from illness or injury; (2) 9,600 deaths from unsafe products; (3) 35,000 deaths from environmental pollution; and (4) 12,000 deaths from unnecessary surgery. Adding these figures together, 113,025 people a year die from corporate and professional crime and misconduct (p. 388).

Moreover, the U.S. Public Health Service now defines violence as one of the top health concerns facing Americans. Health initiatives related to crime include reducing drug and alcohol use and the deaths and diseases associated with them, lowering rates of domestic violence, preventing child abuse and neglect, and reducing violence through public health interventions. Finally, it must be noted that crime has mental as well as physical health consequences. For example, violent crime, and particularly rape and sexual assault, are related to post-traumatic stress disorder (see Social and Psychological Costs).

Economic Costs

Conklin (2007, p. 50) suggested that the financial costs of crime can be classified into at least six categories. First are *direct losses* from crime, such as the destruction of buildings through arson, of private property through vandalism, and of the environment by polluters. In 2009, the average dollar loss of destroyed or damaged property as a result of arson was $17,612 (FBI 2011). Second are costs associated with the *transferring of property*. Bank robbers, car thieves, and embezzlers have all taken property from its rightful owner at tremendous expense to the victim and society. For example, it is estimated that, in 2010, $4.5 billion was lost as a result of motor vehicle theft; the average value per vehicle at the time of the theft was $6,152 (FBI 2011).

A third major cost of crime is that associated *criminal violence*, including the medical cost of treating crime victims. The National Crime Prevention Council (NCPC 2005) estimates that the average cost for *each* criminal incident of rape or sexual assault is $7,700, including expenses related to medical and mental health care, law enforcement, and victim and social services.

Fourth are the costs associated with the production and sale of illegal goods and services, that is, *illegal expenditures.* The expenditure of money on drugs, gambling, and prostitution diverts funds away from the legitimate economy and enterprises, and lowers property values in high-crime neighborhoods. Fifth is the cost of *prevention and protection*—the billions of dollars spent on locks and safes, surveillance cameras, guard dogs, and the like. It is estimated that Americans spend $65 billion annually on self-protection items (Surgeon General 2002). The cost of a home security system alone can be as much as $25,000 for installation, in addition to monthly fees (Pollack 2007).

Finally, there is the cost of *controlling crime*—billions of dollars, and the costs are escalating. Figure 4.4 indicates, however, that criminal justice spending varies by function, with spending on law enforcement increasing over the years and spending on criminal justice assistance (i.e., victim services) decreasing (Austin 2011). The cost of corrections has also soared, leading Missouri to require that judges are informed of the costs of the sentences they impose. A second-degree robbery conviction, for example, costs $9,000 for six years of probation but would run $50,000 for six years of incarceration followed by parole (Davey 2010).

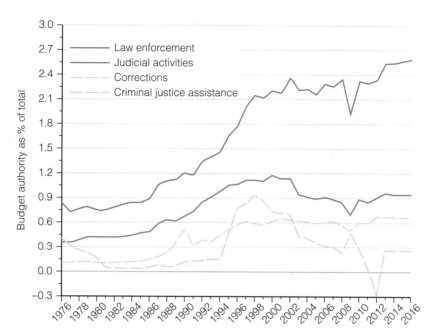

Figure 4.4:

Administration of Justice Costs, by Function, 1976–2012 (projected)*
Source: Austin 2011.
*Discretionary Budget Authority as a Percentage of GDP, FY1976–FY2016; FY2012–FY2016 levels projected.

Although the costs from "street crimes" are staggering, the costs from "crimes in the suites," such as tax evasion, fraud, false advertising, and antitrust violations, are greater than the cost of the FBI index crimes combined (Reiman & Leighton 2010). Further, Barkan (2006), using an FBI estimate, reported that the total cost of property crime and robbery is $17.1 billion annually. This is less than the $44 billion price tag for employee theft alone.

Social and Psychological Costs

Crime entails social and psychological costs as well as economic costs. One such cost—fear—is dependent upon individual perceptions of crime as a problem. For example, surveys since 2001 indicate that Americans' fear of victimization has increased even as violent and property crime rates have decreased (Saad 2010). Such misconceptions are fueled by media presentations that may not accurately reflect the crime picture. For example, Krisberg et al. (2009), when comparing youth crime rates to newspaper coverage of juvenile crime, concluded that the media doesn't "provide a balanced perspective on crime or youth issues," instead focusing on "crime increases and 'crime emergencies'" (p. 7).

Not only do Americans worry about crime at the aggregate level, but they also worry about crime at the individual level. When a random sample of Americans was asked the extent to which they worry about crime, nearly 25 percent of Americans responded that they would be, "afraid to walk alone at night within a mile of their home." As Table 4.4 indicates, fear of crime is not randomly distributed. Both women when compared to men, and low socioeconomic status households compared to higher socioeconomic status households, are more likely to be afraid of criminal victimization (Saad 2010).

How do most Americans deal with their fear of street crime? According to a Gallup Poll, the most common method of dealing with fear of victimization is to "avoid going to certain places/neighborhoods you might otherwise want to go to"

TABLE 4.4 Fear of Walking Alone at Night, by Gender and Household Income, 2010

	AFRAID	NOT AFRAID
	%	%
Gender		
Men	22	78
Women	50	49
Household income		
$75,000 or more	23	76
$30,000 to $49,999	36	64
Less than $30,000	48	52
$75,000 or more by gender		
Men	13	87
Women	36	64
$30,000 to $49,999 by gender		
Men	20	79
Women	52	48
Less than $30,000 by gender		
Men	30	70
Women	59	41

Source: Gallup Poll 2010b. Lydia Saad, "Nearly 4 in 10 Americans Still Fear Walking Alone at Night," 11/05/2010. Copyright © Gallup, Inc. All Rights Reserved. Reprinted by permission.

(Carroll 2007). Such behavioral changes are just one category of social and psychological costs that Shapland and Hall (2007) identified. Others include a sense of shock, a loss of trust, feelings of guilt for being victimized, anger, and a sense of vulnerability. Although these responses vary by type of offense, it should not be concluded that white-collar crimes do not carry a social and psychological toll. In 2009, Bernard Madoff pled guilty to 11 felony counts related to a massive Ponzi scheme run through his investment firm. The scheme to defraud investors of $65 billion took place over a 20-year period and involved thousands of victims. The 71-year-old Madoff was sentenced to 150 years in prison, leaving behind him a trail of misery. Following are excerpts from just seven of the thousands of victims impacted by his crime. They attest to the social and psychological costs of white-collar crime (Victim Statements 2009):

> He robbed of us not only of our money, but of our faith in humanity, and in the systems in place that were supposed to protect us.

> I can't tell you how scattered we feel—it goes beyond financially. It reaches the core and affects your general faith in humanity, our government and basic trust in our financial system.

> I am constantly nervous and anxious about my future. I jump at the slightest noise. I can't sleep and all I do is worry about what will happen to us.

> I don't know which emotion is more destructive, the fear and anxiety or the major depression that I experienced daily.

> . . . when you sentence Madoff . . . I trust that you touch on the loss of money, the loss of dignity, the loss of freedom from financial worries and possible financial ruin. . . .

> At this point . . . we cannot trust anyone.

> How do I live the rest of my life?

Strategies for Action: Crime and Social Control

Clearly, one way to combat crime is to attack the social problems that contribute to its existence. Moreover, when a random sample of Americans were asked which of two views came closer to their own in dealing with the crime problem, increasing law enforcement or resolving social problems, the majority of respondents (65 percent) selected resolving social problems (Gallup Poll 2007).

In addition to policies that address social problems, numerous social programs have been initiated to alleviate the crime problem. These policies and programs include local initiatives, criminal justice policies, legislative action, and international efforts in the fight against crime.

Local Initiatives

Youth programs such as the Boys and Girls Club, and community programs that involve families and schools are an effective "first line of defense" against crime and juvenile delinquency.

Youth Programs Early intervention programs acknowledge that preventing crime is better than "curing" it once it has occurred. *Fight Crime: Invest in Kids* is a nonpartisan, nonprofit anticrime organization made up of more than 5,000 law enforcement leaders and violence survivors (*Fight Crime* 2011). The organization takes a "hard look at the research about what prevents kids from becoming criminals and put[s] that information in the hands of policymakers and the general public" (p. 1). *Fight Crime* advocates four primary initiatives:

- Quality early care and education for young children (birth to 5 years of age)
- Parenting education for at-risk parents of young children
- Effective programming for all children before, during, and after school hours
- Intervention once troubled and delinquent children are identified (*Fight Crime* 2011)

This approach not only benefits society by reducing crime rates, it also saves the taxpayers money. Cohen and Piquero (2009) report that programs that prevent a child from using drugs, dropping out of school, and becoming a career criminal save society more than $2.5 million per person over the course of their lifetime.

The Perry Preschool Project is a good example of an early childhood intervention program. After a sample of 123 African American children was randomly assigned to either a control group or an experimental group, the experimental group members received academically oriented interventions for one to two years, frequent home visits, and weekly parent–teacher conferences. The control and experimental groups were compared on 14 occasions from age 3 to 40. As adults, experimental group members had higher employment and homeownership rates, and significantly lower violent and property crime rates (Schweinhart 2007).

Finally, many youth programs are designed to engage juveniles in noncriminal activities and integrate them into the community. In *Weed and Seed*, a program under the Department of Justice, "law enforcement agencies and prosecutors cooperate in 'weeding out' violent criminals and drug abusers . . . and 'seed' much-needed human services including prevention, intervention, treatment, and neighborhood restoration programs" (*Weed and Seed* 2011, p. 1). As part of the program, "safe havens" are established in, for example, schools, where multiagency services are provided for youth.

Community Programs Neighborhood watch programs involve local residents in crime prevention strategies. For example, MAD DADS (Men Against Destruction—Defending Against Drugs and Social Disorder) patrol the streets in high-crime areas of the city on weekend nights, providing positive adult role models and fun community activities for troubled children. Members also report crime and drug sales to police, paint over gang graffiti, organize gun buyback programs, and counsel incarcerated fathers. At present, 75,000 men, women, and children are in MAD DADS in 60 chapters in 17 states (MAD DADS 2011).

In 2010, 15,000 communities in 50 states—more than 37 million people—participated in "National Night Out," a crime prevention event in which citizens, businesses, neighborhood organizations, and local officials joined together in outdoor activities to heighten awareness of neighborhood problems, promote anti-crime messages, and strengthen community ties (National Night Out 2011).

Mediation and victim–offender dispute resolution programs are also increasing, with thousands of such programs worldwide. The growth of these programs is a reflection of their success rate: Two-thirds of cases referred result in face-to-face meetings, 95 percent of these cases result in a written restitution agreement, and 90 percent of the written restitution agreements are completed within one year (VORP 2009).

Criminal Justice Policy

The criminal justice system is based on the principle of **deterrence**—the use of harm or the threat of harm to prevent unwanted behaviors. The criminal justice system assumes that people rationally choose to commit crime, weighing the rewards and consequences of their actions. Thus "get-tough" measures hold that maximizing punishment will increase deterrence and cause crime rates to go down. Yet, most recently, incarceration rates have declined even as crime has declined, calling into question the principle of deterrence. Further, 30 years of "get-tough" policies have not significantly reduced recidivism rates and have created other criminal justice problems such as prison crowding. For example, in 2011, the U.S. Supreme Court ordered the State of California to reduce its prison population by 30,000 inmates, saying that overcrowded conditions are a violation of the Eighth Amendment's prohibition against cruel and unusual punishment. That said, experts are scratching their collective heads and asking, "what works?"

> . . . programs that prevent a child from using drugs, dropping out of school, and becoming a career criminal save society more than $2.5 million per person over the course of their lifetime.

deterrence The use of harm or the threat of harm to prevent unwanted behaviors.

Law Enforcement Agencies In 2009, the United States had 705,009 full-time law enforcement officers and 308,599 full-time civilian employees (e.g., clerks, meter attendants, correctional guards), yielding an estimated three law enforcement personnel per 1,000 inhabitants (FBI 2011). There are over 18,000 law enforcement agencies in the United States, including municipal (e.g., city police), county (e.g., sheriff's department), state (e.g., highway patrol), and federal agencies (e.g., FBI), often with overlapping jurisdictions (Siegel 2009). Research by Shaw and Brannan (2009) on the public's confidence in law enforcement at the federal and local levels indicates a slight decline in recent years, despite falling crime rates. Confidence in judges and in the U.S. Supreme Court has also declined.

In 2008, the latest year for which national data is available, 17 percent of the U.S. population 16 and older experienced face-to face contact with a police officer. The most common reason given, over 44 percent of the total, was traffic-related. Although white, black, and Hispanic drivers were equally likely to be stopped by the police, blacks and Hispanics were more likely to be searched. Further, 1.4 percent of people having face-to-face police contact reported force or the threat of force being used against them, and a majority of these drivers felt that the force was excessive. (Eith and Durose 2011).

Accusations of racial profiling, police brutality, and discriminatory arrest practices have made police–citizen cooperation in the fight against crime difficult. In response to such trends, the Violent Crime Control and Law Enforcement Act of 1994 established the Office of Community Oriented Policing Services (COPS). Community policing

> . . . emphasizes proactive problem solving in a systematic and routine fashion. Rather than responding to crime only after it occurs, community policing encourages agencies to proactively develop solutions to the immediate underlying conditions contributing to public safety problems. Problem solving must be infused into all police operations and guide decision-making efforts (COPS 2009, p. 12).

Such an approach often includes "problem-oriented policing," a model that includes analyzing the underlying causes of crime, looking for solutions, and actively seeking out alternatives to standard law enforcement practices.

Rehabilitation versus Incapacitation An important debate concerns the primary purpose of the criminal justice system: Is it to rehabilitate offenders or to incapacitate them through incarceration? Both rehabilitation and incapacitation are concerned with **recidivism** rates, or the extent to which criminals commit another crime. Advocates of **rehabilitation** believe that recidivism can be reduced by changing the criminal, whereas proponents of **incapacitation** think that recidivism can best be reduced by placing offenders in prison so that they are unable to commit further crimes against the general public.

Fear of crime has led to a public emphasis on incapacitation and a demand for tougher mandatory sentences, a reduction in the use of probation and parole, support of a "three strikes and you're out" policy, and truth-in-sentencing laws. However, these tough measures have recently come under attack for three reasons. First, research indicates that incarceration may not deter crime. Recent data indicate that,

recidivism A return to criminal behavior by a former inmate, most often measured by re-arrest, re-conviction, or re-conviction

rehabilitation A criminal justice philosophy that argues that recidivism can be reduced by changing the criminal through such programs as substance abuse counseling, job training, education, and so on.

incapacitation A criminal justice philosophy that argues that recidivism can be reduced by placing offenders in prison so that they are unable to commit further crimes against the general public.

. . . despite the massive increase in corrections spending, in many states there has been little improvement in the performance of corrections systems. If more than four out of 10 adult American offenders still return to prison within three years of their release, the system designed to deter them from continued criminal behavior clearly is falling short (Pew 2011a, p. 3).

Second is the accusation that get-tough measures, such as California's "three strikes and you're out" policy, are not equally applied. Chen's (2008) analysis of over 170,000 California inmates indicates that African Americans compared to whites and Hispanics are more likely to receive "third-strike sentences," with the greatest racial disparities being for property and drug offenses. Similarly, males are more likely to receive third-strike sentencing than females.

Finally, in an environment of budget deficits and legislative cuts, states simply can no longer afford the policies of decades ago. At a total cost of $52 billion in 2010, state corrections spending "quadrupled during the past two decades, making it the second fastest-growing area of state budgets, trailing only Medicaid" (Pew 2011a, p. 1). As a response to the economic downturn and concerns over the effectiveness of get-tough policies, many states are rethinking correctional policies—closing prisons, eliminating mandatory sentencing, replacing jail time with community-based programs, commuting sentences, expanding parole, and providing treatment rather than punishment for nonserious drug offenders (Steinhauer 2009; Davey 2010).

AP Photo/Ted S. Warren

According to the U.S. Bureau of Justice Statistics, 2,374,140 prisoners were in federal or state prisons and local jails at year-end 2009. Typically, individuals convicted of felonies are confined in prisons, whereas people who have committed misdemeanors and have sentences of one year or less are confined in jails.

Clearly, sentencing more offenders for longer periods of time to confinement enhances incapacitation. However, faced with budget cuts, states—as well as the Obama administration—are revisiting the ideals of rehabilitation. Rehabilitation assumes that criminal behavior is caused by sociological, psychological, and/or biological forces rather than being solely a product of free will. If such forces can be identified, the necessary change can be instituted. Many rehabilitation programs focus on helping the inmate reenter society. In 2008, President Bush signed the Second Chance Act, which supports reentry programs in the hopes of reducing recidivism (Greenblatt 2008). This chapter's Photo Essay highlights successful rehabilitation programs, including reentry initiatives.

Corrections In 2010, 5 million people were under court supervisions (e.g., probation), and 2.3 million were incarcerated (Porter 2011). An examination of global rates reveals that the United States has the highest incarceration rate in the world—743 per 100,000 population. The U.S. rate exceeds many times over those of other countries; for example, the rate in Russia is 585, China's rate is 120, the rate for England and Wales is 150, for Canada 117, and France and Germany, 96 and 88, respectively (Lambert 2011).

The U.S. incarceration rate has grown at an alarming rate—700 percent between 1950 and 2005, and, despite a general decrease in crime, it is expected to continue to grow, with the greatest increases being in the West, the South, and the Midwest (Pew 2007). Growth means more needed funds and, at an average prisoner cost of $29,000 a year, states are relying more and more on community alternatives, including probation and parole (Moore 2009).

In 2010, 5 million people were under court supervisions (e.g., probation), and 2.3 million were incarcerated . . . the highest incarceration rate in the world—743 prisoners per 100,000 population.

Photo Essay

Prison Programs That Work

Millions of men and women are behind bars, and recidivism is at an all-time high. The number of female inmates is estimated to increase by 16 percent, and there will be more elderly, both groups adding to the already skyrocketing cost of corrections. There are alternatives, however, and many of them have proven to be successful. Each of the described correctional practices has been empirically evaluated and found to be associated with positive changes in the inmate participants. From reducing recidivism rates and enhancing self-esteem, to lowering aggression and increasing the likelihood of post-release employment, these programs not only are cost-effective, but they are also humane.

Shaul Schwarz/Getty Images

Prison University Project

Thousands of inmates across the United States participate in postsecondary education programs. Research documents their benefits: reduced recidivism, enhanced problem-solving skills, safer prison conditions, a more marketable post-release inmate, and taxpayer savings (Correctional Association 2009). San Quentin's Prison University Project (PUP) is just such a program. Taught by college professors, graduate students, and volunteers, PUP provides 12 college courses each semester—classes in the humanities, liberal arts, social sciences, math, and science—leading to an associate of arts degree (PUP 2011).

Heather Rowley/Prison University Project

Indiana Canine Assistant and Adolescent Network

Programs such as Puppies Behind Bars (see photo), Puppies in Prison, Pen Pals, Project Pooch, and Prison Pet Partnership have been instrumental in changing the lives of inmates, breeder and shelter dogs, and the beneficiaries of the inmate–trainers' months of hard work and discipline. The men and women in the Indiana Canine Assistant and Adolescent Network (ICAAN) program train service and therapy dogs. An empirical evaluation of ICAAN documents the positive impact on the rehabilitation of participating offenders— higher self-esteem and better communication

◄The goals of the Prison University Program are to "educate and challenge students intellectually; to prepare them to lead thoughtful and productive lives inside and outside of prison; to provide them with skills needed to obtain meaningful employment and economic stability post-release; and to prepare them to become providers, leaders, and examples for their families and communities" (PUP 2011, p. 1). Pictured, a San Quentin inmate writing a paper for a class.

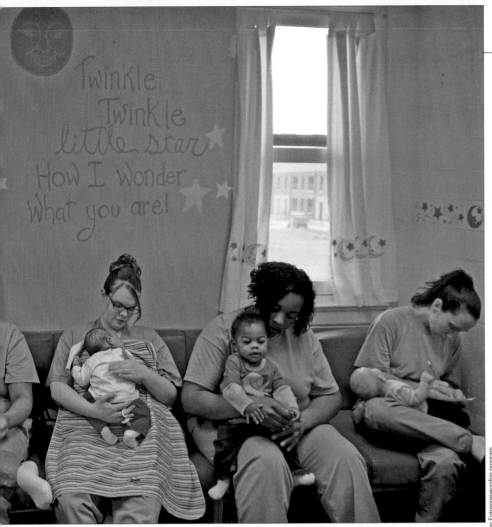

◀ Here, inmates wait for a parenting class at Indiana Women's Prison. Only nine states presently provide nursery facilities for female prisoners.

▲ Since its creation in 2004, the Prison Entrepreneurship Program (PEP) has been instrumental in helping inmates transition to successful businessmen. Here, PEP students talk with community business leaders as part of their five-month series of classes (PEP 2009).

skills, and a marked improvement in patience and trust, leading to better inmate–prison employee relations (Turner 2007).

Nebraska's Prison Nursery Program

There are over 200,000 women incarcerated in the United States (WPA 2011). With about 4 percent of female inmates pregnant at the time of incarceration, there has been a growing trend to provide prison nursery programs (see photo). Nebraska's Prison Nursery Program provides inmates with prenatal care education, parenting skills, information on child development, "hands-on training" for new and expectant mothers, and community resources upon release. After 10 years of operation, an evaluation of the program revealed lower misconduct and recidivism rates when compared to inmates who were required to give up their infants (Carlson 2009).

Prison Entrepreneurship Program

When Catherine Rohr, a Wall Street investor, toured a Texas prison, she had an epiphany—criminals and people in business are a lot alike. They both assess risks, live by their instincts, share profits, network, and compete with one another. It was then she founded the Prison Entrepreneurship Program (PEP). Today, corporate leaders and faculty volunteers teach business skills to PEP participants—former drugs dealers, gang leaders, hustlers, and felons—by equipping them with the tools for success. Over 90 percent of the 440 graduates have found jobs within four weeks of being released, 57 have started their own businesses, and recidivism rates are as low as 5 percent (PEP 2009; Beiser 2009).

◀ In the Puppies Behind Bars (PBB) program, puppies live in a prison cell for 16 months with their inmate-trainer who teaches them basic obedience skills. At the end of that time, the dogs are assessed for suitability as service dogs for the disabled or as explosive detection dogs for law enforcement agencies. In 2006, PBB began a program called Dog Tags, whereby service dogs are donated to injured soldiers coming home from Iraq and Afghanistan (PBB 2009).

Figure 4.5: **Growth of Correctional Population Over the Last 20 Years** Source: Pew 2009. From *One in 31: The Long Reach of American Corrections*. Copyright © 2009 The Pew Charitable Trusts. All Rights Reserved. Reprinted with permission.

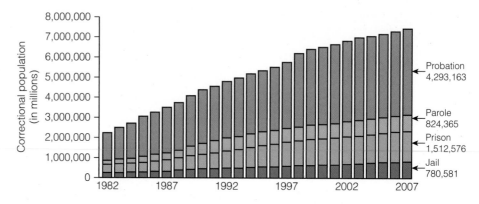

NOTE: Due to offenders with dual status, the sum of these four correctional categories slightly overstates the total correctional population.

Probation entails the conditional release of an offender who, for a specific time period and subject to certain conditions, remains under court supervision in the community. **Parole** entails release from prison, for a specific time period and subject to certain conditions, before the inmate's sentence is finished. Although varying by race, age, and gender, over 5 million people were on probation or parole in the United States in 2009 (BJS 2010b) (see Figure 4.5). Some argue that new technologies (e.g., global positioning systems) coupled with research-based treatment and reentry programs "can produce double-digit reductions in recidivism and save states money along the way" Pew 2009, p. 2).

> **What Do You Think?** Christian Longo, 37, admits to brutally killing his wife and three children, and was convicted of murder and sentenced to death. Now he wants to donate his organs to save the lives of some of the over 100,000 people waiting for transplants but the State of Oregon will not allow it (Hanrahan 2011; UNOS 2011). There are concerns that some of the lethal injection drugs used in the execution might damage his organs, and that if permitted, prisoners would start selling their organs to the highest bidder. Says Mr. Longo, "If I donated all of my organs today, I could clear nearly 1 percent of my state's organ waiting list. I am 37 years old and healthy; throwing my organs away after I am executed is nothing but a waste." (Hanrahan 2011, p. 1) What do you think?

probation The conditional release of an offender who, for a specific time period and subject to certain conditions, remains under court supervision in the community.

parole Parole entails release from prison, for a specific time period and subject to certain conditions, before the inmate's sentence is finished.

capital punishment The state (the federal government or a state) takes the life of a person as punishment for a crime.

Capital Punishment With **capital punishment**, the state (the federal government or a state) takes the life of a person as punishment for a crime. In 2008, China, Iran, Saudi Arabia, Pakistan, and the United States were responsible for 93 percent of all executions worldwide and, in 2010, more than two-thirds of the world's countries had abolished capital punishment (Amnesty International 2009; 2011b). The United States is the only western industrialized nation in the world to retain the death penalty (see Figure 4.6).

In 2010, 46 inmates were executed in the United States, one less than in 2009 (BJS 2011c). Of the 34 states that have the death penalty, the majority and the federal government almost exclusively use lethal injection as the method of execution. Two concerns, however, have been raised with the use of lethal injection, leading some states to halt executions. First is the question of whether or not death by lethal injection violates the Eighth Amendment's prohibition against cruel and unusual punishment. In 2007, a district court judge held that Tennessee's lethal injection procedures "present a substantial risk of unnecessary pain" that could "result in a terrifying, excruciating death" (Schelzig 2007, p. 1).

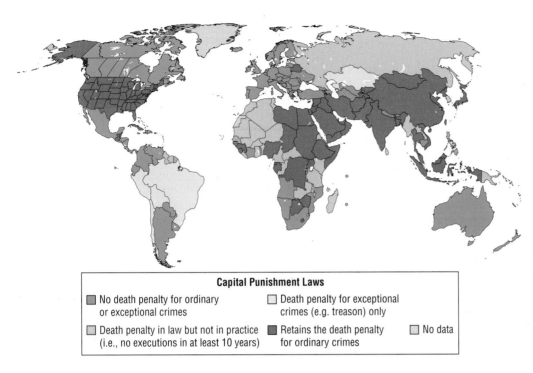

Figure 4.6: **Capital Punishment around the World, 2011**
Source: Amnesty International 2011a.

Capital Punishment Laws

☐ No death penalty for ordinary or exceptional crimes

☐ Death penalty for exceptional crimes (e.g. treason) only

☐ Death penalty in law but not in practice (i.e., no executions in at least 10 years)

☐ Retains the death penalty for ordinary crimes

☐ No data

However, in 2008, the U.S. Supreme Court held that Kentucky's use of lethal injection was not a violation of the Eighth Amendment (*Baze v. Rees* 2008).

The second issue concerns the role of physicians in state executions. According to Vu (2007, p. 1), the "American Medical Association is adamant that it is a violation of medical ethics for doctors to participate in, or even be present at, executions." In 2010, the American Board of Anesthesiology (ABA) adopted the position that participation in capital punishment would prohibit a physician from becoming certified by the ABA, arguing that physicians "should not be expected to act in ways that violate the ethics of medical practice, even if these acts are legal" (ABA 2011).

Proponents of capital punishment argue that executions of convicted murderers are necessary to convey public disapproval and intolerance for such heinous crimes. Those against capital punishment believe that no one, including the state, has the right to take another person's life and that putting convicted murderers behind bars for life is a "social death" that conveys the necessary societal disapproval.

Proponents of capital punishment also argue that it deters individuals from committing murder. Critics of capital punishment hold, however, that because most homicides are situational and are not planned, offenders do not consider the consequences of their actions before they commit the offense. Critics also point out that the United States has a higher murder rate than many western European nations that do not practice capital punishment, and that death sentences are racially discriminatory. A study of capital punishment in the United States, between 1973 and 2002, found that ". . . minority death row inmates convicted of killing whites face higher execution probabilities than other capital offenders" (Jacobs et al. 2007, p. 610). In 2010, 64 percent of Americans favored the death penalty for a person convicted of murder, 29 percent did not, and 6 percent were unsure (Gallup Poll 2010a).

Capital punishment advocates suggest that executing a convicted murderer relieves taxpayers of the costs involved in housing, feeding, guarding, and providing medical care for inmates. Opponents of capital punishment argue that financial considerations should not determine the principles that decide life and death issues. In addition, taking care of convicted murderers for life may actually be less costly than sentencing them to death because of the lengthy and costly appeals process for capital punishment cases. As of 2011, 16 states have abolished the death penalty including Illinois, New York, Michigan, and Wisconsin. Note, however, no Southern state to date has repealed their death penalty law (DPIC 2011). California is considering abolishing the death penalty. Former attorney general of California, John Van de Kamp (2009) argues:

> Confinement on death row (with all the attendant security requirements) adds $90,000 per inmate per year to the normal cost of incarceration. Appeals and habeas corpus proceedings add tens of thousands more. In all, it costs [California] $125 million a year more to prosecute and defend death penalty cases and to keep inmates on death row than it would simply to put all those people in prison for life without parole.

Nevertheless, those in favor of capital punishment argue that it protects society by preventing convicted individuals from committing another crime, including the murder of another inmate or prison official. One study of the deterrent effect of capital punishment concluded that each execution is associated with at least eight fewer homicides (Rubin 2002). Opponents contend that capital punishment may result in innocent people being sentenced to death. According to the Innocence Project, there have been 250 post-conviction exonerations using DNA evidence since 1989. The most common reasons for wrongful convictions, in order, are (1) eyewitness misidentification, (2) faulty forensic science, (3) false confessions or admissions, (4) government misconduct, (5) informants (e.g., paid informants lying), and (6) inadequate legal counsel (Innocence Project 2011).

Legislative Action

Legislative action is one of the most powerful methods of fighting crime. Federal and state legislatures establish criminal justice policy by the laws they pass, the funds they allocate, and the programs they embrace.

Gun Control In 2010, when a national sample of U.S. adults were asked about gun control, 50 percent responded that controlling gun ownership was more important than protecting the rights of gun owners, 47 percent the reverse. Those concerned with protecting the rights of gun owners were more likely to be Republican, non-Hispanic, and men with high school degrees living in a southern state than their respective counterparts (Pew 2011b).

Those against gun control argue that not only do citizens have a constitutional right to own guns but also that more guns may actually lead to less crime as would-be offenders retreat in self-defense when confronted (Lott 2003). Advocates of gun control, however, insist that the 250 million privately owned firearms in the United States, one-third of them handguns (Conklin 2007), significantly contribute to the violent crime rate in the United States and distinguish the country from other industrialized nations.

After a seven-year battle with the National Rifle Association (NRA), gun control advocates achieved a small victory in 1993, when Congress passed the Brady Bill. The law initially required a five-day waiting period on handgun purchases so that sellers can screen buyers for criminal records or mental instability. The law was amended in 1998, to include an instant check of buyers and their suitability for gun ownership. Today, the law requires background checks of not just handgun users but also those who purchase rifles and shotguns.

In addition to federal regulations, cities and states can create other restrictions. In Washington, D.C., gun ownership has virtually been banned for the last 30 years (Williams 2007). Only residents with permits such as police, security guards, and the like, can possess handguns and rifles. In what has been called the most important gun control case in decades, in 2008, the U.S. Supreme Court in *District of Columbia v. Heller* affirmed the right of an individual to own a firearm for private use (*District of Columbia v. Heller 2008*). The recent killings in Tucson, concern over the illegal flow of weapons to Mexico, and the possible change in administrations may, however, signal a renewed interest in gun legislation.

Other Legislation Major legislative initiatives have been passed in recent years, including the 1994 Violent Crime Control and Law Enforcement Act, which created community policing, "three strikes and you're out," and truth-in-sentencing laws. A sample of significant crime-related legislation presently before Congress includes the following bills (U.S. Congress 2011):

- *Medicare Fraud Enforcement and Prevention Act.* If passed, this act would increase civil fines, criminal penalties, and prison sentences for fraud and abuse under the Medicare program.
- *Preventing Gun Violence Act.* This act, if passed, would prohibit anyone convicted of a violent juvenile crime from possessing a firearm.
- *Safety from Sex Offenders Act.* This legislation would forbid a property manager from allowing a sex offender to access a residence, and prohibit a sex offender from possessing such means of accessing a residence (i.e., a key) while acting on behalf of a property manager.
- *Nicole's Law.* This bill, if passed, would require that state protections similar to those for victims of domestic violence be extended to victims of a sex offense who are not in a "familiar or dating relationship" with the perpetrator of the offense.
- *Child Gun Safety and Gun Access Prevention Act.* If passed, this bill would amend the Brady Bill by raising the age of handgun eligibility from 18 to 21, and by banning anyone under the age of 21 from possessing semi-automatic assault weapons.

International Efforts in the Fight against Crime

Europol is the European law enforcement organization that handles criminal intelligence. Unlike the FBI, Europol officers do not have the power to arrest; they predominantly provide support services for law enforcement agencies of counties that are members of the European Union. For example, Europol coordinates the dissemination of information, provides operational analysis and technical support, and generates strategic reports (Europol 2011). Europol, in conjunction with law enforcement agencies in member states, fights against transnational crimes such as illicit drug trafficking, child pornography, human trafficking, money laundering, illegal immigration, and counterfeiting of the euro.

INTERPOL, the International Criminal Police Organization, was established in 1923, and is the world's largest international police organization, with 188 member countries (INTERPOL 2011b). Similar to Europol, INTERPOL provides support services for law enforcement agencies of member nations. It has four core functions (INTERPOL 2011b). First, INTERPOL operates a worldwide police communications network that operates 24 hours a day, 7 days a week. Second, INTERPOL's extensive databases (e.g., DNA profiles, fingerprints, suspected terrorists) ensure that police get the information they need to investigate existing crime and prevent new crime from occurring. Third, INTERPOL provides emergency support services and operational activities to law enforcement personnel in the field. Finally, INTERPOL provides police training and development to help member states better fight the increasingly complex and globalized nature of crime.

Finally, the International Center for the Prevention of Crime (ICPC) is a consortium of policy makers, academicians, police, governmental officials, and nongovernmental agencies from all over the world (ICPC 2009). Located in Montreal, Canada, ICPC is a ". . . unique international forum and resource center dedicated to the exchange of ideas and knowledge on crime prevention and community safety" (p. 1). In fulfilling such a task, the ICPC seeks to (1) raise awareness of and access to crime prevention knowledge; (2) enhance community safety; (3) facilitate the sharing of crime prevention information between countries, cities, and justice systems; and (4) respond to calls for technical assistance.

Understanding Crime and Social Control

What can we conclude from the information presented in this chapter? Research on crime and violence supports the contentions of both structural functionalists and conflict theorists. Inequality in society, along with the emphasis on material well-being and corporate profit, produces societal strains and individual frustrations. Poverty, unemployment,

INTERPOL The largest international police organization in the world.

urban decay, and substandard schools—the symptoms of social inequality—in turn lead to the development of criminal subcultures and conditions favorable to law violation. Furthermore, criminal behavior is encouraged by the continued weakening of social bonds among members of society and between individuals and society as a whole, the labeling of some acts and actors as "deviant," and the differential treatment of minority groups by the criminal justice system.

Recently, there has been a general decline in crime, making it tempting to conclude that get-tough criminal justice policies are responsible for the reductions. Other valid explanations exist and are likely to have contributed to the falling rates: changing demographics, community policing, stricter gun control, and a reduction in the use of crack cocaine.

Concerns over the cost of "nail 'em and jail 'em" policies, overcrowded prisons, and high recidivism rates have some policy makers looking elsewhere. Several states are already expanding the use of community-based initiatives and developing evidence-based reentry programs. Further, in 2011, the National Criminal Justice Commission Act of 2011 was reintroduced into the U.S. Senate. If passed, the act would "create a blue-ribbon commission to look at every aspect of our criminal justice system with an eye toward reshaping the criminal justice system from top to bottom" (Webb 2011).

Rather than getting tough on crime after the fact, some advocate getting serious about prevention. Prevention programs are not only preferable to dealing with the wreckage crime leaves behind, but they are also cost-effective. For example, the Perry Preschool Project, as discussed earlier, cost $15,166 per participant but produced savings of $258,888 per participant. Of that savings, 88 percent was associated with a reduction in costs related to criminal justice (Schweinhart 2007).

Lastly, the movement toward **restorative justice,** a philosophy primarily concerned with repairing the victim–offender–community relation, is in direct response to the concerns of an adversarial criminal justice system that encourages offenders to deny, justify, or otherwise avoid taking responsibility for their actions.

Restorative justice holds that the justice system, rather than relying on "punishment, stigma, and disgrace" (Siegel 2006, p. 275), should "repair the harm" (Sherman 2003, p. 10). Key components of restorative justice include restitution to the victim, remedying the harm to the community, and mediation. In a meta-analysis of 35 restorative justice programs, offenders in restorative justice programs were less likely to become recidivists and more likely to meet restitution obligations than offenders not in restorative justice programs (Latimer et al. 2005).

restorative justice A philosophy primarily concerned with reconciling conflict between the victim, the offender, and the community.

CHAPTER REVIEW

- **Are there any similarities between crime in the United States and crime in other countries?**
  All societies have crime and have a process by which they deal with crime and criminals; that is, they have police, courts, and correctional facilities. Worldwide, most offenders are young males, and the most common offense is theft; the least common offense is murder.

- **How can we measure crime?**
  There are three primary sources of crime statistics. First are official statistics, for example, the FBI's Uniform Crime Reports, which are published annually. Second are victimization surveys designed to get at the "dark figure" of crime, crime that official statistics miss. Finally, self-report studies have all the problems of any survey research. Investigators must be cautious about whom they survey and how they ask the questions.

- **What sociological theory of criminal behavior blames the schism between the culture and structure of society for crime?**
  Strain theory was developed by Robert Merton (1957) and uses Durkheim's concept of *anomie*, or normlessness. Merton argued that, when the structure of society limits legitimate means (e.g., a job) of acquiring culturally defined goals (e.g., money), the resulting strain may lead to crime. Individuals, then, must adapt to the inconsistency between means and goals in a society that socializes everyone into wanting the same thing but provides opportunities for only some.

- **What are index offenses?**
  Index offenses, as defined by the FBI, include two categories of crime: violent crime and property crime. Violent crimes include murder, robbery, assault, and rape; property crimes include larceny, car theft, burglary, and arson. Property crimes, although less serious than violent crimes, are the most numerous.

- **What is meant by white-collar crime?**
  White-collar crime includes two categories: occupational crime, that is, crime committed in the course of one's occupation; and corporate crime, in which corporations violate the law in the interest of maximizing profits. In occupational crime, the motivation is individual gain.

- **How do social class and race affect the likelihood of criminal behavior?**
  Official statistics indicate that minorities are disproportionately represented in the offender population. Nevertheless, it is inaccurate to conclude that race and crime are causally related. First, official statistics reflect the behaviors and policies of criminal justice actors. Thus, the high rate of arrests, conviction, and incarceration of minorities may be a consequence of individual and institutional bias against minorities. Second, race and social class are closely related in that nonwhites are overrepresented in the lower classes. Because lower-class members lack legitimate means to acquire material goods, they may turn to instrumental, or economically motivated, crimes. Thus, the apparent relationship between race and crime may, in part, be a consequence of the relationship between these variables and social class.

- **What are some of the economic costs of crime?**
  First are direct losses from crime, such as the destruction of buildings through arson or of the environment by polluters. Second are costs associated with the transferring of property (e.g., embezzlement). A third major cost of crime is that associated with criminal violence (e.g., the medical cost of treating crime victims). Fourth are the costs associated with the production and sale of illegal goods and services. Fifth is the cost of prevention and protection. Finally, there is the cost of the criminal justice system, law enforcement, litigation and judicial activities, corrections, and victims' assistance.

- **What is the present legal status of capital punishment in this country?**
  Of the 34 states that have the death penalty, the majority and the federal government almost exclusively use lethal injection as the method of execution. Questions concerning the constitutionality of lethal injection and the role of physicians in state executions have led to several court cases.

TEST YOURSELF

1. The United States has the highest violent crime rate in the world.
   a. True
   b. False
2. The Uniform Crime Reports is a compilation of data from
   a. the U.S. Census Bureau
   b. law enforcement agencies
   c. victimization surveys
   d. the Department of Justice
3. According to ___, crime results from the absence of legitimate opportunities as limited by the social structure of society.
   a. Hirschi
   b. Marx
   c. Merton
   d. Becker
4. Which of the following is not an index offense?
   a. Drug possession
   b. Homicide
   c. Rape
   d. Burglary
5. The economic costs of white-collar crime outweigh the costs of traditional street crime.
   a. True
   b. False
6. Women everywhere commit less crime than men.
   a. True
   b. False
7. Probation entails
   a. early release from prison
   b. a suspended sentence
   c. court supervision in the community in lieu of incarceration
   d. incapacitation of the offender
8. Europol is an advisory and support law enforcement agency for European Union members.
   a. True
   b. False
9. Lethal injection
   a. violates the Eighth Amendment of the U.S. Constitution
   b. has been determined to be painless
   c. by physicians is approved by the American Medical Association
   d. is the most common method of execution in the United States
10. The United States has the highest incarceration rate in the world.
    a. True
    b. False

Answers: 1: b; 2: b; 3: c; 4: a; 5: a; 6: b; 7: c; 8: a; 9: d; 10: a.

KEY TERMS

acquaintance rape 108
capital punishment 128
classic rape 108
clearance rate 102
computer crime 114
corporate violence 113
crime 102
crime rate 102
deterrence 123
feminist criminology 116

identity theft 114
incapacitation 124
index offenses 107
INTERPOL 131
insider trading 113
larceny 109
organized crime 111
parole 128
primary deviance 106
probation 128

racial profiling 117
recidivism 124
rehabilitation 124
restorative justice 132
secondary deviance 106
transnational crime 101
victimless crimes 111
white-collar crime 112

MEDIA RESOURCES

Turning to Video

Watch the video on *The Criminal Justice System: Plea Bargaining* (running time 3:13), available through **CengageBrain.com.** This video describes the difficulty in managing the criminal justice system and the use of conflicting policies–three strikes to deter people from committing crime and plea bargaining which calls into question the integrity of the foundation of deterrence–if you do the crime, you'll do the time. Think about how each of the policies discussed in the video serves society, or not.

Online Study Resources

Log in to **www.cengagebrain.com** to access the resources your instructor has assigned. For this book, you can access:

CourseMate

Access chapter-specific learning tools including learning objectives, practice quizzes, videos, Internet exercises, flash cards, and glossaries, as well as web links, and more in your Sociology CourseMate.

Family Problems

"We must recognize that there are healthy as well as unhealthy ways to be single or to be divorced, just as there are healthy and unhealthy ways to be married."

—Stephanie Coontz, Family historian

Roy Morsch/Surf/CORBIS

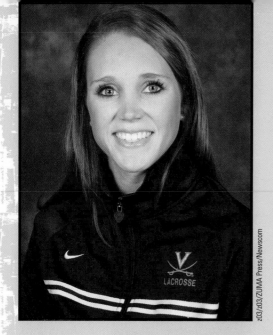

Yeardley Love, 22-year-old senior at the University of Virginia, was murdered in her off-campus apartment in 2010. Her ex-boyfriend was arrested for the murder.

IN 2010, YEARDLEY LOVE, a 22-year-old senior lacrosse player at the University of Virginia, was beaten to death. Her ex-boyfriend, George Huguely, also a 22-year-old senior lacrosse player at UVA, was arrested and charged with her murder. This highly publicized murder brought increased attention to the problem of intimate partner violence.

In this chapter, we turn our attention to family problems, focusing on violence and abuse in intimate and family relationships, and problems associated with divorce and its aftermath. Note that many of the problems families face, such as health problems, poverty, job-related problems, drug and alcohol abuse, discrimination, and military deployment of a spouse are dealt with in other chapters in this text. We begin by examining the diversity in family life across the globe and the changing patterns in U.S. families.

The Global Context: Families of the World

family A kinship system of all relatives living together or recognized as a social unit, including adopted people.

monogamy Marriage between two partners; the only legal form of marriage in the United States.

serial monogamy A succession of marriages in which a person has more than one spouse over a lifetime but is legally married to only one person at a time.

polygamy A form of marriage in which one person may have two or more spouses.

polygyny A form of marriage in which one husband has more than one wife.

polyandry The concurrent marriage of one woman to two or more men.

bigamy The criminal offense in the United States of marrying one person while still legally married to another.

The U.S. Census Bureau defines *family* as a group of two or more people related by blood, marriage, or adoption. Sociology offers a broader definition of family: A **family** is a kinship system of all relatives living together, or recognized as a social unit, including adopted people. This broader definition recognizes foster families, unmarried same-sex and opposite-sex couples with or without children, and any relationships that function and feel like a family. As we describe in the following section, family forms and patterns vary worldwide.

Monogamy and Polygamy In many countries, including the United States, the only legal form of marriage is **monogamy**—a marriage between two partners. A common variation of monogamy is **serial monogamy**—a succession of marriages in which a person has more than one spouse over a lifetime but is legally married to only one person at a time.

Polygamy—a form of marriage in which one person may have two or more spouses—is practiced on all continents throughout the world (Zeitzen 2008). The most common form of polygamy, known as **polygyny**, involves one husband having more than one wife. A less common form of polygamy is **polyandry**—the concurrent marriage of one woman to two or more men.

In the United States, Congress outlawed polygamy in 1892; thus, being married to more than one spouse is a crime referred to as **bigamy**. Although the Mormon Church has officially banned polygamy, it is still practiced among members of the Fundamentalist Church of Jesus Christ Latter-Day Saints (FLDS)—a Mormon splinter group with about 10,000 members who believe that a man must marry at least three wives to go to heaven (Zeitzen 2008).

Polygamy in the United States also occurs among some immigrants who come from countries where polygamy is accepted, such as Mali and Ghana and other West African countries. It is estimated, for example, that thousands of New Yorkers are involved in polygamous marriages (Bernstein 2007). Immigrants who practice polygamy generally keep their lifestyle a secret because polygamy is grounds for deportation under U.S. immigration law.

The issue of greatest concern regarding polygamy in the United States is the forcing of underage girls into polygamous marriages. In 2008, Texas state officials raided a FLDS ranch in Eldorado, Texas, and removed more than 400 children, placing them in temporary state custody to protect them from allegedly abusive conditions. Although 31 of these children were girls aged 14 to 17 who had children or were pregnant, a Court of Appeals ruled that the state did not have sufficient evidence of imminent danger to remove the children, and the court ordered the return of the children.

HBO / THE KOBAL COLLECTION/Picture Desk

The HBO series *Big Love* gave visibility to the illegal practice of polygamy among some religious fundamentalist groups.

What Do You Think? Fundamentalist Mormons, who practice polygamy as an extension of their religious beliefs, have attempted to justify polygamy by referring to the Constitution's First Amendment right to freedom of religion. In 1879, the Supreme Court in *Reynolds v. United States* ruled that, although people are free to believe in their religious principles, religious belief is not justification for acting against the law. Do you think that the criminalization of polygamy violates the right to freedom of religion? Should the government be tolerant of polygamy?

Division of Power in the Family In many societies, male dominance in the larger society is reflected in the dominance of husbands over wives in the family (see also Chapter 10, "Gender Inequality"). In many societies, for example, men make decisions about their wife's health care, when their wife may visit relatives, and/or daily and larger household purchases (Population Reference Bureau 2011). In sub-Saharan Africa, male dominance is linked to widespread wife abuse. According to Nigeria's minister for women's affairs, "It is like it is a normal thing for women to be treated by their husbands as punching bags. . . . The Nigerian man thinks that a woman is his inferior" (quoted by LaFraniere 2005, p. A1). In some societies, many women and men believe that it is acceptable for a man to beat his wife if she argues with him and/or refuses to have sex with him (Population Reference Bureau 2011).

In developed Western countries, although gender inequality persists (see Chapter 10), marriages tend to be more egalitarian, which means women and men view each other as equal partners who share decision making and assign family roles based on choice rather than on traditional beliefs about gender (Lindsey 2005). A study by the Pew Research Center (2008) found that nearly a third (31 percent) of U.S. couples reported joint decision making for most decisions; 43 percent of couples reported that the woman makes decisions in more areas than the man, and 26 percent of couples reported that men make more of the decisions in the family.

Social Norms Related to Childbearing In less developed societies, where social expectations for women to have children are strong, women on average have four to five children in their lifetime, and begin having them at an early age, with more than a third of women getting married by age 18 (Population Reference Bureau 2011). In developed societies, women may view having children as optional—as a personal choice.

These women and children of the Fundamentalist Church of Jesus Christ of Latter-Day Saints were removed from a compound in Eldorado, Texas, where polygamy is practiced, after allegations of abuse were reported.

Among U.S. young adults ages 18 to 29, although 74 percent say they want children, a significant minority aren't sure they want to be parents (19 percent) or say they don't want to have children (7 percent) (Wang & Taylor 2011).

Norms about childbirth out of wedlock also vary across the globe. In many poor countries, nonmarital childbirth is rare because girls often marry at a young age (before they are 18). In India, it is almost unheard of for a Hindu woman to have a child outside marriage; unwed childbearing would bring great shame to a woman and her family (Laungani 2005).

In the United States, 4 in 10 births in 2009 were to unmarried women (Hamilton et al. 2010). Yet, compared with some countries, the United States has a low proportion of births outside of marriage. Two-thirds of births in Iceland and more than half of births in France, Slovenia, Mexico, Estonia, Norway, and Sweden are out of wedlock. At the other end of the spectrum, less than 10 percent of births in Korea, Japan, and Greece are out of wedlock (OECD 2010).

Same-Sex Couples Norms and policies concerning same-sex intimate relationships also vary around the world. In some countries, homosexuality is punishable by imprisonment or even death. In a handful of countries, and in some U.S. states, same-sex couples are granted legal rights to marry. Several countries, and some U.S. states, grant same-sex couples legal rights and protections that are more limited than marriage. In the United States, support for gay marriage reached a milestone in 2011, when a poll of registered voters found that, for the first time, the majority (53 percent) support gay marriage (Langer 2011). And a 2011 Gallup poll found that the majority of U.S. adults (56 percent) view gay or lesbian relations as "morally acceptable" (Saad 2011). See Chapter 11, "Sexual Orientation and the Struggle for Equality," for in-depth information about same-sex families.

Looking at families from a global perspective underscores the fact that families are shaped by the social and cultural context in which they exist. As we discuss the family problems related to violence and abuse and divorce, we refer to social and cultural forces that shape these events and the attitudes surrounding them. Next, we look at changing patterns and structures of U.S. families and households.

Changing Patterns in U.S. Families

Some of the significant changes in U.S. families and households that have occurred over the past several decades include the following:

- *Increased singlehood and older age at first marriage.* U.S. women and men are staying single longer. From 1960 to 2010, the median age at first marriage increased from about 20 to 26 for women and from about 23 to 28 for men. Today, 13.8 percent of women and 20.4 percent of men ages 40 to 44 have never been married—the highest figures in this nation's history (U.S. Census Bureau 2010).

- *Increased heterosexual and same-sex cohabitation.* Many U.S. adults who are technically "single" are in long-term committed cohabiting relationships with a partner. From 1960 to 2010, the number of cohabiting unmarried couples skyrocketed (see Figure 5.1). The percentage of people who cohabited with their spouses before marriage more than doubled between 1980 and 2000, rising from 16 percent to 41 percent (Amato et al. 2007).

Figure 5.1: Number of Cohabiting, Unmarried U.S. Couples of the Opposite Sex, by Year
Source: U.S. Census Bureau 2010 and earlier reports in the same series.

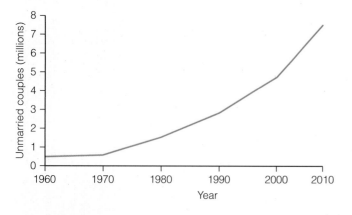

Couples live together to assess their relationship, to reduce or share expenses, or to avoid losing pensions or alimony from prior spouses. For some, including many same-sex couples who live in states where they cannot legally marry, cohabitation is an alternative to marriage. Some heterosexual couples do not marry out of solidarity with gays and lesbians, who can't legally wed in most states. Brad Pitt said that he and Angelina Jolie will "consider tying the knot when everyone . . . who wants to be married is legally able" (quoted in Davis 2009, p. 58).

Increased cohabitation among adults means that children are increasingly living in families that may function as two-parent families but do not have the social or legal recognition that married-couple families have. More than half of opposite-sex unmarried partner households have biological children in the home and about one-quarter of all same-sex households are raising children (ABC News 2011; U.S. Census Bureau 2010). When children are denied a legal relationship to both parents because of the parents' unmarried status, they may be denied Social Security survivor benefits, health care insurance, or the ability to have either parent authorize medical treatment in an emergency, among other protections. Some states, cities, counties, and employers allow unmarried partners (same-sex and/or heterosexual partners) to apply for a **domestic partnership** designation, which grants them some legal entitlements, such as health insurance benefits and inheritance rights that have traditionally been reserved for married couples.

Brad Pitt implied solidarity with gays and lesbians when he said that he and Angelina Jolie will "consider tying the knot when everyone . . . who wants to be married is legally able."

- *A new family form: Living apart together.* Some couples live apart in different cities or states because of their employment situation. Known as "commuter marriages," these couples generally would prefer to live together, but their jobs require them to live apart. However, other couples (married or unmarried) live apart in separate residences out of choice. Family scholars have identified this arrangement as an emerging family form known as **living apart together (LAT) relationships**. Couples may choose this family form for a number of reasons, including the desire to maintain a measure of independence and to avoid problems that may arise from living together. This new social phenomenon has been observed in several western European countries as well as in the United States (Lara 2005; Levin 2004).

- *Increased births to unmarried women.* The percentage of births to unmarried women rose to historic levels in 2009: 4 in 10 U.S. births in 2009 were to unmarried women (Hamilton et al. 2010). The highest rates of nonmarital births are among blacks, Native Americans/Alaskan natives, and Hispanics (see Figure 5.2).

domestic partnership A status that some states, counties, cities, and workplaces grant to unmarried couples, including gay and lesbian couples, which conveys various rights and responsibilities.

living apart together (LAT) relationships An emerging family form in which couples—married or unmarried—live apart in separate residences.

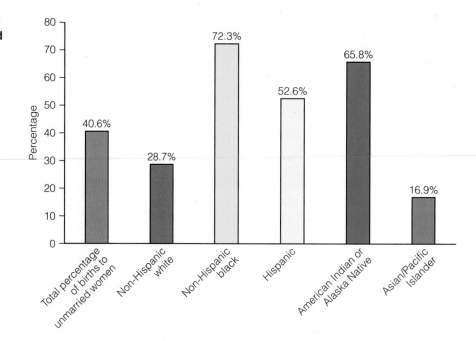

Figure 5.2: Percentage of All Births to Unmarried U.S. Women by Race and Hispanic Origin, 2009
Source: Hamilton et al. 2010.

Having a baby outside marriage has become more socially acceptable, with more than half (54 percent) of U.S. adults saying that having a baby outside of marriage is "morally acceptable." However, a significant minority (41 percent) view nonmarital childbearing as "morally wrong" (Saad 2011).

The percentage of births to unmarried women rose to historic levels in 2009: 4 in 10 U.S. births in 2009 were to unmarried women.

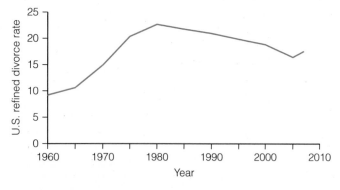

Figure 5.3: U.S. Refined Divorce Rate,* by Year
Source: The National Marriage Project 2009.
*Number of divorces per 1,000 married women

refined divorce rate The number of divorces per 1,000 married women.

Family scholar Stephanie Coontz (1997) emphasized the fact that we must be careful not to overdramatize the increase because "much illegitimacy was covered up in the past" (p. 29). In addition, not all unwed mothers are single parents; many unwed mothers are cohabiting with their partners when their children are born. Nevertheless, compared with children living in families with two parents, children in single-parent families tend to have more physical and mental health problems, higher drug use, and earlier onset of sexual behavior (Carr & Springer 2010).

- *Increased divorce and blended families.* The **refined divorce rate**—the number of divorces per 1,000 married women—increased dramatically from 1960, to its peak around 1980, then decreased until 2005, before increasing again (see Figure 5.3) (The National Marriage Project 2009). Between 40 and 50 percent of new U.S. marriages will end in divorce (Cherlin 2010). Most divorced individuals remarry and create blended families, traditionally referred to as stepfamilies. More than 4 in 10 U.S. adults have at least one step-relative in their family—either a stepparent, a step- or half sibling or a stepchild (see Figure 5.4) (Parker 2011). Seven in 10 adults who have at least one step-relative say they are very satisfied with their family life. Those who don't have any step-relatives report slightly higher levels of family satisfaction (78 percent very satisfied) (Parker 2011).

Stepparents and stepchildren do not have the same legal rights and responsibilities as biological/adopted children and biological/adoptive parents. Stepchildren do not automatically inherit from their stepparents, and courts have been reluctant to give stepparents legal access to stepchildren in the event of a divorce. In general, U.S. law does not recognize stepparents' roles, rights, and obligations regarding their stepchildren. In case of divorce, stepparents have little or no rights to their stepchildren (Sweeney 2010).

Percentage of U.S. adults* who have....

At least one step-relative	42%
A step- or half sibling	30%
A living step parent	18%
A step child	13%

0 5 10 15 20 25 30 35 40 45

*based on national sample of 2,691 adults

Figure 5.4: Percentage of U.S. Adults with Step-Relatives
Source: Parker 2011. From "A Portrait of Step Families", January 13, 2011, Social and Demographic Trends, a project of Pew Research Center. Reprinted with permission.

- *Increased employment of mothers.* Employment of married women with children under age 18 rose from 24 percent in 1950, to 71 percent in 2010 (Gilbert 2003; Bureau of Labor Statistics 2011). In the majority of married-couple families with children under 18, both parents are employed (58 percent in 2010) (Bureau of Labor Statistics 2011). Yet work, school, and medical care in the United States tend to be organized around the expectation that every household has a full-time parent at home who is available to transport children to medical appointments, pick up children from school on early dismissal days, and stay home when a child is sick (Coontz 1992).

Public Attitudes toward Changes in Family Life

A national survey found that U.S. adults are sharply divided in their judgments about the changes in U.S. families over the past several decades. A Pew Research Center survey asked a national sample of 2,691 adults whether they considered the following seven trends to be good, bad, or of no consequence to society: more unmarried couples raising children; more gay and lesbian couples raising children; more single women having children without a male partner to help raise them; more people living together without getting married; more mothers of young children working outside the home; more people of different races marrying each other; and more women not ever having children (Morin 2011).

The survey results found that about a third (31 percent) accept the changes, with most saying that changes in family structure either make no difference to society or are good. A similar share (32 percent) rejects the changing trends in family life, saying that five out of the seven trends are bad for society. The only trends they generally accepted are interracial marriage and fewer women having children. Finally, about a third (37 percent) is skeptical about the changes in American families, as they generally are tolerant of the trends, but also express concern about the impact of these trends on society. The trend they are most concerned about is single mothers raising children.

Getty Images Entertainment/Getty Images

Not all nonmarital births imply the absence of a father. Actress Goldie Hawn and actor Kurt Russell have been in a committed cohabiting relationship for more than 20 years. Although their child Wyatt was born "out of wedlock," he has been raised in a stable, loving family with his mother and his father. Kurt Russell also helped Goldie raise her two children (Kate and Oliver) from a previous marriage.

Marital Decline? Or Marital Resiliency?

Do the recent transformations in American families signify a collapse of marriage and family in the United States? Does the trend toward diversification of family forms mean that marriage and family are disintegrating, falling apart, or even disappearing? Or has family simply undergone transformations in response to changes in socioeconomic conditions, gender roles, and cultural values? The answers to these questions depend on whether we adopt the marital decline perspective or the marital resilience perspective.

According to the **marital decline perspective,** (1) personal happiness has become more important than marital commitment and family obligations, and (2) the decline in lifelong marriage and the increase in single-parent families have contributed to a variety of social problems, such as poverty, delinquency, substance abuse, violence, and the erosion of neighborhoods and communities (Amato 2004). According to the **marital resiliency perspective,** "poverty, unemployment, poorly funded schools, discrimination, and the lack of basic services (such as health insurance and child care) represent more serious threats to the well-being of children and adults than does the decline in married two-parent families" (Amato 2004, p. 960). According to this perspective, many marriages in the past were troubled, but because divorce was not socially acceptable, these problematic marriages remained intact. Rather than the view of divorce as a sign of the decline of marriage, divorce provides adults and children an escape from dysfunctional home environments.

Although the high rate of marital dissolution seems to suggest a weakening of marriage, divorce may also be viewed as resulting from placing a high value on marriage, such that a less than satisfactory marriage is unacceptable. In effect, people who divorce may be viewed not as incapable of commitment but as those who would not settle for a bad marriage. Indeed, the expectations that young women and men have of marriage have changed. Whereas once the main purpose of marriage was to have and raise children, today women and men want marriage to provide adult intimacy and companionship (Coontz 2000).

> Whereas once the main purpose of marriage was to have and raise children, today women and men want marriage to provide adult intimacy and companionship.

The high rate of childbirths out of wedlock and single parenting is also not necessarily indicative of a decline in the value of marriage. In interviews with a sample of low-income single women with children, most women said they would like to be married but just have not found "Mr. Right" (Edin 2000). Low-income single mothers in Edin's study were reluctant to marry the father of their children because these men had low economic status, traditional notions of male domination in household and parental decisions, and patterns of untrustworthiness and even violent behavior. Given the low level of trust these mothers have of men and given their view that husbands want more control than the women are willing to give them, women realize that a marriage that is also economically strained is likely to be conflictual and short-lived. "Interestingly, mothers say they reject entering into economically risky marital unions out of respect for the institution of marriage, rather than because of a rejection of the marriage norm" (Edin 2000, p. 130).

Is the well-being of a family measured by the degree to which that family conforms to the idealized married, two-parent, stay-at-home mom model of the 1950s? Or is family well-being measured by function rather than form? As suggested by family scholars Mason et al. (2003), "the important question to ask about American families . . . is not how much they conform to a particular image of the family, but rather how well do they function—what kind of love, care, and nurturance do they provide?" (p. 2).

It is also important to have a perspective that takes into account the historical realities of families. Family historian Stephanie Coontz (2004) explained:

> . . . many things that seem new in family life are actually quite traditional. Two-provider families, for example, were the norm through most of history. Stepfamilies were more numerous in much of history than they are today. There have been several times and places when cohabitation, out-of-wedlock births, or nonmarital sex were more widespread than they are today. (p. 974)

marital decline perspective A pessimistic view of the current state of marriage that includes the beliefs that (1) personal happiness has become more important than marital commitment and family obligations, and (2) the decline in lifelong marriage and the increase in single-parent families have contributed to a variety of social problems.

marital resiliency perspective A view of the current state of marriage that includes the beliefs that (1) poverty, unemployment, poorly funded schools, discrimination, and the lack of basic services (such as health insurance and child care) represent more serious threats to the well-being of children and adults than does the decline in married two-parent families, and (2) divorce provides adults and children an escape from dysfunctional home environments.

In *Alone Together: How Marriage in America Is Changing*, Paul Amato and colleagues (2007) examined survey data to see how marital quality changed between 1980 and 2000. One of the key questions the authors attempted to answer is: Has marriage become less satisfying and stable? Or has marriage become stronger and more satisfying for couples today compared to in the past?

Sample and Methods

Data for this study came from two national random samples of married adults 55 years and younger. Response rates for both the 1980 and the 2000 sample were over 60 percent. Data were collected through telephone interviews designed to measure five dimensions of marital quality: (1) marital happiness; (2) marital interaction; (3) marital conflict; (4) marital problems; and (5) divorce proneness. To assess whether marital quality changed between 1980 and 2000, the researchers compared the responses from 1980 with those from 2000. They also looked at gender differences to see whether marital quality for wives and husbands differed.

Selected Findings and Discussion

Analysis of the 1980 and 2000 data suggests that marriages have become stronger and more satisfying in some respects and weaker and less satisfying in other respects. In the following selected findings, note that two dimensions of marital quality improved (conflict and problems), one dimension deteriorated (interaction), and two dimensions were unchanged (happiness and divorce proneness).

1. *Marital happiness:* The percentage of spouses who rated their marriage as better than average was nearly equivalent in 1980 (74 percent) as in 2000 (68 percent).
2. *Marital interaction:* All five measures of marital interaction showed a decline from 1980 to 2000. Compared with spouses in 1980, spouses in 2000 were less likely to report that their husband or wife almost always accompanied them while visiting friends, shopping, eating their main meal of the day, going out for leisure, and working around the home.
3. *Marital conflict:* Of the five items that measure marital conflict, three declined significantly between 1980 and 2000: reports of disagreeing with one's spouse "often" or "very often," reports of violence ever occurring in the marriage, and reports of marital violence within the previous three years. No change was observed in the percentage of spouses who reported arguments over the division of household labor or in the number of serious quarrels in the previous two months.
4. *Marital problems:* Compared with spouses in 1980, spouses in 2000 were less likely to report marital problems in the following four areas: getting angry easily, having feelings that are hurt easily, experiencing jealousy, and being domineering. There was little change in the percentage of spouses who reported that getting in trouble with police, drinking or drugs, or extramarital sex were problems in their marriage.
5. *Divorce proneness:* The average level of divorce proneness was stable between 1980 and 2000. However, the percentage of spouses who reported either a low or a high degree of divorce proneness increased from 1980 to 2000, whereas the percentage reporting moderate divorce proneness declined.
6. *Gender differences:* Responses to items measuring marital interaction, marital problems, and divorce proneness were similar for husbands and wives. Two items measuring marital happiness revealed a gender difference: Between 1980 and 2000, wives (but not husbands) reported greater happiness with the amount of understanding received from their spouses. Wives also reported more happiness with their husbands' work around the house, whereas husbands reported less happiness with their wives' work around the house. Regarding the measure of marital conflict, both husbands and wives reported declines in violence.

In sum, this research suggests that between 1980 and 2000, "marriages became more peaceful, with fewer disagreements, less aggression, and fewer interpersonal sources of tension between spouses" (Amato et al. 2007, p. 68). Although overall levels of marital satisfaction did not seem to change from 1980 to 2000, the results for divorce proneness were mixed, with an increase in the proportion of both stable and unstable marriages. The finding that most concerned the researchers was that, between 1980 and 2000, the lives of husbands and wives became more separate, as spouses shared fewer activities. Citing previous research that has found that spouses who spend less time together are less happy in their marriage and more likely to divorce, Amato et al. suggested that "it is possible that the gradual decline in marital interaction between 1980 and 2000 will erode future marital happiness and increase subsequent levels of marital instability" (p. 69).

In sum, it is clear that the institution of marriage and family has undergone significant changes in the last few generations. What is not as clear is whether these changes are for the better or for the worse. As this chapter's *Social Problems Research Up Close* feature discusses, changes in marriage and family may be viewed as neither all good nor all bad, but are perhaps a more complex mix. Coontz (2005a) noted, "Marriage has become more joyful, more loving, and more satisfying for many couples than ever before in history. At the same time it has become optional and more brittle" (p. 306).

Sociological Theories of Family Problems

Three major sociological theories—structural functionalism, conflict theory, and symbolic interactionism—help to explain different aspects of the family institution and the problems in families today.

Structural-Functionalist Perspective

The structural-functionalist perspective views the family as a social institution that performs important functions for society, including producing and socializing new members, regulating sexual activity and procreation, and providing physical and emotional care for family members. When the family does not perform these functions adequately, social problems such as crime, ill health, and drug abuse may occur. This perspective views the high rate of divorce and the rising number of single-parent households as constituting a "breakdown" of the family institution. This supposed breakdown of the family is considered a primary social problem that leads to secondary social problems such as crime, poverty, and substance abuse.

According to the structural-functionalist perspective, traditional gender roles contribute to family functioning: Women perform the "expressive" role of managing household tasks and providing emotional care and nurturing to family members, and men perform the "instrumental" role of earning income and making major family decisions. According to this view, families have been disrupted and weakened by the change in gender roles, particularly women's participation in the workforce—an assertion that has been criticized and refuted, but is still supported by some "family values" scholars and supporters who advocate a return to traditional gender roles in the family.

Structural-functionalist explanations of family problems examine how changes in other social institutions contribute to family problems. For example, a structural-functionalist view of divorce examines how changes in the economy (such as more dual-earner marriages) and in the legal system (such as the adoption of "no-fault" divorce) contribute to high rates of divorce. Changes in the economic institution, specifically falling wages among unskilled and semiskilled men, also contribute to both intimate partner abuse and the rise in female-headed single-parent households (Edin 2000).

Conflict and Feminist Perspectives

Conflict theory focuses on how capitalism, social class, and power influence marriages and families. Feminist theory is concerned with how gender inequalities influence and are influenced by marriages and families. Feminists are critical of the traditional male domination of families—a system known as **patriarchy**—that is reflected in the tradition of wives taking their husband's last name and children taking their father's name. Patriarchy implies that wives and children are the property of husbands and fathers.

The overlap between conflict and feminist perspectives is evident in views on how industrialism and capitalism have contributed to gender inequality. With the onset of factory production during industrialization, workers—mainly men—left the home to earn incomes and women stayed home to do unpaid child care and domestic work. This arrangement resulted in families founded on what Friedrich Engels calls "domestic slavery of the wife" (quoted by Carrington 2002, p. 32). Modern society, according to Engels, rests on gender-based slavery, with women doing household labor for which they receive neither income nor status, whereas men leave the home to earn an income. Times have certainly changed since Engels made his observations, with most wives today leaving the home to earn incomes. However, wives employed full-time still do the bulk of unpaid domestic labor, and women are more likely than men to compromise their occupational achievement to take on child care and other domestic responsibilities. The continuing unequal distribution of wealth that favors men contributes to inequities in power and fosters economic dependence of wives on husbands. When wives do earn more money than their husbands, the divorce rate is higher—the women can afford to leave abusive or inequitable relationships (Jalovaara 2003).

patriarchy A male-dominated family system that is reflected in the tradition of wives taking their husband's last name and children taking their father's name.

Economic factors have also influenced norms concerning monogamy. In societies in which women and men are expected to be monogamous within marriage, there is a double standard that grants men considerably more tolerance for being nonmonogamous. Engels explained that monogamy arose from the concentration of wealth in the hands of a single individual—a man—and from the need to bequeath this wealth to children of his own, which required that his wife be monogamous. The "sole exclusive aims of monogamous marriage were to make the man supreme in the family and to propagate, as the future heirs to his wealth, children indisputably his own" (quoted by Carrington 2002, p. 32).

Feminist and conflict perspectives on domestic violence suggest that the unequal distribution of power among women and men and the historical view of women as the property of men contribute to wife battering. When wives violate or challenge the male head of household's authority, the male may react by "disciplining" his wife or using anger and violence to reassert his position of power in the family (see Figure 5.5).

Although modern gender relations within families and within society at large are more egalitarian than in the past, male domination persists, even if less obvious. Lloyd and Emery (2000) noted that "one of the primary ways that power disguises itself in courtship and marriage is through the 'myth of equality between the sexes.' . . . The widespread discourse on 'marriage between equals' serves as a cover for the presence of male domination in intimate relationships . . . and allows couples to create an illusion of equality that masks the inequities in their relationships" (pp. 25–26).

Conflict theorists emphasize that powerful and wealthy segments of society largely shape social programs and policies that affect families. The interests of corporations and businesses are often in conflict with the needs of families. Corporations and businesses strenuously fought the passage of the 1993 Family and Medical Leave Act, which gives people employed full-time for at least 12 months in companies with at least 50 employees up to 12 weeks of unpaid time off for parenting leave, illness or death of a family member, and elder care. Government, which corporate interests largely influence through lobbying and political financial contributions, enacts policies and laws that serve the interests of for-profit corporations rather than families.

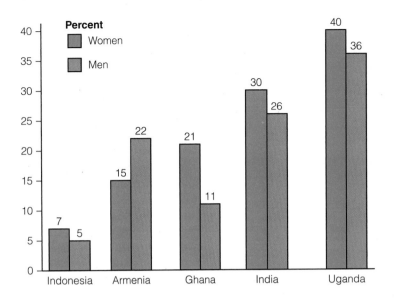

Figure 5.5a: Percent of Women and Men Who Agree That Wife Beating Is Acceptable if a Wife Argues with Her Husband

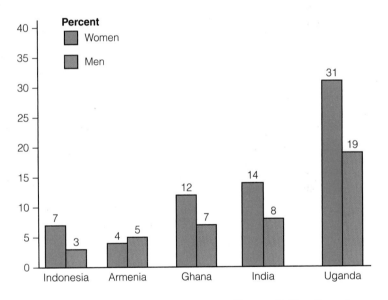

Figure 5.5b: Percent of Women and Men Who Agree That Wife Beating Is Acceptable if a Wife Refuses Sex with Her Husband
Source: Population Reference Bureau 2011.

Symbolic Interactionist Perspective

The symbolic interactionist perspective is concerned with how labels affect meaning and behavior. For example, when a noncustodial divorced parent (usually a father) is awarded "visitation" rights, he may view himself as a visitor in his children's lives. The meaning attached to the visitor status can be an obstacle to the father's involvement because the label *visitor* minimizes the importance of the noncustodial parent's role (Pasley & Minton 2001). Fathers' rights advocates suggest replacing the term *visitation* with terms such as *parenting plan* or *time-sharing arrangement,* because these terms do not minimize either parent's role.

Symbolic interactionists also point to the effects of interaction on one's self-concept, especially the self-concept of children. In a process called the "looking-glass self," individuals form a self-concept based on how others interact with them. Family members, such as parents, grandparents, siblings, and spouses, have a powerful effect on our self-concepts. For example, negative self-concepts may result from verbal abuse in the family, whereas positive self-concepts may develop in families in which interactions are supportive and loving. The importance of social interaction in children's developing self-concept suggests a compelling reason for society to accept rather than stigmatize nontraditional family forms. Imagine the effect on children who are called "illegitimate" or who are teased for having two moms or dads.

The symbolic interactionist perspective is useful in understanding the dynamics of domestic violence and abuse. For example, some abusers and their victims learn to define intimate partner violence as an expression of love (Lloyd 2000). Emotional abuse often involves using negative labels (e.g., *stupid, whore,* or *bad*) to define a partner or family member. Such labels negatively affect the self-concept of abuse victims, often convincing them that they deserve the abuse. In the next section, we examine violence and abuse in intimate and family relationships.

Violence and Abuse in Intimate and Family Relationships

In U.S. society, people are more likely to be physically assaulted, abused and neglected, sexually assaulted and molested, or killed in their own homes rather than anywhere else, and by other family members rather than by anyone else (Gelles 2000). Before reading further, you may want to take the Abusive Behavior Inventory in this chapter's *Self and Society* feature.

> In U.S. society, people are more likely to be physically assaulted, abused and neglected, sexually assaulted and molested, or killed in their own homes rather than anywhere else, and by other family members rather than by anyone else.

Intimate Partner Violence and Abuse

Abuse in relationships can take many forms, including emotional and psychological abuse, physical violence, and sexual abuse. **Intimate partner violence (IPV)** refers to actual or threatened violent crimes committed against individuals by their current or former spouses, cohabiting partners, boyfriends, or girlfriends. Emotional abuse can take many forms including yelling and screaming, withholding physical contact, isolating someone from their family and friends, belittling or insulting a person, restricting a person's activities, controlling a person's behavior (e.g., telling them what to wear, what to eat, where they can go, what time they have to be home, and so on), and exhibiting unreasonable jealousy (Follingstad & Edmundson 2010).

Globally, one woman in every three has been subjected to violence in an intimate relationship (United Nations Development Programme 2009). In the United States, women are more than four times more likely to be victims of IPV than are men (Bureau of Justice Statistics 2011). When women assault their male partners, these assaults tend to be acts of retaliation or self-defense (Johnson 2001; Swan et al. 2008).

Johnson and Ferraro (2003) identified the following four patterns of partner violence:

1. *Common couple violence* refers to occasional acts of violence arising from arguments that get "out of hand." Common couple violence usually does not escalate into serious or life-threatening violence.
2. *Intimate terrorism* is motivated by a wish to control one's partner and involves not only violence but also economic subordination, threats, isolation, verbal and emotional abuse, and other control tactics. Intimate terrorism is almost entirely

intimate partner violence (IPV) Actual or threatened violent crimes committed against individuals by their current or former spouses, cohabiting partners, boyfriends, or girlfriends.

Circle the number that best represents your closest estimate of how often each of the behaviors happened in your relationship with your partner or former partner during the previous six months (1, never; 2, rarely; 3, occasionally; 4, frequently; 5, very frequently).

1. Called you a name and/or criticized you   1 2 3 4 5
2. Tried to keep you from doing something you wanted to do (e.g., going out with friends, going to meetings)   1 2 3 4 5
3. Gave you angry stares or looks   1 2 3 4 5
4. Prevented you from having money for your own use   1 2 3 4 5
5. Ended a discussion with you and made the decision himself/herself   1 2 3 4 5
6. Threatened to hit or throw something at you   1 2 3 4 5
7. Pushed, grabbed, or shoved you   1 2 3 4 5
8. Put down your family and friends   1 2 3 4 5
9. Accused you of paying too much attention to someone or something else   1 2 3 4 5
10. Put you on an allowance   1 2 3 4 5
11. Used your children to threaten you (e.g., told you that you would lose custody, said he/she would leave town with the children)   1 2 3 4 5
12. Became very upset with you because dinner, housework, or laundry was not done when he/she wanted it done or done the way he/she thought it should be   1 2 3 4 5
13. Said things to scare you (e.g., told you something "bad" would happen, threatened to commit suicide)   1 2 3 4 5
14. Slapped, hit, or punched you   1 2 3 4 5
15. Made you do something humiliating or degrading (e.g., begging for forgiveness, having to ask permission to use the car or to do something)   1 2 3 4 5
16. Checked up on you (e.g., listened to your phone calls, checked the mileage on your car, called you repeatedly at work)   1 2 3 4 5
17. Drove recklessly when you were in the car   1 2 3 4 5
18. Pressured you to have sex in a way you didn't like or want   1 2 3 4 5
19. Refused to do housework or child care   1 2 3 4 5
20. Threatened you with a knife, gun, or other weapon   1 2 3 4 5
21. Spanked you   1 2 3 4 5
22. Told you that you were a bad parent   1 2 3 4 5
23. Stopped you or tried to stop you from going to work or school   1 2 3 4 5
24. Threw, hit, kicked, or smashed something   1 2 3 4 5
25. Kicked you   1 2 3 4 5
26. Physically forced you to have sex   1 2 3 4 5
27. Threw you around   1 2 3 4 5
28. Physically attacked the sexual parts of your body   1 2 3 4 5
29. Choked or strangled you   1 2 3 4 5
30. Used a knife, gun, or other weapon against you   1 2 3 4 5

Scoring

Add the numbers you circled and divide the total by 30 to find your score. The higher your score, the more abusive your relationship.

Source: Shepard, Melanie F., and James A. Campbell. 1992 (September). "The Abusive Behavior Inventory: A Measure of Psychological and Physical Abuse." *Journal of Interpersonal Violence* 7(3):291–305. Copyright © 1992 by Sage Publications. Used by permission.

---

perpetrated by men and is more likely to escalate over time and to involve serious injury.

3. *Violent resistance* refers to acts of violence that are committed in self-defense, usually by women against a male partner.

4. *Mutual violent control* is a rare pattern of abuse "that could be viewed as two intimate terrorists battling for control" (Johnson & Ferraro 2003, p. 169).

Intimate partner abuse also takes the form of sexual aggression, which refers to sexual interaction that occurs against one's will through use of physical force, threat of force, pressure, use of alcohol or drugs, or use of position of authority. In 2010, 73 percent of female rape or sexual assault victims (12 and older) reported that the offender was an intimate partner, friend, or acquaintance (Truman 2011).

A survey of college students found that, in the past 12 months, students were much more likely to have experienced an emotionally abusive relationship than a physically or sexually abusive relationship (see Table 5.1) (American College Health Association 2011).

TABLE 5.1 Abusive Relationships Reported by College Students

PERCENT OF COLLEGE STUDENTS REPORTING ABUSIVE RELATIONSHIP IN PAST 12 MONTHS

TYPE OF ABUSE	MALE	FEMALE
Emotional	7	12
Physical	2	2
Sexual	1	2

Source: American College Health Association 2011.

In 2009, music artist Chris Brown assaulted his then girlfriend, pop singer Rihanna, leaving her with a broken nose and lip and bruises on her face.

DAILY NEWS
MILLION READERS EVERY DAY   NYDailyNews.com
50¢

BRUTAL DETAILED ACCOUNT OF STAR'S BEATING

In just minutes, he pounded Rihanna's face to a pulp

WIREIMAGE

NY Daily News / Getty Images

Three Types of Male Perpetrators of Intimate Partner Violence Researchers have identified three types of male abusers (Fowler & Westen 2011):

1. The *psychopathic abuser* is generally violent, impulsive, and lacks empathy and remorse. Men in this subgroup often experienced high rates of childhood abuse and neglect.
2. The *hostile/controlling abuser* is suspicious, hypersensitive to perceived criticism, holds grudges, feels misunderstood, and has few friends. He is a "controlling" person who, instead of taking responsibility for his actions, blames and attacks his partner.
3. The *borderline/dependent abuser* is unhappy, depressed, and prone to emotions that spiral out of control. This type of abuser suffers from deep fears of abandonment and lashes out at the person he loves and needs the most. Men in this subgroup appear more fragile than violent, and after abusive episodes, they tend to feel self-loathing and guilt and are likely to beg for forgiveness.

Effects of Intimate Partner Violence and Abuse IPV results in bruises, broken bones, traumatic brain injury, cuts, burns, and other physical injuries. Nearly a third (30 percent) of female murder victims and 5 percent of male murder victims in the United States are killed by intimate partners (Federal Bureau of Investigation 2010). In 2009, more than 1,300 individuals (1,081 women and 279 men) were murdered by a partner or ex-partner (FBI 2010). The abuse of pregnant women often results in miscarriage or birth defects. Psychological consequences for victims of intimate partner violence can include depression, anxiety, suicidal thoughts and attempts, lowered self-esteem, inability to trust men, fear of intimacy, and substance abuse.

Battering also interferes with women's employment. Some abusers prohibit their partners from working. Other abusers "deliberately undermine women's employment by depriving them of transportation, harassing them at work, turning off alarm clocks, beating them before job interviews, and disappearing when they promise to provide child care" (Johnson & Ferraro 2003, p. 508). Battering also undermines employment by causing repeated absences, impairing women's ability to concentrate, and lowering their self-esteem and aspirations.

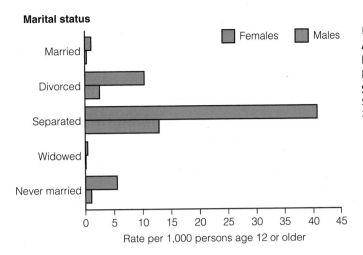

Marital status

Figure 5.6: **Average Annual Nonfatal Intimate Partner Victimization Rate, by Gender and Marital Status, 2001–2005** Source: Bureau of Justice Statistics 2011.

Abuse is also a factor in many divorces, which often results in a loss of economic resources. Women who flee an abusive home and who have no economic resources may find themselves homeless.

Children who witness domestic violence are at risk for emotional, behavioral, and academic problems as well as future violence in their own adult relationships (Parker et al. 2000; Kitzmann et al. 2003). Children may also commit violent acts against a parent's abusing partner.

Why Do Some Adults Stay in Abusive Relationships? Adult victims of abuse are commonly blamed for tolerating abusive relationships and for not leaving the relationship as soon as the abuse begins. However, from the point of view of the victims, there are compelling reasons to stay, including economic dependency, emotional attachment, commitment to the relationship, hope that things will get better, the view that violence is legitimate because they "deserve" it, guilt, and fear. Having children complicates the decision to leave an abusive relationship. Although some abuse victims leave to protect the children from harm by the abuser, others may stay because they can't support and/or raise the children alone, or may fear losing custody of the children if they leave (Lacey et al. 2011). Many victims of intimate partner abuse stay because they fear retribution from their abusive partner if they leave. Indeed, the rate of intimate partner violence is highest among separated couples (see Figure 5.6). Some victims also delay leaving a violent home because they fear the abuser will hurt or neglect a family pet (Fogle 2003).

Victims also stay because abuse in relationships is usually not ongoing and constant but rather occurs in cycles. The **cycle of abuse** involves a violent or abusive episode followed by a makeup period when the abuser expresses sorrow and asks for forgiveness and "one more chance." The makeup period may last for days, weeks, or even months before the next violent outburst occurs. Nevertheless, research suggests that most abused women eventually leave their abusers for good, although leaving tends to be a process rather than a one-time event, and the process often involves leaving and returning multiple times (Lacey et al. 2011).

What Do You Think? A major reason why abused women stay in abusive relationships is economic dependency. Yet, in a national study of abused women, researchers found that among women of color, those with higher socioeconomic status were more likely to stay in abusive relationships than those with lower socioeconomic status (Lacey et al. 2011). How might this unexpected finding be explained?

cycle of abuse A pattern of abuse in which a violent or abusive episode is followed by a makeup period when the abuser expresses sorrow and asks for forgiveness and "one more chance," before another instance of abuse occurs.

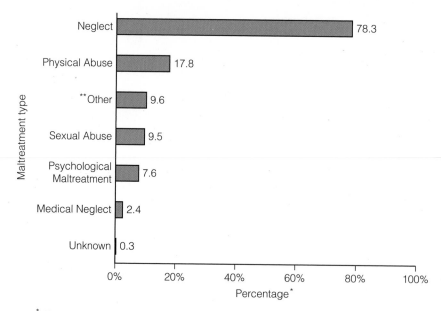

Figure 5.7: Types of Child Maltreatment, 2009
Source: U.S. Department of Health and Human Services, Administration on Children, Youth, and Families 2010.

Maltreatment type

Neglect — 78.3
Physical Abuse — 17.8
**Other — 9.6
Sexual Abuse — 9.5
Psychological Maltreatment — 7.6
Medical Neglect — 2.4
Unknown — 0.3

0%   20%   40%   60%   80%   100%
*Percentage**

* Percentages sum to more than 100% because children may experience more than one type of maltreatment.

** Includes "abandonment," "threats of harm," and "congenital drug addiction."

Child Abuse

Child abuse refers to the physical or mental injury, sexual abuse, negligent treatment, or maltreatment of a child under the age of 18 by a person who is responsible for the child's welfare. The most common form of child maltreatment is **neglect**—the caregiver's failure to provide adequate attention and supervision, food and nutrition, hygiene, medical care, and a safe and clean living environment (see Figure 5.7).

The highest rates of victimization are for the youngest children (birth to 1 year), and for children with disabilities. Although half of child abuse and neglect victims are white, rates of victimization are higher among some minority children (see Figure 5.8). Children reared in lesbian families are less likely to be abused by a parent or caregiver than are children reared in other family contexts (Gartrell et al. 2010).

Perpetrators of child abuse are most often the parents of the victim. Abuse is more likely to occur in families where there is a lot of stress that can result from a family history of violence, drug or alcohol abuse, poverty, chronic health problems, and social isolation (Centers for Disease Control and Prevention 2010).

Effects of Child Abuse Physical injuries sustained by child abuse cause pain, disfigurement, scarring, physical disability, and death. In 2009, an estimated 1,770 U.S. children died of abuse or neglect (U.S. Department of Health and Human Services, Administration on Children, Youth, and Families 2010). Most of these children were younger than age 4; nearly half were younger than one year; and most child deaths were caused by a parent or other family member. Head injury is the leading cause of death in abused children (Rubin et al. 2003). **Shaken baby syndrome**, whereby a caregiver shakes a baby to the point of causing the child to experience brain or retinal hemorrhage, most often occurs in response to a baby, typically younger than 6 months, who will not stop crying (Ricci et al. 2003; Smith 2003).

Adults who were abused as children have an increased risk of a number of problems, including depression, smoking, alcohol and drug abuse, eating disorders, obesity, high-risk sexual behavior, and suicide (Centers for Disease Control and Prevention 2010). Sexual abuse of young girls is associated with decreased self-esteem, increased levels of depression, running away from home, alcohol and drug use, and multiple sexual partners (Jasinski et al.

child abuse The physical or mental injury, sexual abuse, negligent treatment, or maltreatment of a child younger than age 18 by a person who is responsible for the child's welfare.

neglect A form of abuse involving the failure to provide adequate attention, supervision, nutrition, hygiene, health care, and a safe and clean living environment for a minor child or a dependent elderly individual.

shaken baby syndrome A form of child abuse whereby the caretaker shakes a baby to the point of causing the child to experience brain or retinal hemorrhage.

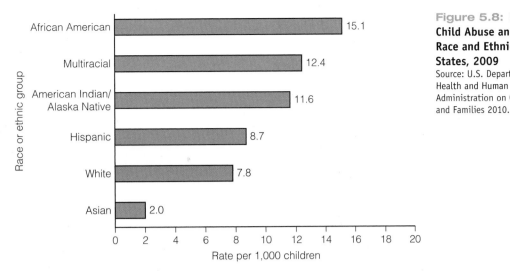

Figure 5.8: Rates of Child Abuse and Neglect by Race and Ethnicity: United States, 2009
Source: U.S. Department of Health and Human Services, Administration on Children, Youth, and Families 2010.

2000; Whiffen et al. 2000). A review of the research suggests that sexual abuse of boys produces many of the same reactions that sexually abused girls experience, including depression, sexual dysfunction, anger, self-blame, suicidal feelings, guilt, and flashbacks (Daniel 2005). Married adults who were physically and sexually abused as children report lower marital satisfaction, higher stress, and lower family cohesion than married adults with no abuse history (Nelson & Wampler 2000). Contrary to prior research findings, a recent study found that physical abuse in childhood is not associated with future violent delinquency, whereas sexual abuse and neglect are strongly related to future violent delinquent behavior (Yun et al. 2011).

Elder, Parent, Sibling, and Pet Abuse

Domestic violence and abuse may involve adults abusing their elderly parents or grandparents, children abusing their parents, and siblings abusing each other. Pets (or companion animals) are also victimized by domestic violence.

Elder Abuse **Elder abuse** includes physical abuse, sexual abuse, psychological abuse, financial abuse (such as improper use of the elder person's financial resources), and neglect. The most common form of elder abuse is neglect—failure to provide basic health and hygiene needs, such as clean clothes, doctor visits, medication, and adequate nutrition. Neglect also involves unreasonable confinement, isolation of elderly family members, lack of supervision, and abandonment.

Older women are far more likely than older men to suffer from abuse or neglect. Two out of every three cases of elder abuse reported to state adult protective services involve women. Although elder abuse also occurs in nursing homes, most cases of elder abuse occur in a domestic setting. The most likely perpetrators are adult children, followed by other family members and spouses or intimate partners (Teaster et al. 2006).

Parent Abuse Some parents are victimized by their children's violence, ranging from hitting, kicking, and biting to pushing a parent down the stairs and using a weapon to inflict serious injury to or even kill a parent. More violence is directed against mothers than against fathers, and sons tend to be more violent toward parents than are daughters

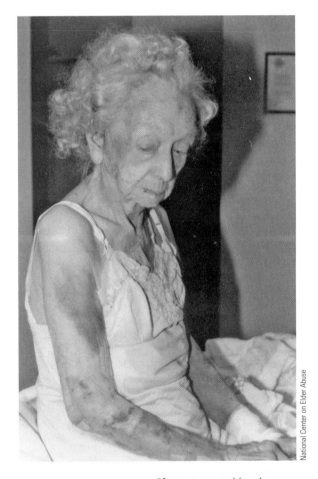

National Center on Elder Abuse

If you suspect elder abuse or are concerned about the well-being of an older person, call your state abuse hotline immediately.

elder abuse The physical or psychological abuse, financial exploitation, or medical abuse or neglect of the elderly.

In 2005, an 81-year-old woman walked nearly a mile from her house to a restaurant, pleading for help to escape the abuse she endured. She said her son repeatedly beat her and her dog. The restaurant owner called the local sheriff's office, and a deputy arrested the son. The woman relocated, but could not take her dog, a 2-year-old 60-pound Catahoula, with her. Kathy Cornwell, one of the victim advocates who helped the woman, adopted the dog. Kathy and the dog were in the bedroom when her husband came home. At the sight of a strange man, the dog jumped onto the bed and laid his body across Kathy. When Kathy later spoke with the abuse victim, she learned the dog had always tried to protect the woman from her abusive son, covering her body and absorbing the blows. As a result of the beatings, the dog had permanent nerve damage in his eyes and must wear goggles when he is outside because his pupils can't dilate. The dog, known as Little Horatio, goes to universities and other public forums on abuse. Cromwell explains that Horatio gives abuse victims hope: "This dog has gone through so much. And look at how happy he is. He's not stuck in the past. It shows people that it's hard, but you can move on" (quoted in Sullivan 2010, n.p.).

Domestic violence and animal abuse are linked. In a review of the literature, DeGue (2009) noted that between 47 and 71 percent of women seeking services from domestic violence shelters reported that a male abuser had threatened, hurt, or killed their companion animals. One study that compared the reports of women in shelters with a sample of non-abused women found that abused women were 11 times more likely to report that their partner had hurt or killed a pet (54 percent versus 5 percent) (DeGue 2009). Numerous studies have reported that fear of their companion animals being harmed or killed creates a significant barrier that prevents abused women from escaping their violent situations (Ascione 2007). Research has also

Little Horatio was a victim of domestic violence.

documented that children exposed to domestic violence are more likely than nonexposed children to have abused animals (Ascione & Shapiro 2009).

Because animal cruelty and family violence commonly co-occur, cross-reporting and cross-training have been instituted in many communities to teach human service personnel how to recognize and report cases of animal abuse and, conversely, to teach animal welfare personnel to recognize and report child, spousal, and elder abuse. Florida and San Diego County, California, mandate child protective personnel to report suspected animal abuse to animal welfare agencies, and four states require animal care and control personnel to report possible child abuse to child welfare agencies. Several states have laws permitting pets to be included in domestic violence protective orders (Ascione & Shapiro 2009). Finally, some programs provide procedures for housing pets in domestic violence situations at animal

shelters or with animal rescue groups. The American Humane Association PAWS (Pets and Women's Shelters) program provides grants to shelters to create housing where abuse victims and their pets can be together. Carol Wick, CEO of Harbor House, a domestic violence shelter in Orange County, Florida, describes the benefits of providing shelters for abuse victims and their pets:

> Every week we receive calls from domestic violence victims who want to leave their situation but are fearful of what will happen to their pets if they are left behind. . . . The new on-site kennel will allow women and children to safely leave an abusive situation and bring their pet with them—so no one gets left behind. (quoted in American Humane Association 2010)

(Ulman 2003). In most cases of children being violent toward their parents, the parents had been violent toward the children.

Sibling Abuse The most prevalent form of abuse in families is sibling abuse. Ninety-eight percent of the females and 89 percent of the males in one study reported having been emotionally abused by a sibling, and 88 percent of the females and 71 percent of the males reported having been physically abused by a sibling (Simonelli et al. 2002). Sexual abuse also occurs in sibling relationships.

Pet Abuse As pets are often viewed as "members of the family," abused pets (or companion animals) can be considered victims of family violence (see also this chapter's *Animals and Society* feature). Abusers threaten to hurt pets to control their victims, and all too often their threats are carried out. A student in one of the author's classes revealed that her abusive boyfriend warned her not to go out with her friends or she would "be sorry." The student went out with her friends, and when she came home, her cat was on her bed, dead. Her boyfriend had killed the cat by putting it in the dryer.

Factors Contributing to Intimate Partner and Family Violence and Abuse

Various individual, family, and social factors contribute to domestic violence and abuse.

Individual and Family Factors Several factors are associated with a higher risk that a person will become abusive to an intimate partner or family member. These risk factors include having witnessed or been a victim of abuse as a child, past violent or aggressive behavior, lack of employment and other stressful life events or circumstances, and drug and alcohol use (National Center for Injury Prevention and Control 2011). Alcohol use is involved in 50 percent to 75 percent of incidents of physical and sexual aggression in intimate relationships (Lloyd & Emery 2000). Alcohol and other drugs increase aggression in some individuals and enable the offender to avoid responsibility by blaming the violent behavior on drugs or alcohol.

Although abuse in adult relationships occurs among all socioeconomic groups, it is more prevalent among the poor. However, most poor people do not maltreat their children, and poverty, per se, does not cause abuse and neglect; the correlates of poverty, including stress, drug abuse, and inadequate resources for food and medical care, increase the likelihood of maltreatment (Kaufman & Zigler 1992).

Gender Inequality and Gender Socialization In the United States before the late 19th century, a married woman was considered the property of her husband. A husband had a legal right and marital obligation to discipline and control his wife through the use of physical force. This traditional view of women as property may contribute to men's doing with their "property" as they wish. In a study of men in battering intervention programs, about half of the men viewed battering as acceptable in certain situations (Jackson et al. 2003).

The view of women and children as property also explains marital rape and father-daughter incest. Historically, the penalties for rape were based on property rights laws designed to protect a man's property—his wife or daughter—from rape by other men (Russell 1990). Although a husband or father "taking" his own property in the past was not considered rape, today, marital rape and incest is considered a crime in all 50 states. Acquaintance and date rape can be explained in part by the fact that, in U.S. culture, "men receive support, even permission, from each other to sexually assault women through their direct encouragement or by ignoring problematic behavior" (Foubert et al. 2010, p. 2,238).

Traditional male gender roles have taught men to be aggressive and to be dominant in male–female relationships. "Because it is so clearly associated with masculinity in American culture, violence is a social practice that enables men to express a masculine identity" (Anderson 1997, p. 667). Traditional male gender socialization also discourages men from verbally expressing their feelings, which increases the potential for violence and abusive behavior (Umberson et al. 2003). Traditional female gender roles have also taught some women to be submissive to their male partner's control.

Acceptance of Corporal Punishment **Corporal punishment**—the intentional infliction of pain intended to change or control behavior—is widely accepted as a parenting practice. In a review of research on corporal punishment, Straus (2010) concluded that, in the United States, corporal punishment is (1) almost universal—94 percent of toddlers are spanked; (2) chronic—toddlers are often spanked three or more times a week; (3) often severe, with more than one in four parents using an object such as a paddle or belt to punish their children; and (4) of long duration—13 years for a third of U.S. children, 17 years for 14 percent of U.S. children. Parents of black children are more likely to use corporal punishment than parents of white or Latino children (Grogan-Kaylor & Otis 2007). A study of 35 low- and middle-income countries found that 17 percent of children, on average, experienced severe physical punishment, such as hitting the child on the head, ears, or face, or hitting the child hard and repeatedly (UNICEF 2010).

Although not everyone agrees that all instances of corporal punishment constitute abuse, some episodes of parental "discipline" are undoubtedly abusive. Many mental health and child development specialists advise against physical punishment, arguing that it is ineffective and potentially harmful to the child (Straus 2000). Yet acceptance of spanking as a method of child discipline is deeply ingrained in the United States and other societies. In an examination of 10 child-development textbooks with 2009 or 2010 copyrights, coverage of the topic of corporal punishment was minimal, and not one of the books concluded that children should never be spanked (Straus 2010).

Inaccessible or Unaffordable Community Services Many cases of child and elder abuse and neglect are linked to inaccessible or unaffordable health care, day care, elder care, and respite care facilities. Failure to provide medical care to children and elderly family members (a form of neglect) is sometimes a result of the lack of accessible or affordable health care services in the community. Failure to provide supervision for children and adults may result from inaccessible day care and elder care services. Without elder care and respite care facilities, socially isolated families may not have any help with the stresses of caring for elderly family members and children with special needs.

Strategies for Action: Preventing and Responding to Domestic Violence and Abuse

Next, we look at strategies for preventing and responding to violence and abuse in intimate and family relationships.

Prevention Strategies

Abuse-prevention strategies include public education and media campaigns, which may help to reduce domestic violence by conveying the criminal nature of domestic assault and offering ways to prevent abuse. Other abuse-prevention efforts focus on parent education to teach parents realistic expectations about child behavior and methods of child discipline that do not involve corporal punishment. For example, Mental Health America (2003) distributes a fact sheet on alternatives to spanking (see Table 5.2).

corporal punishment The intentional infliction of pain for a perceived misbehavior.

Another abuse-prevention strategy involves reducing violence-provoking stress by reducing poverty and unemployment and providing adequate housing, child care programs and facilities, nutrition, medical care, and educational opportunities.

Strengthening the supports for poor families with children reduces violence-provoking stress and minimizes neglect that results from inaccessible or unaffordable community services.

Responding to Domestic Violence and Abuse

After domestic violence and abuse has occurred, victims may seek safety in a shelter or safe house and take legal action against the abuser. Children may be placed in foster care, and the abuser may participate in mandatory or voluntary treatment.

Shelters and Safe Houses Between 1993 and 2004, about 21 percent of female victims and 10 percent of male victims of nonfatal IPV contacted a private or government agency for assistance (Catalano 2006). The National Domestic Violence Hotline (1-800-799-SAFE) is a 24-hour, toll-free service that provides crisis assistance and local domestic violence shelter and safe house referrals for callers across the country.

Shelters provide abused women and their children with housing, food, and counseling services. Safe houses are private homes of individuals who volunteer to provide temporary housing to abused people who decide to leave their violent homes. Some communities have abuse shelters for victims of elder abuse. Because one in four victims reports a delay in leaving dangerous domestic situations because of concerns over the safety of a pet, some programs offer a safe shelter for pets of victims of domestic violence (Fogle 2003).

TABLE 5.2 Effective Discipline Techniques for Parents: Alternatives to Spanking

Punishment is a "penalty" for misbehavior, but discipline is a method of teaching a child right from wrong. Alternatives to physical discipline include the following:

1. Be a positive role model.

Children learn behaviors by observing their parents' actions, so parents must model the ways in which they want their children to behave. If a parent yells or hits, the child is likely to do the same.

2. Set rules and consequences.

Make rules that are fair, realistic, and appropriate to a child's level of development. Explain the rules to children along with the consequences of not following them. If children are old enough, they can be included in establishing the rules and consequences for breaking them.

3. Encourage and reward good behavior.

When children are behaving appropriately, give them verbal praise and occasionally reward them with tangible objects, privileges, or increased responsibility.

4. Create charts.

Charts to monitor and reward behavior can help children learn appropriate behavior. Charts should be simple and should focus on one behavior at a time, for a certain length of time.

5. Give time-outs.

A "time-out" involves removing a child from a situation following a negative behavior. This can help the child calm down, end the inappropriate behavior, and reenter the situation in a positive way. Explain what the inappropriate behavior is, why the time-out is needed, when the time-out will begin, and how long it will last. Set an appropriate length of time for the time-out based on age and level of development, usually just a few minutes.

Source: Based on Mental Health America 2003. *Effective Discipline Techniques for Parents: Alternatives to Spanking.* Strengthening Families Fact Sheet. http://www.nmha.org/. Reprinted with permission.

Arrest and Restraining Order Domestic violence and abuse are crimes for which individuals can be arrested, jailed, and/or ordered to leave the home or enter a treatment program. About half of the states and Washington, D.C., now have mandatory arrest policies that require police to arrest abusers, even if the victim does not want to press charges. Abuse victims may obtain a restraining order that prohibits the perpetrator from going near the abused partner. However, many victims of intimate partner violence do not report the violence to law enforcement authorities because victims (1) believe that such violence is a private or personal matter, (2) fear retaliation, (3) view the violence as a "minor" crime, (4) want to protect the offender, and (5) believe that the police will not help or will be ineffective (Catalano 2006). In 2010, only half of rape or sexual assault incidents were reported to police (Truman 2011).

Foster Care Placement Children who are abused in the family may be removed from their homes and placed in government-supervised foster care. Foster care placements include other family members, certified foster parents, group homes, and other institutional facilities. More than 700,000 children are in foster care, waiting to be reunited with their families or adopted.

Due to the economic recession, more prospective adoptive parents are considering adopting foster children because they cannot afford private adoptions. However, there continues to be a shortage of people willing to adopt foster children because foster children tend to be older and are more likely to have emotional or physical problems (Koch 2009). Every year, more than 20,000 children who have not been adopted or reunited with their families must leave foster care because they turn 18. In some states, youth who age out of the foster care system can receive government aid such as housing assistance and Medicaid. However, many youth who age out of foster care struggle to fend for themselves, and many become homeless.

Another problem that plagues the foster care system is that, although it is intended to protect children from abuse, foster parents or caregivers sometimes abuse the children. Excluding extreme cases of abuse and neglect, some evidence suggests that children whose families are investigated for abuse or neglect are better off staying with their families than entering foster care, as children in foster care are more likely to drop out of school, commit crimes, abuse drugs, and become teen parents (Doyle 2007).

Treatment for Abusers Treatment for abusers typically involves group and/or individual counseling, substance abuse counseling, and/or training in communication, conflict resolution, and anger management. Most men in batterer intervention programs are referred by courts, although some may enter treatment voluntarily (Carter 2010). Treatment for men who sexually abuse children typically involves cognitive-behavior therapy (changing the thoughts that lead to sex abuse) and medication to reduce the sex drive (Stone 2004). Men who stop abusing their partners learn to take responsibility for their abusive behavior, develop empathy for their partner's victimization, reduce their dependency on their partners, and improve their communication skills (Scott & Wolfe 2000).

Problems Associated with Divorce

The United States has the highest divorce rate among Western nations. The lifetime probability of divorce among couples getting married today is between 40 and 50 percent (Cherlin 2010). This does not include the "informal divorces" that occur when long-term cohabitation relationships end (Amato 2010). In a sample of 2,922 undergraduates, 26 percent reported that their parents were divorced (Knox & Hall 2010).

The United States has the highest divorce rate among Western nations.

Divorce is considered problematic because of the negative effects it has on children as well as the difficulties it causes for adults. However, in some societies, legal and social barriers to divorce are considered problematic because such barriers limit the options of spouses in unhappy and abusive marriages. Ireland did not allow divorce under any condition until 1995, and Chile did not allow divorce until 2004.

Even when divorce is a legal option, social barriers often prevent spouses from divorcing. Hindu women, for example, experience great difficulty leaving a marriage, even when the husband is abusive, because divorce leads to loss of status, possible loss of custody of the children, homelessness, poverty, and being labeled a "loose" woman (Laungani 2005, p. 88).

Social Causes of Divorce

When we think of why a particular couple gets divorced, we typically think of a number of individual and relationship factors that might have contributed to the marital breakup: incompatibility in values or goals, poor communication, lack of conflict resolution skills, sexual incompatibility, extramarital relationships, substance abuse, emotional or physical abuse or neglect, boredom, jealousy, and difficulty coping with change or stress related to parenting, employment, finances, in-laws, and illness. However, understanding the high rate of divorce in U.S. society requires awareness of how the following social and cultural factors contribute to marital breakup:

No couple is immune from divorce. Al and Tipper Gore, who wrote a book called *Joined at the Heart*, shocked the U.S. public when they announced they were getting divorced.

1. *Changing function of marriage.* Before the Industrial Revolution, marriage functioned as a unit of economic production and consumption that was largely organized around producing, socializing, and educating children. However, the institution of marriage has changed over the last few generations:

   > Marriage changed from a formal institution that meets the needs of the larger society to a companionate relationship that meets the needs of the couple and their children and then to a private pact that meets the psychological needs of individual spouses. (Amato et al. 2007, p. 70)

   When spouses do not feel that their psychological needs—for emotional support, intimacy, affection, love, or personal growth—are being met in the marriage, they may consider divorce with the hope of finding a new partner to fulfill these needs.

2. *Increased economic autonomy of women.* Before 1940, most wives were not employed outside the home and depended on their husband's income. Today, the majority of married women are in the labor force. A wife who is unhappy in her marriage is more likely to leave the marriage if she has the economic means to support herself (Jalovaara 2003). An unhappy husband may also be more likely to leave a marriage if his wife is self-sufficient and can contribute to the support of the children.

3. *Increased work demands and economic stress.* Another factor influencing divorce is increased work demands and the stresses of balancing work and family roles. Some workers are putting in longer hours, often working overtime or taking second jobs, while others face job loss and unemployment. As discussed in Chapters 6 and 7, many families struggle to earn enough money to pay for rising housing, health care, and child care costs. Financial stress can cause marital problems. Couples with an annual income under $25,000 are 30 percent more likely to divorce than couples with incomes over $50,000 (see Table 5.3).

TABLE 5.3 Factors That Decrease Women's Risk of Separation or Divorce during the First 10 Years of Marriage	
FACTOR	PERCENT DECREASE IN RISK OF DIVORCE OR SEPARATION
Annual income over $50,000 (versus under $25,000)	30
Having a baby 7 months or more after marriage (versus before marriage)	24
Marrying over 25 years of age (versus under 18)	24
Having an intact family of origin (versus having divorced parents)	14
Religious affiliation (versus none)	14
Some college (versus high school dropout)	13

Source: National Marriage Project 2009.

Wives tend to be less happy in marriage than husbands when they perceive the division of household labor to be unfair.

Luc Ubaghs / Shutterstock.com

Women want to be equal partners in their marriages, not just in earning income but also in sharing the work of household chores, child rearing, marital communication, and in making decisions for the family.

second shift The household work and child care that employed parents (usually women) do when they return home from their jobs.

no-fault divorce A divorce that is granted based on the claim that there are irreconcilable differences within a marriage (as opposed to one spouse being legally at fault for the marital breakup).

individualism The tendency to focus on one's individual self-interests and personal happiness rather than on the interests of one's family and community.

4. *Inequality in marital division of labor.* Many employed parents, particularly mothers, come home to work a **second shift**—the work involved in caring for children and household chores (Hochschild 1989). Wives are more likely than husbands to perceive the marital division of labor as unfair (Nock 1995). This perception of unfairness can lead to marital tension and resentment, as reflected in the following excerpt:

> My husband's a great help watching our baby. But as far as doing housework or even taking the baby when I'm at home, no. He figures he works five days a week; he's not going to come home and clean.

But he doesn't stop to think that I work seven days a week. Why should I have to come home and do the housework without help from anybody else? My husband and I have been through this over and over again. Even if he would just pick up from the kitchen table and stack the dishes for me, that would make a big difference. He does nothing. . . . He'll help out if I'm not here, but the minute I am, all the work at home is mine. (quoted by Hochschild 1997, pp. 37–38)

Women want to be equal partners in their marriages, not just in earning income but also in sharing the work of household chores, child rearing, marital communication, and in making decisions for the family. Frustrated by men's lack of participation in marital work, women who desire relationship egalitarianism may see divorce as the lesser of two evils (Hackstaff 2003).

5. *Liberalized divorce laws.* Before 1970, the law required a couple who wanted a divorce to prove that one of the spouses was at fault and had committed an act defined by the state as grounds for divorce—adultery, cruelty, or desertion. In 1969, California became the first state to initiate **no-fault divorce,** which permitted a divorce based on the claim that there were "irreconcilable differences" in the marriage. Today, all 50 states recognize some form of no-fault divorce. No-fault divorce law has contributed to the U.S. divorce rate by making divorce easier to obtain.

6. *Increased individualism.* U.S. society is characterized by **individualism**—the tendency to focus on one's individual self-interests and personal happiness rather than

on the interests of one's family and community. "Marital commitment lasts only as long as people are happy and feel that their own needs are being met" (Amato 2004, p. 960). Belief in the right to be happy, even if it means getting divorced, is reflected in social attitudes toward divorce: More than two-thirds (69 percent) of U.S. adults report that divorce is morally acceptable (Saad 2011).

7. *Weak social ties.* Couples with strong social ties have a network of family and friends who provide support during difficult times, and who express disapproval for behavior that threatens the stability of the marriage. Couples who live in the same community for a long time have the opportunity to develop and maintain strong social ties. But many Americans move from place to place during their adult years, more so than people in other countries, which may help explain why the U.S. divorce rate is higher than in other countries. Cherlin (2009) explains that "moving from one community to another could affect marriages because it disrupts social ties. Migration can separate people from friends and relatives who could help them through family crises" (pp. 148–149).

8. *Increased life expectancy.* Finally, more marriages today end in divorce, in part, because people live longer than they did in previous generations and "till death do us part" involves a longer commitment than it once did. Indeed, one can argue that "marriage once was as unstable as it is today, but it was cut short by death not divorce" (Emery 1999, p. 7).

Consequences of Divorce

When parents have bitter and unresolved conflict and/or if one parent is abusing a child or the other parent, divorce may offer a solution to family problems. However, divorce often has negative effects for ex-spouses and their children and also contributes to problems that affect society as a whole.

Physical and Mental Health Consequences Numerous studies show that divorced individuals have more health problems and a higher risk of mortality than married individuals; divorced individuals also experience lower levels of psychological well-being, including more unhappiness, depression, anxiety, and poorer self-concepts (Amato 2003). Both divorced and never-married individuals are, on average, more distressed than married people because unmarried people are more likely than married people to have low social attachment, low emotional support, and increased economic hardship (Walker 2001). Some research suggests that divorce leads to higher levels of depressive symptoms for women, but not for men (Kalmijn & Monden 2006), especially when young children are in the family (Williams & Dunne-Bryant 2006). This finding is probably due to the increased financial and parenting strains experienced by divorced mothers who have custody of young children.

However, some studies have found that divorced individuals report higher levels of autonomy and personal growth than married individuals do (Amato 2003). For example, many divorced mothers report improvements in career opportunities, social lives, and happiness after divorce; some divorced women report more self-confidence, and some men report more interpersonal skills and a greater willingness to self-disclose. For people in a poor-quality marriage, divorce has a less negative or even a positive effect on well-being (Amato 2003; Kalmijn & Monden 2006). However, leaving a bad marriage does not always result in increased well-being because "divorce is a trigger for even more problems after the divorce" (Kalmijn & Monden 2006, p. 1210). In sum, some men and women experience a decline in well-being after divorce; others experience an improvement.

> . . . Some men and women experience a decline in well-being after divorce; others experience an improvement.

Economic Consequences Following divorce, there tends to be a dramatic drop in women's income and a slight drop in men's income (Gadalla 2009). Compared with married individuals, divorced individuals have a lower standard of living, have less wealth, and experience greater economic hardship, although this difference is considerably greater for women than for men (Amato 2003). The economic costs of divorce are often greater for women and children because women tend to earn less than men (see Chapter 10) and because mothers devote substantially more time to household and child care tasks than fathers do. The time women invest in this unpaid labor restricts their educational and job opportunities as well as their income. Men are less likely than women to be economically disadvantaged after divorce because they continue to profit from earlier investments in education and career.

After divorce, both parents are responsible for providing economic resources to their children. However, some nonresident parents fail to provide child support. In some cases, failure to pay child support is not due to fathers being "deadbeats" but rather to the fact that many fathers are "dead broke." Fathers who are unemployed or who have low-wage jobs may be unable to make child support payments.

Effects on Children and Young Adults Parental divorce is a stressful event for children and is often accompanied by a variety of stressors, such as continuing conflict between parents, a decline in the standard of living, moving and perhaps changing schools, separation from the noncustodial parent (usually the father), and parental remarriage. These stressors place children of divorce at higher risk for a variety of emotional and behavioral problems. Reviews of research on the consequences of divorce for children have found that children with divorced parents score lower on measures of academic success, psychological adjustment, self-concept, social competence, and long-term health; they also have higher levels of aggressive behavior and depression (Amato 2003; Wallerstein 2003).

Many of the negative effects of divorce on children are related to the economic hardship associated with divorce. Economic hardship is associated with less effective and less supportive parenting, inconsistent and harsh discipline, and emotional distress in children (Demo et al. 2000). Despite the adverse effects of divorce on children, research findings suggest that "most children from divorced families are resilient, that is, they do not suffer from serious psychological problems" (Emery et al. 2005, p. 24). Other researchers conclude that "most offspring with divorced parents develop into well-adjusted adults," despite the pain they feel associated with the divorce (Amato & Cheadle 2005, p. 191).

Divorce can also have positive consequences for children and young adults. In highly conflictual marriages, divorce may actually improve the emotional well-being of children relative to staying in a conflicted home environment (Jekielek 1998). In interviews with 173 grown children whose parents divorced years earlier, Ahrons (2004) found that most of the young adults reported positive outcomes for their parents and for themselves. Although many young adults who have divorced parents fear that they too will have an unhappy marriage (Dennison & Koerner 2008), such a fear can also lead young adults to think carefully about their choices regarding marriage.

Effects on Father-Child Relationships Children who live with their mothers may have a damaged relationship with their nonresidential father, especially if he becomes disengaged from their lives. Some research has found that young adults whose parents divorced are less likely to report having a close relationship with their father compared with children whose parents are together (DeCuzzi et al. 2004). However, in another study of 173 adult children of divorce, more than half felt that their relationships with their fathers improved after the divorce (Ahrons 2004). Children may benefit from having more quality time with their fathers after parental divorce. Some fathers report that they became more active in the role of father after divorce.

One study has found that the older the child is at the time of parental separation, the greater the amount of time children spend with their fathers (Swiss & Bourdais 2009). This study also found that fathers earning $50,000 or more per year are likely to see their children more often than fathers who earn less than $30,000 per year.

Women who have primary custody of the children serve as gatekeepers for the relationship their children have with their fathers (Trinder 2008). As gatekeepers, custodial mothers can either keep the gate open and encourage contact between the children and their father, or make every effort to keep the gate closed, cutting off the children's contact with their father. Some noncustodial divorced fathers discontinue contact with their children as a coping strategy for managing emotional pain (Pasley & Minton 2001). Many divorced fathers are overwhelmed with feelings of failure, guilt, anger, and sadness over the separation from their children (Knox 1998). Hewlett and West (1998) explained that "visiting their children only serves to remind these men of their painful loss, and they respond to this feeling by withdrawing completely" (p. 69). Divorced fathers commonly experience the legal system as favoring the mother in child-related matters. One divorced father commented:

> I believe that the system [judges, attorneys, etc.] have [sic] little or no consideration for the father. At some point the system creates an environment where the father loses any natural desire to see his children because it becomes so difficult, both financially and emotionally. At that point, he convinces himself that the best thing to do is wait until they are older. (quoted by Pasley & Minton 2001, p. 242)

Parental Alienation **Parental alienation** refers to the intentional efforts of one parent to turn a child against the other parent and essentially destroy any positive relationship a child has with the other parent (Baker & Chambers 2011). When parents influence their children to hate or fear the other parent, both the alienated parent and child may be said to be victims of *parental alienation syndrome* (PAS)—a controversial diagnosis that the American Psychiatric Association is considering including in the next revision of the *Diagnostic and Statistical Manual of Mental Disorders.* Parental alienation has been described as a form of emotional abuse involving "psychological kidnapping," whereby one parent manipulates children to make them hate and reject the other parent (see this chapter's *The Human Side* feature). Parents may alienate their child from the other parent by engaging in a variety of behaviors (see Table 5.4) (Baker & Chambers 2011).

Long-term effects of parental alienation on children can include low self-esteem, depression, drug and alcohol problems, mistrust, and divorce (Baker 2007). The effects on the rejected parent are equally devastating.

The alienating parent can be the mother, or the father, and although parental alienation is probably more common in divorced families, it also occurs in intact families. Extended family of the alienating parent often play a role in maintaining and supporting the alienation.

Some lawyers, judges, therapists, and academics challenge the validity of parental alienation, claiming that it does not really exist. Baker (2006) explains,

Perhaps such skeptics hold the belief that a parent must have done *something* to warrant their child's rejection and/or the other parent's animosity. This is the double victimization of PAS: the shame and frustration of being misunderstood in addition to the grief and anger associated with being powerless to prevent the alienation in the first place. (p. 192)

TABLE 5.4 Parental Alienation Behaviors
limiting the child's contact with the other parent
interfering with communication between the child and the other parent
destroying gifts or memorabilia from the other parent
failing to display any positive interest in the child's activities or experiences with the other parent
expressing disapproval or dislike of the child spending time with the other parent
limiting mention and photographs of the other parent
withdrawal of love or expressions of anger if the child indicates positive feelings for the other parent
telling the child that the other parent does not love him or her
forcing the child to choose between his or her parents
creating the impression that the other parent is dangerous (when the parent is not) forcing the child to reject the other parent
asking the child to spy on the other parent
asking the child to keep secrets from the other parent
referring to the other parent by his or her first name
referring to a stepparent as "Mom" or "Dad" and encouraging the child to do the same
withholding medical, social, or academic information from the other parent and keeping the other parent's name off of such records
changing the child's name to remove association with the other parent
denigration of the other parent, the other parent's extended family, and any new partner/spouse

parental alienation The intentional efforts of one parent to turn a child against the other parent and essentially destroy any positive relationship a child has with the other parent.

Amy Baker (2007) conducted interviews with 40 individuals who believed that they had been turned against one parent by the other parent. This *The Human Side* feature presents excerpts from these interviews, providing a glimpse into the experience of parental alienation from the child's perspective.

Larissa (speaking about her mother): *She'd always made both my brother and me feel that our father was somehow to blame for everything. Every day there'd be some attempt by her, some tale she'd tell me, to turn me against my father, so many incidents, its simply impossible to list them all. . . . I grew to detest him, with a truly visceral hate. I couldn't stand to be in the same room with him, or to even talk to him or have him talk to me.* (pp. 30–31)

Bonnie: *My dad tried several different tactics to talk to me but my mom foiled his plans, I guess you could say. At the time I sort of knew what he was doing. . . . It felt like he was trying to contact me but she would stop it. For example, when I was in middle school in 6th grade I think, my dad went up to the school to have lunch with me and she had told the school that if he would ever go up there they were to immediately call the police and I remember it was a big production. They came and got me out of class and they put me in the principal's office and they locked the door. I found out later he was escorted off the grounds by the police.* (p. 126)

Maria: *Basically I was a chess piece between the two of them, and it was very hard. . . . I started picking up on signals from my mother that showing any kind of affection or love for my father could be a problem for me. If I showed any positiveness toward him, it was, "How could you do this to me? . . . You are betraying me." When I go to visit him and I come back, to this day, my mother still says, "How can you do this to me? You are betraying me. You are slapping me in the face any time you have anything to do with this man. How could you do this to me?"* (p. 139) *. . . . My dad and I were very close and then after a time and all of this stuff from my mother and then I think I took on a lot of her anger toward him, feeling responsible for her because she was so, "Oh poor me. I am so broken and so wounded." . . . And even if I would have desired anything to do with him, I would not have acted on it because of how my mother would have reacted.* (p. 141)

As we have seen, the effects of divorce on adults and children are mixed and variable. In a review of research on the consequences of divorce for children and adults, Amato (2003) concluded that "divorce benefits some individuals, leads others to experience temporary decrements in well-being that improve over time, and forces others on a downward cycle from which they might never fully recover" (p. 206).

Strategies for Action: Strengthening Marriage and Alleviating Problems of Divorce

Two general strategies for responding to the problems of divorce are those that prevent divorce by strengthening marriages and those that strengthen postdivorce families.

Strategies to Strengthen Marriage and Prevent Divorce

A growing "marriage movement" involves efforts to strengthen marriage and prevent divorce through a number of strategies, including premarital and marriage education, covenant marriage and divorce law reform, and provision of workplace and economic supports. Amato et al. (2007) explained,

> Policies to strengthen marital quality and stability are based on consistent evidence that happy and stable marriages promote the health, psychological well-being, and financial security of adults . . . as well as children. . . . Moreover, recent research suggests that a large proportion of marriages that end in divorce are not deeply troubled, and that many of these marriages might be salvaged if spouses sought assistance for relationship problems . . . and stayed the course through difficult times. (pp. 245–46)

Marriage Education Marriage education, also known as family life education, includes various types of workshops and classes that (1) teach relationship skills, communication, and problem solving; (2) convey the idea that sustaining healthy marriages requires effort; and (3) convey the importance of having realistic expectations of marriage, commitment, and a willingness to make personal sacrifices (Hawkins et al. 2004). An alternative or supplement to face-to-face family life education is web-based education, such as the Forever Families website, a faith-based family education website (Steimle & Duncan 2004).

The Healthy Marriage Initiative provides federal funds to support research and programs to encourage healthy marriages and promote involved and responsible fatherhood. Funds may be used for a variety of activities, including marriage and premarital education, public advertising campaigns that promote healthy marriage, high school programs on the value of marriage, marriage mentoring programs, and parenting skills programs. Some states have passed or considered legislation that requires marriage education in high schools or that provides incentives (such as marriage license fee reductions) to couples who complete a marriage education program.

Covenant Marriage and Divorce Law Reform With the passing of the 1996 Covenant Marriage Act, Louisiana became the first state to offer two types of marriage contracts: (1) the standard marriage contract that allows a no-fault divorce (after a six-month separation), or (2) a **covenant marriage**, which permits divorce only under condition of fault (e.g., abuse, adultery, or felony conviction) or after a two-year separation. Couples who choose a covenant marriage must also get premarital counseling. Variations of the covenant marriage have also been adopted in Arizona and Arkansas. Only 3 percent of couples in states with covenant marriage laws have chosen the covenant marriage option (Coontz 2005b).

The covenant marriage option was designed to strengthen marriages and decrease divorce. However, critics argue that covenant marriage may increase family problems by making it more emotionally and financially difficult to terminate a problematic marriage and by prolonging the exposure of children to parental conflict (Applewhite 2003).

What Do You Think? Several states have considered reforming divorce laws to make divorce harder to obtain by extending the waiting period required before a divorce is granted or requiring proof of fault (e.g., adultery or abuse). Do you think that making divorce more difficult to obtain would help keep families together? Or would it hurt families by increasing the conflict between divorcing spouses (which harms the children as well as the adults involved) and the legal costs of getting a divorce (which leaves less money to support any children)?

Workplace and Economic Supports The most important pro-marriage and divorce-prevention measures may be those that maximize employment and earnings. Given that research finds a link between financial hardship and marital quality, policies to strengthen marriage should include a focus on the economic well-being of poor and near-poor couples and families (Amato et al. 2007).

> The most important pro-marriage and divorce-prevention measures may be those that maximize employment and earnings.

"Policy makers should recognize that any initiative that improves the financial security and well-being of married couples is a pro-marriage policy" (Amato et al. 2007, p. 256). Supports such as job training, employment assistance, flexible workplace policies that decrease work–family conflict, affordable child care, and economic support, such as the earned income tax credit are discussed in Chapters 6 and 7. In addition, policy makers should take a hard look at policies that penalize poor couples for marrying. Poor couples who marry are often penalized by losing Medicaid benefits, food stamps, and other forms of assistance.

Strategies to Strengthen Families During and After Divorce

When one or both marriage partners decide to divorce, what can the couple do to minimize the negative consequences of divorce for themselves and their children? According to Ahrons (2004), the postdivorce conflict between parents and not the divorce itself is most traumatic for children. A review of the literature on the effects of parental conflict

covenant marriage A type of marriage (offered in a few states) that requires premarital counseling and that permits divorce only under condition of fault or after a marital separation of more than two years.

on children suggests that children who are exposed to high levels of parental conflict are at risk for anxiety, depression, and disruptive behavior; they are more likely to be abusive toward romantic partners in adolescence and adulthood and are likely to have higher rates of divorce and maladjustment in adulthood (Grych 2005). Next, we discuss ways in which divorced (or divorcing) parents can minimize the conflict with their ex-spouse and develop a cooperative parenting relationship.

Forgiveness Research suggests that divorced parents who forgive each other are more likely to have positive and cooperative co-parenting after divorce (Bonach 2009). Forgiveness does not mean accepting, condoning, or excusing the offender. Rather, forgiveness involves making a choice to think, feel, and behave less negatively toward someone who has hurt or offended you and to act with good will toward the offender (Fincham et al. 2006).

Divorce Mediation In **divorce mediation,** divorcing couples meet with a neutral third party, a mediator, who helps them resolve issues of property division, child support, child custody, and spousal support (i.e., alimony) in a way that minimizes conflict and encourages cooperation. In a longitudinal study, researchers compared two groups of divorcing parents who were petitioning for a court custody hearing: parents who were randomly assigned to try mediation and those who were randomly assigned to continue the adversarial court process (Emery et al. 2005). If mediation did not work, the parents in the mediation group could still go to court to resolve their case. The parents who participated in mediation were much more likely to settle their custody dispute outside court than the parents who did not. The researchers found that mediation can not only speed settlement, save money on attorney and court fees, and increase compliance, it can also result in improved relationships between nonresidential parents and children, as well as between divorced parents 12 years after the dispute settlement. An increasing number of jurisdictions and states have mandatory child custody mediation programs, whereby parents in a custody or visitation dispute must attempt to resolve their dispute through mediation before a court will hear the case.

Divorce Education Programs Divorce education programs are designed to help parents who are divorced or planning to divorce reduce parental conflict and educate them about the factors that affect their children's adjustment. Parents are taught how to respond to their children's reactions to divorce, and how to cooperate in co-parenting. An experimental research study on the effectiveness of a court-ordered education program for divorcing parents found that participation in the program had a significant positive effect on co-parenting skills and parent relationship for both mothers and fathers (Whitehurst et al. 2008).

What Do You Think? Many counties and some states (e.g., Arizona and Hawaii) require divorcing spouses to attend a divorce education program, whereas it is optional in other jurisdictions. Do you think that parents of minor children should be required to complete a divorce education program before they can get a divorce? Why or why not?

divorce mediation A process in which divorcing couples meet with a neutral third party (mediator) who assists the individuals in resolving issues such as property division, child custody, child support, and spousal support in a way that minimizes conflict and encourages cooperation.

Understanding Family Problems

Family problems can best be understood within the context of the society and culture in which they occur. Although domestic violence and divorce may appear to result from individual decisions, myriad social and cultural forces influence these decisions.

The impact of family problems, including divorce and abuse, is felt not only by family members but also by society at large. Family members experience life difficulties such as poverty, school failure, low self-esteem, and mental and physical health problems.

Each of these difficulties contributes to a cycle of family problems in the next generation. The impact on society includes public expenditures to assist single-parent families and victims of domestic violence and neglect, increased rates of juvenile delinquency, and lower worker productivity.

For some, the solution to family problems implies encouraging marriage and discouraging other family forms, such as single parenting, cohabitation, and same-sex unions. However, many family scholars argue that the fundamental issue is making sure that children are well cared for, regardless of their parents' marital status or sexual orientation. Some even suggest that marriage is part of the problem, not part of the solution. Martha Fineman of Cornell Law School said, "This obsession with marriage prevents us from looking at our social problems and addressing them. . . . Marriage is nothing more than a piece of paper, and yet we rely on marriage to do a lot of work in this society: It becomes our family policy, our policy in regard to welfare and children, the cure for poverty" (quoted by Lewin 2000, p. 2).

Strengthening marriage is a worthy goal because strong marriages offer many benefits to individuals and their children. However, "strengthening marriage does not have to mean a return to the patriarchal family of an earlier era. . . . Indeed, greater marital stability will only come about when men are willing to share power, as well as housework and child care, equally with women" (Amato 1999, p. 184). Strengthening marriage does not mean that other family forms should not also be supported. The reality is that the postmodern family comes in many forms, each with its strengths, needs, and challenges. Given the diversity of families today, social historian Stephanie Coontz (2004) suggested that "the appropriate question . . . is not what single family form or marriage arrangement we would prefer in the abstract, but how we can help people in a wide array of different committed relationships minimize their shortcomings and maximize their solidarities" (p. 979). She further argued that

> If we withdrew our social acceptance of alternatives to marriage, marriage itself might suffer. . . . The same personal freedoms that allow people to expect more from their married lives also allow them to get more out of staying single and give them more choice than ever before in history about whether or not to remain together. (Coontz 2005a, p. 310)

The family problems emphasized in this chapter—domestic violence and abuse, and problems of divorce—have something in common: Economic hardship and poverty can be a contributing factor and a consequence of each of these problems. In the next chapter, we turn our attention to poverty and economic inequality, problems that are at the heart of many other social ills.

CHAPTER REVIEW

- **What are some examples of diversity in families around the world?**
  Some societies recognize monogamy as the only legal form of marriage, whereas other societies permit polygamy. Societies also vary in their policies regarding same-sex couples and their norms regarding childbearing and the roles of women and men in the family.

- **What are some of the major changes in U.S. families that have occurred in the past several decades?**
  Some of the major changes in U.S. families that have occurred in recent decades include increased singlehood and older age at first marriage, increased heterosexual and same-sex cohabitation, the emergence of living apart

together (LAT) relationships, increased births to unmarried women, increased divorce and blended families, and increased employment of married mothers.

- **What is the marital decline perspective? What is the marital resiliency perspective?**
  According to the marital decline perspective, the recent transformations in American families signify a collapse of marriage and family in the United States. According to the marital resiliency perspective, poverty, unemployment, poorly funded schools, discrimination, and the lack of basic services (such as health insurance and child care) are more harmful to the well-being of children and adults than is the decline in married two-parent families.

- **Feminist theories of family are most similar to which of the three main sociological theories: structural functionalism, conflict theory, or symbolic interactionism?**

  Feminist theories of family are most aligned with conflict theory. Both feminist and conflict theories are concerned with how gender inequality influences and results from family patterns.

- **What are the four patterns of partner violence that Johnson and Ferraro (2003) identified?**

  The four patterns of partner violence are (1) common couple violence (occasional acts of violence arising from arguments that get "out of hand"); (2) intimate terrorism (violence that is motivated by a wish to control one's partner); (3) violent resistance (acts of violence that are committed in self-defense); and (4) mutual violent control (both partners battling for control).

- **Why do some abused adults stay in abusive relationships?**

  Adult victims of abuse are commonly blamed for choosing to stay in their abusive relationships. From the point of view of the victim, reasons to stay in the relationship include economic dependency, emotional attachment, commitment to the relationship, hope that things will get better, the view that violence is legitimate because they "deserve" it, guilt, and fear. Some victims with children leave to protect the children, but others stay because they need help raising the children or fear losing custody of them if they leave. Some victims stay because they fear the abuser will abuse or neglect a pet.

- **What is the most common form of child abuse?**

  The most common form of child abuse is neglect.

- **What are some of the effects of divorce on children?**

  Reviews of recent research on the consequences of divorce for children find that children with divorced parents score lower on measures of academic success, psychological adjustment, self-concept, social competence, and long-term health, and that they have higher levels of aggressive behavior and depression. Such effects are related to the economic hardship associated with divorce, the reduced parental supervision resulting from divorce, and parental conflict during and after divorce. In highly conflictual marriages, divorce may actually improve the emotional well-being of children relative to staying in a conflicted home environment.

- **What is divorce mediation?**

  In divorce mediation, divorcing couples meet with a neutral third party, a mediator, who helps them resolve issues of property division, child custody, child support, and spousal support in a way that minimizes conflict and encourages cooperation. In some states, counties, and jurisdictions, divorcing couples who are disputing child custody issues are required to participate in divorce mediation before their case can be heard in court.

TEST YOURSELF

1. The United States has the highest nonmarital birthrate of any country in the world.
   a. True
   b. False
2. Two perspectives on the state of marriage in the United States are the marital decline perspective and the marital ___ perspective.
   a. health
   b. resiliency
   c. incline
   d. stability
3. A study on how marriage has changed found that, between 1980 to 2000, marriages
   a. have become more prone to divorce
   b. have become more conflictual
   c. involve less interaction between husband and wife
   d. all of the above
4. In the United States, people are more likely to be physically assaulted, sexually assaulted and molested, or killed by ___ than by anyone else.
   a. a family member
   b. an employee
   c. a stranger
   d. a friend
5. A survey of college students found that, in the past 12 months, students were much more likely to have experienced an emotionally abusive relationship than a physically or sexually abusive relationship.
   a. True
   b. False
6. Which of the following is the most prevalent form of abuse in families?
   a. Sexual abuse by a father
   b. Sexual abuse by an uncle or cousin
   c. Verbal abuse by a mother
   d. Sibling abuse
7. How many states recognize no-fault divorce?
   a. None
   b. 5
   c. 25 and the District of Columbia
   d. All 50
8. Most U.S. adults believe that divorce is morally acceptable.
   a. True
   b. False
9. Parental alienation occurs in both divorced and intact families.
   a. True
   b. False
10. The purpose of divorce mediation is to help couples who are considering divorce repair their relationship and stay together.
    a. True
    b. False

Answers: 1: b; 2: b; 3: c; 4: a; 5: a; 6: d; 7: d; 8: a; 9: a; 10: b.

KEY TERMS

bigamy   136
child abuse   150
corporal punishment   154
covenant marriage   163
cycle of abuse   149
divorce mediation   164
domestic partnership   139
elder abuse   151
family   136

individualism   158
intimate partner violence (IPV)   146
living apart together (LAT)
    relationships   139
marital decline perspective   142
marital resiliency perspective   142
monogamy   136
neglect   150
no-fault divorce   158

parental alienation   161
patriarchy   144
polyandry   136
polygamy   136
polygyny   136
refined divorce rate   140
second shift   158
serial monogamy   136
shaken baby syndrome   150

MEDIA RESOURCES

Turning to Video

▶ ❚❚   Watch the BBC video *Divorced Dads* (running time 1:59), available through **CengageBrain.com.** This video features a father who explains how getting divorced has made him a better parent. As you watch the video, think about this: What factors help divorced dads be good parents? What hinders them?

Online Study Resources

Log in to **www. cengagebrain.com** to access the resources your instructor has assigned. For this book, you can access:

CourseMate

Access chapter-specific learning tools including learning objectives, practice quizzes, videos, Internet exercises, flash cards, and glossaries, as well as web links, and more in your Sociology CourseMate.

© Irving Olson, 2007

6

Poverty and Economic Inequality

"We are the first generation that can look extreme poverty in the eye, and say this and mean it—we have the cash, we have the drugs, we have the science. Do we have the will to make poverty history?"

—Bono, U2 (rock music group)

Bettmann/CORBIS

Washington, D.C., the capital of one of the wealthiest nations in the world, has one of the highest rates of poverty in the United States.

NOT FAR FROM the Capitol Building, 60-year-old John Treece pondered his life in deep poverty as he left a local food pantry with two bags of free groceries. Plagued by arthritis and back problems from years of manual labor, Treece has been unable to find a full-time job for 15 years. He's tried to get Social Security disability benefits, but the Social Security Administration disputes his injuries and work history. Treece earns a little more than $5,000 a year doing odd jobs. His clothes are tattered and he lives in a $450-a-month room in a boarding house in a high-crime neighborhood. Treece does not go hungry, thanks to food stamps, the food pantry, and help from relatives. But items that require cash, such as toothpaste, soap, and toilet paper, are harder to come by. "Sometimes it makes you want to do the wrong thing, you know," said Treece, referring to crime. "But I ain't a kid no more. I can't do no time. At this point, I ain't got a lotta years left." Despite his poor circumstances, Treece is positive and grateful for what he has. "I don't ask for nothing. . . . I just thank the Lord for this day and ask that tomorrow be just as blessed" (quoted in Pugh 2007).

In this chapter, we examine the extent of poverty globally and in the United States, focusing on the consequences of poverty for individuals, families, and societies. We present theories of poverty and economic inequality and consider strategies for rectifying economic inequality and poverty.

The Global Context: Poverty and Economic Inequality around the World

Who are the poor? The answer depends on how we define and measure poverty.

Defining and Measuring Poverty

Absolute poverty refers to the lack of resources necessary for well-being—most importantly, food and water, but also housing, sanitation, education, and health care. In contrast, **relative poverty** refers to the lack of material and economic resources compared with some other population. If you are a struggling college student living on a limited budget, you may feel as though you are "poor" compared with the middle- or upper-middle-class lifestyle to which you may aspire. However, if you have a roof over your head; access to clean water, toilets, and medical care; and enough to eat, you are not absolutely poor; indeed, you have a level of well-being that millions of people living in absolute poverty may never achieve.

Measures of Poverty The World Bank sets a "poverty threshold" of $1.25 per day for the least developed countries, and $2 per day for middle-income countries and regions, including Latin America and eastern Europe. In 2010, 15.8 percent of the world's population lived in **extreme poverty**—less than $1.25 a day (Chandy & Gertz 2011).

A measure of relative poverty is based on comparing the income of a household to the median household income in a specific country. According to this relative poverty measure, members of a household are considered poor if their household income is less than 50 percent of the median household income in that country.

absolute poverty The lack of resources necessary for material well-being—most importantly, food and water, but also housing, sanitation, education, and health care.

relative poverty The lack of material and economic resources compared with some other population.

extreme poverty Living on less than $1.25 a day.

People in this tented village in New Delhi, India, live in extreme poverty.

Low income is only one indicator of impoverishment. The **Multidimensional Poverty Index** is a measure of serious deprivation in the dimensions of health, education, and living standards that combines the number of deprived and the intensity of their deprivation (see Table 6.1). Data from 104 countries (with 78 percent of the world's population) revealed that a third of the world's population (1.7 billion) live in multidimensional poverty (UNDP 2010). Half of the world's poor, as measured by MPI, live in South Asia, though rates are highest in sub-Saharan Africa. In Niger, the country with the highest rate of multidimensional poverty, 93 percent of the people are multidimensionally poor.

Another measure of poverty is the proportion of people who consume less than a minimum amount of calories to maintain health—who are chronically hungry or undernourished. About one billion people do not consume adequate calories. Of these, only one percent lives in developed countries; 63 percent are in Asia and the Pacific, and 26 percent are in sub-Saharan Africa (UNDP 2010).

TABLE 6.1 Three Dimensions of Multidimensional Poverty
1. Health (nutrition, child mortality)
2. Education (years of schooling, children enrolled)
3. Standard of living (cooking fuel, toilet, water, electricity, floor, assets)

What Do You Think? The next section describes how poverty is measured in the United States by comparing the annual pretax income of a household to the official U.S. poverty line—a dollar amount that determines who is considered poor. Before reading further, answer these questions: How much annual income do you think a household with one adult needs to earn to avoid living in poverty? What about a household with two adults? One adult and one child? Two adults and one child? Compare your answers with the official poverty thresholds in Table 6.2.

U.S. Measures of Poverty In 1964, the Social Security Administration devised a poverty index based on data that indicated that families spent about one-third of their income on food. The official poverty level was set by multiplying food costs by three. Since then, the poverty level has been updated annually for inflation, and differs by the number of adults and children in a household and by the age of the head of household, but is the same across the continental United States (see Table 6.2). Anyone living in a household with pretax income below the official poverty line is considered "poor." Individuals living in households with incomes that are above the poverty line, but not very much above it, are classified as "near-poor," and those living in households with income below 50 percent of the poverty line live in "deep poverty," also referred to as "severe poverty." A common working definition of "low-income" households is households with incomes that are between 100 percent and 200 percent of the federal poverty line or up to twice the poverty level.

Multidimensional Poverty Index A measure of serious deprivation in the dimensions of health, education, and living standards that combines the number of deprived and the intensity of their deprivation.

The U.S. poverty line has been criticized on several grounds. First, the official poverty line is based on pretax income so tax burdens, as well as tax credits, are disregarded. Family wealth, including savings and property, are also excluded in official poverty calculations, and noncash government benefits that assist low-income families—food stamps, Medicaid, and housing and child care assistance—are not taken into account. In addition, the current poverty measure is a national standard that does not reflect the significant variation in the cost of living from state to state and between urban and rural areas. Finally, the poverty line underestimates the extent of material hardship in the United States because it is based on the assumption that low-income families spend one-third of their household income on food. That was true in the 1950s, but because housing, medical care, child care, and transportation costs have risen more rapidly than food costs, low-income families today spend far less than one-third of their income on food.

When a 2007 Gallup Poll asked the American public to estimate the minimum amount of yearly income a family of four would need "to get along in your local community," the average answer was $52,000 (rounded to the nearest thousand) (Jones 2007). The **Basic Economic Security Tables Index (BEST),** a measure of the basic needs and income workers require for economic security, finds that a family with two working parents (who receive employment-based benefits) and two young children needs to earn $67,920 a year, or about $16 an hour per worker (Wider Opportunities for Women 2010). Table 6.3 presents the annual pretax income required to meet basic needs for different family types. Notice that workers without employment benefits need more income than those with benefits. Table 6.4 provides a breakdown of specific expenses that are considered in calculating BEST values.

The Extent of Global Poverty and Economic Inequality

Between 1980 and 2005, the percentage of people in the developing world who lived in extreme poverty dropped from more than half to about a quarter. Extreme poverty dropped further to about 16 percent in 2010, and is projected to be less than 10 percent by 2015 (see Table 6.5) (Chandy & Gertz 2011). Although the poverty reductions shown in Table 6.5 are not insignificant, keep in mind that these statistics refer to the percent of people in *extreme* poverty—those living on less than $1.25 a day. Many more people live on between $1.25 and $2.00 a day, and then many more live on not much more above that.

Basic Economic Security Tables Index (BEST) A measure of the basic needs and income workers require for economic security.

TABLE 6.2 Poverty Thresholds, 2010 (Householder Younger Than 65 Years)	
HOUSEHOLD MAKEUP	POVERTY THRESHOLD
One adult	$11,344
Two adults	$14,602
One adult, one child	$15,030
Two adults, one child	$17,552
Two adults, two children	$22,113

Source: U.S. Census Bureau 2011a.

TABLE 6.3 Basic Economic Security Tables, 2010

U.S., by Family Type and Receipt of Employment-Based Benefits

1 WORKER		1 WORKER, 1 INFANT		1 WORKER, 1 PRESCHOOLER, 1 SCHOOLCHILD		2 WORKERS, 1 PRESCHOOLER, 1 SCHOOLCHILD	
Workers with Employment-Based Benefits	Workers without Employment-Based Benefits	Workers with Employment-Based Benefits	Workers without Employment-Based Benefits	Workers with Employment-Based Benefits	Workers without Employment-Based Benefits	Workers with Employment-Based Benefits	Workers without Employment-Based Benefits
$30,012	$34,728	$46,438	$53,268	$57,756	$63,012	$67,920	$73,296

Notes: "Benefits" include unemployment insurance and employment-based health insurance and retirement plans.

Source: Wider Opportunities for Women. 2010. *The Basic Economic Security Tables for the United States.* Washington, D.C.: Wider Opportunities for Women.

TABLE 6.4 Expenses Involved in Calculating Basic Economic Security Tables*, 2010

(WORKERS WITH EMPLOYMENT-BASED BENEFITS)

United States
Monthly Expenses for: 1 Worker, 1 infant

Housing	$821
Utilities	$178
Food	$351
Transportation	$536
Child Care	$610
Personal & Household Items	$364
Health Care	$267
Emergency Savings	$116
Retirement Savings	$73
Taxes	$720
Tax Credits	−$172
Monthly Total	**$3,864**
Annual Total	**$46,368**
Hourly Wage	**$21.95**
Additional Asset Building Savings	
Children's Higher Education	$43
Homeownership	$130

*For U.S. workers with employment-based benefits.

Source: Wider Opportunities for Women. 2010. *The Basic Economic Security Tables for the United States.* Washington, D.C.: Wider Opportunities for Women.

TABLE 6.5 Regional and Global Extreme Poverty*, 2005, 2010, 2015

	2005	2010	2015
East Asia	16.8%	7.4%	2.7%
Europe and Central Asia	3.4%	1.8%	0.9%
Latin America and Caribbean	8.4%	6.2%	4.5%
Middle East and North Africa	3.8%	2.5%	1.9%
South Asia	40.2%	20.3%	8.7%
Sub-Saharan Africa	54.5%	46.9%	39.3%
WORLD	25.7%	15.8%	9.9%

*Living on less than $1.25 a day.
Source: Chandy and Gertz 2011.

wealth The total assets of an individual or household minus liabilities.

With so much wealth in the world, it is hard to fathom that so many people live in poverty. Economic inequality is a well-established feature of our social world. In 2000, the average income of the richest 20 countries was 37 times that of the poorest 20 countries—a gap that doubled since 1960 (World Bank 2001). From 1960 to 2002, income per person in the world's poorest countries rose only slightly, from $212 to $267, whereas income in the richest 20 nations tripled from $11,417 to $32,339 (Schifferes 2004).

The United States has the greatest degree of income inequality and the highest rate of poverty of any industrialized nation. From 1979 to 2007, Americans in the top 10 percent of the income distribution received nearly two-thirds of gains to overall incomes; the top one percent alone received 38.7 percent of overall gains in income (Economic Policy Institute 2011). Between 1979 and 2009, real family income rose 49 percent for the top 20 percent (earning $112,540 and up); rose only 11 percent for the middle 20 percent (earning $47,914 to $73,338); and declined 7 percent for the bottom 20 percent (earning less than $26,934) (Hartman 2011).

Another example of economic inequality in the United States is the gap between the compensation (salaries, bonuses, stock options, and so on) of chief executive officers (CEOs) and the average employee. In 2009, CEOs at the top 500 U.S. corporations received an average of nearly $8.5 million—263 times the average compensation of U.S. workers (Anderson et al., 2010). That means that a typical worker would have to work 263 years to earn what a CEO makes in one year! Table 6.6 shows the dramatic increase in the ratio of CEO pay to average worker pay since 1965.

The inequality in the distribution of global wealth is even greater than income inequality. **Wealth** refers to the total assets of an individual or household minus liabilities (mortgages, loans, and debts). Wealth includes the value of a home, investment real estate, cars, unincorporated business, life insurance (cash value), stocks, bonds, mutual funds, trusts, checking and savings accounts, individual retirement accounts (IRAs), and valuable collectibles.

A comprehensive study on the world distribution of household wealth revealed that:

- The richest one percent of adults in the world own 40 percent of global household wealth; the richest 2 percent of adults own more than half of global wealth; and the richest 10 percent of adults own 85 percent of total global wealth.
- The poorest half of the world adult population owns barely one percent of global wealth.
- Households with per-adult assets of $2,200 are in the top half of the world wealth distribution, assets of $61,000 per adult places a household in the top 10 percent, and assets of more than $500,000 per adult places a household in the richest one percent worldwide.

What Do You Think? Researchers asked citizens of various nations about their perceptions of income inequality in their country, and then compared those responses to the actual measurements of income inequality (Forster & d'Ercole 2005). Among 17 advanced countries, U.S. citizens had the largest gap—by a wide margin—between their perception of inequality and its reality. Why do you think Americans perceive the level of inequality in the United States to be much less than what it actually is?

- Although North America has only 6 percent of the world adult population, it accounts for one-third (34 percent) of all household wealth worldwide. More than one-third (37 percent) of the richest one percent of individuals in the world resides in the United States (Davies et al. 2006).

In the United States, the wealthiest one percent of U.S. households owns 35.6 percent of all private wealth—nearly as much wealth as the entire bottom 95 percent (Economic Policy Institute 2011). As shown in Figure 6.1, median net worth (i.e., wealth) is much higher among white Americans than among African Americans and Hispanic Americans.

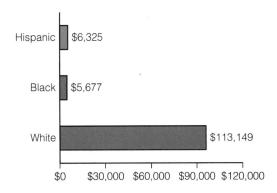

Figure 6.1: Median Household Worth by Race/Hispanic Origin, 2009
Source: Pew Research Center 2011. From "Wealth Gaps Rise to Record Highs Between Whites, Blacks, Hispanics", July 26, 2011, Social and Demographic Trends, a project of Pew Research Center. Reprinted with permission.

Sociological Theories of Poverty and Economic Inequality

The three main theoretical perspectives in sociology—structural functionalism, conflict theory, and symbolic interactionism—offer insights into the nature, causes, and consequences of poverty and economic inequality.

Structural-Functionalist Perspective

According to the structural-functionalist perspective, poverty results from institutional breakdown: economic institutions that fail to provide sufficient jobs and pay, educational institutions that fail to equip members of society with the skills they need for employment, family institutions that do not provide two parents, and government institutions that do not provide sufficient public support. Sociologist William Julius Wilson explains:

> Where jobs are scarce . . . and where there is a disruptive or degraded school life purporting to prepare youngsters for eventual participation in the workforce, many people eventually lose their feeling of connectedness to work in the formal economy; they no longer expect work to be a regular, and regulating, force in their lives. . . . These circumstances also increase the likelihood that the residents will rely on illegitimate sources of income, thereby further weakening their attachment to the legitimate labor market. (Wilson 1996, pp. 52–53)

From a structural-functionalist perspective, economic inequality within a society can be beneficial for society, as a system of unequal pay motivates people to achieve higher levels of training and education and to take on jobs that are more important and difficult by offering higher rewards for higher achievements (Davis & Moore 1945). However, this argument is criticized on the grounds that many important occupational

TABLE 6.6 Ratio of CEO Pay to Average Worker Pay, U.S., 1965–2009	
1965	24:1
1978	35:1
1989	70:1
2000	298:1
2002	143:1
2009	263:1

Sources: Anderson et al., 2010; Mishel et al. 2009.

roles, such as child care workers and nurse assistants, have low salaries, whereas many individuals in nonessential roles (e.g., professional sports stars and entertainers) earn outrageous sums of money. The structural-functionalist argument that CEO pay is high to reward high performance is shattered by the fact that CEOs are paid huge salaries and bonuses even when they contribute to the economic failure of their corporation and/or to the problem of unemployment. Consider that, in 2009, CEOs of the 50 companies that laid off the most workers since the economic crisis earned 42 percent more than the average CEO pay at S&P 500 companies (Anderson et al. 2010).

The structural-functionalist perspective also points to the ways in which poverty is functional for society (Gans 1972). For example, although poverty-level wages hurt workers and their families, such wages allow industries to flourish. Low-wage labor also allows the affluent to hire gardeners, child care workers, and housecleaners. Poverty also provides a market for used goods, such as clothing and household appliances and furnishings that are purchased by or donated to the poor. By providing a market for used goods, the poor help to lessen the burden on landfills.

Conflict Perspective

Karl Marx (1818–1883) proposed that economic inequality results from the domination of the *bourgeoisie* (owners of the factories, or "means of production") over the *proletariat* (workers). The bourgeoisie accumulate wealth as they profit from the labor of the proletariat, who earn wages far below the earnings of the bourgeoisie. Modern conflict theorists recognize that the power to influence economic outcomes comes not only from ownership of the means of production but also from management position, interlocking board memberships, control of media, and financial contributions to politicians.

Following a report that GE paid no federal taxes in 2010, liberal groups rallied for GE CEO Jeffrey R. Immelt to resign from his position as the head of the White House's Council on Jobs and Competitiveness.

mk/mk/Newscom

For example, wealthy corporations use financial political contributions to influence politicians to enact policies that benefit the wealthy. Laws and policies that favor the rich, such as tax breaks that benefit the wealthy, are sometimes referred to as **wealthfare.** For example, tax and accounting loopholes save CEOs a total of $20 billion per year (Anderson et al. 2008). Households with incomes over $1 million in 1961 paid an average of 43.1 percent of their incomes in federal income taxes. Due to tax policy that favors the wealthy, households making $1 million or more in 2010 paid 23 percent (Goldberg et al. 2011).

Corporate welfare refers to laws and policies that benefit corporations, such as low-interest government loans to failing businesses and special subsidies and tax breaks to corporations. For example, in 2010, General Electric (GE) earned $14.2 billion in profits, but not only did it pay no federal income tax, it received a $3.2 billion tax credit (Kocieniewski 2011). Many large, profitable corporations take advantage of tax shelters to avoid paying taxes. Although the official U.S. corporate tax rate is 35 percent, from 2008 to 2010, Chevron paid less than 5 percent a year, Merck paid 5 percent, Hewlett-Packard paid 3 percent, Exxon and IBM paid 2 percent, and Carnival paid 1 percent. In the same three-year period, Verizon, Boeing, Dow, and DuPont made profits, but all paid zero taxes. And Citigroup and Bank of America, with a combined $8 billion of pretax earnings in 2009 and 2010, each paid zero taxes two years in a row (Buchheit 2011). Indeed, 100 of the largest U.S. companies paid less than 10 percent of pretax profits in federal income tax in 2010. If these 100 companies had paid the 35 percent official corporate tax rate, an additional $150 billion would have been collected in federal taxes in one year—an amount approximately equal to the total budget deficits for all 50 states (Buchheit 2011).

Conflict theorists also note that, throughout the world "free market," economic reform policies have been hailed as a solution to poverty. Yet, although such economic reform

wealthfare Laws and policies that benefit the rich.

corporate welfare Laws and policies that benefit corporations.

Being poor is knowing exactly how much everything costs.

Being poor is getting angry at your kids for asking for all the crap they see on TV.

Being poor is having to keep buying $800 cars because they're what you can afford, and then having the cars break down on you, because there's not an $800 car in America that's worth a damn.

Being poor is hoping the toothache goes away.

Being poor is knowing your kid goes to friends' houses but never has friends over to yours.

Being poor is going to the restroom before you get in the school lunch line so your friends will be ahead of you and won't hear you say "I get free lunch" when you get to the cashier.

Being poor is living next to the freeway.

Being poor is coming back to the car with your children in the backseat, clutching that box of Raisin Bran you just bought and trying to think of a way to make the kids understand that the box has to last.

Being poor is a heater in only one room of the house.

Being poor is hoping your kids don't have a growth spurt.

Being poor is Goodwill underwear.

Being poor is your kid's school being the one with the 15-year-old textbooks and no air conditioning.

Being poor is thinking $8 an hour is a really good deal.

Being poor is finding the letter your mom wrote to your dad, begging him for the child support.

Being poor is stopping the car to take a lamp from a stranger's trash.

Being poor is making lunch for your kid when a cockroach skitters over the bread, and you looking over to see if your kid saw.

Being poor is not taking the job because you can't find someone you trust to watch your kids.

Being poor is not talking to that girl because she'll probably just laugh at your clothes.

Being poor is needing that 35-cent raise.

Being poor is six dollars short on the utility bill and no way to close the gap.

Being poor is crying when you drop the mac and cheese on the floor.

Being poor is people surprised to discover you're not actually stupid or lazy.

Being poor is a six-hour wait in an emergency room with a sick child asleep on your lap.

Being poor is knowing you're being judged.

Being poor is a box of crayons and a $1 coloring book from a community center Santa.

Being poor is knowing you really shouldn't spend that buck on a Lotto ticket.

Being poor is a $200 paycheck advance from a company that takes $250 when the paycheck comes in.

Being poor is four years of night classes for an associates of art degree.

Being poor is people who have never been poor wondering why you choose to be so.

Being poor is seeing how few options you have.

Being poor is knowing how hard it is to stop being poor.

Source: Adapted from John Scalzi. 2005 (September 3). "Being Poor." Copyright © 2005 by John Scalzi. Reprinted by permission. www.whatever.scalzi.com

has benefited many wealthy corporations and investors, it has also resulted in increasing levels of global poverty. As companies relocate to countries with abundant supplies of cheap labor, wages decline. Lower wages lead to decreased consumer spending, which leads to more industries closing plants, going bankrupt, and/or laying off workers (downsizing). These actions result in higher unemployment rates and a surplus of workers, enabling employers to lower wages even more.

Symbolic Interactionist Perspective

Symbolic interactionism focuses on how meanings, labels, and definitions affect and are affected by social life. This view calls attention to ways in which wealth and poverty are defined and the consequences of being labeled "poor." Individuals who are viewed as poor—especially those receiving public assistance (i.e., welfare)—are often stigmatized as lazy, irresponsible, and lacking in abilities, motivation, and moral values. Wealthy individuals, on the other hand, tend to be viewed as capable, motivated, hardworking, and deserving of their wealth.

The symbolic interactionist perspective also focuses on the meanings of being poor. A qualitative study of more than 40,000 poor women and men in 50 countries revealed that the experience of poverty involves psychological dimensions such as powerlessness, voicelessness, dependency, shame, and humiliation (Narayan 2000). This chapter's *The Human Side* feature conveys what it means to be poor.

Meanings and definitions of wealth and poverty vary across societies and across time. Although many Americans think of poverty in terms of income level, for millions of people, poverty is not primarily a function of income but of their alienation from sustainable

Photo courtesy of the author

Children are more likely than adults to live in poverty.

patterns of consumption and production. For indigenous women living in the least developed areas of the world, poverty and wealth are determined primarily by access to and control of their natural resources (such as land and water) and traditional knowledge, which are the sources of their livelihoods (Susskind 2005).

By global standards, the Dinka, the largest ethnic group in the sub-Saharan African country of Sudan, are among the poorest of the poor, being among the least modernized peoples of the world. In Dinka culture, wealth is measured in large part by how many cattle a family owns. Although modernized populations might label the Dinka as poor, the Dinka view themselves as wealthy. As one Dinka elder explained, "It is for cattle that we are admired, we, the Dinka. . . . All over the world, people look to us because of cattle . . . because of our great wealth; and our wealth is cattle" (Deng 1998, p. 107).

Patterns of Poverty in the United States

Although poverty is not as widespread or severe in the United States as it is in many other parts of the world, the United States has the highest rate of poverty among wealthy countries belonging to the Organization for Economic Cooperation and Development (OECD 2011a). In 2010, 46.2 million Americans—15.1 percent of the U.S. population— lived below the poverty line (DeNavas-Walt et al. 2011). More than half (58 percent) of Americans between the ages of 20 and 75 will spend at least one year in poverty, and one in three Americans will experience a full year of extreme poverty at some point in adult life (Pugh 2007).

Age and Poverty

Children are more likely than adults to live in poverty (see Table 6.7). More than one-third (35.5 percent) of the U.S. poor population are children (DeNavas-Walt et al. 2011). Compared with other industrialized countries, the United States has the highest child poverty rate.

What Do You Think? In our sociology classes, we introduce the topic of U.S. poverty by asking students to think of an image of a person who represents poverty in America and to draw that imaginary person. Students are asked to give the person a name (to indicate their sex) and to write down the age of the person. Most students draw a picture of a middle-aged man. Yet U.S. poverty statistics reveal that the higher poverty rates are among women, not men, and among youth, not middle-aged adults. Why do you think the most common image of a U.S. poor person is a middle-aged man?

Sex and Poverty

feminization of poverty The disproportionate distribution of poverty among women.

Women are more likely than men to live below the poverty line—a phenomenon referred to as the **feminization of poverty.** The 2010 poverty rates for U.S. women and men were 16.2 and 14.0, respectively (U.S. Census Bureau 2011b). As discussed in

Chapter 10, women are less likely than men to pursue advanced educational degrees and tend to have low-paying jobs, such as service and clerical jobs. However, even with the same level of education and the same occupational role, women still earn significantly less than men. Women who are minorities and/or who are single mothers are at increased risk of being poor.

Education and Poverty

Education is one of the best insurance policies for protecting an individual against living in poverty. In general, the higher a person's level of educational attainment, the less likely that person is to be poor (see Figure 6.2). The relationship between educational attainment and poverty points to the importance of fixing our educational system so that students from all socioeconomic backgrounds have access to quality education (see also Chapter 8). But we also need to consider the fact that many jobs do not require advanced education. Indeed, most job growth (68 percent) through 2018 will involve jobs that do not require a four-year college degree (Wider Opportunities for Women 2010). Wright and Rogers (2011) suggest that "poverty in a rich society does not simply reflect a failure of equal opportunity to acquire a good education; it reflects a social failure in the creation of sufficient jobs to provide an adequate standard of living for all people regardless of their education or levels of skills" (p. 224).

Family Structure and Poverty

Poverty is much more prevalent among female-headed single-parent households than among other types of family structures (see Figure 6.3). In other industrialized countries, poverty rates of female-headed families are lower than those in the United States. Unlike the United States, other developed countries offer a variety of supports for single mothers, such as income supplements, tax breaks, universal child care, national health care, and higher wages for female-dominated occupations.

Lesbian couples and their families are more likely to be poor than heterosexual couples and their families; gay male couple families have the lowest poverty rate of all family structures. Children in same-sex couple families are twice as likely to be poor as children of married heterosexual couples (Albelda et al. 2009).

Race or Ethnicity and Poverty

As displayed in Figure 6.4, poverty rates are higher among blacks, Hispanics, and Asians than among non-Hispanic whites. As discussed in Chapter 9, past and present discrimination has contributed to the persistence of poverty among minorities. Other contributing factors include the loss of manufacturing

TABLE 6.7 U.S. Poverty Rates by Age, 2010	
AGE (YEARS)	POVERTY RATE
Younger than 18	22.0%
18–24	21.9%
25–34	15.2%
35–44	12.6%
45–54	10.6%
55–59	10.1%
60–64	10.1%
65 and older	9.0%
All ages	15.1%

Source: DeNavas-Walt et al. 2011.

Education is one of the best insurance policies for protecting an individual against living in poverty.

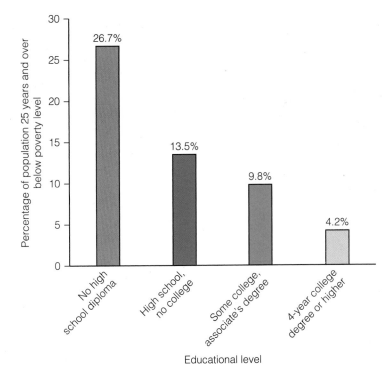

Figure 6.2: **Relationship between Education and Poverty, 2010**
Source: U.S. Census Bureau 2011c.

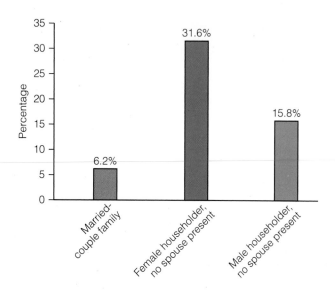

jobs from the inner city, the movement of whites and middle-class blacks out of the inner city, and the resulting concentration of poverty in predominantly minority inner-city neighborhoods (Massey 1991; Wilson 1987, 1996). Finally, blacks and Hispanics are more likely to live in female-headed households with no spouse present—a family structure that is associated with high rates of poverty.

Labor Force Participation and Poverty

A common image of the poor is that they are jobless and unable or unwilling to work. Although the poor in the United States are primarily children and adults who are not in the labor force, many U.S. poor are classified as **working poor**—individuals who spend at least 27 weeks per year in the labor force (working or looking for work), but whose income falls below the official poverty level.

Consequences of Poverty and Economic Inequality

Poverty is associated with health problems and hunger, increased vulnerability from natural disasters, problems in education, problems in families and parenting, and housing problems. These various problems are interrelated and contribute to the perpetuation of poverty across generations, feeding a cycle of intergenerational poverty. In addition, poverty and economic inequality breed social conflict and war.

Health Problems, Hunger, and Poverty

In developing countries, absolute poverty is associated with hunger and malnutrition, high rates of maternal and infant deaths, indoor air pollution from heating and cooking fumes, and unsafe water and sanitation (see this chapter's Photo Essay) (World Health Organization 2002).

In 2010, 925 million people globally were undernourished (FAO 2010). Most of the world's hungry live in Asia and the Pacific (see Figure 6.5). Inadequate nutrition hampers the ability to work and generate income, and can produce irreversible health problems

Figure 6.4: U.S. Poverty Rates by Race and Hispanic Origin, 2010
Source: DeNavas-Walt et al. 2011; DeNavas-Walt et al. 2006.

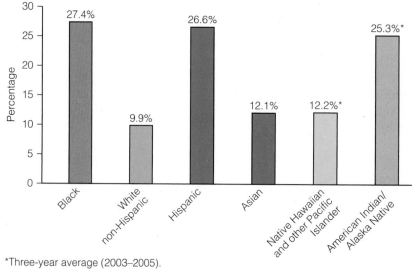

*Three-year average (2003–2005).

working poor Individuals who spend at least 27 weeks per year in the labor force (working or looking for work) but whose income falls below the official poverty level.

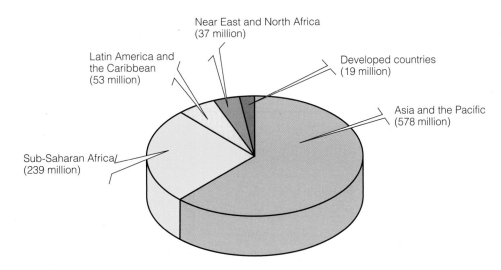

Figure 6.5: **Where Do the World's Hungry Live?** Source: FAO 2010.

Near East and North Africa (37 million)

Latin America and the Caribbean (53 million)

Developed countries (19 million)

Asia and the Pacific (578 million)

Sub-Saharan Africa (239 million)

such as blindness (from vitamin A deficiency) and physical stunting (from protein deficiency) (UNDP 2010).

In the United States, low socioeconomic status is associated with higher incidence and prevalence of health problems, disease, and death (Malatu & Schooler 2002). Hunger in the United States is measured by the percentage of households that are "food insecure," which means that the household had difficulty providing enough food for all its members due to a lack of resources. In 2010, 14.5 percent of U.S. households were food insecure at some time during the year, and 5.4 percent of households had very low food security (some household members had reduced food intake and disrupted normal eating patterns due to lack of resources) (Coleman-Jensen et al. 2011). Assess your own degree of food security in this chapter's *Self and Society* feature.

Poor U.S. children and adults tend to receive inadequate and inferior health care, which exacerbates their health problems. Finally, poverty is linked to higher levels of mental health problems, including stress, depression, and anxiety (Leventhal & Brooks-Gunn 2003).

Economic inequality also affects physical and psychological health. Poor and middle-income adults who live in states with the greatest gap between the rich and the poor are much more likely to rate their own health as poor or fair than people who live in states where income is more equitably distributed (Kennedy et al. 1998). Subjective perceptions of happiness depend more on how an individual's income compares with other people's income than on the actual amount of their income (International Labour Organization 2008a).

Natural Disasters and Poverty

Although natural disasters such as hurricanes, tsunamis, floods, and earthquakes strike indiscriminately—rich and poor alike—the poor are more vulnerable to devastation from such disasters. After a tsunami devastated a large part of South and Southeast Asia in December 2004—killing thousands and tearing apart families, villages, and communities—Oxfam director Barbara Stocking noted that, "it is not mere chance that most of those who died or have been left homeless and destitute were already among the world's poorest. Poor families are always much more severely affected by natural disasters. . . . They live in flimsier homes" (Stocking 2005).

The poor have few to no resources to help them avoid or cope with natural disasters. Many poor people in the path of Hurricane Katrina lacked a means of transportation to evacuate. When the poor lose their homes and their livelihoods in a natural disaster, they do not have the resources to rebuild. For example, when poor fishing communities lost their boats and nets—their very means of survival—in the Asian tsunami, they had no bank accounts or insurance policies to replace their losses.

Although natural disasters such as hurricanes, tsunamis, floods, and earthquakes strike indiscriminately—rich and poor alike—the poor are more vulnerable to devastation from such disasters.

More than one-third of the world's population—2.6 billion people—do not use improved sanitation facilities (which ensure hygienic separation of human excreta from human contact), and 884 million people do not use improved sources of drinking water (water that is likely to be safe to drink) (WHO/UNICEF 2010). Worldwide, 1.1 billion people—17 percent of the world's population—practice open defecation in public places (WHO/ UNICEF 2010). Water-related diseases cause 3 million deaths per year (Oxfam 2006). Each year, 1.5 million children under age 5 die from diarrhea resulting from unsafe water and lack of sanitation (UNICEF 2006). Other diseases associated with lack of clean water and sanitation include trachoma (which causes blindness), typhoid fever, intestinal worms, and guinea worm.

In many poor areas of the world, people drink whatever water is available, and they urinate and defecate in the street, along the roadside, in buckets, or in plastic bags that are tied up and thrown in ditches or along the road. Used in this way, plastic bags are known as "flying toilets."

Three-quarters of the world's rural population obtain water from a communal source, which means family members (usually women and girls) must walk to the water source and carry water back to the family. To collect enough water for drinking, food preparation, personal hygiene, house cleaning, and laundry, a household of five needs at least 32 gallons of water per day (a little more than 6 gallons per person) (Satterthwaite & McGranahan 2007). This is equivalent to carrying six heavy suitcases of water every day.

Cultural norms in many countries require that women not be seen urinating or defecating, which forces them to limit their food and water intake so they can relieve themselves in the dark of night in fields or roadsides. One woman in Bangladesh explained, "Men can answer the call of nature anytime they want . . . but women have to wait until darkness" (United Nations Development Programme 2006, p. 48). Delaying bodily functions can cause liver infection and acute constipation, and going out in darkness to eliminate places women at risk for physical attack.

Improving access to clean water and sanitation for poor populations is one of the most important priorities in the fight against poverty. For every $1 invested in water and sanitation, $3 to $4 are saved in health spending or through increased productivity (Oxfam 2006). More importantly, access to clean water and sanitation is a basic human right that should be enjoyed by every woman, man, and child.

▶ The Musca sorbens fly, which breeds in human feces, causes 2 million new cases of blindness-causing trachoma each year in the developing world. The flies burrow into human eyes, which causes decades of repeat infections. Victims describe the infections as feeling as if they have thorns in their eyes.

Neil Cooper/Alamy

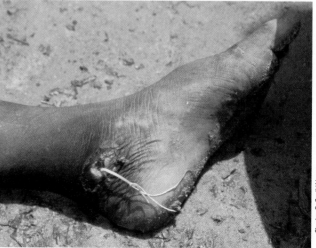

▶ A person gets guinea worm, a parasite, by drinking water contaminated with the larvae. The worm grows up to three feet long in the body and eventually erupts through the skin, causing extremely painful blisters.

Charles O. Cecil/Alamy

◄ Imagine having to share this one toilet with more than 1,000 other people.

Rob Elliott/AFP/Getty Images

▼ Whereas Americans and other wealthy populations take running water for granted, poor populations must leave their homes to get water, often from a water kiosk or standpipe where they may have to wait in a long line.

Caroline Penn/CORBIS

James Pomerantz/INSTITUTE

Ajit Solanki/AP Photo

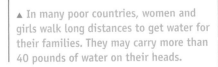

◄ In many poor areas of the world, it is common for residents to defecate in plastic bags that they dump in ditches or throw on the roadside.

▲ In many poor countries, women and girls walk long distances to get water for their families. They may carry more than 40 pounds of water on their heads.

Many of the more than 1,300 people who died in the wake of Hurricane Katrina were poor.

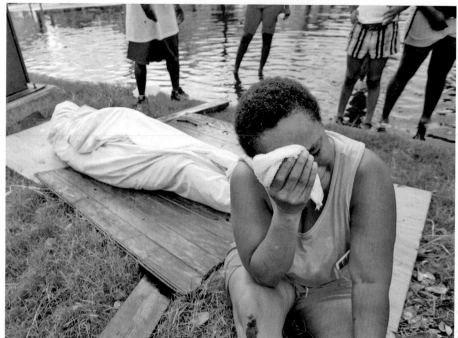

AP Photo/Eric Gay

Poverty also affects natural disaster relief efforts. Because the poorest of the tsunami victims lived in areas with weak or nonexistent infrastructure, hundreds of thousands of tsunami victims were cut off from aid because they did not live near a road or airport. In addition, just as upper-class passengers on the *Titanic* were given priority access to lifeboats, some reports claimed that the tsunami-affected areas that catered to well-off tourists received more assistance than the thousands of poor people who lived in the villages (Roberts 2005). Similarly, in the wake of Hurricane Katrina, some of the local poor residents waited at the back of an evacuation line while 700 guests and employees of the Hyatt Hotel were bused out first (Dowd 2005). Other victims of Hurricane Katrina were stranded for days without food, water, and medical supplies. Five days after the hurricane hit, 20,000 hungry, dehydrated, desperate people stranded in the rain-soaked and sweltering hot Louisiana Superdome in New Orleans, where overflowed toilets forced people to relieve themselves in hallways and stairwells, waited to be evacuated. For days, in some cases a week or more, some hurricane victims waited to be rescued from rooftops or from neck-high floodwater in their attics. For many, rescue came too late. Scores of media interviews and editorials suggested that the government's slow response was at least partly due to the fact that Katrina's victims were predominantly poor. A church pastor lamented, "I think a lot of it has to do with race and class. The people affected were largely poor people. Poor, black people" (quoted by Gonzalez 2005).

What Do You Think? Do you think that the federal response to the disaster left in the wake of Hurricane Katrina would have been different if Katrina had devastated an area of the country where wealthier people resided? Do you think, for example, that residents of Hollywood, California, or Long Island, New York, would have been stranded for days on their rooftops with signs saying, "HELP ME"?

The U.S. Department of Agriculture conducts national surveys to assess the degree to which U.S. households experience food security, food insecurity, and food insecurity with hunger. To assess your own level of food security, respond to the following items and use the scoring key to interpret your results:

1. In the last 12 months, the food that (I/we) bought just didn't last, and (I/we) didn't have money to get more.
   a. Often true
   b. Sometimes true
   c. Never true

2. In the last 12 months, (I/we) couldn't afford to eat balanced meals.
   a. Often true
   b. Sometimes true
   c. Never true

3. In the last 12 months, did you ever cut the size of your meals or skip meals because there wasn't enough money for food?
   a. Yes
   b. No (skip Question 4)

4. If you answered yes to Question 3, how often did this happen in the last 12 months?
   a. Almost every month
   b. Some months but not every month
   c. Only 1 or 2 months

5. In the last 12 months, did you ever eat less than you felt you should because there wasn't enough money to buy food?
   a. Yes
   b. No

6. In the last 12 months, were you ever hungry but didn't eat because you couldn't afford enough food?
   a. Yes
   b. No

Scoring and Interpretation

The answer responses in boldface type indicate affirmative responses. Count the number of affirmative responses you gave to the items, and use the following scoring key to interpret your results.

Number of Affirmative Responses and Interpretation

0 or 1 item: *Food secure* (In the last year, you have had access to enough food for an active, healthy life.)

2, 3, or 4 items: *Food insecure* (In the last year, you have had limited or uncertain availability of food and have been worried or unsure you would get enough to eat.)

5 or 6 items: *Food insecure with hunger evident* (In the last year, you have experienced more than isolated occasions of involuntary hunger as a result of not being able to afford enough food.)

If you scored as food insecure (with or without hunger), you might consider exploring whether you are eligible for public food assistance (e.g., food stamps) or whether there is a local food assistance program (e.g., food pantry or soup kitchen) that you could use.

Source: Based on the short form of the 12-month Food Security Scale found in Bickel et al. 2000.

Educational Problems and Poverty

In many countries, most children from the poorest households have no schooling, and enter their adult lives never having completed the first grade (United Nations Population Fund 2002). In the United States, children living in poverty are more likely to suffer academically than are children who are not poor. "Overall, poor children receive lower grades, receive lower scores on standardized tests, are less likely to finish high school, and are less likely to attend or graduate from college than are nonpoor youth" (Seccombe 2001, p. 323). Health problems associated with childhood poverty, including poorer vision, lead poisoning, asthma, and inadequate nutrition, contribute to poor academic performance (Rothstein 2004). The poor often attend schools that are characterized by lower quality facilities, overcrowded classrooms, and a higher teacher turnover rate (see also Chapter 8). Because poor parents have less schooling on average than do nonpoor parents, they may be less able to encourage and help their children succeed in school. Poor parents have fewer resources to provide their children with books, computers, travel, and other goods and experiences that promote cognitive development and educational achievement (Sobolewski & Amato 2005). With the skyrocketing costs of tuition and other fees, many poor parents cannot afford to send their children to college. Poor adults who want to escape poverty by furthering their education may have to work while attending school or may be unable to attend school because of unaffordable child care, transportation, and/or tuition, fees, and books.

Family Stress and Parenting Problems Associated with Poverty

The stresses associated with low income contribute to substance abuse, domestic violence, child abuse and neglect, divorce, and questionable parenting practices. For example, economic stress is associated with greater marital discord (Sobolewski & Amato 2005), and couples with incomes less than $25,000 are more likely to divorce than couples with incomes greater than $50,000 (Popenoe 2008). Child neglect is more likely to be found with poor parents who are unable to afford child care or medical expenses and leave children at home without adult supervision or fail to provide needed medical care. Poor parents are more likely than other parents to use harsh physical disciplinary techniques, and they are less likely to be nurturing and supportive of their children (Mayer 1997; Seccombe 2001).

Another family problem associated with poverty is teenage pregnancy. Poor adolescent teenagers are at higher risk of having babies than their nonpoor peers. Early childbearing is associated with increased risk of premature babies or babies with low birth weight, dropping out of school, and lower future earning potential as a result of lack of academic achievement. Luker (1996) noted that "the high rate of early childbearing is a measure of how bleak life is for young people who are living in poor communities and who have no obvious arenas for success" (p. 189). For poor teenage women who have been excluded from the American dream and disillusioned with education, "childbearing . . . is one of the few ways . . . such women feel they can make a change in their lives" (p. 182):

> Having a baby is a lottery ticket for many teenagers: it brings with it at least the dream of something better, and if the dream fails, not much is lost. . . . In a few cases it leads to marriage or a stable relationship; in many others it motivates a woman to push herself for her baby's sake; and in still other cases it enhances the woman's self-esteem, since it enables her to do something productive, something nurturing and socially responsible. . . . (Luker 1996, p. 182)

slums Concentrated areas of poverty and poor housing in urban areas.

ghettos Slums in the United States that are occupied primarily by African Americans.

barrios Slums in the United States that are occupied primarily by Latinos.

Housing Problems

Housing problems include substandard housing, homelessness, and "the housing crisis."

Substandard Housing Having a roof over one's head is considered a basic necessity. However, for the poor, that roof may be literally caving in. In addition to having leaky roofs, housing units of the poor often have holes in the floor and open cracks in the walls or ceiling. Low-income housing units often lack central heating and air conditioning, sewer or septic systems, and electric outlets in one or more rooms. Housing for the poor is also often located in areas with high crime rates and high levels of pollution.

Concentrated areas of poverty and poor housing in urban areas are called **slums.** One-third of urban populations in developing regions are living in slums. In sub-Saharan Africa, a staggering two-thirds of urban populations are living in slums (UN-Habitat 2010). In the United States, slums that are occupied primarily by African Americans are called **ghettos,** and those occupied primarily by Latinos are called **barrios.**

Ric Feld/AP Photo

Many Americans would be shocked to see the conditions under which many poor people in this country live.

Homelessness Due to the recent rise in unemployment and foreclosures, more women, men, and children have been pushed into homelessness. The number of homeless nationwide is estimated to be more than one million (Lichtblau 2009). Over the course of a lifetime, an estimated 9 percent to 15 percent of the U.S. population becomes homeless (Hoback & Anderson 2007). Individuals who do not have a home of their own and who stay at the home of family or friends are known as **couch-homeless,** or "couch-surfers" (Hoback & Anderson 2007).

Many factors contribute to homelessness. According to a survey of mayors in major U.S. cities, the primary cause of homelessness is unemployment, followed by lack of affordable housing (City Mayors Society 2011). Housing is considered affordable when a household pays no more than 30 percent of its income on rent. Other causes of homelessness include domestic violence, mental illness and the lack of mental health services, substance abuse and the lack of substance abuse services, low-paying jobs, poverty, and prison release.

Homeless individuals live on the street or outdoors; others live in homeless shelters or makeshift dwellings made of a variety of discarded materials such as pieces of wood and boards, cardboard, mattresses, fabric, and plastic tarps. For homeless people living on the street, every day is a struggle for survival:

> The primary cause of homelessness is unemployment, followed by lack of affordable housing.

> Living on the streets makes you do a lot of things that you wouldn't normally do. . . . Comin' into this environment I've done a lot of things I said I wouldn't do. . . . There was some people that came along in a van and just threw sandwiches on the street and I picked them up and ate them. . . . The guilt almost killed me . . . but my stomach said, "Hey, listen, you better eat this food." (Dordick 1997, pp. 5–6)

In recent years, there has been a surge in unprovoked violent attacks against homeless individuals. In a decade, more than a thousand homeless men, women, and children have been attacked; 291 have been killed (National Coalition for the Homeless 2010). In most cases, the attacks are by teenage and young adult males. In Toms River, New Jersey, five high school students were charged with beating a 50-year-old homeless man nearly to death with pipes and baseball bats as he slept in the woods. In Spokane, Washington, a 22-year-old man was charged with first-degree murder in the case of a one-legged 50-year-old homeless man who was set on fire in his wheelchair on a downtown street; he died of his burns. In Nashville, Tennessee, two young men were charged with shoving a 32-year-old homeless woman who was sleeping on a boat ramp into the river and leaving her to drown (Lewan 2007). Many acts of violence toward the homeless are not reported to the police, so documented cases may be just the tip of the iceberg. During the years he lived homeless on the street, David Pirtle was attacked five times and he did not report the attacks to police. "I was struck on the back, kicked, urinated on, spray-painted. . . . A lot of people who are homeless go through it, and it's just the way it is" (quoted in Dvorak 2009, p. DZ01).

There is also a new fascination with "bum bashing" or "bum fight" videos on *You-Tube*—videos shot by young men and boys who are seen beating the homeless or who pay homeless people a few

couch-homeless Individuals who do not have a home of their own and who stay at the home of family or friends.

Thousands of people in the United States are homeless on any given day.

Elena Rooraid/PhotoEdit

dollars to fight each other. As of May 2010, about 11,300 *YouTube* videos have been tagged with the search phrase "bum fight" (National Coalition for the Homeless 2010). Individuals who find the idea of bum bashing entertaining can also purchase bum fight DVDs and play web-based bum-bashing games.

> **What Do You Think?** Under hate crime laws, violators are subject to harsher legal penalties if their crime is motivated by the victim's race, religion, national origin, or sexual orientation. In 2009, Maryland became the first state to add homeless status to its hate crime law. A few other states (California, Maine, Alaska, Puerto Rico, and Florida), as well as several cities and counties, have also taken measures to recognize homeless status in their laws or procedures. Proposed legislation to add homelessness to the federal hate crime law has not, as of this date, passed. Do you think that violent acts toward homeless individuals should be categorized as hate crimes and subject to harsher penalties? Why or why not?

The "Housing Crisis." The recent housing crisis has caused financial ruin for millions of people in the United States, and has contributed to a global financial crisis that has exacerbated poverty around the world. In the 1990s and early 2000, inflated housing values enabled middle-class homeowners with maxed-out credit cards to keep spending by refinancing their mortgages. At the same time, there was an increase in **subprime mortgages**—high-interest or adjustable-rate mortgages that require little money down and are issued to borrowers with poor credit ratings or limited credit history. Subprime mortgage lending enabled low-wage earners to buy a house, and the increased housing demand raised house values further. When the housing bubble burst and house values fell in 2007/08, millions of people were stuck with "upside-down mortgages," in which the amount owed on a mortgage is more than the value of the property. Many homeowners with subprime mortgages could not make their payments, and foreclosures skyrocketed. In this housing crisis, millions of people have lost their homes and their credit, and renters in foreclosed dwellings lost their lease with little notice.

Intergenerational Poverty

Research finds that nearly half of U.S. children born to low-income parents became low-income adults (Corak 2006). Problems associated with poverty, such as health and educational problems, create a cycle of poverty from one generation to the next. Poverty that is transmitted from one generation to the next is called **intergenerational poverty.**

Intergenerational poverty creates a persistently poor and socially disadvantaged population, referred to as the underclass. Although the underclass is stereotyped as being composed of minorities living in inner-city or ghetto communities, the underclass is a heterogeneous population that includes poor whites living in urban and nonurban communities (Alex-Assensoh 1995).

William Julius Wilson attributes intergenerational poverty and the underclass to a variety of social factors, including the decline in well-paid jobs and their movement out of urban areas, the resultant decline in the availability of marriageable males able to support a family, declining marriage rates and an increase in out-of-wedlock births, the migration of the middle class to the suburbs, and the effect of deteriorating neighborhoods on children and youth (Wilson 1987, 1996).

War and Social Conflict

Poverty and economic inequality are often root causes of conflict and war within and between nations. Poorer countries are more likely than wealthier countries to be involved in civil war, and countries that experience civil war tend to become and/or remain poor. Armed conflict and civil war are generally more likely to occur in countries with extreme and growing inequalities between ethnic groups (United Nations 2005):

subprime mortgages High-interest or adjustable-rate mortgages that require little money down and are issued to borrowers with poor credit ratings or limited credit history.

intergenerational poverty Poverty that is transmitted from one generation to the next.

When inequalities become persistent and some groups are systematically barred from the benefits of growth . . . those at the bottom claim their share of the national income by any means possible. (International Labour Organization 2008b)

In the developing world, most of the people recruited for armed conflict are unemployed. "They don't have education opportunities and they don't really see what the future holds for them other than war and misery" (World Population News Service 2003, p. 4). Tanzania president Benjamin Mkapa said that "countries with impoverished, disadvantaged and desperate populations are breeding grounds for present and future terrorists" (quoted by Schifferes 2004). A United Nations report suggested that "the most effective conflict prevention strategies . . . are those aimed at achieving reductions in poverty and inequality, full and decent employment for all, and complete social integration" (United Nations 2005, p. 94).

Not only does poverty breed conflict and war, but war also contributes to poverty. War devastates infrastructures, homes, businesses, and transportation systems. In the wake of war, populations often experience hunger and homelessness.

In the United States, the widening gap between the rich and the poor may lead to class warfare (hooks 2000). Briggs (1998) asked how long the United States can maintain social order "when increasing numbers of persons are left out of the banquet while a few are allowed to gorge?" (p. 474). Although Karl Marx predicted that the have-nots would revolt against the haves, Briggs did not foresee a revival of Marxism. "The means of surveillance and the methods of suppression by the governments of industrialized states are far too great to offer any prospect of success for such endeavors" (p. 476). Instead, Briggs predicted that American capitalism and its resulting economic inequalities would lead to social anarchy—a state of political disorder and weakening of political authority.

Strategies for Action: Alleviating Poverty

Government programs designed to alleviate poverty include various types of welfare and public assistance, the earned income tax credit, and policies and proposals that involve increasing wages. We also look at faith-based initiatives and international responses to poverty.

Government Public Assistance and Welfare Programs in the United States

Many public assistance programs stipulate that households are not eligible for benefits unless their income and/or assets fall below a specified guideline. Programs that have eligibility requirements based on income are called **means-tested programs.** Government public assistance programs designed to help the poor include Supplemental Security Income, Temporary Assistance for Needy Families, food programs, housing assistance, medical care, educational assistance, child care, child support enforcement, and the earned income tax credit (EITC).

Supplemental Security Income Supplemental Security Income (federal SSI), administered by the Social Security Administration, provides a minimum income to poor people who are age 65 or older, blind, or disabled. Under the 1996 welfare reforms, the definition of disability has been sharply restricted, and the eligibility standards have been tightened.

Temporary Assistance for Needy Families Before 1996, a cash assistance program called Aid to Families with Dependent Children (AFDC) provided single parents (primarily women) and their children with a minimum monthly income. In 1996, Congress passed the Personal Responsibility and Work Opportunity Reconciliation Act (PRWORA), commonly referred to as "welfare reform," which replaced AFDC with a program called **Temporary Assistance for Needy Families (TANF).** Within two years of receiving benefits, adult TANF recipients must be either employed or involved in

means-tested programs Assistance programs that have eligibility requirements based on income.

Temporary Assistance for Needy Families (TANF) A federal cash welfare program that involves work requirements and a five-year lifetime limit.

work-related activities, such as on-the-job training, job search, and vocational education. A federal lifetime limit of five years is set for families receiving benefits, and able-bodied recipients ages 18 to 50 without dependents have a two-year lifetime limit. Some exceptions to these rules are made for individuals with disabilities, victims of domestic violence, residents in high unemployment areas, and those caring for young children. To qualify for TANF benefits, unwed mothers younger than age 18 are required to live in an adult-supervised environment (e.g., with their parents) and to receive education and job training. Legal immigrants who entered the United States before August 22, 1996, can receive TANF, but those who entered after this date can receive services only after they have been in the country for five years.

Food Assistance The largest food assistance program in the United States is the **Supplemental Nutrition Assistance Program (SNAP)** (formerly known as the Food Stamp Program), followed by school meals and the Special Supplemental Nutrition Program for Women, Infants, and Children (WIC). SNAP issues monthly benefits through coupons or a plastic card similar to a credit card. In 2010, the average household receiving SNAP received a monthly benefit of $289.61; the benefit for an individual was $133.79 (USDA Food and Nutrition Service 2011).

To supplement SNAP, school meals, and WIC, many communities have food pantries (which distribute food to poor households), "soup kitchens" (which provide cooked meals on site), and food assistance programs for the elderly population (such as Meals on Wheels). Despite the various forms of food assistance, a significant share of poor U.S. children (18 percent of young children and 12 percent of school-age children) receives no food assistance (Zedlewski & Rader 2005).

Housing Assistance Although media and the public tend to focus attention on rising gas and food prices, the biggest expense for most families is housing. For the poor, housing costs are a major burden: half of the working poor spend at least 50 percent of their income on housing (Grunwald 2006). Federal housing assistance programs include public housing, Section 8 housing, and other private project-based housing.

The **public housing** program, initiated in 1937, provides federally subsidized housing that is owned and operated by local public housing authorities (PHAs). To save costs and avoid public opposition, high-rise public housing units were built in inner-city projects. These have been plagued by poor construction, managerial neglect, inadequate maintenance, and rampant vandalism. Poor-quality public housing has serious costs for its residents and for society:

> Distressed public housing subjects families and children to dangerous and damaging living environments that raise the risks of ill health, school failure, teen parenting, delinquency, and crime—all of which generate long-term costs that taxpayers ultimately bear. . . . These severely distressed developments are not just old, outmoded, or run down. Rather, many have become virtually uninhabitable for all but the most vulnerable and desperate families. (Turner et al. 2005, pp. 1–2)

Section 8 housing involves federal rent subsidies provided either to tenants (in the form of certificates and vouchers) or to private landlords. Unlike public housing that confines low-income families to high-poverty neighborhoods, the aim with Section 8 housing is to disperse low-income families throughout the community. However, because of opposition by residents in middle-class neighborhoods, most Section 8 housing units remain in low-income areas.

The level of housing assistance available is sorely inadequate to meet the housing needs of low-income Americans. For every low-income family that receives federal housing assistance, three eligible families are without it (Grunwald 2006).

Lack of affordable housing is not just a problem for the poor living in urban areas. "The problem has climbed the income ladder and moved to the suburbs, where service workers cram their families into overcrowded apartments, college graduates have to

Supplemental Nutrition Assistance Program (SNAP) The largest U.S. food assistance program.

public housing Federally subsidized housing that is owned and operated by local public housing authorities (PHAs).

Section 8 housing A housing assistance program in which federal rent subsidies are provided either to tenants (in the form of certificates and vouchers) or to private landlords.

crash with their parents, and firefighters, police officers, and teachers can't afford to live in the communities they serve" (Grunwald 2006).

A major barrier to building affordable housing is zoning regulations that set minimum lot size requirements, density restrictions, and other controls. Such zoning regulations serve the interests of upper-middle-class suburbanites who want to maintain their property values and keep out the "riffraff"—the lower-income segment of society who would presumably hurt the character of the community. Thus, one answer to the housing problem is to change zoning regulations that exclude affordable housing. Fairfax County, Virginia, is one of more than 100 communities that have adopted "inclusionary zoning," which requires developers to reserve a percentage of units for affordable housing (Grunwald 2006).

Alleviating Homelessness Programs to temporarily alleviate homelessness include "homeless shelters" that provide emergency shelter beds, and transitional housing programs, which provide time-limited (usually two years) housing and services designed to help individuals gain employment, increase their income, and resolve substance abuse and other health problems. Resolving homelessness requires strategies to *prevent* homelessness from occurring in the first place, such as increasing employment and living wages, providing tax benefits to renters (not just to homeowners), providing more affordable housing, and protecting both homeowners and renters against foreclosures. To reduce the number of people losing their homes to foreclosure, President Obama implemented a program called Making Home Affordable. This program includes a refinancing program that offers refinancing at lower rates for eligible homeowners. Obama also signed the Protecting Tenants at Foreclosure Act of 2009, which gives tenants the right to stay in a foreclosed property at least until the end of the lease, and entitles month-to-month tenants to a 90-day notice before having to move out.

What Do You Think? The number of spaces in homeless shelters is grossly inadequate to accommodate the numbers of homeless individuals, which means that hundreds of thousands of homeless people have no place to be, except in public. Many cities have passed laws that prohibit homeless people from begging as well as sleeping, sitting, and/or "loitering" in public. Do you think that such laws unfairly punish the homeless? Or are these laws necessary to protect the public?

Medicaid The largest U.S. public medical care assistance program is Medicaid, which provides medical services and hospital care for the poor through reimbursements to physicians and hospitals. However, many low-income individuals and families do not qualify for Medicaid and either cannot afford health insurance or cannot pay the deductible and co-payments under their insurance plan.

In the earlier AFDC welfare program, all recipients were automatically entitled to Medicaid. Under the TANF program, states decide who is eligible for Medicaid; eligibility for cash assistance does not automatically convey eligibility for Medicaid. A provision of the 1996 welfare reform legislation guarantees welfare recipients at least one year of transitional Medicaid when leaving welfare for work (see Chapter 2 also).

Educational Assistance Educational assistance includes Head Start and Early Head Start programs and college assistance programs (see also Chapter 8). Head Start and Early Head Start programs provide educational services for disadvantaged infants, toddlers, and preschool-age children and their parents. Evaluations of Head Start and Early Head Start programs indicate that they improve children's cognitive, language, and social-emotional development and strengthen parenting skills (Administration for Children and Families 2002). According to the Children's Defense Fund (2003), every $1 invested in high-quality early childhood care and education saves as much as $7, by increasing the likelihood that children will be literate, go to college, and be employed, and by decreasing the likelihood that they will drop out of schools, be dependent on welfare, or be arrested for criminal activity.

To alleviate economic barriers for low-income individuals wanting to attend college, the federal government offers grants, loans, and work opportunities. The Pell Grant program aids students from low-income families. The guaranteed student loan program enables college students and their families to obtain low-interest loans with deferred interest payments. The federal college work-study program provides jobs for students with "demonstrated need."

Child Care Assistance In the United States, lack of affordable, good child care is a major obstacle to employment for single parents and a tremendous burden on dual-income families and employed single parents. The cost of child care for a 4-year-old child ranges from about $4,000 to nearly $12,000 per year. Child care fees for an infant are even higher, ranging from about $5,000 to more than $15,000 a year.

Some public- and private-sector programs and policies provide limited assistance with child care. The Dependent Care Assistance Plan provisions of the 1981 Economic Recovery Tax Act permit individuals to exclude the value of employer-provided child care services from their gross income. However, few employers provide on-site child care or subsidies for child care. At the same time, Congress increased the amount of the child care tax credit and modified the federal tax code to allow taxpayers to shelter pretax dollars for child care in "flexible spending plans." The Family Support Act of 1988 offered additional funding for child care services for the poor (in conjunction with mandatory work requirements). The Child Care and Development Block Grant, which became law in 1990, targeted child care funds to low-income groups, and the PRWORA appropriated funds for child care. However, child care assistance is inadequate; many states have waiting lists for child care assistance, and many families earn more than the eligibility limit, but not enough to afford child care expenses.

Child Support Enforcement To encourage child support from absent parents, the PRWORA requires states to set up child support enforcement programs, and single parents who receive TANF are required to cooperate with child support enforcement efforts. The welfare reform law established a Federal Case Registry and National Directory of New Hires to track delinquent parents across state lines, increased the use of wage withholding to collect child support, and allowed states to seize assets and to revoke driving licenses, professional licenses, and recreational licenses of parents who fall behind in their child support.

Earned Income Tax Credit The federal **earned income tax credit (EITC),** created in 1975, is a refundable tax credit based on a working family's income and number of children. The EITC is designed to offset Social Security and Medicare payroll taxes on working poor families and to strengthen work incentives. The federal EITC lifts more children out of poverty than any other program (Llobrera & Zahradnik 2004).

Welfare in the United States: Myths and Realities

The majority of U.S. adults—63 percent in 2010—favor more generous government assistance to the poor; 31 percent are opposed (Pew Research Center 2010). Nevertheless, negative attitudes toward welfare assistance and welfare recipients are not uncommon (Epstein 2004). What are some of the common myths about welfare that perpetuate negative images of welfare and welfare recipients?

Myth 1. People who receive welfare are lazy, have no work ethic, and prefer to have a "free ride" on welfare rather than work.

Reality. Most recipients of TANF and SNAP benefits are children and therefore are not expected to work. In 2006, one in five adult TANF recipients worked, earning an average monthly salary amount of $703 (Office of Family Assistance 2009). Unemployed adult

earned income tax credit (EITC) A refundable tax credit based on a working family's income and number of children.

welfare recipients experience a number of barriers that prevent them from working, including poor health, job scarcity, lack of transportation, lack of education, and/or the desire to stay home and care for their children (which often stems from the inability to pay for child care or the lack of trust in child care providers) (Zedlewski 2003). Welfare recipients who stay home to care for children *are* doing very important work: parenting. "Raising children is work. It requires time, skills, and commitment. While we as a society don't place a monetary value on it, it is work that is invaluable—and indeed, essential to the survival of our society" (Albelda & Tilly 1997, p. 111).

It is also important to note that many adults receiving public assistance do earn an income or are participating in work activities, including job training or education, job searches, and employment. However, many do not work because there are not enough jobs available. In fall 2011, there were more than four unemployed workers for every available job (Shierholz 2011).

Finally, most adult welfare recipients would rather be able to support themselves and their families than rely on public assistance. The image of a welfare "freeloader" lounging around enjoying life is far from the reality of the day-to-day struggles and challenges of supporting a household on a monthly TANF check of $372, which was the average monthly cash and cash-equivalent assistance to TANF families in 2006 (Office of Family Assistance 2009).

Myth 2. Most welfare mothers have large families with many children.

Reality. The average number of individuals in TANF families is 2.3, including an average of only 1.8 children (Office of Family Assistance 2009).

Myth 3. Welfare benefits are granted to many people who are not really poor or eligible to receive them.

Reality. Although some people obtain welfare benefits through fraudulent means, it is much more common for people who are eligible to receive welfare not to receive benefits. Fewer than one in five individuals living below the poverty line receive cash assistance and fewer than half receive food stamps (see Table 6.8).

A main reason for not receiving benefits is lack of information; people do not know they are eligible. Many people who are eligible for public assistance do not apply for it because they do not want to be stigmatized as lazy people who just want a "free ride" at the taxpayers' expense—their sense of personal pride prevents them from receiving public assistance. Others have difficulty navigating the administrative process of applying for assistance. One homeless person explained,

TABLE 6.8	Percentage of Individuals Living below the Poverty Level in Households That Receive Means-Tested Assistance, 2009
TYPE OF ASSISTANCE	**PERCENTAGE**
Any type of assistance	72.5
Cash assistance	18.7
Food stamps	45.8
Medicaid	60.2
Public or subsidized housing	14.8

Source: U.S. Census Bureau 2011c.

> I don't get welfare. I just can't . . . do it. I hate those people in there. They make you . . . sit and ask you questions that don't make any sense. . . . You're homeless but you have to have an address. What kind of shit is that? Give me a break. They want you to get so . . . upset that you do get up and walk out. (Dordick 1997, p. 58)

Finally, some individuals who are eligible for public assistance do not receive it because it is not available. In cities across the United States, thousands of eligible low-income households are on waiting lists for public housing assistance because there are not enough public housing units available, and some cities have even stopped accepting housing applications.

Myth 4. Immigrants place a huge burden on our welfare system.

Reality. In 2006, only 5.9 percent of adult recipients of TANF and 1.2 percent of child TANF recipients were qualified immigrants (Office of Family Assistance 2009).

Minimum Wage Increase and "Living Wage" Laws

In 2009, the federal minimum wage increased to $7.25 an hour. Some states have established a minimum wage that is higher than the federal minimum wage. As of 2011, 17 states and the District of Columbia have mandated a minimum wage that is higher than the federal $7.25.

> **What Do You Think?** Just as there is a federal minimum wage, do you think that there should be a federal maximum wage? Why or why not? If you favor the idea of a maximum wage, what should that wage be?

Many cities and counties throughout the United States have **living wage laws** that require state or municipal contractors, recipients of public subsidies or tax breaks, or, in some cases, all businesses to pay employee wages that are significantly above the federal minimum, enabling families to live above the poverty line. Research findings show that businesses that pay their employees a living wage have lower worker turnover and absenteeism, reduced training costs, higher morale and productivity, and a stronger consumer market (Kraut et al. 2000).

As more individuals receiving welfare (TANF) reach their time limits and are forced to enter the job market, it is increasingly important to provide jobs that pay a living wage. As shown in this chapter's *Social Problems Research Up Close* feature, single mothers who work in low-wage jobs often have more hardships than those who are dependent on welfare.

Faith-Based Services for the Poor

In 2001, President G. W. Bush established the Faith-Based and Community Initiative, which provided federal funding for faith-based programs that serve the needy, such as homeless services and food aid programs. President Obama renamed the program **Faith-Based and Neighborhood Partnerships**, expanding the program to include non–faith-based neighborhood programs.

Critics are concerned about the degree to which the faith-based initiative violates the separation of church and state and affects the rights of clients seeking services. "How are the lives of the jobless improved when they are told they won't get work until they first get right with God?" (Boston 2005). Although religious groups are prohibited from using government funding to promote religion, this policy is difficult to enforce.

International Responses to Poverty

In 2000, leaders from 191 United Nations member countries pledged to achieve eight **Millennium Development Goals**—an international agenda for reducing poverty and improving lives. One of the Millennium Development Goals (MDGs) is to halve, between 1990 and 2015, the proportion of people who live in severe poverty and who suffer from hunger. As can be seen in Table 6.9 (page 194), several other MDGs involve alleviating problems related to poverty, such as disease, child and maternal mortality, and lack of access to education. Approaches for achieving poverty reduction throughout the world include promoting economic development, investing in human development, providing financial aid and debt cancellation, and providing microcredit programs that provide loans to poor people.

Another strategy to reduce poverty is to increase public awareness of the magnitude of the problem. For many of our students who have gone on study abroad programs to poor areas in India, Africa, and South America, witnessing poverty firsthand was a transformative experience. Some students came back from the experience with a commitment to raise funds for health care or clean water facilities in a specific village; others changed their career plans to include working with poor populations.

living wage laws Laws that require state or municipal contractors, recipients of public subsidies or tax breaks, or, in some cases, all businesses to pay employees wages that are significantly above the federal minimum, enabling families to live above the poverty line.

Faith-Based and Neighborhood Partnerships A program in which faith-based and other neighborhood organizations receive federal funding for programs that serve the needy, such as homeless services and food aid programs.

Millennium Development Goals Eight goals that comprise an international agenda for reducing poverty and improving lives.

How do individuals in low-income jobs compare with those dependent on welfare in terms of their well-being? And how do both low-wage earners and welfare recipients survive on income that does not meet their basic needs? Researchers Kathryn Edin and Laura Lein (1997) conducted research to answer these questions.

Sample and Methods

The sample consisted of 379 Black American, white, and Mexican American single mothers from four cities (Chicago, San Antonio, Boston, and Charleston, South Carolina).

The mothers either received welfare cash assistance (N = 214) or were nonrecipients who held low-wage jobs earning $5 to $7 an hour between 1988 and 1992 (N = 165). Edin and Lein (1997) used a "snowball sampling" technique in which each mother who was interviewed was asked to refer researchers to one or two friends who might also participate in interviews. Nearly 90 percent of the mothers contacted agreed to be interviewed. Edin and Lein conducted multiple semistructured in-depth interviews with women in the sample on topics including the mothers' income and job experience, types and amount of welfare benefits they received, spending behavior, housing situation, use of medical care and child care, and hardships the women and their children experienced because of lack of financial resources.

Findings and Conclusions

Single mothers earning low wages had a higher monthly reported income than welfare-reliant mothers. However, the expenses of wage-earning mothers were also higher. This is because employed mothers usually have to pay for child care, transportation to work, and additional clothing to wear to work. If newly employed mothers have a federal housing subsidy, some of their new income is spent on the increase in the rent they must pay. Employed mothers are also usually not eligible for Medicaid, which means that they have more out-of-pocket medical expenses and often go uninsured.

The monthly expenses of both groups of women exceeded their reported monthly income, forcing women to use various strategies to make ends meet. Cash welfare and food stamps covered only three-fifths of welfare-reliant mothers' expenses. The main job of mothers earning low wages covered only 63 percent of their expenses. Edin and Lein (1997) found that women relied on three basic strategies to make ends meet: work in the formal, informal, or underground economy; cash assistance from absent fathers, boyfriends, relatives, and friends; and cash assistance or help from agencies, community groups, or charities in paying overdue bills. Welfare recipients had to keep their income-generating activities hidden from their welfare caseworkers and other government officials. Otherwise, their welfare checks would be reduced by nearly the same amount as their earnings. Many of the wage-earning mothers also concealed income generated "on the side" to maintain eligibility for food stamps, housing subsidies, or other benefits that would have been reduced or eliminated if they had reported this additional income.

Most of the single mothers in the study described experiencing serious material hardship during the previous 12 months. Material hardships included not having enough food and clothes, not receiving needed medical care, not having health insurance, having the utilities or phone cut off, not having a phone, and being evicted and/or homeless. An important finding was that wage-reliant mothers experienced more hardship than welfare-reliant mothers. In addition to the increased financial pressures of child care costs, transportation, health care, and work clothing, employed mothers worried about not providing adequate supervision of their children and struggled with balancing work and parenting responsibilities, especially when their children were sick. Nevertheless, almost all the mothers said they would rather work than rely on welfare. They believed that work provided important psychological benefits and increased self-esteem, avoided the stigma of welfare, and enabled them to be good role models for their children.

Source: Based on Edin & Lein 1997.

Non-student travelers are also interested in learning about areas of the world where extreme poverty persists. Some tourists seek out guides to take them through poor slum areas. Although slum tourism can promote greater awareness of poverty, Kennedy Odede, who was born and raised in the Kibera slum of Nairobi, Kenya, notes that, "Slums will not go away because a few dozen Americans or Europeans spent a morning walking around them. There are solutions to our problems—but they won't come about through tours" (Odede 2010, p. A25).

Economic Development One approach to alleviating poverty involves increasing the economic output or the gross domestic product of a country. However, economic development does not always reduce poverty; in some cases, it increases it. Policies that involve cutting government spending, privatizing basic services, liberalizing

TABLE 6.9 The Millennium Development Goals

1. Eradicate extreme poverty and hunger.

2. Achieve universal primary education.

3. Promote gender equality and empower women.

4. Reduce child mortality.

5. Improve maternal health.

6. Combat HIV/AIDS, malaria, and other diseases.

7. Ensure environmental sustainability.

8. Develop a global partnership for development.

Source: U.S. Census Bureau 2006.

trade, and producing goods primarily for export may increase economic growth at the national level, but the wealth ends up in the hands of the political and corporate elite at the expense of the poor. Economic growth does not help poverty reduction when public spending is diverted away from meeting the needs of the poor and instead is used to pay international debt, finance military operations, and support corporations that do not pay workers fair wages. The World Bank lends billions of dollars a year to developing nations to pay primarily for roads, bridges, and industrialized agriculture that mostly benefit corporations. "Relatively little attention or money has been given to developing basic social services, building schools and clinics, and building decent public sanitation and clean water systems in some of the world's poorest countries" (Mann 2000, p. 2).

Another problem with economic development is that the environment and natural resources are often destroyed and depleted in the process of economic growth. Economic development also threatens the lives and cultures of the 370 million indigenous people who live in 70 countries around the world. Indigenous people who live on land that is rich in natural resources are displaced by corporations that want access to the land and its natural resources, and by government forces that help the corporations expand their activities (Ramos et al. 2009). As remote areas are "developed," many indigenous people are forced to give up their traditional ways of life and become assimilated into the dominant culture. Conflict between indigenous people who want to continue their traditional ways of life and governments and multinationals who want to pursue economic development and the profits it produces can become violent. In June 2009, 600 Peruvian police opened fired on thousands of peaceful indigenous protesters who were blocking a road to protest new laws that allow the expansion of mining, oil drilling, logging, and destructive agriculture in their territories in the Amazon rain forest (Robinson 2009).

Human Development Unlike the economic development approach to poverty alleviation, the human development approach views people—not money—as the real wealth of a nation.

> The human development approach views people—not money—as the real wealth of a nation.

The central contention of the human development approach. . . is that well-being is about much more than money. . . . Income is critical but so are having access to education and being able to lead a long and healthy life, to influence the decisions of society and to live in a society that respects and values everyone. (UNDP 2010, p. 114)

In many poor countries, large segments of the population are illiterate and without job skills and/or are malnourished and in poor health. Investments in human development involve programs and policies that provide adequate nutrition, sanitation, housing, health care (including reproductive health care and family planning), and educational and job training.

Providing Financial Aid and Debt Cancellation To increase both economic and human development, poor countries depend on financial aid from wealthier countries. In 2010, development aid from wealthy countries reached a historic high, with the United States being the largest donor (OECD 2011b).

Another way to help poor countries invest in human capital and reduce poverty is to provide debt relief. In poor countries with large debts, money needed for health and education is instead spent on debt repayment. Canceling the debts of 32 of the poorest countries would cost citizens of the richest countries of the world just $2.10 a year for each citizen for 10 years (Oxfam 2005).

Microcredit Programs The old saying "It takes money to make money" explains why many poor people are stuck in poverty: They have no access to financial resources and services. **Microcredit programs** refer to the provision of loans to people who are generally excluded from traditional credit services because of their low socioeconomic status. Microcredit programs give poor people the financial resources they need to become self-sufficient and to contribute to their local economies.

The Grameen Bank in Bangladesh, started in 1976, has become a model for the more than 3,000 microcredit programs that have served millions of poor clients (Roseland & Soots 2007). To get a loan from the Grameen Bank, borrowers must form small groups of five people "to provide mutual, morally binding group guarantees in lieu of the collateral required by conventional banks" (Roseland & Soots 2007, p. 160). Initially, only two of the five group members are allowed to apply for a loan. When the initial loans are repaid, the other group members may apply for loans.

Understanding Poverty and Economic Inequality

As we have seen in this chapter, economic prosperity has not been evenly distributed; the rich have become richer while the poor have become poorer. A common belief among U.S. adults is that the rich are deserving and the poor are failures. Blaming poverty on the individual rather than on structural and cultural factors implies not only that poor individuals are responsible for their plight but also that they are responsible for improving their condition. If we hold individuals accountable for their poverty, we fail to make society accountable for making investments in human development that are necessary to alleviate poverty, such as providing health care, adequate food and housing, education, child care, job training and job opportunities, and living wages. Lastly, blaming the poor for their condition diverts attention away from the recognition that the wealthy—individuals and corporations—receive far more benefits in the form of wealthfare or corporate welfare, without the stigma of welfare.

Ending or reducing poverty begins with the recognition that doing so is a worthy ideal and an attainable goal. Imagine a world where everyone had comfortable shelter, plentiful food, clean water and sanitation, adequate medical care, and education. If this imaginary world were achieved and if absolute poverty were effectively eliminated, what would be the effects on social problems such as crime, drug abuse, family problems (e.g., domestic violence, child abuse, and divorce), health problems, prejudice and racism, and international conflict? In the current global climate of conflict and terrorism, we might consider that "reducing poverty and the hopelessness that comes with human deprivation is perhaps the most effective way of promoting long-term peace and security" (World Bank 2005).

Efforts to alleviate poverty are often motivated by a sense of moral responsibility. But alleviating poverty also makes sense from an economic standpoint. In the United States, the cost of sustained childhood poverty is more than $500 billion (4 percent of the GDP) per year due to increased health care costs, increased crime-related costs, and lowered productivity (vanden Heuvel 2011). According to one source, the cost of eradicating poverty worldwide would be only about one percent of global income—and no more than 2 percent to 3 percent of national income in all but the poorest countries (UNDP 1997). Certainly the costs of allowing poverty to continue are much greater than that.

microcredit programs The provision of loans to people who are generally excluded from traditional credit services because of their low socioeconomic status.

• **What is the difference between absolute poverty and relative poverty?**

Absolute poverty refers to a lack of basic necessities for life, such as food, clean water, shelter, and medical care. In contrast, relative poverty refers to a deficiency in material and economic resources compared with some other population.

• **How is poverty measured?**

The World Bank sets a "poverty threshold" of $1.25 per day for the least developed countries, and $2 per day for middle-income countries and regions, including Latin America and eastern Europe. Another measure of poverty is the proportion of people who consume less than a minimum amount of calories to maintain health—who are chronically hungry or undernourished. According to measures of relative poverty, members of a household are considered poor if their household income is less than 50 percent of the median household income in that country. Each year, the U.S. federal government establishes "poverty thresholds" that differ by the number of adults and children in a family and by the age of the family head of household. Anyone living in a household with pretax income below the official poverty line is considered "poor." To capture the multidimensional nature of poverty, the Multidimensional Poverty Index measures serious deprivation in the dimensions of health, education, and living standards.

• **Which sociological perspective criticizes wealthy corporations for using financial political contributions to influence politicians to enact policies that benefit corporations and the wealthy?**

The conflict perspective is critical of wealthy corporations that use financial political contributions to influence laws and policies that favor corporations and the rich. Such laws and policies, sometimes referred to as wealthfare or corporate welfare, include low-interest government loans to failing businesses and special subsidies and tax breaks to corporations.

• **In the United States, what age group has the highest rate of poverty?**

U.S. children are more likely than adults to live in poverty. More than one-third of the U.S. poor population is children. Child poverty rates are much higher in the United States than in any other industrialized country.

• **What are some of the consequences of poverty and economic inequality for individuals, families, and societies?**

Poverty is associated with health problems and hunger, increased vulnerability from natural disasters, problems in education, problems in families and parenting, and housing problems. These various problems are interrelated and contribute to the perpetuation of poverty across generations, feeding a cycle of intergenerational poverty. In addition, poverty and economic inequality breed social conflict and war.

• **What are some of the U.S. government public assistance programs designed to help the poor?**

Government public assistance programs designed to help the poor include Supplemental Security Income, Temporary Assistance for Needy Families (TANF), food programs (such as school meal programs and SNAP), housing assistance, Medicaid, educational assistance (such as Pell Grants), child care, child support enforcement, and the earned income tax credit (EITC).

• **What are four common myths about welfare and welfare recipients?**

Common myths about welfare and welfare recipients are (1) that welfare recipients are lazy, have no work ethic, and prefer to have a "free ride" on welfare rather than work; (2) that most welfare mothers have large families with many children; (3) that welfare benefits are granted to many people who are not really poor or eligible to receive them; and (4) that immigrants place an enormous burden on our welfare system.

• **What are four general approaches for achieving poverty reduction throughout the world?**

Approaches for achieving poverty reduction throughout the world include promoting economic growth, investing in human development, providing financial aid and debt cancellation to nations, and providing microcredit programs that provide loans to poor people.

TEST YOURSELF

1. The _____ Poverty Index is a measure of serious deprivation in the dimensions of health, education, and living standards that combines the number of deprived and the intensity of their deprivation.
   a. Relative
   b. Human
   c. International
   d. Multidimensional

2. According to the official U.S. poverty threshold guidelines, a single adult earning $12,000 a year is considered "poor."
   a. True
   b. False

3. In 2009, a typical U.S. worker would have to work _____ years to make what a typical CEO of a large U.S. corporation earns in one year.
   a. 15
   b. 42
   c. 111
   d. 263

4. Corporate welfare refers to which of the following?
   a. Taxes corporations pay that provide most of the funding for federal welfare programs for the poor
   b. Tax-deductible contributions that corporations make to charitable organizations

c. Laws and policies that benefit corporations

d. Employee assistance programs offered by corporations to help employees who are struggling with debt

5. What age group in the United States has the highest rate of poverty?

a. Younger than 18

b. 30 to 44

c. 45 to 64

d. Older than 65

6. According to the text, the wealthy are hardest hit by natural disasters because they have more to lose than do the poor.

a. True

b. False

7. Subprime mortgages are mortgages that charge very low interest rates and are available only to people with excellent credit and ample collateral.

a. True

b. False

8. Which federal program lifts more children out of poverty than any other program?

a. Public and Section 8 housing

b. TANF

c. EITC

d. SNAP

9. Nearly half of recipients of welfare in the United States are immigrants.

a. True

b. False

10. Economic development can result in increased poverty and economic inequality in a country.

a. True

b. False

Answer: 1: d; 2: b; 3: d; 4: c; 5: a; 6: b; 7: b; 8: c; 9: b; 10: b

KEY TERMS

absolute poverty   169

barrios   184

Basic Economic Security Tables Index (BEST)   171

corporate welfare   174

couch-homeless   185

earned income tax credit (EITC)   190

extreme poverty   169

Faith-Based and Neighborhood Partnerships   192

feminization of poverty   176

ghettos   184

intergenerational poverty   186

living wage laws   192

means-tested programs   187

microcredit programs   195

Millennium Development Goals   192

Multidimensional Poverty Index (MPI)   170

public housing   188

relative poverty   169

Section 8 housing   188

slums   184

subprime mortgages   186

Supplemental Nutrition Assistance Program (SNAP)   188

Temporary Assistance for Needy Families (TANF)   187

wealth   172

wealthfare   174

working poor   178

MEDIA RESOURCES

Turning to Video

▶❚❚ Watch the ABC video *Growing Up Fast: Children of Poverty* (running time 9:34), available through **CengageBrain.com.** This video describes the obstacles that poor children in Camden, New Jersey, face as they struggle to succeed. After you watch the video, think about how the video has influenced your view of poor people in the United States.

Online Study Resources

Log in to **www.cengagebrain.com** to access the resources your instructor has assigned. For this book, you can access:

CourseMate

Access chapter-specific learning tools including learning objectives, practice quizzes, videos, Internet exercises, flash cards, and glossaries, as well as web links, and more in your Sociology CourseMate.

Fan Xia/ZUMA Press/Newscom

7

Work and Unemployment

"When a man tells you that he got rich through hard work, ask him whose."

—Don Marquis, Journalist

ON APRIL 5, 2010, 29 coal miners were killed in an explosion at the Massey Energy Upper Big Branch Mine in West Virginia—the worst coal mine disaster in the United States in 40 years. Miner Gary Quarles, who worked at the Massey Mine with his son, Gary Wayne Quarles, said, "I worked the day shift that day, same as he did . . . I came home. He didn't" (quoted in Dorell 2011). In May 2011, Quarles attended a meeting with other families of the 29 men killed in the explosion to hear a report about how and why the tragedy happened. Based on an independent investigation, the report found that the mine explosion was primarily the fault of Massey Energy management, who could have prevented the accident by providing adequate ventilation and by limiting explosive coal dust in the mine (Maher 2011). Massey disputes the report's conclusions and claims that the explosion was sparked by an uncontrollable inundation of natural gas. According to records from the Mine Safety and Health Administration (MSHA), the mine had more than 500 safety violations issued against it in the last year. The Massey Mine disaster has raised serious questions about the adequacy of mine safety laws, regulations, and oversight, particularly how a mine with so many documented violations could continue to operate.

The West Virginia coal mine tragedy in 2010 was not the only mine accident in the state. In 2006, 12 miners died after a blast at the Sago Mine trapped them in a toxic air shaft. Mine officials claimed that lightening caused the blast, but others blame faulty equipment. Sago mine had a long list of safety violations and fines, with more than 270 safety citations in the two years prior to the fatal explosion.

Health and safety hazards in the workplace, often due to employers' willful violations of health and safety regulations, are among the work-related problems discussed in this chapter. Other problems we examine include unemployment, slavery and forced labor, child labor, sweatshop labor, alienation, work-life conflict, and declining labor strength and representation. We set the stage with a brief look at the global economy.

The Global Context: The New Global Economy

In recent decades, innovations in communication and information technology have spawned the emergence of a **global economy**—an interconnected network of economic activity that transcends national borders and spans the world. The globalization of economic activity means that our jobs, the products and services we buy, and our nation's economic policies and agendas are influenced by economic activities occurring around the world.

Beginning in 2007, the economic situation around the world took a downward turn, as banks faltered, credit froze, businesses closed, unemployment rates soared, and investments plummeted. This global financial crisis, which originated in the United States and spread throughout the world, illustrates the globalization of the **economic institution.** Some say that the cause of the global economic crisis was the lack of U.S. financial regulatory oversight that enabled financial institutions to engage in predatory and subprime lending during the housing boom in the early 2000s. As the housing boom turned to bust, and adjustable rate mortgages were reset to higher rates, millions of homeowners

global economy An interconnected network of economic activity that transcends national borders.

economic institution The structure and means by which a society produces, distributes, and consumes goods and services.

were unable to keep up with mortgage payments, and foreclosures skyrocketed. Home-owners lost their homes, renters lost their leases, and banks suffered because the fore-closed homes they now owned were often worth less than the mortgages owed on them. As banks lost revenue, they had less money to lend so credit froze, consumer spending plummeted, businesses went bust, and stockholders watched their investments and re-tirement accounts take a nosedive. The whole banking system was faltering; some got bailed out, others (e.g., Bear Stearns) went bust. All this happened in the United States, but in this new global economy, what happens in Vegas does not stay in Vegas; the crisis spread around the world. This is because those risky subprime and adjustable rate mort-gages were packaged and resold as "mortgage-backed securities" to financial institutions around the world.

The U.S. economic crisis also triggered a huge drop in world trade. Between 2000 and 2007, U.S. consumption accounted for more than a third of the growth in global consumption, and much of that consumption was based on borrowed money (Baily & Elliott 2009). When the United States went into recession and consumer spending de-clined, all the countries that depended on U.S. consum-ers to buy their goods and services lost a major source of revenue.

> In this new global economy, what happens in Vegas does not stay in Vegas.

The global economic crisis reignited debate between those who view U.S. capitalism as the cause of economic problems in the world, and those who hail capitalism as "the greatest engine of economic progress and prosperity known to mankind" (Ebeling 2009). After summarizing capitalism and socialism—the two main economic systems in the world—we describe how industrialization and postindustrialization have changed the nature of work, and look at the emergence of free trade agreements and transnational corporations.

> **What Do You Think?** The economic health of a country is commonly measured by how much the country is producing (the total value of goods and services) and how much money consumers are spending on the pur-chase of goods and services. In what ways might high levels of produc-tion and consumption contribute to individual and social ills rather than to health and well-being?

Capitalism and Socialism

Socialism is an economic system characterized by state ownership of the means of pro-duction and distribution of goods and services. In a socialist economy, the government controls income-producing property. Theoretically, goods and services are equitably dis-tributed according to the needs of the citizens. Socialist economic systems emphasize collective well-being rather than individualistic pursuit of profit.

Critics of socialism argue that socialism creates excessive government control, re-duces work incentives and technological development, and lowers the standard of living. A national survey of U.S. adults found that 59 percent say they have a negative reaction to the word *socialism,* compared with 29 percent who have a positive reaction (Pew Re-search Center 2010).

Under **capitalism,** private individuals or groups invest capital (money, technology, machines) to produce goods and services to sell for a profit in a competitive market. Whereas socialism emphasizes social equality, capitalism emphasizes individual free-dom. Capitalism is characterized by economic motivation through profit, the determi-nation of prices and wages primarily through supply and demand, and the absence of government intervention in the economy.

Critics of capitalism argue that it creates too many social evils, including alienated workers, poor working conditions, near-poverty wages, unemployment, a polluted and

socialism An economic system characterized by state owner-ship of the means of produc-tion and distribution of goods and services.

capitalism An economic sys-tem characterized by private ownership of the means of production and distribution of goods and services for profit in a competitive market.

depleted environment, and world conflict over resources. Although capitalistic values are deeply ingrained value in U.S. society, public dissatisfaction with big business (see Table 7.1) has revived debates about how much government regulation is desirable. Still, a slight majority (52 percent) of Americans have a positive reaction to the word *capitalism* whereas 37 percent have a negative reaction (Pew Research Center 2010).

In reality, there are no pure socialist or capitalistic economies. Rather, most countries have mixed economies, incorporating elements of both capitalism and socialism. Most developed countries, for example, have both private-owned and state-owned enterprises, as well as a social welfare system. The U.S. economy is dominated by capitalism, but there are elements of socialism in our welfare system and in government subsidies and low-interest loans to industry, fiscal stimulus money, and bailout money.

TABLE 7.1 Attitudes toward Big Business in the United States	
PERCENT OF U.S. ADULTS AGREEING THAT . . .	
There's too much power concentrated in the hands of a few big companies.	77%
Businesses make too much profit.	62%
Free market needs regulation to best serve public interest needs.	62%

Source: Pew Research Center 2009.

Industrialization, Postindustrialization, and the Changing Nature of Work

The nature of work has been shaped by the Industrial Revolution, the period between the mid-18th century and the early 19th century when the factory system was introduced in England. **Industrialization** dramatically altered the nature of work: Machines replaced hand tools, and steam, gasoline, and electric power replaced human or animal power. Industrialization also led to the development of the assembly line and an increased division of labor as goods began to be mass-produced. The development of factories contributed to the emergence of large cities. Instead of the family-centered economy characteristic of an agricultural society, people began to work outside the home for wages.

Postindustrialization refers to the shift from an industrial economy dominated by manufacturing jobs to an economy dominated by service and information technology jobs. In the global economy, jobs in the service sector outnumber jobs in both agriculture and industry (see Figure 7.1). In developed countries and the European Union, the majority of jobs (73 percent) are in services, followed by jobs in industry (25 percent) and agriculture (4 percent) (ILO 2011).

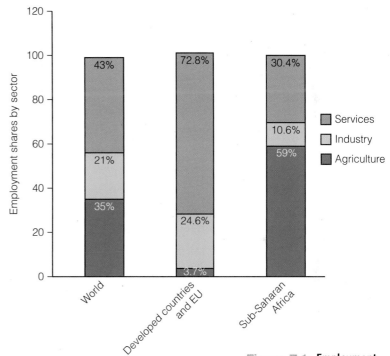

Figure 7.1: **Employment Shares by Sector, World and Selected Regions, 2009**
Source: ILO 2011.

industrialization The replacement of hand tools, human labor, and animal labor with machines run by steam, gasoline, and electric power.

postindustrialization The shift from an industrial economy dominated by manufacturing jobs to an economy dominated by service-oriented, information-intensive occupations.

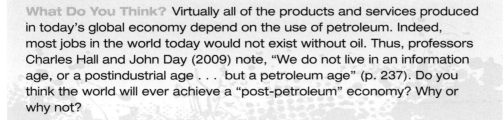

What Do You Think? Virtually all of the products and services produced in today's global economy depend on the use of petroleum. Indeed, most jobs in the world today would not exist without oil. Thus, professors Charles Hall and John Day (2009) note, "We do not live in an information age, or a postindustrial age . . . but a petroleum age" (p. 237). Do you think the world will ever achieve a "post-petroleum" economy? Why or why not?

McFlexible

We have shift patterns to suit the lifestyles of all our Crew.

Not bad for a McJob

We know it is important to have a life outside of work, so many of our Crew work shift patterns that allow them to pursue their hobbies and interests.
They also enjoy some great discounts as well as thorough training, in addition to access to our free private healthcare scheme after they've been with us for 3 years.
www.mcdonalds.co.uk

© 2006 McDonald's. The Golden Arches and the McD if logos are trademarks of McDonald's Corporation and affiliates.

This poster is part of McDonalds' advertising campaign to discredit the negative image implied by the term *McJob*.

McDonalds/PA Photos/Landov

McDonaldization of the Workplace

Sociologist George Ritzer (1995) coined the term **McDonaldization** to refer to the process by which the principles of the fast-food industry are being applied to more and more sectors of society, particularly the workplace. McDonaldization involves four principles:

1. *Efficiency.* Tasks are completed in the most efficient way possible by following prescribed steps in a process overseen by managers.
2. *Calculability.* Quantitative aspects of products and services (such as portion size, cost, and the time it takes to serve the product) are emphasized over quality.
3. *Predictability.* Products and services are uniform and standardized. A Big Mac in Albany is the same as a Big Mac in Tucson. Workers behave in predictable ways. For example, servers at McDonald's learn to follow a script when interacting with customers.
4. *Control through technology.* Automation and mechanization are used in the workplace to replace human labor.

What are the effects of McDonaldization on workers? In a McDonaldized workplace, employees are not permitted to use their full capabilities, be creative, or engage in genuine human interaction. Workers are not paid to think, just to follow a predetermined set of procedures. Because human interactions are unpredictable and inefficient (they waste time), "we're left with either no interaction at all, such as at ATMs, or a 'false fraternization.' Rule number 17 for Burger King workers is to smile at all times" (Ritzer, quoted by Jensen 2002, p. 41). Workers may also feel that they are merely extensions of the machines they operate. The alienation that workers feel—the powerlessness and meaninglessness that characterize a "McJob"—may lead to dissatisfaction with one's job and, more generally, with one's life.

What Do You Think? The slang term *McJob* is found in several dictionaries. For example, the *Oxford English Dictionary* defines McJob as "an unstimulating, low-paid job with few prospects, especially one created by the expansion of the service sector." *Merriam-Webster* defines McJob as "a low-paying job that requires little skill and provides little opportunity for advancement." Not surprisingly, McDonald's is opposed to such characterizations of their employment conditions and has fought back with an advertising campaign in which posters at more than 1,200 restaurants play up "the positive aspects of working for McDonald's," and include the phrase, "Not bad for a McJob." Do you think that the *Oxford* and *Merriam-Webster* dictionary definitions of McJob are accurate and fair?

McDonaldization The process by which principles of the fast-food industry (efficiency, calculability, predictability, and control through technology) are being applied to more sectors of society, particularly the workplace.

The Globalization of Trade and Free Trade Agreements

Just as industrialization and postindustrialization changed the nature of economic life, so has the globalization of trade—the expansion of trade of raw materials, manufactured goods, and agricultural products across national and hemispheric borders. The first set of global trade rules were adopted through the General Agreement on Tariffs and Trade (GATT) in 1947. GATT members met periodically to revise trade agreements in negotiations called "rounds." In 1995, the World Trade Organization (WTO) replaced GATT as the organization overseeing the multilateral trading system.

In the 1980s and early 1990s, U.S. officials began negotiating regional free trade agreements that would open doors to U.S. goods in neighboring countries and reduce the massive U.S. trade deficit, which had grown from $25.3 billion in 1980, to $122 billion in 1985 (Schaeffer 2003). **Free trade agreements** are pacts between two countries or among a group of countries that make it easier to trade goods across national boundaries. Free trade agreements reduce or eliminate foreign restrictions on exports, reduce or eliminate tariffs (or taxes) on imported goods, and prevent U.S. technology from being copied and used by competitors through protection of "intellectual property rights." Treaties such as the Canada–U.S. Free Trade Agreement, the North American Free Trade Agreement (NAFTA), the Free Trade Area of the Americas (FTAA), and the Central America Free Trade Agreement (CAFTA) are designed to accomplish these trade goals.

U.S. officials have also used Section 301 of the Trade Acts of 1984 and 1988 to force trade negotiations with individual countries. If U.S. trade officials determine that other countries have denied U.S. corporations "reasonable" access to domestic markets, sold their goods in the United States at below-market prices, or failed to protect the patents and copyrights of U.S. companies, Section 301 allows the United States to impose retaliatory sanctions and tariffs on goods from these countries (Schaeffer 2003).

Through GATT and the WTO, free trade agreements, and Section 301, U.S. trade officials have expanded trading opportunities, benefiting large export manufacturing and service industries in the global north, specifically aircraft, auto, computer, pharmaceutical, and entertainment industries in western Europe, the United States, and Japan. But trade globalization also hurt the U.S. steel and textile-apparel industries and the workers employed in them, small businesses that cannot compete with large retail chain stores, supermarkets, franchises, and small farmers (Schaeffer 2003). Since NAFTA was signed in 1993, the growth of U.S. exports supported 1 million U.S. jobs, but the growth of imports from Mexico and Canada displaced production that would have supported 2 million jobs. Thus, NAFTA has resulted in the loss of 1 million U.S. jobs, two-thirds of which are in the manufacturing industries (Scott & Ratner 2005).

Foreign workers have also been hurt by trade agreements. Since NAFTA took effect, imports of highly subsidized U.S. and Canadian grain and other agricultural products undercut Mexico's agricultural economy and put over 2 million family farmers out of business, many of whom have come to the United States to find work. After NAFTA, annual illegal immigration from Mexico doubled (Faux 2008). Free trade agreements also undermine the ability of national, state, and local governments to implement environmental and food or product safety policies.

Transnational Corporations

Although free trade agreements have increased business competition around the world, resulting in lower prices for consumers for some goods, they have also opened markets to monopolies (and higher prices) because they have facilitated the development of large-scale transnational corporations. **Transnational corporations,** also known as *multinational corporations*, are corporations that have their home base in one country and branches, or affiliates, in other countries. The number of transnational corporations more than doubled from about 35,000 in 1990, to about 75,000 in 2005 (Roach 2007). Among the world's largest economies, 29 are companies, rather than countries (Roach 2007).

Transnational corporations provide jobs for U.S. managers, secure profits for U.S. investors, and help the United States compete in the global economy. Transnational

free trade agreements Pacts between two countries or among a group of countries that make it easier to trade goods across national boundaries by reducing or eliminating restrictions on exports and tariffs (or taxes) on imported goods and protecting intellectual property rights.

transnational corporations Also known as multinational corporations, corporations that have their home base in one country and branches, or affiliates, in other countries.

corporations benefit from increased access to raw materials, cheap foreign labor, and the avoidance of government regulations:

> By moving production plants abroad, business managers may be able to work foreign employees for long hours under dangerous conditions at low pay, pollute the environment with impunity, and pretty much have their way with local communities. Then the business may be able to ship its goods back to its home country at lower costs and bigger profits. (Caston 1998, pp. 274–275)

Transnational companies can also avoid or reduce tax liabilities by moving their headquarters to a "tax haven." And the savings that big business reaps from cheap labor abroad are not passed on to consumers. "Corporations do not outsource to far-off regions so that U.S. consumers can save money. They outsource in order to increase their margin of profit" (Parenti 2007). For example, shoes made by Indonesian children working 12-hour days for 13 cents an hour cost only $2.60 but are still sold for $100 or more in the United States. In 2010, the wages and salaries of U.S. employees accounted for less than half of national income—an all-time low—while the share of national income that went to corporate profits was at its highest level ever (Norris 2011).

Transnational corporations contribute to the trade deficit in that more goods are produced and exported from outside the United States than from within. Transnational corporations also contribute to the budget deficit, because the United States does not get tax income from U.S. corporations abroad, yet transnational corporations pressure the government to protect their foreign interests; as a result, military spending increases. Third, transnational corporations contribute to U.S. unemployment by letting workers in other countries perform labor that U.S. employees could perform. Finally, transnational corporations are implicated in an array of other social problems, such as poverty resulting from fewer jobs, urban decline resulting from factories moving away, and racial and ethnic tensions resulting from competition for jobs.

Sociological Theories of Work and the Economy

In sociology, structural functionalism, conflict theory, and symbolic interactionism serve as theoretical lenses through which we may better understand work and economic issues and activities.

Structural-Functionalist Perspective

According to the structural-functionalist perspective, the economic institution is one of the most important of all social institutions. By providing the basic necessities common to all human societies, including food, clothing, and shelter, the economic institution contributes to social stability. After the basic survival needs of a society are met, surplus materials and wealth may be allocated to other social uses, such as maintaining military protection from enemies, supporting political and religious leaders, providing formal education, supporting an expanding population, and providing entertainment and recreational activities. Societal development is dependent on an economic surplus in a society (Lenski & Lenski 1987).

The economic institution can also be dysfunctional when it fails to provide members with the goods and services they need, when the distribution of goods and services is grossly unequal, and when the production, distribution, and consumption of goods and services depletes and pollutes the environment.

The structural-functionalist perspective is also concerned with how changes in one aspect of society affect other aspects. For example, when unemployment rates rise, college

Jmhite/iStock photo

Cities across the United States have cancelled the traditional fireworks show on Independence Day to save money.

enrollments go up, crime increases, and tax revenues decrease (unemployed people pay less in income tax and sales tax), which hurts the government's ability to pay for services such as education, garbage pickup, police and fire services, and road repairs.

What Do You Think? Over the last few years, many cities across the country have cancelled their annual Fourth of July fireworks to save money. Mayor Bill Cervenik, of Euclid, Ohio, said, "It came down to this: Did we want to spend $150,000 on something that would be over in a few hours? Or did we want to use that money to keep city workers employed?" In the Los Angeles suburb of Montebello, the City Council voted to use its $39,000 fireworks budget on donations to local food banks. Mayor Rosemarie Vasquez explained, "We figured that, instead of burning the money in the air, why not give it to people who need it?" (Huffstutter 2009). Protestors against the cancellation of fireworks shows argued that they are an important American tradition. If you were on the City Council, and the issue of whether or not to cancel the Fourth of July fireworks was being considered, what would your vote be?

Conflict Perspective

According to the conflict perspective, the ruling class controls the economic system for its own benefit and exploits and oppresses the working masses. The conflict perspective is critical of ways that the government caters to the interests of big business at the expense of workers, consumers, and the public interest. This system of government that serves the interests of corporations—known as **corporatocracy**—involves ties between government and business. For example, in Chapter 2, we discussed how the pharmaceutical and health insurance industries influence politicians on matters related to health care.

Corporate interests find their way into politics through large political contributions and "soft money," which is money that flows through a loophole to provide political parties, candidates, and contributors a means to evade federal limits on political contributions. Critics of this system of campaign financing argue that corporations and interest groups purchase political influence through financial contributions and lobbying. For example, in 2008, Congress passed the Emergency Economic Stabilization Act, which created a $700 billion Troubled Assets Relief Program (TARP), allowing the government to purchase failing bank assets that resulted largely from the subprime mortgage crisis. Companies that received taxpayers' money from the TARP bailout program had spent $77 million on lobbying and $37 million on federal campaign contributions in 2008. Companies that spent the most on lobbying and political campaign contributions, including General Motors, Bank of America, and American International Group (AIG) are also the ones that received the most bailout money (Center for Responsive Politics 2009).

A survey of business leaders' views on political fund-raising found that the main reasons U.S. corporations make political contributions are fear of retribution and to buy access to lawmakers (*Multinational Monitor* 2000). Although 75 percent of the surveyed business leaders said that political donations give them an advantage in shaping legislation, nearly three-quarters (74 percent) said that business leaders are pressured to make large political donations. Half of the executives said that their colleagues "fear adverse consequences for themselves or their industry if they turn down requests" for contributions.

The pervasive influence of corporate power in government exists worldwide. The policies of the International Monetary Fund (IMF) and the World Bank pressure developing countries to open their economies to foreign corporations, promoting export production at the expense of local consumption, encouraging the exploitation of labor as a means of attracting foreign investment, and hastening the degradation of natural

corporatocracy A system of government that serves the interests of corporations and that involves ties between government and business.

During much of the 1990s, the U.S. economy flourished—incomes rose, unemployment and poverty rates fell, and stock market values soared. In the early 2000s, these economic trends changed, and unemployment rates went back up. But unlike earlier job loss patterns of the 1980s that largely affected blue-collar workers, more recent job losses have increasingly involved middle-class professionals. Research presented here investigates how job loss affected the self-concepts and views about employment among a sample of unemployed professionals in midlife (Mendenhall et al. 2008).

Sample and Methods

The sample consisted of 77 men and women who were recruited through two Chicago-area networking groups that help unemployed managers and executives find employment, and also through announcements at local churches and posted flyers at cafés and public libraries. Participants had to meet four eligibility criteria: They had to (1) have been unemployed for at least three months during the past year; (2) have been married at the time of their job loss; (3) have children between the ages of 12 and 18 living at home; and (4) live in the Chicago area. Most participants were male (83 percent) and white (80 percent). Respondents had been unemployed for an average of 15 months at the time of their first interview. More than half had earned annual salaries of $100,000 or more; no one earned less than $50,000, and most had jobs with generous benefits.

The researchers interviewed participants at libraries, cafés, or on the University of Chicago campus. Interview topics included the job loss event, how the job loss affected family relationships, and educational plans for their children. Participants also completed a survey about their job loss experience, family economic circumstances, family relationships, and their own health and well-being. About a year later, participants were interviewed again, and completed another questionnaire.

Selected Findings

This study revealed some interesting patterns in how job loss among professionals in midlife (1) affected their self-concept, (2) influenced their job-seeking strategies, and (3) shaped the messages about employment they conveyed to their children. The participants viewed their job termination as evidence of a lack of employer loyalty and a change from a lifetime employment contract to one in which even high-level employees can be terminated without warning—a shift in which participants came to view themselves as free agents who effectively "rent" their services to employers. As free agents, there is no expectation of permanent employment or loyalty in the employee-employer relationship.

More than half of the participants 50 years and older said they perceived age discrimination in the job market, in that employers signaled that they were looking for entry-level employees, which is code for "young." In response, participants often de-emphasized their age and experience by omitting graduation dates and some of their work history on their resumes—a phenomenon known as "deprofessionalization." A former CEO explained that, in the first several months of job searching, he listed the date of his college graduation on his resume and got no responses. Then he deleted the date, and got a half dozen responses.

Many participants viewed their job loss experience as providing an opportunity to teach real-life lessons to their children about the world of work. A 50-year-old project manager said,

I think [that my son getting a good glimpse into the reality of life is] a positive because . . . if he goes out there with rose-tinted glasses, he's going to get smacked upside the head real hard someday. (p. 203)

One 47-year-old information technology consultant used his job loss experience to teach his children to prepare for a job market in which neither employee nor employer expects a lifetime employment contract. He told his children,

For most jobs now, you need to view [it] like the movie industry where . . . you're rented . . . you're contracted to do that movie, whether you're the lights or the cameraman or whatever, and then you're out of work again. And that's more the way most jobs are now where you should view a job as just your job until you're out of work again. (p. 204)

Respondents advised their children to take specific steps to prevent being overly dependent on employers. Some advised their children to own their own businesses; others urged their children to develop skills that could be "transferable" to a new employer. Others encouraged their children to choose careers where their potential client base was diversified:

I've started talking to them about how . . . they should start thinking about [whether] they want to work for big companies or . . . go into a field where their client base is very much diversified, which is lawyers, doctors, CPAs, therapists. . . . Because then if someone fires you, you don't care because you have another 100 [clients] whereas when you work for a company, and . . . your boss . . . fires you . . . you're out of a job. (p. 204)

Source: Mendenhall et al. 2008.

---

resources as countries sell their forests and minerals to earn money to pay back loans. In his book *Confessions of an Economic Hit Man*, John Perkins (2004) described his prior job as an "economic hit man"—a highly paid professional who would convince leaders of poor countries to accept huge loans (primarily from the World Bank) that were much bigger than the country could possibly repay. The loans would be used to help develop the country by paying for needed infrastructure, such as roads, electrical plants, airports, shipping ports, and industrial plants. One of the conditions of the loan was that the borrowing country had to give 90 percent of the loan back to U.S.

companies (such as Halliburton or Bechtel) to build the infrastructure. The result: The wealthiest families in the country benefit from additional infrastructure and the poor masses are stuck with a debt they cannot repay. The United States uses the debt as leverage to ask for "favors," such as land for a military base or access to natural resources such as oil. According to Perkins, large corporations want "control over the entire world and its resources, along with a military that enforces that control" (quoted by MacEnulty 2005, p. 10).

Symbolic Interactionist Perspective

According to symbolic interactionism, the work role is a central part of a person's self-concept and social identity. When making a new social acquaintance, one of the first questions we usually ask is, "What do you do?" The answer largely defines for us who that person is. An individual's occupation is one of the person's most important statuses; for many, it represents a "master status," that is, the most significant status in a person's social identity. This chapter's *Social Problems Research Up Close* feature describes a study that looks at how job loss of white-collar professionals in midlife affects their self-concepts and attitudes about work and unemployment.

Symbolic interactionism emphasizes the fact that attitudes and behavior are influenced by interaction with others. The applications of symbolic interactionism in the workplace are numerous: Employers and managers use interpersonal interaction techniques to elicit the attitudes and behaviors they want from their employees; union organizers use interpersonal interaction techniques to persuade workers to unionize. And, as noted in the Social Problems Research Up Close feature, parents teach their young adult children important lessons about work and unemployment through interaction with them.

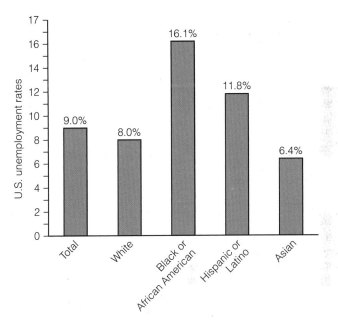

Figure 7.2: U.S. Unemployment Rates by Race and Hispanic Origin, April 2011
Source: Bureau of Labor Statistics 2011a.

Problems of Work and Unemployment

In this section, we examine unemployment and other problems associated with work. Poverty, minimum wage and living wage issues, workplace discrimination, and retirement and employment concerns of older adults are discussed in other chapters. Here, we discuss problems concerning unemployment and underemployment, child labor, forced labor, sweatshop labor, health and safety hazards in the workplace, alienation, work-life conflict, and labor unions and the struggle for workers' rights.

Unemployment and Underemployment

In 2010, 205 million people worldwide—6.2 percent of the global labor force—were unemployed, with the highest rates being in the Middle East (10.3 percent) and in North Africa (9.8 percent) (ILO 2011). Measures of **unemployment** in the United States consider individuals to be unemployed if they are currently without employment, are actively seeking employment, and are available for employment. In 2000, the U.S. unemployment rate dipped to a 31-year low of 4 percent. During the economic recession that began in 2007, mass layoffs pushed the unemployment rate to 10 percent in the last quarter of 2009, as companies went out of business and plants closed. A **recession** refers to a significant decline in economic activity spread across the economy and lasting for at least six months. In communities hardest hit by the recession, unemployment rates have been over 20 percent. Rates of unemployment are higher among racial and ethnic minorities (see Figure 7.2) and among those with lower levels of education (see Chapter 8).

unemployment To be currently without employment, actively seeking employment, and available for employment, according to U.S. measures of unemployment.

recession A significant decline in economic activity spread across the economy and lasting for at least six months.

GOING OUT OF BUSINESS

Circuit City

When Circuit City closed its 567 stores nationwide, some 30,000 employees lost their jobs.

job exportation The relocation of jobs to other countries where products can be produced more cheaply.

outsourcing A practice in which a business subcontracts with a third party to provide business services.

automation The replacement of human labor with machinery and equipment.

long-term unemployment rate The share of the unemployed who have been out of work for 27 weeks or more.

underemployment Unemployed workers as well as (1) those working part-time but who wish to work full-time, (2) those who want to work but have been discouraged from searching by their lack of success, and (3) others who are neither working nor seeking work but who want and are available to work and have looked for employment in the last year. Also refers to the employment of workers with high skills and/or educational attainment working in low-skill or low-wage jobs.

Causes of Unemployment A primary cause of unemployment is lack of available jobs. In December 2000, jobs were plentiful: For every job opening, there was 1.1 job seeker. In summer 2009, for every job opening, there were *more than six unemployed U.S. workers*! By March 2011, the ratio of jobs to job seekers was 1 to 4.3, marking more than two years of there being more than four job seekers for every available job (Shierholz 2011).

Another cause of U.S. unemployment is **job exportation,** or the relocation of jobs to other countries where products can be produced more cheaply. **Outsourcing,** which involves a business subcontracting with a third party to provide business services, saves companies money as they pay lower salaries and no benefits to those who provide outsourced services. Many commonly outsourced jobs, including accounting, web development, information technology, telemarketing, and customer support, are outsourced to non-U.S. workers. **Automation,** or the replacement of human labor with machinery and equipment, also contributes to unemployment.

Another cause of unemployment is increased global and domestic competition. Mass layoffs in the U.S. automobile industry have occurred, in part, due to competition from makers of foreign cars who, unlike U.S. automakers, do not have the burden of providing health insurance for their employees. Finally, unemployment results from mass layoffs that occur when plants close and companies downsize or go out of business. In 2009, Circuit City went out of business after revenues steadily declined due, in part, to competition from Best Buy and Wal-Mart. Circuit City was also hurt by the recession that started in late 2007, as consumers cut back on spending.

Long-Term Unemployment The **long-term unemployment rate** refers to the share of the unemployed who have been out of work for 27 weeks or more. In April 2011, more than four in 10 unemployed Americans had been jobless for 27 weeks or more (Bureau of Labor Statistics 2011a).

Underemployment Unemployment figures do not include "discouraged" workers, who have given up on finding a job and are no longer looking for employment. In 2010, unemployed people gave up looking for work and left the labor force after about 20 weeks of being jobless (Ilg 2011). Unemployment figures also do not count those working part-time but who wish to work full-time ("involuntary" part-timers). Measures of **underemployment** include unemployed workers as well as discouraged workers and "involuntary" part-timers. The underemployment rate is always higher than the unemployment rate, which means that official unemployment rates—frequently reported in the media—undercount those whose employment needs are not being met. For example, although the official unemployment rate in April 2011 was 9.5 percent, the underemployment rate was 16.6 percent (Bureau of Labor Statistics 2011a).

Effects of Unemployment on Individuals, Families, and Societies A recent study found that poor-quality jobs—those with high demands, low control over decision making, high job insecurity, and low pay had more negative effects on mental health than having no job at all (Butterworth et al. 2011). Nevertheless, unemployment has been linked to depression, low-self esteem, and increased mortality rates (Turner & Irons 2009). One study found that, among workers with no preexisting health conditions, losing a job due to a business closure increased the risk for a new health problem by 85 percent, with the most common health problems including hypertension, heart disease, and arthritis (Strully 2009). For workers who receive health insurance through their employer, losing a job can mean losing health insurance coverage (see also Chapter 2).

Long-term unemployment can have lasting effects, such as increased debt, diminished retirement and savings accounts (which are depleted to meet living expenses), home foreclosure, and/or relocation from secure housing and communities to unfamiliar places to find a job. But, even when individuals who are laid off find another job in one or two weeks, they still suffer damage to their self-esteem from having been told they are no longer wanted or needed at their workplace. And being fired affects worker trust and loyalty in future jobs. Employees who are not fired during a mass layoff are also affected, as they worry that "their job could be next" (Uchitelle 2006).

A recent study found that poor-quality jobs—those with high demands, low control over decision making, high job insecurity, and low pay had more negative effects on mental health than having no job at all.

Unemployment is a risk factor for homelessness, substance abuse, and crime, as some unemployed individuals turn to illegitimate, criminal sources of income, such as theft, drug dealing, and prostitution. Globally, the high numbers of young adults without jobs create a risk for crime, violence, and political conflict (United Nations 2005). As discussed in other chapters, unemployed young adults are targeted for recruitment into terrorist groups.

In families, unemployment is also a risk factor for child and spousal abuse and marital instability. When an adult is unemployed, other family members are often compelled to work more hours to keep the family afloat. And unemployed noncustodial parents, usually fathers, fall behind on their child support payments.

Plant closings and large-scale layoffs affect communities by lowering property values and depressing community living standards. High numbers of unemployed adults create a drain on societies that provide support to those without jobs. And early and mid-career job displacement is linked to lower rates of social participation in church groups, charitable organizations, youth and community groups, and civic and neighborhood groups (Brand & Burgard 2008).

Unemployment and underemployment create a vicious cycle: The un- and underemployed (as well as those who fear job loss) cut back on spending, which hurts businesses that then must cut jobs to stay afloat. During times of economic stress or uncertainty, have you cut back on everyday spending? See this chapter's *Self and Society* feature.

Employment Concerns of Recent College Grads

Most college students seek a college degree to improve their job opportunities. But having a college degree is no guarantee of employment in a struggling economy. The unemployment rate for young recent college graduates (adults under age 24 and younger with a bachelor's degree or higher) was 9.4 percent in 2010—the highest since the Labor Department began keeping records in 1985 (Isidore 2011).

Due to the high unemployment rate, high cost of living, and college debts averaging $20,000 per graduate, many college graduates end up living with their parents instead of on their own. Almost one-third of recent college grads are living with their parents, 17 percent say they are financially dependent on their parents, and nearly one in four say they are in debt (Isidore 2011). Some new college graduates are glad that jobs are scarce. After graduation, some new grads are taking time off to relax and to travel instead of seeking employment (Weiss 2009). Other college grads question whether going to college was worth the expense and the effort. One survey of 4 million college graduates found that less than half were working in jobs that required a college degree (Pugh 2009). Another survey found that nearly one in five (18 percent) young recent college grads resorted to taking a job outside their field of study (Isidore 2011).

Forced Labor and Slavery

Forced labor, also known as *slavery*, refers to any work that is performed under the threat of punishment and is undertaken involuntarily. There are more slaves in the world today than at any other time in history—27 million (Skinner 2008). A resurgence of slavery

forced labor Also known as slavery, any work that is performed under the threat of punishment and is undertaken involuntarily.

During the recent economic crisis, Americans bought fewer big-ticket items such as computers and new cars. In addition, many Americans cut back on their everyday spending. Difficult economic times make people reconsider what is important for them to have in their daily lives, and what they can live without or change to save money. How do your spending habits change in hard economic times? For each of the following spending categories, indicate how your spending has changed in response to tough economic times.

	Have Done	Have Considered	Have Not Done or Considered	n/a
1. Buy more generic brands	___	___	___	___
2. Take lunch from home instead of buying lunch out	___	___	___	___
3. Go to hairdresser/barber less often	___	___	___	___
4. Stopped purchasing bottled water; switched to refillable water bottle	___	___	___	___
5. Cancelled one or more magazine subscriptions	___	___	___	___
6. Cancelled or cut back on cable service	___	___	___	___
7. Stopped buying coffee in the morning	___	___	___	___
8. Cut down on dry cleaning	___	___	___	___
9. Changed or canceled cell phone service	___	___	___	___
10. Cancelled landline phone service and only use cell phone	___	___	___	___
11. Cancelled a newspaper subscription	___	___	___	___
12. Begun carpooling or using mass transit	___	___	___	___

Comparison data: Table 1 presents the percentages of responses to each of the 12 previous items from a 2010 online Harris Poll of 3,084 U.S. adults. The most common changes in spending behavior include purchasing more generic brands and taking lunch from home instead of buying lunch out.

TABLE 1 Spending/Savings over Past Six Months, in Percentages

"Have you done or considered doing any of the following over the past six months to save money?"

	HAVE DONE	HAVE CONSIDERED	HAVE NOT DONE OR CONSIDERED	N/A
1.	62	12	17	9
2.	45	9	13	32
3.	37	8	30	25
4.	37	10	22	31
5.	27	8	23	41
6.	22	21	40	17
7.	22	7	19	53
8.	21	5	16	57
9.	17	16	50	17
10.	17	21	40	21
11.	17	11	27	44
12.	14	9	31	45

Source: Harris Poll 2010.

around the world is linked to three main factors: (1) rapid growth in population, especially in the developing world; (2) social and economic changes that have displaced many rural dwellers to urban centers and their outskirts, where people are powerless and jobless and are vulnerable to exploitation and slavery; and (3) government corruption that allows slavery to go unpunished, even though it is illegal in every country (Bales 1999).

> There are more slaves in the world today than at any other time in history.

Forced labor exists all over the world but is most prevalent in India, Pakistan, Bangladesh, and Nepal. Most forced laborers work in agriculture, mining, prostitution, and factories. Forced laborers produce goods we use every day, including sugar from the Dominican Republic, chocolate from the Ivory Coast, paper clips from China, carpets from Nepal, toys from China, and clothing from India.

Forms of Slavery and Forced Labor The form of slavery most people are familiar with is **chattel slavery,** in which slaves are considered property that can be bought and sold. In the past, the high cost of purchasing a slave (about $40,000 in today's money) gave the master incentive to provide a minimum standard of care to ensure that the slave would be healthy enough to work and generate profit for the long term. Today, slaves are cheap, costing an average of $100 or less. In Port-au-Prince, Haiti, a 10-year-old girl can be bought for $50.00 (Skinner 2008). Because they are so cheap and abundant, slaves are no longer a major investment worth maintaining. If slaves become ill or injured, too old to work, or troublesome to the slaveholder, they are dumped or killed and replaced with another slave (Cernasky 2003).

Although chattel slavery still exists in some areas, most forced laborers today are not "owned" but are rather controlled by violence, the threat of violence, and/or debt. The most common form of forced labor today is called *bonded labor.* Bonded laborers are usually illiterate, landless, rural, poor individuals who take out a loan simply to survive or to pay for a wedding, funeral, medicines, fertilizer, or other necessities. Debtors must work for the creditor to pay back the loan, but often they are unable to repay it. Creditors can keep debtors in bondage indefinitely by charging the debtors illegal fines (for workplace "violations" or for poorly performed work) or charge laborers for food, tools, and transportation to the work site while keeping wages too low for the debt to ever be repaid. Alternatively, creditors can claim that all the labor the debtor performs is collateral for the debt and cannot be used to reduce it (Miers 2003).

Mark Peterson/CORBIS

Forced prison labor is a type of forced labor that is controlled by the state. Forced prison labor is particularly widespread in China.

Another common form of forced labor involves luring individuals with the promise of a good job and instead holding them captive and forcing them to work. Migrant workers are particularly vulnerable because, if they try to escape and report their abuse, they risk deportation. Organized crime rings are sometimes involved in the international trafficking of human beings, which often flows from developing nations to the West. A form of forced labor most common in South Asia is sex slavery, in which girls are forced into prostitution by their own husbands, fathers, and brothers to earn money to pay family debts. Other girls are lured by offers of good jobs and then are forced to work in brothels under the threat of violence.

Another type of forced labor is conducted by the state or military. Forced military service, which has been reported in parts of Africa, and forced prison labor (common in China) are examples of state and military forced labor.

chattel slavery A form of slavery in which slaves are considered property that can be bought and sold.

Forced Labor in the United States Each year, 14,000 to 17,000 people are trafficked into the United States and forced into slavery, most commonly in domestic work, farm labor, and the sex industry (Skinner 2008). Migrant workers are tricked into working for little or no pay as a means of repaying debts from their transport across the U.S. border, similar to debt bondage in South Asia. Traffickers posing as employment agents lure women into the United States with the promise of good jobs and education but then place them in "jobs" where they are forced to do domestic or sex work.

Sweatshop Labor

Millions of people worldwide work in **sweatshops**—work environments that are characterized by less-than-minimum wage pay, excessively long hours of work (often without overtime pay), unsafe or inhumane working conditions, abusive treatment of workers by employers, and/or the lack of worker organizations aimed at negotiating better working conditions. Sweatshop labor conditions occur in a wide variety of industries, including garment production, manufacturing, mining, and agriculture.

Sweatshop labor commonly occurs in the garment industry.

More than 97 percent of all clothing purchased in the United States is imported, often made under sweatshop conditions (Institute for Global Labour & Human Rights 2011). Garment workers at a factory in Ha-Meem, Bangladesh, work 12 to 14 hour shifts, 7 days a week, with one day off a month. Young women making garments for Gap, J.C. Penney, Phillips-Van Heusen, Target, and Abercrombie & Fitch make 20 to 26 cents an hour (Institute for Global Labour & Human Rights 2011). In 2010, 29 workers died in a fire at the Ha-Meem Factory; they could not escape because the emergency exits were illegally locked.

Many products in the U.S. consumer market are made under sweatshop conditions. An investigative report on working conditions in five Chinese factories that produce products for Disney, Wal-Mart, Kmart, Mattel, and McDonald's revealed sweatshop conditions that violate Chinese labor laws (Students and Scholars Against Corporate Misbehavior 2005). Workers are forced to work grueling 12- to 15-hour days, earning just 33 cents to 41 cents an hour. Workers are housed in overcrowded dorm rooms and fed horrible food at the factory canteen. They are charged for the housing and food provided at the factory (even if they live and eat elsewhere), which often costs them one-fifth to one-third of their monthly wages. Some factories have no fans and become oppressively hot. Workers often faint from exhaustion and the unbearably stifling heat. Some workers are exposed to strong-smelling gases from working with glue, with no protective masks or ventilation system. Crushed fingers and other injuries are common in some factory departments. Workers have no health insurance, no pension, and no right to freedom of association or to organize.

Sweatshop Labor in the United States Sweatshop conditions in overseas industries have been widely publicized. However, many Americans do not realize the extent to which sweatshops exist in the United States. The Department of Labor estimates that more than half of the country's 22,000 sewing shops violate minimum wage and overtime laws, and that 75 percent violate safety and health laws ("The Garment Industry" 2001). Most garment workers in the United States are immigrant women who

sweatshops Work environments that are characterized by less-than-minimum wage pay, excessively long hours of work (often without overtime pay), unsafe or inhumane working conditions, abusive treatment of workers by employers, and/or the lack of worker organizations aimed to negotiate better working conditions.

typically work 60 to 80 hours a week, often earning less than minimum wage with no overtime, and many face verbal and physical abuse.

Immigrant farm workers, who process 85 percent of the fruits and vegetables grown in the United States, also work under sweatshop conditions. Many live in substandard and crowded housing provided by their employer and lack access to safe drinking water as well as bathing and sanitary toilet facilities. Farm workers commonly suffer from heat exhaustion, back and muscle strains, injuries resulting from the use of sharp and heavy farm equipment, and illness resulting from pesticide exposure (Austin 2002). Working 12-hour days under hazardous conditions, farm workers have the lowest annual family incomes of any U.S. wage and salary workers, and more than 60 percent of them live in poverty (Thompson 2002). Problems associated with immigrant labor are discussed further in Chapter 9.

Child Labor

Child labor involves a child performing work that is hazardous, that interferes with a child's education, or that harms a child's health or physical, mental, social, or moral development. Even though virtually every country in the world has laws that prohibit or limit the extent to which children can be employed, child labor persists throughout the world. According to the most recent data available, 215 million school-age children are engaged in child labor; 115 million are involved in hazardous work (ILO 2010).

Child labor is involved in many of the products we buy, wear, use, and eat. Child laborers work in factories, workshops, construction sites, mines, quarries, fields, and on fishing boats. Child laborers make bricks, shoes, soccer balls, fireworks and matches, furniture, toys, rugs, and clothing. They work in the manufacturing of brass, leather goods, and glass. They tend livestock and pick crops. Tens of thousands of children in at least 24 countries and territories are recruited by armed forces where children are used in combat or for sexual exploitation (UNICEF 2009).

Child laborers work long hours with few (or no) breaks or days off, often in unsafe conditions where they are exposed to toxic chemicals and/or excessive heat, and they endure beatings and other forms of mistreatment from their employers, all for as little as a dollar a day. A former textile worker in Bangladesh described conditions at a company called Harvest Rich, where clothing is sewn for the U.S. firms including Wal-Mart and J. C. Penney. She testified that hundreds of children, some as young as 11 years old, were illegally working at Harvest Rich, sometimes for up to 20 hours a day: "Before clothing shipments had to leave for the United States, there are often mandatory 19- to 20-hour shifts from 8:00 a.m. to 3:00 or 4:00 a.m. . . . The workers would sleep on the factory floor for a few hours before getting up for their next shift in the morning. If they did anything wrong, they were beaten every day." Workers had two days off a month and were paid $3.20 a week (Tate 2007). At another garment factory that made fleece jackets for Wal-Mart, 14- or 15-year-old kids worked 18- or 20-hour shifts, from 8:00 a.m. to midnight or 4:00 a.m., seven days a week. When they passed out, rulers struck them to wake them up. Some of the girls were raped by management (Tate 2007).

Child labor also exists in the United States in restaurants, grocery stores, meatpacking plants, garment factories, and agriculture. Despite federal prohibitions, U.S. youth employed in service and retail jobs are exposed to harmful conditions and dangerous equipment (e.g., paper balers, box crushers, and dough mixers) (Runyan et al. 2007). One of the most dangerous forms of child labor in the United States is agricultural work. This chapter's Photo Essay looks at child labor in U.S. agriculture.

child labor Involves a child performing work that is hazardous, that interferes with a child's education, or that harms a child's health or physical, mental, social, or moral development.

Health and Safety in the U.S. Workplace

Although many workplaces are safer today than in generations past, fatal and disabling occupational injuries and illnesses still occur in troubling numbers. In 2009, 4,551 U.S. workers—most of whom were men—died of fatal work-related injuries (Bureau of

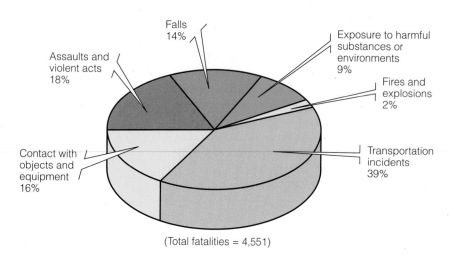

Falls
14%

Assaults and
violent acts
18%

Exposure to harmful
substances or
environments
9%

Fires and
explosions
2%

Contact with
objects and
equipment
16%

Transportation
incidents
39%

(Total fatalities = 4,551)

Figure 7.3: Causes of Workplace Fatalities, 2009
Source: Bureau of Labor Statistics 2011b.

Labor Statistics 2011b). The most common type of job-related fatality involves transportation accidents (see Figure 7.3). Although the highest number of fatal injuries occurred in construction, the highest rates (per 100,000 workers) occurred in agriculture/forestry/fishing/hunting, transportation and warehousing, and mining.

Nonfatal occupational injuries in U.S. workplaces are not uncommon—more than 4 million cases were reported in 2009 (AFL-CIO 2011). The most common nonfatal injuries are sprains, strains, and tears (Bureau of Labor Statistics 2010). Workers who do repeated motions such as typing or assembly line work are prone to repetitive strain injuries (see this chapter's *The Human Side* feature).

The incidence of illnesses resulting from hazardous working conditions is probably much higher than the reported statistics show, because long-term latent illnesses caused by, for example, exposure to carcinogens often are difficult to relate to the workplace and are not adequately recognized and reported. In addition, employees don't always report workplace injuries or illnesses for fear of losing their jobs, or in the case of undocumented immigrants, being deported. And companies don't always maintain accurate records of workplace injuries and illnesses, to avoid scrutiny and fines for possible violations of health and safety regulations.

In many cases, workplace injuries, illnesses, and deaths are due to employers' willful violations of health and safety regulations. The Occupational Safety and Health Administration (OSHA), created in 1970, develops, monitors, and enforces health and safety regulations in the workplace. But inadequate funding leaves OSHA unable to hire enough workplace inspectors to do the job effectively. Federal and state OSHA have a total of (in 2010) 2,218 inspectors to inspect about 8 million workplaces, which means that federal OSHA can inspect workplaces an average of once every 129 years; state OSHA can inspect workplaces once every 67 years (AFL-CIO 2011). The International Labour Office recommends that industrial economies have one inspector per 10,000 workers. To comply with this standard, the U.S. would have needed (in 2010) 12,792 inspectors, but instead had only 2,218 inspectors (AFL-CIO 2011). Nevertheless, in 2010, OSHA inspectors documented more than 90,000 violations of health and safety regulations in the workplace, most of which were classified as "serious" (AFL-CIO 2011).

What Do You Think? Suppose that a corporation is guilty of a serious violation of health and safety laws, in which "serious violation" is defined as one that poses a substantial probability of death or serious physical harm to workers. What penalty do you think that corporations should pay for such a violation? Serious violations of workplace health and safety laws in 2010 carried an average federal penalty of $1,052 or a state penalty of $858 (AFL-CIO 2011). Under federal law, willful violation that results in a worker's death is considered a misdemeanor and carries a maximum prison sentence of only six months. Contrast this with the crime of harassing a wild burro on federal lands, which is punishable by one year in prison (Barstow & Bergman 2003). Why do you think penalties for violating workplace health and safety laws are so weak?

Celia Lopez felt lucky when she was hired at the House of Raeford Farms turkey plant in Raeford, North Carolina. But after six years, the 44-year-old mother of three said she feared the "hands that take care of my family" are ruined. She has had carpal tunnel surgery on both hands. At the request of the *Charlotte Observer* newspaper, Lopez recounted a typical day:

6:45 a.m. Lopez walks through the gate of the sprawling plant. She's struck by the pungent smell of ammonia. She punches her time card and puts on her gear—rubber boots, apron, hairnet, and two pairs of gloves. She rushes to position. Workers must be at their posts before the production line starts. No excuses.

7 a.m. The line starts. Lopez begins by grabbing and placing turkey breasts on plates to be weighed. Each plate must weigh between 6 and 6 ½ pounds. She grabs meat with her right hand and uses her left to hold the plate, then pushes the turkey along the line. She'll repeat this process hundreds of times an hour.

9:30 a.m. If Lopez needs a bathroom break, she must wait until a supervisor finds someone to replace her on the line. This can take minutes or hours—if approved at all. "Bathroom breaks are a privilege, not a necessity," she said her bosses told her. If granted, she has 10 minutes to remove her gear, use the facilities, and return.

11 a.m. — Lunch.

11:30 a.m. Back on the line. She has processed hundreds of pounds of meat. The line is moving fast; workers struggle to keep pace, she says. Conversation is minimal.

2 p.m. Break. She looks for a wall to press her back against and stretch her muscles.

2:30 p.m. The next two hours are the hardest—the piles of meat seem endless, she says. Her back cramps, pain spreading to her shoulder, arms, and hands. She is exhausted from standing. Sometimes she feels dizzy.

4 p.m. She punches out. She changes out of her work clothes, washes her face, and leaves.

4:30 p.m. She arrives home and takes a shower. "The meat smell gets stuck in your skin," she says.

About 7 p.m. She helps cook dinner for her family. Grasping a spoon is hard, she says. She uses two hands to carry a dinner plate. Basic tasks take longer because of the pain. "It's like ants crawling through my hands, up my arms," she says.

9 p.m. She takes two ibuprofen pills before rubbing her hands with alcohol and lotion—a nightly routine.

9:30 p.m. She goes to bed.

Midnight —2 a.m. Lopez frequently wakes up, hands cramping. She squeezes her fists and rubs her fingers to get blood flowing. She may wake up four times a night, each time the pain is worse. She swallows more ibuprofen.

6 a.m. Her alarm sounds. The line starts in two hours. "Sometimes I cry. I just pray to God that he will show me the way."

Source: Adapted from Franco Ordonez "A Worker's Grueling Day." *The Charlotte Observer*, 6/25/2010. Copyright © 2010 by the Charlotte Observer. Reprinted by permission.

Job Stress Another work-related health problem is job stress. A Gallup poll found that one-third of respondents is dissatisfied with the amount of stress in their jobs, making job stress workers' most common complaint (Saad 2010). Workers can be stressed by the number of hours they work, as the 40-hour workweek has become, for many workers, a 60-hour (and more) workweek. As the following excerpt reveals, employers are asking more of their employees, expecting them to work harder and faster, and employees fear that they could lose their jobs if they don't work harder:

> With many companies shedding one group of workers after another, the employees who survive often have more and more responsibilities heaped on them. Many workers feel internal pressures to perform above expectations, fearing that if they don't, they might be next to lose their jobs to downsizing or offshoring. (Greenhouse 2008, p. 186)

Prolonged job stress, also known as **job burnout,** can cause or contribute to physical and mental health problems, such as high blood pressure, ulcers, headaches, anxiety, and depression. Taking time off to heal and "recharge one's batteries" is not an option for many workers: One-half of the U.S. workforce has no paid sick leave, and one-fourth has no paid vacation (Watkins 2002). The United States is the only advanced nation that does not mandate a minimum number of vacation days; in the 27 European Union countries, employers are required to give workers at least four weeks of vacation each year (some countries mandate five or six weeks of vacation) (Greenhouse 2008). And, even when the workday is officially over, workers are connected to their jobs by their Blackberry, cell phones, e-mail accounts, and laptops. One in five employees works on a laptop computer while on vacation, 40 percent check office e-mail, and 50 percent check voicemail (Fram 2007).

The United States is the only advanced nation that does not mandate a minimum number of vacation days.

job burnout Prolonged job stress that can cause or contribute to high blood pressure, ulcers, headaches, anxiety, depression, and other health problems.

In most jobs, U.S. law requires children to be at least 16 years old to work, with a few exceptions, such as that 14- and 15-year-olds can work as cashiers, grocery baggers, and car washers. But in agricultural jobs, U.S. law allows children of any age to work on small farms as long as they have a parent's permission. Children ages 12 and older may be hired on any farm with their parent's consent, or if they work on the same farm with their parents. U.S. law allows children at age 14 to work on any farm, even without parental permission.

Hundreds of thousands of U.S. child workers labor on commercial farms (Human Rights Watch 2010). Child farmworkers in the United States are typically of Hispanic origin; many are U.S. citizens.

Farm work is considered the most dangerous form of child labor in the United States. Child farm workers are exposed to chemicals, sun, and temperature extremes; they work with sharp tools and heavy machinery, climb tall ladders, and carry heavy buckets and sacks. They frequently work 12-hour days without adequate access to drinking water, toilets, or hand-washing facilities. In nonagricultural jobs, U.S. law requires workers engaged in hazardous work to be at least 18 years old. But in agricultural jobs, children are permitted to do work that the U.S. Department of Labor deems "particularly hazardous" at age 16, and at any age on farms owned or operated by their parents.

The Children's Act for Responsible Employment (CARE Act) was introduced into Congress in 2009. This bill proposes to raise the minimum age of farmworkers from 12 to 14 (except for children working on their parents' farm). The bill also contains provisions to protect young farmworkers from hazardous work and exposure to pesticides. As of this writing, the CARE Act has not passed.

Melissa Farlow/National Geographic/Getty Images

◄Girls who work on U.S. farms are sometimes victims of sexual harassment.

◄ Child farmworkers are frequently exposed to pesticides sprayed on crops.

▼ Child laborers working on a tobacco farm are especially vulnerable to "green tobacco sickness" a poisoning that occurs when workers absorb tobacco through the skin when they come in contact with the leaves. Symptoms include nausea, vomiting, weakness, dizziness, abdominal pain, diarrhea, and shortness of breath.

Melissa Farlow/National Geographic/Getty Images

Jim Sugar/CORBIS

◄ This child farmworker uses a sharp saw to cut trees at this Christmas tree farm.

▼ Children are more susceptible to heat stroke than adults. Working long hours in the hot sun places child farmworkers at the risk of heat stroke and dehydration.

Stacy Barnett/Shutterstock.com

LOWELL GEORGIA/National Geographic Stock

Work-Life Conflict

A major source of stress for U.S. workers is the day-to-day struggle to meet the demands of work and other life responsibilities, including family and education, simultaneously. Work-life conflicts are common among U.S. workers: Seventy percent of U.S. women and men report some interference between work and nonwork responsibilities (Schieman et al. 2009).

Employed spouses strategize ways to coordinate their work schedules to have vacation time together, or simply to have meals together. Employed parents with young children must find and manage arrangements for child care and negotiate with their employers about taking time off to care for a sick child, or to attend a child's school event or extracurricular activity. Some employers post their workers' weekly schedules only a few days in advance, making it difficult to arrange child care. In some two-parent households, spouses or partners work different shifts so that one adult can be home with the children. However, working different shifts strains marriage relationships, because the partners rarely have time off together. Some employed parents who cannot find or afford child care leave their children with no adult supervision.

Employees with elderly and/or ill parents worry about how they will provide care for their parents, or arrange for and monitor their care, while putting in a 40-hour (or more) workweek. About two in 10 employees are caregivers providing care to a person over age 50 (Council of Economic Advisors 2010).

Women tend to experience their jobs as interfering with their nonwork lives more so than do men; however, the level of work-life conflict women experience has remained stable over the past three decades, whereas men's reported level of work-life conflict has increased, probably due to the increased involvement of men in child care and household responsibilities (Galinsky et al. 2009).

Balancing the responsibilities of a job with the demands of school is also a challenge for many adults. Millions of college students, both traditional age (18 to 24) and over age 25 work either part-time or full-time jobs while pursuing their college degree.

Alienation

Work in industrialized societies is characterized by a high degree of division of labor and specialization of work roles. As a result, workers' tasks are repetitive and monotonous and often involve little or no creativity. Limited to specific tasks by their work roles, workers are unable to express and utilize their full potential—intellectual, emotional, and physical. According to Marx, when workers are merely cogs in a machine, they become estranged from their work, the product they create, other human beings, and themselves. Marx called this estrangement "alienation." As we discussed earlier, the McDonaldization of the workplace also contributes to alienation.

Alienation has four components: powerlessness, meaninglessness, normlessness, and self-estrangement. Powerlessness results from working in an environment in which one has little or no control over the decisions that affect one's work. Meaninglessness results when workers do not find fulfillment in their work. Workers may experience normlessness if workplace norms are unclear or conflicting. For example, many companies that have family leave policies informally discourage workers from using them, or workplaces that officially promote nondiscrimination in reality practice discrimination. Alienation also involves a feeling of self-estrangement, which stems from the workers' inability to realize their full human potential in their work roles.

Labor Unions and the Struggle for Workers' Rights

In times of high unemployment, many people with jobs are thankful that they are employed. But having a job is no guarantee of having favorable working conditions and receiving decent pay and benefits. **Labor unions** are worker advocacy organizations that developed to protect workers and represent them at negotiations between management and labor.

labor unions Worker advocacy organizations that developed to protect workers and represent them at negotiations between management and labor.

Benefits and Disadvantages of Labor Unions to Workers Labor unions have played an important role in fighting for fair wages and benefits, healthy and safe work environments, and other forms of worker advocacy. Compared with nonunion workers, unionized workers tend to have higher earnings, better insurance and pension benefits, and more paid time off.

Labor unions are also influential in achieving better working conditions. For example, the United Food and Commercial Workers (UFCW), the country's largest union representing poultry-processing workers, was instrumental in the formation of an OSHA rule that established a federal workplace "potty" policy governing when employees can use the bathroom while on the job. Prior to this rule, workers in food processing industries were denied the right to go to the bathroom when needed, and often had no other choice but to relieve themselves while standing on the assembly line because their boss would not let them leave their work station ("New OSHA Policy Relieves Employees" 1998). Now, OSHA mandates that employers must make toilet facilities available so that employees can use them when they need to.

One of the disadvantages of unions is that members must pay dues and other fees, and these dues have been rising in recent years. Union members resent the high salaries that many union leaders make. Another disadvantage for unionized workers is the loss of individuality. Unionized workers are members of an overall bargaining unit in which the majority rules. Decisions made by the majority may conflict with individual employees' specific employment needs.

Declining Union Density The strength and membership of unions in the United States have declined over the last several decades. **Union density**—the percentage of workers who belong to unions—grew in the 1930s and peaked in the 1940s and 1950s, when 35 percent of U.S. workers were unionized. In 2010, the percentage of U.S. workers belonging to unions had fallen to just under 12 percent (Bureau of Labor Statistics 2011c).

One reason for the decline in union representation is the loss of manufacturing jobs, which tend to have higher rates of unionization than other industries. Job growth has occurred in high technology and financial services, where unions have little presence. In addition, globalization has led to layoffs and plant closings at many unionized work sites as a result of companies moving to other countries to find cheaper labor. A major reason why union representation has declined is that corporations take active measures to keep workers from unionizing, and weak U.S. labor laws fail to support and protect unionization.

Corporate Antiunion Activities In the 1960s and 1970s, U.S. corporations mounted an offensive attack on labor unions, "aiming to tame them or maim them" (Gordon 1996, p. 207). Corporations hired management consultants to help them develop and implement antiunion campaigns. They threatened unions with decertification, fired union leaders and organizers, and threatened to relocate their plants unless the unions and their members "behaved":

> One management consultant firm . . . was unusually blunt in broadcasting its methods. A late-1970s blurb promoting its manual promised: "We will show you how to screw your employees (before they screw you)—how to keep them smiling on low pay—how to maneuver them into low-pay jobs they are afraid to walk away from." (Gordon 1996, p. 208)

At least 23,000 workers each year are fired or discriminated against at their workplace because of involvement in union-related activity (Bonior 2006). Some employers faced with organizing campaigns have engaged in the following antiunion strategies: (1) firing pro-union workers, (2) threatening to close a work site when workers try to form a union, (3) coercing workers into opposing unions with bribery or favoritism, (4) hiring high-priced union-busting consultants to fight union-organizing drives, and (5) forcing employees to attend one-on-one antiunion meetings with their supervisors (Bonior 2006; Mehta & Theodore 2005).

union density The percentage of workers who belong to unions.

Antiunion Legislation In March 2011, Governor Scott Walker (Wisconsin) signed legislation to weaken unions representing state and local government employees. The legislation prohibited collective bargaining by most public workers for issues beyond wages, required unions to hold annual votes on whether they should remain in existence, and increased workers' contributions for pensions and health care. Although Judge Marilyn Sumi struck down the legislation on procedural grounds, the Wisconsin Supreme Court later upheld the new law. Opposition to the antiunion bill resulted in special elections due to demands that legislators who supported Governor Walker's bill be removed from office.

Weak U.S. Labor Laws The 1935 National Labor Relations Act (NLRA) is the primary federal labor law in the United States. The NLRA guarantees the right to unionize, bargain collectively, and to strike against private-sector employees. However, in addition to excluding public-sector workers, the law excludes agricultural and domestic workers, supervisors, railroad and airline employees, and independent contractors. As a result, millions of workers do not have the right under U.S. law to negotiate their wages, hours, or employment terms.

> Penalties for violating U.S. labor law are so weak that many employers consider them as a cost of doing business and a small price to pay for defeating workers' attempts to organize.

In addition, changes in U.S. labor law over the years have eroded workers' rights to freedom of association. Originally, labor law required employers to grant a demand for union recognition if a majority of workers signed a card indicating they wanted a union. But since 1947, employers can reject workers' demand for unionization and force a National Labor Relations Board (NLRB) election, which requires about one-third of workers to petition for the board to hold the election. "The company then uses the time leading up to the election to focus its campaign against union formation, while disallowing opportunities for opposing views" (Human Rights Watch 2007, p. 18).

In the United States, the NLRB and the courts play an important role in upholding workers' rights to unionize and sanctioning employers who violate these rights. The NLRB has the authority to issue job reinstatement and "back pay" orders or other remedial orders to workers wrongfully fired or demoted for participating in union-related activities. Although it is illegal to fire workers for engaging in union activities, there are few consequences for employers that do so. Penalties for violating U.S. labor law are so weak that many employers consider them as a cost of doing business and a small price to pay for defeating workers' attempts to organize (Human Rights Watch 2009).

If you get fired for trying to organize, for example, you can apply to the NLRB. If the NLRB finds that you were illegally fired, the employer has to give you back pay for the time you were fired—minus any money that you may have earned at another job. As you can imagine, most people who are fired for trying to organize will, in fact, get another job somewhere, so there's no compensation for them at all. Then all the employer has to do is post on a bulletin board at the work site that they won't do it again. So there are effectively no sanctions, and it is in the employers' interest to fire people. They really don't suffer many consequences for doing so, and firing a leading union supporter sends a very powerful message to the rest of the employees. The message is: If you too try to lead an organizing campaign, you are going to lose your job; and if you vote for a union, you could lose your job (Bonior 2006).

In addition, there is a backlog of thousands of cases of unfair labor practices by employers, and workers often wait years from the filing of a charge until the NLRB resolves a case, discouraging many workers from filing charges (Greenhouse 2008).

Labor Union Struggles Around the World In 1949, the International Labour Office established the Convention on the Right to Organize and Collective Bargaining. About half of the world's workforce lives in countries that have not ratified this convention, including China, India, Mexico, Canada, and the United States. The International

Confederation of Free Trade Unions (ICFTU) publishes an *Annual Survey of Violations of Trade Union Rights* that describes severe abuses of workers' rights in countries around the world. This annual survey is frightening documentation of the lack of workers' rights around the world and of the abuses of governments and employers in the continued suppression of workers' rights. The survey results in the last few years showed that between 75 and several hundred trade unionists are killed each year. Several thousands more are imprisoned, beaten in demonstrations, tortured by security forces or others, and often sentenced to long prison terms. And each year, hundreds of thousands of workers lose their jobs merely for attempting to organize a trade union.

Colombia remains the most dangerous place in the world to be a trade unionist. Of the 90 documented murders of trade unionists in 2010, 49 occurred in Columbia (International Confederation of Free Trade Unions 2011). In some cases, the families of the unionist murder victims were also killed. Most reported cases of trade unionist assassinations are not properly investigated, and the murderers are not caught or punished.

Both employers and governments violate trade union rights. Governments in numerous countries hamper trade union activity or strike action and fail to enforce existing national and international laws that protect workers' rights. Some countries, such as Oman, Saudi Arabia, and Burma, do not recognize the right to form trade unions. Other countries, such as China, Egypt, and Syria, impose a trade union monopoly. And some governments, such as those in Belarus and Moldova (countries in eastern Europe), try to coerce workers into joining the government-supported union. Eager to secure financial benefits from participating in the global market, governments see trade unions as an obstacle to their economic development.

Strategies for Action: Responses to Problems of Work and Unemployment

Government, private business, human rights organizations, labor organizations, college student activists, and consumers play important roles in responding to problems of work and unemployment.

Reducing Unemployment

Efforts to reduce unemployment include (1) workforce development programs, and (2) programs and policies that create and save jobs.

Workforce Development The 1998 **Workforce Investment Act (WIA)** provides a wide array of programs and services designed to assist individuals to prepare for and find employment. These include the following: assessment of skills and abilities; access to job vacancy listings; job search and placement assistance; individual career planning and counseling; resume preparation; English as a second language instruction; computer literacy; wage subsidies for on-the-job training; and support services such as transportation and child care to enable individuals to participate in WIA programs. Some workforce development programs focus on strategies to improve the employability of hard-to-employ individuals through providing targeted interventions such as substance abuse treatment, domestic violence services, prison release reintegration assistance, mental health services, and homelessness services, in combination with employment services (Martinson & Holcomb 2007).

Efforts to prepare high school students for work include the establishment of technical and vocational high schools and high school programs and school-to-work programs. School-to-work programs involve partnerships between business, labor, government, education, and community organizations that allow high school students to explore different careers, and provide job skill training and work-based learning experiences (Bassi & Ludwig 2000; Leonard 1996).

Educational attainment is often touted as the path to employment and economic security. Although education and skill development are important for individuals to realize

Workforce Investment Act (WIA) Legislation passed in 1998 that provides a wide array of programs and services designed to assist individuals to prepare for and find employment.

their potential and to become better-informed and more productive citizens, education alone is not the answer to unemployment and economic insecurity. More than two-thirds of jobs in the United States require little education (Martinson & Holcomb 2007); thus, it is important to ensure that all jobs pay a minimum "living wage" (see also Chapter 6). But, as long as our economy allows people who work full-time to earn poverty-level wages, having a job is not necessarily the answer to economic self-sufficiency. As David Shipler (2005) explained in his book *The Working Poor:*

> A job alone is not enough. Medical insurance alone is not enough. Good housing alone is not enough. Reliable transportation, careful family budgeting, effective parenting, effective schooling are not enough when each is achieved in isolation from the rest. There is no single variable that can be altered to help working people move away from the edge of poverty. Only where the full array of factors is attacked can America fulfill its promise. (p. 11)

Job Creation and Preservation In a national poll, one in four Americans said that the best way to create more U.S. jobs is to keep manufacturing jobs in the United States and stop sending work overseas (Newport 2011). Although both Democrat and Republican respondents cited keeping manufacturing jobs in the United States as the best way of creating jobs, Republicans also favored job-creation strategies that involve reducing taxes and limiting government involvement in regulating business, whereas Democrats were more likely to favor using government to create jobs through infrastructure projects.

In response to spiraling unemployment rates, President Obama signed into law the **American Recovery and Reinvestment Act of 2009,** which provided an economic stimulus package of $787 billion to create and save jobs and to reinvigorate the U.S. economy. About $280 billion of these monies were administered by states and localities, largely for health, education, and transportation programs. The stimulus package also included measures to increase or extend benefits under Medicaid, **unemployment insurance,** and nutrition assistance programs to help those most affected by the economic downturn.

In 2010, Obama signed into law the Small Business Jobs Act, which provides tax breaks and better access to credit for small businesses so that small businesses can create new jobs. The 2010 Hiring Incentives to Restore Employment (HIRE) Act provided tax incentives to companies that hired employees who had been looking for work for 60 days or more. Obama also launched the National Export Initiative with a goal of supporting new jobs through doubling exports.

Efforts to End Slavery and Child Labor

More than 50 years ago, the United Nations stated in Article 4 of its Universal Declaration of Human Rights, that "no one shall be held in slavery or servitude; slavery and the slave trade shall be prohibited in all their forms." Yet slavery persists throughout the world. The international community has drafted treaties on slavery, but many countries have yet to ratify and implement the different treaties.

One strategy to fight slavery is punishment. In at least 25 countries, slave trafficking is actively prosecuted and treated as a serious crime. However, slave traffickers often avoid punishment because, as a former official of the U.S. Agency for International Development explained, "government officials in dozens of countries assist, overlook, or actively collude with traffickers" (quoted by Cockburn 2003, p. 16). In many countries, the justice system is more likely to jail or expel sex slaves than to punish traffickers ("Sex Trade Enslaves Millions of Women, Youth" 2003).

In the United States, the Victims of Trafficking and Violence Protection Act, passed by Congress in 2000, protects slaves against deportation if they testify against their former owners. Convicted slave traffickers in the United States are subject to prison sentences, as shown in the following examples:

- Louisa Satia and Kevin Waton Nanji each received nine years for luring a 14-year-old girl from Cameroon with promises of schooling and then isolating her in their

American Recovery and Reinvestment Act of 2009 Legislation that provided an economic stimulus package of $787 billion to create and save jobs and to reinvigorate the U.S. economy.

unemployment insurance A federal-state program that temporarily provides laid-off workers with a portion of their paychecks.

Maryland home, raping her, and forcing her to work as their domestic servant for three years.

- Sardar and Nadira Gasanov were sentenced to five years each for recruiting women from Uzbekistan with promises of jobs, taking their passports, and forcing them to work in strip clubs and bars in Texas.
- Juan, Ramiro, and Jose Ramos each received 10 to 12 years for transporting Mexicans to Florida and forcing them to work as fruit pickers. (Cockburn 2003)

U.S. corporations are also being held accountable for enterprises that involve forced labor and other human rights and labor violations. In 2003, the Unocal oil company became the first corporation in history to stand trial in the United States for human rights violations abroad (George 2003). Unocal was accused of involvement in a pipeline project that used Myanmar (formerly Burma) military personnel to provide "security" for a natural gas pipeline project in the remote Yadana region near the Thai border. According to the Ninth U.S. Circuit Court of Appeals, the soldiers' true role was to force villagers in the pipeline region to work without pay. The military also forced villagers living along the pipeline route to relocate without compensation, raped and assaulted villagers, and imprisoned and/or executed those who opposed them. In 2004, Unocal announced that it had reached a settlement with the parties who alleged that Unocal was complicit in the human rights violations committed by the Myanmar military. In 2005, Chevron/Texaco purchased Unocal.

In 1989, the General Assembly of the United Nations adopted the Convention on the Rights of the Child, which asserts the right that children should not be engaged in work deemed to be "hazardous or to interfere with the child's education, or to be harmful to the child's health." The International Labour Office has taken a leading role in enforcing these rights, leading efforts to prevent and eliminate child labor. Although almost every country has laws prohibiting the employment of children below a certain age, some countries exempt certain sectors—often the very sectors where the highest numbers of child laborers are found. Efforts to prevent child labor also focus on increasing penalties for violating child labor laws, which are often weak and poorly enforced.

Education is a primary means to combat child labor (International Labour Office 2010). Children with no access to education have little choice but to enter the labor market. In most countries, primary education is not free and parents must pay for costs such as uniforms and books. Parents who cannot afford these fees are also likely to view their children's labor as a necessary source of income to the household. Promoting Education for All—a global movement advocating that every child has access to free, basic education (primary school plus two or three years of secondary school)—is critical in the effort to reduce child labor.

Responses to Sweatshop Labor

The Fair Labor Association (FLA), established in 1996, is a coalition of companies, universities, and nongovernmental organizations (NGOs) that works to promote adherence to international labor standards and improve working conditions worldwide. In 2009, 30 leading companies that sell brand-name apparel, footwear, and other goods voluntarily participated in FLA's monitoring system, which inspects their overseas factories and requires them to meet minimum labor standards, such as not requiring workers to work more than 60 hours a week. In addition, more than 200 colleges and universities require their collegiate licensees (companies that manufacture logo-carrying goods for colleges and universities) to participate in FLA's monitoring system.

In its first few years of operation, the FLA was criticized for allowing firms to select and directly pay their own monitors and to have a say in which factories were audited. In 2002, the FLA responded to these criticisms by taking much more control over external monitoring, with the FLA staff selecting factories for evaluation, choosing the monitoring organization, and requiring that inspections be unannounced (O'Rourke 2003). The FLA continues to be criticized, however, for having low standards in allowing below-poverty wages and excessive overtime and for requiring that only a small percentage of a manufacturer's supplier factories be inspected each year. Critics also suggest that companies use their participation in the FLA as a marketing tool. Once "certified" by the FLA, companies can sew a label into their products saying that the products were made under fair working conditions (Benjamin 1998). In 2006, United Students Against Sweatshops created "FLA Watch" to "expose the truth about the Fair Labor Association . . . and . . . the FLA's ongoing failure to defend the rights of workers" (FLA Watch 2007a). The home page of the FLA Watch website explains, "The FLA purports to be an 'independent' monitor of working conditions in the apparel industry. But the organization is funded and controlled by the very corporations that have been repeatedly found to be sweatshop violators." FLA Watch further accuses the Fair Labor Association of being "nothing more than a public relations mouthpiece for the apparel industry. Created, funded, and controlled by Nike, Adidas, and other leading sweatshop abusers, the FLA is a classic case of the 'fox guarding the hen house'" (FLA Watch 2007b).

Kirschner/United Students Against Sweatshops

College student groups across the country have participated in boycotts against Coca-Cola in protest of the violence against union leaders at Colombian Coca-Cola plants.

Student Activism United Students Against Sweatshops (USAS), formed in 1997, is a grassroots organization of youth and students who fight against labor abuses and for the rights of workers around the world, particularly campus workers and garment workers who make collegiate licensed apparel. USAS student activists have influenced more than 180 colleges and universities to affiliate with the Worker Rights Consortium (WRC), which investigates factories that produce clothing and other goods with school logos to make sure that the factory meets the code of conduct developed by each school. A typical code of conduct includes fair wages, a safe working environment, a ban on child labor, and the right to be represented by a union or other form of employee representation. If the WRC investigation finds that a factory fails to meet the code of conduct, the companies—often well-known international brands—who purchase items from that factory are warned that their contract with the school will be terminated if working conditions at the factory do not improve.

What Do You Think? Do you know where the clothing with your college or university logo is made? Do you think most students care if the college or university logo clothing or products they buy are made under sweatshop conditions?

Sodexo—a profitable multinational food service company—serves food to more U.S. college students than any other company. In 2009, United Students Against Sweatshops began a *Kick Out Sodexo* campaign after discovering that Sodexo paid its campus food service and janitorial workers poverty-level wages and interfered with employees' attempts to form a union. Students who are active in the Kick Out Sodexo campaign have written letters to Sodexo, expressing their concerns about the company's labor practices, and have campaigned to convince college and university administrators to cancel their food service contracts with Sodexo and to do business instead with a food service company that meets certain labor standards.

Legislation Perhaps the most effective strategy against sweatshop work conditions is legislation. In the United States, as of 2011, 9 states, 40 cities, 15 counties, and 118 school districts have passed "sweatfree" procurement laws that prohibit public entities (such as schools, police, and fire departments) from purchasing uniforms and apparel made under sweatshop conditions (SweatFree Communities n.d.).

In 2006, the Decent Working Conditions and Fair Competition Act was introduced in U.S. Congress. This legislation, which proposed to prohibit the import, export, or sale of sweatshop goods in the United States, died in committee.

Establishing and enforcing labor laws to protect workers from sweatshop labor conditions is difficult in a political climate that offers more protections to corporations than it does to workers. Companies have demanded and won all sorts of intellectual property and copyright laws to defend their corporate trademarks, labels, and products. Yet, the corporations have long said that extending similar laws to protect the human rights of the 16-year-old girl in Bangladesh who sews the garment would be "an impediment to free trade." Under this distorted sense of values, the label is protected, but not the human being, the worker who makes the product (National Labor Committee 2007).

Responses to Workplace Health and Safety Concerns

A series of horrific workplace tragedies in 2010—the explosion at the Massey Energy Mine in West Virginia that killed 29 coal miners; the explosion at the Kleen Energy Plant in Middletown, Connecticut, that killed six workers and another at the Tesoro Refinery in Washington State that killed seven workers; and the BP/Transocean Gulf Coast oil rig explosion that killed 11 workers—highlight the need for improvements in workplace safety. In 2009, the Protecting America's Workers Act (PAWA) was introduced in Congress, but as of this writing, has not passed. The PAWA would strengthen OSHA by extending coverage to uncovered workers, including state and local public employees; enhance whistleblower protections so that workers who report workplace safety violations would have job protection; and increase penalties for serious and willful violations and in cases of worker death. Other proposed legislation would strengthen OSHA's authority to shut down operations that pose an imminent danger to workers; require large corporate employers to provide regular reports to OSHA on work-related injuries, illnesses, and fatalities; mandate OSHA to issue a standard on safe patient handling to protect health care workers from injuries; and issue a standard to protect workers from explosions and fires (AFL-CIO 2011). Efforts have also been made to strengthen the Mine Safety and Health Administration (MSHA) to give it more authority to close down mines that are in violation of safety standards and to prevent "black lung disease" in miners by establishing stricter rules on the amount of coal dust miners inhale. These and other initiatives to strengthen workplace health and safety are opposed by industry groups, and Republicans in Congress who are against government regulations argue that excessive regulation is burdensome to business, hampers investment, and hurts job creation.

One victory in worker health legislation occurred in December 2010, when Congress passed the James Zadroga 9/11 Health and Compensation Act. First introduced in 2004, this legislation establishes a health monitoring, treatment, and compensation program for the tens of thousands of 9/11 responders and others who became ill as a result of exposures at the World Trade Center.

In developing countries, governments fear that strict enforcement of workplace regulations will discourage foreign investment. Investment in workplace safety in developing countries, whether by domestic firms or foreign multinationals, is far below that in the rich countries. Unless global standards of worker safety are implemented and enforced in all countries, millions of workers throughout the world will continue to suffer under hazardous work conditions. Low unionization rates and workers' fears of losing their jobs—or their lives—if they demand health and safety protections leave most workers powerless to improve their working conditions.

Behavior-Based Safety Programs A controversial health and safety strategy used by business management is behavior-based safety programs. Instead of examining how work processes and conditions compromise health and safety on the job, **behavior-based safety programs** direct attention to workers themselves as the problem. Behavior-based safety programs claim that most job injuries and illnesses are caused by workers' own carelessness and unsafe acts (Frederick & Lessin 2000). These programs focus on teaching employees and managers to identify, "discipline," and change unsafe worker behaviors that cause accidents and encourage a work culture that recognizes and rewards safe behaviors.

Critics contend that behavior-based safety programs divert attention away from the employers' failures to provide safe working conditions. They also say that the real goal of behavior-based safety programs is to discourage workers from reporting illness and injuries. Workers whose employers have implemented behavior-based safety programs describe an atmosphere of fear in the workplace, such that workers are reluctant to report injuries and illnesses for fear of being labeled "unsafe workers."

Work-Life Policies and Programs

Policies that help women and men balance their work and family responsibilities are referred to by a number of terms, including *work-family, work-life,* and *family-friendly* policies. As shown in Table 7.2, the United States lags far behind many other countries in national work-family provisions.

Federal and State Family and Medical Leave Initiatives In 1993, President Clinton signed into law the first national policy designed to help workers meet the dual demands of work and family. **The Family and Medical Leave Act (FMLA)** requires all public agencies and private-sector employers (with 50 or more employees who worked at least 1,250 hours in the preceding year) to provide up to 12 weeks of job-protected, *unpaid* leave so that they can care for a seriously ill child, spouse, or parent; stay home to care for their newborn, newly adopted, or newly placed foster child; or take time off when they are seriously ill. A 2008 amendment to the FMLA requires employers to provide up to 26 weeks of unpaid leave to employees to care for a seriously ill or injured family member who is in the armed forces, including the National Guard or Reserves. However, nearly 40 percent of employees are not eligible for the FMLA benefit because they work for companies with fewer than 50 employees or they work part-time (National Partnership for Women and Families 2008). Some employers do not comply with the FMLA either because they are unaware of their responsibilities under FMLA, or because they are deliberately violating the law (Galinsky et al. 2008). And some eligible

behavior-based safety programs A strategy used by business management that attributes health and safety problems in the workplace to workers' behavior, rather than to work processes and conditions.

Family and Medical Leave Act (FMLA) A federal law that requires public agencies and companies with 50 or more employees to provide eligible workers with up to 12 weeks of job-protected, unpaid leave so that they can care for an ill child, spouse, or parent; stay home to care for their newborn, newly adopted, or newly placed child; or take time off when they are seriously ill, and up to 26 weeks of unpaid leave to care for a seriously ill or injured family member who is in the armed forces, including the National Guard or Reserves.

TABLE 7.2 How Do U.S. Work Policies Compare with Other Countries?

POLICY	UNITED STATES	OTHER COUNTRIES
Paid childbirth leave	No federal policy	168 countries offer paid leave to women; 98 countries offer 14 or more weeks of paid leave
Right to breast-feed at work	No federal policy	107 countries protect working women's right to breast-feed
Paid sick leave	No federal policy	145 countries provide paid sick leave
Paid annual leave	No federal policy	137 countries require employers to provide paid annual leave; 121 countries guarantee 2 weeks or more
Guaranteed leave for major family events (e.g., weddings, funerals)	No federal policy	49 countries guarantee leave for major family events (leave is paid in 40 countries)

Source: Based on Heymann et al. 2007.

employees do not take advantage of the FMLA benefit because of lack of awareness: In a survey of employees covered by the FMLA, only 38 percent correctly reported that the FMLA applied to them and about half said they did not know whether it did (Cantor et al. 2001). Finally, many eligible workers do not use their FMLA benefit because they cannot afford to take leave without pay, and/or they fear they will lose their job if they take time off (Cantor et al. 2001).

Only a handful of states offers paid leave programs that provide eligible workers a family leave benefit with a portion of their salary for up to six weeks. The proposed federal Family Leave Insurance Act would provide 8 to 12 weeks of partially paid leave for FMLA purposes, paid for by employer and worker contributions.

As shown in Table 7.2, U.S. employers are not required to provide workers with any paid sick leave, and nearly half (47 percent) of the private-sector workforce has no paid sick leave (Galinsky et al. 2008). The proposed Healthy Families Act would require employers with at least 15 employees to provide 7 days of paid sick leave annually for full-time employees who work at least 30 hours a week. Employees could take the benefit if they or a family member is sick. Supporters of the Healthy Families Act point out that, when sick employees go to work because they cannot afford to take unpaid sick days, or fear losing their job by taking a sick day, they risk spreading infectious diseases at the workplace. And sending sick children to school because their parents cannot afford to take unpaid sick days to stay home with them risks spreading infectious diseases at school. The Healthy Families Act is a bill proposing to establish a federally mandated paid sick day policy.

Employer-Based Work-Life Policies Aside from government-mandated work-family policies, some corporations and employers have "family-friendly" work policies and programs, including unpaid or paid family and medical leave, child care assistance, assistance with elderly parent care, and flexible work options such as **flextime, compressed workweek,** and **telecommuting.** Studies show that flexible work arrangements reduce work-life conflict and increase job satisfaction. For example, Best Buy Co., Inc. developed a Results-Only Work Environment (ROWE), which allows employees and managers to control when and where they work as long as they get the job done. A controlled study found that employees participating in ROWE benefitted from reduced work-family conflict (Kelly et al. 2011).

Flexible work arrangements also benefit employers in reducing absenteeism, lowering turnover, improving the health of workers, and increasing productivity. Yet, less than one-third of full-time workers report having flexible work hours and only 15 percent report working from home at least once a week (Council of Economic Advisors 2010). National surveys of more than 1,000 employers (with 100 or more employees) found that, between 1998 and 2008, some employer-provided work-life benefits increased, whereas others decreased (see Table 7.3).

> **What Do You Think?** A national survey found that more than half of employers (52 percent) that offer maternity leave provide at least partial pay to employees on maternity leave, whereas only 16 percent provide any pay for paternity leave (Galinsky et al. 2008). Do you think this is fair to fathers? Is this discrepancy a form of discrimination against men?

Efforts to Strengthen Labor

A Pew Research Center (2011) poll found that about half (47 percent) of U.S. adults say they have a favorable opinion of labor unions, compared with 39 percent who say they have an unfavorable opinion of unions. Although efforts to strengthen labor are viewed as problematic to corporations, employers, and some governments, such efforts have the potential to remedy many of the problems facing workers.

flextime A work arrangement that allows employees to begin and end the workday at different times so long as 40 hours per week are maintained.

compressed workweek A work arrangement that allows employees to condense their work into fewer days (e.g., four 10-hour days each week).

telecommuting A work arrangement involving the use of information technology that allows employees to work part- or full-time at home or at a satellite office.

TABLE 7.3 Employer-Based Work-Life Benefits and Policies, U.S., 1998–2008*

BENEFIT OR POLICY	PERCENTAGE OF EMPLOYERS* THAT PROVIDE BENEFIT/POLICY	
	1998	2008
Full pay for maternity leave	27%	16%
Partial pay for maternity leave	60%	67%
Private space for breast-feeding	37%	53%
Private space and storage facilities for women who are nursing to express milk	49%	66%
Some paid time off for paternity leave	13%	16%
Health insurance for unmarried partners	14%	31%
Allow some employees to periodically change their arrival and departure time	68%	79%
Provide information about elder care services	23%	39%
Child care at or near worksite	9%**	9%
Dependent Care Assistance Plans that help employees pay for child care with pretax dollars	46%**	46%

Source: Galinsky et al. 2008.

*Employers with 100 or more employees

**Report did not specify 1998 value, but did report that the 1998 value was not statistically different from the 2008 value.

In an effort to strengthen their power, some labor unions have merged with one another. Labor union mergers result in higher membership numbers, thereby increasing the unions' financial resources, which are needed to recruit new members and to withstand long strikes. Because workers must fight for labor protections within a globalized economic system, their unions must cross national boundaries to build international cooperation and solidarity. Otherwise, employers can play working and poor people in different countries against each other.

Strengthening labor unions requires combating the threats and violence against workers who attempt to organize or who join unions. One way to do this is to pressure governments to apprehend and punish the perpetrators of such violence. Although about 3,500 trade unionists were murdered between 1990 and 2005, only 600 cases were investigated, resulting in just six convictions (Moloney 2005). Another tactic is to stop doing business with countries where government-sponsored violations of free trade union rights occur.

Proposed legislation called the Employee Free Choice Act would allow workers to sign a card stating that they want to be represented by a union. If a majority of the employees in any workplace sign such a card, the company would then have to recognize the union and bargain over terms and conditions of employment. If the company does not negotiate a first contract in a timely manner after workers unionize, the Employee Free Choice Act requires binding arbitration. This legislation would also strengthen U.S. labor law enforcement by increasing penalties for violations. In 2007, the U.S. House of Representatives passed the Employee Free Choice Act and, as of this writing, the bill is pending in the U.S. Senate.

The Global Jobs Pact Improving workers' lives and the economy as a whole requires a coordinated and comprehensive effort that combines a number of goals and strategies. In 2009, the International Labour Organization adopted a Global Jobs Pact designed to guide national and international policies aimed at stimulating the economy, creating jobs, and providing protection to workers and their families.

The Global Jobs Pact calls for a wide range of measures to retain workers, sustain businesses, create jobs, and provide social protections to workers and the unemployed. The pact also calls for more stringent supervision and regulation of the financial industry so that it better serves the economy and protects individuals' savings and pensions. The pact urges a number of measures, including (1) a shift to a low-carbon, environmentally friendly economy that will help create new jobs; (2) investments in public infrastructure; and (3) increases in social protection and minimum wages (to reduce poverty, increase demand for goods, and stimulate the economy). The Global Jobs Pact provides a vision for a healthier economy that meets the needs of workers and consumers. The hard work of translating the Global Jobs Pact into reality falls to employers, trade unions, and especially governments.

Understanding Work and Unemployment

On December 10, 1948, the General Assembly of the United Nations adopted and proclaimed the Universal Declaration of Human Rights. Among the articles of that declaration are the following:

Article 23. Everyone has the right to work, to free choice of employment, to just and favourable conditions of work and to protection against unemployment.

Everyone, without any discrimination, has the right to equal pay for equal work.

Everyone who works has the right to just and favourable remuneration ensuring for himself and his family an existence worthy of human dignity, and supplemented, if necessary, by other means of social protection.

Everyone has the right to form and to join trade unions for the protection of his interests.

Article 24. Everyone has the right to rest and leisure, including reasonable limitation of working hours and periodic holidays with pay.

More than a half century later, workers around the world are still fighting for these basic rights as proclaimed in the Universal Declaration of Human Rights.

To understand the social problems associated with work and unemployment, we must first recognize that corporatocracy—the ties between government and corporations—serves the interests of corporations over the needs of workers. We must also be aware of the roles that technological developments and postindustrialization play on what we produce, how we produce it, where we produce it, and who does the producing. With regard to what we produce, the United States has moved away from producing manufactured goods to producing services. With regard to production methods, the labor-intensive blue-collar assembly line has declined in importance, and information-intensive white-collar occupations have increased. Although some people argue that the growth of multinational corporations brings economic growth, jobs, lower prices, and quality products to consumers throughout the world, others view global corporations as exploiting workers, harming the environment, dominating public policy, and degrading cultural values.

Decisions made by U.S. corporations about what and where to invest influence the quantity and quality of jobs available in the United States. As conflict theorists argue, such investment decisions are motivated by profit, which is part of a capitalist system. Profit is also a driving factor in deciding how and when technological devices will be used to replace workers and increase productivity. If goods and services are produced too efficiently, however, workers are laid off and high unemployment results. When people have no money to buy products, sales slump, recession ensues, and social welfare programs are needed to support the unemployed. When the government increases spending to pay for its social programs, it expands the deficit and increases the national debt. Deficit spending and a large national debt make it difficult to recover from the recession, and the cycle continues.

What can be done to break the cycle? Those adhering to the classic view of capitalism argue for limited government intervention on the premise that business will regulate itself by means of an "invisible hand" or "market forces." But Americans are growing increasingly skeptical of the notion that big business, which has caused many of the economic problems in our country and throughout the world, can solve the very problems it creates. As this book goes to press, a social movement known as "Occupy Wall Street," which began with a few dozen demonstrations in front of the New York Stock Exchange, has spread to cities across the country. Participants in the Wall Street protests include students, the unemployed, union members, professionals, and others who are fed up with corporate greed and the widening gap between the rich and the poor. They are frustrated by high unemployment and the erosion of workers' salaries, benefits, and rights. Although this growing movement has not, as of this writing, formulated any specific demands, what is certain is that Americans want change in the economy and in the workplace.

CHAPTER REVIEW

- **According to a Pew Research survey, what do the majority of U.S. adults believe about big business?**
  The majority of U.S. adults believe that (1) there's too much power concentrated in the hands of a few big companies; (2) businesses make too much profit; and (3) free market needs regulation to best serve public interest needs.

- **The United States is described as a "postindustrialized" society. What does that mean?**
  Postindustrialization refers to the shift from an industrial economy dominated by manufacturing jobs to an economy dominated by service-oriented, information-intensive occupations.

- **What are transnational corporations?**
  Transnational corporations are corporations that have their home base in one country and branches, or affiliates, in other countries.

- **What are the four principles of McDonaldization?**
  The four principles of McDonaldization are (1) efficiency, (2) predictability, (3) calculability, and (4) control through technology.

- **What are some of the causes of unemployment?**
  Causes of unemployment include not enough jobs, job exportation (the relocation of jobs to other countries), automation (the replacement of human labor with machinery and equipment), increased global competition, and mass layoffs as plants close and companies downsize or go out of business.

- **Does slavery still exist today? If so, where?**
  Forced labor, commonly known as slavery, exists today all over the world, including the United States, but it is most prevalent in India, Pakistan, Bangladesh, and Nepal. Most forced laborers work in agriculture, mining, prostitution, and factories.

- **What is the most common cause of job-related fatality and nonfatal job-related illness or injury?**
  The most common type of job-related fatality involves transportation accidents. Sprains and strains are the most

common nonfatal occupational injury or illness involving days away from work.

- **According to a Gallup Poll, what is the most common complaint among U.S. workers?**
  A Gallup Poll found that one-third of respondents is dissatisfied with the amount of stress in their jobs, making job stress the most common complaint of workers.

- **What are some of the challenges that workers face in balancing work and family?**
  Workers often struggle to find the time and energy to care for elderly parents and children and to have time with their families.

- **Why has union membership in the United States declined over the last several decades?**
  Union membership has declined for several reasons: the loss of manufacturing jobs, which tend to have higher rates of unionization than other industries; layoffs and plant closings at many unionized work sites as a result of companies moving to other countries to find cheaper labor; active measures by corporations to discourage unionization; and weak U.S. labor laws that fail to support and protect unionization.

- **What is the federal Family and Medical Leave Act?**
  In 1993, President Clinton signed into law the Family and Medical Leave Act (FMLA), which requires all companies with 50 or more employees to provide eligible workers with up to 12 weeks of job-protected, unpaid leave so that they can care for a seriously ill child, spouse, or parent; stay home to care for their newborn, newly adopted, or newly placed child; or take time off when they are seriously ill. The United States is the only advanced nation that does not mandate paid leave for family and medical reasons.

TEST YOURSELF

1. The global financial crisis that began in 2007 started in what country?
   a. China
   b. India
   c. United States
   d. Mexico

2. According to your text, corporations outsource labor to other countries so that U.S. consumers can save money on cheaper products.
   a. True
   b. False

3. Under federal law, willful violation that results in a worker's death is considered which of the following?
    a. A misdemeanor with a maximum prison sentence of 6 months
    b. A misdemeanor with a minimum prison sentence of 6 years
    c. A felony with a maximum prison sentence of 26 years
    d. A felony with a minimum prison sentence of 6 years
4. The most common type of job-related fatality involves which of the following?
    a. Mining accidents
    b. Construction accidents
    c. Transportation accidents
    d. Homicide
5. The United States is the only advanced nation that does not mandate a minimum number of vacation days.
    a. True
    b. False
6. The 1935 National Labor Relations Act (NLRA) gives all U.S. workers the right to unionize, bargain collectively, and to strike.
    a. True
    b. False

7. Which is the most dangerous country in the world to be involved in trade union activities?
    a. China
    b. Ireland
    c. Egypt
    d. Colombia
8. The Decent Working Conditions and Fair Competition Act is the first proposed federal legislation in the United States aimed at which of the following?
    a. Combating sweatshop labor
    b. Preventing outsourcing of jobs
    c. Strengthening unions
    d. Prohibiting age discrimination in the workplace
9. The United States is the only country in the world that protects working women's right to breast-feed at work.
    a. True
    b. False
10. The majority of U.S. adults have an unfavorable opinion of labor unions.
    a. True
    b. False

Answers: 1: c; 2: b; 3: a; 4: c; 5: a; 6: b; 7: d; 8: a; 9: b; 10: b.

KEY TERMS

American Recovery and Reinvestment Act of 2009 222
automation 208
behavior-based safety programs 226
capitalism 200
chattel slavery 211
child labor 213
compressed workweek 227
corporatocracy 205
economic institution 199
Family and Medical Leave Act (FMLA) 226

flextime 227
forced labor 209
free trade agreements 203
global economy 199
industrialization 201
job burnout 215
job exportation 208
labor unions 218
long-term unemployment rate 208
McDonaldization 202
outsourcing 208
postindustrialization 201

recession 207
socialism 200
sweatshops 212
telecommuting 227
transnational corporations 203
underemployment 208
unemployment 207
unemployment insurance 222
union density 219
Workforce Investment Act (WIA) 221

MEDIA RESOURCES

Turning to Video

▶❚❚ Watch the BBC video *Recession and Small Towns* (running time 1:54), available through **CengageBrain. com**. This video provides a glimpse into the causes and consequences of unemployment in small-town America. To what degree has media coverage of job loss and unemployment affected your choice of major in college?

Online Study Resources

Log in to **www.cengagebrain.com** to access the resources your instructor has assigned. For this book, you can access:

CourseMate

Access chapter-specific learning tools, including learning objectives, practice quizzes, videos, Internet exercises, flash cards, and glossaries, as well as web links, and more in your Sociology CourseMate.

Bloomberg/Getty Images

8

Problems in Education

"At the present time, public education is in peril. Efforts to reform public education are, ironically, diminishing its quality and endangering its very survival."

—Diane Ravitch, Education historian
and former U.S. Assistant Secretary of
Education

JEFFERY SLATTERY, A Baltimore public school teacher enjoyed teaching at one of the city's alternative high schools. Throughout his career, he had regularly told students to get to their classes during "announcements"—that they could not be in the hallways. But this time, his verbal instructions and attempt at redirecting a 14-year-old former student back to his classroom resulted in a broken jaw and major surgery to repair it. It was not the first time this student had acted aggressively toward Mr. Slattery, but it was the first time anything was done about it. The boy was arrested and charged with first-degree assault. Understandably, Mr. Slattery is angry, but not just at the young man. He's angry that no one intervened sooner, before the student reacted so violently. "The thing is . . ." said Slattery, "this child needs help."

School violence is just one of the many issues that must be addressed in today's schools. Students continue to graduate from high school unable to read, work simple mathematics problems, or write grammatically correct sentences. Graduates discover that they are ill prepared for corporations that demand literate, articulate, informed employees. Teachers leave the profession because of uncontrollable discipline problems, inadequate pay, and overcrowded classrooms. Students and teachers alike are "dumbing down"—lowering their standards, expectations, and role performances to fit increasingly undemanding and unresponsive systems of learning.

Yet, education is often claimed as a panacea—the cure-all for poverty and prejudice, drugs and violence, war and hatred, and the like. Can one institution, riddled with problems, be a solution to other social problems? In this chapter, we focus on this question and on what is being called an educational crisis. We begin with a look at education around the world.

The Global Context: Cross-Cultural Variations in Education

Looking only at the American educational system might lead one to conclude that most societies have developed some method of formal instruction for their members.

After all, the United States has more than 140,000 schools, 4.6 million primary and secondary school teachers and college faculty, 5.4 million administrators and support staff, and 75.9 million students (NCES 2011a). In reality, many societies have no formal mechanism for educating the masses. One in five adults cannot read or write—776 million people worldwide—two-thirds of them women (UNESCO 2011).

Education at a Glance, a publication of the Organization for Economic Cooperation and Development (OECD), reports education statistics on over 35 countries (OECD 2010). Some interesting findings are revealed. First, in general, educational levels are rising. For example, many countries saw an increase in the proportion of young people attending colleges and universities. However, large numbers of people do not graduate from tertiary institutions. On the average, only 38 percent of all adults in OECD-participating countries have completed postsecondary education; rates range from less than 10 percent in Luxembourg to over 60 percent in Finland.

> One in five adults cannot read or write—776 million people worldwide—the majority of them women.

Second, there is a clear link between education and income, and between education and employment. In general, across the OECD-participating countries, the more education people have, the higher their income and the greater the likelihood they will be employed. In 21 of the 31 OECD-participating countries for which data were available, investment in a college degree has an "earnings premium" of over 50 percent (OECD 2010).

Third, across all OECD-participating countries, an average of $9,195 is spent per student each year they are in school from elementary school to college. Spending, although increasing in general, varies widely by country. In Mexico, Chile, and Brazil, less than $4,000 is spent per student per year. However, the United States spends over $14,000, on the average, per student per year, the highest per-pupil spending of any of the participating countries (OECD 2010).

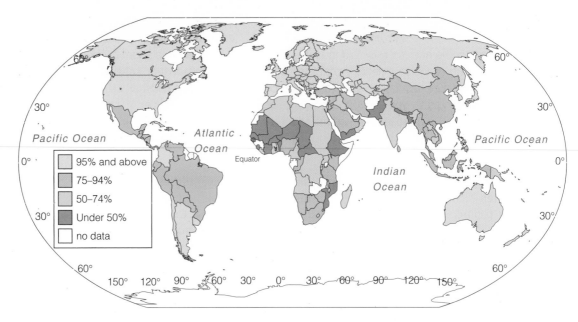

Figure 8.1: **Literacy Percentages around the World, 15 Years Old and Older, 2010**
Source: Copyright © 2011 Maps.com. Reprinted by permission.

Fourth, in reference to teachers, the average student-teacher ratio in elementary schools in OECD-participating countries is 22:1. Teachers' salaries in almost all countries have increased over the last decade, with the greatest increases in Finland, Hungary, and Mexico (OECD 2010). Despite the fact that public school teachers' salaries in the United States averaged $53,910 in 2009 (NCES 2010), near the OECD average, teachers in the United States work more hours than in any of the other OECD-participating countries.

Globally, few countries have the quality of schools or accessibility to education as enjoyed in the United States. Here, students in Lahtora, India, have outdoor lessons because their classroom is too crowded and too cold to have classes inside.

Fifth, educational attainment is increasing worldwide (HDR 2010). In almost all of the OECD countries studied, having a high school degree has become the norm among younger cohorts. For example, the percent of 25- to 34-year-olds who have earned a high school degree is 22 percent higher than the percent of 55- to 64-year-olds that have completed a high school degree (OECD 2010).

Finally, the *Programme for International Student Assessment* (PISA) measures "how well students who are nearing the end of compulsory education are prepared to meet the challenges of today's knowledge societies—what PISA refers to as 'literacy'" (OECD 2009, p. 80). A random sample of over 400,000 15-year-old students in 65 countries participates in the PISA. The results indicate that, of the 31 OECD participating countries, 12 had average science scores above the OECD average. Those with the highest average science scores were Finland, Japan, the Republic of Korea, New Zealand, and Canada. These same five countries earned the highest reading scores and, with the exception of New Zealand, were also the top performers in mathematics literacy. The United States scored below the OECD average in science and in mathematics, but slightly above the OECD average in reading (OECD 2010).

Sociological Theories of Education

The three major sociological perspectives—structural functionalism, conflict theory, and symbolic interactionism—are important in explaining different aspects of American education.

Structural-Functionalist Perspective

According to structural functionalism, the educational institution serves important tasks for society, including instruction, socialization, the sorting of individuals into various statuses, and the provision of custodial care (Sadovnik 2004). Many social problems, such as unemployment, crime and delinquency, and poverty, can be linked to the failure of the educational institution to fulfill these basic functions (see Chapters 4, 6, and 7). Structural functionalists also examine the reciprocal influences of the educational institution and other social institutions, including the family, political institution, and economic institution.

Instruction A major function of education is to teach students the knowledge and skills that are necessary for future occupational roles, self-development, and social functioning. Although some parents teach their children basic knowledge and skills at home, most parents rely on schools to teach their children to read, spell, write, tell time, count money, and use computers. As discussed later, many U.S. students display a low level of academic achievement. The failure of schools to instruct students in basic knowledge and skills both causes and results from many other social problems.

Socialization The socialization function of education involves teaching students to respect authority—behavior that is essential for social organization (Merton 1968). Students learn to respond to authority by asking permission to leave the classroom, sitting quietly at their desks, and raising their hands before asking a question. Students who do not learn to respect and obey teachers may later disrespect and disobey employers, police officers, and judges.

The educational institution also socializes youth into the dominant culture. Schools attempt to instill and maintain the norms, values, traditions, and symbols of the culture in a variety of ways, such as celebrating holidays (e.g., Martin Luther King Jr. Day and Thanksgiving); requiring students to speak and write in standard English; displaying the American flag; and discouraging violence, drug use, and cheating.

As the number and size of racial and ethnic minority groups have increased, American schools are faced with a dilemma: Should public schools promote only one common culture, or should they emphasize the cultural diversity reflected in the U.S. population? Some evidence suggests that most Americans believe that schools should do both—they should promote one common culture and emphasize diverse cultural traditions. **Multicultural education**—that is, education that includes all racial and ethnic groups in the school curriculum—promotes awareness and appreciation for cultural diversity (also see Chapter 9).

Sorting Individuals into Statuses Schools sort individuals into statuses by providing credentials for individuals who achieve various levels of education at various schools within the system. These credentials sort people into different statuses—for example, "high school graduate," "Harvard alumna," and "English major." In addition, schools sort individuals into professional statuses by awarding degrees in fields such as medicine, engineering, and law. The significance of such statuses lies in their association with occupational prestige and income—the higher one's education, the higher one's income. Furthermore, unemployment rates and earnings are tied to educational status as seen in Figure 8.2.

Custodial Care The educational system also serves the function of providing custodial care by providing supervision and care for children and adolescents until they are 18 years old (Merton 1968). Despite 12 years and almost 13,000 hours of instruction, some school districts are increasing the number of school hours and/or days beyond the "traditional" schedule. In 2006, Massachusetts began the Expanded Learning Time Initiative, which adds an additional 300 hours of learning over the course of a school year by having longer and/or more school days (ELT 2009). Concerns over mandatory year-round schedules led a parents' group to sue the local school district. In 2009, the North Carolina Supreme Court held that a school system does not need parental permission to assign students to year-round schools (Bowens 2009).

multicultural education Education that includes all racial and ethnic groups in the school curriculum, thereby promoting awareness and appreciation for cultural diversity.

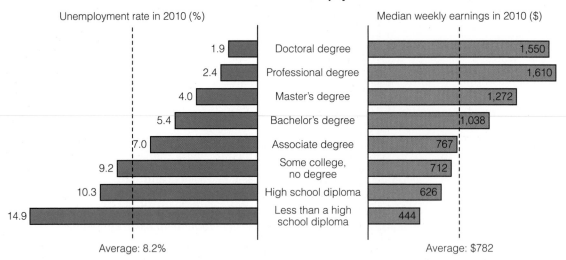

Education pays:

Unemployment rate in 2010 (%) | | Median weekly earnings in 2010 ($)

Unemployment rate	Level of Education	Median weekly earnings
1.9	Doctoral degree	1,550
2.4	Professional degree	1,610
4.0	Master's degree	1,272
5.4	Bachelor's degree	1,038
7.0	Associate degree	767
9.2	Some college, no degree	712
10.3	High school diploma	626
14.9	Less than a high school diploma	444

Average: 8.2%   Average: $782

Note: Data are 2010 annual averages for persons age 25 and over. Earnings are for full-time wage and salary workers.

Figure 8.2: **Unemployment Rate and Median Weekly Earnings for Individuals 25 and Over by Highest Level of Education, 2010**
Source: BLS 2011.

Conflict Perspective

Conflict theorists emphasize that the educational institution solidifies the class positions of groups and allows the elite to control the masses. Although the official goal of education in society is to provide a universal mechanism for achievement, in reality, educational opportunities and the quality of education are not equally distributed.

Conflict theorists point out that the socialization function of education is really indoctrination into a capitalist ideology (Sadovnik 2004). In essence, students are socialized to value the interests of the state and to function to sustain it. Such indoctrination begins in kindergarten. Rosabeth Moss Kanter (1972) coined the term the *organization child* to refer to the child in nursery school who is most comfortable with supervision, guidance, and adult control. Teachers cultivate the organization child by providing daily routines and rewarding those who conform. In essence, teachers train future bureaucrats to be obedient to authority.

In addition, to conflict theorists, education serves as a mechanism for **cultural imperialism,** or the indoctrination into the dominant culture of a society. When cultural imperialism exists, the norms, values, traditions, and languages of minorities are systematically ignored. A Mexican American student recalls his feelings about being required to speak English (Rodriguez 1990):

To cover financial costs associated with the increasing needs of educational programs, school systems often find it necessary to contract with major corporations such as Coca-Cola.

When I became a student, I was literally "remade"; neither I nor my teachers considered anything I had known before as relevant. I had to forget most of what my culture had provided, because to remember it was a disadvantage. The past and its cultural values became detachable, like a piece of clothing grown heavy on a warm day and finally put away. (p. 203)

Conflict theorists are also quick to note that learning is increasingly a commercial enterprise as corporations anxious to bombard students with advertising and other pro-capitalist messages fund necessary financial support for equipment, laboratories, and

cultural imperialism The indoctrination into the dominant culture of a society.

technological upgrades. Moreover, as Molnar et al. observe (2010) ". . . the economic recession of the last several years appears to have resulted in intensified corporate marketing efforts in schools, as parents, teachers and administrators welcome 'partnerships' that they think may help avoid program cuts" (p. 1).

Finally, the conflict perspective focuses on what Kozol (1991) called the "savage inequalities" in education that perpetuate racial disparities. Kozol documented gross inequities in the quality of education in poorer districts, largely composed of minorities, compared with districts that serve predominantly white middle-class and upper-middle-class families. Kozol revealed that schools in poor districts tend to receive less funding and to have inadequate facilities, books, materials, equipment, and personnel. For example, disadvantaged children are more likely to attend schools with fewer certified or experienced teachers and to have fewer advanced placement courses than schools attended by more advantaged children (Viadero 2006).

Symbolic Interactionist Perspective

Whereas structural functionalism and conflict theory focus on macro-level issues, such as institutional influences and power relations, symbolic interactionism examines education from a micro-level perspective. This perspective is concerned with individual and small-group issues, such as teacher-student interactions and the self-fulfilling prophecy.

Teacher-Student Interactions Symbolic interactionists have examined the ways that students and teachers view and relate to one another. For example, children from economically advantaged homes may be more likely to bring to the classroom social and verbal skills that elicit approval from teachers. From the teachers' point of view, middle-class children are easy and fun to teach. They grasp the material quickly, do their homework, and are more likely to "value" the educational process. Children from economically disadvantaged homes often bring fewer social and verbal skills to those same middle-class teachers, who may, inadvertently, hold up social mirrors of disapproval. Teacher disapproval contributes to lower self-esteem among disadvantaged youth.

Self-Fulfilling Prophecy The **self-fulfilling prophecy** occurs when people act in a manner consistent with the expectations of others. For example, a teacher who defines a student as a slow learner may be less likely to call on that student or to encourage the student to pursue difficult subjects. The teacher may also be more likely to assign the student to lower ability groups or curriculum tracks (Riehl 2004). As a consequence of the teacher's behavior, the student is more likely to perform at a lower level.

A classic study by Rosenthal and Jacobson (1968) provided empirical evidence of the self-fulfilling prophecy in the public school system. Five elementary school students in a San Francisco school were selected at random and identified for their teachers as "spurters." Such a label implied that they had superior intelligence and academic ability. In reality, they were no different from the other students in their classes. At the end of the school year, however, these five students scored higher on their intelligence quotient (IQ) tests and made higher grades than their classmates who were not labeled as spurters. In addition, the teachers rated the spurters as more curious, interesting, and happy, and more likely to succeed than the non-spurters. Because the teachers expected the spurters to do well, they treated the students in a way that encouraged better school performance.

Who Succeeds? The Inequality of Educational Attainment

Figure 8.3 shows the extent of the variation in highest level of education attained by individuals 25 years of age and older in the United States. As noted earlier, conflict theory focuses on such variations in discussions of educational inequalities. Educational inequality is based on social class and family background, race and ethnicity, and gender. Each of these factors influences who succeeds in school.

self-fulfilling prophecy A concept referring to the tendency for people to act in a manner consistent with the expectations of others.

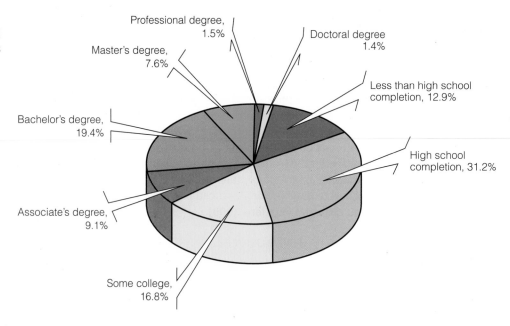

Figure 8.3: Highest Level of Education Attained by Individuals Age 25 Years Old and Older, 2010*
*Percentages do not sum to 100 percent because of rounding.
Source: NCES 2011a.

Professional degree, 1.5%

Doctoral degree 1.4%

Master's degree, 7.6%

Less than high school completion, 12.9%

Bachelor's degree, 19.4%

High school completion, 31.2%

Associate's degree, 9.1%

Some college, 16.8%

One of the best predictors of educational success and attainment is socioeconomic status. Children whose families are in middle and upper socioeconomic brackets are more likely to perform better in school and to complete more years of education than children from lower socioeconomic class families.

Social Class and Family Background

One of the best predictors of educational success and attainment is socioeconomic status. Children whose families are in middle and upper socioeconomic brackets are more likely to perform better in school and to complete more years of education than children from lower socioeconomic class families. Using father's education as a proxy for socioeconomic status, global statistics indicate that students are significantly more likely to attend college if their fathers graduated from college—more than twice as likely in Austria, Germany, France, and England (OECD 2009).

Socioeconomic status also predicts academic achievement. On standardized tests such as the SAT and the ACT, "children from the lowest-income families have the lowest average test scores, with an incremental rise in family income associated with a rise in test scores" (Corbett et al. 2008, p. 3). Muller and Schiller (2000) reported that students from higher socioeconomic backgrounds are more likely to enroll in advanced courses for mathematics credit and to graduate from high school—two indicators of future educational and occupational success. In addition, a recent report to the U.S. Congress on educational inequality acknowledged that students from high-income families compared to low- and middle-income families are not only more likely to attend college, but they are also more likely to complete it (ACSFA 2010).

> **What Do You Think?** Research suggests that poor children who attend high-quality preschools are "less likely to drop out of school, repeat grades, or need special education, [and] as adults, are less likely to commit crimes, more likely to be employed, and likely to have higher earnings" (Olson 2007, p. 30). However, access to preschools varies by family income—the wealthier the family, the higher the probability a child will attend preschool. Do you think the federal government should establish a nationalized program of preschools for all 3- and 4-year-olds?

Families with low incomes have fewer resources to commit to educational purposes—less money to buy books or computers or to pay for tutors. Disadvantaged parents are less involved in learning activities. As parental education and income levels increase, the likelihood of a parent taking a child to a library, play, concert or other live show, art gallery, museum, or historical site also increases. Further, parents who have less education and lower income levels are less likely to be involved in school-related activities, for example, attending a school event or volunteering at a school (NCES 2011a).

Head Start and Early Head Start In 1965, Project **Head Start** began to help preschool children from the most disadvantaged homes. Head Start provides an integrated program of health care, parental involvement, education, and social services. Today, over 904,000 preschoolers (3- and 4-year-olds) in 50 states participate in the program that, in 2009, had an average cost of $7,600 per child (Office of Head Start 2011). Graduates of Head Start "score better on intelligence and achievement tests, their health status is better, and they have the socio-emotional traits to help them adjust to school" (Zigler et al. 2004, p. 341).

Early Head Start, a program for infants and toddlers from low-income families, has also been evaluated. A seven-year national evaluation of Early Head Start programs culminated in a 2004 report to Congress. In summarizing the report, the authors conclude that "participating children perform significantly better in cognitive, language, and social-emotional development than their peers who do not participate. The program also had important impacts on many aspects of parenting and the home environment, and supported parents' progress toward economic self-sufficiency" (MPR 2007, p. 1).

Realizing the benefits of Head Start and Early Head Start, the final 2011 federal budget included increased funding to compensate for revenue losses as monies allocated from the 2009 American Recovery and Reinvestment Act (ARRA) run out. Federal stimulus dollars, however, cannot entirely compensate for the loss of federal and state expenditures on education. Although there are signs of economic recovery, most state revenues remain significantly lower than in pre-recession years (McNichol, Oliff, and Johnson 2011). The result is cuts in state education budgets. For example, several districts in Nevada have gone to a four-day school week, and Washington State suspended professional development programs for teachers (ERC 2011).

Further, local dollars, which make up 47 percent of school funding, varies by socioeconomic status of the district (NCES 2011b). For example, local expenditures on schools come from taxes, usually property taxes, and as housing prices decline, property taxes decline. The U.S. system of decentralized funding for schools has several additional consequences:

- School districts with low socioeconomic status are more likely to be in urban areas and in inner cities where the value of older and dilapidated houses have depreciated; less desirable neighborhoods are hurt by "white flight," with the result that the tax base for local schools is lower in deprived areas.
- School districts with low socioeconomic status are less likely to have businesses or retail outlets where revenues are generated; such businesses have closed or moved away.
- Because of their proximity to the downtown area, school districts with low socioeconomic status are more likely to include hospitals, museums, and art galleries, all of which are tax-free facilities. These properties do not generate revenues.
- Neighborhoods with low socioeconomic status are often in need of the greatest share of city services; fire and police protection, sanitation, and public housing consume the bulk of the available revenues. Precious little is left over for education in these districts.
- In school districts with low socioeconomic status, a disproportionate amount of the money has to be spent on maintaining the school facilities, which are old and in need of repair, and on free or reduced-priced lunches, so less is available for the children themselves.

Head Start Begun in 1965 to help preschool children from the most disadvantaged homes, Head Start provides an integrated program of health care, parental involvement, education, and social services for qualifying children.

Many poor children have to depend on donations for much-needed school supplies. Here, a first grader smiles as he opens his box of school supplies, including a backpack from "Kits for Kidz" of Chicago.

Although states provide additional funding to supplement local taxes, this funding is not always enough to lift schools in poorer districts to a level that even approximates the funding available to schools in wealthier districts. For example, in *Leandro v. State* (1997), the North Carolina Supreme Court held that the state constitution required that all schools must provide adequate resources to fully educate disadvantaged students (i.e., those who are poor, in special education, and have limited English proficiency). Yet, over a decade later, at least one county in North Carolina is doing such a poor job of educating its disadvantaged students that a judge has called it "academic genocide" (Waggoner 2009). Further, in 2011, attorneys for several poor counties in North Carolina argued that legislative budget cuts in education would prevent North Carolina children from getting a "basic education" in violation of the *Leandro* decision (Stancill 2011).

Race and Ethnicity

Between 1989 and 2009, the percentage of racial and ethnic minorities in public schools increased from 32 percent to 45 percent (NCES 2011b). To a large extent, this demographic shift in the racial and ethnic makeup of public school students is a result of the growth in the number of Hispanic students—the fastest-growing minority in the United States.

In comparison to whites, Hispanics and blacks are less likely to succeed in school at almost every level. As early as the start of kindergarten, Hispanics and blacks have lower mean achievement scores than other racial or ethnic groups in reading and mathematics (NCES 2011b). By fourth grade, over 85 percent of blacks and Hispanics are reading below grade level compared to 59 percent of whites. Similarly, by eighth grade, over 83 percent of black and Hispanic students compared to 57 percent of white students are below grade level in mathematics (CDF 2010). Further, as Table 8.1 indicates, although educational attainment has increased over time, racial and ethnic disparities remain. Note that, in general, the high school graduation gap between racial and ethnic groups is narrowing; the college graduation gap, however, is getting wider—whites and Asians on one side and Hispanics and Black Americans on the other.

It is important to note, however, that socioeconomic status interacts with race and ethnicity. Because race and ethnicity are so closely tied to socioeconomic status, i.e., a disproportionate number of racial and ethnic minorities are poor, it *appears* that race or ethnicity determines school success. Although race and ethnicity may have an independent effect on educational achievement, their relationship is largely a result of the association between race and ethnicity and socioeconomic status.

In addition to the socioeconomic variables, there are several reasons why minority students have academic difficulty. First, minority children may be English language

Although race and ethnicity may have an independent effect on educational achievement, their relationship is largely a result of the association between race and ethnicity and socioeconomic status.

learners (ELL). Although ELL students come from over 400 different language backgrounds, over 75 percent are Spanish speakers (Goldenberg 2008). Now imagine, as Goldenberg (2008) suggests,

> you are in second grade and don't speak English very well and are expected to learn in one year . . . irregular spelling patterns, diphthongs, syllabication rules, regular and irregular plurals, common prefixes and suffixes, antonyms and synonyms; how to follow written instructions, interpret words with multiple meanings, locate information in expository texts . . . read fluently and correctly at least 80 words per minute, add approximately 3,000 words to your vocabulary . . . and write narratives and friendly letters using appropriate forms, organization, critical elements, capitalization, and punctuation, revising as needed. (p. 8)

Not surprisingly, ELL students score significantly below non-ELL students on standardized tests in both reading and mathematics. To help ELL students, some educators advocate **bilingual education,** teaching children in both English and their non-English native language.

Advocates claim that bilingual education results in better academic performance of minority students, enriches all students by exposing them to different languages and cultures, and enhances the self-esteem of minority students. Critics argue that bilingual education limits minority students and places them at a disadvantage when they compete outside the classroom, reduces the English skills of minorities, costs money, and leads to hostility with other minorities who are also competing for scarce resources.

TABLE 8.1 Educational Attainments by Race, Ethnicity, and Sex, 1970 and 2009

	1970		2009	
	MALES	FEMALES	MALES	FEMALES
Four years of high school or more				
White	54.0	55.0	86.5	87.7
Black	30.1	32.5	84.0	84.1
Hispanic	37.9	34.2	60.6	63.7
Asian and Pacific Islander	61.3	63.1	90.4	86.2
Total	51.9	52.8	86.2	87.1
College graduate or more				
White	14.4	8.4	30.6	29.3
Black	4.2	4.6	17.8	20.6
Hispanic	7.8	4.3	12.5	14.0
Asian and Pacific Islander	23.5	17.3	55.7	49.3
Total	13.5	8.1	30.1	29.1

Source: U.S. Census Bureau 2011.

What Do You Think? Despite the many critics of bilingual education, the regional government in Madrid, Spain, has made about a third of its elementary schools bilingual and hopes to increase that number to 50 percent by 2015, specifically to increase job competitiveness (Minder 2011). Keeping globalization in mind, do you think U.S. schools that have English-only policies are hurting or helping their students?

The second reason why racial and ethnic minorities don't perform well in school, and compounding the difficulty ELL students have, is that many of the tests used to assess academic achievement and ability are biased against minorities. Questions on standardized tests often require students to have knowledge that is specific to the white middle-class majority culture, knowledge that racial and ethnic minorities may not have.

A third factor that hinders minority students' academic achievement is overt racism and discrimination. Discrimination against minority students takes the form of unequal funding, as discussed earlier, as well as racial profiling and school segregation.

Studies indicate that minority students, and specifically black students, may be the victims of what is being called "learning while black" (Morse 2002). The allegation is often supported by an examination of school discipline statistics. In 2007, 43 percent of black students in sixth through twelfth grades were suspended compared to 16 percent of white students and 22 percent of Hispanic students (NCES 2011b). Although the debate is likely to continue, it must be noted that differences in discipline patterns are not necessarily a consequence of racism. They may reflect differences in behavior.

bilingual education In the United States, teaching children in both English and their non-English native language.

School Desegregation In 1954, the U.S. Supreme Court ruled in *Brown v. Board of Education* that segregated education was unconstitutional because it was inherently unequal. In 1966, a landmark study titled *Equality of Educational Opportunity* (Coleman et al. 1966) revealed that almost 80 percent of all U.S. schools attended by whites contained 10 percent or fewer blacks and that, with the exception of Asian Americans, whites outperformed minorities on standardized tests. Coleman and colleagues emphasized that the only way to achieve quality education for all racial groups was to desegregate the schools. This recommendation, known as the **integration hypothesis**, advocated busing to achieve racial balance.

Increasingly, school classrooms will be characterized by racial and ethnic diversity. By 2040, less than half of all school-age children will be non-Hispanic whites.

Despite the Coleman report, court-ordered busing, and a societal emphasis on the equality of education, public schools remain largely segregated. Most black and Hispanic U.S. students attend schools that are predominantly minority in enrollment. For example, in 2007, over half of all Black American and Hispanic public school students attended schools that had a combined minority enrollment of 75 percent or more (NCES 2011b). Commenting on the "restoration of apartheid schooling in America" Jonathan Kozol (2005) notes that "[s]chools that were already deeply segregated 25 or 30 years ago . . . are no less segregated now, while thousands of other schools that had been integrated either voluntarily or by force of law have since been rapidly re-segregating . . ." (p. 18). Attending a desegregated school, although having no negative achievement effect on white students, tends to have positive effects on black and Hispanic students in terms of learning and graduation rates (Orfield & Lee 2006).

In 2007, the U.S. Supreme Court held that public school systems "cannot seek to achieve or maintain integration through measures that take explicit account of a student's race" (Greenhouse 2007, p. 1). The court's decision reflects a general trend toward using socioeconomic or income-based integration rather than race-based integration variables. Kahlenberg (2006) advocated this approach for several reasons. First, "socioeconomic integration more directly and effectively achieves the first aim of racial integration: raising the achievement of students" (Kahlenberg 2006, p. 10). Second, socioeconomic integration, because of the relationship between race and income, achieves racial integration, and racial integration, in turn, fosters racial tolerance and social cohesion. Lastly, unlike race-based integration that is subject to "strict scrutiny" by the government, school assignments based on socioeconomic status are perfectly legal. Nonetheless, several school districts throughout the United States remain under federal oversight as required by decades-old racial desegregation cases.

Despite a half-century movement toward the racial desegregation of schools, some school districts are reverting, at least in part, to traditional methods of student assignment, i.e., neighborhood schools. In North Carolina, for example, the Wake County School Board, ". . . abolished the policy behind one of the nation's most celebrated integration efforts" (McCrummen 2011, p. 1). Fears that the new policy will lead to racial resegregation has led the National Association for the Advancement of Colored People (NAACP) to file a civil suit against the school district in an attempt to prevent the Republican-backed board from instituting the new policy.

Gender

Not only do women comprise two-thirds of the world's illiterate, girls comprise more than 70 percent of the 125 million children who don't attend school (Save the Children 2011). Although progress in reducing the education gender gap has been made, gender parity in primary and secondary schools has not been achieved. Globally, of the 176 countries with data available, only two-thirds have achieved gender parity in primary schools, and less than 40 percent have achieved gender parity in secondary schools (EFA Global Monitoring Report 2009).

Historically, U.S. schools have discriminated against women. Before the 1830s, U.S. colleges accepted only male students. In 1833, Oberlin College in Ohio became the first college to admit women. Even so, in 1833, female students at Oberlin were required to

integration hypothesis A theory that the only way to achieve quality education for all racial and ethnic groups is to desegregate the schools.

wash male students' clothes, clean their rooms, and serve their meals and were forbidden to speak at public assemblies (Fletcher 1943; Flexner 1972).

In the 1960s, the women's movement sought to end sexism in education. Title IX of the Education Amendments of 1972 states that no person shall be discriminated against on the basis of sex in any educational program receiving federal funds. These guidelines were designed to end sexism in the hiring and promoting of teachers and administrators. Title IX also sought to end sex discrimination in granting admission to college and awarding financial aid. Finally, the guidelines called for an increase in opportunities for female athletes by making more funds available to their programs.

Although gender inequality in education continues to be a problem worldwide, the push toward equality has had considerable effect in the United States. For example, in 1970, nearly twice as many men as women had four years of college or more—8.1 percent compared with 13.5 percent. By 2009, 30.1 percent of men and 29.1 percent of women had four years of college or more (see Table 8.1). Further, scores on the National Assessment of Educational Progress (NAEP) exam indicate that the gender gap in both mathematics and reading scores has decreased over the last few decades. Where differences exist, for example, in a specific grade level, they generally follow academic stereotypes—boys outscore girls in mathematics, girls outscore boys in reading. Similarly, on the SAT college admissions test, males, on the average, outperform females, with the largest achievement gap on the mathematics portion of the exam (Corbett et al. 2008).

Traditionally, the gender gap in achievement has been explained by differences in gender role socialization (see Chapter 10). Recently, however, popular portrayals of male and female differences in achievement have been attributed to "hardwiring," as though there are " pink and blue brains." As neuroscientist Lise Eliot (2010) notes

. . . a closer look reveals that the gaps vary considerably by age, ethnicity, and nationality. For example, among the countries participating in PISA, the reading gap is more than twice as large in some countries (Iceland, Norway, and Austria) as in others (Japan, Mexico, and Korea); for math, the gap ranges from a large male advantage in certain countries (Korea and Greece) to essentially no gap in other countries—or even reversed in girls' favor (Iceland and Thailand). What's more, a recent analysis of PISA data found that higher female performance in math correlates with higher levels of gender equity in individual nations with higher levels of gender equity in individual nations. (p. 32)

Most of the research on gender inequality in the schools focuses on how female students are disadvantaged in the educational system. But what about male students? Although often achieving higher scores on standardized tests (Corbett et al. 2008), boys are more likely to lag behind girls in the classroom, be diagnosed with attention-deficit/hyperactivity disorder (ADHD), have learning disabilities, feel alienated from the learning process, and drop out or be expelled from school (see this chapter's *Self and Society* feature and assess your own school alienation rate) (Dobbs 2005; Mead 2006; Tyre 2008). Further, black males compared to white males score lower on the NAEP, are less likely to be in advanced placement (AP) classes, are more likely to be in special education classes, and are less likely to graduate from high school or college (Schott Report 2010). Thus, the problems boys have in school may indeed require schools to devote more resources and attention to them (e.g., recruit male teachers). Unfortunately, in a time of economic crisis, that may not be possible.

Problems in the American Educational System

When a random sample of adults was asked what grade they would give the nation's public schools, over half the respondents gave the nation's public schools a C (Bushaw & Lopez 2010). The implication that American schools are just "average"

Indicate your agreement with each statement by selecting one of the responses provided:

1. It is hard to know what is right and wrong because the world is changing so fast.
   ___ Strongly agree ___ Agree ___ Disagree ___ Strongly disagree
2. I am pretty sure my life will work out the way I want it to.
   ___ Strongly agree ___ Agree ___ Disagree ___ Strongly disagree
3. I like the rules of my school because I know what to expect.
   ___ Strongly agree ___ Agree ___ Disagree ___ Strongly disagree
4. School is important in building social relationships.
   ___ Strongly agree ___ Agree ___ Disagree ___ Strongly disagree
5. School will get me a good job.
   ___ Strongly agree ___ Agree ___ Disagree ___ Strongly disagree
6. It is all right to break the law as long as you do not get caught.
   ___ Strongly agree ___ Agree ___ Disagree ___ Strongly disagree
7. I go to ball games and other sports activities at school.
   ___ Always ___ Most of the time ___ Some of the time ___ Never
8. School is teaching me what I want to learn.
   ___ Always ___ Most of the time ___ Some of the time ___ Never
9. I go to school parties, dances, and other school activities.
   ___ Always ___ Most of the time ___ Some of the time ___ Never
10. A student has the right to cheat if it will keep him or her from failing.
   ___ Always ___ Most of the time ___ Some of the time ___ Never
11. I feel like I do not have anyone to reach out to.
   ___ Always ___ Most of the time ___ Some of the time ___ Never
12. I feel that I am wasting my time in school.
   ___ Always ___ Most of the time ___ Some of the time ___ Never
13. I do not know anyone that I can confide in.
   ___ Strongly agree ___ Agree ___ Disagree ___ Strongly disagree
14. It is important to act and dress for the occasion.
   ___ Always ___ Most of the time ___ Some of the time ___ Never
15. It is no use to vote because one vote does not count very much.
   ___ Strongly agree ___ Agree ___ Disagree ___ Strongly disagree
16. When I am unhappy, there are people I can turn to for support.
   ___ Always ___ Most of the time ___ Some of the time ___ Never
17. School is helping me get ready for what I want to do after college.
   ___ Strongly agree ___ Agree ___ Disagree ___ Strongly disagree
18. When I am troubled, I keep things to myself.
   ___ Always ___ Most of the time ___ Some of the time ___ Never
19. I am not interested in adjusting to American society.
   ___ Strongly agree ___ Agree ___ Disagree ___ Strongly disagree
20. I feel close to my family.
   ___ Always ___ Most of the time ___ Some of the time ___ Never
21. Everything is relative and there just aren't any rules to live by.
   ___ Strongly agree ___ Agree ___ Disagree ___ Strongly disagree
22. The problems of life are sometimes too big for me.
   ___ Always ___ Most of the time ___ Some of the time ___ Never
23. I have lots of friends.
   ___ Always ___ Most of the time ___ Some of the time ___ Never
24. I belong to different social groups.
   ___ Strongly agree ___ Agree ___ Disagree ___ Strongly disagree

Interpretation

This scale measures four aspects of alienation: (1) powerlessness, or the sense that high goals (e.g., straight As) are unattainable; (2) meaninglessness, or lack of connectedness between the present (e.g., school) and the future (e.g., job); (3) normlessness, or the feeling that socially disapproved behavior (e.g., cheating) is necessary to achieve goals (e.g., high grades); (4) and social estrangement, or lack of connectedness to others (e.g., being a "loner"). For items 1, 6, 10, 11, 12, 13, 15, 18, 19, 21, and 22, the response indicating the greatest degree of alienation is "strongly agree" or "always. " For all other items, the response indicating the greatest degree of alienation is "strongly disagree" or "never. "

Source: Mau 1992. Used by permission of Libra Publishers, Inc., 3089 Clairemont Drive, Suite 383, San Diego, California 92117.

is strengthened by the Educational Research Center's (ERC) annual assessment of education in the United States. Analyzing data across states in several substantive areas (e.g., learning, teachers), the center assigned an overall score to the nation's schools of 76.3 percent—a C (ERC 2011). Not surprisingly, the present administration hopes to address many of the problems—low academic achievement, high dropout rates, questionable teacher training, school violence, and the challenges of higher education—that the ERC and other experts identified (see Figure 8.4).

Low Levels of Academic Achievement

The Educational Research Center uses three indicators to measure achievement in public elementary and secondary schools: current levels of performance, improvement over time, and achievement gap between poor and nonpoor learners (ERC 2011). Based on a 100-point scale, the achievement average for the nation was 68.7 (D+), ranging from a high of 85.0 (B) for Massachusetts and a low of 55.3 (F) for Mississippi (ERC 2011).

One way to measure performance is to look at the results of the National Assessment of Education Progress, the "nation's report card" on public and private school student performances. Although varying by race, ethnicity, and gender, in general, mathematics scores of 9-year-olds (fourth graders), 13-year-olds (eighth graders), and 17-year olds (twelfth graders) have increased since 2005 (NAEP 2010). Reading scores have also increased for each of the three age groups.

It is important to note that even statistically significant increases, although a step in the right direction, may mask poor performances. For example, as noted, reading scores increased significantly for 9-, 13-, and 17-year-olds between 2005 and 2009. However, 33 percent of 9-year-olds and 25 percent of 13-year-olds scored below the "basic" reading level in 2009. Similarly, 38 percent of 17-year-olds scored below the "basic" mathematics level in 2009 (NAEP 2010).

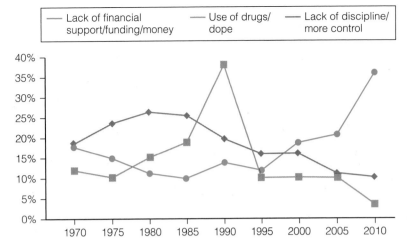

Figure 8.4: Responses by a sample of Americans 18 years and older when asked, "What do you think are the biggest problems the public schools of your community must deal with?"
Source: Bushaw & Lopez 2010. From *The 42nd Annual Phi Delta Kappa/Gallup Poll of the Public's Attitudes Toward the Public Schools*, September 2010. Reprinted with permission of Phi Delta Kappa International, www.pdkintl.org. All rights reserved.

U.S. students are also outperformed by many of their foreign counterparts—something particularly troubling in a knowledge-based global economy. Based upon the *Program for International Student Assessment* (PISA), U.S. 15-year-olds perform below the OECD country averages in both science and mathematics. The U.S. average high school graduation rate is also below the average high school graduation rate for OECD-participating countries (OECD 2010). Perhaps most troubling, the international achievement gap "widens the longer children are in school" and is "not merely an issue for poor children attending schools in poor neighborhoods; instead, it affects most children in most schools" (McKinsey & Company 2009, p. 7). The international achievement gap impacts individuals and communities and costs the economy millions of dollars in lost human potential (McKinsey & Company 2009).

School Dropouts

The *status dropout rate* is the percentage of 16- to 24-year-olds that are not in school and have not earned a high school degree or its equivalent. In the last several decades, the status dropout rate has significantly declined, dropping from 14 percent in 1980 to 8 percent in 2009 (NCES 2011b). In each of those years, status dropout rates were higher for Hispanics than for blacks and whites. As Figure 8.5 indicates, dropout rates vary

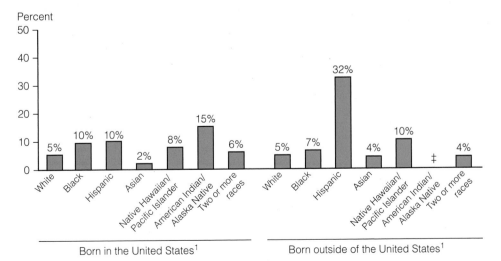

Figure 8.5: Status Dropout Rates of 16- to 24-Year-Olds, by Race/ Ethnicity and Nativity, 2009
Source: NCES 2011b.

‡Reporting standards not met (too few cases).

[1]United States refers to the 50 states and the District of Columbia.

considerably by race and ethnicity, and by nativity. They also vary by gender, with males having higher dropout rates than females, 9 percent and 7 percent, respectively.

Students from lower socioeconomic backgrounds who have only one parent in the home and who have changed schools frequently are more likely to drop out (Barton 2005; Tyler & Lofstrom 2009; Orthner et al. 2009). Further, a study funded by the Bill and Melinda Gates Foundation, using focus groups and a survey of more than 500 dropouts, reports that respondents identified five major reasons for dropping out of high school: (1) "classes were not interesting"; (2) "missed too many days and can't catch up"; (3) "spent time with people who were not interested in school"; (4) "had too much freedom and not enough rules"; and (5) "was failing in school" (Bridgeland et al. 2006, p. 3).

The former students were also asked what could be done to improve a student's chances of remaining in school. Eighty-one percent responded "opportunities for real-world learning (internships, service-learning, etc.) to make classrooms more relevant" and "better teachers who keep classes interesting" (Bridgeland et al. 2006, p. 13). Other top responses included smaller classes and more individual attention, better communication between parents and schools, and increased supervision at home and at school to ensure students attend classes.

The economic and social consequences of dropping out of school are significant. A million children drop out of high school every year, reducing tax revenues and increasing societal costs for public assistance, crime, and health care (Tyler & Lofstrom 2009). Some programs are successful in reducing the dropout rate. A review of dropout prevention programs by Tyler and Lofstrom (2009) suggests that successful programs share five common elements: (1) management of students, (2) mentoring, (3) family involvement, (4) curricular reform, and (5) contending with out-of-school problems.

Second-chance initiatives such as General Educational Development (GED) certification allow students to complete their high school requirements. Alternatively, early or middle-school college programs allow dropouts to enroll in community colleges or, in some cases, four-year degree programs. There, they receive a secondary school education, earn a high school diploma, and often accrue college credits (Manzo 2005).

What Do You Think? California, as roughly half of all states, requires an exit exam for high school graduation. Researchers found that after the exam was put into place in California, (1) lower-achieving students had significantly higher dropout rates, and (2) the negative effect of the exit exam was stronger for minority and female students than for nonminority and male students, even when levels of academic achievement were held constant (Reardon et al. 2009). These findings support the **"stereotype threat"** hypothesis, i.e., the tendency for minorities and women to perform poorly on high-stakes tests because of the anxiety created by fear that a poor performance will validate negative societal stereotypes. Do you think high schools should or should not have exit exams as a requirement for graduation? Do you think exit exams unfairly target minorities and females?

Crime, Violence, and School Discipline

stereotype threat The tendency of minorities and women to perform poorly on high-stakes tests because of the anxiety created by the fear that a negative performance will validate societal stereotypes about one's member group.

Despite the horrors of high-publicity school killings such as those at Columbine and Virginia Tech, the chance of a student dying at school is quite rare—about one homicide or suicide per 2.5 million students (NCES 2011a). The unlikelihood of such an event is reflected in students' perceptions of safety. In 2008, 93 percent of students said they felt very or somewhat safe while at school, a number that has changed little over the last 15 years (MetLife 2008).

In the 2007–2008 school year, the latest term for which data are available, 85 percent of schools recorded at least one violent crime. As in the nation as a whole, the number of school-related nonviolent offenses exceeded the number of school-related violent offenses. Nearly half of all schools reported that a theft had taken place during the school year—nearly one million thefts in all (NCES 2011c). Schools that experienced higher-than-average rates of school-related violent crime, whether against a student or teacher, were disproportionately public versus private schools, in poor, urban neighborhoods, and were more likely to have gang-related activity (NCES 2011c).

Discipline problems such as verbal abuse of teachers, disorder in classrooms, disrespect for teachers, fighting, insubordination, and the use of drugs or alcohol are also a concern. Of the disciplinary actions available to school personnel, suspension lasting more than five days, expulsion, and transfer to an alternative school are the most serious. In the 2007–2008 school years, 46 percent of public school principals reported using at least one serious disciplinary action against a student (NCES 2011c).

Of late, one particular type of school-related problem, bullying, has become the focus of research and legislative action. For example, the Safe Schools Improvement Act of 2011, if passed, would require states and school districts to clearly articulate policies that prohibit bullying, develop bullying prevention programs, and maintain a bullying database of recorded bullying incidents. Much of the impetus for new policies comes from the recent suicides of several students, including Phoebe Prince, a first-year high school student in Massachusetts. Three of her tormentors were charged with and found guilty of criminal harassment.

Bullying is characterized by an "imbalance of power that exists over a long period of time between two individuals, two groups, or a group and an individual in which the more powerful intimidate or belittle others" (Hurst 2005, p. 1) (see this chapter's *Social Problems Research Up Close* feature). Bullying may be direct (e.g., hitting someone) or indirect (e.g., spreading rumors) and may be considered a type of aggression (Wong 2009). Sometimes called cyber-bullying, bullying can also take place remotely, i.e., through electronic communication devices (e.g., cell phone) (see also Chapters 11 and 14).

Research by Wong (2009), AASA (2009), and NCES (2011c), indicates that:

- In 2007, about a third of students between 12 and 18 reported being bullied in the last year.
- Students who bully often perform poorly academically, have high dropout rates, and are more likely to get into fights, drink alcohol, vandalize property, and be truant.
- Victims of bullying report, in order of frequency, being the subject of rumors, being pushed, shoved, tripped, spit on, being threatened with harm, and being intentionally excluded from activities.
- Most bullying takes place inside the school building and is more often directed at females than males.
- Being bullied is associated with low self-esteem, anxiety, depression, alcohol and drug use, running away, and suicide.

In response to crime, violence, and other disciplinary problems, schools throughout the country have police officers patrolling the halls, require students to pass through metal detectors before entering school, and conduct random locker searches. Video cameras set up in classrooms, cafeterias, halls, and buses purportedly deter some school violence. A relatively new way to deal with school "troublemakers" is the *alternative* school. In theory, such schools deter disciplinary problems while providing low-achieving students with the remedial help they need. Concerns about alternative schools include the disproportionate number of minorities who are transferred to such facilities for discretionary rather than mandated reasons (Booker & Mitchell 2011).

Inadequate School Facilities and Programs

A report by the American Federation of Teachers (AFT) has documented many of the troubling conditions that exist in U.S. schools:

> . . . rodent infestation, mice droppings, fallen ceiling tiles, poor lighting, mold that has caused mushrooms to grow, crumbling exterior walls, asbestos, severely overcrowded classrooms and hallways, freezing rooms in the winter and extreme heat in the summer, old carpeting, clogged bathroom toilets and no stall doors, inadequate circuit breakers causing frequent outages, and poor ventilation. (AFT 2006, p. 4)

bullying Bullying "entails an imbalance of power that exists over a long period of time in which the more powerful intimidate or belittle others" (Hurst 2005, p. 1).

Bullying and Victimization among Black and Hispanic Adolescents

Researchers often study the variables associated with being a student bully or victim. The hope is that, if we can identify the characteristics of a student bully or victim early on, we can develop successful interventions that will reduce the prevalence of the behavior. To that end, Peskin et al. (2006) investigate variables associated with being a bully, a victim, or both, in a sample of middle and high school students.

Sample and Methods

Students from eight predominantly black and Hispanic secondary schools located in a large urban school district in Texas were selected for study. Classes were sampled by grade, resulting in a sample size of 1,413 respondents and a response rate of 52 percent. Nearly 60 percent of the sample was females. Middle school students (sixth, seventh, and eighth graders) comprised 56 percent of the sample, ninth graders comprised 11 percent

of the sample, and tenth through twelfth graders comprised 32 percent of the sample. Sixty-four percent of the sample described themselves as Hispanic, with the remainder self-identifying as African American.

Students were asked about their participation in bullying and their rates of victimization in the last 30 days. Response options included two categories: 0 to 2 times, and 3 or more times. Students were asked the frequency of their bullying behavior: (1) upsetting other students for the fun of it, (2) group teasing, (3) harassing, (4) teasing, (5) rumor spreading, (6) starting arguments, (7) getting others to fight, and (8) excluding others. There were four student victimization variables: (1) called names, (2) picked on by others, (3) made fun of, and (4) got hit and pushed.

In addition to measuring the frequency of bullying and victimization events, students were classified into one of four categories:

. . . a student was classified as a bully if he/she participated in at least two of the "bullying" behaviors at least three times in the last 30 days. Victims were classified as those students who reported that at least one of the "victim" behaviors happened to them at least three times in the last 30 days. Four mutually exclusive categories were constructed: (1) bullies; (2) victims; (3) those who reported both bullying and being a victim (bully-victim); and (4) students reporting neither behavior (p. 471)

Findings and Conclusions

Seven percent of the sample was defined as bullies, 12 percent as victims, and 5 percent as bully-victims. Grade level was found to be significantly related to bully or victim status. The prevalence of bullying is highest in the ninth grade (11.5 percent) and lowest in the sixth grade (4.9 percent) and

It is estimated that state governments need over $254 billion for school infrastructure repairs and maintenance, renovations and new construction, retrofitting, major improvements, and existing building additions (AFT 2009). Older schools have greater needs than newer ones, and are more often located in disadvantaged neighborhoods. Courts have consistently held that the quality of building facilities is part and parcel of equal educational opportunities resulting in government monitoring of state spending on infrastructure needs to ensure equitable distributions of funds (AFT 2009).

There is a considerable body of evidence that documents the relationship between school environment and academic achievement. Milkie and Warner (2011), after interviewing over 10,000 parents and teachers of first grade students, conclude that students' stress levels are negatively impacted by deteriorated school facilities. Tanner (2008) found a significant relationship between school environment (e.g., space, movement patterns, light, etc.) and academic achievement of third graders even when controlling for socioeconomic status of the school. Air quality, noise, overcrowding, inadequate space, and lighting

Mark Richards/PhotoEdit

More school buildings and facilities are in need of repair. Mold, defective ventilation systems, faulty plumbing, and the like are not uncommon. Still, quality education is expected to continue in the classrooms despite such deplorable conditions.

tenth grade (6.2 percent). The highest rate of victimization is in the sixth grade, with 20.8 percent of sixth graders meeting the criteria for being a victim. The lowest rate of being a victim was among eleventh and twelfth graders (7.5 percent). Bully-victims also varied by grade level, with the highest levels occurring in the eleventh and twelfth grades (7.9 percent) and lowest levels in the sixth grade (3.8 percent) and ninth grade (3.8 percent).

There were no significant relationships between the dependent variable and gender. Males and females were equally likely to report being bullies, victims, and bully-victims. However, race and ethnicity were significantly associated with the dependent variable. Blacks, compared with Hispanics, were more likely to be bullies, victims, and bully-victims. For example, although only 3.7 percent of Hispanics reported being bully-victims, 8.6 percent of blacks reported being bully-victims.

When specific bullying behaviors are examined, "upsetting students for the fun of it" was the most common kind of bullying activity, followed by teasing, group teasing, and starting arguments. The lowest prevalence of bullying behaviors was getting others to fight. Males were significantly more likely to participate in harassing and teasing behaviors than females and, with the exception of rumor spreading, African Americans were more likely to be involved in bullying behaviors than Hispanics.

Victims report that the most common form of bullying is name calling, followed by being made fun of. Males were significantly more likely to be hit and pushed than females, and blacks were significantly more likely to be made fun of, called names, or be hit and pushed than their Hispanic counterparts. Further, eleventh and twelfth graders were statistically more likely to report being picked on or made fun of than students in other class levels.

The authors conclude that the results of the study suggest the need for early intervention and the direction it should take:

While steps to decrease physical types of bullying may be targeted largely at males, steps to reduce verbal and relational types should be targeted at all students......... Interventions should be developed in middle school as the prevalence of these behaviors seems to peak as students begin high school..... . ... In our study, teasing and name calling were most prevalent; thus, targeted actions for the reduction of these behaviors may be a focus for intervention activities. (p. 479)

Finally, the authors note that future research should concentrate on the role of race and ethnicity in the prevalence of bullying and victimization. Rather than simply noting racial and ethnic differences, however, research must begin to articulate the way in which racial and ethnic differences impact "the content" of bullying behavior. Only then can we fully appreciate the role of social factors "in the development of bullying and victimization problems" (p. 480).

Source: Peskin et al. 2006.

---

all affect a child's ability to learn, a teacher's ability to teach, and a staff member's ability to be effective.

Green schools are "education buildings that operate in harmony with the natural environment" (AFT 2009, p. 4). Green schools reduce energy use by using natural and solar light. They conserve water and use recycled materials, reduce air pollution and excessive noise, and enhance indoor climate control. Collectively, such innovations reduce absenteeism, increase student performance and teacher retention, and, ultimately, save money.

Special education programs pose yet another problem for school systems. Before 1997, when the Individuals with Disabilities Education Act (IDEA) was adopted, more than 1 million students with disabilities were excluded from public schools; 3.5 million were not receiving appropriate educational services (NCD 2000). With the implementation of IDEA, which was reaffirmed in 2004, and is presently being considered for reauthorization, more than 6.5 million children and young people with disabilities, ages 3 through 21, qualified for educational interventions under IDEA (NCES 2011b). Pursuant to IDEA mandates, public school special education programs are now required to provide guaranteed access to education for disabled children. These specially designed instructional programs are structured to meet the unique needs of each child at no cost to the parent. As part of the program, if the school district cannot meet the unique needs of the child, by law, it becomes the school system's fiscal responsibility to pay for the child's education in an appropriate private school placement even if the child never received "special education or related services" through the public schools (Walsh & Robelen 2009).

Although the number of special education students in private settings is estimated to be just over one percent, financially strapped school districts may find it difficult to cover the high costs of private school placements for these students. There is also concern that Black American students, particularly Black American males, are disproportionately

green schools Schools that are "in harmony with the natural environment."

placed in special education programs that contribute to their comparatively low high school graduation rates (Schott Report 2010). Finally, many teachers are opting to leave special education classrooms for traditional classrooms or to pursue different fields entirely. With a lack of certified teachers in this area, a critical shortage has occurred nationwide, affecting the quality of education taking place in special education classrooms.

> **What Do You Think?** All students, whether in regular classrooms or special education classes, are required by law to have a free and appropriate education. However, because of the increased costs associated with educating a child with special needs, this often means that students of special education are receiving a higher-quality, more individualized education than mainstream students who are in overcrowded classrooms in poor school districts (Berger 2007). How do you balance the unique needs of the students in special education with the basic needs of the mainstream students who are poor? How would you allocate funds?

Recruitment and Retention of Quality Teachers

School districts with inadequate funding and facilities, low salaries, lack of community support, and minimal professional development have difficulty attracting and retaining qualified school personnel. According to the U.S. Department of Education, the number of schoolteachers leaving the profession has increased over the last 20 years, reaching 8 percent in 2009 (NCES 2011b). Teachers leaving the profession tended to be at either end of the experience continuum, i.e., they had less than 3 years experience or more than 20 years experience. High teacher turnover is problematic in a number of ways (Boyd et al. 2009). First, newer teachers are less experienced and often less effective. Second, teacher turnover contributes to a lack of continuity in programs and educational reforms. Finally, recruiting and training expenses in addition to the time and effort devoted to replacing teachers who have left the profession is considerable.

> School districts with inadequate funding and facilities, low salaries, lack of community support, and minimal professional development have difficulty attracting and retaining qualified school personnel.

A report by the Bill and Melinda Gates Foundation concludes that the key to a good education hinges on the quality of the teachers in the classroom (Blankinship 2009). Unfortunately, in the United States, qualified and experienced teachers are not equally distributed across all school districts. Because teacher salaries are the largest component of a school district's costs, poor school districts have less money to compete for qualified teachers (McKinsey & Company 2009). Thus, they are more likely to employ beginning teachers with less than three years of experience, and are more likely to assign these teachers to areas outside their specialty. Knowledge of subject taught is one of the key characteristics of an effective teacher (Stotsky 2009). Further, students in poorer school districts are twice as likely to be taught by substitute teachers, and to have less effective teachers, when compared to students in more affluent districts (Barton 2004; Sawchuk 2010).

Recruiting and retaining quality teachers in poverty-level schools is critical to the success of its students. For example, Hanushek et al. (2005) report that, if a child from a poor family has a good teacher for five consecutive years, the achievement gap between that child and a child from a higher-income family would be closed. Because minority students disproportionately populate poor school districts, it is also important to recruit and retain teachers who meet the needs of children from diverse backgrounds and of varying abilities. The number of minority teachers who can serve as role models, have similar life experiences, and have similar language and cultural backgrounds is far too few for the number of minority students.

Recruiting and retaining quality teachers may be made more difficult with the recent emphasis on accountability and **value-added measurement (VAM).** VAM is the use of student achievement data to assess a teacher's effectiveness. States are required to use VAM to receive Race to the Top funds from the federal government (see National Standards in the Strategies for Action section). Although some would argue that quantitative measures of accountability are a necessary evil in a time of budget shortfalls, critics are quick to note that assessing teachers based on student performance assumes all else constant, and ignores the reality of student differences in such nonschool factors as family life, poverty, emotional and physical obstacles, and the like (Mitchell 2010). Further, there are concerns that teachers, fearing for their jobs and concerned about merit-based pay, may begin "teaching to the test."

What Do You Think? School districts have begun publishing teachers' evaluations, known as VAM scores, both online and in print. In 2010, Rigoberto Ruelas Jr., a fifth grade teacher, committed suicide after his "less effective" overall evaluation was published in the *Los Angeles Times*. Recently, New York City teachers have gone to court to prevent their scores from being released to the public (Gabriel & Lester 2011). Given that 60 percent of parents believe that the primary purpose of teachers' evaluations is to "help them improve their ability to teach" (Bushaw & Lopez 2010, p. 15), do you think VAM scores should be made public?

Even seasoned teachers may not be competent or effective. There is evidence that those who choose teaching as a career, on average, have lower college entrance exam scores than the average college student. Further, there is a documented relationship between college entrance scores and the likelihood of continued teaching—the *lower* the college entrance scores, the *higher* the probability of still teaching 10 years postbaccalaureate (NCES 2007). Nonetheless, there is some evidence that the quality of teachers is improving. Both teachers and principals, when asked about the qualifications and competence of new teachers, were more likely to rate them as excellent compared with teachers' and principals' ratings 20 years ago (MetLife 2008).

To place quality teachers in the classroom, many states have implemented mandatory competency testing (e.g., the Praxis Series). The need for teachers who are officially classified as "highly qualified" is tied to federal mandates that place an emphasis on the importance of having licensed teachers in the classroom. Additionally, teachers who have a bachelor's degree and have been in the classroom for three or more years are also eligible for national board certification. Some studies indicate that students of "highly qualified" teachers and/or board-certified teachers perform better on standardized tests and have shown greater testing gains than students of teachers who are not "highly qualified" and/ or board-certified (Viadero 2005; NBPTS 2007). It should be noted, however, that there is no single variable that consistently predicts success in the classroom (Sawchuk 2010).

In an effort to meet the demands of placing teachers in classrooms while facing teacher shortages due to baby boomer retirements, states are now allowing skilled professionals who have an interest in teaching but did not receive a teaching degree to enter the teaching profession. Called lateral entry by some states, the program allows the person to obtain a lateral entry teaching license while actually teaching in the classroom. In addition, more than half of the states have adopted **alternative certification programs,** whereby college graduates with degrees in fields other than education can become certified if they have "life experience" in industry, the military, or other relevant jobs. Teach for America (TFA), a program originally conceived by a Princeton University student in an undergraduate honors thesis, is an alternative teacher education program with the aim of recruiting liberal arts graduates into teaching positions in economically deprived and socially disadvantaged schools.

value-added measurement (VAM). VAM is the use of student achievement data to assess teacher effectiveness.

alternative certification programs Programs whereby college graduates with degrees in fields other than education can become certified if they have "life experience" in industry, the military, or other relevant jobs.

Critics argue that the program may place unprepared personnel in schools. However, an analysis of TFA teachers versus traditional teachers concludes that TFA teachers are more effective in the classroom than traditional teachers as measured by student achievement (Xu et al. 2007). In 2010, 8,000 TFA teachers made up 3 percent of the total number of new U.S. teaching hires (Darling-Hammond 2011).

The Challenges of Higher Education in America

Although there are many types of postsecondary education, higher education usually refers to two- or four-year, public or private, degree-granting institutions. Between 1994 and 2008, enrollment in degree-granting institutions increased by 34 percent. Because of unemployment rates and the increase in the college-age population, enrollments are expected to grow another 17 percent by 2019 (NCES 2011d). Of the 19 million undergraduate and graduate students presently enrolled in colleges and universities, full-time students, women, younger students, minorities, and students enrolled at four-year schools have disproportionately contributed to the dramatic enrollment growth (NCES 2011a).

Higher education employs an estimated 2.6 million professional staff (e.g., administrators, faculty, nonteaching professional staff, etc.) and 0.9 million nonprofessional staff (e.g., clerical, service, maintenance, etc.) (NCES 2011a). Over the last decade, there has been a significant decrease in full-time tenured or tenure track faculty, once the "core" of academia, and significant increases in noninstructional staff (e.g., administrators) and nontenure track, part- or full-time instructors. Interestingly, a 2011 survey of college and university presidents reveals that only 24 percent would prefer that most of their faculty be tenured, the majority wishing to rely rather on annual or long-term contracts (Pew 2011).

As the number of students attending college has increased, so have the costs associated with getting a college degree (see Figure 8.6). In 2010, the annual average expense for tuition and fees, books and materials, and money for living expenses for a full-time undergraduate student at a four-year public institution was $19,300 (NCES 2011a). About 65 percent of all undergraduates receive some type of student aid; in 2008, the average total amount of financial aid was $9,100 (Wei et al. 2009). Full-time students attending four-year private doctoral-granting institutions were the most likely to receive student aid, and federal and state grants and student loans comprise the largest proportion of student aid. Despite budget deficits and cutbacks to education, in 2010, President Obama signed into law a bill that eliminates fees paid to banks as student loan intermediaries and, with the $68 billion savings, expands the number and maximum amount of Pell Grants (Baker & Herszenhorn 2010).

Access to and completion of higher education among minority and/or low-income students is particularly problematic. Only 14 percent of four-year degree-granting institution students are black, and only 12 percent are Hispanic. Minority representation at two year institutions, 34 percent in total, is significantly higher than at four year institutions (NCES 2011a). It should be noted, however, that a fairly high proportion of all black students, approximately 12 percent, attend "historically black universities and colleges" (HBCU) (NCES 2011b).

Racial and ethnic minorities, although varying by state, are also less likely to graduate from college than their white counterparts. In part, their lower levels of graduation rates

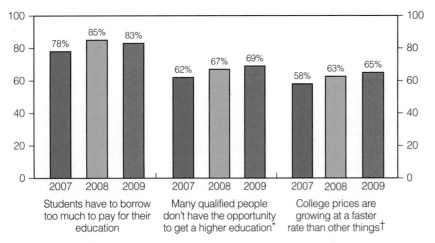

*Versus those who say, "The vast majority of people who are qualified to go to college have the opportunity to do so."

†Versus those who say, "Slower or at the same rate"

Figure 8.6: Survey Results of 1,031 Randomly Selected Americans, Percent that Agree, 2009
Source: NCPPHE 2010.

can be explain by parents' education, one of the best predictors of student outcomes, including educational achievement and attainment. Moreover, college completion rates at all degree-granting institutions have decreased over time, leading to concerns over U.S. competitiveness in global markets, particularly labor markets (NCPPHE 2009). Of particular concern is whether the United States is training a sufficient number of graduates in science, technology, engineering, and mathematics (STEM) occupations (see Chapter 14).

To address some of the issues in higher education, a number of policy changes have been adopted. Community colleges, for example—the nation's "unsung heroes of the American education system" (President Obama quoted in Adams 2011, p. 1)—are now awarding bachelor's degrees in 17 states (Lewin 2009). Some bachelor degree–granting institutions, on the other hand, are now offering three-year degree programs to attract students. Critics argue that economic considerations should not drive educational policy; proponents argue that an economic crisis must be met with innovation (Strauss 2009).

> College completion rates at all degree-granting institutions have decreased over time, leading to concerns over U.S. competitiveness in global markets, particularly labor markets.

Finally, the present administration has made a commitment to several higher education programs. The federal stimulus package included $15.6 billion in Pell Grant student aid, and $13.9 billion over 10 years in higher education tax cuts. Further, the 2012 budget proposal requests $123 million for the *First in the World* program, which will help ensure that the United States has the highest proportion of college graduates of any country in the world by 2020. The 2012 budget also requests a total of $340 million to strengthen HBCU, and $117.4 million to "enhance the academic quality, institutional management, fiscal stability, and self-sufficiency of colleges and universities that enroll large percentages of Hispanic students" (U.S. Department of Education 2011a, p. 1).

Strategies for Action: Trends and Innovations in American Education

Americans consistently rank improving education as one of their top priorities. Recent attempts to improve schools include raising graduation requirements, barring students from participating in extracurricular activities if they are failing academic subjects, lengthening the school day, prohibiting dropouts from obtaining driver's licenses, implementing year-round schooling, and extending the number of years permitted to complete a high school degree. However, educational reformers on both sides of the political aisle continue to call for changes that go beyond these get-tough policies that some say simply maintain the status quo.

National Educational Policy

The challenges facing national educational policies are considerable. In 2011, President Obama speaking at Kenmore Middle School in Arlington, Virginia, stated that ". . . unfortunately too many students aren't getting a world-class education today. As many as a quarter of American students aren't finishing high school. The quality of our math and science education lags behind many other nations. And America has fallen to 9th in the proportion of young people with a college degree. Understand, we used to be first, and we now rank 9th. That's not acceptable" (quoted in Adams 2011, p. 1). Although it may be difficult to predict the totality of his educational policy and its impact, there is evidence of both significant changes and "things as they were" politics (see this chapter's *The Human Side*).

No Child Left Behind The No Child Left Behind (NCLB) Act of 2001 was signed into law in January 2002. The federally funded plan was organized around four principles: *accountability* for learning outcomes, *flexibility* in funding, *expanding school*

Paul Karrer is a fifth grade teacher in North Monterey, California. Here, in a "A Letter to My President—the One I Voted for . . . ," Mr. Karrer expresses his frustration over national educational policy that continues to emphasize standardized testing and teacher accountability. Although there may be many who disagree with his position, the impact of social forces on students' lives that he describes is undeniable.

Dear President Obama:

I mean this with all respect. I'm on my knees here, and there's a knife in my back, and the prints on it kinda match yours. I think you don't get it.

Your Race to the Top is killing the wrong guys. You're hitting the good guys with friendly fire. I'm teaching in a barrio in California. I had 32 kids in my class last year. I love them to tears. They're 5th graders. That means they're 10 years old, mostly. Six of them were 11 because they were retained. Five more were in special education, and two more should have been. I stopped using the word *parents* with my kids because so many of them don't have them. Amanda's mom died in October. She lives with her 30-year-old brother. (A thousand blessings on him.) Seven kids live with their "Grams," six with their dads. A few rotate between parents. So "parents" is out as a descriptor.

Here's the kicker: Fifty percent of my students have set foot in a jail or prison to visit a family member.

Do you and your Secretary of Education, Arne Duncan, understand the significance of that? I'm afraid not. It's not bad teaching that got things to the current state of affairs. It's pure, raw poverty. We don't teach in failing schools. We teach in failing communities. It's called the Zip Code Quandary. If the kids live in a wealthy zip code, they have high scores; if they live in a zip code that's entombed with poverty, guess how they do?

We also have massive teacher turnover at my school. Now, we have no money. We haven't had an art or music teacher in 10 years. We have a nurse twice a week. And because of the No Child Left Behind Act, struggling public schools like mine are held to impossible standards and punished brutally when they don't meet them. Did you know that 100 percent of our students have to be on grade level, or else we could face oversight by an outside agency? That's like saying you have to achieve 100 percent of your policy objectives every year. It's not bad teaching that got things to the current state of affairs. It's pure, raw poverty.

You lived in Indonesia, so you know what conditions are like in the rest of the world. President Obama, I swear that conditions in my school are akin to those in the third world. We had a test when I taught in the Peace Corps. We had to describe a glass filled to the middle. (We were supposed to say it was half full.) Too many of my kids *don't even have the glass!*

Next, gangs. Gangs eat my kids, their parents, and the neighborhood. One of my former students stuffed an AK-47 down his pants at a local bank and was shot dead by the police. Another one of my favorites has been incarcerated since he was 13. He'll be 27 in November. I've been writing to him for 10 years and visiting him in the maximum-security section of Salinas Valley State Prison. He's a major gangster.

Do you get that it's tough here? Charter schools and voucher schools aren't the solution. They are an excuse not to fix the real issues. You promised us so much. And you want to give us merit pay? Anyway, I think we really need to talk. Oh, and can you pull the knife out while you're standing behind me? It really hurts.

Sincerely yours,

Paul Karrer

options for parents, and the use of sound *teaching methods,* including the use of only "highly qualified" teachers in the classroom by 2006.

However, soon after the act was signed, it became clear that the implementation of NCLB was problematic. For example, accountability efforts required that, to make adequate yearly progress (AYP), the factor by which a school is measured, all student groups (e.g., students who are black, white, Native American, of limited English proficiency, disabled, or others) must attain a set level of achievement in both reading and mathematics. According to the act, if one student group does not reach the set levels, the entire school receives a failing grade and is in danger of being severely sanctioned (McNeil 2011a). Because of the tremendous disparities between student groups, the original NCLB regulations were changed to allow for alternative testing of students who were disabled or had limited English proficiency; deadlines for making AYP in reading and mathematics were also extended to 2014 for all states (McNeil 2011a).

In addition to concerns about accountability, critics of the law also argue that it unfairly burdens the states, which must absorb the financial cost of its provisions. The National Education Association, the largest organization of teachers in the nation, joined schools in several districts to bring the first federal lawsuit against the U.S. Department of Education for failing to provide funding for NCLB initiatives. The U.S. Court of

Appeals held in favor of the government, finding that "whether one is a school district, teacher, principal, or state education agency, one cannot lawfully refuse to comply with NCLB . . . by arguing there are insufficient federal funds to meet the requirements of the law" (Talbert 2010, p. 1).

Further, empirical tests of NCLB are mixed. For example, a study by Hall and Kennedy (2006) indicates that, overall, achievement gains were found in elementary schools where reading scores showed gains in 27 of 31 states and mathematics scores improved in 29 of 32 states. However, the picture in middle schools and high schools was less clear, with a combination of gains and losses in both reading and mathematics. Although minority and low-income students showed achievement gains, whites and higher-income students, in general, showed more dramatic gains, thus increasing the achievement gap between the two groups. Braun, Chapman, and Vezzu (2010), after comparing eighth grade mathematics scores across 10 states over time, conclude that the black-white achievement gap remains considerably large and has only modestly been reduced by the mandated high-stakes testing of NCLB. Further, Demarest (2011) observes that, over the decade in which NCLB was law, "[T]est scores in the United States have not significantly improved in recent years, the achievement gap has not closed, and other developed nations have continued to build their educational capacity and surpass U.S. performance on international assessments" (p. 1).

In 2010, President Obama issued his "blueprint" for educational reform, which includes reauthorization of the Elementary and Secondary Education Act (ESEA). A comparison of NCLB to the proposed reauthorization of ESEA indicates that ESEA, if passed, will have:

- Greater flexibility for state and local education leaders to institute needed changes;
- Less of a punitive focus on testing and accountability, and a greater emphasis on student growth and school progress;
- Greater acknowledgement, support, and investment in teachers;
- More attention to funding equity and investment in low-performing schools;
- Greater use of data-driven, evidence-based interventions and instructional models; and
- Increased standards to ensure a "complete education" for all children. (U.S. Department of Education 2011b)

National Standards The National Governors Association and the Council of Chief State School Officers have led the move to adopt a common set of academic standards. Presently, 47 states have adopted the Common Core State Standards. The initiative was motivated by concerns that students in different states were not equally being prepared for postsecondary education and/or jobs in the global workforce.

States that have adopted the common core standards approach are eligible to receive funds from the Race to the Top initiative, part of a funding scheme under the federal stimulus package. Member states of the National Common Core State Standards Initiative are presently drafting English and mathematics standards, and two consortia are developing assessment methods to be used across participating states' school districts beginning in the 2014–2015 school year (CCSSI 2011; SBAC 2011).

Fiscal Stabilization Although the American Recovery and Revitalization Act of 2009 provided needed funds for "fiscal stabilization," they were not without strings. To receive the funds states had to ". . . show improvements in four specific areas: low-performing schools, data systems, teacher effectiveness, and standards and assessments" (McNeil 2011b, p. 1). These four areas are called the **four assurances.**

The first annual report concerning fiscal stabilization funds reveals several interesting results. First, taken as a whole, school districts spent most of their stabilization funds on salaries and employee benefits (69 percent). The second largest pool of stabilization fund spending was classified as "other expenses," including everything from paying utility bills to buying textbooks and computers (20 percent). The remaining two categories included carrying over funds to the following school year (11 percent), and construction,

four assurances The four assurances refer to a state's commitment to improving teacher quality, raising academic standards, intervening in failing schools, and developing assessment databases in return for federal dollars.

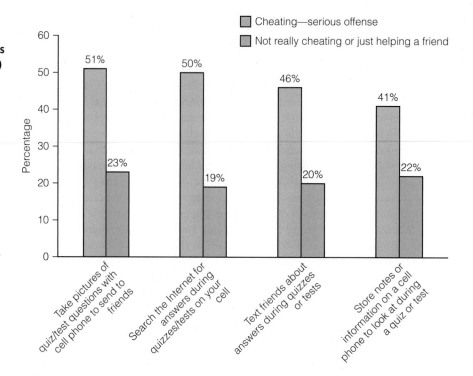

Character Education and Service Learning

Research indicates that cheating is a fairly common event among students in the United States. A 2010 survey on the values and ethical behavior of more than 43,000 high school students indicates that 18 percent admitted to stealing from a friend, 80 percent admitted lying to their parents about something important, 59 percent admitted to cheating on a test in the previous year, and 33 percent admitted to plagiarizing an assignment from the Internet (Josephson Institute of Ethics 2011). To many educators, results like these signify the need for **character education.**

Character education entails teaching students to act morally and ethically, including the ability to "develop just and caring relationships, contribute to community, and assume the responsibilities of democratic citizenship" (Lickona & Davidson 2005). Despite most schools' emphasis on academic achievement, knowledge without character is potentially devastating. Sanford McDonnell (2009), former CEO of McDonnell Aircraft Corporation and chairman emeritus of the Character Education Partnership, recounts a letter written by a principal and former concentration camp survivor to his teachers at the start of a new school year:

> My eyes saw what no person should witness: gas chambers built by learned engineers, children poisoned by educated physicians, infants killed by trained nurses, women and babies shot and burned by high school and college graduates. Your efforts must never produce learned monsters and skilled psychopaths. Reading, writing, and arithmetic are important only if they serve to make our children more humane. (p. 1)

President Obama's educational reform policy includes additional dollars for expanding and redefining character education. Government analyses of research on established

character education Education that emphasizes the moral and ethical aspects of an individual.

Above the Character Education section:

modernization, and infrastructure repairs (1 percent) (SFSF 2011). Despite stabilization efforts, budget shortfalls in many states and school districts have lead to teacher lay-offs, increased class sizes, canceled capital improvement projects, and the elimination of needed programs.

character education programs indicate that classroom activities designed to teach "core values" are relatively ineffective in producing the desired student outcomes (IES 2010). Keeping with the administration's emphasis on evidence-based programs, federal funds have been redirected to a more broadly conceived notion of character education administered through the Safe and Drug-Free Schools and Communities National Programs.

Service learning programs, one type of character education, are increasingly popular at universities and colleges nationwide. Service learning programs are community-based initiatives in which students volunteer in the community and receive academic credit for doing so. Service learning programs have been found to increase academic performance, promote "hands-on" learning, enhance civic engagement and moral reasoning, reduce the likelihood of risky behaviors, and increase the self-esteem of program participants (Independent Sector 2002; Ramierz-Valles & Brown 2003; CNCS 2008).

The Edward M. Kennedy Serve America Act, signed in 2009, creates several programs designed to increase student participation in service activities, and is hoped to increase the participation of volunteers in AmeriCorps and similar programs (Zehr 2009). Examples of new programs include Summer of Service, which pays middle and high school students $500 toward college tuition for 100 hours of community service, and Youth Engagement Zones, which partners schools and community organizations to facilitate community service by secondary school students (CNCS 2010).

Despite most schools' emphasis on academic achievement, knowledge without character is potentially devastating.

Use of Computer Technology and E-Learning

Computers in the classroom allow students to access large amounts of information. The proliferation of computers both in school and at home may mean that teachers will become facilitators and coaches rather than the sole providers of information. Not only do computers enable students to access enormous amounts of information, they also allow students to progress at their own pace. However, computer technology is not equally accessible to all students and varies dramatically by parents' education, parents' income, and race and ethnicity. Although there are lingering concerns over the "digital divide" (see Chapter 14) and its impact on students' achievement, some research indicates that having a computer in the home decreases student math and reading scores (Vigdor & Ladd 2010).

The Obama administration has issued a proposal entitled, "Transforming American Education: Learning Powered by Technology," which outlines his administration's national education technology plan. The goals of "ed-tech," are to increase student achievement through the use of technology, and to create technologically literate students by the end of eighth grade. The plan also advocates integrating technology into the classroom

Tim Shaffer/Microsoft via Getty Images

Students work on their laptops at the Philadelphia School of the Future. Sponsored as part of a public/private partnership between the City of Philadelphia and Microsoft, this model for future schools opened on September 6, 2006.

Figure 8.8: Percentage Summary of Responses to the Question: How Important Do You Believe the Following Reasons Are for a School District to Offer Fully Online or Blended Learning Courses?
Source: Picciano & Seaman 2009. From *K-12 Online Learning: a 2008 Follow Up of the Survey of U.S. School District Administrators*. Copyright © 2009 by Sloan-C™ All rights reserved. Reprinted with permission.

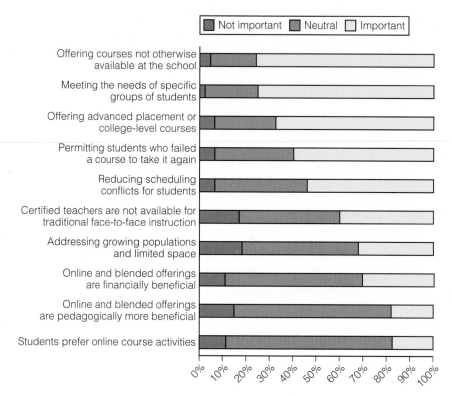

through ". . . both teacher training and curriculum development" to establish "innovative, research-based instructional methods that can be widely implemented" (U.S. Department of Education 2010, p. 1.).

E-learning separates the student and the teacher by time and/or place. They are, however, connected by some communication technology (e.g., videoconferencing, e-mail, real-time chat room, or closed-circuit television). Some classes have blended learning, a mix of online and traditional face-to-face learning. Based on sampled school districts, a report examining e-learning in the nation's elementary and secondary schools found the following:

- In 2008, there were over a million K–12 students taking online courses.
- Seventy-five percent of the school districts offered online or blended courses.
- School districts depended on multiple online service providers (e.g., private, postsecondary institutions).
- Online learning meets specific needs of students (e.g., additional help).
- Online learning provides courses that simply would not be available to students otherwise. (Picciano & Seaman 2009)

Figure 8.8 graphically portrays additional benefits of online education and their relative importance. Although low in rank, financial considerations were identified as a benefit of online education and may increasingly be so in the face of funding cuts.

Online education often serves a segment of the population that would not otherwise be able to attend school—older, married, full-time employees, and people from remote areas. Thus, for example, the number of public school online course offerings is higher in rural areas than in towns, cities, or urban areas (Bausell & Klemick 2007). Further, some research suggests that online learning benefits those who have historically been disadvantaged in the classroom. A study by DeNeui and Dodge (2006) indicates that females in a blended course were more likely to use a learning management system (e.g., Blackboard) than males, and to significantly outperform them as measured by final grades in an introductory psychology class.

e-learning Learning in which, by time or place, the learner is separated from the teacher.

The Debate over School Choice

Traditionally, children have gone to school in the district where they live. School vouchers, charter schools, private schools, and home schooling provide parents with alternative school choices for their children. **School vouchers** are tax credits that are transferred to the public or private school that parents select for their child. For example, the Washington, D.C. voucher program provides $9,500 to students in kindergarten through eighth grade, and $11,000 for students in grades 9 through 12 who are from disadvantaged families. The families can then use the funds to send their children to the school of their choice (Mead 2011). The program, however, as vouchers in general, is a contentious issue and one the Obama administration opposes.

Proponents of the voucher system argue that it increases the quality of schools by creating competition for students. Those who oppose the voucher system argue that it will drain needed funds and the best students away from public schools. After examining 17 empirical studies on the impact of vouchers on public schools, Forster (2009) concluded that 16 of the studies "find that vouchers improve public schools and one—the only one examining a program that insulates public schools from voucher competition—finds no visible difference. No empirical studies find that vouchers harm public schools" (Forster 2009, p. 34).

Opponents also argue that vouchers increase segregation because white parents use the vouchers to send their children to private schools with few minorities, and that the use of vouchers for religious schools violates the constitutional guarantee of separation of church and state. However, the U.S. Supreme Court, in reviewing the voucher program in Cleveland, Ohio, held that the use of tax dollars for enrollment in religious schools is not unconstitutional (NSBA 2007).

Vouchers can be used for charter schools. **Charter schools** originate in contracts, or charters, which articulate a plan of instruction that local or state authorities must approve. Although foundations, universities, private benefactors, and entrepreneurs can fund charter schools, many are supported by tax dollars. Between 2000 and 2009, the number of students attending charter schools increased from 340,000 to 1.4 million. Over the same time period, the number of charter schools whose student bodies were at least 75 percent poor increased from 13 percent to 30 percent. In contrast to traditional schools, over half of charter schools are in large cities (NCES 2011b).

Charter schools, like school vouchers, were designed to expand schooling options and to increase the quality of education through competition. Like vouchers, charter schools have come under heavy criticism for increasing school segregation, reducing public school resources, and "stealing away" top students. Proponents, including President Obama and Secretary of Education Arne Duncan, argue that charter schools encourage innovation and reform, and increase student learning outcomes. However, a recent analysis of New York City charter schools indicates that, when compared to traditional public schools, charter schools disproportionately underserved Hispanic and immigrant children, special education students, English language learners, and children from the lowest socioeconomic classes (UFT 2010). Because charter schools in New York City can be run by for-profit firms, the report also notes that management fees and charges far exceed those for traditional public schools. The authors conclude that "growth of the charter sector will only exacerbate existing inequities" (p. 3).

Another school choice parents can make is to send their children to a private school. In the 2009–2010 school year, private school enrollment in elementary and secondary schools numbered nearly 4.7 million, about 10 percent of the total K–12 student population, with over 400,000 full-time teachers (Broughman, Swaim, and Hryczaniuk 2011). The primary reason parents send their children to private schools is for religious instruction. The second most common reason is the belief that private schools are superior to public schools in terms of academic achievement. Contrary to expectations, however, there is evidence that public school students fair as well or better academically as private school students. For example, Lubienski and Lubienski (2006), using National Assessment of Educational Progress (NAEP) data, report that mathematics scores for

school vouchers Tax credits that are transferred to the public or private school that parents select for their child.

charter schools Schools that originate in contracts, or charters, which articulate a plan of instruction that local or state authorities must approve.

public school students were higher than mathematics scores for private school students. Parents also choose private schools for their children to have greater control over school policy, to avoid busing, or to obtain a specific course of instruction, such as dance or music.

Some parents choose not to send their children to school at all but to teach them at home. In 2007, the latest year for which data are available, 1.5 million were homeschooled, the majority of whom, 77 percent, were white (NCES 2011a). Reasons for homeschooling children vary. The three most common reasons given by parents for **homeschooling** include the need for religious and moral instruction, avoidance of the negative environment of public schools (e.g., drugs), and concerns over the quality of instruction in public schools (NCES 2011a). How does being schooled at home instead of attending public school affect children? Some evidence suggests that homeschooled children perform as well as or better than their institutionally schooled counterparts (Winters 2001).

What Do You Think? Using survey data, transcripts, and standardized tests administered to college students in their first semester and again at the end of their second year, Arum and Roksa (2011) conclude that almost half of the 2,300 students studied made no significant improvement in a variety of academic skills including critical thinking, complex reasoning, and writing. Although the authors suggest several reasons for their results, what do you think are the reasons so many college students fail to learn?

Understanding Problems in Education

What can we conclude about the educational crisis in the United States? Any criticism of education must take into account the fact that just over a century ago, the United States had no systematic public education system at all. Many American children did not receive even a primary school education. Instead, they worked in factories and on farms to help support their families. Whatever education they received came from the family or the religious institution. In the mid-1800s, educational reformer Horace Mann advocated mandatory education for all U.S. children. In 1852, the first compulsory education laws in the United States were passed, requiring school-aged children to attend 12 weeks of school each year. By World War I, every state mandated primary school education and, by World War II, secondary education was compulsory as well.

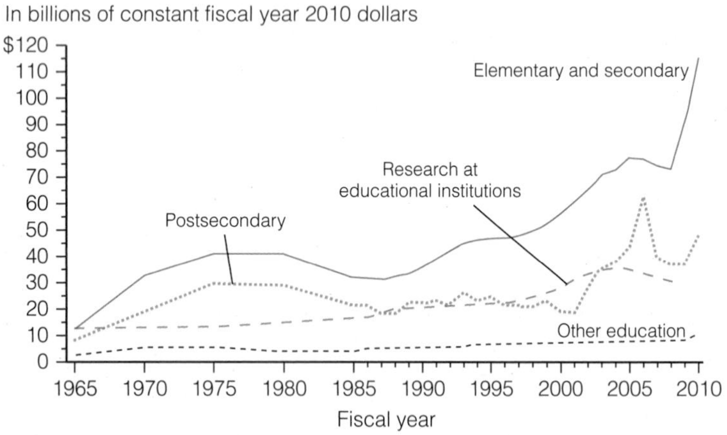

In billions of constant fiscal year 2010 dollars

Figure 8.9: Federal On-Budget Funds for Education, by Level or Other Educational Purpose: Selected Years, 1965 through 2010
Source: NCES 2011a.

homeschooling The education of children at home instead of in a public or private school.

Today, the United States spends more money per student, at every level of education, than any other industrialized nation in the world (OECD 2010), and federal spending on education continues to escalate (see Figure 8.9). Nonetheless, U.S. students are outperformed at almost every level by their international peers. Significant educational reform is needed to meet the needs of a global economy in the 21st century and, perhaps more importantly, to fulfill Horace Mann's dream of education as the "balanced wheel of social machinery," equalizing social differences among members of an immigrant nation.

First, we must invest in teacher education and in teaching practices that have been empirically documented to work in raising student outcomes. Teachers' salaries also need to better reflect the priority Americans place on children, education, and the education of children, and should not be tied to student performance. After all, there are lingering doubts that standardized tests accurately measure student performance, let alone a teacher's performance (Ravitch 2010).

Second, the "savage inequalities" in education, primarily based on race, ethnicity, and socioeconomic status, must be addressed. Segregation, rather than decreasing is increasing, a reflection of housing patterns, local school districts' heavy reliance on property taxes, and immigration patterns. Public schools should provide all U.S. children with the academic and social foundations necessary to participate in society in a productive and meaningful way; however, for many children, schools perpetuate an endless downward cycle of failure, alienation, and hopelessness.

Third, the general public needs to become involved, not just in their children's education but also in the *institution* of education. An uneducated, and perhaps more importantly, unthinking populace hurts all of society, particularly in terms of global competitiveness. As Kohn (2011) observes, children from low-income families continue to be taught "the pedagogy of poverty" (Haberman 1991) or what more recently has been called the "McEducation of the Negro" (Hopkinson 2011). Like Big Macs, children are being packaged in one-size-fits-all wrappers—with learning how to think, explore, question, and debate being replaced by worksheets and standardized tests. Sadly, as education historian Diane Ravitch notes, the present reform movement that "once was an effort to improve the quality of education [has] turned into an accounting strategy" (Ravitch 2010, p.16).

Paul Burns

Research suggests that, for many children in poor, often urban schools, ". . . learning [is about following] the rules and following directions. Not critical thinking. Not creativity. It's about how to correctly eliminate three out of four bubbles" (Hopkinson 2011, p. 1).

Like Big Macs, children are being packaged in one-size-fits-all wrappers—with learning to think, explore, question, and debate being replaced by worksheets and standardized tests.

Finally, as a society, we must attend to and be cognizant of the importance of early childhood development. Poliakoff (2006) notes

Children's physical, emotional, and cognitive development are profoundly shaped by the circumstances of their preschool years. Before some children are even born, birth weight, lead poisoning, and nutrition have taken a toll on their capacity for academic achievement. Other factors—excessive television watching, little exposure to conversation or books, parents who are absent or distracted, inadequate nutrition—further compromise their early development. (p. 10)

We must provide support to families so that children grow up in healthy, safe, and nurturing environments. Children are the future of our nation and of the world. Whatever resources we provide to improve the lives and education of children are sure to be wise investments in our collective future.

- **Do all countries educate their citizens?**
No. Many societies have no formal mechanism for educating the masses. As a result, there are millions of illiterate adults around the world. The problem of illiteracy is greater in developing countries than in developed nations and, worldwide, disproportionately affects women more than men (see Figure 8.1).

- **According to the structural-functionalist perspective, what are the functions of education?**
Education has four major functions. The first is instruction—that is, teaching students knowledge and skills. The second is socialization that, for example, teaches students to respect authority. The third is sorting individuals into statuses by providing them with credentials. The fourth function is custodial care—a babysitting agency of sorts.

- **What is a self-fulfilling prophecy?**
A self-fulfilling prophecy occurs when people act in a manner consistent with the expectations of others.

- **What variables predict school success?**
Three variables tend to predict school success. Socioeconomic status predicts school success: The higher the socioeconomic status, the higher the likelihood of school success. Race predicts school success, with nonwhites and Hispanics having more academic difficulty than whites and non-Hispanics. Gender also predicts success, although it varies by grade level.

- **What are the three reasons given as to why Black and Hispanic Americans, in general, do not perform as well in school as their white and Asian counterparts?**
First, because race and ethnicity are so closely tied to socioeconomic status, it appears that race or ethnicity alone can determine school success when, in fact, it may be socioeconomic status. Second, many minorities are not native English speakers, making academic achievement significantly more difficult. And third, racial and ethnic minorities may be the victims of racism and discrimination.

- **What are some of the conclusions of the study summarized in the Social Problems Research Up Close feature?**
The results of the study indicate that there are more victims of bullying than bullies, and that bullying behavior is highest in the ninth grade and lowest in the sixth and tenth grades. The highest rate of victimization is in the sixth grade. Males and females were equally likely to report being bullies, victims, and bully-victims. Blacks, compared to Hispanics, were more likely to be bullies, victims, and bully-victims. "Upsetting students for the fun of it" was the most common kind of bullying activity, followed by teasing, group teasing, and starting arguments.

- **What are some of the problems associated with the American school system?**
One of the main problems is the lack of student achievement in our schools—particularly when U.S. data are compared with data from other industrialized countries. Minority dropout rates are high, and school violence, crime, and discipline problems continue to be a threat. School facilities are in need of repair and renovations, and personnel, including teachers, have been found to be deficient. Higher education must also address several challenges.

- **What is meant by value-added measurement (VAM) and why are there concerns about its use?**
VAM is the use of student achievement data to assess a teacher's effectiveness. Critics of VAM argue that assessing teachers based on student performance assumes all else constant, and ignores the reality of student differences in such nonschool factors as family life, poverty, emotional and physical obstacles, and the like.

- **What are the arguments for and against school choice?**
Proponents of school choice programs argue that they reduce segregation and that schools that have to compete with one another will be of a higher quality. Opponents argue that school choice programs increase segregation and treat disadvantaged students unfairly. Low-income students cannot afford to go to private schools, even with vouchers. Furthermore, those opposed to school choice are quick to note that using government vouchers to help pay for religious schools is unconstitutional.

TEST YOURSELF

1. All societies have some formal mechanism to educate their citizenry.
    a. True
    b. False
2. According to structural functionalists, which of the following is not a major function of education?
    a. Teach students knowledge and skills
    b. Socialize students into the dominant culture
    c. Indoctrinate students into the capitalist ideology
    d. Provide custodial care for children
3. NCLB has been demonstrated to consistently increase student performances of all children regardless of gender, race, and ethnicity.
    a. True
    b. False
4. In a study supported by the Bill and Melinda Gates Foundation, researchers identified five major reasons why

students drop out of high school. Which of the following is a reason given for dropping out of school?

    a. Classes were not interesting

    b. Had to drop out to work

    c. Pregnancy

    d. Hated school

5. Which of the following statements about bullying is not true?

    a. Student bullies tend to be marginal students.

    b. More females are bullied than males.

    c. There are serious consequences for the student victims of bullying.

    d. Most bullying takes place outside so it is undetected.

6. Public school special education programs are required to provide guaranteed access to education for disabled children.

    a. True

    b. False

7. Over the next decade, the need for teachers is likely to

    a. decrease because of the baby boom retirements

    b. increase because of immigration patterns

    c. remain stable

    d. decrease because of the aging population

8. Which of the following is not part of President Obama's "blueprint" for educational reform?

    a. More directives and less discretion for state and local education leaders

    b. Less of a punitive focus on testing and accountability

    c. More use of data-driven evidence to inform educational interventions

    d. Increased standards to ensure a "complete education" for all children

9. Computer software in the classroom has been shown to significantly increase academic achievement.

    a. True

    b. False

10. School vouchers are tax credits that are transferred to the public or private school that parents select for their child.

    a. True

    b. False

Answers: 1: b; 2: c; 3: b; 4: a; 5: d; 6: a; 7: b; 8: a; 9: b; 10: a.

KEY TERMS

MEDIA RESOURCES

Turning to Video

▶❚❚ Watch the BBC video *Teacher Shortages* (running time 1:46), available through **CengageBrain.com**. This video discusses the impact of state budget shortfalls on education and, specifically, teacher layoffs. Has your school been affected by teacher layoffs? Do you think that class size is associated with student outcomes?

Online Study Resources

Log in to **www.cengagebrain.com** to access the resources your instructor has assigned. For this book, you can access:

CourseMate

Access chapter-specific learning tools, including learning objectives, practice quizzes, videos, Internet exercises, flash cards, and glossaries, as well as web links, and more in your Sociology CourseMate.

AP Photo/Kiichiro Sato

9

Race, Ethnicity, and Immigration

"The 21st century will be the century in which we redefine ourselves as the first country in world history which is literally made up of every part of the world."

—Kenneth Prewitt, Census Bureau director

IN 2009, A New Orleans justice of the peace, Keith Bardwell, refused to issue a marriage license to Terence McKay and Beth Humphrey—an interracial couple in their early 30s. Bardwell explained, "I am not a racist. I just don't believe in the mixing of races that way. . . . I have plenty of black friends. They come to my home, I marry them, they use my bathroom. I treat them like everyone else" (quoted in Dawkins 2010). Bardwell claims that his refusal to marry interracial couples stems from his concern for the well-being of children born to mixed marriages. He explained, "There is a problem with both groups accepting a child from such a marriage. . . . I think those children suffer, and I won't help put them through it" (quoted in Deslatte 2009). In response to public outcry over this blatant act of discrimination and violation of civil rights, public officials, including Louisiana Governor Bobby Jindal, called for the removal of Bardwell from the Office of the Justice of the Peace. In November 2009, Bardwell resigned.

AP Photo/Bill Haber

Beth Humphrey (in photo above) and her fiancé, Terrance McKay, wanted to be married by Louisiana Justice of Peace Keith Bardwell, but Bardwell refused to marry them because Beth and Terrance are an interracial couple.

The election of Barack Obama represents a significant milestone in U.S. history, and has ushered in a new era of hope for racial and ethnic minorities. Another milestone for minorities was achieved when Sonia Sotomayor was sworn in as the first Hispanic Supreme Court Justice in 2009. But despite these milestones, the United States continues to be a nation where racial and ethnic minorities, as well as mixed couples, are treated unfairly.

A **minority group** is a category of people who have unequal access to positions of power, prestige, and wealth in a society and who tend to be targets of prejudice and discrimination. Minority status is not based on numerical representation in society but rather on social status. For example, although Hispanic individuals outnumber non-Hispanic whites in California, Texas, and New Mexico, they are considered a "minority" because they are underrepresented in positions of power prestige, and wealth, and because they are targets of prejudice and discrimination.

In this chapter, we focus on prejudice and discrimination, their consequences for racial and ethnic minorities, and the strategies designed to reduce these problems. We also examine issues related to U.S. immigration, because immigrants often bear the double burden of being minorities *and* foreigners who are not welcomed by many native-born Americans. We begin by examining racial and ethnic diversity worldwide and in the United States, emphasizing first that the concept of race is based on social rather than biological definitions.

The Global Context: Diversity Worldwide

A first grade teacher asked the class, "What is the color of apples?" Most of the children answered red. A few said green. One boy raised his hand and said, "white." The teacher tried to explain that apples could be red, green, or sometimes golden, but never white. The boy insisted his answer was right and finally said, referring to the apple, "Look inside" (Goldstein 1999). Like apples, human beings may be similar on the "inside," but they are often classified into categories according to external

minority group A category of people who have unequal access to positions of power, prestige, and wealth in a society and who tend to be targets of prejudice and discrimination.

appearance. After examining the social construction of race and ethnicity, we review patterns of interaction among racial and ethnic groups and examine racial and ethnic diversity in the United States.

The Social Construction of Race and Ethnicity

Martin Marger (2012), a researcher and writer on race and ethnic relations, says that "race is one of the most misunderstood, misused, and often dangerous concepts of the modern world" (p. 12). Marger explains that the term *race* has been used to describe people of a particular nationality (the Mexican "race"), religion (the Jewish "race"), skin color (the white "race"), and even the entire human species (the human "race"). Confusion around the term *race* stems from the fact that it has both biological and social meanings.

Race as a Biological Concept As a biological concept, race refers to a classification of people based on hereditary physical characteristics such as skin color, hair texture, and the size and shape of the eyes, lips, and nose. But there is no scientific basis—no blood test or genetic test—that reveals a person's race. There are also no clear guidelines for distinguishing racial categories on the basis of visible traits. Skin color is not black or white but rather ranges from dark to light with many gradations of shades. Noses are not either broad or narrow but come in a range of shapes. Physical traits come in an infinite number of combinations. For example, a person with dark skin can have a broad nose (a common combination in West Africa), a narrow nose (a common combination in East Africa), or even blond hair (a combination found in Australia and New Guinea). Further, skin color, hair texture, and facial features are only a few of the many traits that vary among human beings.

> What if we classified people into racial categories based on eye color instead of skin color?

Another problem with race as a biological concept is that the physical traits used to mark a person's race are arbitrary. What if we classified people into racial categories based on eye color instead of skin color? Or hair color? Or blood type? What if all dark-haired individuals were considered to belong to one race, and all light-haired people to another race? Is there any scientific reason for selecting certain traits over others in determining racial categories? The answer is "No." As a biological concept, "races are not scientifically valid because there are no objective, reliable, meaningful criteria scientists can use to construct or identify racial groupings" (Mukhopadhyay et al. 2007, p. 5).

The science of genetics also challenges the biological notion of race. Geneticists have discovered that the genes of any two unrelated people, chosen at random from around the globe, are 99.9 percent alike (Ossorio & Duster 2005). Furthermore, "most human genetic variation—approximately 85 percent—can be found between any two individuals from the same group (racial, ethnic, religious, etc.). Thus, the vast majority of variation is within-group variation" (Ossorio & Duster 2005, p. 117). Finally, classifying people into different races fails to recognize that, over the course of human history, migration and intermarriage have resulted in the blending of genetically transmitted traits.

> To summarize, races are unstable, unreliable, arbitrary, culturally created divisions of humanity. This is why scientists . . . have concluded that race, as scientifically valid biological divisions of the human species, is fiction not fact. (Mukhopadhyay et al. 2007, p. 14)

Race as a Social Concept The idea that race is socially created is one of the most important lessons in understanding race from a sociological perspective. The social construction of race means that "the actual meaning of race lies not in people's physical characteristics, but in the historical treatment of different groups and the significance that society gives to what is believed to differentiate so-called racial groups" (Higginbotham & Andersen 2012, p. 3). The concept of race grew out of social institutions and practices in which groups defined as "races" have been enslaved or otherwise exploited.

People learn to perceive others according to whatever racial classification system exists in their culture. Systems of racial classification vary across societies and change over time. For example, as late as the 1920s, U.S. Italians, Greeks, Jews, the Irish, and other "white" ethnic groups were not considered to be white. Over time, the category of "white" changed so that it included these groups. As an example of cross-cultural variation in racial categories, Brazilians use dozens of terms to racially categorize people based on various combinations of physical characteristics, although officially, the major racial categories in Brazil are *branco* (white), *pardos* (brown or mulatto), *pretos* (black), and *amarelos* (yellow).

Incorporating both biological and social meanings of race, we define **race** as a category of people who are perceived to share distinct physical characteristics that are deemed socially significant. The significance of race is not biological but social and political, because race is used to separate "us" from "them" and becomes a basis for unequal treatment of one group by another. Despite the increasing acceptance that "there is no biological justification for the concept of 'race'" (Brace 2005, p. 4), its social significance continues to be evident throughout the world.

TABLE 9.1 Who's Hispanic?

THE U.S. CENSUS BUREAU APPROACH TO DEFINING WHO IS HISPANIC:

Q. *I immigrated to Phoenix from Mexico. Am I Hispanic?*

A. You are if you say so.

Q. *My parents moved to New York from Puerto Rico. Am I Hispanic?*

A. You are if you say so.

Q. *My grandparents were born in Spain but I grew up in California. Am I Hispanic?*

A. You are if you say so.

Q. *I was born in Maryland and married an immigrant from El Salvador. Am I Hispanic?*

A. You are if you say so.

Q. *My mom is from Chile and my dad is from Iowa. I was born in Des Moines. Am I Hispanic?*

A. You are if you say so.

Q. *I was born in Argentina but grew up in Texas. I don't consider myself Hispanic. Does the Census count me as Hispanic?*

A. Not if you say you aren't.

Source: Passel & Taylor 2009.

What Do You Think? Do you think the time will ever come when a racial classification system will no longer be used? Why or why not? What arguments can be made for discontinuing racial classification? What arguments can be made for continuing it?

Ethnicity as a Social Construction **Ethnicity,** which refers to a shared cultural heritage, nationality, or lineage is also socially constructed in part. Ethnicity can be distinguished on the basis of language, forms of family structures and roles of family members, religious beliefs and practices, dietary customs, forms of artistic expression such as music and dance, as well as national origin or origin of one's parents.

Although the Census Bureau defines Hispanic or Latino as "a person of Cuban, Mexican, Puerto Rican, South or Central American, or other Spanish culture or origin regardless of race" (Ennis, Rios-Vargas, & Albert 2011, p. 2), when it comes down to collecting census data on the U.S. population, a person is Hispanic if they say they are Hispanic (see Table 9.1). Hence, ethnicity is socially constructed.

Patterns of Racial and Ethnic Group Interaction

When two or more racial or ethnic groups come into contact, one of several patterns of interaction occurs; these include genocide, expulsion, segregation, acculturation, pluralism, and assimilation. Although not all patterns of interaction between racial and ethnic groups are destructive, author and Mayan shaman Martin Prechtel reminded us, "Every human on this earth, whether from Africa, Asia, Europe, or the Americas, has ancestors

race A category of people who are believed to share distinct physical characteristics that are deemed socially significant.

ethnicity Shared cultural heritage or nationality.

In 1994, Hutus in Rwanda committed genocide against the Tutsis, resulting in 800,000 deaths.

whose stories, rituals, ingenuity, language, and life ways were taken away, enslaved, banned, exploited, twisted, or destroyed" (quoted by Jensen 2001, p. 13).

Genocide refers to the deliberate, systematic annihilation of an entire nation or people. The European invasion of the Americas, beginning in the 16th century, resulted in the decimation of most of the original inhabitants of North and South America. Some native groups were intentionally killed; others fell victim to diseases brought by the Europeans. In the 20th century, Hitler led the Nazi extermination of 12 million people, including 6 million Jews, in what is known as the Holocaust. In the early 1990s, ethnic Serbs attempted to eliminate Muslims from parts of Bosnia—a process they called "ethnic cleansing." In 1994, genocide took place in Rwanda when Hutus slaughtered hundreds of thousands of Tutsis—an event depicted in the 2004 film *Hotel Rwanda*. Genocide is continuing in the Darfur region of Sudan, where the Sudanese government, using Arab *Janjaweed* militias, its air force, and organized starvation, is systematically killing the African Muslim communities because some among them have challenged the authoritarian rule of the Sudanese government (see also Chapter 15).

Expulsion occurs when a dominant group forces a subordinate group to leave the country or to live only in designated areas of the country. The 1830 Indian Removal Act called for the relocation of eastern tribes to land west of the Mississippi River. The movement, lasting more than a decade, has been called the Trail of Tears because tribes were forced to leave their ancestral lands and endure harsh conditions of inadequate supplies and epidemics that caused illness and death. After Japan's attack on Pearl Harbor in 1941, 120,000 Japanese Americans, who became viewed as threats to national security, were forced from their homes and into evacuation camps surrounded by barbed wire.

Segregation refers to the physical separation of two groups in residence, workplace, and social functions. Segregation can be *de jure* (Latin meaning "by law") or *de facto* ("in fact"). Between 1890 and 1910, a series of U.S. laws, which came to be known as *Jim Crow laws,* were enacted to separate blacks from whites by prohibiting blacks from using "white" buses, hotels, restaurants, and drinking fountains. In 1896, the U.S. Supreme Court (in *Plessy v. Ferguson*) supported de jure segregation of blacks and whites by declaring that "separate but equal" facilities were constitutional. Blacks were forced to live in separate neighborhoods and attend separate schools. Beginning in the 1950s, various rulings overturned these Jim Crow laws, making it illegal to enforce racial segregation. Although de jure segregation is illegal in the United States, de facto segregation still exists in the tendency for racial and ethnic groups to live and go to school in segregated neighborhoods.

Acculturation refers to adopting the culture of a group different from the one in which a person was originally raised. Acculturation may involve learning the dominant language, adopting new values and behaviors, and changing the spelling of the family name. In some instances, acculturation may be forced. For decades, the Australian government removed aboriginal children from their families and placed them in missions or foster families, forcing them to abandon their language and traditional aboriginal culture. Authorities targeted children of mixed descent—what they referred to as "half-caste"—because they thought these children could be more easily acculturated into white society. Today, these individuals are known as the "stolen generation" because they had been stolen from their families and their culture.

genocide The deliberate, systematic annihilation of an entire nation or people.

expulsion Occurs when a dominant group forces a subordinate group to leave the country or to live only in designated areas of the country.

segregation The physical separation of two groups in residence, workplace, and social functions.

acculturation The process of adopting the culture of a group different from the one in which a person was originally raised.

Pluralism refers to a state in which racial and ethnic groups maintain their distinctness but respect each other and have equal access to social resources. In Switzerland, for example, four ethnic groups—French, Italians, Swiss Germans, and Romansch—maintain their distinct cultural heritage and group identity in an atmosphere of mutual respect and social equality.

Assimilation is the process by which formerly distinct and separate groups merge and become integrated as one. *Primary assimilation* occurs when members of different racial or ethnic groups are integrated in personal, intimate associations, as with friends, family, and spouses. *Secondary assimilation* occurs when different groups become integrated in public areas and in social institutions, such as neighborhoods, schools, the workplace, and in government.

Assimilation is sometimes referred to as the "melting pot," whereby different groups come together and contribute equally to a new, common culture. Although the United States has been referred to as a melting pot, in reality, many minorities have been excluded or limited in their cultural contributions to the predominant white Anglo-Saxon Protestant culture.

This is a scene from the 2002 Australian film *Rabbit-Proof Fence,* which tells the story of three aboriginal girls who were taken from their family by the Australian government as part of a program to force Aborigines to adopt the white culture.

Racial and Ethnic Group Diversity in the United States

The first census in 1790 divided the U.S. population into four groups: free white males, free white females, slaves, and other people (including free blacks and Indians). To increase the size of the slave population, the *one-drop rule* specified that even one drop of "negroid" blood defined a person as black and therefore eligible for slavery. The "one-drop rule" is still operative today: Biracial individuals are typically seen as a member of whichever group has the lowest status (Wise 2009).

In 1960, the census recognized only two categories: white and nonwhite. In 1970, the census categories consisted of white, black, and "other" (Hodgkinson 1995). In 1990, the U.S. Census Bureau recognized four racial classifications: (1) white, (2) black, (3) American Indian, Aleut, or Eskimo, and (4) Asian or Pacific Islander. The 1990 census also included the category of "other." Beginning with the 2000 census, racial categories expanded to include Native Hawaiian or other Pacific Islander, and also allowed individuals the option of identifying themselves as being more than one race rather than checking only one racial category (see Figure 9.1).

Although U.S. citizens come from a variety of ethnic backgrounds, the largest ethnic population in the United States is of Hispanic origin. The Census Bureau began collecting data on the U.S. Hispanic population in 1970.

As you read the following section on U.S. census data on race and Hispanic origin, keep in mind that the use of racial and ethnic labels is often misleading and imprecise. The ethnic classification of "Hispanic/Latino," for example, lumps together disparate groups such as Puerto Ricans, Mexicans, Cubans, Venezuelans, Colombians, and others from Latin American countries. The racial term *American Indian* includes more than 300 separate tribal groups that differ enormously in language, tradition, and social structure. The racial label *Asian* includes individuals from China, Japan, Korea, India, the Philippines, or one of the countries of Southeast Asia. And what about people who are Asian but who live in the United States? The term *Asian American* is used to describe people with Asian racial features who are born in the United States, as well as those who immigrate to the United States. Columbia University professor Derald Wing Sue, an Asian American who was born in the United States, said, "When I get out of a cab after having a conversation with a white cab driver, they'll say something like, 'Boy, you speak excellent English. . . . From their perspective, that's meant as a compliment, but another hidden

pluralism A state in which racial and ethnic groups maintain their distinctness but respect each other and have equal access to social resources.

assimilation The process by which formerly distinct and separate groups merge and become integrated as one.

Figure 9.1: 2010 Census Questions

Source: Humes, Jones, & Ramirez 2011.

→ **NOTE: Please answer BOTH Question 5 about Hispanic origin and Question 6 about race. For this census, Hispanic origins are not races.**

5. **Is this person of Hispanic, Latino, or Spanish origin?**

☐ **No,** not of Hispanic, Latino, or Spanish origin
☐ Yes, Mexican, Mexican Am., Chicano
☐ Yes, Puerto Rican
☐ Yes, Cuban
☐ Yes, another Hispanic, Latino, or Spanish origin — *Print origin, for example, Argentinean, Colombian, Dominican, Nicaraguan, Salvadoran, Spaniard, and so on.* ↙

6. **What is this person's race?** Mark ☒ one or more boxes.

☐ White
☐ Black, African Am., or Negro
☐ American Indian or Alaska Native — *Print name of enrolled or principal tribe.* ↙

☐ Asian Indian   ☐ Japanese   ☐ Native Hawaiian
☐ Chinese   ☐ Korean   ☐ Guamanian or Chamorro
☐ Filipino   ☐ Vietnamese   ☐ Samoan
☐ Other Asian — *Print race, for example, Hmong, Laotian, Thai, Pakistani, Cambodian, and so on.* ↙   ☐ Other Pacific Islander — *Print race, for example, Fijian, Tongan, and so on.* ↙

☐ Some other race — *Print race.* ↙

meaning is being communicated, and that is that I am a perpetual foreigner in my own land" (quoted in Tanneeru 2007).

U.S. Census Data on Race and Hispanic Origin

Census data show that the U.S. population is becoming increasingly diverse: From 2000 to 2010, the percentage of the population that is non-Hispanic white declined from 69 percent to 64 percent of the total population (Humes, Jones, & Ramirez 2011). Figure 9.2 shows that from 2000 to 2010, the percent increase in minority populations has increased significantly more than the increase in the white population.

In 2010, 16 percent of the U.S. population was Hispanic, up from 13 percent in 2000, with the majority being Mexican (see Figure 9.3). More than half of the growth in the total U.S. population between 2000 and 2010 was due to the increase in the Hispanic population (Humes, Jones, & Ramirez 2011). More than half of the U.S. Hispanic population lives in just three states: California, Texas, and Florida.

The current Census Bureau classification system does not allow people of mixed Hispanic or Latino ethnicity to identify themselves as such. Individuals with one Hispanic and one non-Hispanic parent still must say that they are either Hispanic or not Hispanic.

Among the general public, one of the most common misunderstandings about Hispanic origin is the belief that "Hispanic" is a race. According to the federal government, Hispanic origin is NOT a race—it is an ethnicity (Ennis, Rios-Vargas, & Albert 2011). In the 2010 census, more than half of Hispanics identified their race as "white." Despite

> One of the most common misunderstandings about Hispanic origin is the belief that "Hispanic" is a race. . . . Hispanic origin is NOT a race—it is an ethnicity.

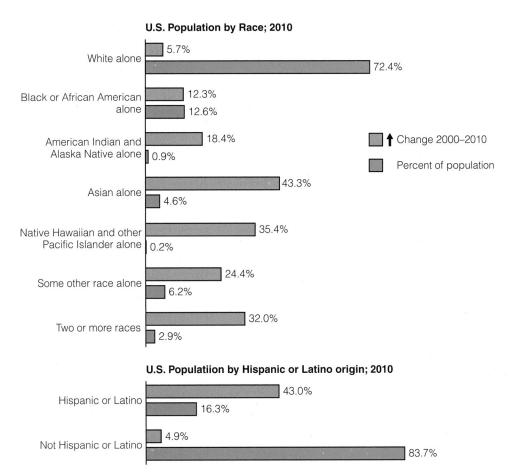

U.S. Population by Race; 2010

White alone — 5.7% / 72.4%

Black or African American alone — 12.3% / 12.6%

American Indian and Alaska Native alone — 18.4% / 0.9%

Asian alone — 43.3% / 4.6%

Native Hawaiian and other Pacific Islander alone — 35.4% / 0.2%

Some other race alone — 24.4% / 6.2%

Two or more races — 32.0% / 2.9%

↑ Change 2000–2010

Percent of population

U.S. Populatiion by Hispanic or Latino origin; 2010

Hispanic or Latino — 43.0% / 16.3%

Not Hispanic or Latino — 4.9% / 83.7%

Figure 9.2: U.S. Population by Race and Hispanic/Latino Origin, 2010
Source: U.S. Census Bureau 2011a.

the fact that the 2010 census included a new instruction that stated, "for this census, Hispanic origins are not races," many Hispanics identified their race as "Latino," "Mexican," "Puerto Rican," "Salvadoran," or other ethnicity or national origin. Responses to the census race question that reflected Hispanic origin were classified in the category of "Some Other Race" (see Table 9.2).

Currently, racial and ethnic minority populations outnumber non-Hispanic whites in four states—California, New Mexico, Hawaii, and Texas. In these states, Hispanics are the largest minority group, except for Hawaii, where the largest minority group is

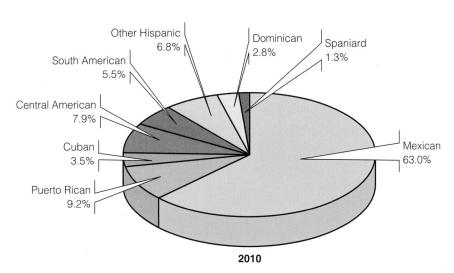

Other Hispanic 6.8%

South American 5.5%

Central American 7.9%

Cuban 3.5%

Puerto Rican 9.2%

Dominican 2.8%

Spaniard 1.3%

Mexican 63.0%

2010

Figure 9.3: Percent Distribution of U.S. Hispanics by Type, 2010
Source: Ennis, Rios-Vargas, & Albert 2011.

TABLE 9.2 Racial Identification of Hispanics/Latinos in the United States, 2010

White	53%
Black	2.5%
American Indian/Alaskan Native	1.4%
Asian	0.4%
Native Hawaiian/Pacific Islander	0.1%
Some Other Race	36.7%
Two or More Races	6.0%

Source: Ennis, Rios-Vargas, & Albert 2011.

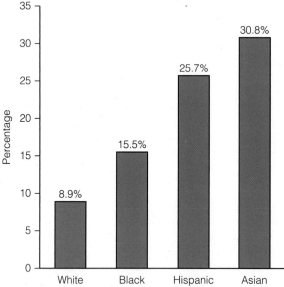

% of newlyweds in 2008 who married someone of a different race/Hispanic ethnicity

White 8.9%
Black 15.5%
Hispanic 25.7%
Asian 30.8%

Figure 9.4: U.S. Intermarriage Rates, by Race and Hispanic Origin
Source: Passel, Wang, and Taylor 2010.

Asian American (Humes, Jones, & Ramirez 2011). And in 10 states, racial and ethnic minority youth outnumber non-Hispanic white youth (Tavernise 2011).

Mixed-Race Identity

As shown in Figure 9.2, 2010 census data found that only a small percentage (2.9 percent) of the U.S. population identify themselves as being of more than one race. But in the 10-year period between 2000 and 2010, the mixed-race U.S. population has grown 32 percent. And among U.S. children, the multiracial population has grown almost 50 percent, making it the fastest-growing youth group in the country (Saulny 2011).

The multiracial population has grown as mixed-race marriages have increased over recent years. Until 1967, 19 states had **antimiscegenation laws** banning interracial marriage. In 1967, the Supreme Court (in *Loving v. Virginia*) declared these laws unconstitutional. This percentage will likely increase as the rate of interracial marriage increases. In 2008, one in seven new marriages in the United States was between spouses with different racial or ethnic identities—a rate double that from 1980, and six times that from 1960 (Passel, Wang, & Tyler 2010). This dramatic increase is due to the weakening of cultural taboos against mixed marriages, the increased independence that adult children have from their parents and communities of origin, and the influx of Latin American and Asian immigrants over the last several decades (Passel et al. 2010; Rosenfeld & Kim 2005) (see Figure 9.4). Although most Americans—more than six in ten—say it "would be fine" with them if a family member married outside of his or her race or ethnicity (Passel et al. 2010), 15 percent of U.S. adults disapprove of marriage between blacks and whites (Gallup Organization 2011a).

What Do You Think? Do you mostly think of Barack Obama as a black person or mostly as a person of mixed race? A Pew Research (2010) poll found that only 24 percent of whites and 23 percent of Hispanics view Obama as black, whereas more than half of blacks (55 percent) view Obama as black. Why do you think blacks are more likely than whites or Hispanics to view Obama as black?

Race and Ethnic Group Relations in the United States

Despite significant improvements over the last two centuries, race and ethnic group relations continue to be problematic. The racial divide in the United States sharpened in 2005, in the wake of Hurricane Katrina, which left victims—who were predominantly black and poor—waiting for days to be rescued from their flooded attics or rooftops or to be evacuated from overcrowded "shelters" with no food, water, medical supplies, or working toilets. Following this crisis, a national survey found that most blacks (77 percent), compared with only 17 percent of whites, believe that the

antimiscegenation laws Laws banning interracial marriage until 1967, when the Supreme Court (in *Loving v. Virginia*) declared these laws unconstitutional.

government's response to the disaster would have been faster if most of Katrina's victims had been white (Pew Research Center 2005).

In response to a question asking whether relations between blacks and whites will always be a problem for the United States or whether a solution will eventually be worked out, more than a third (38 percent) of U.S. adults said that race relations will always be a problem (Gallup Organization 2011a). A Gallup Poll asked a national sample of U.S. adults to rate relations between various groups in the United States. Table 9.3 presents results of this survey. Relations between various racial and ethnic groups are influenced by prejudice and discrimination (discussed later in this chapter). Race and ethnic relations are also complicated by issues concerning immigration—the topic we turn to next.

TABLE 9.3 Perceptions of Race and Ethnic Relations in the United States, 2008

A national sample of U.S. adults was asked to rate relations between various groups in the United States. The results are depicted in this table.

	VERY OR SOMEWHAT GOOD	VERY OR SOMEWHAT BAD
Whites and blacks	68%	30%
Whites and Hispanics	65%	33%
Blacks and Hispanics	49%	40%
Whites and Asians	81%	12%

Note: Percentages do not add up to 100 because of "no opinion" responses.
Source: Gallup Organization 2011a.

What Do You Think? A Gallup Organization (2011a) poll asked U.S. adults, "If blacks and whites honestly expressed their true feelings about race relations, do you think this would do more to bring races together or cause greater racial division?" More than half (56 percent) replied "bring races together," and about a third (37 percent) said "cause greater division" (7 percent had no opinion). What would your answer be? Why?

Immigrants in the United States

The growing racial and ethnic diversity of the United States is largely the result of immigration as well as the higher average birthrates among many minority groups. Immigration generally results from a combination of "push" and "pull" factors. Adverse social, economic, and/or political conditions in a given country "push" some individuals to leave that country, whereas favorable social, economic, and/or political conditions in other countries "pull" some individuals to those countries. Before reading further, you may want to assess your attitudes toward U.S. immigrants and immigration in this chapter's Self and Society feature.

U.S. Immigration: A Historical Perspective

For the first 100 years of U.S. history, all immigrants were allowed to enter and become permanent residents. The continuing influx of immigrants, especially those coming from nonwhite, non-European countries, created fear and resentment among native-born Americans, who competed with immigrants for jobs and who held racist views toward some racial and ethnic immigrant populations. America's open-door policy on immigration ended in 1882 with the Chinese Exclusion Act, which suspended the entrance of the Chinese to the United States for 10 years and declared Chinese ineligible for U.S. citizenship. The Immigration Act of 1917 required all immigrants to pass a literacy test before entering the United States. And in 1921, the Johnson Act introduced a limit on the number of immigrants who could enter the country in a single year, with stricter limitations

© AP Photo/Alan Diaz

Although Barack Obama's background is biracial—he is the son of a black Kenyan immigrant father and a white Kansas native mother—he identifies as a black African American.

For each of the following questions, choose the answer that best reflects your attitudes. After answering all the questions, compare your answers with the results of a national Gallup Poll sample of U.S. adults.

1. On the whole, do you think immigration is a good thing or a bad thing for this country?
   a. Good thing
   b. Bad thing
2. In your view, should immigration be kept at its present level? Increased? Or decreased?
   a. Present level
   b. Increased
   c. Decreased
3. Which of the following comes closer to your point of view? (a) Illegal immigrants in the long run become productive citizens and pay their fair share of taxes, or (b) illegal immigrants cost the taxpayers too much by using government services like public education and medical services.
   a. Pay fair share
   b. Cost too much

4. Which of the following comes closer to your point of view? (a) Illegal immigrants mostly take jobs that American workers want, or (b) illegal immigrants mostly take low-paying jobs Americans don't want.
   a. Take jobs that American workers want
   b. Take jobs that American workers do not want

Comparison Data

Gallup Poll data revealed the following attitudes toward U.S. immigrants and immigration:

1. Bad thing: 37 percent; Good thing: 59 percent; Mixed or no opinion: 5 percent
2. Present level: 35 percent; Increased: 18 percent; Decreased: 43 percent; No opinion: 4 percent
3. Pay fair share of taxes: 32 percent; Cost taxpayers too much: 62 percent; Neither/both/no opinion: 7 percent
4. Take jobs Americans want: 15 percent; Take low-paying jobs Americans don't want: 79 percent; Neither/both/no opinion: 6 percent.

Source: Gallup Organization 2011b.

for certain countries (including those in Africa and the Near East). The 1924 Immigration Act further limited the number of immigrants allowed into the United States and completely excluded the Japanese. Other federal immigration laws include the 1943 repeal of the Chinese Exclusion Act, the 1948 Displaced Persons Act (which permitted refugees from Europe), and the 1952 Immigration and Nationality Act (which permitted a quota of Japanese immigrants).

In the 1960s, most immigrants were from Europe, but most immigrants today are from Latin America (predominantly Mexico) or Asia (see Figure 9.5). In 2009, one in eight U.S. residents (12.5 percent) was born in a foreign country (Grieco & Trevelyan 2010) (see Table 9.4).

Guest Worker Program

The United States has two guest worker programs that allow employers to import unskilled labor for temporary or seasonal work: the H-2A program for agricultural work and the H-2B program for non-agricultural work. H-2 visas generally do not permit guest workers to bring their families to the United States.

A Southern Poverty Law Center report revealed that the guest worker program constitutes a "modern-day system of indentured servitude" (SPLC 2007, p. 2). Guest workers often incur debts ranging from $500 to more than $10,000 to pay for visas, travel costs, and recruiter fees. When they arrive at their jobs, their employers often take their identity documents (passports and Social Security cards) to ensure that workers do not leave before their contract is fulfilled. Guest workers are often paid substantially less than the minimum wage and are rarely paid overtime pay, despite working well over 40 hours a week. Although guest workers perform some of the most difficult and dangerous jobs in America, many who are injured on the job are unable to obtain medical treatment

Following Hurricane Katrina in 2005, a national survey found that the majority of blacks believe that the government's response to the crisis would have been faster if most of Katrina's victims had been white.

and workers' compensation benefits. Although employers hiring H-2A workers are required to provide them with free housing, the quality of housing is often substandard, even dangerous, and located in isolated rural areas where workers are dependent on their employers for transportation to work, grocery stores, and banks (for which employers often charge exorbitant fees). In some cases, guest workers are literally locked up in their living quarters. Immigrant women working at low-wage jobs are often targets of sexual violence; one survey found that three out of four Latinas in the South view sexual harassment as a major workplace problem (SPLC 2009).

Unable to obtain legal assistance, guest workers who are abused and denied their legal rights must endure the abuse or try to escape in a foreign land without passports, money, or tickets home.

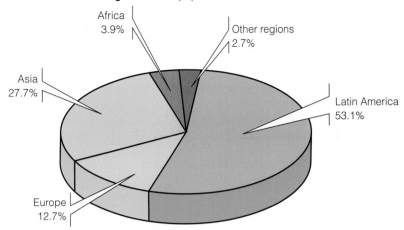

Total foreign-born U.S. population in 2009: 38.5 million

Africa 3.9%
Other regions 2.7%
Asia 27.7%
Latin America 53.1%
Europe 12.7%

Figure 9.5: U.S. Foreign-Born Residents by Region of Birth, 2009
Source: Grieco and Trevelyan 2010.

Illegal Immigration

Illegal immigration occurs when immigrants enter the United States without going through legal channels such as the H-2 visa program, and when immigrants who were admitted legally stay past the date they were required to leave. Although the term *illegal immigrants* is commonly used, we prefer the terms *unauthorized immigrants* and *undocumented immigrants,* which we use interchangeably.

In January 2010, an estimated 10.8 million unauthorized immigrants were living in the United States, down from the peak of 11.8 million in 2007 (Hoefer, Rytina, & Baker 2011). The majority of undocumented immigrants (62 percent) are from Mexico, followed by El Salvador, Guatemala, Honduras, and the Philippines.

TABLE 9.4 Foreign-Born Population and Percentage of Total Population for the United States, 1890–2009

YEAR	NUMBER (IN MILLIONS)	PERCENTAGE OF TOTAL
2009	38.5	12.5
1990	19.8	7.9
1970	9.6	4.7
1950	10.3	6.9
1930	14.2	11.6
1890	9.2	14.8

Source: Grieco and Trevelyan 2010; and Lollock 2001.

Border Crossing U.S. Customs and Border Protection, an agency within the Department of Homeland Security, has more than 20,000 border patrol agents who are responsible for patrolling 6,000 miles of Canadian and Mexican borders and more than 2,000 miles of coastal waters around Florida and Puerto Rico (U.S. Customs and Border Protection 2011). These agents, along with a "fence" at the U.S.–Mexico border, are not able to prevent people from illegally crossing the border to get into the United States.

Some people cross (or attempt to cross) the U.S.–Mexican border with the help of a *coyote*—a hired guide who typically charges $3,000 to $5,000 to lead people across the border (Maril 2011). Crossing the border illegally involves a number of risks, including death from drowning (e.g., while trying to cross the Rio Grande) or dehydration. Border crossers also risk encounters with members of **nativist extremist groups**—organizations that not only advocate restrictive immigration policy, but also encourage their members to use vigilante tactics to confront or harass suspected undocumented immigrants. In 2009, there were 319 nativist extremist groups around the country, including such groups as the Minuteman Civil Defense Corps (MCDC) and the Federal Immigration Reform and Enforcement Coalition (FIRE) (Beirich 2011).

In 2003, a nativist extremist group called Ranch Rescue tried to stop undocumented Latinos from crossing a U.S. rancher's land. Fatima Leiva and Edwin Mancia, Salvadorans,

nativist extremist groups
Organizations that not only advocate restrictive immigration policy, but also encourage their members to use vigilante tactics to confront or harass suspected undocumented immigrants.

Chris Simoux is one of a growing number of Americans who patrol the U.S.-Mexico border looking for undocumented immigrants. Such self-appointed border patrol guards sometimes use violence when they encounter suspected undocumented immigrants.

Reuters/Francisco Medina/Landov

were among a group of immigrants traveling on foot when members of Ranch Rescue detained them. During the detention, Mancia was struck on the back of the head and attacked by a Rottweiler dog owned by a Ranch Rescue member. Although Ranch Rescue purports to be protecting the United States from illegal immigration, the underlying motive for their anti-immigrant actions is hate. The president and spokesperson for Ranch Rescue has described Mexicans as "dog turds" who are "ignorant, uneducated and desperate for a life in a decent nation because the one [they] live in is nothing but a pile of dog [excrement] made up of millions of little dog turds" (quoted in SPLC 2005, p. 3).

Undocumented Immigrants in the Workforce In 2010, there were 8 million undocumented immigrants in the U.S. labor force, comprising 5.2 percent of the U.S. workforce (Passel & Cohn 2011). Sociologist Robert Maril (2004) noted, "The vast majority of illegal immigrants leave their home countries to work hard, save their money, then return to their homeland. . . . These individuals do not travel their difficult and dangerous journeys searching for a welfare handout; they immigrate to work" (pp. 11–12). Virtually all undocumented men are in the labor force. Their labor force participation exceeds that of men who are legal immigrants or who are U.S. citizens because undocumented men are less likely to be disabled, retired, or in school. Undocumented women are less likely to be in the labor force than undocumented men because they are more likely to be stay-at-home mothers.

Undocumented workers often do work that U.S. workers are unwilling to do—they routinely work 60 or more hours per week and earn less than the minimum wage, with no paid overtime and no benefits.

Undocumented workers often do work that U.S. workers are unwilling to do—they routinely work 60 or more hours per week and earn less than the minimum wage, with no paid overtime and no benefits.

Policies Regarding Illegal Immigration In 1986, Congress approved the Immigration Reform and Control Act, which made hiring illegal immigrants an illegal act punishable by fines and even prison sentences. In 2005, Wal-Mart agreed to pay a record $11 million to settle charges that it used hundreds of illegal immigrants to clean its stores (Greenhouse 2005).

Recent concerns about terrorism and drug trafficking have led to stepped-up efforts to apprehend "illegal aliens." The Secure Fence Act of 2006 authorized the construction of a 700-mile "fence" along the U.S.-Mexico border to prevent unauthorized immigrant workers, as well as drug dealers and terrorists, from entering the United States. Critics of the fence argue that this barrier—miles of double chain link and barbed wire fences with light and infrared camera poles—is expensive and ineffective in stopping illegal immigration; disrupts the environment and harms wildlife along the border; increases the risk and danger to immigrants trying to cross the border; and damages diplomatic relations with Mexico.

Several states, cities, and counties have passed various laws and ordinances related to illegal immigration. In 2006, Hazelton, Pennsylvania, became the first city to take business licenses away from employers who hire undocumented workers and to fine landlords $1,000 a day for renting to illegal immigrants. In 2010, Arizona passed the Support Our Law Enforcement and Safe Neighborhoods Act, known as SB 1070—one of the toughest illegal immigration bills in the country. Federal law requires that legal immigrants carry registration papers with them at all times. Arizona's SB 1070 makes failure to carry registration documents a state crime, and requires police to verify the legal status of a person during traffic stops, detentions, or arrests if the police suspect that person is in the country illegally. In response to criticism that the law would lead to racial profiling among police, the Arizona law was modified to specify that law enforcement officials may not use race, color, or national origin as the basis of implementing the law. SB 1070 also criminalizes anyone who transports undocumented immigrants, which, according to the law, constitutes "smuggling of human beings." This provision not only affects employers who transport undocumented day laborers to the worksite, but also church members who transport immigrants to church. Thus, a local church that transports an undocumented immigrant for Sunday service could be charged with a misdemeanor and have their church van impounded. A church volunteer who transports a group of 10 immigrants could be classified as a felon and incur a $1,000 fine for each immigrant in the van (Hwang 2010).

On May 1, 2006, thousands of immigrant workers did not show up for work to show the nation how valuable immigrant labor is for the U.S. economy. Across the country, many businesses closed for the day; others experienced high rates of employee absenteeism.

Although about six in ten U.S. adults approve of Arizona's strict immigration law (Pew Research Center 2011a), fierce public opposition to SB 1070 is evidenced by the thousands of protests around the country and the calls for boycotts against Arizona. Several lawsuits have been filed against Arizona over issues of constitutionality and compliance with civil rights laws. A lawsuit filed by the U.S. Department of Justice argued that SB 1070 interferes with federal immigration regulations. U.S. District Court Judge Susan Bolton granted an injunction blocking key provisions of SB 1070 from going into effect, including the mandate requiring police to check the immigration status of those apprehended or arrested. Arizona has tried, unsuccessfully to date, to reverse the injunction

On May 1, 2010, tens of thousands of people protested in over 70 cities against Arizona's immigration law.

in federal appeals court and, in February 2011, Arizona sued the federal government for failing to secure the border against the continued influx of illegal immigrants. Other states have considered or have enacted similar legislation. For example, in 2011, South Carolina enacted SB 20, which states that if a law enforcement officer stops, detains, arrests, or investigates someone for a criminal offense and has a "reasonable suspicion" that the person is not in the United States legally, the officer must ask for proof of citizenship in the form of identification or documentation. Arizona's and South Carolina's immigration laws require local and state police to play an increased role in immigration enforcement. According to Tramonte (2011), "most state and local police do not want to be put in the position of identifying non-criminal immigrants for deportation because they believe doing so would make it more difficult for them to earn the trust of immigrant residents and protect the entire community from criminals" (p. 4). In contrast to states that have taken a hard-line approach to illegal immigration, more than 70 cities and states have policies that prohibit police from asking local residents who have not been arrested to prove their legal immigration status (Tramonte 2011).

What Do You Think? Guatemalan immigrant Danny Sigui helped convict a murderer in Providence, Rhode Island, by providing critical testimony against the accused. The state attorney general's office learned that Sigui was an undocumented immigrant, and reported him to the U.S. Department of Homeland Security. Following the trial in 2003, Sigui was deported. "When asked whether he would have come forward again, knowing that doing so would result in his deportation, Sigui replied: 'If I had known they would take my liberty, that they would take my children away from me, that they would put me [in immigration detention], I would not do this'" (quoted in Tramonte 2011, p. 4). Sigui's deportation and the publicity surrounding it will likely discourage other undocumented immigrants from reporting crimes to police and testifying in trials. But fear of deportation did not stop Antonio Diaz Chacon from rescuing a 6-year-old girl from a suspected kidnapper. In 2011, Chacon, an undocumented immigrant who lived in the United States for four years and who was married to an American, saw a man scoop up a young girl and put her in his van. Chacon chased the van in his pickup truck. When the van crashed, the suspected kidnapper fled the scene and Chacon rescued the girl.

What do you think U.S. policy should be regarding undocumented immigrants who report crimes, help in criminal investigations and prosecutions, or commit acts of bravery to save the lives of others?

Churches, synagogues, and other religious institutions have also played a role in assisting undocumented immigrants. In August 2006, Elvira Arellano, a Mexican citizen who lived illegally in the United States, took refuge in the Adalberto United Methodist Church in an attempt to avoid deportation by U.S. Immigration and Customs Enforcement. Arellano, whose young son is a U.S. citizen, said she should not have to choose between leaving her son in the states and taking him to Mexico. Arellano was arrested in August 2007, when she left the church and traveled to California and was deported to Mexico.

Elvira Arellano's deportation ignited a debate concerning immigration policies involving illegal immigrants whose children are U.S. citizens. Worksite raids and home raids by immigration officials have resulted in the arrest, detention, and deportation of thousands of immigrants, leaving tens of thousands of children, including children who are U.S. citizens, separated from their parents, or effectively deported with their families (Kremer et al. 2009). A report documenting the harm these children suffer states that,

"In a country that emphasizes the importance of family unity in the socialization and upbringing of its children, an immigration system that promotes family separation is a broken system" (Kremer et al. 2009, p. 5). Some parents even lose legal custody of their children. In 2007, Encarnación Bail Romero, a Guatemalan immigrant, was arrested along with 136 undocumented immigrants at a Missouri processing plant. While serving time in jail for presenting false identification to the police, a county court terminated her rights to her 2-year-old son Carlos on grounds of abandonment, and the child was adopted by a local couple (Thompson 2009).

After taking refuge in a Chicago church for one year to avoid deportation to Mexico, Elvira Arellano was arrested in California and deported to Mexico, leaving behind her 8-year-old son, a U.S. citizen.

The Development, Relief and Education for Alien Minors Act (DREAM Act), introduced in Congress in 2009, would provide a path to legal status for undocumented immigrants who were brought to the United States as children. If passed, the DREAM Act would permit certain immigrant students who have grown up in the United States to apply for temporary legal status and to eventually obtain permanent status and become eligible for U.S. citizenship if they go to college or serve in the U.S. military. The law would also eliminate a federal provision that penalizes states that provide in-state tuition to undocumented immigrants.

What Do You Think? Although the Supreme Court ruled in 1982 that states must educate undocumented immigrants through the twelfth grade, states are divided in their laws regarding supporting undocumented immigrants through in-state college tuition. In 2011, Connecticut became the twelfth state to grant in-state tuition to undocumented immigrants who are working toward legal status (Associated Press 2011). Immigrants' rights advocates argue that it is in the best interest of the United States to help undocumented immigrant youth achieve a college education. But opponents worry that granting in-state tuition to undocumented immigrants would (1) be a financial burden to taxpayers, (2) make it harder for legal citizens to get accepted into state universities, and (3) open the door to granting other benefits now reserved for state residents. Do you think that undocumented immigrants should qualify for in-state college tuition? Why or why not?

Immigration policies will continue to be hotly debated. Although most Americans (78 percent) support stronger enforcement of immigration laws and border security, the majority (72 percent) also favors creating a path to citizenship for illegal immigrants currently in the country (Pew Research Center 2011b).

Becoming a U.S. Citizen

In 2009, of all foreign-born U.S. residents, 44 percent were **naturalized citizens** (immigrants who applied and met the requirements for U.S. citizenship) (Gryn & Larsen 2010). Requirements to become a U.S. citizen for most immigrants include (1) having resided

naturalized citizens Immigrants who apply for and meet the requirements for U.S. citizenship.

continuously as a lawful permanent U.S. resident for at least five years (three years for a spouse of a U.S. resident); (2) being able to read, write, speak, and understand basic English (certain exemptions apply); (3) being "a person of good moral character" (can't have a record of criminal offenses such as prostitution, illegal gambling, failure to pay child support, drug violations, and violent crime); (4) demonstrating willingness to support and defend the U.S. Constitution by taking the Oath of Allegiance; and (5) passing an examination on English (speaking, reading, and writing) and U.S. government and history (U.S. Citizenship and Immigration Services 2011).

Myths about Immigration and Immigrants

Many foreign-born U.S. residents work hard to succeed educationally and occupationally. The percentage of foreign-born adults (age 25 and older) with at least a bachelor's degree (27 percent) matches that of native-born U.S. adults (28 percent) (Kandel 2011). Despite the achievements and contributions of immigrants, many myths about immigration and immigrants persist, largely perpetuated by anti-immigrant groups and campaigns:

Myth 1. Immigrants increase unemployment and lower wages among native workers.
Reality: Most academic economists agree that immigration has a small but positive impact on the wages of native-born workers because although new immigrant workers add to the labor supply, they also consume goods and services, which creates more jobs (Shierholz 2010). Immigrants also start their own businesses at a higher rate than native U.S. residents, which increases demand for business-related supplies (such as computers and office furniture) and service providers (such as accountants and lawyers) (Pollin 2011).

Myth 2. Immigrants drain the public welfare system and our public schools.
Reality: Unauthorized and temporary immigrants are ineligible for major federal benefit programs, and even legal immigrants may face eligibility restrictions. Two benefit programs that do not have restrictions against unauthorized immigrants are the Special Supplemental Nutrition Program for Women, Infants, and Children (WIC) and the National School Lunch Program.

Regarding public education, a 1982 Supreme Court case (*Plyer v. Doe*) held that states cannot deny students access to public education, even if they are not legal U.S. residents. The court ruled that denying public education could impose a lifetime of hardship "on a discrete class of children not accountable for their disabling status" (Armario 2011). Children of unauthorized immigrants, 73 percent of whom are U.S. citizens, comprise only 6.8 percent of students in elementary and secondary schools (Passel & Cohn 2009), although the percentage is much higher in communities with large immigrant populations.

Although the states bear the cost of education, social services, and medical services for the immigrant population, research suggests that the economic benefits that immigrants provide for the states outweigh the costs associated with supporting them. For example, a study of immigrants in North Carolina found that, over the prior 10 years, Latino immigrants had cost the state $61 million in a variety of benefits—but were responsible for more than $9 billion in state economic growth (Beirich 2007). Half to three-fourths of undocumented immigrants pay federal, state, and local taxes, including Social Security taxes for benefits they will never receive (*Teaching Tolerance* 2011).

Myth 3. Immigrants do not want to learn English.
Reality: The demand for English classes for immigrants exceeds their availability; out of 176 providers of English as a Second Language (ESL) classes, more than half reported waiting lists ranging from a few weeks to more than three years (Bauer & Reynolds 2009). Even so, the majority of both naturalized U.S. citizens (83 percent) and noncitizens (59 percent) speak English "well" or "very well" and, among immigrant youth, nine out of ten report that they speak English "well" or "very well" (Kandel 2011).

Myth 4. Undocumented immigrants have children in the United States as a means of gaining legal status.

Reality: Under the Fourteenth Amendment of the U.S. Constitution, any child born in the United States is automatically granted U.S. citizenship. Of the 5.5 million children of undocumented immigrants in the United States, 4.5 million are U.S. born and therefore citizens of the United States (Passel & Cohn 2011). But having children who are U.S. citizens does not provide immigrants with a means of gaining legal status in the United States. Children under 21 are not allowed to petition for their parents' U.S. citizenship. Nevertheless, some legislators have called for ending birthright citizenship by amending the Constitution or enacting state law to limit citizenship to children who have at least one authorized parent (Dwyer 2011).

Myth 5. Immigrants have high rates of criminal behavior.

Reality: Immigrants are less likely than natives to commit crimes. Because they risk deportation, undocumented immigrants have a strong motivation to avoid involvement with the law. In 2000, the U.S. incarceration rate for native-born men aged between 18 and 39 was 3.5 percent—five times greater than that of their foreign-born counterparts (Bauer & Reynolds 2009). El Paso, Texas, a city with a high immigrant population, is among the safest big cities in America. Criminologist Jack Levin said, "If you want to find a safe city, first determine the size of the immigrant population. If the immigrant community represents a large proportion of the population, you're likely in one of the country's safer cities" (quoted in Balko 2009).

> Immigrants are less likely than natives to commit crimes. Because they risk deportation, undocumented immigrants have a strong motivation to avoid involvement with the law.

Myth 6. Undocumented immigrants should have just "gotten in line" to gain legal entry into the United States.

Reality: Despite the demand of U.S. business for immigrant labor, only 5,000 permanent visas for lawful entry of less-skilled workers are available per year. Temporary work visas are also limited, which means that "the avenues for lawful entry into the U.S. by the lower-skilled, lower-educated immigrant that makes up the vast majority of the undocumented population are virtually nonexistent" (Kremer et al. 2009, pp. 2–3).

Sociological Theories of Race and Ethnic Relations

Some theories of race and ethnic relations suggest that individuals with certain personality types are more likely to be prejudiced or to direct hostility toward minority group members. Sociologists, however, concentrate on the impact of the structure and culture of society on race and ethnic relations. Three major sociological theories lend insight into the continued subordination of minorities.

Structural-Functionalist Perspective

Structural functionalists emphasize that each component of society affects the stability of the whole. From this perspective, racial and ethnic inequality is dysfunctional for society (Williams & Morris 1993). A society that practices discrimination fails to develop and utilize the resources of minority members. Prejudice and discrimination aggravate social problems, such as crime and violence, war, unemployment and poverty, health problems, family problems, urban decay, and drug use—problems that cause human suffering as well as impose financial burdens on individuals and society.

Picca and Feagin (2007) explained how "the system of racial oppression in the United States affects not only Americans of color but white Americans and society as a whole" (p. 271):

> Whites lose when they have to pay huge taxes to keep people of color in prisons because they are not willing to remedy patterns of unjust enrichment and . . . to

pay to expand education, jobs, or drug-treatment programs that would be less costly. They lose by driving long commutes so they do not have to live next to people of color in cities. . . . They lose when white politicians use racist ideas and arguments to keep from passing legislation that would improve the social welfare of all Americans. Most of all, whites lose . . . by not having in practice the democracy that they often celebrate to the world in their personal and public rhetoric. (p. 271)

The structural-functionalist analysis of manifest and latent functions also sheds light on issues of race and ethnic relations. For example, the manifest function of the civil rights legislation in the 1960s was to improve conditions for racial minorities. However, civil rights legislation produced an unexpected negative consequence, or latent dysfunction. Because civil rights legislation supposedly ended racial discrimination, whites were more likely to blame blacks for their social disadvantages and thus perpetuate negative stereotypes such as "blacks lack motivation" and "blacks have less ability" (Schuman & Krysan 1999).

Conflict Perspective

The conflict perspective examines how competition over wealth, power, and prestige contributes to racial and ethnic group tensions. Consistent with this perspective, the "racial threat" hypothesis views white racism as a response to perceived or actual threats to whites' economic well-being or cultural dominance by minorities.

For example, between 1840 and 1870, large numbers of Chinese immigrants came to the United States to work in mining (the California Gold Rush of 1848), railroads (the transcontinental railroad, completed in 1869), and construction. As Chinese workers displaced whites, anti-Chinese sentiment rose, resulting in increased prejudice and discrimination and the eventual passage of the Chinese Exclusion Act of 1882, which restricted Chinese immigration until 1924. More recently, white support for Proposition 209—a 1996 resolution passed in California that ended state affirmative action programs—was higher in areas with larger Latino, African American, or Asian American populations, even after controlling for other factors (Tolbert & Grummel 2003). In other words, opposition to affirmative action programs that help minorities was higher in areas with greater racial and ethnic diversity, suggesting that whites living in diverse areas felt more threatened by the minorities.

In another study, researchers interviewed individuals in white racist Internet chat rooms to examine the extent to which people would advocate interracial violence in response to alleged economic and cultural threats (Glaser et al. 2002). The researchers posed three scenarios that might be perceived as threatening: interracial marriage, minority in-migration (i.e., blacks moving into one's neighborhood), and job competition (i.e., competing with a black person for a job). Respondents' reactions to interracial marriage were the most volatile, followed by in-migration. The researchers concluded that violent ideation among white racists stems from perceived threats to white cultural dominance and separateness rather than from perceived economic threats.

Furthermore, conflict theorists suggest that capitalists profit by maintaining a surplus labor force, that is, by having more workers than are needed. A surplus labor force ensures that wages will remain low because someone is always available to take a disgruntled worker's place. Minorities who are disproportionately unemployed serve the interests of the business owners by providing surplus labor, keeping wages low, and, consequently, enabling them to maximize profits.

Conflict theorists also argue that the wealthy and powerful elite foster negative attitudes toward minorities to maintain racial and ethnic tensions among workers. So long as workers are divided along racial and ethnic lines, they are less likely to join forces to advance their own interests at the expense of the capitalists. In addition, the "haves" perpetuate racial and ethnic tensions among the "have-nots" to deflect attention away from their own greed and exploitation of workers.

Symbolic Interactionist Perspective

The symbolic interactionist perspective focuses on the social construction of race and ethnicity—how we learn conceptions and meanings of racial and ethnic distinctions through interaction with others—and how meanings, labels, and definitions affect racial and ethnic groups. We have already explained that contemporary race scholars agree that there is no scientific, biological basis for racial categorizations. However, people have learned to think of racial categories as real, and, as the *Thomas theorem* suggests, if things are defined as real, they are real in their consequences. Ossorio and Duster (2005) explain:

> People often interact with each other on the basis of their beliefs that race reflects physical, intellectual, moral, or spiritual superiority or inferiority. . . . By acting on their beliefs about race, people create a society in which individuals of one group have greater access to the goods of society—such as high-status jobs, good schooling, good housing, and good medical care—than do individuals of another group. (p. 119)

The labeling perspective directs us to consider the role that negative stereotypes play in race and ethnicity. **Stereotypes** are exaggerations or generalizations about the characteristics and behavior of a particular group. Negative stereotyping of minorities leads to a self-fulfilling prophecy—a process in which a false definition of a situation leads to behavior that, in turn, makes the originally falsely defined situation come true (see also Chapter 8). Marger (2012) provides the following example of how the negative stereotype of blacks as less intelligent than whites can lead to a self-fulfilling prophecy:

> If blacks are considered inherently less intelligent, fewer community resources will be used to support schools attended primarily by blacks on the assumption that such support would only be wasted. Poorer-quality schools, then, will inevitably turn out less capable students, who will score lower on intelligence tests. The poorer performance on these tests will "confirm" the original belief about black inferiority. Hence, the self-fulfilling prophecy. (p. 15)

Even stereotypes that appear to be positive can have negative effects. The view of Asian Americans as a "model minority" involves the stereotypes of Asian Americans as excelling in academics and occupational success. These stereotypes mask the struggles and discrimination that many Asian Americans experience and also put enormous pressure on Asian American youth to live up to the social expectation of being a high academic achiever (Tanneeru 2007).

The symbolic interactionist perspective is concerned with how individuals learn negative stereotypes and prejudicial attitudes through language. Different connotations of the colors white and black, for example, may contribute to negative attitudes toward people of color. The white knight is good, and the black knight is evil; angel food cake is white, and devil's food cake is black. Other negative terms associated with black include *black sheep, black plague, black magic, black mass, blackballed,* and *blacklisted.* The continued use of derogatory terms such as *Jap, gook, spic, frog, kraut, coon, chink, wop, towel-head, kike,* and *mick* also confirms the power of language in perpetuating negative attitudes toward minority group members. Similarly, advocates for immigrant rights suggest that the term *illegal aliens* is derogatory; they prefer the term *undocumented* or *unauthorized immigrant* as a more neutral term.

What Do You Think? Are racial and ethnic jokes insulting and harmful to minority group members? Or are such jokes innocent and playful forms of fun? Have you ever felt offended by hearing a racial or ethnic joke? How did you respond to the person telling the joke?

In the next section, we explore the concepts of prejudice and racism in more depth and discuss ways in which socialization and the media perpetuate negative stereotypes and prejudicial attitudes toward racial and ethnic groups.

stereotypes Exaggerations or generalizations about the characteristics and behavior of a particular group.

In a book titled *Two-Faced Racism*, sociologists Leslie Picca and Joe Feagin present research on white college students' encounters with racism in their everyday lives.

Sample and Methods
With the help of many college and university professors, Picca and Feagin (2007) recruited 934 college students to participate in research that required students to keep a diary or journal of "everyday events and conversations that deal with racial issues, images, and understandings" (p. 31). Students did not write their name on the journals, but did indicate their gender, racial group, and age. Instructors who recruited students for this research used the journal writing as either a class assignment or as extra credit. The majority of students (63 percent) were from colleges and universities in the South, 19 percent were from the Midwest, 14 percent were from the West, and 4 percent were from the Northeast. Most of the participating students were between 18 and 25 years of age, although many were in their late 20s and a few were older. Rates of participation varied from 0 percent to 80 percent (15 percent to 30 percent average). The results presented

here are based on roughly 9,000 journal entries written by 626 white students (68 percent women and 32 percent men).

Selected Findings
About three-quarters of the students' journal entries described racist events. In a few of these accounts, students described their own racist actions or thoughts.

> **Kristi:** When I went to pick up the laundry, I saw a young black man sitting in the driver's side of a minivan with the engine running. My first thought was that he was waiting for a friend to rob the store and he was the getaway driver. . . . I am so embarrassed and saddened by my thinking. . . . (p. 1)

More frequently, students described the racist comments or actions of friends, family members, acquaintances, and strangers. Students wrote about racist accounts targeted at African Americans, Latinos, Asian Americans, Native Americans, Jewish Americans, and Middle Easterners. The most frequently targeted minority group was black Americans. A few journal accounts described antiracist actions by whites.

Picca and Feagin used sociologist Erving Goffman's theatrical metaphors to describe and understand the racist behavior described in the journals. Goffman suggested that when people are "backstage" in private situations, they act differently than they do when they are in "frontstage" public situations. The expression of racist attitudes has gone backstage to private settings where whites are among other whites, especially family and friends because "many . . . whites realize it is no longer socially acceptable to be blatantly racist in frontstage areas" (p. xi). Consider the following diary entry:

> **Hannah:** Three of my friends (a white girl and two white boys) and I went back to my house to drink a little more before we ended the night. My one friend, Dylan, started telling jokes. . . . Dylan said: "What's the most confusing day of the year in Harlem?" "Father's Day . . . who's your daddy?" Dylan also referred to black people as "porch monkeys." Everyone laughed a little, but it was obvious that we all felt a little less comfortable when he was telling jokes like that. My friend Dylan is not a racist person. He has more black friends than I do, that's why I was

Prejudice and Racism

Prejudice refers to negative attitudes and feelings toward or about an entire category of people. Prejudice can be directed toward individuals of a particular religion, sexual orientation, political affiliation, age, social class, sex, race, or ethnicity. **Racism** is the belief that race accounts for differences in human character and ability and that a particular race is superior to others. In a national sample of U.S. adults, twice as many blacks as whites said that racism is a "big problem" in our society today (see Figure 9.6) (*Washington Post* 2009). This chapter's Social Problems Research Up Close feature suggests that racism in the United States is more common than many people realize.

Forms of Racism

Compared with traditional, "old-fashioned" prejudice, which is blatant, direct, and conscious, contemporary forms of prejudice are often subtle, indirect, and unconscious. Three variants of these more subtle forms of prejudice are aversive racism, modern racism, and "Racism 2.0."

Aversive Racism **Aversive racism** represents a subtle, often unintentional, form of prejudice exhibited by many well-intentioned white Americans who possess strong egalitarian values and who view themselves as unprejudiced. The negative feelings that aversive racists have toward blacks and other minority groups are not feelings of hostility or hate but rather feelings of discomfort, uneasiness, disgust,

prejudice Negative attitudes and feelings toward or about an entire category of people.

racism The belief that race accounts for differences in human character and ability and that a particular race is superior to others.

aversive racism A subtle form of prejudice that involves feelings of discomfort, uneasiness, disgust, fear, and pro-white attitudes.

surprised he so freely said something like that. Dylan would never have said something like that around anyone who was a minority. . . . It is this sort of "joking" that helps to keep racism alive today. People know the places they have to be politically correct and most people will be. However, until this sort of "behind-the-scenes" racism comes to an end, people will always harbor those stereotypical views that are so prevalent in our country. This kind of joking really does bother me, but I don't know what to do about it. I know that I should probably stand up and say I feel uncomfortable when my friends tell jokes like that, but I know my friends would just get annoyed with me and say that they obviously don't mean anything by it. (pp. 17–18)

Hannah's journal entry illustrates the social norm that defines expressions of racism as acceptable in private "backstage" settings, and inappropriate in public "frontstage" settings. Hannah's journal entry also exemplifies a theme of "white innocence," whereby white students offer excuses for and minimize the significance of racist events. Whites often view racist comments not as "real racism," but as harmless remarks that are not to be taken seriously.

Although white women also display racist behavior, the journal entries in this research suggest that "white men disproportionately make racist jokes and similar racially barbed comments" (p. 133):

Carissa: The white men got on the subject of so-called nicknames for black people. Some mentioned were porch monkeys, jig-aboos, tree swingers, etc. The one thing I took notice of was that not one girl made a comment. (p. 133)

Some students made comments in their journals or on the cover sheet that point to a general lack of awareness of racism. One student, for example, wrote, "This assignment made me realize how many racial remarks are said every single day and I usually never catch any of them or pay close attention" (p. 39).

Many students were offended by the racist remarks they recorded in their journals, but did not express their disapproval and, so, participated in racism as a "bystander." Some students indicated that, after participating in the journal writing assignment, they are more aware of the need to intervene in social situations where racism is being expressed:

Kyle: As my last entry in this journal, I would like to express what I have gained

out of this assignment. I watched my friends and companions with open eyes. I was seeing things that I didn't realize were actually there. By having a reason to pick out the racial comments and actions, I was made aware of what is really out there. Although I noticed that I wasn't partaking in any of the racist actions or comments, I did notice that I wasn't stopping them either. I am now in a position to where I can take a stand and try to intervene in many of the situations. (p. 275)

Conclusion

Based on the research presented in *Two-Faced Racism*, Picca and Feagin suggest that "the majority of whites still participate in openly racist performances in the backstage arena . . . and do not define such performances as problematic and deserving of action aimed at eradication" (p. 22). The researchers further note that "most of our college student diarists did not, or could not, see the connection between the everyday racial performances they recorded and the unjust discrimination and widespread suffering endured by people of color in society generally" (p. 28).

Source: Picca & Feagin 2007.

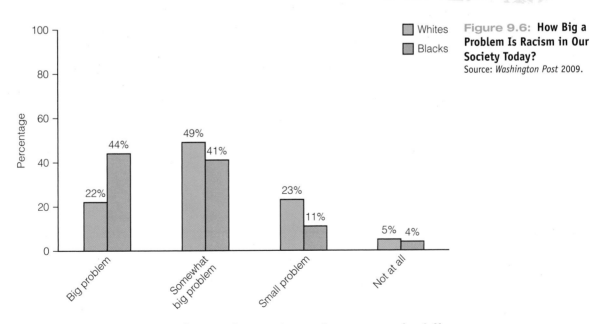

Figure 9.6: **How Big a Problem Is Racism in Our Society Today?**
Source: *Washington Post* 2009.

and sometimes fear (Gaertner & Dovidio 2000). Aversive racists may not be fully aware that they harbor these negative racial feelings; indeed, they disapprove of individuals who are prejudiced and would feel falsely accused if they were labeled prejudiced. "Aversive racists find blacks 'aversive,' while at the same time find any suggestion that they might be prejudiced 'aversive' as well" (Gaertner & Dovidio 2000, p. 14).

Another aspect of aversive racism is the presence of pro-white attitudes, as opposed to anti-black attitudes. In several studies, respondents did not indicate that blacks were worse than whites, only that whites were better than blacks (Gaertner & Dovidio 2000). For example, blacks were not rated as being lazier than whites, but whites were rated as being more ambitious than blacks. Gaertner and Dovidio (2000) explain that "aversive racists would not characterize blacks more negatively than whites because that response could readily be interpreted by others or oneself to reflect racial prejudice" (p. 27). Compared with anti-black attitudes, pro-white attitudes reflect a more subtle prejudice that, although less overtly negative, is still racial bias.

Modern Racism Like aversive racism, **modern racism** involves the rejection of traditional racist beliefs, but a modern racist displaces negative racial feelings onto more abstract social and political issues. The modern racist believes that serious discrimination in the United States no longer exists, that any continuing racial inequality is the fault of minority group members, and that demands for affirmative action for minorities are unfair and unjustified. "Modern racism tends to 'blame the victim' and places the responsibility for change and improvements on the minority groups, not on the larger society" (Healey 1997, p. 55). Like aversive racists, modern racists tend to be unaware of their negative racial feelings and do not view themselves as prejudiced.

Racism 2.0 The election of Barack Obama has unveiled a new form of racism: Racism 2.0. According to Tim Wise (2009), **Racism 2.0** is a type of racism that "allows for and even celebrates the achievements of individuals of color, but only because those individuals generally are seen as different from a less appealing, even pathological black or brown rule" (p. 9). The fact that Barack Obama's supporters and admirers include a huge segment of the white population does not necessarily imply that these white Obama supporters are nonracist:

> If whites come to like, respect, and even vote for persons of color like Barack Obama, but only because they view them as having "transcended" their blackness in some way, to claim that the success of such candidates proves the demise of racism makes no sense at all. (Wise 2009, p. 9)

What Do You Think? A national poll (*Washington Post* 2009) asked a sample of U.S. adults: "If you honestly assessed yourself, would you say that you have at least some feelings of racial prejudice?" How would you answer that question? Would it surprise you to learn that 34 percent of whites and 38 percent of blacks answered "Yes"?

modern racism A subtle form of racism that involves the belief that serious discrimination in America no longer exists, that any continuing racial inequality is the fault of minority group members, and that the demands for affirmative action for minorities are unfair and unjustified.

Racism 2.0 A form of racism that allows for and celebrates the achievements of individuals of color who are viewed as having "transcended" their minority status.

Learning to Be Prejudiced: The Role of Socialization and the Media

Psychological theories of prejudice focus on forces within the individual that give rise to prejudice. For example, the frustration-aggression theory of prejudice (also known as the scapegoating theory) suggests that prejudice is a form of hostility that results from frustration. According to this theory, minority groups serve as convenient targets of displaced aggression. The authoritarian-personality theory of prejudice suggests that prejudice arises in people with a certain personality type. According to this theory, people with an authoritarian personality—who are highly conformist, intolerant, cynical, and preoccupied with power—are prone to being prejudiced.

Rather than focus on the individual, sociologists focus on social forces that contribute to prejudice. Earlier we explained how intergroup conflict over wealth, power, and prestige gives rise to negative feelings and attitudes that serve to protect and enhance dominant group interests. Prejudice is also learned through socialization and the media.

Learning Prejudice through Socialization

Although most researchers agree that the majority of children learn conceptions of racial and ethnic distinctions by the time they are about 6 years old, Van Ausdale and Feagin (2001) suggest that children as young as 3 years old have acquired prejudicial attitudes:

> Well before they can speak clearly, children are exposed to racial and ethnic ideas through their immersion in and observation of the large social world. Since racism exists at all levels of society and is interwoven in all aspects of American social life, it is virtually impossible for alert young children either to miss or ignore it. . . . Children are inundated with it from the moment they enter society. (pp. 189–190)

Byron Calvert operates Panzerfaust Records, one of the nation's largest "white power" music labels.

In the socialization process, individuals adopt the values, beliefs, and perceptions of their family, peers, culture, and social groups. Prejudice is taught and learned through socialization, although it need not be taught directly and intentionally. Parents who teach their children to not be prejudiced yet live in an all-white neighborhood, attend an all-white church, and have only white friends may be indirectly teaching negative racial attitudes to their children. Socialization can also be direct, as in the case of parents who use racial slurs in the presence of their children or who forbid their children from playing with children from a certain racial or ethnic background. Children can also learn prejudicial attitudes from their peers. The telling of racial and ethnic jokes among friends, for example, perpetuates stereotypes that foster negative racial and ethnic attitudes.

What Do You Think? In Canada, a 7-year-old girl went to school with a swastika drawn on her arm. Alerted by the school, Child and Family Services investigated the child's home and, after finding neo-Nazi symbols and flags, they removed the girl and her 2-year-old brother from the home and placed them in temporary custody with their aunt. In a custody hearing, a social worker that interviewed the girl said that the girl spoke about how people of other races should be dead because this is a white man's world, and she provided graphic descriptions of how to kill people. The girl told the social worker that she watched skinhead videos with her parents and that her parents belonged to a skinhead website (CBCNews. ca 2009). This case has drawn international attention and sparked debate over whether the government has the right to protect children from their parents' racist views. What do you think?

Prejudice and the Media

Media—including television, radio, and the Internet—plays a role in perpetuating prejudice and hate. Television often portrays minorities and immigrants in negative and stereotypical ways. An analysis of character portrayals in the 2003–2004 prime-time television season revealed that Latinos were more likely than any other group to be cast in low-status occupations, such as domestic worker, and as criminals. Nearly half (46 percent) of Arabs and Middle Easterners were cast as criminals (Glaubke & Heintz-Krowles 2004).

Another form of media that is used to promote hatred toward minorities and to recruit young people to the white power movement is "white power music," which contains anti-Semitic, racist, and homophobic lyrics. A CD that was distributed to thousands of middle and high school students by the neo-Nazi record label Panzerfaust Records contains the following lyrics from a group called the Bully Boys: "Whiskey bottles/baseball bats/pickup trucks/and rebel flags/we're going on the town tonight/hit and run/let's have some fun/we've got jigaboos on the run." And in a song called "Wrecking Ball," the band H8Machine advises kids to "destroy all your enemies," promising that "the best things come to those who hate" (SPLC 2004). Hate groups also use social networking sites such as Facebook, MySpace, and YouTube to spread their message of hate and to recruit new members (*Turn It Down* 2009).

Discrimination against Racial and Ethnic Minorities

Whereas prejudice refers to attitudes, **discrimination** refers to actions or practices that result in differential treatment of categories of individuals. Although prejudicial attitudes often accompany discriminatory behavior or practices, one can be evident without the other.

Individual versus Institutional Discrimination

Individual discrimination occurs when individuals treat other individuals unfairly or unequally because of their group membership. Individual discrimination can be overt or adaptive. In **overt discrimination,** individuals discriminate because of their own prejudicial attitudes. For example, a white landlord may refuse to rent to a Mexican American family because of a prejudice against Mexican Americans. Or a Taiwanese American college student who shares a dorm room with an African American student may request a roommate reassignment from the student housing office because of prejudice against blacks.

Suppose that a Cuban American family wants to rent an apartment in a predominantly non-Hispanic neighborhood. If the landlord is prejudiced against Cubans and does not allow the family to rent the apartment, that landlord has engaged in overt discrimination. But what if the landlord is not prejudiced against Cubans but still refuses to rent to a Cuban family? Perhaps that landlord is engaging in **adaptive discrimination,** or discrimination that is based on the prejudice of others. In this example, the landlord may fear that other renters who are prejudiced against Cubans may move out of the building or neighborhood and leave the landlord with unrented apartments. Overt and adaptive individual discrimination can coexist.

Institutional discrimination refers to institutional policies and procedures that result in unequal treatment of and opportunities for minorities. Institutional discrimination is covert and insidious and maintains the subordinate position of minorities in society. For example, when schools use standard intelligence tests to decide which children will be placed in college preparatory tracks, they are limiting the educational advancement of minorities whose intelligence is not fairly measured by culturally biased tests developed from white middle-class experiences. And the funding of public schools through local tax dollars results in less funding for schools in poor and largely minority school districts.

Institutional discrimination is also found in the criminal justice system, which more heavily penalizes crimes that minorities are more likely to commit. Until recently, the penalties for crack cocaine, more often used by minorities, were higher than those for other forms of cocaine use, even though the same prohibited chemical substance is involved. As conflict theorists emphasize, majority group members make rules that favor their own group.

Employment Discrimination

When a national sample of U.S. adults in 2008 was asked, "Do you feel that racial minorities in this country have equal job opportunities as whites?" nearly half (46 percent) said "No" (Gallup Organization 2011a). Despite laws against it, discrimination against

discrimination Actions or practices that result in differential treatment of categories of individuals.

individual discrimination The unfair or unequal treatment of individuals because of their group membership.

overt discrimination Discrimination that occurs because of an individual's own prejudicial attitudes.

adaptive discrimination Discrimination that is based on the prejudice of others.

institutional discrimination Discrimination in which institutional policies and procedures result in unequal treatment of and opportunities for minorities.

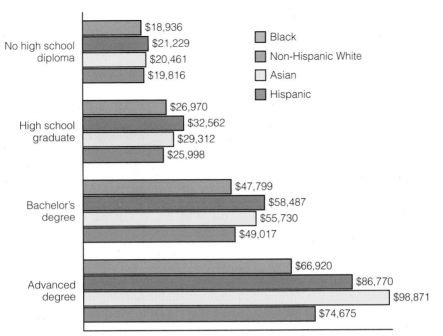

Figure 9.7: **Mean Earnings of Workers 18 Years and Over, by Educational Attainment, Race, and Hispanic Origin, 2009**
Source: U.S. Census Bureau 2011b.

minorities occurs today in all phases of the employment process, from recruitment to interview, job offer, salary, promotion, and firing decisions.

A sociologist at Northwestern University studied employers' treatment of job applicants in Milwaukee, Wisconsin, by dividing job applicant "testers" into four groups: blacks with a criminal record, blacks without a criminal record, whites with a criminal record, and whites without a criminal record (Pager 2003). Applicant testers, none of whom actually had a criminal record, were trained to behave similarly in the application process and were sent with comparable résumés to the same set of employers. The study found that white applicants with no criminal record were the most likely to be called back for an interview (34 percent) and that black applicants with a criminal record were the least likely to be called back (5 percent). But surprisingly, white applicants with a criminal record (17 percent) were more likely to be called back for a job interview than were black applicants *without* a criminal record (14 percent). The researcher concluded that "the powerful effects of race thus continue to direct employment decisions in ways that contribute to persisting racial inequality" (Pager 2003, p. 960).

Discrimination in hiring may be unintended. For example, many businesses rely on their existing employees to refer new recruits when a position opens up. Word-of-mouth recruitment is inexpensive and efficient; some companies offer bonuses to employees who bring in new recruits. But this traditional recruitment practice tends to exclude minority workers, because they often do not have a network of friends and family members in higher positions of employment who can recruit them (Schiller 2004).

Employment discrimination contributes to the higher rates of unemployment and lower incomes of blacks and Hispanics compared with those of whites (see Chapters 6 and 7). Lower levels of educational attainment among minority groups account for some, but not all, of the disadvantages they experience in employment and income. As shown in Figure 9.7, mean earnings of whites are higher than those for blacks, Hispanics, and Asians at most levels of educational attainment.

Workplace discrimination also includes unfair treatment and harassment. African American employees working for A. C. Widenhouse—a North Carolina–based freight trucking company—were repeatedly subjected to derogatory racial comments and slurs by employees and managers. These insulting comments and slurs included "n----r," "monkey," and "boy" (EEOC 2011). One employee was approached by a co-worker with a noose who said, "This is for you. Do you want to hang from the family tree?" A company manager allegedly told an employee, "We are going coon hunting, are you going to be the coon?"

Housing Discrimination and Segregation

Before the 1968 Fair Housing Act and the 1974 Equal Credit Opportunity Act, discrimination against minorities in housing and mortgage lending was as rampant as it was blatant. Banks and mortgage companies commonly engaged in "redlining"—the practice of denying mortgage loans in minority neighborhoods on the premise that the financial risk was too great, and the ethical standards of the National Association of Real Estate Boards prohibited its members from introducing minorities into white neighborhoods. Instead, realtors practiced "geographic steering," whereby they discouraged minorities from moving into certain areas by showing them homes only in minority neighborhoods.

Although housing discrimination is illegal today, it is not uncommon. To assess discrimination in housing, researchers use a method called "paired testing." In a paired test, two individuals—one minority and the other nonminority—are trained to pose as home seekers, and they interact with real estate agents, landlords, rental agents, and mortgage lenders to see how they are treated. The testers are assigned comparable or identical income, assets, and debt as well as comparable or identical housing preferences, family circumstances, education, and job characteristics. A paired testing study of housing discrimination in 23 metropolitan areas found that whites in the rental market were more likely to receive information about available housing units and had more opportunities to inspect available units than did blacks and Hispanics (Turner et al. 2002). The incidence of discrimination was greater for Hispanic renters than for black renters. The same study found that, in the home sales market, white home buyers were more likely to be able to inspect available homes and to be shown homes in more predominantly non-Hispanic white neighborhoods than were comparable black and Hispanic buyers. Whites were also more likely to receive information and assistance with financing.

In a study of housing discrimination in the Philadelphia area, Massey and Lundy (2001) found that, compared with whites, African Americans were less likely to have a rental agent return their calls, less likely to be told that a unit was available, more likely to pay application fees, and more likely to have credit mentioned as a potential problem in qualifying for a lease. Sex and class exacerbated these racial effects. Lower-class blacks experienced less access to rental housing than middle-class blacks, and black females experienced less access than black males. Lower-class black females were the most disadvantaged group. They experienced the lowest probability of contacting and speaking to a rental agent and, even if they did make contact, they faced the lowest probability of being told of a housing unit's availability. Lower-class black females also faced the highest chance of paying an application fee. On average, lower-class black females were assessed $32 more per application than white middle-class males.

Despite continued housing discrimination, homeownership rates among minorities and low-income groups increased substantially in the 1990s, reaching record rates in many central cities. However, minority and low-income homeowner rates still lag behind the overall homeownership rate. Also, many of the gains in minority and low-income homeownership rates are due to increases in *subprime lending*—loans with higher fees and higher interest rates that are offered to borrowers who have poor (or nonexistent) credit records (Williams et al. 2005).

Residential segregation of racial and ethnic groups also persists. Almost a quarter of all census tracts within the largest U.S. metropolitan areas are more than 90 percent white, and 12 percent are more than 90 percent minority (Turner & Fortuny 2009).

Educational Discrimination and Segregation

Both institutional discrimination and individual discrimination in education negatively affect racial and ethnic minorities and help to explain why minorities (with the exception of Asian Americans) tend to achieve lower levels of academic attainment and success (see also Chapter 8). Institutional discrimination is evidenced by inequalities in school funding—a practice that disproportionately hurts minority students (Kozol 1991). Nearly half of school funding comes from local taxes. In 2002, the federal government supplied 8 percent of educational expenditures, and state government contributed 48 percent;

local governments provided the remainder (Schiller 2004). Because minorities are more likely than whites to live in economically disadvantaged areas, they are more likely to go to schools that receive inadequate funding. Inner-city schools, which serve primarily minority students, receive less funding per student than do schools in more affluent, primarily white areas.

Another institutional education policy that is advantageous to whites is the policy that gives preference to college applicants whose parents or grandparents are alumni. The overwhelming majority of alumni at the highest-ranked universities and colleges are white. Thus, white college applicants are the primary beneficiaries of these so-called legacy admissions policies. About 10 percent to 15 percent of students in most Ivy League colleges and universities are children of alumni. Harvard University accepts about 11 percent of its overall applicant pool, but the admission rate is 40 percent for legacy applicants (Schmidt 2004). As a result of pressure from state lawmakers and minority rights activists, in 2004, Texas A&M University became the first public college to abandon its legacy admittance policy.

Minorities also experience individual discrimination in the schools as a result of continuing prejudice among teachers. In a survey conducted by the Southern Poverty Law Center, 1,100 educators were asked whether they had heard racist comments from their colleagues in the past year. More than one-fourth of survey respondents answered yes ("Hear and Now" 2000). It is likely that teachers who are prejudiced against minorities discriminate against them, giving them less teaching attention and less encouragement.

Racial and ethnic minorities are also treated unfairly in educational materials, such as textbooks, which often distort the history and heritages of people of color (King 2000). For example, Zinn (1993) observed, "To emphasize the heroism of Columbus and his successors as navigators and discoverers, and to deemphasize their genocide, is not a technical necessity but an ideological choice. It serves, unwittingly—to justify what was done" (p. 355).

Finally, racial and ethnic minorities are largely isolated from whites in a largely segregated school system. U.S. schools in the 2000–2001 school year were more segregated than they were in 1970 (Orfield 2001). School segregation is largely due to the persistence of housing segregation and the termination of court-ordered desegregation plans. Court-mandated busing became a means to achieve equality of education and school integration in the early 1970s, after the Supreme Court (in *Swann v. Charlotte-Mecklenburg*) endorsed busing to desegregate schools. But in the 1990s, lower courts lifted desegregation orders in dozens of school districts (Winter 2003a). And in 2007, the United States Supreme Court issued a landmark ruling, in a bitterly divided 5-to-4 vote, that race cannot be a factor in the assignment of children to public schools. The decision jeopardizes similar plans in hundreds of districts nationwide, and it further restricts how public school systems may achieve racial diversity. Recent data suggest that racial and ethnic segregation in U.S. suburban schools has declined slightly, as the percentage of minority students in suburban school districts has increased from 28 percent in the 1993–94 school year to 41 percent in the 2006–07 school year (Fry 2009).

Hate Crimes

In June 2011, 49-year-old James Craig Anderson, an African American auto plant worker, was robbed, beaten, and run over by a truck near a hotel in Jackson, Mississippi. Deryl Dedmon, the 18-year-old who drove the truck that ran over Anderson, was charged with murder. Dedmon and a group of other white teenagers were reportedly out looking for a black person to assault when they found Anderson. During the brutal attack, which was captured on hotel security video, racial slurs were used, and later Dedmon bragged that he "just ran that n----- over" (Associated Press 2011).

As this book goes to press, the FBI is investigating the murder of James Craig Anderson to determine if it is a **hate crime**—an unlawful act of violence motivated by prejudice or bias. Examples of hate crimes, also known as "bias-motivated crimes," include intimidation (e.g., threats), destruction of or damage to property, physical assault, and murder.

hate crime An unlawful act of violence motivated by prejudice or bias.

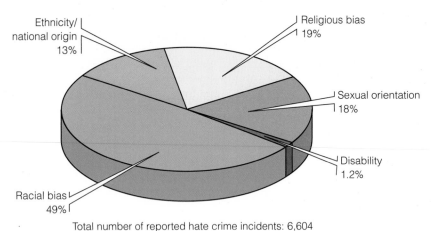

Ethnicity/
national origin
13%

Religious bias
19%

Sexual orientation
18%

Disability
1.2%

Racial bias
49%

Total number of reported hate crime incidents: 6,604

Figure 9.8: Hate Crime Incidence by Category of Bias, 2009 (rounded to nearest percent)
Source: FBI 2010.

The Federal Bureau of Investigation (FBI) reported that, in 2009, there were 6,604 hate crime incidents in the United States (FBI 2010). However, FBI hate crime data undercount the actual number of hate crimes because (1) not all U.S. jurisdictions report hate crimes to the FBI (reporting is voluntary); (2) it is difficult to prove that crimes are motivated by hate or prejudice; (3) law enforcement agencies shy away from classifying crimes as hate crimes because it makes their community "look bad"; and (4) victims are often reluctant to report hate crimes to the authorities.

From the first year that FBI hate crime data were published in 1992, most hate crimes have been based on racial bias (see Figure 9.8). In 2009, most (72 percent) of the reported race-based hate crimes were directed against blacks, followed by whites (17 percent). "From lynching, to burning crosses and churches, to murdering a man by chaining him to a truck and dragging him down a road for three miles, anti-black violence has been and still remains the prototypical hate crime, intended not only to injure and kill individuals but to terrorize an entire group of people" (Leadership Council on Civil Rights Education Fund 2009, p. 25).

In 2009, most hate crimes motivated by religious bias (72 percent) were based on anti-Jewish bias, and the majority of hate crimes based on ethnic bias (62 percent) were anti-Hispanic (FBI 2010). Hate crimes against Hispanic individuals are largely motivated by hate toward immigrants. This chapter's *The Human Side* feature describes the hate crime victimization of a Hispanic immigrant in Georgia.

After the terrorist attacks of September 11, 2001, hate crimes against individuals perceived to be Muslim or Middle Eastern increased significantly. More than half of America's 7 million Muslims (individuals who adhere to the religion of Islam) said they experienced bias or discrimination after September 11 (Morrison 2002). Anti-Muslim and

This public demonstration in New York was staged to show support for the construction of an Islamic cultural center near Ground Zero.

Domingo Lopez Vargas left his dirt-poor Guatemalan farm village to come to the United States, where he hoped to earn decent money for his wife and nine children. After picking oranges in Florida, he moved to Georgia, where the booming construction business lured immigrant workers. Unlike many of his compadres, Vargas had legal status, which helped him find steady work hanging doors and windows. When work dried up, Vargas joined the more than 100,000 jornaleros—day laborers—who wait for landscaping and construction jobs on street corners and in front of convenience stores all across Georgia. Usually there are plenty of pickup trucks that swing by, offering $8 to $12 an hour for digging, planting, painting, or hammering. But, this day, nada. By late afternoon, Vargas had tired of waiting in the cold, so he walked up the street to pick up a few things at a grocery store.

"I got milk, shampoo, and toothpaste," Vargas recalled. "When I was leaving the store, this truck stopped right in front of me and said, 'Do you want to work?'. . . I said, yes, how much? They said nine dollars an hour. I didn't ask what kind of job. I just wanted to work, so I said yes."

Until that afternoon, Vargas said, "Americans had always been very nice to me"—which might explain why he wasn't concerned that the four guys in the pickup truck looked awfully young to be contractors. Or why he didn't think twice about

Domingo Lopez Vargas, an immigrant day laborer, was brutally beaten in a hate crime in Canton, Georgia.

being picked up so close to sunset. "I took the offer because I know sometimes people don't stop working until 9 at night," he says.

The four young men, all high school students, drove Vargas to a remote spot strewn with trash. "They told me to pick up some plastic bags that were on the ground. I thought that was my job, to clean up the trash. But when I bent over to pick it up,

I felt somebody hit me from behind with a piece of wood, on my back." It was just the start of a 30-minute pummeling that left Vargas bruised and bloody from his thighs to his neck. "I thought I was dying," he said. "I tried to stand up but I couldn't." Finally, after he handed over all the cash in his wallet, $260, along with his Virgin Mary pendant, the teenagers sped away.

As a result of injuries incurred by the beating, Vargas could not work for four months, and he was left with $4,500 in medical bills. Sometimes he still puzzles over his attackers' motives. "They were young," he speculates, "and maybe they didn't have enough education. Or maybe their families . . . taught them to kill people, and that is what they have learned."

Having sworn off day labor, Vargas works night shifts now, cutting up chickens at the nearby Tyson plant, wincing through the pain that shoots up his right arm when he lowers the boom on a bird. But it's only temporary, he says. "I called my wife and told her what happened. She told me to move back to Guatemala. I wanted to, but I didn't have enough money to go back and the police officers told me not to move out of the country because they will still need me to work on the case. After the case is finished, I want to go back to my family."

Source: Moser, Bob, "The Battle of 'Georgiafornia,'" *Southern Poverty Law Center's Intelligence Report* 116:40–50. Copyright © 2004. Reprinted by permission of Southern Poverty Law Center.

anti-Islam bias—commonly known as **Islamophobia**—was rekindled after the proposed building of an Islamic cultural center (Park51) in New York City, near Ground Zero. Islamophobia is largely based on ignorance and misunderstanding of Islam. Georgetown professor John L. Esposito (2011) explains,

> Mainstream American Muslims have too often been equated inaccurately with terrorists and people who reject democracy. Muslim Americans cherish the freedoms guaranteed by the American Constitution as much as others and . . . majorities of Muslims globally desire democracy and freedom and fear and reject religious extremism and terrorism.

Islamophobia is largely based on ignorance and misunderstanding of Islam.

Motivations for Hate Crimes Levin and McDevitt (1995) found that the motivations for hate crimes were of three distinct types: thrill, defensive, and mission. Thrill hate crimes are committed by offenders who are looking for excitement and attack victims for the "fun of it." Defensive hate crimes involve offenders who view their attacks as necessary to protect their community, workplace, or college campus from "outsiders" or to protect their racial and cultural purity from being "contaminated" by interracial

Islamophobia Anti-Muslim and anti-Islam bias.

TABLE 9.5 Hate Groups in the United States, 2010	
HATE GROUP	NUMBER OF CHAPTERS
Ku Klux Klan	221
Neo-Nazi	170
Black separatist	149
White nationalist	136
Racist skinhead	136
Neo-Confederate	42
Christian identity	26
General hate (including anti-gay, anti-immigrant, anti-Muslim, holocaust denial, racist music, and other)	122
Total	1,002

marriage and childbearing. Mission hate crimes are perpetrated by white supremacist group members or other offenders who have dedicated their lives to bigotry. Hate groups known to engage in violent crimes include the Ku Klux Klan, the Identity Church Movement, the neo-Nazis, and the skinheads. The Southern Poverty Law Center identified 1,002 hate group chapters in 2010—a significant increase since 2000 when there were 602 hate group chapters in the United States (see Table 9.5) (Potok 2011). Although some members of hate groups have visible features, such as tattoos and armbands, that identify their hate group membership, many hate group members are not so easily identifiable. Criminologist Jack Levin, who studies hate crimes, noted that "many white supremacist groups are going more mainstream . . . they are eliminating the sheets and armbands. . . . The groups realize if they want to be attractive to middle-class types, they need to look middle class" (quoted in Leadership Council on Civil Rights Education Fund 2009, p. 19). Some anti-immigrant hate groups, such as Federation for American Immigration Reform (FAIR), the Center for Immigration Studies (CIS), and Numbers USA portray themselves as legitimate mainstream advocates against illegal immigration, but "some of these organizations have disturbing links to or relationships with extremists in the anti-immigrant movement" (p. 14).

Hate on Campus Karl Nichols, a white residence hall director at the University of Mississippi, learned about racism in college, but not in the classroom and not from a textbook. Two chunks of asphalt were hurled through Nichols's dormitory room window along with a note warning, "You're going to get it, you Godforsaken nigger-lover" (SPLC 2000, p. 10). The next night, someone attempted to set Nichols's door on fire. According to the university's investigation of these incidents, Nichols "may have violated racist taboos . . . by openly displaying his affinity for African American individuals and black culture, by dating black women, by playing black music . . . and by promoting diversity" in his dormitory (p. 10).

Karl Nichols is not alone. At Brown University in Rhode Island, a black student was beaten by three white students who told her she was a "quota" who did not belong at a university. At the State University of New York in Binghamton, an Asian American student was left with a fractured skull after a racially motivated assault by three students. Two students at the University of Kentucky—one white, one black—were crossing the street just off campus when 10 white men attacked them. The attackers yelled racist slurs at the black student and choked him until he could not speak or move. The assailants called the white student a "nigger lover" as they broke his hand and nose. "I definitely thought I was going to lose my life," the black student said later. The white student was shocked by the incident, commenting, "I didn't know that much hate existed" (SPLC 2000, p. 7).

According to the FBI, more than 1 in 10 hate crimes occur at schools or colleges (FBI 2010). A survey of 2,612 undergraduates on 10 college campuses found that more than half personally experienced or witnessed bias incidents—verbal insults, threats, assaults, or graffiti—targeting individuals based on their group identity (Campus Tolerance Foundation 2009).

Hate Group Members in the Military In December 1995, two members of the 82nd Airborne at Fort Bragg, North Carolina, who belonged to a white supremacist skinhead gang, shot and killed a black couple in a random, racially motivated double murder that shocked the nation. These hate crime murders led to congressional hearings and a major investigation of extremism in the military. The killers were sentenced to life in prison, and 19 other members of the 82nd Airborne were dishonorably discharged for neo-Nazi gang activities.

The Fort Bragg murders were not the first instance of military personnel involvement in hate groups. In the late 1980s and early 1990s, a number of cases came to light in which extremists in the military were caught diverting stolen firearms and explosives to neo-Nazi and white supremacist organizations, conducting guerrilla training for paramilitary racist militias, and murdering nonwhite civilians. According to investigations by the Southern Poverty Law Center, the military's tough stance against hate group affiliations among military personnel has relaxed since the recent war in Iraq and Afghanistan and the pressure to maintain enlistment numbers. An FBI report titled "White Supremacist Recruitment of Military Personnel since 9/11" revealed that racist extremists were taking advantage of lowered wartime recruiting standards to enlist in the military (FBI 2008). Department of Defense investigator Scott Barfield said,

> Recruiters are knowingly allowing neo-Nazis and white supremacists to join the armed forces, and commanders don't remove them from the military even after we positively identify them as extremists or gang members. . . . Last year, for the first time, they didn't make their recruiting goals. They don't want to start making a big deal again about neo-Nazis in the military, because then parents who are already worried about their kids signing up and dying in Iraq are going to be even more reluctant about their kids enlisting if they feel they'll be exposed to gangs and white supremacists. (quoted in Holthouse 2006)

In one year, Barfield, who is based at Fort Lewis, identified and submitted evidence on 320 extremists there in the past year, but only two were discharged.

Strategies for Action: Responding to Prejudice, Racism, and Discrimination

Because racial and ethnic tensions exist worldwide, strategies for combating prejudice, racism, and discrimination globally require international cooperation and commitment. The World Conference against Racism, Racial Discrimination, Xenophobia, and Related Intolerance, held in Durban, South Africa, in 2001, exemplifies international efforts to reduce racial and ethnic tensions and inequalities and to increase harmony among the various racial and ethnic populations of the world. Unfortunately, the U.S. delegation to this conference withdrew because of the expectation that hateful language would be used against Israel (because of the Israeli-Palestinian conflict).

In the following sections, we discuss the Equal Employment Opportunity Commission's role in responding to employment discrimination and examine the issue of affirmative action in the United States. We also discuss educational strategies to promote diversity and multicultural awareness and appreciation in schools. Finally, we look at apologies and reparations as a means of achieving racial reconciliation.

The Equal Employment Opportunity Commission

The **Equal Employment Opportunity Commission (EEOC),** a U.S. federal agency charged with ending employment discrimination in the United States, is responsible for enforcing laws against discrimination, including Title VII of the 1964 Civil Rights Act that prohibits employment discrimination on the basis of race, color, religion, sex, or national origin. The EEOC investigates, mediates, and may file lawsuits against private employers on behalf of alleged victims of discrimination. For example, after receiving and investigating complaints from across the country that Walgreens discriminated against African American retail management and pharmacy employees in promotion, compensation, and assignment, the EEOC filed a discrimination suit against Walgreens that was settled for $20 million in monetary relief for an estimated 10,000 class members (EEOC 2007). The most frequently filed claims with the EEOC are allegations of race discrimination, racial harassment, or retaliation from opposition to racial discrimination.

In 2007, the EEOC launched a national initiative to combat racial discrimination in the workplace. The goals of this initiative, called E-RACE (Eradicating Racism And

Equal Employment Opportunity Commission (EEOC) A U.S. federal agency charged with ending employment discrimination in the United States that is responsible for enforcing laws against discrimination, including Title VII of the 1964 Civil Rights Act that prohibits employment discrimination on the basis of race, color, religion, sex, or national origin.

Colorism from Employment), are to (1) identify factors that contribute to race and color discrimination, (2) explore strategies to improve the administrative processing and litigation of race and color discrimination cases, and (3) increase public awareness of race and color discrimination in employment.

Affirmative Action

Affirmative action refers to a broad range of policies and practices in the workplace and educational institutions to promote equal opportunity as well as diversity. Affirmative action is an attempt to compensate for the effects of past discrimination and prevent current discrimination against women and racial and ethnic minorities. Vietnam veterans and people with disabilities may also qualify under affirmative action policies.

Federal Affirmative Action Affirmative action policies developed in the 1960s, from federal legislation that required any employer (universities as well as businesses) who received contracts from the federal government to make "good faith efforts" to increase the pool of qualified minorities and women (U.S. Department of Labor 2002). Such efforts can be made by expanding recruitment and training programs. Hiring decisions are to be made on a nondiscriminatory basis.

Affirmative Action in Higher Education The Supreme Court's 1974 ruling in *Regents of the University of California v. Bakke* marked the beginning of the decline of affirmative action. Allan Bakke, a white male, had applied to the University of California at Davis medical school and was rejected, even though his grade point average and score on the medical school admissions test were higher than those of several minority applicants who had been admitted. The medical school had established fixed racial quotas, guaranteeing admission to 16 minority applicants regardless of their qualifications. Bakke claimed that such quotas discriminated against him as a white male and that the University of California had violated his Fourteenth Amendment right to equal protection under the law. The Supreme Court ruled in Bakke's favor by a 5-to-4 vote (showing how split the court was), concluding that the University of California unwittingly engaged in "reverse discrimination," which was unconstitutional. Affirmative action programs, the court ruled, could not use fixed quotas in admission, hiring, or promotion policies. However, the court affirmed the right for universities and employers to consider race as a factor in admission, hiring, and promotion to achieve diversity.

Since the *Bakke* case, numerous legal battles have challenged affirmative action (Olson 2003). In *Hopwood v. Texas,* Cheryl Hopwood and three other white individuals who had been denied admission to the University of Texas School of Law claimed that their Fourteenth Amendment rights to equal protection under the law had been violated by the university's affirmative action admission policies. In 1997, the Fifth Circuit Court of Appeals decided that the University of Texas could no longer use race as a factor in awarding financial aid, admitting students, and hiring and promoting faculty. Higher courts refused to review the decision. But in 2003, the U.S. Supreme Court, after hearing the appeals from two white applicants who applied but were not accepted to the University of Michigan, affirmed the right of colleges to consider race in admissions, but the court rejected Michigan's use of a point system to do so. According to the University of Michigan undergraduate admissions procedure, minority status provided 20 points on a 150-point scale for admission. The court found that the point system was problematic in that it turned race into the decisive factor for some applicants instead of just one of many factors (Winter 2003b). In response, the University of Michigan abandoned its point system and created an undergraduate admissions policy similar to its law school admissions policy, which may serve as a model for how other universities can achieve a diverse student body while following court guidelines. In the new "holistic review" approach, the university considers the unique circumstances of each student, prioritizing academics and treating all other factors, including race, equally. Applicants are now required to write more essays, including one on cultural diversity. Other universities,

affirmative action A broad range of policies and practices in the workplace and educational institutions to promote equal opportunity as well as diversity.

including the University of Wisconsin, University of Washington, and University of California, have also adopted "holistic admissions" procedures to increase their minority student populations.

Some universities, such as the University of Texas at Austin, use the "10 percent plan" as a way to maintain minority enrollment. Graduates in the top 10 percent of their high school classes are admitted automatically to the public college or university of their choice; standardized test scores and other factors are not considered.

In 2006, Michigan voters decided to ban affirmative action in university admissions, but in 2011, a federal appeals court struck down Michigan's ban on affirmative action, ruling that the ban burdens minorities and is unconstitutional (Jaschik 2011). States with similar bans include Arizona, California, Nebraska, and Washington State. Legal battles over affirmative action in higher education will likely continue.

Attitudes toward Affirmative Action Affirmative action remains a divisive issue among Americans. Among first-year college students, half (49.6 percent) agreed that "affirmative action in college admissions should be abolished" (Pryor et al. 2010). In an NBC News/*Wall Street Journal* survey, a sample of U.S. adults was asked whether (a) "Affirmative action programs are still needed to counteract the effects of discrimination against minorities, and are a good idea as long as there are no rigid quotas"; or (b) "Affirmative action programs have gone too far in favoring minorities, and should be ended because they unfairly discriminate against whites." Nearly half (49 percent) said affirmative action programs are still needed, 43 percent said they should be ended, and 8 percent were unsure.

Public opinion poll results are influenced by how survey questions are worded and framed. Survey questions that ask whether respondents favor "affirmative action programs for women and minorities" elicit more favorable responses than questions that ask about affirmative action for minorities only (Paul 2003). In addition, terms such as *affirmative action, equal,* and *opportunity* in survey questions yield more support for affirmative action policies, whereas terms such as *special preferences, preferential treatment,* and *quotas* tend to lessen support.

Supporters of affirmative action suggest that such policies have many social benefits. In a review of more than 200 scientific studies of affirmative action, Holzer and Neumark (2000) concluded that these policies produce benefits for women, minorities, and the overall economy. Holzer and Neumark (2000) found that employers who adopt affirmative action increase the relative number of women and minorities in the workplace by an average of 10 percent to 15 percent. Since the early 1960s, affirmative action in education has contributed to an increase in the percentage of blacks attending college by a factor of 3 and the percentage of blacks enrolled in medical school by a factor of 4. Black doctors choose more often to practice medicine in inner cities and rural areas serving poor or minority patients than their white medical school classmates do (Holzer & Neumark 2000). Increasing the numbers of minorities in educational and professional positions also provides positive role models for other, especially younger, minorities.

Opponents of affirmative action suggest that such programs constitute reverse discrimination, which hurts whites. In 2004, 20 firefighters who are white (one is also Hispanic) sued the city of New Haven, Connecticut, after the city threw out the results of an examination given to determine promotion to positions of captain and lieutenant. Of the 118 candidates who took the test, 27 were black. None of the black candidates scored high enough to qualify for promotion, whereas all 20 of the plaintiffs qualified. The city decided to scrap the test results and promote no one, leading the 20 white firefighters who passed the test to sue the city for reverse discrimination. The hotly debated case reached the Supreme Court; in June 2009, the court ruled that the 20 white New Haven firefighters who were denied promotion were victims of illegal racial discrimination (Mahony & Kovner 2009).

During times of high unemployment rates, it is tempting for some whites to blame unemployment on affirmative action policies (as well as on immigrant labor). However, the main causes of unemployment among the white population are corporate downsizing,

computerization and automation, factory relocations outside the United States, and other macroeconomic conditions, not affirmative action.

Some critics of affirmative action argue that it undermines the self-esteem of women and minorities. Although affirmative action may have this effect in rare cases, affirmative action can raise the self-esteem of women and minorities in many cases, by providing them with opportunities for educational advancement and employment (Plous 2003). Another criticism of affirmative action is that it fails to help the most impoverished of minorities—those whose deep and persistent poverty impairs their ability to compete not only with whites but also with other more advantaged minorities (Wilson 1987).

Attorney Karen Torre, center, stands with the firefighters she represented in a reverse discrimination case that made it all the way to the Supreme Court.

Educational Strategies

Schools and universities play an important role in whether minorities succeed in school and in the job market. One way to improve minorities' chances of academic success is to reduce or eliminate disparities in school funding. As noted earlier, schools in poor districts—which predominantly serve minority students—have traditionally received less funding per pupil than do schools in middle- and upper-class districts (which predominantly serve white students). Other educational strategies focus on reducing prejudice, racism, and discrimination and fostering awareness and appreciation of racial and ethnic diversity. These strategies include multicultural education, "whiteness studies," and efforts to increase diversity among student populations.

Multicultural Education in Schools and Communities In schools across the nation, multicultural education, which encompasses a broad range of programs and strategies, works to dispel myths, stereotypes, and ignorance about minorities; to promote tolerance and appreciation of diversity; and to include minority groups in the school curriculum (see also Chapter 8). With multicultural education, the school curriculum reflects the diversity of U.S. society and fosters an awareness and appreciation of the contributions of different racial and ethnic groups to U.S. culture. The Southern Poverty Law Center's program Teaching Tolerance publishes and distributes materials and videos designed to promote better human relations among diverse groups. These materials are sent to schools, colleges, religious organizations, and a variety of community groups across the nation.

Many colleges and universities have made efforts to promote awareness and appreciation of diversity by offering courses and degree programs in racial and ethnic studies and by sponsoring multicultural events and student organizations. Many colleges mandate that students take a certain number of required diversity courses to graduate. Positive outcomes for students who take college diversity courses include increased racial understanding and cultural awareness, increased social interaction with students who have backgrounds different from their own, improved cognitive development, increased support for efforts to achieve educational equity, and higher satisfaction with their college experience (Humphreys 1999).

Positive outcomes for students who take college diversity courses include increased racial understanding and cultural awareness, increased social interaction with students who have backgrounds different from their own, improved cognitive development, increased support for efforts to achieve educational equity, and higher satisfaction with their college experience.

   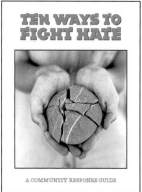

Whiteness Studies Traditionally, studies of and courses on race have focused on the social disadvantages that racial minorities experience, while ignoring or minimizing the other side of the racial inequality equation—the social advantages conferred upon whites. Courses in Whiteness Studies, which are being offered in many colleges and universities, focus on increasing awareness of white privilege—an awareness that is limited among white students. "White privilege is so fundamental as to be largely invisible, expected, and normalized" (Picca & Feagin 2007, p. 243). In an often-cited paper called "White Privilege: Unpacking the Invisible Knapsack," Peggy McIntosh (1990) likened white privilege to an "invisible weightless knapsack" that whites carry with them without awareness of the many benefits inside the knapsack. Just a few of the many benefits McIntosh associates with being white include being able to

- Go shopping without being followed or harassed.
- Be assured that my skin color will not convey that I am financially unreliable when I use checks or credit cards.
- Swear, or dress in secondhand clothes, or not answer letters without having people attribute these choices to the bad morals, poverty, or the illiteracy of my race.
- Avoid ever being asked to speak for all the people of my racial group.
- Be sure that, if a traffic cop pulls me over or if the IRS audits my tax return, it is not because I have been singled out because of my race.
- Take a job with an affirmative action employer without having coworkers on the job suspect that I got it because of race.

Whiteness studies "serve to rectify something wrong with the way we study race in America: By traditionally focusing on minority groups, the implicit message that scholarship projects is that nonwhites are 'deviant,' that's why they are studied" (Conley, 2002). As explained by Professor Gregory Jay (2005):

> Whiteness Studies attempts to trace the economic and political history behind the invention of "whiteness," to challenge the privileges given to so-called "whites," and to analyze the cultural practices (in art, music, literature, and popular media) that create and perpetuate the fiction of "whiteness.". . . "Whiteness Studies" is an attempt to think critically about how white skin preference has operated systematically, structurally, and sometimes unconsciously as a dominant force in American—and indeed in global—society and culture. (n.p.)

Diversification of College Student Populations Recruiting and admitting racial and ethnic minorities in institutions of higher education can foster positive relationships among diverse groups and enrich the educational experience of all students—minority and nonminority alike (American Council on Education and American Association of University Professors 2000). Psychologist Gordon Allport's (1954) "contact hypothesis" suggested that contact between groups is necessary for the reduction of prejudice between group members. Ensuring a diverse student population

More than 600,000 teachers receive *Teaching Tolerance* magazine—a free resource for tolerance education in the classroom (Dees 2000). Other free materials available from the Southern Poverty Law Center (www.splcenter .org) include *101 Tools for Tolerance, Responding to Hate at School,* and *Ten Ways to Fight Hate.*

can provide students with opportunities for contact with different groups and can thereby reduce prejudice. In a study of 2,000 UCLA students, researchers found that students who were randomly assigned to roommates of a different race or ethnicity developed more favorable attitudes toward students of different backgrounds (Sidanius et al. 2010). Another study found that students with the most exposure to diverse populations during college had the most cross-racial interactions five years after leaving college (Gurin 1999).

Retrospective Justice Initiatives: Apologies and Reparations

In 2003, Brown University President Ruth J. Simmons appointed a Steering Committee on Slavery and Justice to investigate and issue a public report on the university's historical relationship to slavery and the trans-Atlantic slave trade. The steering committee's research concluded that "there is no question that many of the assets that underwrote the University's creation and growth derived, directly and indirectly, from slavery and the slave trade" (Brown University Steering Committee on Slavery and Justice 2007, p. 13). The committee's final report recommended that Brown University should acknowledge and make amends for its past ties to the slave trade.

Various governments around the world have issued official apologies for racial and ethnic oppression. After World War II, West Germany signed a reparations agreement with Israel in which West Germany agreed to pay Israel for the enslavement and persecution of Jews during the Holocaust and to compensate for Jewish property that the Nazis stole. In 2008, the Australian government issued a formal apology for the treatment of the country's aboriginal people, specifically, for the decades during which the government removed aboriginal children from their families in a forced acculturation program.

In the United States, President Gerald Ford and Congress apologized to Japanese Americans for their internment during World War II, and reparations of $20,000 were granted to each surviving internee who was a U.S. citizen or legal resident alien at time of internment. In 1993, President Bill Clinton apologized to native Hawaiians for overthrowing the government of their nation. In 1994, the state of Florida offered monetary compensation to the survivors and descendents of the 1923 "Rosewood massacre," in which a white mob attacked and murdered black residents in Rosewood, Florida, and set fire to the town. And in 1997, the U.S. government offered monetary reparations to surviving victims of the Tuskegee syphilis study, in which blacks suffering from syphilis were denied medical treatment.

Although various forms of reparations were offered to Native American tribes to compensate for land that had been taken by force or deception, it wasn't until 2008 that the Senate Committee on Indian Affairs passed a resolution apologizing to all Indian tribes for the mistreatment and violence committed against them. In 2008, the U.S. House of Representatives issued a formal apology to African Americans for slavery, and in 2009, the U.S. Senate passed a resolution apologizing for slavery, adding the stipulation that the official apology cannot be used to support claims for restitution. The resolution "acknowledges the fundamental injustice, cruelty, brutality, and inhumanity of slavery and Jim Crow laws," and "apologizes to African Americans . . . for the wrongs committed against them and their ancestors who suffered under slavery and Jim Crow laws" (CNN 2009).

> **What Do You Think?** Some scholars argue that achieving black/white racial reconciliation in the United States requires that the U.S. government not only issue an official apology but also provide substantial monetary and other reparations to African Americans (Brooks 2004). Do you think African Americans should receive monetary or other reparations? Why or why not?

The growing movement to redress past large-scale violations of human rights is based on the moral principles of taking responsibility for and attempting to rectify past wrongdoings. Supporters of the reparative justice movement believe that the granting of apologies and reparations to groups that have been mistreated promotes dialogue and healing, increases awareness of present inequalities, and stimulates political action to remedy current injustices. Some who are opposed to this movement claim that "preoccupation with past injustice is a distraction from the challenge of present injustice" (Brown University Steering Committee on Slavery and Justice 2007, p. 39).

The Brown University Steering Committee on Slavery and Justice (2007) examined "retrospective justice initiatives" from around the world and concluded that the most successful generally combined three elements: (1) formal acknowledgment of an offense; (2) a commitment to truth telling, to ensure that the relevant facts are uncovered, discussed, and properly memorialized; and (3) the making of some form of amends in the present to give material substance to expressions of regret and responsibility. In the committee's view, "reparative justice is not an invitation to 'wallow in the past' but a way for societies to come to terms with painful histories and move forward" (Brown University Steering Committee on Slavery and Justice 2007, p. 39).

Understanding Race, Ethnicity, and Immigration

After considering the material presented in this chapter, what understanding about race and ethnic relations are we left with? First, we have seen that racial and ethnic categories are socially constructed; they are largely arbitrary, imprecise, and misleading. Although some scholars suggest that we abandon racial and ethnic labels, others advocate adding new categories—multiethnic and multiracial—to reflect the identities of a growing segment of the U.S. and world population.

Conflict theorists and structural functionalists agree that prejudice, discrimination, and racism have benefited certain groups in society. But racial and ethnic disharmony has created tensions that disrupt social equilibrium. Symbolic interactionists note that negative labeling of minority group members, which is learned through interaction with others, contributes to the subordinate position of minorities.

Prejudice, racism, and discrimination are debilitating forces in the lives of minorities and immigrants. Despite these negative forces, many minority group members succeed in living productive, meaningful, and prosperous lives. But many others cannot overcome the social disadvantages associated with their minority status and become victims of a cycle of poverty (see Chapter 6). Minorities are disproportionately poor, receive inferior education and health care, and, with continued discrimination in the workplace, have difficulty improving their standard of living.

Achieving racial and ethnic equality requires alterations in the structure of society that increase opportunities for minorities—in education, employment and income, and political participation. In addition, policy makers concerned with racial and ethnic equality must find ways to reduce the racial and ethnic wealth gap and to foster wealth accumulation among minorities (Conley 1999). Social class is a central issue in race and ethnic relations. Professor and activist bell hooks (2000) (who spells her name in all lowercase) warned that focusing on issues of race and gender can deflect attention away from the larger issue of class division that increasingly separates the haves from the have-nots. Addressing class inequality must, suggests hooks, be part of any meaningful strategy to reduce inequalities that minority groups suffer. Civil rights activist Lani Guinier, in an interview with Paula Zahn on the *CBS Evening News* on July 18, 1998, suggested that

Reverend Joseph Lowery, giving the benediction at President Obama's inauguration in January 2009.

Jason Reed/Reuters/Landov

"the real challenge is to . . . use race as a window on issues of class, issues of gender, and issues of fundamental fairness, not just to talk about race as if it's a question of individual bigotry or individual prejudice. The issue is more than about making friends—it's about making change." But, as Shipler (1998) noted, making change requires members of society to recognize that change is necessary, that there is a problem that needs rectifying:

> One has to perceive the problem to embrace the solutions. If you think racism isn't harmful unless it wears sheets or burns crosses or bars blacks from motels and restaurants, you will support only the crudest antidiscrimination laws and not the more refined methods of affirmative action and diversity training. (p. 2)

The historic election of the first African American U.S. president and the appointment of the first Hispanic woman to the U.S. Supreme Court have raised hopes for a new era in race and ethnic relations in the United States. We end this chapter with an excerpt from the benediction that Dr. Joseph Lowery delivered at President Obama's inauguration:

> Lord . . . we ask you to help us work for that day when black will not be asked to get back, when brown can stick around—and when yellow will be mellow—when the red man can get ahead, man—and when white will embrace what's right. (quoted in Kovac 2009)

CHAPTER REVIEW

- **What is a minority group?**
A minority group is a category of people who have un-equal access to positions of power, prestige, and wealth in a society and who tend to be targets of prejudice and discrimination. Minority status is not based on numerical representation in society but rather on social status.

- **What is meant by the idea that race is socially constructed?**
The concept of race refers to a category of people who are perceived to share distinct physical characteristics that are deemed socially significant. The significance of race is not biological but social and political, because race is used to separate "us" from "them" and becomes a basis for unequal treatment of one group by another. Races are cultural and social inventions; they are not scientifically valid because there are no objective, reliable, meaningful criteria scien-tists can use to identify racial groupings. Different societ-ies construct different systems of racial classification, and these systems change over time.

- **What are the various patterns of interaction that may occur when two or more racial or ethnic groups come into contact?**
When two or more racial or ethnic groups come into con-tact, one of several patterns of interaction occurs, including genocide, expulsion, segregation, acculturation, pluralism, and assimilation.

- **Beginning with the 2000 census, what are the five race cat-egories used to identify the race composition of the United States?**
Beginning with the 2000 census, the five race categories are (1) white, (2) black or African American, (3) American Indian or Alaska Native, (4) Asian, and (5) Native Hawai-ian or other Pacific Islander. In addition, respondents to federal surveys and the census have the option of officially identifying themselves as being of more than one race, rather than checking only one racial category.

- **What is an ethnic group?**
An ethnic group is a population that has a shared cul-tural heritage, nationality, or lineage. Ethnic groups can be distinguished on the basis of language, forms of family structures and roles of family members, religious beliefs and practices, dietary customs, forms of artistic expression such as music and dance, and national origin. The larg-est ethnic population in the United States is Hispanics or Latinos.

- **What percentage of the U.S. population (in 2009) was born outside the United States?**
More than 1 in 10 U.S. residents (12.5 percent) were born in a foreign country.

- **What were the manifest function and latent dysfunction of the civil rights movement?**
The manifest function of the civil rights legislation in the 1960s was to improve conditions for racial minorities. However, civil rights legislation produced an unexpected consequence, or latent dysfunction. Because civil rights legislation supposedly ended racial discrimination, whites were more likely to blame blacks for their social disadvantages and thus perpetuate negative stereotypes such as "blacks lack motivation" and "blacks have less ability."

- **How does contemporary prejudice differ from more tradi-tional, "old-fashioned" prejudice?**
Traditional, old-fashioned prejudice is easy to recognize, because it is blatant, direct, and conscious. More contem-porary forms of prejudice are often subtle, indirect, and

unconscious. In addition, racist expressions have gone "backstage" to private social settings.

- **Is it possible for an individual to discriminate without being prejudiced?**
Yes. In overt discrimination, individuals discriminate because of their own prejudicial attitudes. But sometimes individuals who are not prejudiced discriminate because of someone else's prejudice. For example, a store clerk may watch black customers more closely because the store manager is prejudiced against blacks and has instructed the employee to follow black customers in the store closely. Discrimination based on someone else's prejudice is called adaptive discrimination.

- **Are U.S. schools segregated?**
Racial and ethnic minorities are largely isolated from whites in an increasingly segregated school system. One study found that U.S. schools in the 2000–2001 school year were more segregated than they were in 1970. The upward trend in school segregation is due to large increases in minority student enrollment, continuing white flight from urban areas, the persistence of housing segregation, and the termination of court-ordered desegregation plans.

- **According to FBI data, the majority of hate crimes are motivated by what kind of bias?**
Since the FBI began publishing hate crime data in 1992, the majority of hate crimes have been based on racial bias.

- **What is the role of the Equal Employment Opportunity Commission (EEOC) in combating employment discrimination?**
The Equal Employment Opportunity Commission (EEOC) is responsible for enforcing laws against discrimination, including Title VII of the 1964 Civil Rights Act that prohibits employment discrimination on the basis of race, color, religion, sex, or national origin. The EEOC investigates, mediates, and may file lawsuits against private employers on behalf of alleged victims of discrimination.

- **What group constitutes the largest beneficiary of affirmative action policies?**
Affirmative action policies are designed to benefit racial and ethnic minorities, women, and, in some cases, Vietnam veterans and people with disabilities. The largest category of affirmative action beneficiaries is women.

- **What are Whiteness Studies?**
Courses in Whiteness Studies, which are being offered in many colleges and universities, focus on increasing awareness of white privilege—an awareness that is limited among white students.

- **According to the Brown University Steering Committee on Slavery and Justice, successful retrospective justice initiatives contain what three elements?**
The Brown University Steering Committee on Slavery and Justice examined retrospective justice initiatives from around the world and concluded that the most successful generally combined three elements: (1) formal acknowledgment of an offense; (2) a commitment to truth telling, to ensure that the relevant facts are uncovered, discussed, and properly memorialized; and (3) the making of some form of amends in the present to give material substance to expressions of regret and responsibility.

TEST YOURSELF

1. The U.S. federal government considers "Hispanic" origin to be a race.
   a. True
   b. False
2. Which of the following occurs when a person adopts the culture of a group different from the one in which that person was originally raised?
   a. Secondary assimilation
   b. Acculturation
   c. Genocide
   d. Pluralism
3. More than half of the growth in the total U.S. population between 2000 and 2010 was due to the increase in which of the following populations?
   a. Non-Hispanic whites
   b. Blacks
   c. Hispanics
   d. Mixed-race
4. A Southern Poverty Law Center report reveals that the guest worker program constitutes a "modern-day system of ___."
   a. amnesty
   b. colonialism
   c. paid apprenticeship
   d. indentured servitude
5. Which minority group in the United States is considered a "model minority"?
   a. Women
   b. Black women
   c. Scandinavian immigrants
   d. Asian Americans
6. Which of the following has/have moved "backstage"?
   a. Racist behavior
   b. Undocumented immigrants
   c. Multicultural education
   d. Affirmative action
7. The number of hate groups in the United States has been steadily declining over recent years.
   a. True
   b. False
8. Which of the following is responsible for enforcing laws against discrimination?
   a. Local police departments
   b. Equal Employment Opportunity Commission
   c. U.S. Department of Labor
   d. The Supreme Court

9. Colleges and universities are prohibited from requiring students to take diversity courses as a graduation requirement.
   a. True
   b. False

10. The U.S. Congress has issued an official apology to African Americans for slavery and Jim Crow laws.
    a. True
    b. False

Answers: 1: b; 2: b; 3: c; 4: d; 5: d; 6: a; 7: b; 8: b; 9: b; 10: a.

KEY TERMS

acculturation 268
adaptive discrimination 288
affirmative action 296
antimiscegenation laws 272
assimilation 269
aversive racism 284
discrimination 288
ethnicity 267
Equal Employment Opportunity
   Commission (EEOC) 295

expulsion 268
genocide 268
hate crime 291
individual discrimination 288
institutional discrimination 288
Islamophobia 293
minority group 265
modern racism 286
nativist extremist groups 275
naturalized citizens 279

overt discrimination 288
pluralism 269
prejudice 284
race 267
racism 284
Racism 2.0 286
segregation 268
stereotypes 283

MEDIA RESOURCES

Turning to Video

 Watch the BBC video *Multiracial Americans* (running time 2:02), available through **CengageBrain.com.**
This video discusses the growing multiracial U.S. population and the challenges of being multiracial. As you watch the video, think about this: Do you think that the multiracial population will ever outnumber single-race populations in the United States? Why or why not?

Online Study Resources

Log in to **www.cengagebrain.com** to access the resources your instructor has assigned. For this book, you can access:

CourseMate

Access chapter-specific learning tools, including learning objectives, practice quizzes, videos, Internet exercises, flash cards, and glossaries, as well as web links, and more in your Sociology CourseMate.

© Kaadaa/Getty Images

Gender Inequality

"Gender equality is more than a goal in itself. It is a precondition for meeting the challenge of reducing poverty, promoting sustainable development and building good governance."

—Kofi A. Annan, Nobel Prize winner, 7th Secretary-General of the United Nations

10

STORM STOCKER IS a beautiful, chubby cheeked, red-haired little baby. But unlike so many other children whose proud parents announce the birth of their "little girl" or "little boy" with pink and blue, Storm's arrival was accompanied by an e-mail message: "We've decided not to share Storm's sex for now — a tribute to freedom and choice in place of limitation, a stand up to what the world could become in Storm's lifetime . . ." (quoted in Davis 2011).

The idea to keep Storm's sex private came from her big brother. He wondered out loud if, ". . . people would respond differently if they didn't know the baby's sex. What gifts would they bring? If Storm was a boy, would he be allowed to wear dresses?" (Witterick 2011, p. 1). And so, as his mother writes, "There are these moments as a parent when you wish your child could bring a different issue to the table — but there it is, plop! And if you really mean what you say about being kind, honouring difference, having an open mind and placing limits thoughtfully where they help children develop competencies and be safe, then you better walk the talk. We agreed to keep the sex of our new baby private" (Witterick 2011, p. 1).

The response to Kathy Witterick and her husband David Stocker's decision, most of it quite negative, attests to the importance of sex and gender in society. **Sex** refers to one's biological classification, whereas **gender** refers to the social definitions and expectations associated with being female or male. Today, however, researchers, educators, and parents alike are challenging the binary concepts of sex and gender and, just as our evolving notion of sexual orientation (see Chapter 11), have led to feelings of uneasiness for some. For example, words like *transgender* and *transsexual, gender variant, androgyny, intersexed, boi and birl, metrosexual, two-spirited, gender neutrality, a third sex,* and *gender bender* were not part of the American lexicon just decades ago.

> . . . researchers, educators, and parents alike are challenging the binary concepts of sex and gender and, just as our evolving notion of sexual orientation, have led to feelings of uneasiness for some.

In most Western cultures, we take for granted that there are two categories of gender. However, in many other societies, three and four genders have been recognized. For example, in parts of Mexico, *muxes* (pronounced MOO-shays) live "between two genders" (Lacey 2008, p. 1), and many Polynesian cultures recognize the *mahu (*pronounced MAH-ho*)*—individuals who take on the work roles of members of the opposite sex (Nanda 2000).

A **transgender individual** (sometimes called "trans" or "transgender") is a person whose sense of gender identity—masculine or feminine—is inconsistent with their birth (sometimes called chromosomal) sex. Transgender is not a sexual orientation and transgender individuals may have any sexual orientation—heterosexual, homosexual, or bisexual. Transsexuals are transgender individuals "who have changed or are in the process of changing [their] . . . physical sex to conform to [their] . . . internal sense of gender identity" (HRC 2005, p. 7).

Although there is documented movement away from a binary emphasis on sex and gender (see this chapter's photo essay, *The Gender Continuum*), most Americans still think in terms of females and males, and women and men, much the same way they continue to use such false dichotomies as black and white, gay and straight, and young and old. Although empirically inaccurate, this *social shorthand* makes conversation easier and fulfills our need to "know" and respond "appropriately." Thus, upon meeting someone, we quickly attach the label of gay, white male, or elderly African American female, though knowing little about their social biographies. Similarly, this chapter emphasizes **sexism** and gender inequality as traditionally defined i.e., as the inequality between women and men but, wherever possible, information on transgender individuals will be included in the discussion. Note that civil rights issues for transgender individuals are discussed in Chapter 11 (see this chapter's *The Human Side*).

sex A person's biological classification as male or female.

gender The social definitions and expectations associated with being female or male.

transgender individual A transgender individual is a person whose sense of gender identity—masculine or feminine—is inconsistent with their birth (sometimes called chromosomal) sex (male or female).

sexism The belief that innate psychological, behavioral, and/or intellectual differences exist between women and men and that these differences connote the superiority of one group and the inferiority of the other.

The following is a speech given to the National Press Club by Dr. Jennifer Finney Boylan, a professor of English at Colby College in Maine. She is the author of 12 books and numerous short stories, is a regular contributor to the New York Times, and has appeared on many television shows including the Oprah Winfrey Show. The speech was delivered in May, 2007 as Dr. Boylan and other transgender individuals lobbied for the passage of the Employment Non-discrimination Act and the Hate Crimes Bill.

Thank you. Look at you all. It's very cool to see all of you gathered in one spot, all these trans people and their allies.

There are a lot of things I don't know, but I know this: Tomorrow is going to be a great day.

One night recently, my children and my partner and I were talking about the usual stuff at dinner—about whether bloodhounds drool too much, about who would win, The Incredible Hulk, or Abraham Lincoln? At one point we even fell into the classic discussion of what makes the best superpower? While I argued for super-speed, my children tried to make the case for Time-Travel, and Flying, and something they called Super-stickiness, which might be the thing that enables Spiderman to climb walls, or which might be something else entirely.

My son Sean was doing a book report on Martin Luther King at the time. And in the midst of our conversation, Sean suddenly looked up at Grace and me and said, "Why did Martin Luther King say he wanted to dream?"

And we said, well, it's good to dream.

My son said he understood that. But why, he asked, didn't Martin Luther King want to wake up? And step out into a world where those dreams are at last coming true?

As I think about all of us—transgendered Americans in this room and across the country, I can't help but think that my son is right. While our dreams give us courage and hope, it is also surely time that we all wake up, and enjoy our rights as American citizens, in a country that respects our diversity, our courage, and our strength.

And so I say to you:

I want to wake up in a country where transgendered people are seen as human, where our curiously gendered lives are seen as one more variation in the rich tapestry of experience, as something not to be shocked by, but as something to be celebrated, and honored, and understood.

I want to wake up in a country where Americans understand that transgender people come in all shapes and sizes and embodiments, where to be a cross dresser or a transsexual or a drag queen or trans man or genderqueer is seen as simply another way of being human, a person endowed by the creator with certain inalienable rights, and that among these rights are life, liberty and the pursuit of happiness.

I want to wake up tomorrow.

I want to wake up in a country where coming out as transgender is not seen as the end of the world, but as a beginning, where the lives of people such as ourselves are celebrated, where we are seen as precious, vital parts of a democracy, where we have the right to earn a living without fear of being fired for what we are, where we have the right to get married to the people we love, where the President of the United States will reach out and shake our hands and say that he is proud of everything we bring to the American experience. I want to wake up.

I want to wake up in a country where qualified, hardworking Americans will never be denied job opportunities because of the sexual orientation or their gender identity or expression, a country where every individual will have a fundamental right under Federal Law, to be protected from discrimination. I want to wake up in a country in which the thirty-three states at present where a person can be fired because of her sexuality have to change their laws. I want to wake up in a country in which the forty-two states in which a person can be fired because of her gender identity have to change their laws. I want to wake up in a country in which men and women are judged not by what they are wearing, or whom they love, but by the content of their characters. I want to wake up.

I want to wake up to a county in which crimes against transgender people will never be excused by anybody, ever, for any reason.

Tomorrow morning, when you open your eyes, you will wake up into a country which is changing, one human face at a time. And in so doing, you will also answer for my family another one of those questions we ask around the dinner table, namely, who is the best superhero? Wolverine? Spiderman? Thomas Jefferson? And what exactly does it mean, in the end, to be a hero?

If you ask me, the best superheroes are the transgender people in this room and all across America. In your grace, your courage, in your unquenchable desire to make this a better country, you are all heroes.

It is an honor to be here with you all, fighting this fight. With all our super powers tomorrow—super love, super compassion, and yes, even a little bit of super-stickiness—I know that in the morning, we are all going to wake up to a better country, and to a better future.

Tomorrow is going to be a great day. Thank you.

Source: Boylan 2007. Copyright © Jennifer Finney Boylan. Reprinted by permission of the author.

What Do You Think? Ceara Lynn Sturgis graduated from high school in 2010 (NSBA 2011). When having her senior picture taken, feeling uncomfortable in the "scoop-neck drape" worn by women in her class, she asked and received permission to wear a tuxedo as were her male classmates. However, when Ceara, who identifies as a female, received her yearbook, her picture had been excluded. She is suing for damages based on sex discrimination and civil rights violations. Do you think she should be awarded damages?

Photo Essay

Race and ethnicity, sex and gender, and age are considered core demographic variables because of their effectiveness in predicting life chances—education, occupation, income, and the like. Yet, over the last several decades, the meanings we associate with each of these variables have evolved. Race is no longer black, white, and other in the U.S. Census, and being middle aged extends into the 60s as life expectancy increases. Such is the case with sex and gender. As a society, we are beginning to embrace those who don't fit neatly into the female–male, woman–man binary. For example, in 2011, the governor of Connecticut signed into law a "bill prohibiting discrimination on the basis of gender identity and expression," thereby extending legal protection to transgender and gender nonconforming students in school, employees in the workplace, and would-be homeowners (Warbelow 2011, p. 1).

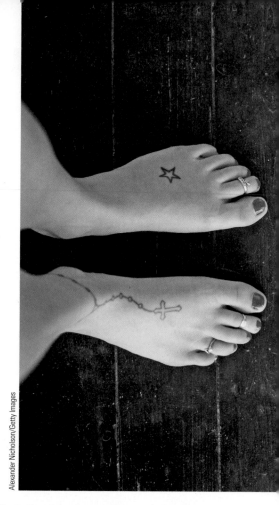

Alexander Nicholson/Getty Images

Pep Roig/Alamy

▲ In India's 2011 national census, the census form had three options under the heading of "sex"—male, female, and other (Haub 2011). Similarly, Nepal, after a ruling by their highest court which held that lesbian, gay, bisexual, and transgendered (LGBT) persons needed greater protection, also used the three gender option in their census (Cohn 2011). Pictured here is a *hijra*, the Indian word for "third sex," institutionalized in India for thousands of years.

▼ In Native American culture, prior to the invasion of the Europeans, there existed three genders—male, female, and male-female. Berdache, the Native American word used to describe the male-female gender, was translated into English as "the two spirited person" (McGill 2011). Rather than being ostracized, the two-spirited person was embraced, revered, and envied as a person who had the "privilege to house both male and female spirits in their body." (p.1). Here a We-Wa Zuni man is dressed as a woman, weaving a belt on a waist-high loom with a reed heddle (National Archives 2010).

National Archives

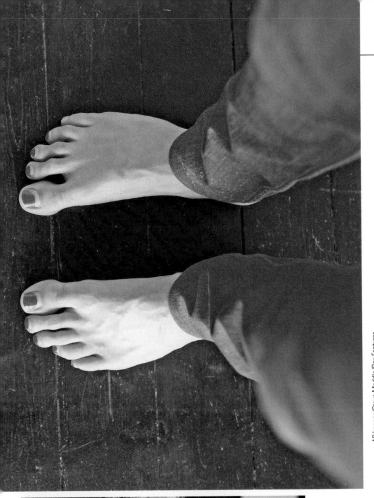

◀ In 2011, an advertisement for J. Crew showing the store's creative director painting her son's toenails sparked both accolades and indignation. *The Daily Mail*, a British newspaper, noting that the photo created "gender identity debates across the U.S.," published photographs of young boys with nail polish, including the sons of Gwen Stefani, Jennifer Lopez, and Christina Aguilera (Abraham & Madison 2011). The paper also noted that singer Justin Bieber launched his own brand of nail polish in 2010, which sold over 2 million bottles in the first few months.

AP Images/Steve Meddle/Rex Features

▲ Research indicates that in classes where teachers call attention to gender differences, even in seemingly harmless ways such as saying "Good morning boys and girls" or by dividing the class by sex for a spelling bee, children are less likely to interact with opposite-sex children and more likely to use gender stereotypes (Moskowitz 2010). Schools are beginning to acknowledge that sex and gender are not categorical and that some children may identify as female, male, both, or neither. Pictured here is Liivy, a 10-year-old male to female transgender. When Livvy returned to school for the first time as a female, there was a special assembly to explain Livvy's situation to the other children. The mood at school was one of acceptance until some concerned parent went to the newspaper and the little girls story when global.

▼ Whenever social change takes place, requests for interviews with the agents, victims, or beneficiaries of change increase as media outlets search for the latest "news." Transgendered children have recently made appearances on *Oprah, 20/20, Dr. Oz, Tyra Banks, National Public Radio (NPR)*, and the *Barbara Walters Show*. There are also emerging support groups, conferences, professional workshops, and the like. Here, Chaz Bono (third from the right), child star of Cher and the late Sonny Bono, poses with his cast mates from the popular television show *Dancing with the Stars*.

REUTERS/Benoit Tessie

▲ Andrej Pejic, a transgender model, states that "Sometimes I feel like more of a woman, other times I feel male" (quoted in Daily Mail 2011, p. 1). The 19-year-old Serbian who relocated to Australia is ranked the 11th top male model in the world. He's been described as the "poster boy for fashion androgyny" (Models 2011, p. 1).

Getty Images

The Global Context: The Status of Women and Men

There is no country in the world in which women and men have equal status. Although much progress has been made in closing the gender gap in areas such as education, health care, employment, and government, gender inequality is still prevalent throughout the world.

The World Economic Forum assessed the gender gap in 134 countries by measuring the extent to which women have achieved equality with men in four areas: economic participation and opportunity, educational attainment, health and survival, and political empowerment (Hausmann et al. 2010). Table 10.1 presents the overall scores and rankings

TABLE 10.1 Gender Gap Rankings: Top and Bottom 10 Countries and the United States, 2010

COUNTRY	OVERALL SCORE*	RANKINGS				
		OVERALL	ECONOMIC PARTICIPATION AND OPPORTUNITY	EDUCATIONAL ATTAINMENT	HEALTH AND SURVIVAL	POLITICAL EMPOWERMENT
TOP 10 COUNTRIES						
Iceland	0.8496	1	18	1	96	1
Norway	0.8404	2	3	1	91	3
Finland	0.8260	3	16	28	1	2
Sweden	0.8024	4	11	41	80	4
New Zealand	0.7808	5	9	1	91	8
Ireland	0.7773	6	25	1	89	7
Denmark	0.7719	7	23	1	68	10
Lesotho	0.7678	8	1	1	1	34
Philippines	0.7654	9	13	1	1	17
Switzerland	0.7562	10	30	71	74	13
United States	**0.7411**	**19**	**6**	**1**	**38**	**40**
BOTTOM 10 COUNTRIES						
Egypt	0.5899	125	121	110	52	125
Turkey	0.5876	126	131	109	61	99
Morocco	0.5767	127	127	116	85	103
Benin	0.5719	128	85	133	110	100
Saudi Arabia	0.5713	129	132	92	53	131
Cote d'Ivoire	0.5691	130	132	130	1	104
Mali	0.5713	131	106	131	55	81
Pakistan	0.5691	132	133	127	122	52
Chad	0.5330	133	77	134	110	122
Yemen	0.4603	134	134	132	81	130

Source: Hausmann et al. 2010

*All overall scores are reported on a scale of 0 to 1, with 1 representing maximum gender equality.

of (1) the 10 countries with the smallest gender gap (i.e., the least gender inequality); (2) the 10 countries with the largest gender gap (i.e., the most gender inequality); and (3) the United States, which did not rank in the top 10 or the bottom 10 but ranked number 19 of the 134 countries studied. Ties were possible. For example, Finland, the Philippines, and Lesotho tied for first place on the composite measure of health and survival. Further, note that the overall score approximates the proportion of the gender gap a country has *closed*—in the United States, 0.7411 or 74.11 percent.

Gender inequality varies across cultures, not only in its extent or degree but also in its forms. For example, in the United States, gender inequality in family roles commonly takes the form of an unequal division of household labor and child care, with women bearing the heavier responsibility for these tasks. In other countries, forms of gender inequality in the family include the expectation that wives ask their husbands for permission to use birth control (see Chapter 12), unequal penalties for spouses who commit adultery, with wives receiving harsher punishment, and the practice of aborting female fetuses in cultures that value male children over female children. In a recent book entitled, *Unnatural Selection*, author Mara Hvistendahl (2011) documents how medical technology, and specifically the increased availability of ultrasounds, has made sex-selection abortions commonplace around the world. Today, there are over 163 million "missing" females, more than the entire female population of the United States.

A global perspective on gender inequality must also take into account the different ways in which such inequality is viewed. For example, many non-Muslims view the practice of Muslim women wearing a headscarf in public as a symbol of female subordination and oppression. To Muslims who embrace this practice (and not all Muslims do), wearing a headscarf reflects the high status of women and represents the view that women should be respected and not treated as sexual objects.

Similarly, cultures differ in how they view the practice of female genital mutilation (FGM/C), also known as female genital cutting or female circumcision. There are several forms of FGM/C, ranging from a symbolic nicking of the clitoris to removal of the clitoris and labia and partial closure of the vaginal opening by stitching the two sides of the vulva together, leaving only a small opening for the passage of urine and menstrual blood. After marriage, the sealed opening is reopened to permit intercourse and childbearing.

Nonmedical personnel perform most FGM/C procedures using unsterilized blades or string. Health risks associated with FGM/C include pain, hemorrhage, infection, shock, scarring, and infertility. Worldwide, over 70 million girls and women are estimated to have experienced FGM/C, and an additional 3 million are at risk each year (WHO 2008; UNICEF 2011).

People from countries in which FGM/C is not the norm generally view this practice as a barbaric form of violence against women. For example, an Ethiopian immigrant in the United States was convicted of aggravated battery and cruelty to children for using a pair of scissors to remove his daughter's clitoris. He was sentenced to 10 years in prison (Haines 2006). In countries where it commonly occurs, FGM/C is viewed as an important and useful practice. In some countries, it is considered a rite of passage that enhances a woman's status. In other countries, it is aesthetically pleasing. For others, FGM/C is a moral imperative based on religious beliefs (Yoder et al. 2004; WHO 2008).

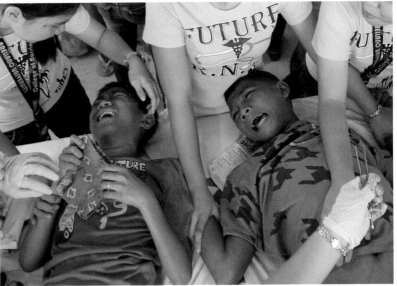

© REUTERS/Cheryl Ravelo

In 2011, in Manila, 1,500 boys were forced to take place in a mass circumcision as the organizers of the event tried to get into the *Guinness Book of World Records* for the most circumcisions performed on boys 9 and older. Of late, the medical procedure has come under fire as questions arise about its safety, necessity, and, similar to FGM/C, abuse of a child's anatomical integrity.

Inequality in the United States

Although attitudes toward gender equality are becoming increasingly liberal, the United States has a long history of gender inequality. Women have had to fight for equality: the right to vote, equal pay for comparable work, quality education, entrance into male-dominated occupations, and legal equality. As shown in Table 10.1, the World Economic Forum (Hausmann et al. 2010)—based on its assessment of women's economic participation and opportunities, political empowerment, educational attainment, and health and survival—ranks the United States 19th in the world in terms of gender equality. Most U.S. citizens agree that American society does not treat women and men equally: Women have lower incomes, hold fewer prestigious jobs, earn fewer graduate degrees, and are more likely than men to live in poverty.

Men are also victims of gender inequality. In 1963, sociologist Erving Goffman wrote that in the United States there is only

> one complete unblushing male . . . a young, married, white, urban, northern heterosexual, Protestant father of college education, fully employed, of good complexion, weight and height, and a recent record in sports. . . . Any male who fails to qualify in one of these ways is likely to view himself . . . as unworthy, incomplete, and inferior. (p. 128)

Although standards of masculinity have relaxed, Williams (2000) argued that masculinity is still based on "success"—at work, on the athletic field, on the streets, and at home. Similarly, Vandello et al. (2008) concluded that ". . . manhood is tenuous, and therefore requires public proof, is consistent with research across multiple areas" (p. 1,326).

When U.S. college students were asked to list the best and worst things about being the opposite sex, the same qualities, although in opposite categories, emerged (Cohen 2001). For example, what males list as the best thing about being female (e.g., free to be emotional), females list as the worst thing about being male (e.g., not free to be emotional). Similarly, what females list as the best thing about being male (e.g., higher pay), males listed as the worst thing about being female (e.g., lower pay). As Cohen (2001) noted, although "some differences are exaggerated or oversimplified . . . we identif[ied] a host of ways in which we 'win' or 'lose' simply because we are male or female" (p. 3).

Sociological Theories of Gender Inequality

Both structural functionalism and conflict theory concentrate on how the structure of society and, specifically, its institutions contribute to gender inequality. However, these two theoretical perspectives offer opposing views of the development and maintenance of gender inequality. Symbolic interactionism, on the other hand, focuses on the culture of society and how gender roles are learned through the socialization process.

Structural-Functionalist Perspective

Structural functionalists argue that preindustrial society required a division of labor based on gender. Women, out of biological necessity, remained in the home performing functions such as bearing, nursing, and caring for children. Men, who were physically stronger and could be away from home for long periods of time, were responsible for providing food, clothing, and shelter for their families. This division of labor was functional for society and became defined as both normal and natural over time.

Industrialization rendered the traditional division of labor less functional, although remnants of the supporting belief system still persist. With increased control over reproduction (e.g., contraception), declining birth rate, and fewer jobs dependent upon physical size and strength, women's opportunities for education and workforce participation increased (Wood & Eagly 2002). Thus, modern conceptions of the family have, to some extent, replaced traditional ones—families have evolved from extended to nuclear, authority is more egalitarian, more women work outside the home, and greater role variation exists in the division of labor. Structural functionalists argue, therefore, that, as the needs of society change, the associated institutional arrangements also change.

Conflict Perspective

Many conflict theorists hold that male dominance and female subordination are shaped by the relationships men and women have to the production process. During the hunting-and-gathering stage of development, males and females were economic equals, both controlling their own labor and producing needed subsistence. As society evolved to agricultural and industrial modes of production, private property developed and men gained control of the modes of production, whereas women remained in the home to bear and care for children. Inheritance laws that ensured that ownership would remain in their hands furthered male domination. Laws that regarded women as property ensured that women would remain confined to the home.

As industrialization continued and the production of goods and services moved away from the home, the gaps between females and males continued to grow—women had less education, lower incomes, fewer occupational skills, and were rarely owners. World War II necessitated the entry of a large number of women into the labor force, but, in contrast with previous periods, many of them did not return to the home at the end of the war. They had established their own place in the workforce and, facilitated by the changing nature of work and technological advances, now competed directly with men for jobs and wages.

Conflict theorists also argue that continued domination by males requires a belief system that supports gender inequality. Two such beliefs are (1) that women are inferior outside the home (e.g., they are less intelligent, less reliable, and less rational); and (2) that women are more valuable in the home (e.g., they have maternal instincts and are naturally nurturing). Thus, unlike structural functionalists, conflict theorists hold that the subordinate position of women in society is a consequence of social inducement rather than biological differences that led to the traditional division of labor.

> Thus, unlike structural functionalists, conflict theorists hold that the subordinate position of women in society is a consequence of social inducement rather than biological differences that led to the traditional division of labor.

Symbolic Interactionist Perspective

Although some scientists argue that gender differences are innate, symbolic interactionists emphasize that, through the socialization process, both females and males are taught the meanings associated with being feminine and masculine. Gender assignment begins at birth as a child is classified as either female or male. However, the learning of gender roles is a lifelong process whereby individuals acquire society's definitions of appropriate and inappropriate gender behavior.

Gender roles are taught by the family, in the school, in peer groups, and by media presentations of girls and boys and women and men (see the discussion on the social construction of gender roles later in this chapter). Most importantly, however, gender roles are learned through symbolic interaction as the messages that others send us reaffirm or challenge our gender performances. In an examination of parent–child interactions, Tenenbaum (2009) found that discussions regarding course selections for high school followed gender-stereotyped patterns. Here, a father talks to his fifth grade daughter (how the conversation was coded by the researchers is in brackets):

> But you know, spelling is English, right? That's what English is. For the most part generally speaking, girls do better with

In traditional Muslim societies, women are forbidden to show their faces or other parts of their bodies when in public. Muslim women wear a veil to cover their faces and a chador, a floor-length loose-fitting garment, to cover themselves from head to toe. Although some women adhere to this norm out of fear of repercussions, many others believe veiling was first imposed on Muhammad's wives out of respect for women and the desire to protect them from unwanted advances. More than half a million Muslims are living in the United States.

© Tom Stoddart /Reportage/Getty Images

those kind of skills, they have a harder time with math, generally, you know? [Code: Lack of ability.] Now, of course, you know it's nice to do the things you're really good at too, and you like to do. But, sometimes when you're trying to be, when you grow up to be someone in life, you also gotta take classes that are, not really, how do I say? You're not really good at, you have to put more practice in, right? [Code: Lack of ability.] So, I picked, I selected algebra, cause that's kinda, high school, mathematics. (pp. 458–459)

Although the father encouraged his daughter to take mathematics, he twice conveyed to his daughter that she is (and girls in general are) not very good in math.

Feminist theory, although also consistent with a conflict perspective, incorporates many aspects of symbolic interactionism. Feminists argue that conceptions of gender are socially constructed as societal expectations dictate what it means to be female or what it means to be male. Thus, women are generally socialized into **expressive roles** (i.e., nurturing and emotionally supportive roles), and males are more often socialized into **instrumental roles** (i.e., task-oriented roles). These roles are then acted out in countless daily interactions as boss and secretary, doctor and nurse, football player and cheerleader "do gender."

Feminists also hold that gender "is a central organizing factor in the social world and so must be included as a fundamental category of analysis in sociological research" (Renzetti & Curran 2003, p. 8). Noting that the impact of the structure and culture of society is not the same for different groups of women and men, feminists encourage research on gender that takes into consideration the differential effects of age, race and ethnicity, and sexual orientation.

What Do You Think? In Afghanistan, little girls in families with no male children often assume the persona of little boys—short hair, traditional male clothing, and an appropriate male name (Nordberg 2010). In addition to the social pressure to have a boy child, the decision to raise a female child as a "bacha posh" is pragmatic—such a child can help "his" father at work, get a better education, and does not have to be chaperoned. The problem is that when it's time to return to being female, usually at puberty, some have difficulty transitioning. Said one young woman, "Nothing in me feels like a girl. . . . For always I want to be a boy. . . ." (Nordberg 2010, p. 6). Do you think gender is a consequence of nature or nurture?

Gender Stratification: Structural Sexism

As structural functionalists and conflict theorists agree, the social structure underlies and perpetuates much of the sexism in society. **Structural sexism,** also known as institutional sexism, refers to the ways the organization of society and specifically its institutions subordinate individuals and groups based on their sex classification. Structural sexism has resulted in significant differences in the education and income levels, occupational and political involvement, and civil rights of women and men.

Education and Structural Sexism

Literacy rates worldwide indicate that women are less likely than men to be able to read and write, with millions of women being denied access to even the most basic education. Worldwide, for example, on average, 87 girls attend elementary school for every 100 boys (UNESCO 2011).

expressive roles Roles into which women are traditionally socialized (i.e., nurturing and emotionally supportive roles).

instrumental roles Roles into which men are traditionally socialized (i.e., task-oriented roles).

structural sexism The ways in which the organization of society, and specifically its institutions, subordinate individuals and groups based on their sex classification.

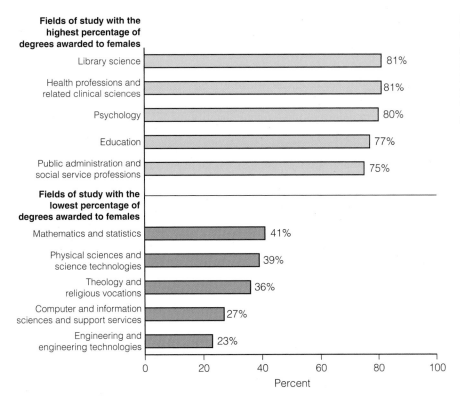

Fields of study with the highest percentage of degrees awarded to females

Library science	81%
Health professions and related clinical sciences	81%
Psychology	80%
Education	77%
Public administration and social service professions	75%

Fields of study with the lowest percentage of degrees awarded to females

Mathematics and statistics	41%
Physical sciences and science technologies	39%
Theology and religious vocations	36%
Computer and information sciences and support services	27%
Engineering and engineering technologies	23%

Percent

Figure 10.1: Percentage of Master's Degrees Awarded to Females by Degree-Granting Institutions in Selected Fields of Study, Academic Year 2008–2009
Source: NCES 2011.

Because children born to educated mothers are less likely to die at a young age, there is an **education dividend** associated with educating women. For example, if universal primary education was available for all girls in sub-Saharan Africa, 200,000 fewer children would die each year, and universal secondary education for all girls would save as many as 1.8 million lives annually (UNESCO 2011).

In 2009, few differences existed between men and women in their completion rates of high school and college degrees (NCES 2011). In fact, in recent years, most U.S. colleges and universities have had a higher percentage of women than men enrolling directly from high school (see Chapter 8). This trend is causing some concern that many young American men may not have the education they need to compete in today's global economy. Although differences are narrowing, men are still more likely to go on to complete a graduate or professional degree than are women. As Figure 10.1 indicates, dramatic differences also appear in the types of advanced degrees that women and men earn. For example, women earn 81 percent of master's degrees in library science but only 23 percent of master's degrees in engineering and engineering technologies.

Concern over the continued lack of women in STEM (science, technology, engineering, and mathematics) is justified. Despite making progress in medicine, women remain wholly underrepresented in STEM majors, degrees, and occupations. Reasons for the STEM gender disparity include reliance on gender stereotyping, ("boys are better in math and science than girls!"), a lack of female STEM role models, little encouragement to follow STEM pursuits, and a lack of awareness about women in STEM fields (AAUW 2011a).

Further, women earn fewer advanced degrees than men because they are socialized to choose marriage and motherhood over long-term career preparation. From an early age, women are exposed to images and models of femininity that stress the importance of domestic family life. Ceci et al. (2009) report that math-proficient women compared to math-proficient men are less likely to enter math-intensive fields such as computer science, engineering, and physics and, if they do, are more likely to leave such careers than their male counterparts. Although the reasons are complex, the greater priority women place on home over career explains much of the gender variation.

education dividend The additional benefit of universal education for women is that it reduces the death rate of children under 5 years of age.

In higher education, there are also structural limitations that discourage women from advancing. For example, women seeking academic careers may find that securing a tenure-track position is more difficult for them than it is for men, and that having children pre-tenure negatively impacts the likelihood of getting tenure, that is, there is a "pregnancy penalty" (Ceci et al. 2009). Similarly, in an analysis of data from the *National Study of Postsecondary Faculty*, Leslie (2007) reports that, as the number of children increases, the number of hours a female faculty member works decreases, and the number of hours a male faculty member works increases (see the discussion on household division of labor later in this chapter). Finally, research also indicates gender stereotyping in letters of recommendation. Women are more often described as socio-emotive (e.g., helpful) rather than active/assertive (e.g., ambitious) and, thus, are evaluated less positively (Madera et al. 2009). Similarly, female faculty are disproportionately assigned service duties (e.g., advising) when compared to their male counterparts, which places them at a disadvantage when being considered for tenure and/or promotion (Misra et al. 2011).

Work and Structural Sexism

According to the International Labour Organization (ILO), in 2008, women made up 40.4 percent of the world's *total* labor force (ILO 2009). Globally, women are disproportionately employed in the agricultural and service sectors, and in vulnerable employment (e.g., unpaid family workers). For example, unpaid work on family-owned farms accounts for 85 percent of all Egyptian women's informal employment (ILO 2011a).

Women are also more likely to be unemployed when compared to men. Increasing rates of female unemployment are, in part, a consequence of the global economic crisis, particularly in developing countries (Social Watch 2011). For example, because women are disproportionately employed in the agricultural sector, as food prices increase as a result of the recession, food exports drop, leading to higher rates of female unemployment (ITUC 2009).

> Women work 66 percent of the total number of labor hours worldwide, yet receive just 10 percent of the world's income.

Worldwide, women tend to work in jobs that have little prestige and low or no pay, where no product is produced, and where they are the facilitators for others. Women are also more likely to hold positions of little or no authority within the work environment and to receive lower wages than men (WHO 2006; ILO 2009; IPC 2008). Women work 66 percent of the total number of labor hours worldwide, yet receive just 10 percent of the world's income (Foerstel 2008).

No matter what the job, if a woman does it, it is likely to be valued less than if a man does it. For example, in the early 1800s, 90 percent of all clerks were men and being a clerk was a prestigious profession. As the job became more routine, in part because of the advent of the typewriter, the pay and prestige of the job declined and the number of female clerks increased. Today, 92.2 percent of clerks are female (U.S. Census Bureau 2011), and the position is one of relatively low pay and prestige.

The concentration of women in certain occupations and men in other occupations is referred to as **occupational sex segregation.** Although occupational sex segregation remains high, as indicated by the first column in Table 10.2, it has decreased in recent years for some occupations. For example, between 1983 and 2009, the percentage of female physicians nearly doubled from 16 percent to 32 percent, female dentists increased from 7 percent to 30 percent, and female clergy increased from 6 percent to 17 percent (U.S. Census Bureau 2009, 2011).

Although the pace is slower, increasingly men are applying for jobs that women traditionally held. Spurred by the loss of jobs in the manufacturing sector and the recent economic crisis, many jobs traditionally defined as male (e.g., auto worker, construction worker) have been lost. Thus, over the last 20 years, there has been a significant increase in the number of men in traditionally held female jobs: for example, a 50 percent increase in male telephone operators, 45 percent increase in male tellers, and 40 percent increase in male preschool and kindergarten teachers (Bourin & Blakemore 2008). Some

occupational sex segregation The concentration of women in certain occupations and men in other occupations.

TABLE 10.2 The Wage Gap in the 10 Most Common Occupations for Women and Men (Full-Time Workers Only), 2010

	PERCENT OF WORKERS IN OCCUPATION THAT ARE FEMALE	MEDIAN WEEKLY EARNINGS FOR WOMEN	MEDIAN WEEKLY EARNINGS FOR MEN	WOMEN'S EARNINGS AS PERCENT OF MEN'S
	44.7%	$669	$824	81.2%
10 MOST COMMON OCCUPATIONS FOR WOMEN				
All female workers (44,472,000)				
Secretaries and administrative assistants	95.7%	$657	$725	90.6%
Registered nurses	90.5%	$1,039	$1,201	86.5%
Elementary and middle school teachers	80.9%	$931	$1,024	90.9%
Nursing, psychiatric, and home health aides	87.0%	$427	$488	87.5%
Customer service representatives	66.2%	$586	$614	95.4%
First-line supervisors/managers of retail sales workers	45.5%	$578	$782	73.9%
Cashiers	71.5%	$366	$400	91.5%
First-line supervisors/managers of office and administrative workers	66.9%	$726	$890	81.6%
Receptionists and information clerks	92.5%	$529	$547	96.7%
Accountants and auditors	59.1%	$953	$1,273	74.9%
10 MOST COMMON OCCUPATIONS FOR MEN				
All male workers (55,059,000)				
Driver/sales workers and truck drivers	3.3%	$492	$691	71.2%
Managers, all other	36.6%	$1,045	$1,395	74.9%
First-line supervisors/managers of retail sales workers	45.5%	$578	$782	73.9%
Janitors and building cleaners	28.5%	$400	$494	81.0%
Retail salespeople	42.1%	$421	$651	64.7%
Laborers and freight, stock, and material movers, hand	15.8%	$419	$508	82.5%
Construction laborers	2.2%	*	$569	*
Sales representatives, wholesale and manufacturing	24.0%	$842	$983	85.7%
Computer software engineers	20.6%	$1,445	$1,590	90.9%
Chief executives	25.6%	$1,598	$2,217	72.1%

Source: Hegewisch et al. 2011.

evidence suggests that men in traditionally held female jobs have an advantage in hiring, promotion, and salaries called the **glass escalator effect** (Williams 2007). Table 10.2, for example, indicates that secretaries and administrative assistants are overwhelmingly female—95.7 percent. Yet the average median weekly salary for male secretaries is $725 compared to $657 for female secretaries. Despite the increase of men into traditionally held female occupations, women are still heavily represented in low-prestige, low-wage, **pink-collar jobs** that offer few benefits.

Sex segregation in occupations continues for several reasons. First, cultural beliefs about what is an "appropriate" job for a man or a woman still exist. Snyder and Green's (2008) analysis of nurses in the United States is a case in point. Using survey data and

glass escalator effect The tendency for men seeking or working in traditionally female occupations to benefit from their minority status.

pink-collar jobs Jobs that offer few benefits, often have low prestige, and are disproportionately held by women.

TABLE 10.3 Median Gender Pay Gap, 43 countries, 2009*

	MEDIAN GROSS HOURLY WAGE ($)		
	MALE	FEMALE	MEDIAN GENDER PAY GAP %**
All respondents	20.0	14.8	26.3
With children	23.1	15.7	31.9
Without children	17.3	13.9	19.9
Has full-time hours	20.2	15.3	24.3
Does not have full-time hours	15.2	12.2	19.7

*There may be a slight sample bias in that younger and more educated individuals are more likely to complete an online survey.

**For example, for all respondents, males were paid 26.3 percent higher gross hourly wages than females.

Source: ITUC 2011. From *Decisions for Work: An Examination of the Factors Influencing Women's Decisions For Work.* Copyright © 2010. Reprinted with permission.

in-depth interviews, the researchers identified patterns of sex segregation. Over 88 percent of all patient-care nurses were in sex-specific specialties (e.g., intensive care and psychiatry for male nurses, and labor or delivery and outpatient services for female nurses). Interestingly, although women rarely mentioned gender as a reason for their choice of specialty, male nurses frequently did so, acknowledging the "process of gender affirmation that led them to seek out 'masculine' positions within what was otherwise construed to be a women's profession" (p. 291).

Second, opportunity structures for men and women differ. For example, women and men, upon career entry, are often channeled by employers into gender-specific jobs that carry different wages and promotion opportunities. However, even women in higher-paying jobs may be victimized by a **glass ceiling**—an often invisible barrier that prevents women and other minorities from moving into top corporate positions. For example, women and minorities have different social networks than do white men, which contributes to this barrier. White men in high-paying jobs are more likely to have interpersonal connections with individuals in positions of authority (Padavic & Reskin 2002). In addition, women often find that their opportunities for career advancement are adversely affected after returning from family leave. Female lawyers returning from maternity leave found their career mobility stalled after being reassigned to less prestigious cases (Williams 2000).

Finally, there is evidence that working mothers pay a price for motherhood. Using an experimental design, Correll et al. (2007) report that, even when qualifications, background, and work experience were held constant, "evaluators rated mothers as less competent and committed to paid work than non-mothers" (p. 1,332). Other examples of the **"motherhood penalty"** include women who feel pressured to choose professions that permit flexible hours and career paths, sometimes known as mommy tracks (Moen & Yu 2000). Thus, women dominate the field of elementary education, which permits them to be home when their children are not in school. Nursing, also dominated by women, often offers flexible hours. Although the type of career pursued may be the woman's choice, it is a **structured choice**—a choice among limited options as a result of the structure of society (see Table 10.3).

Income and Structural Sexism

Worldwide, on the average, women earn between 25 and 50 percent of what men earn (see Table 10.3) (Social Watch 2009). In 2010, full-time working women in the United States earned, on the average, 81 percent of the weekly median earnings of full-time working men, a 1.0 percent increase from the previous year (IWPR 2011a). The gender pay gap not only varies over time, it also varies by state. In Rhode Island, a female college graduate over the age of 25 working full-time earns, on the average, 79 percent of what her male counterpart earns. In Texas, she earns just 68 percent of what a full-time working man earns (AAUW 2011b).

Racial differences also exist. Although women in general earn 81 percent as much as men, Black American and Hispanic American women earn just 69.9 percent and 59.8 percent, respectively, of white men's salaries (IWPR 2011a). Even among celebrities, a significant income gap exists. According to *Forbes* magazine's Celebrity 100 issue, the top female athlete (Maria Sharapova) earned less than 50 percent of what the top male athlete (Tiger Woods) earned in 2010 (Basenhausen 2011).

There are several arguments as to why the gender pay gap exists. One, the **human capital hypothesis** holds that pay differences between females and males are a function of differences in women's and men's levels of education, skills, training, work experience,

glass ceiling An invisible barrier that prevents women and other minorities from moving into top corporate positions.

motherhood penalty The tendency for women with children, particularly young children, to be disadvantaged in hiring, wages, and the like compared to women without children.

structured choice Choices that are limited by the structure of society.

human capital hypothesis The hypothesis that pay differences between females and males are a function of differences in women's and men's levels of education, skills, training, and work experience.

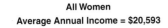

All Women
Average Annual Income = $20,593

Other
1.8%
($364)

Earnings
18.9%
($3,887)

Assets
14.3%
($2,939)

Pensions
17.6%
($3,620)

Social
Security
47.5%
($9,782)

All Men
Average Annual Income = $38,350

Earnings
31.1%
($11,915)

Other
1.3%
($449)

Pensions
23.1%
($8,863)

Assets
11.5%
($4,404)

Social
Security
33.0%
($12,669)

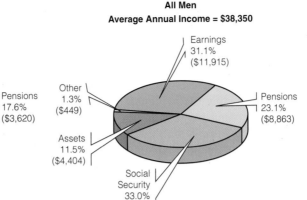

Figure 10.2: **Average Income Amounts by Source for Older Women and Men (Aged 65 and Older)**
Source: Gault 2010.

Note: In 2004 constant dollars.

and the like. For example, Rose and Hartman (2008), using a longitudinal data set, found that over a 15-year period, women worked fewer years than men and, when they worked, they worked fewer hours per year. Rose and Hartman concluded that over ". . . the 15 years, the more likely a woman is to have dependent children and be married, the more likely she is to be a low earner and have fewer hours in the labor market" (Rose & Hartman 2008, p. 1). Bertrand et al. (2009) reported similar findings, concluding that the "presence of children is associated with less accumulation of job experience, more career interruptions, and shorter work hours for female MBAs but not for male MBAs" (p. 24). Based on their analysis, the authors concluded that a decade after graduation, female MBAs earn an average annual salary of $243,481 and male MBAs earn an average annual salary of $442,353. Lower incomes over time create a significant deficit later in life. This is particularly true given a woman's higher life expectancy and the exhaustion of household savings when her husband becomes ill (see Figure 10.2).

One variation of the human capital hypothesis is called the *life-cycle human capital hypothesis*. Here it is argued that women have less incentive to invest in education and marketable skills because they know that they will be working less than their male counterparts as wives and mothers, and that their careers will be interrupted by family responsibilities. Alternatively, men's incentives to acquire marketable skills increase with greater family responsibilities and it is, or so it is argued, this human capital difference that is responsible for the female–male pay gap (Polachek 2006).

Human capital theorists also argue that women make educational choices (e.g., school attended, major, etc.) that limit their occupational opportunities and future earnings. Women, for example, are more likely to major in the humanities, education, or the social sciences rather than science and engineering, which results in reduced incomes (York 2008). Ironically, the relatively low rate of women majoring in science and engineering is a function of the lack of female faculty in science and engineering departments (Sonnert et al. 2007; AAUW 2011a).

Female–male human capital differences are a result of structural constraints (e.g., no national system of child care) as well as expectations that women should remain in the home. The results of a survey indicate that few working mothers (11 percent) or stay-at-home mothers (10 percent) believe that a full-time working mother is the "ideal situation for a child" (Pew Research Center 2007). Further, 72 percent of men with children under 18 years of age responded that working full-time was the ideal situation for them compared to only 20 percent of women with children under the age of 18.

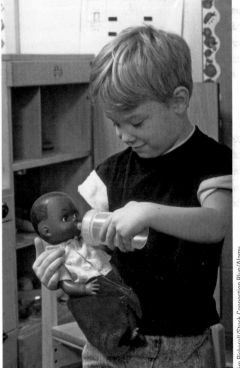

Societal definitions of the appropriateness of gender roles have traditionally restricted women and men in terms of educational, occupational, and leisure-time pursuits. Fifty years ago, little boys playing with dolls and displaying nurturing behaviors would have been unheard of.

The second explanation for the gender gap is called the **devaluation hypothesis.** It argues that women are paid less because the work they perform is socially defined as less valuable than the work men perform. Guy and Newman (2004) argued that these jobs are undervalued in part because they include a significant amount of **emotion work**—that is, work that involves caring, negotiating, and empathizing with people, which is rarely specified in job descriptions or performance evaluations.

Finally, there is evidence that, even when women and men have equal education and experience (and, therefore, not a matter of human capital differences) and are in the same occupations (and, therefore, not a matter of women's work being devalued), pay differentials remain. Table 10.2 indicates that men make more than women in each of the 10 most sex-segregated occupations for females and males. Even among elementary and middle school teachers, a profession that is 81 percent female, women earn a weekly median salary of $931 and men earn a weekly median salary of $1,024—a difference of $4,650 annually. An examination of high- and low-wage professions is also telling. For example, the profession with the highest median weekly earning for women is physician, and about a third of all physicians are women. Yet female physicians earn just 71 percent of what male physicians earn (IWPR 2011b).

Clearly, the human capital differences and the devaluation hypotheses may explain some of the variation in women's and men's earnings. But women's and men's earnings vary for many different reasons including discrimination in education, and in hiring, promotions, and salaries in the workplace. For example, the American Association of University Women (2007) reported that:

> . . . just one year after college graduation, women earn only 80 percent of what their male counterparts earn. Ten years after graduation, women fall further behind, earning only 69 percent of what men earn. Even when controlling for hours, occupation, parenthood, and other factors known to affect earnings, the research indicates that one-quarter of the pay gap remains unexplained and is likely due to sex discrimination. (p. 1)

What Do You Think? In 2010, Ana Lopez-Rodriguez was fired from Area Temps, a northeastern Ohio temporary job agency for not participating in a cover-up. She just "couldn't hide it anymore," she told authorities. The company had been using code words (e.g., "chocolate cupcake" for young black women, "figure skater" for white women, and "basketball player" for black men) to fulfill the discriminatory hiring requests of their clients (EEOC 2011a). Although the company was fined $650,000, the clients were never punished. Do you think they should have been?

devaluation hypothesis The hypothesis that women are paid less because the work they perform is socially defined as less valuable than the work men perform.

emotion work Work that involves caring for, negotiating, and empathizing with people.

comparable worth The belief that individuals in occupations, even in different occupations, should be paid equally if the job requires "comparable" levels of education, training, and responsibility.

Similarly, when the Government Accountability Office examined the gender pay gap between female and male federal employees, some of it was explained by female–male differences in occupation, education, and work experience. However, that portion of the pay gap that could not be explained by such variables grew over the time period studied—1988 to 2007. The report concludes that the portion of the pay gap that cannot be explained "neither confirms nor refutes the presence of discriminatory practices" in government wages (Sherrill 2009, p. 2).

Comparable worth refers to the belief that individuals in occupations, even in different occupations, should be paid equally if the job requires "comparable" levels of education, training, and responsibility. Is it fair that "maids and housecleaners," who are 87 percent women, earn $3,000 a year less than "janitors and building cleaners" (Billitteri 2008)? Does the pay differential reflect differences in the requirements of the job or the traditionally low esteem women's work has held? Advocates of comparable worth, consistent with the devaluation hypothesis, argue that comparable worth policies are the only way to ensure pay equity. Critics argue that comparable worth policies undermine our system of market-based wages and are demeaning to women (Billitteri 2008).

Politics and Structural Sexism

Women received the right to vote in the United States in 1920, with the passage of the Nineteenth Amendment. Even though this amendment went into effect more than 90 years ago, women still play a rather minor role in the political arena. In general, the more important the political office is, the lower the probability that a woman will hold it. Although women constitute 52 percent of the population, the United States has never had a woman president or vice president and, until 2010, when Justice Elena Kagan was appointed, had only three female Supreme Court Justices in the history of the Court. The highest-ranking women ever to serve in the U.S. government have been former Secretaries of State Madeleine Albright and Condoleezza Rice, and current Secretary of State Hillary Clinton. In 2011, women represented only 12 percent of all governors and held only 16.6 percent of all U.S. Congressional seats. Further, women held an average of 23.6 percent of all state legislative seats, with Colorado having the highest female representation at 41

Shortly after Prime Minister José Luis Rodríguez Zapatero appointed Spain's first female Minister of Defense, Carmé Chacón, she gave birth to a son. Chacón's 2008 appointment was part of Spain's "commitment to gender parity" in a "traditionally macho society whose new equality laws are among the most progressive in Europe" (Burnett 2008, p. 1). Today, she officiates over 125,000 soldiers and is in charge of a $12.1 billion defense budget (Zuber 2010).

percent (CAWP 2011a). Worldwide, the percentage of women in legislative posts varies by region of the world (see Figure 10.3—international representation).

In response to the underrepresentation of women in the political arena, some countries have instituted electoral quotas. There are 74 countries "where quotas have been

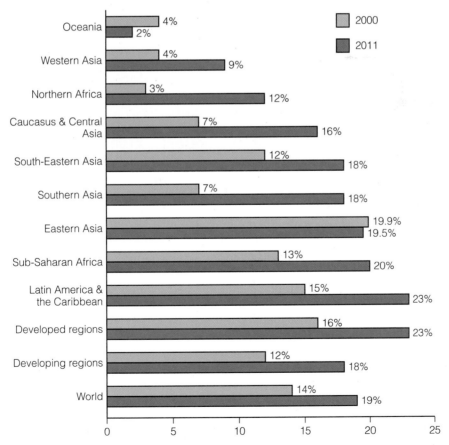

Figure 10.3: Proportion of Seats Held by Women in National Legislatures, 2000 and 2011
Source: United Nations 2011. From *The Millennium Development Goals Report 2011* (page 26). Copyright © 2011 United Nations. All rights reserved. Reprinted with permission.

implemented in the constitution, regulations, and laws, or where political parties have implemented their own internal quotas" (IDEA 2009, p. 1). Quotas are particularly useful in countries where women are underrepresented and their candidacy is threatened by long-standing patriarchal traditions. In Iraq, for example a new law requires that 25 percent of the seats in the 18 provincial councils in Iraq be allocated to women. However, female candidates are rarely taken seriously, often described as "political décor" (Allam 2010).

The relative absence of women in politics, as in higher education and in high-paying, high-prestige jobs in general, is a consequence of structural limitations. Running for office requires large sums of money, the political backing of powerful individuals and interest groups, and a willingness of the voting public to elect women. Disproportionately lacking these resources, minority women have even greater structural barriers to election and, not surprisingly, represent an even smaller percentage of elected officials. Of the 535 U.S. congressional representatives, women of color represent just 6 percent, and no minority woman is presently a member of the U.S. Senate (CAWP 2011b).

> Running for office requires large sums of money, the political backing of powerful individuals and interest groups, and a willingness of the voting public to elect women. Disproportionately lacking these resources, minority women have even greater structural barriers to election and, not surprisingly, represent an even smaller percentage of elected officials.

There is also evidence of gender discrimination against female candidates. In an experiment where two congressional candidates' credentials were presented to a sample of respondents—in one case as Ann Clark and in the other as Andrew Clark—Republican respondents were significantly more likely to say they would vote for a father with young children rather than a mother with young children. They were also more likely to vote for women without small children than women with small children—additional evidence of the "motherhood penalty." The opposite pattern was detected for Democrat respondents (Morin & Taylor 2008).

Finally, gender bias and sexist commentaries dominate the political area. In a gender analysis of the 2008 election, Pozner (2009) notes that Sarah Palin's vice presidential candidacy was characterized by sexual innuendos and trivializing taunts. Palin was described as "by the far the best-looking woman ever to rise to such heights" and "the first indisputably fertile female to dare dance with the big dogs." Donny Deutsch of MSNBC stated that Palin was divisive because "this is the first woman in power with sexual appeal. . . . We're used to seeing a woman in power as nonthreatening." The logical conclusion of such a statement is not lost on Pozner, who asks, "So, Palin's polarizing because she's . . . hot?" (p. 1).

Finally, arguing that part of the American ethic of politics is whether or not you are "man enough to serve," Faludi (2008) contends that John McCain, who survived years in a prisoner of war camp, typifies the American ideal of what it means to be a man. On the other hand, Don Imus, as Faludi notes, "never to be outdone in the sexual slur department, dubbed Mr. Obama as a 'sissy boy'" (pp. 1–2). Questioning a candidates' gender identity or sexual orientation is not uncommon. After Elena Kagan's nomination to the U.S. Supreme Court, conservative talk show host Rush Limbaugh, commented that in Kagan's nomination, "Obama has chosen himself a different gender" (quoted in Media Matters 2010, p.1). Thus, to some, a candidate's gender performance, or at least the perception of it, should play a role in an individual's voting behavior.

Civil Rights, the Law, and Structural Sexism

In many countries, victims of gender discrimination cannot bring their cases to court. This is not true in the United States. The 1963 Equal Pay Act and Title VII of the 1964 Civil Rights Act make it illegal for employers to discriminate in wages or employment on the basis of sex. Nevertheless, such discrimination still occurs as evidenced by the thousands of grievances filed each year with the Equal Employment Opportunity Commission (EEOC)—29,029 grievances in 2010 (EEOC 2011b).

Some child brides are as young as 5 years old and rarely know the ramifications of what's happening to them. Here, a 16-year-old screams in protest as she is carried in a cart to her new husband's village.

Stephanie Sinclair / VII

One technique employers use to justify differences in pay is the use of different job titles for the same type of work. Repeatedly the courts have ruled, however, that jobs that are "substantially equal," regardless of title, must result in equal pay. For example, in one of the largest employment discrimination suits ever filed, Wal-Mart Stores, Inc. has been named in a class-action suit in which 2 million former and current female employees allege that management discriminated against them by (1) paying male employees more and (2) denying promotions to women. In court, Wal-Mart attorneys argued that because the stores were independently owned, there could be no company-wide policy or practice of discrimination in place (Kravets 2007). In 2011, the U.S. Supreme Court held for Wal-Mart Stores, Inc., finding that, according to Justice Antonin Scalia, the workers "provided no convincing proof of a company-wide discriminatory pay and promotion policy (Stohr 2011, p.1).

Discrimination, although illegal, takes place at both the institutional and the individual levels (see Chapter 9). Institutional discrimination includes screening devices designed for men, hiring preferences for veterans, the practice of promoting from within an organization based on seniority, and male-dominated recruiting networks (Reskin & McBrier 2000). For example, the Augusta National Golf Club, home of the Masters Golf Tournament and a virtual "who's who of the corporate world," has refused to change its policy, forbidding women from joining, despite years of political pressure from women's groups and negative publicity (Rudnanski 2011). One of the most blatant forms of individual discrimination is sexual harassment, discussed later in this chapter.

In the United States, women often have difficulty obtaining home mortgages or rental property because they have lower incomes, shorter work histories, and less collateral. Until fairly recently, husbands who raped their wives were exempt from prosecution. Even today, some states require a legal separation agreement and/or separate residences for a wife who has been raped to receive full protection under the law. Women in the military have traditionally been restricted in the duties they can perform, and, finally, since the U.S. Supreme Court's 1973 *Roe v. Wade* decision, which made abortion legal, the right of a woman to obtain an abortion has steadily been limited and narrowed by subsequent legislative acts and judicial decisions (Connolly 2005; Cosgrove 2011). The debate has continued with several recent court decisions (see Chapter 14).

What Do You Think? Throughout the world, as many as 10 to 12 million young girls (Gorney 2011)—some as young as age 5—have been forced into marriage by their parents or as a result of human trafficking. Consequently, they have little or no education, bear children long before they are physically mature and, in many cases, die in childbirth. Do you think that all countries where this tradition exists should have anti–child marriage laws? What if such laws contradict religious beliefs?

The Social Construction of Gender Roles: Cultural Sexism

As social constructionists note, structural sexism is supported by a system of cultural sexism that perpetuates beliefs about the differences between women and men. **Cultural sexism** refers to the ways the culture of society—its norms, values, beliefs, and symbols—perpetuate the subordination of an individual or group because of the sex classification of that individual or group.

Cultural sexism takes place in a variety of settings, including the family, the school, and the media, as well as in everyday interactions.

Family Relations and Cultural Sexism

From birth, males and females are treated differently. The toys that male and female children receive convey different messages about appropriate gender roles. **Gender roles** are patterns of socially defined behaviors and expectations, associated with being female or male. For example, a study by Professor Becky Francis of Roehampton University in England, examined the impact of educational toys on 3- to 5-year-olds' learning. When parents were asked their child's favorite toys, boys' toys "involved action, construction, and machinery," whereas girls' toys were more often "dolls and perceived 'feminine' interests, such as hairdressing" (Lepkowska 2008, p. 1). After purchasing and analyzing the toys parents selected, Francis concluded that girls' toys had limited "learning potential" whereas boys' toys "were far more diverse" and propelled boys "into a world of action as well as technology" designed to "be exciting and stimulating" (quoted in Lepkowska 2008, p. 1).

Household Division of Labor Globally, women and girls continue to be responsible for household maintenance including cooking, gathering firewood and fetching water, and taking care of younger siblings (IPC 2008). In a study of household labor in 10 Western countries, Bittman and Wajcman (2000) report that "women continue to be responsible for the majority of hours of unpaid labor," ranging from a low of 70 percent in Sweden to a high of 88 percent in Italy (p. 173). A study in Mexico found that women in the paid labor force performed household chores for an additional 33 hours of work each week; men contributed 6 hours a week to domestic chores (UNICEF 2007).

In the United States, girls and boys work within the home in approximately equal amounts until the age of 18 when girls' household labor begins to increase. Although men's share of the household labor has more than doubled in the last 25 years, men's share of inside chores is still only between 25 percent and 33 percent of the total (Coltrane & Adams 2008). The fact that women, even when working full-time, contribute significantly more hours to home care than men is known as the "second shift" (Hochschild 1989). According to Coltrane and Adams (2008), husbands of working wives are most likely to do child care, meal preparation, cleanup after meals (e.g., washing dishes), and house cleaning, and least likely to do laundry. The two researchers predict that, as the hours worked outside the home by employed husbands and wives converge, there will be a continued trend toward egalitarian households.

Although some changes have occurred, the traditional division of labor remains to a large extent, even in terms of what one "expects" after marriage (Askari et al., 2010). Three explanations emerge from the literature. The first explanation is the *time-availability approach*. Consistent with the structural-functionalist perspective, this position claims that role performance is a function of who has the time to accomplish certain tasks. Because women are more likely to be at home, they are more likely to perform domestic chores.

A second explanation is the *relative resources approach*. This explanation, consistent with a conflict perspective, suggests that the spouse with the least power is relegated the most unrewarding tasks. Because men, on average, have more education, higher incomes, and more prestigious occupations, they are less responsible for domestic labor.

cultural sexism The ways in which the culture of society perpetuates the subordination of an individual or group based on the sex classification of that individual or group.

gender roles Patterns of socially defined behaviors and expectations associated with being female or male.

Thus, for example, because women earn less money than men do, on average, they turn down overtime and other work opportunities to take care of children and household responsibilities, which subsequently reduces their earnings potential even further (Williams 2000).

Gender role ideology, the final explanation, is consistent with a symbolic interactionist perspective. It argues that the division of labor is a consequence of traditional socialization and the accompanying attitudes and beliefs. Women and men have been socialized to perform various roles and to expect their partners to perform other complementary roles. Women typically take care of the house, and men take care of the yard. These patterns begin with household chores assigned to girls and boys and are learned through the media, schools, books, and toys. A test of the three positions found that, although all three had some support, gender role ideology was the weakest of the three in predicting work allocation (Bianchi et al. 2000). It should be noted that the three positions are not mutually exclusive. For example, Mannino and Deutsch (2007), in their longitudinal study of 81 married women, found support for the relative resources and gender role ideology perspectives.

© Jade Brookbank/Getty Images

According to Hochschild (1989), women are expected to work "second shifts" by having gainful outside employment as well as performing household chores and child care once they arrive home after a day's work. Although research indicates that women contribute significantly more hours to home and child care than their male counterparts, males are increasingly contributing to household labor.

The School Experience and Cultural Sexism

Sexism is also evident in schools (see Chapter 8). The bulk of research on gender images in books and other instructional materials documents the way males and females are portrayed stereotypically. For example, in a study of 200 "top-selling" children's picture books, women and girls were significantly underrepresented, with twice as many male title and main characters. Males were also more likely to be in the illustrations, to be pictured in the outdoors and, if an adult, to be visibly portrayed as employed outside of the home. Both men and women were over nine times more likely to be pictured in traditional rather than in nontraditional occupations, and "female main characters . . . were more than three times more likely than were male main characters . . . to perform nurturing or caring behaviors" (Hamilton et al. 2006, p. 761). Further, children's coloring books are stereotypical. In an analysis of 889 characters in 54 contemporary American coloring books, Fitzpatrick and McPherson (2010) report that males were significantly more likely to be pictured and, when pictured, significantly more likely to be active, older, and more powerful.

The differing expectations and/or encouragement that females and males receive contribute to their varying abilities, as measured by standardized tests, in disciplines such as reading, math, and science. Are such differences a matter of aptitude? Social science research would indicate otherwise. For example, in an experiment at the University of Waterloo, male and female college students, all of whom were good in math, were shown either gender-stereotyped or gender-neutral advertisements. When, subsequently, female students who had seen the female-stereotyped advertisements took a math test, they performed significantly lower than women who had seen the gender-neutral advertisements (Begley 2000).

Further, the results of a recent study at the University of Virginia document the pervasiveness of gender stereotypes and their impact on science achievement (Nosek et al. 2009). Over 500,000 respondents from 34 countries participated in the research designed to measure *implicit bias*—bias that we are unaware of. About 70 percent of the respondents were found to have stereotypical views of the relationship between gender and science in the predictable direction. In countries where gender stereotypes were the strongest, that is, being male was associated with science but being female was not, males performed better and females performed worse on standardized math and science tests. The authors suggest that "implicit stereotypes and sex differences in science participation and performance are mutually reinforcing, contributing to the persistent gender gap in science engagement" (Nosek et al. 2009, p. 10,593).

As discussed in Chapter 8, Title IX of the 1972 Educational Amendments Act prohibits sex discrimination in educational programs and activities that receive federal financial assistance (NCES 2009). An evaluation of Title IX suggests that sex discrimination continues at many levels. Women are underrepresented in administrative positions at all levels of education; women's and girls' athletic programs lag behind boys' and men's programs in terms of participation, resources, and coaching; despite significant gains, women and girls continue to perform behind boys and men in science and math; and career and technical education remain sex-segregated, in part, because of stereotyped career counseling in high school (NCWGE 2009).

> **What Do You Think?** In 2009, Amber Parker, coach of Franklin County girl's varsity basketball team and mother of one of the players, sued the county alleging violations of Title IX. Specifically, the plaintiffs argued that scheduling differences (i.e., the boys games were scheduled Friday and Saturday nights, and the girls on weekday nights) created a hardship on the female basketball players in that it "unfairly placed academic burdens on the girls, who competed on school nights, [and] were deprived of crowd support, and were made to feel like second-class athletes" (AAUW 2011c, p. 1). Do you think female and male athletic teams should have rotating schedules?

Additionally, in 2006, the U.S. Department of Education granted school districts permission to expand single-sex education if it can be shown to accomplish a specific educational objective. The assumption behind single-sex education is that boys and girls learn differently—have "pink and blue brains," if you will—a belief for which there is very little scientific support (see Chapter 8). Single-sex education also furthers sex and gender binaries and ignores that, "[i]n a world of ever-increasing visibility of gender diversity . . . single-sex schooling is an anachronism—one that has the potential to take us back to a time when females and males who behaved outside gender norms were perceived as 'problems' instead of as people" (Jackson 2010, p. 237).

Sexism is also reflected in the way that teachers treat their students. Millions of young girls are subjected to sexual harassment by male teachers, who then fail them when they refuse the teachers' sexual advances (Quist-Areton 2003). There is also convincing evidence that elementary and secondary school teachers pay more attention to boys than to girls—talking to them more, asking them more questions, listening to them more, counseling them more, giving them more extended directions, and criticizing and rewarding them more frequently. In *Still Failing at Fairness,* Sadker and Zittleman (2009) recount a fifth grade teacher's instructions to her students. "There are too many of us here to all shout out at once. I want you to raise your hands, and then I'll call on you. If you shout out, I'll pick somebody else." The discussion on presidents continues, with Stephen calling out (pp. 65–66):

Stephen: I think Lincoln was the best president. He held the country together during a war.
Teacher: A lot of historians would agree with you.
Kelvin (seeing that nothing happened to Stephen, calls out): I don't. Lincoln was OK but my Dad liked Reagan. He always said Reagan was a great president.
David (calling out): Reagan? Are you kidding?

Teacher: Who do you think our best president was, David?

David: FDR. He saved us from the Depression.

Max (calling out): I don't think it's right to pick one best president. There were a lot of good ones.

Teacher: That's interesting.

Rebecca (calling out): I don't think the presidents today are as good as the ones we used to have.

Teacher: Ok, Rebecca. But you forgot the rule. You're supposed to raise your hand.

Gender-based interactions are not the only things that affect student outcomes. Dee (2006) tested the effect of teachers' gender on school performance by using the National Education Longitudinal Survey. Analyzing data from a nationally representative sample of eighth grade boys and girls and their teachers, he found that (1) when the teacher is a female, boys are more likely to be seen as disruptive; (2) when the teacher is a female, girls are less likely to be seen as either disruptive or inattentive; (3) when the teacher is male, female students are less likely to perceive a class as useful, to ask questions in class, or to look forward to the class; and (4) when the teacher is female, regardless of the subject, boys are less likely to report looking forward to the class. In conclusion, the author noted that, "simply put, girls have better educational outcomes when taught by women, and boys are better off when taught by men" (p. 71) (see also Chapter 8).

Media, Language, and Cultural Sexism

Another concern social scientists voice is the extent to which the media portray females and males in a limited and stereotypical fashion and the impact of such portrayals. For example, Levin and Kilbourne (2009), in *So Sexy So Soon,* document the sexualizing of young girls and boys. Advertising, books, cartoons, songs, toys, and television shows create:

> [a] narrow definition of femininity and sexuality [that] encourages girls to focus heavily on appearance and sex appeal. They learn at a very young age that their value is determined by how beautiful, thin, "hot," and sexy they are. And boys, who get a very narrow definition of masculinity that promotes insensitivity and macho behavior, are taught to judge girls based on how close they come to an artificial, impossible, and shallow ideal. (p. 2)

Girls and women are often invisible. For example, a study of Sunday morning television talk shows revealed that of all Sunday morning talk show guests, only 13.5 percent were female (Foser 2010).

Men are also victimized by media images. A study of 1,000 adults found that two-thirds of the respondents thought that women in television advertisements were pictured as "intelligent, assertive, and caring," whereas men were portrayed as "pathetic and silly" (Abernathy 2003). In a study of beer and liquor ads, Messner and Montez de Oca (2005) concluded that, although beer ads of the 1950s and 1960s focused on men in their work roles and only depicted women as a reflection of men's home lives, present-day ads portray young men as "bumblers" and "losers" and women as "hotties" and "bitches."

One of the largest studies of women and men in the media ever conducted is the Global Media Monitoring Project (GMMP 2010). Nearly 16,750 news stories on television, the radio, and in newspapers were monitored by 20,767 news personnel in 108 countries. The results indicate that

- Women are dramatically underrepresented as news subjects (24 percent), and, when they are the subject of news, they are most often ordinary women rather than figures of authority or experts.
- The number of stories reported by women increased up to 2005; however, between 2005 and 2010, the percentage of stories reported by women remained constant at just 37 percent.

Please use the following definition in completing this exercise: Sexist language includes words, phrases, and expressions that unnecessarily differentiate between females and males or exclude, trivialize, or diminish either gender.

For each of the following statements, choose the descriptor (1 = strongly disagree; 2 = tend to disagree; 3 = undecided; 4 = tend to agree; 5 = strongly agree) that most closely corresponds with your beliefs about language, and enter the number in the blank following each statement.

1. Women who think that being called a "chairman" is sexist are misinterpreting the word *chairman*. ___
2. We should not change the way the English language has traditionally been written and spoken. ___
3. Worrying about sexist language is a trivial activity. ___
4. If the original meaning of the word *he* was "person," we should continue to use *he* to refer to both males and females today. ___
5. When people use the term *man and wife*, the expression is not sexist if the users don't mean it to be. ___
6. The English language will never be changed because it is too deeply ingrained in the culture. ___
7. The elimination of sexist language is an important goal. ___
8. Most publication guidelines require newspaper writers to avoid using ethnic and racial slurs. So, these guidelines should also require writers to avoid sexist language. ___
9. Sexist language is related to sexist treatment of people in society. ___
10. When teachers talk about the history of the United States, they should change expressions, such as "our forefathers," to expressions that include women. ___
11. Teachers who require students to use nonsexist language are unfairly forcing their political views upon their students. ___
12. Although change is difficult, we still should try to eliminate sexist language. ___

Scoring

All items are scored on a 5-point Likert-type scale. Items 1, 2, 3, 4, 5, 6, and 11 are reverse-scored. High scores (4 to 5) indicate a positive attitude toward inclusive language; low scores (1 to 2) indicate a negative attitude toward inclusive language. A score of 3 indicates neutrality or uncertainty. Each respondent's score on the instrument is the total of all the items.

Interpretation

The range of possible total scores on the 12-item IASNL is 12 to 60. Across the 12 items, total scores between 42.1 and 60 reflect a supportive attitude toward nonsexist language; total scores between 12 and 30 reflect a negative attitude toward nonsexist language; and total scores between 30.1 and 42 reflect a neutral attitude.

Sources: Parks & Roberton 2000; Parks & Roberton 2001.

- Older women have achieved gender parity with their male counterparts, each presenting about half of all television news stories presented by 50- to 64-year-olds.
- Male journalists are more likely to work on the "hard" news (e.g., politics), whereas women are more often assigned "soft stories" (e.g., social issues).
- Just 13 percent of all news stories focus specifically on women, and just over 5 percent of all news stories focus on gender equality or inequality.

As with media images, both the words we use and the way we use them can reflect gender inequality. Radio talk show host Rush Limbaugh, referring to female reporters as "infobabes" and stating that they have "chickified the news," is demeaning to women and men (Media Matters 2009). The term *nurse* carries the meaning of "a woman who . . ." and the term *engineer* suggests "a man who. . . ." Terms such as *broad, old maid,* and *spinster* have no male counterparts. Sexually active teenage females are described by terms carrying negative connotations, whereas terms for equally sexually active male teenagers are considered complimentary.

Language is so gender-stereotyped that the placement of male or female before titles is sometimes necessary, as in the case of "female police officer" or "male prostitute." Furthermore, as symbolic interactionists note, the embedded meanings of words carry expectations of behavior. This chapter's *Self and Society* assesses your attitudes toward language and gender.

> Language is so gender-stereotyped that the placement of female or male before titles is sometimes necessary, as in the case of "female police officer" or "male prostitute."

Religion and Cultural Sexism

Most Americans claim membership in a church, synagogue, or mosque. Research indicates that women attend religious services more often, rate religion as more important to their lives, and are more likely to believe in an afterlife than are men (Davis et al. 2002; Smith 2006; Adams 2007). In general, religious teachings have tended to promote traditional conceptions of gender. For example, in 2009, the Vatican began investigating the religious and secular lives of U.S. nuns (Goodstein 2009). Many nuns fear that the investigation is motivated by the church's patriarchal need to control them, some of whom have stopped wearing the traditional habit, live outside of convents, have joined advocacy groups, work in academia, and advocate ordination of female priests.

Gallagher (2004) found that most evangelical Christians believe that marriage should be considered an equal partnership, but also believe that the male is the head of the household. Female Orthodox Jews are not counted as part of the *minyan* (i.e., a quorum required at prayer services), are not allowed to read from the Torah, and are required to sit separately from men at religious services. In addition, Roman Catholic Church doctrine forbids the use of artificial forms of contraception, and Muslim women are required to be veiled in public at all times (Renzetti & Curran 2003). Women cannot serve as ordained religious leaders in the Roman Catholic Church, in Orthodox Jewish synagogues, or in Islamic temples.

Despite this "stained glass ceiling" (Adams 2007), women are increasingly active in leadership positions in churches and temples across the nation. For example, in 2009, the Lutheran Church in Great Britain ordained a woman bishop—the first female bishop in Great Britain (Lutheran World Federation 2009), and Alysa Stanton became the first U.S. Black American female rabbi (Kaufman 2009). Nonetheless, today, 83 percent of clergy in the United States are male (U.S. Census Bureau 2011), and even in denominations that allow for the ordination of women, female clergy often do not hold the same status as their male counterparts and are often limited in their duties (Renzetti & Curran 2003).

However, religious teachings are not all traditional in their beliefs about women and men. Quaker women have been referred to as the "mothers of feminism" because of their active role in the early feminist movement. Reform Judaism has allowed ordination of women as rabbis for more than 25 years, and gays and lesbians for more than 15 years. In addition, the women-church movement—a coalition of feminist faith-sharing groups composed primarily of Roman Catholic women—offers feminist interpretations of Christian teachings. Within many other religious denominations, individual congregations choose to interpret their religious teachings from an inclusive perspective by replacing masculine pronouns in hymns, the Bible, and other religious readings (Anderson 1997; Renzetti & Curran 2003).

Social Problems and Traditional Gender Role Socialization

One of the fundamental questions in the sociology of gender is the extent to which observed differences are a consequence of nature (i.e., innate) or nurture (i.e., environmental). It is likely, as author and neuroscientist Lise Eliot states, "Sex differences are real and some are probably present at birth, but then social factors magnify them." So, she continues, "if we, as a society, feel that gender divisions do more harm than good, it would be valuable to break them down" (quoted in Weeks 2011, p. 1).

The recent acknowledgement that gender is not binary but is best conceived of as a continuum suggests that traditional gender roles are changing. However, to a large extent, the social definitions of what it means to be a woman and what it means to be a man have varied little over the decades. These definitions, in turn, are associated with several social problems, including the feminization of poverty, social-psychological costs, death and illness, conflict in relationships, and gendered violence.

The Feminization of Poverty

Of the 1.4 billion people living in poverty, 70 percent are women and girls (*Soroptimist International* 2010). Further, women are more likely to be unemployed than men. In 2010, the world unemployment rate for men was 6.0 percent; for women, 6.5 percent

(ILO 2011b). Women are also disproportionately affected by the global economic crisis, leading labor market specialists to predict that up to an additional 22 million women may become unemployed over the next several years. Jane Hodges, Bureau Director for Gender Equality at the International Labour Organization, explained this continued trend toward the feminization of poverty: "Women's lower employment rates, weaker control over property and resources, concentration in informal and vulnerable forms of employment with lower earnings, and less social protection, all place women in a weaker position than men to weather crises" (*ILO News* 2009, p. 1).

Women in the United States make up the majority of minimum wage workers and are significantly more likely to live in poverty than men (ITUC 2009). Household incomes of transgender individuals are also low. Recent research indicates that nearly four times (4 percent) as many transgender households live on less than $10,000 a year when compared to the general population (15 percent) (Grant et al. 2011). Transgender individuals also have twice the unemployment rate than the general population.

Two groups of women are the most likely to be poor in the United States—heads of households with dependent children and women over the age of 65 who have outlived their spouses. Women with children are far more likely to work part-time than their male counterparts or women that don't have children (ITUC 2011). Further, 13.8 percent of male-headed households are below the poverty level compared to 28.7 percent of female-headed households. Hispanic (39.2 percent) and black female-headed (37.2 percent) households are the poorest of all families, headed by a single woman (U.S. Census Bureau 2011).

It is often assumed that antipoverty programs designed to address overall economic inequality will reduce the feminization of poverty. However, Brady and Kall's (2008) analysis of poverty in 18 affluent countries suggests that not only is the feminization of poverty universal, it is unique in its origins, tied to the percent of children in single-mother families and the male-to-female ratio of the elderly in a country.

The Social Psychological Costs of Gender Socialization

How we feel about ourselves begins in early childhood. Significant others, through the socialization process, expect certain behaviors and prohibit others based upon our birth sex. Both girls and boys, often as a consequence of these expectations, feel varying degrees of self-esteem, autonomy, depression, and life dissatisfaction. For example, Jose and Brown (2009) studied depression, stress, and rumination (i.e., worrying) in a co-ed sample of 10- to 17-year-olds. Each of the three dependent variables was significantly higher in females when compared to males.

Transgender individuals also suffer from self-esteem issues and depression. In a survey of 6,450 transgender adults, 41 percent reported attempting suicide. The comparable number for the general population is 1.6 percent. (Grant et al. 2011). For transgender individuals, family acceptance is an important buffer against societal prejudice and discrimination. Not surprisingly, transgender family members who are accepted and supported by their families have lower rates of risky behavior and negative experiences (see Figure 10.4).

Adolescent girls are more likely to be dissatisfied with their looks, including physical attractiveness, appearance, and body weight than adolescent boys. Canadian nonprofit group Media Awareness Network (Mnet) summarizes the present state of research on the "cult of thinness" (Mnet 2009): Research indicates that 25 percent of college-aged women have tried to lose weight using unhealthy means (e.g., vomiting, laxative abuse, fasting). Women's magazines have 10 times the advertisements and articles promoting weight loss than men's magazines, and 75 percent of the covers of women's magazines contain at least one message about women changing their appearance (e.g., cosmetic surgery, diet).

According to Mnet, media messages that perpetuate insecurities in women are simply a matter of economics. If you liked the way you looked, you wouldn't buy anything. The diet industry alone makes billions of dollars a year selling products that don't work. The social-psychological costs are high. Research on the impact of media images "indicates that

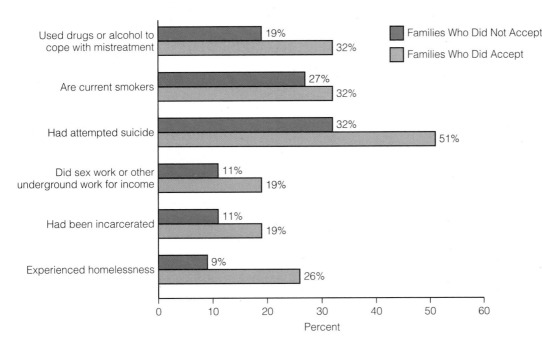

Figure 10.4: **Impact of Family Acceptance on Transgender Life Experiences**
Source: Grant et al. 2011. From Grant, Jaime M. et al., *Injustice at Every Turn: A Report of the National TransGender Discrimination Survey.* Copyright © 2011. Reprinted with permission.

(Bar chart data:)
Legend: Families Who Did Not Accept / Families Who Did Accept

- Used drugs or alcohol to cope with mistreatment: 19% / 32%
- Are current smokers: 27% / 32%
- Had attempted suicide: 32% / 51%
- Did sex work or other underground work for income: 11% / 19%
- Had been incarcerated: 11% / 19%
- Experienced homelessness: 9% / 26%

x-axis: Percent (0, 10, 20, 30, 40, 50, 60)

exposure to images of thin, young, air-brushed female bodies is linked to depression, loss of self-esteem, and the development of unhealthy eating habits in women and girls" (Mnet 2009, p. 1).

Boys too, from an early age, are concerned about body image, and as adults, their self-esteem is also linked to body shape and weight (Grogan 2008). Men are also affected by traditional gender socialization, which places enormous cultural pressure to be successful in their work and to earn high incomes. Sanchez and Crocker (2005) found that, among college-age women *and* men, the more participants were invested in traditional ideals of gender, the *lower* their self-concept and psychological well-being. Traditional male socialization also discourages males from expressing emotion and asking for help—part of what William Pollack (2000a) calls the **boy code.** Much of the boy code continues into adulthood as described in journalist Norah Vincent's 2006 book *Self-Made Man* (see this chapter's *Social Problems Research Up Close* feature).

David C. Ellis/Taxi/Getty Images

According to sociologist Michael Kimmel, the residents of Guyland, from the book of the same name (2008), represent over 22 million young men between the ages of 16 and 26, and are a major consumer group. They are targeted by Hollywood, magazines such as *Stuff, Men's Health,* and *Maxim,* by television sports channels, and by a variety of technology-related products such as computers, video games, software, and CDs.

What Do You Think? The good news is that fathers who live with their children are increasingly becoming involved in their lives, spending 6.5 hours a week with them compared to 2.6 hours a week in 1965 (Livingston & Parker 2011). The bad news is that more and more children live apart from their fathers—27 percent in 2010. What do you think accounts for the increased time spent between fathers and children who live together? Why are fewer fathers living with their children?

In a recent book, sociologist Michael Kimmel (2008) describes the lives of nearly 400 predominantly white, college-educated males between 16 and 21 years of age—the residents of "Guyland." The young men of Guyland live with male friends (e.g., in dormitories, apartments, fraternity houses) or with parents who "just don't get it." They are unmarried, and work in low-paying entry-level jobs, leaving plenty of free time to be "guys." What do the guys of Guyland do? They hook up with women; play video games; listen to, watch, and talk about sports; drink, sometimes nightly; gamble online and watch pornography; live by the adage "Bros Over Hos"; and are misogynistic, homophobic, and, sometimes, violent.

boy code A set of societal expectations that discourages males from expressing emotion, weakness, or vulnerability, or asking for help.

Journalist Norah Vincent wanted to know what it is like to be a man. So, for 18 months, she observed and participated in the world of men, not as Norah Vincent, but as "Ned"—the male persona she created for the purpose of her research. Norah Vincent describes her journey into manhood in the book *Self-Made Man* (2006).

Sample and Methods

Norah Vincent used a method of research known as participant observation—a method in which the researcher participates in the phenomenon being studied to obtain an insider's perspective of the people and/or behavior being observed. She describes her book as "a travelogue as much as anything else . . . a six-city tour of an entire continent, a woman's-eye view of one guy's approximated life, not an authoritative guide to the whole vast and variegated terrain of manhood in America" (p. 17).

To participate in and observe the world of men, Norah sought the help of a makeup artist to transform herself into Ned. She learned how to apply artificial beard stubble, wore masculine-looking glasses, had her hair cut into a man's hairstyle, wore men's clothing, and stuffed her pants with a prosthetic penis known as a "packable softie." Norah bulked up her upper body muscle through six months of weight lifting and increased protein consumption, and she flattened her breasts by wearing a flat-front sports bra two sizes too small. Norah, who already had a deep voice, hired a voice coach to help her speak like a man.

Having achieved a male identity, Ned spent 18 months exploring what it means to be a man by (1) joining a men's bowling team, (2) going to men's strip clubs, (3) dating women, (4) living at a monastery, (5) working at a sales job, and (6) attending a men's movement group.

Findings and Conclusion

Examples of Ned's experiences and observations include the following:

1. *Bowling with the guys*. One of Ned's first outings was to the local bowling alley, where he played in a Monday night men's league. It is here that he was introduced to male interaction cues:

   Our evenings together always started out slowly with a few grunted hellos that among women would have been interpreted as rude. This made my female antennae twitch a little. Were they pissed off at me about something? But among these guys no interpretation was necessary. . . . If they were pissed at you, you'd know it. These gruff greetings . . . were coming from long, wearing workdays, usually filled with hard physical labor and the slow, soul-deadening deprecation that comes of being told what to do all day. (pp. 29–30)

2. *Strip clubs*. Ned went to strip clubs to get a glimpse of "a substratum of the male sexual psyche that most women either don't know about, don't want to know about, or both" (p. 65). After weeks of observing and interacting with various strippers and other patrons, Ned offers the following reflections:

   I'd been inside a part of the male world that most women and even a lot of men never see, and I'd seen it as just another one of the boys. In those places male sexuality felt like something you weren't supposed to feel but did, like something heavy you were carrying around and had nowhere to unload except in the lap of some damaged stranger, and then only for five minutes. . . . It wasn't nearly so simple as men objectifying women. . . . Nobody won . . . nobody was more or less victimized than anyone else. The girls got money. The men got an approximation of sex and flirtation. But in the end everyone was equally debased by the experience. (pp. 90–91)

3. *Dating*. Ned found that most of the women he dated "were carrying the baggage of previous hurts at the hands of men, which in many cases had prejudiced

---

The boys and men of Guyland, as Kimmel describes them, are in "suspended animation," somewhere between boyhood and adulthood, lost in a world they don't understand and have difficulty maneuvering, frustrated by "thwarted entitlement"—the belief that minorities, women, and gays are encroaching on their territory. With compassion and concern for today's young men, Kimmel warns that bullying and hazing, binge drinking and self-medication, sexual violence, gay bashing, and suicides will continue unless, as one reviewer states, "we enable young men to chart their own paths, to stay true to themselves, and to travel safely through *Guyland*, emerging as responsible and fully formed men of integrity and honor" (HarperCollins 2008, p. 1).

The Impact of Gender Socialization on Death and Illness

Men are less likely to go to a doctor than women for a variety of structural and cultural reasons (WHO 2010). Men, for example, work longer hours than women and are more likely to be working full-time, making it difficult to see a doctor or attend preventive medicine programs that are often only available during the day. Men are also less likely to have a regular physician or to go to the hospital, even if time permits (White & Witty 2009).

At every stage of life, "American males have poorer health and a higher risk of mortality than females" (Gupta 2003, p. 84). On average, men in the United States die about three years earlier than women, although gender differences in life expectancy have

them unfairly against the male sex" (p. 100). Ned described that, as a "man" dating women . . . , I often felt attacked, judged, on the defensive. Whereas with the men I met and befriended as Ned there was a presumption of innocence—that is, you're a good guy until you prove otherwise—with women there was quite often a presumption of guilt: you're a cad like every other guy until you prove otherwise. (p. 101)

Ned experienced the difficulty that heterosexual men have in living up to the ideals of women, observing that "women wanted a take-control man, at the same time, they wanted a man who was vulnerable to them, a man who would show his colors and open his doors, someone expressive, intuitive, attuned" (p. 111).

4. *A Catholic monastery.* In the weeks Ned spent at a Catholic monastery, he observed men in a context in which women and sexual expression were, at least in principle, excluded from the male experience. Even in this context, however, male gender role expectations were evident. Ned was at the dinner table with the monks and he told Father Richard that he looked very good for his age:

As soon as the remark came out of my mouth, everyone at the table stopped eating mid-forkful and looked at me as if I had three heads. Father Richard . . . said a very

suspicious, squint-eyed "thank you," and looked away, clearly embarrassed. But the implication from other quarters was clear: "What the hell's wrong with you, kid? Don't you know that properly socialized males don't behave that way with each other?" (pp. 144–145)

5. *Work.* Ned interviewed for several jobs and landed a job as a salesman. One of his coworkers advised him, "You're a man. . . . You gotta pitch like a man." Ned learned that, in sales work, women and men had different strategies:

Girls pitched differently. They flirted. They cajoled and smiled and eased their way into the sales underhandedly, which was exactly how I'd started out trying to do it. I'd tried initially to ask for the sale the way I asked for food in a restaurant as a woman, or the way I asked for help at a gas station—pleadingly. But coming from a man this was off-color. It didn't work. It bred contempt in both men and women. . . . People see weakness in a woman and they want to help. They see weakness in a man and they want to stamp it out. (p. 213)

6. *Men's movement group.* Ned attended a men's movement group in which 25 to 30 guys met once a month. He noted that hugging was a central part of the group:

Most men don't tend to share much physical affection with their male friends,

so here the guys made a point of hugging each other long and hard at every possible opportunity as a way of offsetting what the world had long deprived them of, and what they in turn had been socialized to disallow themselves. (p. 232)

In the concluding chapter of *Self-Made Man,* Norah explains that the experience of being a guy was not what she had expected:

I had thought that by being a guy I would get to do all the things I didn't get to do as a woman, things I'd always envied about boyhood when I was a child: the perceived freedoms of being unafraid in the world. . . . But when it actually came to the business of being Ned I rarely felt free at all. (pp. 275–276)

As Ned, Norah found that "somebody is always evaluating your manhood. Whether it's other men, other women, even children. And everybody is always on the lookout for your weakness or your inadequacy" (p. 276). Norah described the male role as a "straightjacket . . . that is no less constrictive than its feminine counterpart. You're not allowed to be a complete human being. Instead you get to be a coached jumble of stoic poses. You get to be what's expected of you" (p. 276).

Source: Vincent 2006.

been shrinking (U.S. Census Bureau 2011). Traditionally defined gender roles for men are linked to high rates of cirrhosis of the liver (e.g., alcohol consumption), many cancers (e.g., tobacco use), and cardiovascular diseases (e.g., stress). Men also engage in self-destructive behaviors—poor diets, lack of exercise, higher drug use, refusal to ask for help or wear a seatbelt, and stress-related activities—more often than women. Being married improves men's health more than it does women's (Williams & Umberson 2004), in large part because wives encourage their husbands to take better care of themselves.

Women's health is also gendered. Although men have higher rates of HIV/AIDS worldwide, the disease disproportionately affects women in many areas of the world (see Chapter 2). For example, among young people with HIV, 80 percent live in sub-Saharan Africa and, of that number, about 66 percent are female (WHO 2011). Women's inequality contributes to the spread of the disease. First, in many of these societies, "women lack the power in relationships to refuse sex or negotiate protected sex" (Heyzer 2003, p. 1). Second, women are often the victims of rape and sexual assault, with little social or legal recourse. Third, gender norms often dictate that men have more sexual partners than women, putting women at greater risk.

The World Health Organization (2009, 2011) has identified additional ways traditional definitions of gender impact the health and well-being of women and girls. For example, every day, over 1,600 women die from preventable complications during pregnancy and childbirth. Primarily responsible for household duties, women are exposed to hundreds of pollutants as they cook that contribute to their disproportionately high rates of death from chronic obstructive pulmonary disease (COPD). Deaths from lung cancer and other tobacco-related illnesses are expected to rise as the tobacco industry targets women in developing countries. Finally, many women and girls throughout the world have a higher probability of suffering or dying from a variety of diseases because of gender: They are more likely to be poor, less likely to be seen as worthy of care when resources are short, and in many countries, they are forbidden to travel unaccompanied by a male, making access to a hospital difficult.

Gendered Violence

Men are more likely than women to be involved in violence—to kill and be killed; to wage war and die both as combatants and noncombatants; to take their own lives, usually with the use of a firearm; to engage in violent crimes of all types; to bully, harass, and abuse. Although the most serious of violent acts are exceptions rather than the norm, Pollack argued that male violence is a consequence of gender socialization and definitions of masculinity that hold "as long as nobody is seriously hurt, no lethal weapons are employed, and especially within the framework of sports and games—football, soccer, boxing, wrestling—aggression and violence are widely accepted and even encouraged in boys" (Pollack 2000b, p. 40).

Similarly, sociologist Michael Kimmel (2011) notes that violence against transgender and gay youth is rooted in notions of masculinity and the threat that gender nonconformity poses to "real men":

> Calling someone gay or a fag has become so universal that it's become synonymous with dumb, stupid or wrong. . . . And its "dumb" or "wrong" because it isn't masculine enough. To the "that's-so-gay" chorus, homosexuality is about gender nonconformity, not being a "real man," and so anti-gay sentiments become a shorthand method of gender policing.

Not surprisingly, a high proportion of transgender adults recount being harassed (78 percent), and physically (35 percent) and sexually (12 percent) assaulted while in school. In some cases, the victimization was so severe that 15 percent of the respondents report having to leave school (Grant et al. 2011).

Women and girls are also the victims of male violence. Worldwide, as many as 71 percent of women will be physically or sexually abused in their lifetime (see Chapter 5) (WHO 2009). Attacks on women's and girls' bodies routinely take place as they are beaten, raped, and killed in the name of religion, war, and honor. Over 5,000 women and girls are killed each year in **honor killings**—murders, often public, as a result of a female dishonoring, or being perceived to have dishonored, her family or community (Foerstel 2008). In 2007, Du'a Khalil Aswad, 17, was brutally beaten, kicked, stoned, and hit from behind with a cement block, as she struggled to sit up. Her crime? She had fallen in love with someone outside her religious sect. The murder, posted on YouTube, became an international outrage. In 2010, four men were arrested and sentenced to death for killing Du'a Khalil Aswad. Whether or not the death penalty will be imposed now rests with the Iraq Supreme Court (Cockburn 2010).

Violence against women is "rising every day. . . . [B]eheadings, rapes, beatings, suicides through self-immolation, genital mutilation, trafficking, and child abuse masquerading as marriage of girls as young as 9 are all on the increase" (Judd 2008, p. 1). Violence against women and girls, a problem of pandemic proportions, is rooted in gender inequality and the lingering notion that women and girls are property—a belief tied to ancient law and many of the world's religions. Yet, in the United States, being female is not part of the federal hate crime statutes despite the fact that, as one advocate testified before Congress, "women and girls . . . are exposed to terror, brutality, serious injury, and even death because of their sex" (LCCREF 2009, p. 33).

> Attacks on women's and girls' bodies routinely take place as they are beaten, raped, and killed in the name of religion, war, and honor.

honor killings Murders, often public, as a result of a female dishonoring, or being perceived to have dishonored, her family or community.

Strategies for Action: Toward Gender Equality

In recent decades, there has been a growing awareness of the need to increase gender equality throughout the world. Strategies to achieve this end have focused on empowering women in social, educational, economic, and political spheres and improving women's access to education, nutrition, health care, and basic human rights. But as we will see in the following section on grassroots movements, there is also a men's movement that is concerned with gender inequities and the issues facing men.

Grassroots Movements

Efforts to achieve gender equality in the United States have been largely fueled by the feminist movement. Despite a conservative backlash, and to a lesser extent men's activist groups, feminists have made some gains in reducing structural and cultural sexism in the workplace and in the political arena.

Feminism and the Women's Movement **Feminism** is the belief that women and men should have equal rights and responsibilities. The U.S. feminist movement began in Seneca Falls, New York, in 1848, when a group of women wrote and adopted a women's rights manifesto modeled after the Declaration of Independence. Although many of the early feminists were primarily concerned with suffrage, feminism has its "political origins . . . in the abolitionist movement of the 1830s," when women learned to question the assumption of "natural superiority" (Anderson 1997, p. 305). Early feminists were also involved in the temperance movement, which advocated restricting the sale and consumption of alcohol, although their greatest success was the passing of the Nineteenth Amendment in 1920, which recognized women's right to vote.

The rebirth of feminism almost 50 years later was facilitated by a number of interacting forces: an increase in the number of women in the labor force, the publication of Betty Friedan's book *The Feminine Mystique,* an escalating divorce rate, the socially and politically liberal climate of the 1960s, student activism, and the establishment of the Commission on the Status of Women by John F. Kennedy. The National Organization for Women (NOW) was established in 1966, and remains one of the largest feminist organizations in the United States, with more than 500,000 members in 550 chapters across the country.

One of NOW's hardest fought battles is the struggle to win ratification of the **equal rights amendment** (ERA), which states that "equality of rights under the law shall not be denied or abridged by the United States, or by any state, on account of sex." The proposed 28th amendment to the Constitution passed both the House of Representatives and the Senate in 1972, but failed to be ratified by the required 38 states by the 1979 deadline, which was later extended to 1982. With the exception of 2008, when presidential politics took precedence, the bill has been reintroduced into Congress every year since 1982 (ERA 2011). Six states are presently considering ratification. If three of the six ratify the amendment, it could become law within two years of reaching the 38-state requirement

feminism The belief that men and women should have equal rights and responsibilities.

equal rights amendment (ERA) The proposed 28th amendment to the Constitution, which states that "equality of rights under the law shall not be denied or abridged by the United States, or by any state, on account of sex."

(Cook 2009; ERA 2011). Presently, 22 states (Alaska, California, Colorado, Connecticut, Florida, Hawaii, Illinois, Iowa, Louisiana, Maryland, Massachusetts, Montana, New Hampshire, New Jersey, New Mexico, Pennsylvania, Rhode Island, Texas, Utah, Virginia, Washington, and Wyoming) include partial or inclusive guarantees of equal rights on the basis of sex with their state constitutions (ERA 2011).

Proponents of the ERA argue that its opponents used scare tactics—saying that the ERA would lead to unisex bathrooms, mothers losing custody of their children, and mandatory military service for women—to create a conservative backlash. However, Susan Faludi, in *Backlash: The Undeclared War against American Women* (1991), contends that contemporary arguments against feminism are the same as those levied against the movement 100 years ago and that the negative consequences predicted by opponents of feminism (e.g., women unfulfilled and children suffering) have no empirical support. Proponents also argue that "without the explicit wording and intention of women's rights documented in the principles of our government, women remain second-class citizens" (Cook 2009, p. 1).

Today, a new wave of feminism is being led by young women and men who grew up with the benefits their mothers won, but who are shocked by the stoning to death of a woman accused of adultery in Afghanistan, the continuing practice of female genital mutilation, the fact that millions of women throughout the world lack access to modern contraception, and that, even in one of the "freest" countries in the world, U.S. women face increasing restrictions on abortion, can be denied prescription contraception by a pharmacist who morally objects, earn less than men earn for doing the same job, and experience alarming rates of date rape and intimate partner abuse. These young feminists are more inclusive than their predecessors, welcoming all who champion the cause of global equality.

The Men's Movement As a consequence of the women's rights movement, men began to reevaluate their own gender status. As with any grassroots movement, the men's movement has a variety of factions. One of the early branches of the men's movement is known as the mythopoetic men's movement, which began after the publication of Robert Bly's (1990) *Iron John*—a fairy tale about men's wounded masculinity that was on the *New York Times* bestseller list for more than 60 weeks (Zakrzewski 2005). Participants in the men's mythopoetic movement met in men-only workshops and retreats to explore their internal masculine nature, male identity, and emotional experiences through the use of stories, drumming, dance, music, and discussion.

The ManKind Project (MKP), founded in 1985, grew out of the men's mythopoetic movement. It is an international organization with over 40,000 members. Their motto, "changing the world one man at a time," reflects the stated core values of the organization: authenticity, integrity, community, service, and inclusivity (MKP 2010a). MKP conducts The New Warrior Training Adventure which, according to their website, "is a singular type of life affirming event, honoring the best in what men have to offer the planet" (MKP 2010b, p. 1).

In addition to enthusiastic supporters, there are also critics. Some experts argue that MKP "practices therapy without a license; targets vulnerable members of 12-step recovery groups; purposefully withholds the details of the program, thus keeping potential participants from making a fully informed decision whether or not to attend; and does not screen applicants who may be too emotionally frail for the rigors of the program" (Vogel 2008, p. 1).

The men's movement also includes men's organizations that advocate gender equality and work to make men more accountable for sexism, violence, and homophobia. For example, the National Organization for Men Against Sexism (NOMAS) was founded in 1975, and "advocates a perspective that is pro-feminist, gay affirmative, antiracist, dedicated to enhancing men's lives, and committed to justice on a broad range of social issues including class, age, religion, and physical abilities" (NOMAS 2009, p. 1).

Some men's groups focus on issues concerning children's and fathers' rights. Groups such as the American Coalition for Fathers and Children, Dads Against Discrimination, and Fathers4Justice are attempting to change the social and legal bias against men in divorce and child custody decisions, which tend to favor women. Members of the National Coalition of Free Men (NCFM) are also concerned with issues surrounding fatherhood, concentrating on "how gender-based expectations limit men legally, socially, and psychologically" (NCFM 2009, p. 1).

Other concerns on the agenda of men's rights groups include the domestic violence committed against men by women, false allegations of child sexual abuse, wrongful paternity suits, and the oppressive nature of restrictive masculine gender norms. Just as women have fought against being oppressed by expectations to conform to traditional gender stereotypes, men are beginning to want the same freedom from traditional gender expectations. For example, men who enter nontraditional work roles, such as nurse and primary school teacher, are often stigmatized for participating in "feminine" work. A study of men in nontraditional work roles found that these men commonly experience embarrassment, discomfort, shame, and disapproval from friends and peers (Simpson 2005; Sayman 2007).

Other concerns on the agenda of men's rights groups include the domestic violence committed against men by women, false allegations of child sexual abuse, wrongful paternity suits, and the oppressive nature of restrictive masculine gender norms.

U.S. National Policy

A number of important federal statutes have been passed to help reduce gender inequality. They include the Equal Pay Act of 1963, Title VII of the Civil Rights Act of 1964, Title IX of the Education Amendments of 1972, the Victims of Trafficking and Violence Protection Act of 2000, and the Violence against Women Reauthorization Act of 2005. In 2009, President Obama signed the Ledbetter Fair Pay Act, which reversed the 2007 U.S. Supreme Court decision that held that victims of pay discrimination had 180 days to file a grievance after the act of discrimination. The act now defines each paycheck as a separate act of discrimination (Mehmood 2009). In 2011, the Paycheck Fairness Act was reintroduced and presently awaits legislative approval. If passed, it would close loopholes in the Equal Pay Act of 1963, by requiring that employers "demonstrate that wage differences between men and women doing the same work stem from factors other than sex" (ACLU 2011a, p. 1).

Of late, there has been increased concern over sexual violence in schools. It is estimated that 20 percent of female undergraduates and 6 percent of male undergraduates will be the victims of a sexual assault or attempted sexual assault while in college. In 2011, President Obama announced new federal guidelines that require schools to take immediate action once an incident is known, even if no campus complaint is filed and/or a criminal investigation is underway (Mianecki 2011). Further, the Campus Sexual Violence Elimination (SaVE) Act of 2011, if passed, would require campuses to: (1) record incidents of intimate partner violence (e.g., stalking, dating violence) as well as sexual assault; (2) support victims through various accommodations (e.g., move from residence hall), if so desired; (3) inform both the victim and the accused of their due process rights including a summary of disciplinary procedures; and (4) establish sexual assault awareness and prevention programs (SOC 2011).

What Do You Think? After Jessica Gonzales's estranged husband abducted their three little girls, she called the police nine times within a 10-hour period, called 911, and went to the police department asking that the restraining order against her husband be enforced. When her husband drove to the police department and opened fire, police fatally shot him soon after discovering the slain bodies of Ms. Gonzales's three daughters. That was over 10 year ago. After a 2005 U.S. Supreme Court decision determined that there is no constitutional right to enforce a restraining order, Ms. Gonzales, the first individual to do so, sued the United States of America, alleging an international human rights violation (*Gonzales v. The United States of America* 2009; ACLU 2011b). Do you think that the failure to protect domestic violence victims is a human rights violation?

Presently, two hotly debated public policies concern sexual harassment and affirmative action.

Sexual Harassment Sexual harassment is a form of sex discrimination that violates Title VII of the 1964 Civil Rights Act. The U.S. EEOC (2009) defines **sexual harassment** in the workplace as "unwelcome sexual advances, requests for sexual favors, and other verbal or physical conduct of a sexual nature . . . when this conduct explicitly or implicitly affects an individual's employment, unreasonably interferes with an individual's work performance, or creates an intimidating, hostile, or offensive work environment" (p. 1).

Any individual, female, male, or transgender, can be a victim of sexual harassment. There are two types of sexual harassment: (1) *quid pro quo,* in which an employer requires sexual favors in exchange for a promotion, salary increase, or any other employee benefit, and (2) the existence of a hostile environment that unreasonably interferes with job performance, as in the case of sexually explicit comments or insults being made to an employee. Common examples of sexual harassment include unwanted touching, the invasion of personal space, making sexual comments about a person's body or attire, and telling sexual jokes (Uggen & Blackstone 2004). Sexual harassment occurs in a variety of settings, including the workplace, public schools, military academies, and college campuses. Women who work in male-dominated occupations and blue-collar jobs, who are young, financially dependent, and single or divorced are the most likely to experience sexual harassment (Jackson & Newman 2004; ILO 2011a).

Affirmative Action As discussed in Chapter 9, **affirmative action** refers to a broad range of policies and practices to promote equal opportunity as well as diversity in the workplace and on campuses. Affirmative action policies, developed in the 1960s from federal legislation, require that any employer (universities as well as businesses) that receives contracts from the federal government must make "good faith efforts" to increase the number of female and other minority applicants. Such efforts can be made through expanding recruitment and training programs and by making hiring decisions on a nondiscriminatory basis.

However, a 1996 California ballot initiative, the first of its kind, abolished race and sex preferences in government programs, which included state colleges and universities. Over the next three years, several other states followed suit. In 2003, the U.S. Supreme Court held that universities have a "compelling interest" in a diverse student population and therefore may take minority status into consideration when making admissions decisions. Since that time, several states have addressed the issue of affirmative action in government programs with varying results. For example, in Nebraska, voters outlawed affirmative action programs; in Colorado, voters turned down a ban of affirmative action programs; and in Missouri and Oklahoma, anti-affirmative action initiatives were stopped before they reached the ballot. In 2010, Arizona banned "the consideration of race, ethnicity or gender by units of state government, including public colleges and universities" (Jaschik 2010 p. 1).

International Efforts

International efforts to address problems of gender inequality date back to the 1979 Convention on the Elimination of All Forms of Discrimination Against Women (CEDAW), often referred to as the international women's bill of rights, adopted by the United Nations in 1979. Although 185 countries have ratified the women's bill of rights, including every country in Europe and South and Central America, 70 have filed "reservations," meaning they have been exempt from portions of the agreement (Foerstel 2008). The United States is the only industrialized country in the world that has not ratified the document.

Another significant international effort occurred in 1995, when representatives from 189 countries adopted the Beijing Declaration and Platform for Action at the Fourth World Conference on Women sponsored by the United Nations. The platform reflects an international commitment to the goals of equality, development, and peace for women everywhere. The platform identifies strategies to address critical areas of concern related to women and girls, including poverty, education, health, violence, armed conflict, and human rights. Today, over a decade after the platform was adopted, there are concerns that many of the stated goals have not been met (Foerstal 2008).

In addition to the CEDAW and the Beijing Platform, in 2000, all of the members of the United Nations adopted the Millennium Declaration. One of the eight Millennium

sexual harassment In reference to workplace harassment, when an employer requires sexual favors in exchange for a promotion, salary increase, or any other employee benefit and/or the existence of a hostile environment that unreasonably interferes with job performance.

affirmative action A broad range of policies and practices to promote equal opportunity as well as diversity in the workplace and on campuses.

Gender-Based Violence & Harassment:
Your School, Your Rights

WOMEN'S RIGHTS PROJECT

Does any law protect me at school from gender-based violence and harassment?

- Title IX is a federal civil rights law that prohibits discrimination on the basis of sex, including on the basis of sex stereotypes, in education programs and activities. All public schools and any private schools receiving federal funds must comply with Title IX.

- Under Title IX, discrimination on the basis of sex can include sexual harassment or sexual violence such as rape, sexual assault, sexual battery, sexual coercion, or dating violence.

Gender-based violence and harassment are behaviors that are committed because of a person's gender or sex. They can be carried out by a boyfriend or girlfriend, a date, other kids, or adults. If someone does any of the following to you because of gender or sex, it *may* constitute gender-based violence or harassment. Someone:

- follows you around, always wants to know where you are and who you are with, or stalks you
- pressures you to perform sexual acts
- touches you sexually against your will
- forces you to have sex
- interferes with your birth control
- verbally abuses you using anti-gay or sex-based insults

- sends you repeated and unwanted texts, IMs, online messages, and/or phone calls that harass you
- hits, punches, kicks, slaps, or chokes you
- verbally or physically threatens you

Under Title IX, your school is obligated to do something about gender-based violence and harassment IF:

- these behaviors are so severe (for example, even a single incident of rape) or happen so often (for example, numerous harassing texts) that the acts would deprive a student of equal access to education, or to an educational activity like being on a team or in the band, AND

- your school has authority over the person or people committing the violent or harassing behavior, and over the environment where the behavior is happening.

Students can contact the Department of Education's Office for Civil Rights to report sexual harassment by writing a letter or filing a complaint form, available at www.ed.gov/about/offices/list/ocr/complaintintro.html.

For more information you can contact:
Women's Rights Project
American Civil Liberties Union
125 Broad Street, 18th Floor
New York, NY 10004
(212) 549-2644
womensrights@aclu.org
www.aclu.org/sexualassault

Find the local ACLU affiliate office in your state at:
www.aclu.org/affiliates

MAY 2011 © ACLU – American Civil Liberties Union

Development Goals, as stated in the Millennium Declaration, is the promotion of gender equality and women's empowerment by 2015. Progress has been slow. In an evaluation of this goal, a United Nations' Report (2008) concludes the following:

> Ensuring gender equality and empowering women in all respects—desirable objectives in themselves—are required to combat poverty, hunger and disease, and to ensure sustainable development. The limited progress in empowering women and achieving gender equality is a pervasive shortcoming that extends beyond the goal itself. Relative neglect of, and de facto bias against women and girls continues to prevail in most countries. (p. 5)

Further, in 2010, the U.N. General Assembly established the U.N. Entity for Gender Equality and the Empowerment of Women. UN Women, as it is called, combined four existing women-centered offices into one, thereby heightening the effectiveness of each. One of the goals of UN Women is to "hold the UN system accountable for its own commitments on gender equality, including regular monitoring of system-wide progress" (UN Women 2011).

Finally, also in 2010, the European Commission (EC) adopted the European Union strategy for equality between women and men, a five-year action plan with six priority areas. The priority areas include (1) equal economic independence; (2) equal pay for equal work and work of equal value; (3) equality in decision making; (4) dignity, integrity, and an end to gender-based violence; (5) gender equality in external actions; and (6) gender equality beyond European Union states (European Commission 2011).

In addition to international efforts, individual countries have instituted programs or policies designed to combat sexism and gender inequality. For example, for the first time in history, men and women came together to address gender issues in Brazil at the Global Symposium on Engaging Men and Boys in Achieving Gender Equality (Moreno 2009). The government of the United Kingdom has proposed legislation that would require "gender pay audits" of businesses to ensure equality of treatment in wages between women and men (Bennett 2009). However, Afghanistan, a country known for its patriarchal mores, passed legislation that applies to the Shiite minority—a religious denomination of Islam. The objectionable provisions state that a woman has no legal right to refuse her husband's sexual advances, that she must get her husband's permission to go to school or work outside the home, and she must "make herself up" or "dress up" if her husband so desires (Filkins 2009, p. 1). Finally, it must be noted, that China's one-child policy remains intact despite international outrage at the resulting **gendercide,** the disproportionate aborting of female fetuses (Rauhala 2011).

Understanding Gender Inequality

Gender roles and the social inequality they create are ingrained in our social and cultural ideologies and institutions and are therefore difficult to alter as indicated by the persistence of discriminatory attitudes. Nevertheless, as we have seen in this chapter, growing attention to gender issues in social life has spurred change. The traditional gender roles of men and women, and the binary concepts of sex and gender, are weakening. Women who have traditionally been expected to give domestic life first priority are now finding it more acceptable to seek a career outside the home; the gender pay gap is narrowing; and significant improvements in educational disparities between men and women, boys and girls, have been made. Most of these improvements, however, are in developed countries. Globally, millions of women and girls continue to be victimized by poverty, gendered violence, illiteracy, and limited legal rights and political representation.

Eliminating gender stereotypes and redefining gender in terms of equality does not simply mean liberating women but liberating men as well as society. Men are also victimized by discrimination and gender stereotypes that define what they "should" do rather than what they are capable, interested, and willing to do. The National Coalition of Free Men (NCFM 2011) has incorporated this view into their position statement:

> We have heard in some detail from the women's movement how such sex-stereotyping has limited the potential of women. More recently, men have become increasingly aware that they too are assigned limiting roles which they are expected to fulfill regardless of their individual abilities, interests, physical/emotional constitutions, or needs. Men have few or no effective choices in many critical areas of life. They face injustices under the law. And typically they have been handicapped by socially defined "shoulds" in expressing themselves in other than stereotypical ways.

Increasingly, people are embracing **androgyny**—the blending of traditionally defined masculine and feminine characteristics. The concept of androgyny implies that both masculine and feminine characteristics and roles are *equally valued*. However, "achieving gender equality . . . is a grindingly slow process, since it challenges one of the most deeply entrenched of all human attitudes" (Lopez-Claros & Zahidi 2005, p. 1).

An international strategy for achieving gender equality, *gender mainstreaming* is "the process of assessing the implications for women and men of any planned action, including legislation, policies or programmes, in all areas and at all levels" (IASC 2009, p. 7). Difficult? Yes. But, regardless of whether traditional gender roles emerged out of

gendercide The disproportionate aborting of female fetuses.

androgyny Having both traditionally defined feminine and masculine characteristics.

biological necessity as the structural functionalists argue, or out of economic oppression as the conflict theorists hold, or both, it is clear that, today, gender inequality carries a high price: poverty, loss of human capital, feelings of worthlessness, violence, physical and mental illness, and death. Perhaps we have reached a time when we need to ask ourselves, are the costs of traditional gender roles too high to continue to pay?

CHAPTER REVIEW

- **Does gender inequality exist worldwide?**

There is no country in the world in which men and women are treated equally. Although women suffer in terms of income, education, and occupational prestige, men are more likely to suffer in terms of mental and physical health, mortality, and the quality of their relationships.

- **How do the three major sociological theories view gender inequality?**

Structural functionalists argue that the traditional division of labor was functional for preindustrial society and has become defined as both normal and natural over time. Today, however, modern conceptions of the family have replaced traditional ones to some extent. Conflict theorists hold that male dominance and female subordination evolved in relation to the means of production—from hunting-and-gathering societies in which females and males were economic equals to industrial societies in which females were subordinate to males. Symbolic interactionists emphasize that, through the socialization process, both females and males are taught the meanings associated with being feminine and masculine.

- **What is meant by the terms *structural sexism* and *cultural sexism*?**

Structural sexism refers to the ways in which the organization of society, and specifically its institutions, subordinate individuals and groups based on their sex classification. Structural sexism has resulted in significant differences between education and income levels, occupational and political involvement, and civil rights of women and men. Structural sexism is supported by a system of cultural sexism that perpetuates beliefs about the differences between women and men. Cultural sexism refers to the ways the culture of society—its norms, values, beliefs, and symbols—perpetuate the subordination of an individual or group because of the sex classification of that individual or group.

- **What is the difference between the glass ceiling and the glass escalator?**

The glass ceiling is an often invisible barrier that prevents women and other minorities from moving into top corporate positions. The glass escalator, on the other hand, refers to the tendency for men seeking traditionally female jobs to have an edge in hiring and promotion practices.

- **What are some of the problems caused by traditional gender roles?**

First is the feminization of poverty. Women are socialized to put family ahead of education and careers, a belief that is reflected in their less prestigious occupations and lower incomes. Second are social-psychological costs. Women and transgender individuals tend to have lower self-esteem and higher rates of depression than men. Men and boys, on the other hand, are often subject to the emotional restrictions of the "boy code." Third, traditional gender roles carry health costs in terms of death and illness. For example, many traditionally defined male behaviors shorten life expectancy, and women in many regions of the world disproportionately suffer from HIV/AIDS. Finally, gendered violence is responsible for the deaths of men, women, and transgender individuals.

- **What strategies can be used to end gender inequality?**

Grassroots movements, such as feminism and the women's rights movement and the men's rights movement, have made significant inroads in the fight against gender inequality. Their accomplishments, in part, have been the result of successful lobbying for passage of laws concerning sex discrimination, sexual harassment, and affirmative action. Besides these national efforts, international efforts continue as well. One of the most important is the Convention on the Elimination of All Forms of Discrimination Against Women (CEDAW), also known as the international women's bill of rights, which the United Nations adopted in 1979.

TEST YOURSELF

1. The United States is the most gender-equal nation in the world.
   a. True
   b. False

2. Symbolic interactionists argue that
   a. male domination is a consequence of men's relationship to the production process
   b. gender inequality is functional for society

c. gender roles are learned in the family, in the school, in peer groups, and in the media

d. women are more valuable in the home than in the workplace

3. Recent trends indicate that more males enter college from high school than females.
   a. True
   b. False

4. The devaluation hypothesis argues that female and male pay differences are a function of women's and men's different levels of education, skills, training, and work experience.
   a. True
   b. False

5. The glass ceiling is an often invisible barrier that prevents women and other minorities from moving into top corporate positions.
   a. True
   b. False

6. Which of the following statements is true about women and U.S. politics?
   a. There are more female than male governors.
   b. In the history of the U.S. Supreme Court, there have only been four female justices.
   c. The highest ranking woman in the U.S. government is vice-president Hillary Clinton.
   d. Women received the right to vote with the passage of the 21st Amendment.

7. The Global Media Monitoring Project found that
   a. the number of female reporters has decreased in recent years
   b. women are dramatically underrepresented as news subjects
   c. older women were more likely to appear on television than younger women
   d. male journalists are more likely to work on stories dealing with social issues than women journalists

8. The feminization of poverty refers to
   a. the tendency for women to be caretakers of the poor
   b. feminists' criticism of public policy on poverty
   c. the disproportionate number of women who are poor
   d. gender role socialization of poor women

9. *Quid pro quo* sexual harassment refers to the existence of a hostile working environment.
   a. True
   b. False

10. Feminists would argue which of the following?
    a. The ERA should be part of the Constitution.
    b. The passage of the Nineteenth Amendment was a mistake.
    c. Occupational sex segregation benefits everyone.
    d. NOMAS is a radical, sexist organization.

Answers: 1: b; 2: c; 3: b; 4: b; 5: a; 6: b; 7: b; 8: c; 9: b; 10: a.

KEY TERMS

affirmative action   338
androgyny   340
boy code   331
comparable worth   320
cultural sexism   324
devaluation hypothesis   320
education dividend   315
emotion work   320
equal rights amendment (ERA)   335
expressive roles   314

feminism   335
gender   306
gender roles   324
gendercide   340
glass ceiling   318
glass escalator effect   317
honor killings   334
human capital hypothesis   318
instrumental roles   314
motherhood penalty   318

occupational sex segregation   316
pink-collar jobs   317
sex   306
sexism   306
sexual harassment   338
structured choice   318
structural sexism   314
transgender individual   306

MEDIA RESOURCES

Turning to Video

▶❚❚ Watch the ABC video *Are Men Smarter? Sexism in the Classroom* (running time 4:36), available through **CengageBrain.com.** This video describes the unique experiences of neuroscientist Dr. Ben Barres who has endured sexism as both a man and a woman. After you watch the video, think about whether you or any of your friends have been the victim of gender inequality.

Online Study Resources

Log in to **www.cengagebrain.com** to access the resources your instructor has assigned. For this book, you can access:

CourseMate

Access chapter-specific learning tools, including learning objectives, practice quizzes, videos, Internet exercises, flash cards, and glossaries, as well as web links, and more in your Sociology CourseMate.

Michael S. Williamson/The Washington Post/Getty Images

Sexual Orientation and the Struggle for Equality

"Homophobia alienates mothers and fathers from sons and daughters, friend from friend, neighbor from neighbor, Americans from one another. So long as it is legitimated by society, religion, and politics, homophobia will spawn hatred, contempt, and violence, and it will remain our last acceptable prejudice."

—Byrne Fone, Author, Emeritus Professor

AS MOST HIGH SCHOOL seniors, Constance McMillen was eager to attend her prom. But high school officials told her that if she and her girlfriend tried to attend, they would be ejected from the event if there were any complaints. After several meetings, school officials made the decision to cancel the prom for everyone rather than let the two young women attend. The American Civil Liberties Union (ACLU) filed a lawsuit against the school district, requesting that the court reinstate the prom, and accusing the school district of violating Constance's First Amendment right to freedom of expression. The judge ruled that the district could not bar Constance and her date from attending the prom. However, the judge did not direct the school district to proceed with the event because a privately organized prom to which Constance and her date were invited had already been arranged. Constance was sent to a "fake prom" at a local country club along with a handful of students and chaperones, while the rest of her class took part in a "real" prom organized by parents at a secret location (Broverman 2010).

Grand Marshal Constance McMillen attends the 2010 New York City Gay Pride March on the streets of Manhattan on June 27, 2010, in New York City. Ms. McMillen was denied access to her senior prom because her intended date was a young woman.

sexual orientation A person's emotional and sexual attractions, relationships, self-identity, and behavior.

heterosexuality The predominance of emotional, cognitive, and sexual attraction to individuals of the other sex.

homosexuality The predominance of emotional, cognitive, and sexual attraction to individuals of the same sex.

bisexuality The emotional, cognitive, and sexual attraction to members of both sexes.

lesbian A woman who is attracted to same-sex partners.

gay A term that can refer to either women or men who are attracted to same-sex partners.

LGBT, LGBTQ, and **LGBTQI** Terms used to refer collectively to lesbian, gay, bisexual, transgender, questioning or "queer," and/or intersexed individuals.

Despite the progress that has been made with respect to equal rights for same-sex-attracted individuals in recent years, the fight for equality on the basis of sexual orientation and gender identity continues to be met with opposition. The term **sexual orientation** refers to a person's emotional and sexual attractions, relationships, self-identity, and behavior. **Heterosexuality** refers to the predominance of emotional, cognitive, and sexual attraction to individuals of the other sex. **Homosexuality** refers to the predominance of emotional, cognitive, and sexual attraction to individuals of the same sex, and **bisexuality** is the emotional, cognitive, and sexual attraction to members of both sexes. The term **lesbian** refers to women who are attracted to same-sex partners, and the term **gay** can refer to either women or men who are attracted to same-sex partners. Much of the current literature on the treatment, and political and social agendas of individuals who are gay, lesbian, and bisexual includes transgender individuals. As discussed in Chapter 10, a *transgender individual* (sometimes called "*trans*") is a person whose sense of gender identity—masculine or feminine—is inconsistent with their birth (sometimes called chromosomal) sex.

The acronyms **LGBT, LGBTQ,** and **LGBTQI**, as well as an assortment of other variations, are used to refer collectively to individuals who are lesbian, gay, bisexual, transgender, questioning or "queer," and *intersexed*. Sometimes, one will encounter the acronym accompanied by the letter "A," which stands for *allies*.

This chapter focuses on sexual orientation. However, it is impossible to discuss the struggle for same-sex equality without acknowledging transgender individuals, who have been central to the larger LGBT civil rights movement. Thus, some of the research reported in this chapter is on lesbians, gays, and bisexuals, whereas other studies include transgender people as participants in addition to lesbians, gays, and bisexuals. Further, many laws and policies discussed in this chapter do not specifically refer to bisexuals but would, by default, apply if they were in a same-sex relationship.

In this chapter, we focus primarily on Western conceptions of diversity in sexual orientation. It is beyond the scope of this chapter to explore in depth how sexual orientation and its cultural meanings vary throughout the world. The global legal status of lesbians and gay men, however, will be summarized. We also discuss the prevalence of homosexuality, heterosexuality, and bisexuality in the United States, the beliefs about the origins of sexual orientation, and then apply sociological theories to better understand societal reactions to non-heterosexuals. After detailing the ways in which non-heterosexuals are victimized by prejudice and discrimination, we then end the chapter with a discussion of strategies to reduce antigay prejudice and discrimination.

The Global Context: A World View of the Status of Homosexuality

Homosexuality has existed throughout human history and in most, perhaps all, human societies (Joannides 2011; Kirkpatrick 2000). A global perspective on laws and social attitudes regarding homosexuality reveals that countries vary tremendously in their treatment of same-sex sexual behavior—from intolerance and criminalization to acceptance and legal protection. In 79 out of 242 countries throughout the world, homosexuality among males is illegal; in 45 of these countries, homosexuality among females is illegal. Legal penalties vary for violating laws that prohibit homosexual acts. In some countries, homosexuality is punishable by prison sentences and/or corporal punishment, such as whipping or lashing, and in five nations—Iran, Mauritania, Saudi Arabia, Yemen, and Sudan—people found guilty of engaging in same-sex sexual behavior may receive the death penalty (Ottosson 2010).

In general, countries throughout the world are moving toward increased legal protection of non-heterosexuals, as discrimination on the basis of sexual orientation has become part of a broad international human rights agenda. In fact, "global trends towards more approval of same-gender sexual behavior have largely paralleled changes in the United States" (Smith 2011). In 1996, South Africa became the first country in the world to include in its constitution a clause banning discrimination based on sexual orientation. However, days after that nation's Justice Ministry set up a task force to address hate crimes against LGBT South Africans, a 24-year-old woman was sexually assaulted and stabbed to death in a case of "corrective rape" due to living openly as a lesbian (Karimi 2011).

Following South Africa's lead, other countries began amending their constitutions. To date, at least 20 countries—from Ecuador to Greece—also have national constitutions banning discrimination based on sexual orientation. Others have laws prohibiting discrimination in some areas of the country, but not in others, as is the case in Brazil. Finally, many countries have laws prohibiting discrimination based on sexual orientation specific to the workplace, or regarding access to social services (Bruce-Jones & Itaborahy 2011).

A global perspective on laws and social attitudes regarding homosexuality reveals that countries vary tremendously in their treatment of same-sex sexual behavior—from intolerance and criminalization to acceptance and legal protection.

In 2009, the United States joined the majority of the United Nations member states when President Barack Obama supported the decriminalization of homosexuality and the international expansion of human rights protections for those who are non-heterosexual or **gender non-conforming**. The resolution includes language prohibiting harassment, discrimination, exclusion, stigmatization, and prejudice against members of the LGBT population. As of 2011, 85 out of the 192 United Nations members supported the resolution, which passed with 23 votes in favor and 19 opposed by the U.N. Human Rights Council (Dougherty 2011).

> **What Do You Think?** In a 2008 documentary entitled, *Be Like Others*, gay and lesbian Iranians who undergo sex reassignment surgeries were brought into the international spotlight (Ellison 2008). In a country where sex reassignment procedures such as hormone therapy and surgeries are legal, while homosexuality is punished by death, gay men and women have few other alternatives. Why do you think gays and lesbians, who are not transgender, undergo such extreme procedures?

The growing legal recognition of same-sex relationships provides evidence of the changing status of homosexuality throughout the world, yet it remains a complicated issue that has divided jurisdictions in many countries. In 2001, the Netherlands became the first country in the world to offer full legal marriage to same-sex couples. In 2003, Belgium legalized same-sex marriages. In 2005, Spain, Canada, and South Africa became the third, fourth, and fifth countries, respectively, to legalize same-sex marriage. Since then, Argentina, Iceland, Norway, Portugal, Spain, and Sweden have provided legal recognition of same-sex marriages. In Mexico City, same-sex marriages became legal in 2010 (Bruce-Jones and Itaborahy 2011).

Other countries recognize same-sex **registered partnerships** or "civil unions," which are federally recognized relationships that convey most but not all the rights of marriage. Some countries also offer registered partnerships or civil unions to opposite-sex couples. Registered partnerships are also referred to as "domestic partnerships." In 1989, Denmark paved the way for legal recognition of any kind for same-sex relationships. Legally recognized registered partnerships for same-sex couples are also available in 25 countries in Europe, the Americas, and as specific jurisdictions of Mexico, Australia, and Venezuela (Bruce-Jones & Itaborahy 2011). Debates over same-sex marriage and civil union legislation continue in many countries throughout the world.

Homosexuality and Bisexuality in the United States: A Demographic Overview

Before looking at demographic data concerning homosexuality and bisexuality in the United States, it is important to understand the ways in which identifying or classifying individuals as homo-, hetero-, and bisexual is problematic.

Sexual Orientation: Problems Associated with Identification and Classification

The classification of individuals into sexual orientation categories (e.g., gay, bisexual, lesbian, or heterosexual) is problematic for a number of reasons. First, distinctions among sexual orientation categories are simply not as clear-cut as many people would believe. Consider the early research on sexual behavior by Alfred Kinsey and his colleagues (1948; 1953), who found that, although 37 percent of men and 13 percent of women had had at least one same-sex sexual experience since adolescence, few of the individuals reported exclusive same-sex behavior. These data led Kinsey to conclude that heterosexuality and

gender non-conforming Often used synonymously with transgender, gender non-conforming (sometimes called gender variant) refers to displays of gender that are inconsistent with society's expectations.

registered partnerships Federally recognized relationships that convey most but not all the rights of marriage.

homosexuality represent two ends of a continuum and that most individuals fall somewhere along this sexuality continuum. In other words, most people are, to some degree, bisexual. Kinsey and his colleagues (1948) aptly wrote:

> The world is not to be divided into sheep and goats. Not all things are black nor all things white. It is a fundamental of taxonomy that nature rarely deals with discrete categories. Only the human mind invents categories and tries to force facts into separated pigeon-holes. The living world is a continuum in each and every one of its aspects. The sooner we learn this concerning human sexual behavior the sooner we shall reach a sound understanding of the realities of sex. (p. 639)

More recent research has also indicated that many individuals are not exclusively heterosexual or homosexual. In a longitudinal study of women, two-thirds changed their sexual identity labels they had claimed at the beginning of the study, while one-third changed labels two or more times. The most commonly adopted final assessment after 10 years was "unlabeled" (Diamond 2008). In a study of 243 undergraduates, Vrangalova and Savin-Williams (2010) found that 84 percent of the heterosexual women and 51 percent of the heterosexual men reported the presence of at least one same-sex quality—sexual attraction, fantasy, and/or behavior.

The second factor that makes classification difficult is that research with same-sex populations has tended to define sexual orientation based on one of three components: sexual/romantic attraction or arousal, sexual behavior, and sexual identity (Savin-Williams 2006). However, sexual orientation involves *multiple dimensions*—sexual attraction, sexual behavior, sexual fantasies, emotional attraction, self-identification, and the interrelations between these dimensions (Herek & Garnets 2007). Many people with same-sex attractions or who have engaged in same-sex sexual behavior do not identify themselves as gay, lesbian, or bisexual (Rothblum 2000; Savin-Williams 2006; Herek & Garnets 2007).

Lastly, a factor complicating the identification and classification of same-sex populations is the social stigma associated with non-heterosexual identities. The social stigma associated with non-heterosexual behavior is, in part, a function of conservative religious and political beliefs, and an adherence to traditional gender roles (Brown & Henriquez 2008). As a result of the stigma associated with being gay, lesbian, or bisexual, many individuals conceal or falsely portray their sexual orientation to protect themselves from prejudice and discrimination (Meyer 2003; Varjas et al. 2008; Röndahl & Innala 2008).

What Do You Think? Findings from recent psychological research have shown that after having established a male gender identity, female-to-male (FTM) transsexuals' attractions toward men became salient, even though most identified as lesbians before their medical transition (Schleifer 2006; Bockting et al. 2009). Why do you think that some FTM transsexuals identify as gay or bisexual men after they have sex reassignment? How does this influence your understanding of sexual orientation and its relationship to gender identity?

The Prevalence of LGBT Individuals and Same-Sex Couple Households in the United States

Reliable estimates of the percentage of the U.S. population that is gay, lesbian, or bisexual (LGB) are scarce. It is estimated that there are more than 8 million LGB adults in the United States, comprising 3.5 percent of the adult population. In addition, there are approximately 700,000 transgender (T) individuals in the United States. This means that approximately 9 million Americans identify as LGBT, a number nearly comparable to the total population of New Jersey (Gates 2011).

"You know, statistically speaking, at least one of these gingerbread men is gay."

Bisexual individuals comprise approximately 1.8 percent of the U.S. population, slightly more than the 1.7 percent who identify as lesbian or gay in the U.S. population. Further, women are substantially more likely than men to identify as bisexual, and estimates of individuals who report any lifetime same-sex behavior or attraction are considerably higher than estimates of those who actually identify as lesbian, gay, or bisexual. An estimated 19 million Americans (8.2 percent) report that they have engaged in same-sex behavior and nearly 25.6 million Americans (11 percent) acknowledge at least some same-sex attraction (Gates 2011). Interestingly, when Americans are asked to estimate the number of lesbians, gays, and bisexual individuals in the United States, the estimate is that 25 percent of the American population is LGB, a considerably higher estimate than research suggests (Morales 2011).

A national study of U.S. college students found that 4.9 percent identified as gay, lesbian, or bisexual, while 1.5 percent stated they were "unsure" (see Table 11.1). An estimated 1 million to 3 million Americans older than age 65 are gay, lesbian, bisexual, or transgender—a number that is expected to double over the next quarter century (Cahill 2007).

TABLE 11.1	Self-Described Sexual & Gender Identity of U.S. College Students*
Heterosexual	93.5%
Gay/lesbian	1.9%
Bisexual	3.0%
Unsure	1.5%
Transgender	0.2%

Source: American College Health Association 2011.

*Based on a sample of 30,093 students at 39 campuses.

Data on the prevalence of Americans who identify as LGBT are important because they can influence laws and policies that affect lesbian, gay, bisexual, and transgender individuals and their families. The 2010 Census did not ask about sexual orientation or gender identity. However, in a reversal of a Bush administration decision, the 2010 Census became the first census to report the number of same-sex partners and the number of same-sex spouses. LGBT people living with a spouse or partner could identify their relationship by checking either the "husband or wife" or "unmarried partner" box. There are an estimated 581,300 same-sex couple households in the United States, 28 percent of which are in legally recognized relationships, including marriages or the state-level equivalent (U.S. Census Bureau 2010). According to the 2010 Census, an estimated one-quarter of all same-sex households are raising children (James 2011).

The Origins of Sexual Orientation

One of the most common questions regarding sexual orientation centers on its origin or "cause." Questions about the causes of sexual orientation are typically concerned with the origins of homosexuality and bisexuality because heterosexuality is considered normative and "natural." A wealth of biomedical and social science research has investigated the possible genetic, hormonal, developmental, social, and cultural influences on sexual orientation, yet no conclusive findings have suggested that sexual orientation is determined by any particular factor or interplay of factors. In the absence of compelling findings, many practitioners and professionals believe that sexual orientation might be determined by the interplay of environmental and biological factors. Further, most people experience little or no sense of choice about their sexual orientation (American Psychological Association 2008).

Aside from what "causes" homosexuality, sociologists are interested in what people *believe* about the "causes" of homosexuality. Research shows that people rely heavily on ideology, religion, and life experience to form their beliefs on the causes of homosexuality (Haider-Markel & Joslyn 2008). Currently, Americans are narrowly at odds over the factors that contribute to someone being gay—42 percent attribute it to one's upbringing and environment, while 40 percent believe that people are born gay or lesbian. Yet, long-term, there has been a significant shift away from the belief that homosexuality is a choice (Jones 2011 (see Figure 11.1)).

In addition, Americans' views on what causes non-heterosexual orientations appear to be the most strongly associated with their support for same-sex rights. Those most supportive of same-sex rights are much more likely to believe people are born gay or lesbian than that it is a product of their environment, controlling for other attitudinal and demographic factors (Jones 2011). The belief that homosexuality is learned, acquired, or a personal choice implies that those with same-sex attractions can control, and therefore are responsible for, their sexuality.

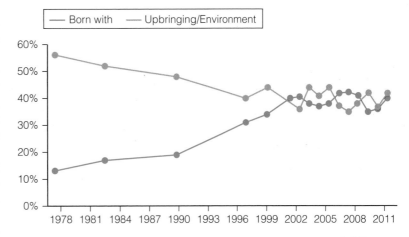

In your view, is being gay or lesbian something a person is born with, or due to factors such as upbringing and environment?

Figure 11.1: Public Opinion on the Causes of Homosexuality, 1978–2011
Source: Jones 2011. Jones, Jeffrey, "Support for Legal Gay Relations Hits New High." Copyright © 2011 by the Gallup Organization. Reprinted with permission.

Can Gays, Lesbians, and Bisexuals Change Their Sexual Orientation?

Those who believe that non-heterosexuals choose to have same-sex attractions tend to think that these individuals can and should change their sexual orientation. Various forms of *reparative therapy, conversion therapy,* and *reorientation therapy* are dedicated to changing the sexual orientation of individuals who are non-heterosexual. The American Psychological Association collectively refers to these "treatments" as **sexual orientation change efforts (SOCE)** (APA 2009).

A wealth of research has investigated the possible genetic, hormonal, developmental, social, and cultural influences on sexual orientation, yet no conclusive findings have suggested that sexual orientation is determined by any particular factor or interplay of factors.

Many conversion and reparative therapy programs view homosexuality as inherently immoral and/or pathological and allegedly achieve "conversion" to heterosexuality through embracing evangelical Christianity and being "born again" (Cianciotto & Cahill 2007). Some "treatments" have gone to unethical extremes, using shame, physical punishment, and other abusive methods to "convert" those with same-sex attractions. Sometimes, against their will, adolescents are sent to camps or group homes by their parents. Exodus International, a leader in the religious ex-gay movement claiming over 240 member ministries across North America and over 300 ministries worldwide (Exodus International 2011), has come under scrutiny after some of its founding members have come forth stating that they were not "cured" (Besen 2010). In 2011, Exodus developed an iPhone application, which was pulled by the Apple Corporation due to the public backlash regarding its offensive nature (Bosker 2011).

The National Association for Research and Therapy of Homosexuality (NARTH) is a secular organization comprised of mental health practitioners who use their training in psychology and related fields to assist clients in ridding themselves of "unwanted" same-sex attractions, behaviors, and identifications (NARTH 2011). Members of NARTH deny that they are "antigay" and assert that they provide non-religious psychosocial services (called *reorientation therapy*) for people who are distressed by their sexual

sexual orientation change efforts (SOCE) Collectively refers to reparative, conversion, and reorientation therapies, according to the APA.

orientation. NARTH claims to have the same success rate for reorientation therapies as for any other presenting problem in therapy (30 to 70 percent), such as anxiety or depression (Hamilton 2011).

Critics of SOCE include the American Medical Association, American Academy of Pediatrics, American Counseling Association, American Psychological Association, American School Health Association, American Psychiatric Association, National Association of Social Workers, National Education Association, and the American Association of School Administrators, who all agree that homosexuality is not a mental disorder, that sexual orientation *cannot* be changed, and that efforts to change sexual orientation may be harmful (Lambda Legal 2011a). In an extensive review of the ex-gay movement, Cianciotto and Cahill (2006) found that depression, social isolation from family and friends, low self-esteem, internalized homophobia, and suicidality can result, rather than a change in sexual orientation.

The most significant statement against SOCE was issued by the American Psychological Association in a review of empirical and peer-reviewed clinical literature. They too came to the conclusion that SOCE are not only unlikely to be successful, but may involve some risk of harm to the individual. Further, individuals who choose to undergo SOCE tend to have strongly conservative religious views that impact their decision to seek to change (APA 2009). According to Lambda Legal, the oldest and largest LGBT legal activist group in the United States, such practitioners can be held liable for the harm they inflict upon clients including malpractice, consumer fraud, false advertising, or child abuse and neglect in the case of minors who are forced to attend an ex-gay program (Cianciotto & Cahill 2007).

> . . . [A]ll agree that homosexuality is not a mental disorder, that sexual orientation *cannot* be changed, and that efforts to change sexual orientation may be harmful.

Sociological Theories of Sexual Orientation Inequality

Sociological theories do not explain the origin or "cause" of sexual orientation diversity; rather, they help to explain societal reactions to homosexuality and bisexuality and ways in which sexual identities are socially constructed.

Structural-Functionalist Perspective

Structural functionalists, consistent with their emphasis on institutions and the functions they fulfill, emphasize the importance of monogamous heterosexual relationships for the reproduction, nurturance, and socialization of children. From a structural-functionalist perspective, homosexual relations, as well as heterosexual non-marital relations, are "deviant" because they do not fulfill the main function of the family institution—producing and rearing children. Clearly, however, this argument is less salient in a society in which (1) other institutions, most notably schools, have supplemented the traditional functions of the family, (2) reducing (rather than increasing) population is a societal goal, and (3) same-sex couples can and do raise children.

Some structural functionalists argue that antagonisms between individuals who are heterosexual and homosexual disrupt the natural state, or equilibrium, of society. Durkheim (1993 [1938]), however, recognized that deviation from society's norms can also be functional. Specifically, the gay rights movement has motivated many people to reexamine their treatment of sexual orientation minorities and has produced a sense of cohesion and solidarity among members of the gay population. Gay activism has also been instrumental in advocating HIV/AIDS prevention strategies and health services that benefit the society as a whole.

The structural-functionalist perspective is concerned with how changes in one part of society affect other aspects. For example, research has shown that the worldwide increase in

Gay cowboys line dance to country and western music at the Windy City Rodeo sponsored by the Illinois Gay Rodeo Association. The number of companies and organizations that support LGBT rights in the United States is steadily rising.

legal and social support of same-sex equality has been influenced by three cultural changes: the rise of individualism, increasing gender equality, and the emergence of a global society in which nations are influenced by international pressures (Frank & McEneaney 1999).

According to Frank and McEneaney (1999), individualism "appears to loosen the tie between sex and procreation, allowing more personal modes of sexual expression" (p. 930). They add:

> Whereas once sex was approved strictly for the purpose of family reproduction, sex increasingly serves to pleasure individualized men and women in society. This shift has involved the casting off of many traditional regulations on sexual behavior, including prohibitions of male-male and female-female sex. (p. 936)

Gender equality involves the breakdown of sharply differentiated sex roles, thereby supporting the varied expressions of male and female sexuality. Globalization permits the international community to influence individual nations. For example, when Zimbabwe president Robert Mugabe pursued antigay policies in the 1990s, 70 U.S. Congress members as well as several international human rights organizations pressured Mugabe to halt his antigay campaign (Gallagher 1995). Currently, international leaders have expressed serious concerns over Uganda's Anti-Homosexuality Bill that was first proposed in 2009. Uganda lawmakers recently removed the death penalty clause from the bill, but nonetheless are hoping to pass the bill, which would criminalize the promotion of homosexuality (Dixon 2011).

The structural-functionalist perspective is also concerned with latent functions, or unintended consequences. A latent function of the gay rights movement is increased opposition to gay rights. For example, after California granted same-sex couples the right to be legally married, a fierce campaign was launched to pass Proposition 8—a voter referendum that, in 2008, amended the state constitution to prohibit same-sex marriage. Following the elections, demonstrations and protests over the passage of Proposition 8 spread across California and throughout the United States. Numerous lawsuits were filed with the California Supreme Court challenging the proposition's validity and effect on previously administered same-sex marriages. In *Strauss v. Horton*, the court upheld Proposition 8, but allowed existing same-sex marriages to stand. On August 4, 2010, in the case of *Perry et al. v. Schwarzenegger*, Judge Vaughn R. Walker overturned Proposition 8, and a federal judge later upheld Vaughn's ruling after questions were raised about his impartiality because he is gay (Mears 2011).

Conflict Perspective

The conflict perspective frames the gay rights movement and the opposition to it as a struggle over power, prestige, and economic resources. This struggle is largely over prestige, or social respect. Sexual minorities want to be recognized as full and decent human beings who are deserving of all the rights and protections entitled to individuals who are heterosexual. A major achievement in gaining acceptance of gay and lesbian individuals occurred in 1973, when the American Psychiatric Association (APA) removed homosexuality from its official list of mental disorders (Bayer 1987).

More recently, gay and lesbian individuals have been waging a political battle to win civil rights protections against employment discrimination and to be allowed to marry a same-sex partner. Conflict theory helps to explain why many business owners and corporate leaders support nondiscrimination policies: It is good for the "bottom line." The majority (89 percent) of Fortune 500 companies have included sexual orientation in their nondiscrimination policies, while 43 percent have also included gender identity. Further, employers are increasingly offering benefits to domestic partners of LGBT employees (HRC 2010). Gay-friendly work policies help employers maintain a competitive edge in recruiting and maintaining a talented and productive workforce. Companies are also competing for gay and lesbian dollars. Many LGBT as well as heterosexual consumers prefer to purchase products and services from businesses that provide workplace protections for LGBT employees. Recent trends toward increased social acceptance of homosexuality may, in part, reflect the corporate world's competition for the gay and lesbian consumer dollar.

Many business leaders also see an economic advantage to supporting same-sex marriage. In support of marriage equality, Levi Strauss & Co., with the support of an LGBT activist group, testified before the Supreme Court of California that, "Ending marriage discrimination will improve businesses' ability to attract the best and the brightest to California and enhance California's reputation as a diverse, inclusive and innovative community, both of which are key factors to continued economic growth and prosperity in this state . . ." (California Supreme Court 2011).

Symbolic Interactionist Perspective

Symbolic interactionism focuses on the meanings of heterosexuality, homosexuality, and bisexuality; how these meanings are socially constructed; and how they influence the social status, self-concepts, and well-being of non-heterosexual individuals. The meanings we associate with same-sex relations are learned from society—from family, peers, religion, and the media. The negative meanings associated with homosexuality are reflected in the current slang use of the phrase "That's so gay" or "You're so gay," which is meant to convey that something or someone is stupid or worthless (Kosciw & Diaz 2005; Burn et al. 2005; Sue 2010; GLSEN 2011a). Sociological and psychological research has shown that the use of such words as *fag* and *queer* by heterosexuals to insult one another facilitates social acceptance by peers, particularly among heterosexual males (Burn 2000; Kimmel 2011). However, this behavior comes at an expense to the well-being of non-heterosexuals (Meyer 2003; Burn et al. 2005; Sue 2010; GLSEN 2011a).

The symbolic interactionist perspective also points to the effects of labeling on individuals. Language is power. Once individuals become identified or labeled as lesbian, gay, or bisexual, that label tends to become their **master status.** In other words, the dominant heterosexual community tends to view "gay," "lesbian," and "bisexual" as the most socially significant statuses of individuals who are identified as such. Esterberg (1997)

> Sociological and psychological research has shown that the use of such words as *fag* and *queer* by heterosexuals to insult one another facilitates social acceptance by peers, particularly among heterosexual males

master status The status that is considered the most significant in a person's social identity.

notes that, "unlike heterosexuals, who are defined by their family structures, communities, occupations, or other aspects of their lives, lesbians, gay men, and bisexuals are often defined primarily by what they do in bed" (p. 377).

Symbolic interactionism draws attention to how social interaction affects our self-concept, behavior, and well-being. When gay and lesbian individuals interact with people who express antigay attitudes, they may develop what is known as **internalized homophobia (or internalized heterosexism)**—the internalization of negative messages about homosexuality by lesbian, gay, and bisexual individuals as a result of direct or indirect social rejection and stigmatization. Internalized homophobia has been linked to increased risk for depression, substance abuse, anxiety, and suicidal thoughts (Gilman et al. 2001; Bobbe 2002; Meyer 2003; Szymanski et al. 2008; Newcomb & Mustanski 2010).

Family members also have a powerful effect on the self-concepts, behavior, and well-being of lesbian, gay, and bisexual youth. For example, a study of young lesbian, gay, and bisexual adults found that higher rates of family rejection during adolescence were significantly associated with poorer health outcomes. Further, young LGBT adults who reported higher levels of family rejection were more likely to have attempted suicide, have high levels of depression, use illegal drugs, and engage in unprotected sex compared to LGBT peers from families that reported no or low levels of family rejection (Ryan et al. 2009). Research also indicates that a positive reaction from parents to a youth's coming out is a protective factor that buffers against stigma-related stress and reduces risk for illegal drug use (Padilla et al. 2010). Although newer research is pointing to the resilience of LGBT individuals in the face of sexual and gender identity prejudice (Stanley 2009), research also indicates that the effects of environmental stress combined with internal stress processes, such as internalized homophobia, makes a person more vulnerable to developing a mental health disorder (Meyer 2003).

Prejudice Against Lesbians, Gays, and Bisexuals

Oppression refers to the use of power to create inequality and limit access to resources, which impedes the physical and/or emotional well-being of individuals or groups of people. A person or group is **privileged** when they have a special advantage or benefits as a result of cultural, economic, societal, legal, and political factors (Guadalupe & Lum 2005). **Heterosexism** is a form of oppression and refers to a belief system that gives power and privilege to heterosexuals, while depriving, oppressing, stigmatizing, and devaluing people who are not heterosexual (Herek 2004; Szymanski et al. 2008).

The belief that heterosexuality is superior to non-heterosexuality results in prejudice and discrimination against gays, lesbians, and bisexuals. **Prejudice** refers to negative attitudes, whereas **discrimination** refers to behavior that denies individuals or groups equality of treatment (see Chapter 9). In turn, this often leads non-heterosexual individuals to question the legitimacy of their same-sex attractions (Harper et al. 2004). Before reading further, you may wish to complete this chapter's *Self and Society* feature, which assesses your knowledge of and attitudes toward lesbians, gays, and bisexual individuals.

Homophobia and Biphobia

The term **homophobia** is commonly used to refer to negative or hostile attitudes directed toward same-sex sexual behavior, a non-heterosexually identified individual, and communities of non-heterosexuals. Homophobia is made possible by heterosexism. Homophobia is not necessarily a clinical phobia, i.e., one involving a compelling desire to avoid the feared object despite recognizing that the fear is unreasonable (e.g., fear of flying). Rather, research indicates that disgust and anger, rather than fear and anxiety, are more involved in heterosexuals' negative attitudes and emotional responses to LGB individuals (Herek 2004). When prejudice is directed toward bisexual individuals, this is called **biphobia** (Rust 2002). Both heterosexuals and non-heterosexuals often reject bisexuals; thus, bisexual men and women can experience "double discrimination." Other

internalized homophobia (or internalized heterosexism) The internalization of negative messages about homosexuality by lesbian, gay, and bisexual individuals as a result of direct or indirect social rejection and stigmatization.

oppression The use of power to create inequality and limit access to resources, which impedes the physical and/or emotional well-being of individuals or groups of people.

privilege When a group has a special advantage or benefits as a result of cultural, economic, societal, legal, and political factors.

heterosexism A form of oppression that refers to a belief system that gives power and privilege to heterosexuals, while depriving, oppressing, stigmatizing, and devaluing people who are not heterosexual.

prejudice Negative attitudes and feelings toward or about an entire category of people.

discrimination Actions or practices that result in differential treatment of categories of individuals.

homophobia Negative or hostile attitudes directed toward non-heterosexual sexual behavior, a non-heterosexually identified individual, and communities of non-heterosexuals.

biphobia When prejudice is directed toward bisexual individuals.

Please use the scale below to respond to the following items. For each item, write the number that indicates the extent to which each statement is characteristic or uncharacteristic of you or your views. Please try to respond to every item.

1	2	3	4	5	6
VERY UNCHARACTERISTIC OF ME OR MY VIEWS				VERY CHARACTERISTIC OF ME OR MY VIEWS	

Please consider the ENTIRE statement when making your rating, as some statements contain two parts.

1. I feel qualified to educate others about how to be affirmative regarding LGB issues. ___
2. I have conflicting attitudes or beliefs about LGB people. ___
3. I can accept LGB people even though I condemn their behavior. ___
4. It is important to me to avoid LGB individuals. ___
5. I could educate others about the history and symbolism behind the "pink triangle." ___
6. I have close friends who are LGB. ___
7. I have difficulty reconciling my religious views with my interest in being accepting of LGB people. ___
8. I would be unsure what to do or say if I met someone who is openly lesbian, gay, or bisexual. ___
9. Hearing about a hate crime against a LGB person would not bother me. ___
10. I am knowledgeable about the significance of the Stonewall Riot to the Gay Liberation Movement. ___
11. I think marriage should be legal for same-sex couples. ___
12. I keep my religious views to myself in order to accept LGB people. ___
13. I conceal my negative views toward LGB people when I am with someone who doesn't share my views. ___
14. I sometimes think about being violent toward LGB people. ___
15. Feeling attracted to another person of the same sex would not make me uncomfortable. ___
16. I am familiar with the work of the National Gay and Lesbian Task Force. ___
17. I would display a symbol of gay pride (pink triangle, rainbow, etc.) to show my support of the LGB community. ___

18. I would feel self-conscious greeting a known LGB person in a public place. ___
19. I have had sexual fantasies about members of my same sex. ___
20. I am knowledgeable about the history and mission of the PFLAG organization. ___
21. I would attend a demonstration to promote LGB civil rights. ___
22. I try not to let my negative beliefs about LGB people harm my relationships with lesbian, gay, or bisexual individuals. ___
23. Hospitals should acknowledge same-sex partners equally to any other next of kin. ___
24. LGB people deserve the hatred they receive. ___
25. It is important to teach children positive attitudes toward LGB people. ___
26. I conceal my positive attitudes toward LGB people when I am with someone who is homophobic. ___
27. Health benefits should be available equally to same-sex partners as to any other couple. ___
28. It is wrong for courts to make child custody decisions based on a parent's sexual orientation. ___

Scoring

HATE = 4, 24, 8, 14, 9, 18
KNOWLEDGE = 20, 10, 16, 5, 1
CIVIL RIGHTS = 27, 23, 11, 28, 25
RELIGIOUS CONFLICT = 26, 12, 22, 7, 3, 13, 2
INTERNALIZED AFFIRMATIVENESS = 19, 15, 17, 6, 21

To calculate your subscale scores, add up all the values of your responses on the items in a subscale (e.g., HATE) and divide by the number of items summed. A lower score on each subscale is associated with a *stronger belief* in that subscale. For example, a low score on the Hate subscale would indicate a high level of hate. A low score on the Internal Affirmativeness subscale would indicate a high level of internal affirmativeness.

NOTE: LGB = Lesbian, Gay, or Bisexual

Source: Worthington, Roger L., Frank R. Dillon, and Ann M. Becker-Shutte. "Development, Reliability, and Validity of the Lesbian, Gay, and Bisexual Knowledge and Attitudes Scale for Heterosexuals (LGB-KASH)" *Journal of Counseling Psychology*, 52: 104–118. Copyright © 2005 by the American Psychological Association. Reprinted by permission.

The term *homophobia* is commonly used to refer to negative or hostile attitudes directed toward same-sex sexual behavior, a non-heterosexually identified individual, and communities of non-heterosexuals.

terms that refer to negative attitudes, beliefs, and emotions toward non-heterosexuals include *homonegativity, bi-negativity, antigay bias,* and *sexual prejudice* (Herek 2000a; Eliason 2001; Szymanski et al. 2008).

Although homosexuality has been decriminalized in the United States, some people still believe that gay and lesbian relations should not be legal. A Gallup Poll revealed that 64 percent of Americans believe that gay and lesbian relations between consenting adults should be legal, the highest since Gallup first asked the question over 30 years ago (Jones 2011). Further, almost 67 percent of respondents in a Gallup Poll said they would vote for a president who is non-heterosexual (Page 2011). The Gay and Lesbian Victory Fund calculates that 107 openly gay candidates were elected into office in 2010, which is a one-third increase from 2008. Table 11.2 displays the percentage of Americans by gender, and by political and religious affiliation who consider homosexuality as "morally acceptable."

The gender gap with respect to attitudes toward homo- and bisexuality has narrowed over time (Saad 2010). This is notable because previous public opinion surveys have indicated that men are more likely than women to have negative attitudes toward gay individuals (Saad 2007). Research that has assessed attitudes toward male versus female homosexuality has found that heterosexual women and men hold similar attitudes toward lesbians, but that men are more negative toward gay men (Louderback & Whitley 1997; Price & Dalecki 1998; Herek 2000b, 2002; Falomir-Pichastor & Mugny 2009). As shown in Table 11.2, another unexpected increase in favorable attitudes was found among Catholics, who demonstrated an increase in 16 percentage points.

TABLE 11.2 Percentage Calling Gay/Lesbian Relations "Morally Acceptable"

	2006	2009	2010	2006 v. 2010
Men	% 39	% 46	% 53	percentage points +14
Women	49	51	51	+2
Conservative	28	30	33	+5
Moderate	50	59	64	+14
Liberal	74	81	78	+4
Protestant	36	36	42	+6
Catholic	46	60	62	+16
Other non-Christian	77	85	84	+7
No religion	74	88	85	+11

Source: Saad 2010.

What Do You Think? Research has looked at the phenomenon of using the terms *gay* or *faggot* as a way of expressing disapproval for something or someone (i.e., "That's so gay!" or "Don't be a fag!") and how it impacts non-heterosexuals (Burn 2000; Burn, Kadlec, & Rexer, 2005; Sue 2010). The Gay, Lesbian, and Straight Education Network (GLSEN) recently partnered with the National Basketball Association (NBA) by issuing public service announcements speaking out against this kind of language (www.thinkb4youspeak.com). Do you think the NBA should be taking a stand on social issues?

Cultural Origins of Anti-LGB Bias

Why do many Americans disapprove of homo- and bisexuality? Anti-LGB bias has its roots in various aspects of American culture including religion, rigid gender roles, and myths and negative stereotypes about non-heterosexuals.

Religion Organized religion has been both a source of comfort and distress for many lesbian, gay, and bisexual Americans. Countless LGB individuals have been forced to leave their faith communities due to the condemnation embedded in doctrine and practice. Research has found that higher levels of religiosity are strongly associated with negative attitudes toward homosexuality (Shackelford & Besser, 2007; Brown & Henriquez 2008; APA 2009), religious fundamentalism (Summers 2010), and more conservative political beliefs (Shackelford & Besser, 2007; Brown & Henriquez 2008; APA 2009). Many religious leaders teach that homosexuality is sinful and prohibited by God. In more extreme cases, such as the Westboro Baptist Church (WBC), Reverend Phelps and his followers picket funerals of American servicemen and servicewomen, saying that U.S. military casualties are God's way of punishing the United States for being "nice" to lesbians and gays. The group also maintains a website called godhatesfags .com (Hagerty 2011).

At the same time, there exists variability within the Judeo-Christian faiths regarding attitudes toward LGB individuals. An increasing number of organized religious groups in America have issued statements officially welcoming LGB members and have been supportive of LGB issues, such as freedom from discrimination, the affirmation of same-sex marriage, and the ordination of openly gay clergy. Reformed Judaism ordains openly lesbian, gay, and bisexual individuals as rabbis (HRC 2011a). As early as the 1970s, the United Church of Christ became the first major Christian church to ordain an openly gay minister, and in 2005, became the largest Christian denomination to endorse same-sex marriages (Fone 2000).

Other Christian denominations, with various levels of inclusiveness, include Lutherans, Episcopalians, Methodists, and Presbyterians (Fone 2000; Goodstein 2010; HRC 2011a). The Association of Welcoming and Affirming Baptists (AWAB) is a network of more than 60 churches and organizations that advocate for the inclusion of LGB individuals within the Baptist community of faith (AWAB 2006). With some exceptions, the majority of religious groups in the United States have been silent with respect to transgender individuals.

Rigid Gender Roles Disapproval of homosexuality also stems from rigid gender roles. When Cooper Thompson (1995) was asked to give a guest presentation on male roles at a suburban high school, male students told him that the most humiliating put-down was being called a "fag." The boys in this school gave Thompson the impression that they were expected to conform to rigid, narrow standards of masculinity to avoid being labeled as a "fag." Kimmel (2011), a sociologist who specializes in research on masculinity, writes about the relationship between perceived masculinity and bullying:

> Why are some students targeted? Because they're gay or even "seem" gay—which may be just as disastrous for a teenage boy. After all, the most common put-down in American high schools today is "that's so gay," or calling someone a "fag." It refers to anything and everything: what kind of sneakers you have on, what you're eating for lunch, some comment you made in class, who your friends are or what sports team you like. . . . Calling someone gay or a fag has become so universal that it's become synonymous with dumb, stupid or wrong. (p. 10)

From a conflict perspective, heterosexual men's subordination and devaluation of gay men reinforces gender inequality. "By devaluing gay men . . . heterosexual men devalue the feminine and anything associated with it" (Price & Dalecki 1998, pp. 155–156). Negative views toward lesbians also reinforce the patriarchal system of male dominance. Social disapproval of lesbians is a form of punishment for women who relinquish traditional female sexual and economic dependence on men. Not surprisingly, research findings suggest that individuals with traditional gender role attitudes tend to hold more negative views toward homosexuality (Louderback & Whitley 1997; Brown & Henriquez 2008).

Myths and Negative Stereotypes The stigma associated with homo- and bisexuality can also stem from myths and negative stereotypes. One negative myth about non-heterosexuals is that they are sexually promiscuous and lack "family values," such as monogamy and commitment to relationships. Although some gays and lesbians do engage in casual sex, as do some heterosexuals, many same-sex couples develop and maintain long-term committed relationships and live together. Between 64 percent and 80 percent of lesbians, and between 46 percent and 60 percent of gay men, report that they are in a committed relationship at any given time (Cahill et al. 2002). Newer research with gay men found that approximately 80 percent of respondents reported that they were not happy being single and that it was not their preference (Hostetler 2009). In addition, research conducted in California indicates that 40 percent of gay males cohabit and 60 percent of lesbians live together, while 62 percent of heterosexual couples live together (Carpenter & Gates 2008).

Another myth is that non-heterosexuals, as a group, are a threat to children—most notably child molestation. In other words, people confuse homosexuality with pedophilia, which refers to the perpetration by an adult against a youth who has not yet reached puberty or has just achieved puberty. Having a non-heterosexual orientation is a separate issue from when an adult acts on inappropriate impulses with a youth who is not legally capable of consenting to sex. Research has *not* demonstrated a connection between an adult's same-sex attraction and an increased likelihood of molesting children and teenagers. In fact, a pedophile does not have an adult sexual orientation because they are fixated on children and teenagers (Herek 2009).

On the contrary, Bergman et al. (2010) found that gay men who became parents had heightened self-esteem as a result of becoming parents and raising children. Thus, the simple act of becoming a father had a very positive outcome on gay men's sense of self-worth. Research has also shown that gay men have been stigmatized by social service employees, such as social workers and managers, as being "perverts" as well as being viewed as presenting "problematic" models of gender expression (e.g., feminine) (Hicks 2006). Most often, child abusers are the father, stepfather, or a heterosexual relative or family friend (Herek 2009).

Discrimination Against Lesbians, Gays, and Bisexuals

In June 2003, a Supreme Court decision in *Lawrence v. Texas* invalidated state laws that criminalize sodomy—oral and anal sexual acts. This historic decision overruled a 1986 Supreme Court case (*Bowers v. Hardwick*), which upheld a Georgia sodomy law as constitutional. The 2003 ruling, which found that sodomy laws were discriminatory and unconstitutional, removed the legal stigma and criminal branding that sodomy laws have long placed on LGB individuals. However, sodomy is still illegal in 14 states, four of which (Kansas, Montana, Oklahoma, and Texas) target only same-sex couples (Murphy 2011). In states that criminalize both same- and opposite-sex sodomy, sodomy laws are primarily used against gay men and lesbians (ACLU 2003; Eskridge 2008).

Like other minority groups in American society, gays, lesbians, and bisexual individuals experience various forms of discrimination. Next, we look at discrimination in the workplace, in marriage and parenting, in violent expressions of hate, and in treatment by police.

Workplace Discrimination and Harassment

Most U.S. adults (89 percent) agree that LGB individuals should have equal job opportunities (Gallup Organization 2009). Yet, it is still legal in 29 states to fire, decline to hire or promote, or otherwise discriminate against employees because of their sexual orientation and, in 35 states, it remains legal to discriminate against an employee for being transgender (HRC 2011b). For example, a poll conducted in Utah found that 43 percent of the LGB respondents and 67 percent of the transgender respondents reported that they have been fired, denied a job, or not promoted due to their sexual orientation or gender identity. Further, nearly 30 percent of LGB respondents and 45 percent of transgender respondents reported that they experienced some form of harassment on a weekly basis during the previous year (Rosky et al. 2011).

As discussed later in this chapter, many workplaces have nondiscrimination policies that cover LGBT employees. But gay-affirmative policies do not ensure friendly attitudes and behaviors from coworkers. In a national probability survey representative of the United States, just one-third of LGBT employees were "out" to anyone at work, and only 25 percent were "out" to everyone at work. An examination of Figure 11.2 reveals why so few LGB employees reveal their sexual orientation in the workplace. Employees who are open about their LGB status are more likely to be harassed, lose their job, and, in general, be discriminated against when compared to employees who are not "out" at work (Sears & Mallory 2011).

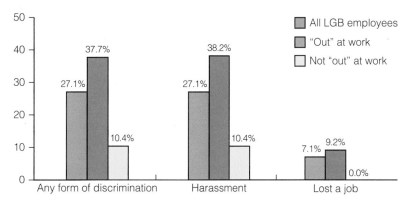

Figure 11.2:
Discrimination Based on Sexual Orientation During the Five Years Prior to the Survey, GSS 2008
Source: Sears & Mallory 2011.

Nearly half of LGBT employees (48 percent) say that they hear coworkers express negative views concerning LGBT issues "at least once in a while," and 61 percent report hearing coworkers tell jokes about LGBT people "at least once in a while." More than one in five respondents said they looked for a new job in the last year because of the uncomfortable working environment in their current job (Fidas 2009). A gay police officer who felt compelled to hide his sexual orientation from his coworkers describes his experience (Carney 2007):

> Can you imagine going to work every day and avoiding any conversations about with whom you had a date . . . or a great weekend . . . or an argument—basically not sharing any part of your personal life for fear of reprisal or being ostracized. I did this in a career that prides itself on integrity, honesty and professionalism—and where a bond with one's colleagues and partner is critical in dangerous and potentially deadly situations.

For many other Americans who risk their lives for the public's welfare, the workplace is the military. Since President Clinton signed a bill instituting a "don't ask, don't tell" (DADT) policy, more than 13,000 servicemen and women have been discharged from the military on the basis of their sexual orientation (Stone 2011). Because of this discriminatory law, countless others have decided not to join the military. DADT does not prohibit lesbians, gays, and bisexuals from serving in the military, but it prohibits recruiters from asking enlistees their sexual orientation and prohibits LGB servicemen and women from revealing their sexual orientation. Not only can LGB servicemen and women lose their jobs under DADT, they are vulnerable to harassment by other service members who threaten to "out" them.

Further, the Williams Institute of the UCLA School of Law estimates that, at a *minimum,* the total cost of DADT was $555 million for the replacement and training of personnel who were discharged as a result of the policy (Frank 2010). In addition, Frank writes that DADT, "infect[s] the morale of the estimated 66,000 gay, lesbian, and bisexual troops and their military peers who must serve in a climate of needless alienation, dishonesty, and fear" (p. 3). On September 22, 2011, the repeal of DADT became official.

What Do You Think? Internationally, other armies have had lesbian, gay, and bisexual recruits without significant problems. Examples include—but are not limited to—Australia, Israel, Uruguay, and South Africa (Palm Center 2009). Why do you think the United States has lagged so far behind other countries on this issue?

Marriage Inequality

Before the 2003 Massachusetts Supreme Court ruling in *Goodridge v. Department of Public Health,* no state had declared that same-sex couples have a constitutional right to be legally married. In response to growing efforts to secure legal recognition of same-sex couples, opponents of same-sex marriage have prompted antigay marriage legislation. For example, in 1996, Congress passed and President Clinton signed the **Defense of Marriage Act (DOMA),** which (1) states that marriage is a "legal union between one man and one woman"; (2) denies federal recognition of same-sex marriage; and (3) allows states to either recognize or not recognize same-sex marriages performed in other states. DOMA prevents any of the over 1,100 federal rights, benefits, and responsibilities of marriage from being afforded to legally married same-sex couples. A public opinion poll suggests that 51 percent of voters oppose DOMA whereas 34 percent favor it (Greenberg et al. 2011).

The proposed Federal Marriage Amendment, also called the *Marriage Protection Amendment,* would amend the U.S. Constitution to define marriage as being a union between a man and a woman. It was first introduced in Congress in 2003, and has been

Defense of Marriage Act (DOMA) Federal legislation that states that marriage is a legal union between one man and one woman and denies federal recognition of same-sex marriage.

defeated every time it has been voted on. For the first time since the Gallup Organization has polled Americans on this issue, a slight majority (53 percent) of Americans are now in favor of same-sex marriage, with the same rights as traditional opposite-sex marriages (Newport 2011).

Currently, state statutes determine whether gays and lesbians can be married or enter into civil unions. Twenty-nine states ban gay marriage through amendments to their constitutions, and an additional 12 states have laws banning same-sex marriage (Schwartz 2011). Many of these states ban not only same-sex marriage, but also other forms of partner recognition such as domestic partnerships and civil unions. Some states ban domestic partnerships for opposite-sex couples as well.

Marriage licenses are granted to same-sex couples in six states and the District of Columbia. Other states recognize same-sex marriages but do not perform them, or offer couples the same or similar state-level spousal rights through domestic partnerships or civil unions (HRC 2011c). In 2008, the campaigns for and against Proposition 8 in California, which defines marriage as between a man and woman, became the most expensive social issues campaign in U.S. history (Matthews 2010). This statistic speaks to just how hotly debated the topic of gay marriage is in America.

Approximately 50,000 same-sex couples have been legally married, according to administrative data from the states of Massachusetts, Connecticut, Iowa, Vermont, and New Hampshire, as well as estimates for same-sex couples marrying in California and the District of Columbia. As many as 30,000 same-sex couples might have married in other countries (e.g., Canada). Another 85,000 same-sex couples have entered civil unions or domestic partnerships in Vermont, California, New Jersey, Oregon, New Hampshire, Washington, and Nevada (Badgett et al. 2011). These numbers do not take into account the number of same-sex marriages that have taken place following the June 2011 New York State ruling that permits same-sex marriages (Confessore & Barbaro 2011).

AP Photo/Gary Kazanjian

A grandson sits next to his grandmother during a Proposition 8 rally in 2008. Although many people define the marriage equality debate as part of the "gay agenda," LGBT individuals have heterosexual brothers and sisters, mothers and fathers, and granddaughters and grandsons.

Arguments Against Same-Sex Marriage Some opponents of same-sex marriage view homosexuality as sick, unnatural, or immoral. They argue that granting legal status to same-sex unions would convey social acceptance of homosexuality and would thus teach youth to view homosexuality as an acceptable lifestyle. Opponents are also concerned that if same-sex marriages are legalized, schools would be pressured to treat LGB individuals as any other minority group resulting in, for example, classes on gay history, gay literature, and the like.

Opponents of same-sex marriages also commonly argue that such marriages would subvert the stability and integrity of the heterosexual family. However, as Sullivan (1997) notes in a now-classic statement, lesbians, gays, and bisexuals are already part of heterosexual families:

> [Homosexuals] are sons and daughters, brothers and sisters, even mothers and
> fathers, of heterosexuals. The distinction between "families" and "homosexuals"
> is, to begin with, empirically false; and the stability of existing families is closely
> linked to how homosexuals are treated within them. (p. 147)

Many opponents of same-sex marriage base their opposition on religious grounds. Notably, 81 percent of Americans who claim no religious affiliation favor legal same-sex marriage, compared to 48 percent support among Catholics, and 33 percent among Protestants (Jones 2010). Research has demonstrated the importance of religion (Olson, Cadge, & Harrison 2006; Whitehead 2010) when predicting attitudes toward same-sex marriage and civil unions, as well as the differences within religions in attitudes. For example, for evangelical Protestants, the question of same-sex marriage elicits a strong, unfavorable response, perhaps because of the belief in the religious sanctity of marriage between a man and a woman (Whitehead 2010).

What Do You Think? Interpretations of the U.S. Constitution and, specifically, the Second Amendment include that there should be a "wall" between the nation-state and organized religion. If the denial of marriage equality to lesbians, gays, and bisexuals is based on religious ideology, is that a violation of the Second Amendment? What do you think?

Arguments in Favor of Same-Sex Marriage Advocates of same-sex marriage argue that banning same-sex marriages or refusing to recognize same-sex marriages granted in other states is a violation of civil rights that denies same-sex couples the countless legal and financial benefits that are granted to heterosexual married couples. For example, married couples have the right to inherit from a spouse who dies without a will, to avoid inheritance taxes between spouses, to make crucial medical decisions for a partner, and to take family leave to care for a partner in the event of the partner's critical injury or illness (Badgett et al. 2011).

Spouses can receive Social Security survivor benefits, and include their partner on health insurance coverage. Other rights bestowed married partners include assumption of a spouse's pension, bereavement leave, burial determination, domestic violence protection, divorce protections, and immunity from testifying against a spouse. Finally, unlike other countries that recognize same-sex couples for immigration purposes, the United States does not recognize same-sex couples in granting immigration status (Caldwell 2011).

Another argument for same-sex marriage is that it would promote relationship stability among gay and lesbian couples. "To the extent that marriage provides status, institutional support, and legitimacy, gay and lesbian couples, if allowed to marry, would likely experience greater relationship stability" (Amato 2004, p. 963). Greater relationship stability benefits not only same-sex couples, but their children as well. Children in same-sex families would gain a range of securities and benefits, including the right to get health insurance coverage and Social Security survivor benefits from a nonbiological parent and the right to continue living with a nonbiological parent should their biological mother or father die (Tobias & Cahill 2003).

> Ironically, the same pro-marriage groups that stress that children are better off in married-couple families disregard the benefits of same-sex marriage to children.

Ironically, the same pro-marriage groups that stress that children are better off in married-couple families disregard the benefits of same-sex marriage to children.

Finally, a cross-cultural and historical view of marriage and family suggests that marriage is a social construction that comes in many forms. In response to supporters of a constitutional amendment banning gay marriage as a threat to civilization, the American Anthropological Association (AAA 2004) released the following statement:

> The results of more than a century of anthropological research on households, kinship relationships, and families, across cultures and through time, provide no support whatsoever for the view that either civilization or viable social orders depend upon marriage as an exclusively heterosexual institution. Rather, anthropological research supports the conclusion that a vast array of family types, including families built upon same-sex partnerships, can contribute to stable and humane societies.

Children and Parental Rights

According to a recent poll, since 2007, the percentage of Americans saying that the increasing number of gay couples raising children is a "bad thing" has fallen from 50 percent to 35 percent (Pew Research Center 2011). Several respected national organizations—including the American Academy of Child and Adolescent Psychiatry, American Academy of Family Physicians, American Academy of Pediatrics, American Bar Association, American Medical Association, American Psychiatric Association, American Psychoanalytic Association, American Psychological Association, Child Welfare League of America, National Adoption Center, National Association of Social Workers, North American Council on Adoptable Children, and Voice for Adoption—have taken the position that a parent's sexual orientation or gender identity has nothing to do with his or her ability to be a good parent (AMA 2011; Court of Appeals 2010; HRC 2011d).

Over 25 years of scholarly research on the children of non-heterosexual parents has yielded clear and consistent results. Irrespective of whether children were from divorced lesbian and gay parents or born to lesbian or gay parents, studies have shown that these children are at least as well adjusted overall as those from opposite-sex parents (Patterson 2009). In addition, children raised in lesbian families actually experience lower rates of physical and sexual abuse than the national norms (Chamberlain et al. 2008).

Nonetheless, across the United States today, the legal contexts for lesbian and gay parents and their children vary significantly from one jurisdiction to another. In Massachusetts, the law recognizes same-sex marriages, stranger adoptions, and second-parent adoptions, and a parent's sexual orientation is considered irrelevant when it concerns foster care, child custody, and visitation proceedings. Yet, in Missouri, the law does not recognize same-sex marriages, and lesbian and gay parents are discriminated against in custody and visitation proceedings. Neither stranger adoptions, nor second-parent adoptions have been reported in Missouri (HRC 2011e).

Laws concerning same-sex adoption vary considerably by state. Here, a multiracial same-sex couple eat dinner with their two adopted Hispanic sons in their kitchen at home in Long Beach, California.

Violence, Hate, and Criminal Victimization

On October 6, 1998, Matthew Shepard, a 21-year-old student at the University of Wyoming, was abducted and brutally beaten. Two motorcyclists who initially thought he was a scarecrow found him tied to a wooden ranch fence. His skull had been smashed, and his head and face had been slashed. The only apparent reason for the attack: Matthew Shepard was gay. On October 12, Matthew died of his injuries. Media coverage of his brutal attack and subsequent death focused nationwide attention on hate crimes against non-heterosexuals.

Anti-LGBT hate crimes are crimes against individuals or their property that are based on bias against the victim because of perceived sexual orientation or gender identity. Such crimes include verbal threats and intimidation, vandalism, sexual assault and rape, physical assault, and murder. According to the Federal Bureau of Investigation (FBI), 18.5 percent of *reported* hate crimes in 2009 were motivated by sexual orientation bias, involving 1,436 victims. Over half (55.6 percent) of the incidents were motivated by anti-gay male bias, whereas 15 percent were motivated specifically by anti-lesbian bias and 1.7 were classified as anti-bisexual bias (FBI 2010). But as discussed in Chapter 9, FBI hate crime statistics underestimate the incidence of hate crimes. The National Coalition of Anti-Violence Programs (NCAVP) documented 2,503 survivors and victims of violence due to perceived non-heterosexuality and gender non-conformity in 2010, including 27 murders.

The following documents an e-mail conversation between "L.R." and advice columnist Dan Savage after the latter was interviewed on the radio. Savage, who has joked that he is the "gay Ann Landers," is also credited with creating the YouTube "It Gets Better Project" following a slew of gay suicides in the fall of 2010. Although some would argue that the listener makes a very valid point, others would be quick to note that Savage's anger is justified.

I was listening to the radio yesterday morning, and I heard an interview with you about your It Gets Better campaign. I was saddened and frustrated with your comments regarding people of faith and their perpetuation of bullying. As someone who loves the Lord and does not support gay marriage, I can honestly say I was heartbroken to hear about the young man who took his own life.

If your message is that we should not judge people based on their sexual preference, how do you justify judging entire groups of people for any other reason (including their faith)? There is no part of me that took any pleasure in what happened to that young man, and I know for a fact that is true of many other people who disagree with your viewpoint.

To that end, to imply that I would somehow encourage my children to mock, hurt, or intimidate another person for any reason is completely unfounded and offensive. Being a follower of Christ is, above all things, recognition that we are all imperfect, fallible, and in desperate need of a savior. We cannot believe that we are better or more worthy than other people.

Please consider your viewpoint, and please be more careful with your words in the future.

L.R.

Dan Savage replies:

I'm sorry your feelings were hurt by my comments.

No, wait. I'm not. Gay kids are dying. So let's try to keep things in perspective: F— your feelings.

A question: Do you "support" atheist marriage? Interfaith marriage? Divorce and remarriage? All are legal, all go against Christian and/or traditional ideas about marriage, and yet there's no "Christian" movement to deny marriage rights to atheists or people marrying outside their respective faiths or people divorcing and remarrying.

Why the hell not?

Sorry, L.R., but so long as you support the denial of marriage rights to same-sex couples, it's clear that you do believe that some people—straight people—are "better or more worthy" than others.

And—sorry—but you are partly responsible for the bullying and physical violence being visited on vulnerable LGBT children. The kids of people who see gay people as sinful or damaged or disordered and unworthy of full civil equality—even if those people strive to express their bigotry in the politest possible way (at least when they happen to be addressing a gay person)—learn to see gay people as sinful, damaged, disordered, and unworthy. And while there may not be any gay adults or couples where you live, or at your church, or in your workplace, I promise you that there are gay and lesbian children in your schools. And while you can only attack gays and lesbians at the ballot box, nice and impersonally, your children have the option of attacking actual gays and lesbians, in person, in real time.

Real gay and lesbian children. Not political abstractions, not "sinners." Gay and lesbian children.

Try to keep up: The dehumanizing bigotries that fall from the lips of "faithful Christians," and the lies about us that vomit out from the pulpits of churches that "faithful Christians" drag their kids to on Sundays, give your children license to verbally abuse, humiliate, and condemn the gay children they encounter at school. And many of your children—having listened to Mom and Dad talk about how gay marriage is a threat to family and how gay sex makes their magic sky friend Jesus cry—feel justified in physically abusing the LGBT children they encounter in their schools. You don't have to explicitly "encourage [your] children to mock, hurt, or intimidate" queer kids. Your encouragement—along with your hatred and fear—is implicit. It's here, it's clear, and we're seeing the fruits of it: dead children.

Oh, and those same dehumanizing bigotries that fill your straight children with hate? They fill your gay children with suicidal despair. And you have the nerve to ask me to be more careful with my words?

Did that hurt to hear? Good. But it couldn't have hurt nearly as much as what was said and done to Asher Brown and Justin Aaberg and Billy Lucas and Cody Barker and Seth Walsh—day-in, day-out for years—at schools filled with bigoted little monsters created not in the image of a loving God, but in the image of the hateful and false "followers of Christ" they call Mom and Dad.

Source: From Dan Savage, "In Your Image," *Savage Love* 10/14/2010. Copyright © by Dan savage. Reprinted by permission.

Of the 27 reported murder victims, 79 percent were people of color—specifically transgender women or feminine-presenting men. Because it is not uncommon for heterosexual men and women to be mistaken for gay men and lesbians, 10.4 percent of victims of anti-LGBT violence in 2010 were identified as being heterosexual (NCAVP 2011).

Anti-LGB Hate and Harassment in Schools and on Campuses On September 19, 2010, Seth Walsh of Tehachapi, California, was found unconscious following an attempt to hang himself. Seth was on life support until his death on September 28.

He was 13 years old, openly gay, and tormented by years of unrelenting bullying based on his sexual orientation. Two days later, Tyler Clementi, an 18-year-old Rutgers University student, jumped from the George Washington Bridge to his death just days after learning that his roommate and another student broadcast a romantic encounter involving Tyler and another male on the Internet (McKinley 2010). These and other suicides by LGBT youth provoked a nationwide dialogue about bullying, sexual orientation, and gender expression. **Gender expression** refers to how a person presents her- or himself to society as a gendered individual (i.e., masculine, feminine, or androgynous) (Bettencourt 2009). A person may have a gender identity as male but nonetheless present their gender as female for any number of reasons.

AP Photo/Reena Rose Sibayan

People participated in a candlelight vigil for Rutgers University freshman Tyler Clementi on the Rutgers campus in October 2010. Clementi jumped to his death off a bridge a day after two classmates surreptitiously recorded him having sex with a man in his dorm room, and then broadcast it over the Internet.

A national survey of students ages 13 through 18 and their teachers found that actual or perceived sexual orientation is one of the most common reasons that students are harassed by their peers, second only to physical appearance (Harris Interactive & GLSEN 2005). Additional research provides further evidence of the pervasiveness of anti-LGBT bullying in schools (Swearer et al. 2008; Toomey et al. 2010), as well as a link between homophobic language, most notably by male classmates, and bullying of LGBT adolescents (Poteat & DiGiovanni 2010).

The 2009 National School Climate Survey of 7,261 middle and high school students found that nearly 9 out of 10 LGBT students experienced harassment at school in the previous year and nearly two-thirds felt unsafe because of their sexual orientation. Further,

- 84.6 percent of LGBT students reported being verbally harassed, 40.1 percent reported being physically harassed, and 18.8 percent reported being physically assaulted at school in the past year because of their sexual orientation.
- 72.4 percent heard homophobic remarks, such as "faggot" or "dyke," frequently or often at school.
- nearly two-thirds (61.1 percent) of students reported that they felt unsafe in school because of their sexual orientation.
- 30 percent of LGBT students missed at least one day of school in the previous month because of safety concerns (GLSEN 2010).

LGBT bullying is also common among college students. Law enforcement agencies in 2009 reported that 10.1 percent of hate crimes based on sexual orientation occurred in schools or on college campuses (FBI 2010). Take, for example, recent findings from the 2011 College Climate Survey conducted by the Iowa Pride Network. According to the survey (Gardner & Roemerman 2011), sexual and gender minority students face more physical harassment because of sexual orientation or gender expression, and more cyberharassment than their heterosexual peers. Forty-four percent of sexual or gender minority students have experienced some form of harassment while in school. The results also indicate that between 82 and 95 percent of students have heard racist, sexist, homophobic comments or negative comments about gender expression, from students on campus. Participants reported that, when these comments are made, very few professors or students intervene. Of those students who reported being harassed or assaulted, the majority (62.6 percent) never informed campus officials (e.g., campus security).

gender expression The way in which a person presents her- or himself as a gendered individual (i.e., masculine, feminine, or androgynous) in society. A person could, for example, have a gender identity as male but nonetheless present their gender as female.

The Effects of Antigay Harassment on Teenagers and Young Adults

LGBT youth in middle and high school settings undergo harsh treatment by their peers and may feel uncomfortable seeking the support of teachers, school administrators, and family members (see this chapter's *The Human Side* feature). Thus, it is not surprising that the reported grade point average of students who were frequently harassed because of their sexual orientation or gender expression was almost half a grade lower than for students who were less often harassed (2.7 percent compared to 3.1 percent) (GLSEN 2010).

Beyond performance in school, LGBT youth may suffer from mental health problems as a result of minority stress. The **minority stress theory** (Meyer 2003) explains that when an individual experiences the social environment as emotionally or physically threatening due to social stigma, the result is an increased risk for mental health problems. If someone is exposed to considerable hostility, or is aware that it happens to similar others, she or he may develop expectations that discrimination or harassment will happen to them. Internalized homophobia, biphobia, and/or transphobia may also develop.

Consistent with prior research and the minority stress theory (D'Augelli et al., 2006; Friedman et al., 2006; Meyer 2003), Toomey et al. (2010) found that victimization due to LGBT status was significantly associated with negative psychosocial adjustment, such as lower levels of life satisfaction and depression, which carried over into adulthood. One study found lifetime suicide attempts were common among LGBT youth aged 16 to 20 years old (Mustanski et al. 2010). Swearer et al. (2008) found that boys who were bullied because they were thought to be gay experienced greater psychological distress, greater verbal and physical bullying, and more negative perceptions of their school experiences than boys who were bullied for other reasons. Finally, Johnson et al. (2011) found that non-heterosexual female high school students were more likely to report nonphysical bullying and experience more depression than their heterosexual peers.

Police Mistreatment

As with heterosexuals and gender conforming individuals, intimate partner violence and sexual assault occurs in the lives of LGBT individuals. Fearing further victimization, many cases of LGBT violence are *not* reported to the police (NCAVP 2011; Ciarlante & Fountain 2010). Among LGBT victims who do report the crime to police, about half report that the police were "courteous," while the other half describe them as "indifferent," verbally abusive, and/or physically abusive (NCAVP 2008).

One factor that contributes to the underreporting of intimate partner violence within LGBT relationships is the failure of police to identify such incidents as criminal. Research has demonstrated that police often fail to recognize that the incident has occurred in the context of an intimate partnership. Survivors of same-sex intimate partner violence must also contend with the misconception among law enforcement that a determination of domestic violence is based primarily on the sex of the victim. As a consequence, many officers simply assign the label of "mutual abuse" and arrest both parties, resulting in a re-victimization of the survivor. It is estimated that police are 10 to 15 times more likely to make a dual arrest in cases of same-sex domestic violence than in heterosexual domestic violence (Ciarlante & Fountain 2010).

Strategies for Action: Toward Equality for All

As has been highlighted in this chapter, attitudes toward homosexuality in the United States have become more accepting over the years, and support for protecting civil rights of gays and lesbians is slowly increasing. Many of the efforts to change policies and attitudes regarding non-heterosexuals and gender non-conforming individuals have been spearheaded by organizations that specifically advocate for LGBT rights including the Human Rights Campaign (HRC), the National Gay and Lesbian Task Force (NGLTF), the Gay and Lesbian Alliance Against Defamation (GLAAD), the International Lesbian and Gay Association (ILGA), the International Gay and Lesbian Human Rights Commission, the Transgender Law and Policy Institute, Amnesty International, the Lambda Legal Defense and Education Fund, the National Center for Lesbian Rights, Parents, Families, &

minority stress theory Explains that when an individual experiences the social environment as emotionally or physically threatening due to social stigma, the result is an increased risk for mental health problems.

Friends of Lesbians and Gays (PFLAG), and the Gay, Lesbian, and Straight Education Network (GLSEN). Many states have Equality organizations, such as Equality North Carolina and Equality Texas.

But the effort to achieve sexual orientation equality is not a "gay agenda"; it is a human rights agenda that many heterosexuals and mainstream organizations support. For example, former New York Giants defensive end and current Fox NFL analyst Michael Strahan, accompanied by his fiancé, became a recent public figure from the sports world to film a public service announcement (PSA) supporting the legalization of same-sex marriage in the state of New York. In addition, Phoenix Suns point guard Steve Nash and New York Rangers forward Sean Avery have filmed PSAs for the Human Rights Campaign (Coogan 2011).

Next, we look at the important roles that "coming out," media, and the political process

Begun in 1972, *Parents, Families, and Friends of Lesbians and Gays* (PFLAG) is one of the earliest LGBT support groups. Today, PFLAG members would more commonly be called "allies."

play in the struggle for LGBT equality. Then we discuss efforts to reduce employment discrimination against non-heterosexuals, provide recognition and support to same-sex families, and include sexual orientation in hate crime legislation. We also provide an overview of educational policies and programs designed to increase acceptance and support of LGBT students in schools and on campuses.

Gays, Lesbians, and the Media

The media has been instrumental in the lives of LGBT individuals for several reasons. First, it has provided LGBT individuals, most importantly youth, role models for **"coming out,"** a phrase referring to the ongoing process whereby a lesbian, gay, or bisexual individual becomes aware of his or her sexuality, accepts and incorporates it into his or her overall sense of self, and shares that information with others such as family, friends, and coworkers (APA 2008). The phrase "coming out" also applies to transgender individuals who become aware of their gender identity and share this with others (Herman 2009). Coming out happens because heterosexuality and gender conformity are considered normative by society.

National Coming Out Day, October 11, is recognized in many countries as a day to raise awareness of the LGBT population and foster discussion of LGBT rights issues. National Coming Out Day signifies the recognition that coming out is not only an important step in the lives of LGBT individuals, but also a critical component of the fight for equality in the United States. Harvey Milk, who in 1977 became the first openly gay person to be elected to public office as a member of the San Francisco Board of Supervisors, believed that the only way for gays and lesbians to combat homophobia was to be visible—to come out of the "closet":

> Gay people, we will not win our rights by staying quietly in our closets. . . . We are coming out to fight the lies, the myths, the distortions! We are coming out to tell the truth about gays! (quoted in Shilts 1982, p. 365)

Coming out can have positive outcomes in terms of strengthening relationships, instilling hope in LGBTs who conceal their identities, and, in turn, their emotional well-being. However, when LGBT individuals come out, they also risk rejection from friends, coworkers, and family. For example, in a study of fathers coming out to their sons and daughters, several of the children rejected their fathers and, at the time of the research, remained hostile (Tasker et al. 2010). Further, criminal victimization is also a possibility. At the height of Harvey Milk's popularity, he was assassinated.

coming out The ongoing process whereby a lesbian, gay, or bisexual individual becomes aware of his or her sexuality, accepts and incorporates it into his or her overall sense of self, and shares that information with others such as family, friends, and coworkers.

National Coming Out Day Celebrated on October 11, this day is recognized in many countries as a day to raise awareness of the LGBT population and foster discussion of gay rights issues.

Harvey Milk was the first gay person to be elected for office in the United States. In 2008, Sean Penn won an award for best actor for his portrayal of Harvey Milk in the movie, aptly called, *Milk*.

. . . Coming out not only is an important step in the lives of LGBT individuals, but also a critical component of the fight for equality in the United States.

As increasingly more LGBT Americans "come out" to their family, friends, and coworkers, heterosexuals and gender-conforming individuals have more personal contact with LGBT individuals. Psychologist Gordon Allport (1954) asserted that contact between groups is necessary for the reduction of prejudice—an idea known as the **contact hypothesis**. Research has shown that heterosexuals have more favorable attitudes toward gay men and lesbian women if they have had prior contact with or know someone who is gay or lesbian (Mohipp & Morry 2004; Bonds-Raacke et al. 2007). For example, in a national opinion poll, less than half of individuals who personally knew someone who was gay were opposed to marriage equality compared to 72 percent of those who did not personally know someone who was gay (Morales 2009).

Second, in addition to providing role models, LGBT visibility in the media counteracts stereotypes of LGBT individuals (Wilcox & Wolpert 2000) and allows non-LGBT individuals to see them not as abstractions, but as real people. In 1998, Ellen DeGeneres came out on her sitcom *Ellen*. After complaints about the episode, the show was cancelled. Today, she is an Emmy-winning daytime talk show host.

Since Ellen's groundbreaking "coming out" episode, many television viewers have watched shows depicting LGBT characters in more realistic ways and that take a supportive stance on LGBT rights issues, including but not limited to *Buffy the Vampire Slayer, Glee, Queer Eye for the Straight Guy,* the *Big Gay Sketch Show, Chelsea Lately, True Blood, Six Feet Under, RuPaul's Drag Race, Will & Grace, Modern Family, The L Word,* and *The Real L Word.* "Honest, non-stereotyped and diverse portrayals of gays and lesbians in prime time can offer youth a realistic representation of the gay community . . . [and] can offer positive role models for gay and lesbian youth" (Miller et al. 2002, p. 21).

One study found that college students reported lower levels of antigay prejudice after watching television shows with prominent gay characters (e.g., *Six Feet Under* and *Queer Eye for the Straight Guy*) (Schiappa et al. 2005). The researchers propose a *parasocial contact hypothesis,* for example, they hypothesize that contact with individuals through the media may reduce prejudice in the same way Allport proposed contact between individuals reduces prejudice (see this chapter's *Social Problems Research Up Close* feature).

The visibility of famous gay, lesbian, and bisexual individuals has also had a societal impact. Many music and television celebrities, as well as politicians, have come out of the closet over the years, including but not limited to Anderson Cooper, Wanda Sykes, Lance Bass, Neil Patrick Harris, Melissa Etheridge, U.S. Representative Barney Frank, Rufus Wainwright, Jane Lynch, Rosie O'Donnell, David Hyde Pierce, Elton John, Suze Orman, K.D. Lang, Adam Lambert, and Ricky Martin (*Life* 2009).

The media has also provided examples of the increasing social disapproval of LGB prejudice. In 2011, some notable celebrities experienced personal and professional consequences as a result of their use of antigay slurs. Los Angeles Lakers player Kobe Bryant, angered by a technical foul penalty, hurled an offensive antigay epithet at a referee. This was followed by Bryant's swift apology; he also paid a $100,000 fine. Additionally, Atlanta Braves pitching coach Roger McDowell attracted attention by verbalizing antigay slurs and making offensive gestures at a group of fans. He was briefly suspended from playing, and later apologized at a news conference (Mungin 2011). Finally, comedian Tracy Morgan, from *Saturday Night Live* and *30 Rock*, also received a great deal of criticism following a "homophobic rant" during a stand-up routine in Nashville, Tennessee. Morgan subsequently apologized to the public, yet many are not satisfied with Morgan's apology (Brown 2011).

Finally, social media has been important in addressing the needs of LGBT youth who feel the effects of minority stress. Following several gay teenage suicides that appeared

contact hypothesis The idea that contact between groups is necessary for the reduction of prejudice.

Meta-analytic research has demonstrated a strong connection between contact with someone who is LGB (Smith et al. 2009) and the reduction of sexual prejudice against nonheterosexuals. In recent years, celebrities have come out in increasing numbers and LGBT television characters have been portrayed more realistically. The following research (Bonds-Raacke et al. 2007) examines the effect of gay and lesbian television characters on attitudes toward real lesbians and gay men.

Sample and Methods

First, a pilot study was conducted in which participants were asked to recall any memorable gay or lesbian television or film character and complete a survey about their perceptions of the character. Participants were 269 midwestern U.S. college students enrolled in a general psychology course at a large public university. The participants' sexual orientation was not asked in this study. Participants were first asked to think of a memorable gay/lesbian character in a TV show/movie. They then wrote down the name of the character and the TV program or movie in which the character appeared. Next, participants rated their perceptions of this character using 7-point Likert-scales for the dimensions of serious/humorous, likeable/not likeable, mentally ill/mentally stable, safe/dangerous, moral/unmoral, honest/dishonest, responsible/irresponsible, kind/cruel, violent/nonviolent, and bad role model/good role model. Afterward, participants responded to attitudinal statements assessing the extent to which they believed: (1) real-life gays/lesbians were like the character, (2) they would enjoy having this

character as a friend, (3) the character's family accepted their homosexuality well, and (4) the character's friends accepted their homosexuality well.

Following the pilot study, an experiment was conducted to directly assess the effects of thinking about either a positive or negative gay or lesbian character on attitudes toward gay men and lesbians. Two groups of participants were used for this experiment. Group 1 (N = 65) recalled a "positive portrayal" of a gay or lesbian media character (as overall likeable or admirable), while Group 2 (N = 49) recalled a "negative portrayal" (as overall not likeable or unlikely to gain respect from others). Next, participants completed a two-part survey. Part one was identical to the pilot study survey in which participants rated their perceptions of the character on various dimensions using Likert scales and responded to four attitudinal statements. In part two of the survey, participants completed the Attitudes Toward Lesbian and Gay Men (ATLG) Scale (Herek 1988), in which participants indicated their level of agreement or disagreement with the statements on a Likert scale ranging from 1 (strongly agree) to 7 (strongly disagree). Examples of statements included: "female/male homosexuality is an inferior form of sexuality" and "female/male homosexuality is a perversion."

Findings and Discussion

Results from the pilot study indicated that all participants were able to recall notable gays and lesbians in the media. Over two-thirds of participants recalled either Ellen or Will, and evaluative ratings for these characters were generally positive. Participants indicated all characters recalled were relatively

humorous, likeable, mentally stable, safe, honest, kind, responsible, and nonviolent. These characters were judged in the middle of the scale for moral/immoral and good role model/bad role model. Participants agreed that the character's friends accepted their homosexuality well and were ambivalent about whether (1) real gay and lesbian people were like the character, (2) they would enjoy having this character as a friend or acquaintance, and (3) if the character's parents accepted their homosexuality well.

The experiment results indicated that those who recalled a positive character portrayal later showed a more positive attitude toward gay men than those recalling a negative portrayal. Further, thinking of a negative portrayal did not substantially change people's attitudes toward homosexuality. Women also were found to have a more positive attitude overall than men toward gay men and lesbians. Such findings illustrate the importance of positive role models in entertainment media as potential primes of social attitudes.

It is important to note that the characters Will and Jack (from *Will and Grace*) and Ellen Degeneres were the most frequent exemplars in both studies, with Will being the most frequently recalled character. The investigators point out that the participants' recall of the same three characters is consistent with Greenberg's (1988) Drench Hypothesis, meaning that a select few leading characters are highly salient in people's minds (e.g., drenching the viewer), while other supporting characters, even if from popular shows, like Carol from *Friends*, have less impact on the viewing population.

Source: Bonds-Raacke et al. 2007.

in the news during September 2010, Dan Savage and his partner created the *It Gets Better Project*. The *It Gets Better Project* has turned into a viral cyber-movement, inspiring thousands of user-created videos that instill messages of hope and support for LGBT youth who have been bullied, feel that they must hide in shame, or who experience rejection from family members. To date, the project has received submissions from celebrities, organizations, activists, politicians, and media personalities, ranging widely from President Barack Obama and Adam Lambert to staff members of Google and Pixar (*It Gets Better Project* 2011).

Ending Workplace Discrimination

Presently, only discrimination on the bases of race, religion, national origin, sex, age, and disability are protected by federal law. The **Employment Non-Discrimination Act (ENDA)**, a proposed federal bill that would protect LGBT individuals from workplace discrimination, has been debated in Congress since 1994, but has never been signed into law. It would offer individuals basic protections against workplace discrimination on the basis of sexual orientation or gender identity. It was re-introduced in the House and Senate in 2011 (Johnson 2011).

With the absence of federal legislation prohibiting discrimination based on sexual orientation, some state and local governments, as well as private corporations, prohibit employment discrimination based on sexual orientation and extend domestic partner benefits to same-sex couples. Twenty-one states and the District of Columbia have laws banning sexual orientation discrimination in the workplace (Sears & Mallory 2011; HRC 2011b). Additionally, all 21 states and the District of Columbia offer domestic partner benefits for their government employees. The Domestic Partnership Benefits and Obligations Act, currently pending in Congress, would extend domestic partnership benefits to all federal employees and their partners (Badgett 2009). Over 90 percent of Fortune 500 companies include sexual orientation in their equal employment opportunity or nondiscrimination policies, and almost half offer same-sex domestic partner health benefits (Equality Forum 2011). Further, nearly 200 cities and counties prohibit sexual orientation discrimination in the public sector (Human Rights Campaign 2009a).

Not long after a poll revealed that the majority of Americans (67 percent) supported repealing DADT (Morales 2010), DADT was repealed by the Senate and subsequently signed into law by President Obama in December 2010 (O'Keefe 2010). In July 2011, the final step toward enacting DADT took place when the President, Secretary of Defense, and chairman of the Joint Chiefs of Staff certified that the Department of Defense is ready to enact the change without hurting military readiness or effectiveness. Lesbian, gay, and bisexual troops will be able to serve openly 60 days from the certification date, September 22, 2011 (ACLU 2011).

Law and Public Policy

Marriage Equality In 2004, Massachusetts became the first state to allow same-sex couples to be legally married. In June 2011, New York became the sixth state to legalize same-sex marriage, joining Massachusetts, Connecticut, Iowa, Vermont, and New Hampshire, as well as the District of Columbia. Several other states allow same-sex couples many of the same legal rights and responsibilities as married heterosexual couples (see Figure 11.3). For example, in some states, same-sex couples can apply for a civil union license. A **civil union** is a legal status parallel to civil marriage under state law, and entitles same-sex couples to *almost* all of the rights and responsibilities available under state law to opposite-sex married couples.

Federal law does not recognize the rights of partners in same-sex civil unions, so they do not have the more than 1,000 federal protections that go along with civil marriage, nor is their legal status recognized in all other states. Some states, counties, cities, and workplaces allow unmarried couples, including gay couples, to register as **domestic partners** (or "reciprocal beneficiaries" in Hawaii). The rights and responsibilities granted to domestic partners vary from place to place but may include coverage under a partner's health and pension plan, rights of inheritance and community property, tax benefits, access to married student housing, child custody and child and spousal support obligations, and mutual responsibility for debts.

In February 2011, the Attorney General of the U.S. announced that the White House would no longer defend the constitutionality of Section 3 of DOMA. Five months later, the White House went further and formally endorsed the **Respect for Marriage Act (RMA),** which was first introduced into Congress in 2009 (ACLU 2011; Savage & Stolberg 2011). If passed, the RMA would overturn DOMA and grant federal recognition to

Employment Nondiscrimination Act (ENDA) A proposed federal bill that would protect LGBTs from workplace discrimination, has been up for congressional debate on a number of occasions since 1994, but has never been signed into law.

civil union A legal status that entitles same-sex couples who apply for and receive a civil union certificate to nearly all of the benefits available to married couples.

domestic partners A status granted to unmarried couples, including gay and lesbian couples, by some states, counties, cities, and workplaces that conveys various rights and responsibilities.

Respect for Marriage Act (RMA) A bill that, if passed, would overturn DOMA and grant federal recognition to same-sex marriages, regardless of the state laws in which they reside.

same-sex marriages, regardless of the state laws in which the couple reside. The bill would *not* require states to recognize same-sex marriages performed in other states. This is the first time the President of the United States and his administration have formally acknowledged that non-heterosexuals are being discriminated against at the federal level. The legal recognition of same-sex marriages by the federal government would mean that same-sex couples would realize all of the benefits of their opposite-sex counterparts (Badgett et al. 2011).

Parental Rights Large national organizations such as the Child Welfare League of America, the National Association of Social Workers, the American Psychological Association, and the American Bar Association all support the suitability of qualified unmarried and non-heterosexual couples to foster and adopt children (ACLU 2010). Politicians are also working to help end discrimination against non-heterosexual parents. For instance, in 2011, Representative Pete Stark (D-CA) reintroduced the **Every Child Deserves a Family Act** into Congress. The legislation would remove obstacles to non-heterosexual individuals providing loving homes for adoption or foster care (Moulton 2011). If passed, this legislation would also widen the pool of homes for children in the foster care system.

Bill Aron/PhotoEdit

Two gay Jewish men wearing yarmulke stand under a chuppah during their outdoor wedding ceremony. The reformed Jewish community was an early supporter of LGBT equality.

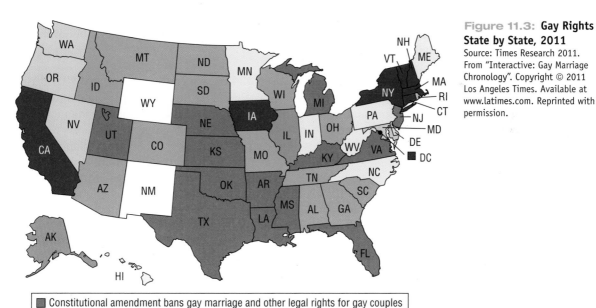

Figure 11.3: Gay Rights State by State, 2011
Source: Times Research 2011. From "Interactive: Gay Marriage Chronology". Copyright © 2011 Los Angeles Times. Available at www.latimes.com. Reprinted with permission.

- ■ Constitutional amendment bans gay marriage and other legal rights for gay couples
- ■ Constitutional amendment bans gay marriage
- □ Legal ban on gay marriage and other legal rights for gay couples
- □ Legal ban on gay marriage
- □ No specific law bans gay marriage
- □ Domestic partnerships legal
- □ Domestic partnerships with comprehensive protections legal
- ■ Gay marriages performed elsewhere recognized
- ■ Civil unions legal
- ■ Gay marriage legal

Every Child Deserves a Family Act This piece of federal legislation would remove obstacles to non-heterosexual (as well as transgender) individuals providing loving homes for adoption or foster care.

As discussed earlier in the chapter, each state has its own laws governing both joint adoptions and second-parent adoptions for same-sex couples. Although some states, such as Utah and Mississippi, prohibit same-sex couples from adopting, there have been legislative advances in others states. In September 2010, Florida's Third District Court of Appeals ruled that a 1977 statute prohibiting "homosexuals" from adopting is unconstitutional. The decision is binding on all trial level courts in Florida (U.S. Court of Appeals 2010). Further, in 2011, the Maryland Department of Health and Mental Hygiene decided to allow a woman to be named as a parent on the Maryland birth certificate of the child born to her same-sex married spouse, without requiring a court order (DHMH 2011).

Hate Crimes Legislation Hate crime laws call for tougher sentencing when prosecutors can prove that the crime committed was a hate crime. According to the most recent data available, as of June 2009, 31 states and the District of Columbia had hate crime laws that include sexual orientation, 14 states had hate crime laws that did not include sexual orientation, and 5 states had no hate crime laws (Human Rights Campaign 2009b; National Gay and Lesbian Task Force 2009).

In October 2009, President Obama signed into law the **Matthew Shepard and James Byrd, Jr. Hate Crimes Prevention Act (HCPA).** This new law expands the original 1969 federal hate crimes law to cover hate crimes based on actual or perceived sexual orientation, gender, gender identity, and disability. The new law was named after Matthew Shepard, a gay Wyoming teenager who died after being severely beaten, and James Byrd Jr., an African American man who was attacked, chained to a vehicle, and dragged to his death in Texas (FBI 2011; CNN 2009).

Educational Strategies and Activism

Educational institutions bear the responsibility of promoting the health and well-being of all students. Thus, they must address the needs and promote acceptance of LGBT youth through certain policies and programs. The strategies for attaining these goals are to include LGBT issues in sex education, include LGBT-affirmative classroom curricula, and promote tolerance in learning environments through policies, education, and activism.

Sex Education and LGBT-Affirmative Classroom Curricula The censorship of LGBT current issues, historical figures and events, and sexual health in both the classroom and in school libraries is a heated debate across the nation (Casey 2011). Whether LGBT themes can be brought into public school classrooms varies considerably between *and* within states. For example, four years ago, one Maryland county school system permitted lessons on non-heterosexuality in the classroom after consulting with the American Academy of Pediatrics on the issue, a bold move at the time (Schemo 2007). Similarly, school officials in one Michigan county are reconsidering their sex education curriculum following a sex education teacher's request to include a discussion of homosexuality in the class (WWMT News 2011).

At the state level, California Governor Jerry Brown signed legislation in July 2011, making California the first state to require that textbooks and history lessons include the contributions of gay, lesbian, bisexual, and transgender Americans (McGreevy 2011). In advocating for the end to censorship of LGBT resources in school libraries, Lambda Legal (2011b) points to the multiple benefits of providing education to students in a way that normalizes and affirms LGBT experiences. This includes promoting greater awareness of human diversity, the reduction of bullying and harassment, and the provision of support, reassurance, and information that non-heterosexual and gender non-conforming students need. Opposition to LGBT curricula topics has garnered momentum in some areas of the country, including the 2011 Tennessee Senate approval of a bill that prohibits teachers from discussing homosexuality in any way with students from kindergarten through the eighth grade. The bill has been termed the "Don't Say Gay Bill" by opponents (Humphrey 2011).

Matthew Shepard and James Byrd, Jr. Hate Crimes Prevention Act (HCPA) This new law expands the original 1969 federal hate crimes law to cover hate crimes based on actual or perceived sexual orientation, gender, gender identity, and disability.

Promotion of Tolerance in Learning Environments Eighteen states and the District of Columbia have laws that prohibit discrimination, harassment, and/or bullying of students based on sexual orientation—16 of those also prohibit prejudice and discrimination based on gender identity. Four states have a school regulation or teachers' ethical code that addresses discrimination, harassment, and/or bullying based on sexual orientation, and 22 states prohibit bullying in schools with no list of protected categories (HRC 2011f). In July 2011, the Miami-Dade County Public School District amended the language of its anti-bullying and harassment policy to explicitly include harm against any student or teacher based on sexual orientation and gender identity (Rapado & Campbell 2011). This is particularly notable given that Florida prohibits bullying in general, but does not specify categories of protection.

Over a number of years, Congresswoman Linda Sanchez from California has spearheaded the effort to have the Safe Schools Improvement Act passed. This is a federal bill that, if passed, would require school districts to adopt policies prohibiting bullying and harassment based on race, gender, religion, sexual orientation, and gender identity, among others. It is pending in Congress as of 2011, when it was reintroduced by Senator Bob Casey of Pennsylvania (Casey 2011).

A number of programs exist that aim to create a "harassment-free" climate and promote understanding and acceptance of sexual orientation and gender diversity in the K–12 school setting. The Gay, Lesbian, and Straight Education Network (GLSEN) is a national organization that collaborates with educators, policy makers, community leaders, and students to protect LGBT students from bullying and harassment, to advance comprehensive safe schools laws and policies, to empower principals to make their schools safer, and to build the skills of educators to teach respect for all people. GLSEN sponsors such initiatives as the "thinkb4youspeak" website, the National Day of Silence in April, No Name-Calling Week in January, and the Training of Trainers Program for teachers and community organizers. GLSEN also supports **gay-straight alliances (GSAs),** which are school-sponsored clubs led by middle or high schools, that strive to address anti-LGBT name-calling and promote respect for all students. One common task of GSAs involves organizing for the **National Day of Silence** at their schools during which students do not speak for an entire day in recognition of the daily harassment that LGBT students endure (GLSEN 2011b).

Campus Programs Many LGBT-affirmative and educational initiatives of middle and high schools also occur on college campuses. Student groups in higher education (i.e., colleges and universities) have been active in the gay liberation movement since the 1960s. In addition to university-wide non-discrimination policies, other measures to support the LGBT college student population include gay and lesbian studies programs, social centers, and support groups, as well as campus events and activities that celebrate diversity. For example, some campuses have a "Lavender Graduation" ceremony in which LGBT graduates are honored and receive rainbow tassels for their caps (Duke University Student Affairs 2011). Many campuses have Safe Zone programs designed to visibly identify students, staff, and faculty who support the LGBT population. Safe Zone programs require a training session that provides a foundation of knowledge needed to be an effective ally to LGBT students and those questioning their sexuality (University of Alabama 2011).

Understanding Sexual Orientation and the Struggle for Equality

In recent years, a growing acceptance of lesbians, gay men, and bisexuals as well as increased legal protection and recognition of these marginalized populations has been witnessed. The advancements in LGB rights are notable and include progress with respect to marriage equality, the repeal of DADT, the growing public support of LGB civil rights, the endorsement of LGB civil rights by mental health and medical organizations, and the growing adoption of nondiscrimination policies covering sexual orientation.

gay-straight alliances (GSAs) School-sponsored clubs led by middle or high schools, that strive to address anti-LGBT name-calling and promote respect for all students.

National Day of Silence A day during which students do not speak in recognition of the daily harassment that LBGT students endure.

Support

SAFE ZONE

Accept · Respect

UAB

University of Alabama at Birmingham, UAB Safe Zone Program

Safe Zone is one way to say that all sexual orientations, gender identities, and expressions are part of our culture and are acknowledged and supported. The Safe Zone Program provides a visible network of volunteers for lesbian, gay, bisexual, transgender, and other students, staff and faculty seeking information and assistance regarding sexual orientation, gender identity and expression.

But political efforts to undermine gay rights and recognition must realize that prejudice and discrimination against individuals based on statuses over which research suggests they have no control hurts everyone.

gay pride Demonstrative and cultural expressions of gay activism that include celebrations, marches, demonstrations, or other cultural activities promoting gay rights.

But the winning of these battles in no way signifies that LGBT individuals have secured equal rights. As evidenced by the years of court maneuvering in the struggle against Proposition 8 in California in 2008, gay rights previously won can be taken away. LGB individuals employed at workplaces with antidiscrimination policies still experience harassment and rejection from their coworkers, and students in schools with policies against bullying are still subjected to antigay taunts. Although many countries worldwide are increasing legal protections for LGBT individuals, homosexuality is formally condemned in some countries, with penalties ranging from fines to imprisonment to death.

Many of the advancements in gay rights have been the result of political action and legislation. Barney Frank (1997), an openly gay U.S. Representative, emphasized the importance of political participation in influencing social outcomes. He noted that demonstrative and cultural expressions of gay activism, such as **"gay pride"** celebrations, marches, demonstrations, or other cultural activities promoting gay rights, are important in organizing gay activists, but cannot be used as substitutes for "conventional, boring, but essential" political participation (p. xi).

As both structural functionalists and conflict theorists note, non-heterosexuality challenges traditional definitions of family, child rearing, and gender roles. Every victory in achieving legal protection and social recognition for non-heterosexuals fuels the backlash against them by groups who are determined to maintain traditional notions of family and gender. Often, this determination is rooted in and derives its strength from religious ideology.

As symbolic interactionists note, the meanings associated with homosexuality are learned. Powerful individuals and groups opposed to gay rights focus their efforts on maintaining the negative meanings of homosexuality to keep the gay, lesbian, and bisexual population marginalized.

But political efforts to undermine gay rights and recognition must realize that prejudice and discrimination against individuals based on statuses over which research suggests they have no control hurts everyone. Using gay epithets as a way of questioning a boy's or man's masculinity threatens heterosexual as well as non-heterosexual males and the community as a whole. Antigay harassment has been identified as a precipitating factor in several school shootings (Pollack 2000a; 2000b).

There is the loss of capital, i.e., the potential contributions LGBT individuals would have made if not for the prevalence of antigay sentiment. For example, fear of reprisals keeps many LGBT employees from sharing information about their personal lives at work. Employees who are "closeted" at work are less satisfied with their jobs, less trusting of their employers, and are more likely to leave their position than their "out" counterparts (Hewlett & Sumberg 2011).

States that don't legally recognize unmarried couples in committed relationships equally ignore the rights of heterosexual and homosexual couples. Heterosexuals who "look" or "act" gay may be the victims of hate crimes. Heterosexual mothers and fathers live in fear that their children will be victimized by antigay prejudice and discrimination—harassed, fired, or even killed.

True, the American public is becoming increasingly supportive of gay rights, and LGBT individuals have been granted legal rights in several states. But, as one scholar notes, "The new confidence and social visibility of homosexuals in American life have by no means conquered homophobia. Indeed it stands as the last acceptable prejudice" (Fone 2000, p. 411).

The arguments for and against LGBT equality are often emotionally charged. Here, crowd control barriers separate gay rights supporters and opponents during a Los Angeles Gay Pride parade.

Colin Young-Wolff/PhotoEdit

CHAPTER REVIEW

- **Are there any countries in which homosexuality is illegal?**

Yes. In 79 out of 242 countries throughout the world, homosexuality among males is illegal; in 45 of these countries, homosexuality among females is illegal. Legal penalties vary, including prison sentences and—in 5 countries—death.

- **Is there any country where same-sex couples can be legally married?**

Yes. In 2001, the Netherlands became the first country in the world to offer full legal marriage to same-sex couples followed by Belgium, Spain, Canada, South Africa, Argentina, Iceland, Norway, Portugal, Spain, Sweden, Mexico City, and six U.S. states.

- **In what ways is the classification of individuals into sexual orientation categories problematic?**

Classifying individuals into sexual orientation categories is problematic for a number of reasons: (1) distinctions among sexual orientation categories are not clear-cut and may better be represented by a continuum; (2) research with same-sex populations has tended to define sexual orientation based on one of three components whereas sexual orientation is complex and multidimensional, and (3) social stigma associated with non-heterosexuality leads people to conceal or falsely portray their sexuality.

- **What is the relationship between beliefs about what "causes" homosexuality and attitudes toward homosexuality?**

Individuals who believe that homo- or bisexuality is biologically based or inborn tend to be more accepting of LGB individuals. In contrast, individuals who believe that

lesbians, gays, and bisexual choose their sexual orientation are less tolerant of LGB individuals.

- **What is the official position of numerous respected professional organizations regarding sexual orientation change efforts (SOCE) for gays and lesbians?**

These organizations agree that sexual orientation cannot be changed and that efforts to change sexual orientation (conversion, reparative, or reorientation therapy) do not work and may, in fact, be harmful.

- **What three cultural changes have influenced the worldwide increase in liberalized national policies on same-sex relations and the gay rights movement?**

The worldwide increase in liberalized national policies on same-sex relations and the gay rights social movement have been influenced by (1) the rise of individualism, (2) increasing gender equality, and (3) the emergence of a global society in which nations are influenced by international pressures.

- **How does social stigma impact the mental health of marginalized individuals, such as LGBT individuals?**

The minority stress theory explains that when an individual experiences the social environment as emotionally or physically threatening due to social stigma, the result is an increased risk for mental health problems. If someone is exposed to considerable hostility, or is aware that it happens to similar others, he or she may develop expectations that discrimination or harassment will happen to them. Internalized homophobia, biphobia, and transphobia may also develop.

- **Is employment discrimination based on sexual orientation illegal in all 50 states?**

As of 2011, it is still legal in 29 states to fire, decline to hire or promote, or otherwise discriminate against employees because of their sexual orientation, and in 37 states, it remains legal to discriminate against an employee for being transgender. As of this writing, a federal bill called the Employment Nondiscrimination Act (ENDA), which would ban employment discrimination against individuals on the basis of sexual orientation, is pending before Congress.

- **What is the "Don't ask, don't tell" (DADT) policy?**
A policy in which recruiting officers are not allowed to ask about sexual orientation, and in which LGB enlistees are encouraged not to volunteer such information. Thousands of service members have been discharged under DADT since it took effect. DADT was repealed by the Senate and signed into law by President Obama in 2010 and put into action months later.

- **Same-sex couples want the same legal and financial benefits that are granted to heterosexual married couples. What are some of these benefits?**
Married couples have the right to inherit from a spouse who dies without a will, to avoid inheritance taxes between spouses, to make crucial medical decisions for a partner and to take family leave to care for a partner in the event of the partner's critical injury or illness, to receive Social Security survivor benefits, and to include a partner in health insurance coverage. Other rights bestowed on married (or once-married) partners include assumption of a spouse's pension, bereavement leave, burial determination, domestic violence protection, reduced rate memberships, divorce protections (such as equitable division of assets and visitation of partner's children), automatic housing lease transfer, immunity from testifying against a spouse, and spousal immigration eligibility.

- **What is Proposition 8 and why is it significant?**
Proposition 8 is a proposed amendment to the California constitution that would define marriage as between a man and a woman. In 2008, the campaigns for and against Proposition 8 became the most expensive social issues campaign in U.S. history. This statistic speaks to just how hotly debated the topic of gay marriage is in America. On August 4, 2010, in the case of *Perry v. Schwarzenegger*, Judge Vaughn R. Walker overturned Proposition 8.

- **What are some reasons for the underreporting of LGBT intimate partner violence and sexual assault to the police?**
Many cases of LGBT violence are not reported to the police for fear of further victimization by police, including indifference or some form of abuse. Additionally, the failure of police to identify such incidents as occurring in the context of an intimate partnership, often erroneously leads to the arrest of both the victim and the perpetrator.

- **Why is the media important in the advancement of LGBT civil rights?**
LGBT visibility in the media counteracts stereotypes of LGBT individuals and allows non-LGBT individuals to see them not as an abstraction, but as real people. This is consistent with Gordon Allport's contact hypothesis that suggests more contact with or exposure to a group results in the reduction of prejudice. The media has also provided LGBT individuals, most importantly youth, with role models for "coming out."

- **What are Safe Zone programs?**
Safe Zone programs are designed to visibly identify students, staff, and faculty who support the LGBT population. Participants in Safe Zone programs display a sign or placard outside their office or residence hall room that identifies them as individuals who are willing to provide a safe haven and support for LGBT people and those struggling with sexual orientation issues.

TEST YOURSELF

1. Some psychological research has indicated that many individuals are not exclusively heterosexual or homosexual and that perhaps sexual orientation can be represented on a continuum.
   a. True
   b. False
2. A national study of U.S. college students found that what percentage identified as gay, lesbian, or bisexual?
   a. 1.2
   b. 3.0
   c. 4.9
   d. 10.1
3. According to the 2010 Census, an estimated one-eighth of all same-sex households are raising children.
   a. True
   b. False

4. NARTH has claimed all but the following:
   a. They are not "antigay."
   b. Collectively, they have a success rate of 75 percent.
   c. They are not a religious group.
   d. They assist those who are distressed by unwanted sexual attractions.
5. Why has Reverend Fred Phelps of the Westboro Baptist Church (Topeka, Kansas) picketed funerals of American servicemen and servicewomen who have died in the Iraq war?
   a. To protest the "Don't ask, don't tell" military policy
   b. To protest efforts to repeal the "Don't ask, don't tell" military policy
   c. To publicize his view that U.S. military casualties are God's way of punishing the United States for being nice to lesbians and gays
   d. To show support of Iraq's harsh penalties for homosexuality

6. Between 2007 and 2011, the percentage of Americans saying that the increasing number of gay couples raising children is a "bad thing" has fallen from 50 percent to 35 percent.
   a. True
   b. False

7. Why was the It Gets Better Project started?
   a. Because same-sex couples continue to encounter obstacles in second parent and joint adoptions
   b. Because LGBT youth continue to be bullied and harassed by their peers, resulting in shame, isolation, and even suicide
   c. Because LGB servicemen and women have been waiting a long time to serve openly in the military
   d. Because American society seems to be growing in its acceptance of the legal recognition of same-sex couples

8. Who was Harvey Milk?
   a. The first openly gay person to be elected to public office as a member of the San Francisco Board of Supervisors
   b. The first openly gay person to be discharged from the military under the "Don't ask don't tell" policy
   c. A 12-year-old boy who committed suicide after being subjected to antigay harassment by his peers
   d. A parent who started the organization Parents, Families, and Friends of Lesbians and Gays (PFLAG) after his son committed suicide due to antigay harassment

9. The "thinkb4youspeak" website created by GLSEN uses public service announcements made by celebrities to help combat biased language (e.g., "You're so gay," etc.).
   a. True
   b. False

10. The first country to legalize same-sex marriage was:
    a. Norway
    b. England
    c. The Netherlands
    d. Argentina

Answers: 1: a; 2: c; 3: b; 4: b; 5: c; 6: a; 7: b; 8: a; 9: a; 10: c.

KEY TERMS

MEDIA RESOURCES

Turning to Video

After watching the ABC video *Forbidden Love: Mormon Gays* (running time 10:01), available through **CengageBrain.com**, what do you think the relationship between religion and being gay should be? Should religion condemn and dictate, accept and embrace, or some combination of the two?

Online Study Resources

Log in to **www.cengagebrain.com** to access the resources your instructor has assigned. For this book, you can access:

CourseMate

Access chapter-specific learning tools, including learning objectives, practice quizzes, videos, Internet exercises, flash cards, and glossaries, as well as web links, and more in your Sociology CourseMate.

12

Population Growth and Aging

"Population may be the key to all the issues that will shape the future: economic growth; environmental security; and the health and well-being of countries, communities, and families."

—Nafis Sadik, former executive
director, United Nations
Population Fund

AS PEOPLE AGE, one of the issues they think about is what they would like done with their body after they die. Although traditions, customs, and laws concerning what to do with human remains vary across cultures and across time, some form of cremation and/or burial is common practice around the world. One of the problems associated with burial of human remains is that this practice requires land. Have you ever driven past a cemetery and wondered how many generations can continue to be buried before we run out of land? Several cemeteries in New York City have already run out of land and have stopped selling burial plots (Santora 2010).

Japan offers an innovative answer to the problem of not enough grave space: a high-tech graveyard in a multistoried building where the ashes of the dead are stored in urns on shelves. Visitors use plastic swipe cards and a touch screen to identify the remains they wish to "visit" and a robotic arm collects the requested urn and delivers it to a "mourning room" where visitors pay tribute to their deceased loved ones (Here and Now 2009). To maximize grave space, some countries allow the practice of vertical burials in which the deceased are wrapped in a biodegradable shroud and buried in a vertical, standing up position (Dunn 2008). In another alternative to burials, the dead can be poured down the drain if the remains are put through a process called alkaline hydrolysis, in which heat and lye are used to transform the body into a nontoxic liquid, leaving behind a small amount of dry bone residue for the family to scatter (Frynes-Clinton 2011). As an alternative to cremation or burial, alkaline hydrolysis is legal in a handful of states, including Kansas, Maryland, and Colorado (Franko 2011).

Running out of land for grave sites is only one of many concerns that stem from the growth in human population over the last century or so. The world's population is not only growing, it is aging. In this chapter, we focus on social problems associated with population growth and aging.

The Global Context: A World View of Population Growth and Aging

Although thousands of years passed before the world's population reached 1 billion around the year 1800, the population exploded from 1 billion to 6 billion in less than 300 years (see Figure 12.1). World population was 1.6 billion when we entered the 20th century, and 6.1 billion when we entered the 21st century. In only 12 years, world population grew from 6 billion (in 1999) to 7 billion (in 2011) (United Nations 2011).

World Population: History, Current Trends, and Future Projections

Humans have existed on this planet for at least 200,000 years. For 99 percent of human history, population growth was restricted by disease and limited food supplies. Around 8000 B.C., the development of agriculture and the domestication of animals led to increased food supplies and population growth, but even then, harsh living conditions and disease still put limits on the rate of growth. This pattern continued until the mid-18th century, when the Industrial Revolution improved the standard of living for much of the world's population. The improvements included better food, cleaner drinking water, improved housing and sanitation, and advances in medical technology, such as antibiotics and vaccinations against infectious diseases—all of which contributed to rapid increases in population.

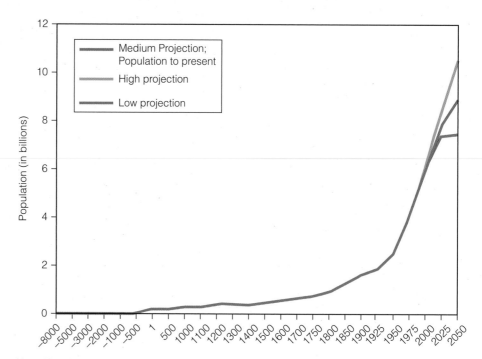

Note: Time is not to scale.

Population **doubling time** is the time required for a population to double from a given base year if the current rate of growth continues. It took several thousand years for the world's population to double to a size of 14 million, but then took only a thousand years to nearly double to 27 million and another thousand to reach 50 million. From there, it took only 500 years to double from 50 million to 100 million, and 400 years for the next doubling to occur. When the Industrial Revolution began around 1750, population growth exploded, taking only 100 years to double. The most recent doubling—from three billion in 1960, to six billion in 1999—took only about 40 years (Weeks 2012). Although world population will continue to grow in the coming decades, it will probably not double in size again.

World population is projected to grow from 6.9 billion in 2010, to 8 billion in 2025, 9.3 billion in 2050, and 10 billion in 2085 (United Nations 2011). Most of the world's population live in less developed countries (see Figure 12.2). The most populated

Figure 12.2: **Distribution of World Population by Level of Development, 2010**
Source: United Nations 2011.

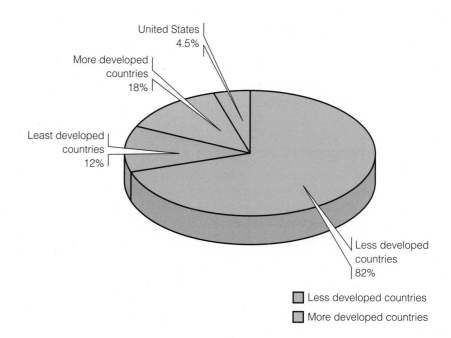

doubling time The time required for a population to double in size from a given base year if the current rate of growth continues.

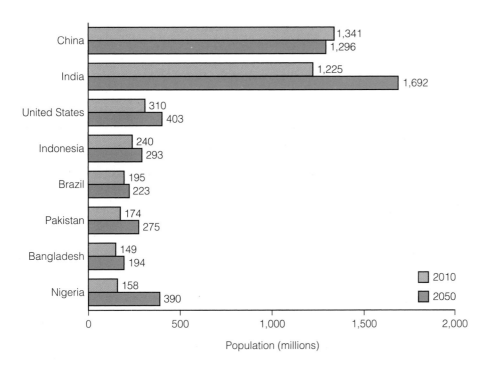

Figure 12.3: **World's Eight Largest Countries in Population, 2010 and 2050 (Projected)**
Source: United Nations 2011.

country in the world today is China, where nearly one in five people on this planet live. By 2050, India will become the most populated country (see Figure 12.3).

Nearly all of world population growth is in less developed countries, (see Figure 12.4), mostly in Africa and Asia. Higher population growth in developing countries is largely due to higher **total fertility rates**—the average lifetime number of births per woman in a population. Although total fertility rates worldwide have declined significantly from nearly 5 in the 1950s to 2.45 between 2010 and 2015, they continue to be highest in the least developed countries of the world. Other factors that affect population size

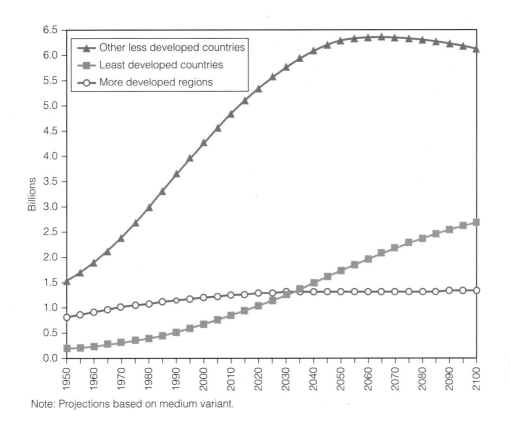

Note: Projections based on medium variant.

Figure 12.4: **World Population Growth by Development Region, 1950–2100**
Source: United Nations 2011.

total fertility rates The average lifetime number of births per woman in a population.

Figure 12.5: **Trends in Aging by World Region**
Source: United Nations 2011.

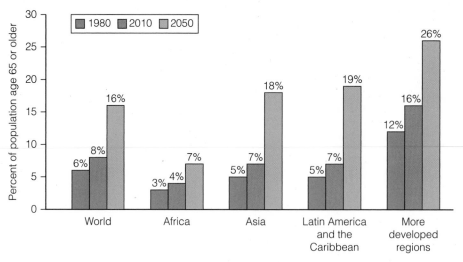

Note: Percentages are rounded to nearest whole percentage.

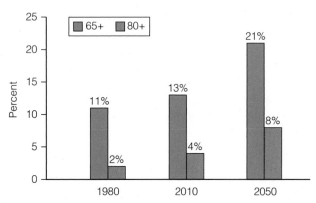

Note: Percentages are rounded to nearest whole percentage.

Figure 12.6: **Percent of U.S. Population Ages 65+ and 80+, 1980–2050**
Source: United Nations 2011.

replacement-level fertility The level of fertility at which a population exactly replaces itself from one generation to the next; currently, the number is 2.1 births per woman (slightly more than 2 because not all female children will live long enough to reach their reproductive years).

population momentum Continued population growth as a result of past high fertility rates that have resulted in a large number of young women who are currently entering their childbearing years.

include migration, armed conflict, economic stagnation, and high rates of disease such as HIV/AIDS.

Will there be an end to the rapid population growth that has occurred in recent decades? Will the population of the world stabilize? Although some predict that population will stabilize around the middle of the 21st century, no one knows for sure. Despite the overall decline in fertility rates, there are 25 countries in the world, primarily in Africa, where women have an average of between 5 and 7 children (United Nations 2011). To reach population stabilization, fertility rates throughout the world would need to achieve what is called "replacement level," whereby births would replace, but not outnumber, deaths. **Replacement-level fertility** is 2.1 births per woman, that is, slightly more than 2 because not all female children will live long enough to reach their reproductive years. The number of countries that have achieved below-replacement fertility rates has grown from 5 between 1950 and 1955, to more than 70 in 2010 and is expected to reach 196 by 2050 (Population Reference Bureau 2010; United Nations 2011). In some of these countries, population will continue to grow for several decades because of **population momentum**—continued population growth as a result of past high fertility rates that have resulted in a large number of young women who are currently entering their childbearing years. But there are 37 countries or areas whose populations are projected to decrease between 2009 and 2050 (United Nations 2009). The U.S. population has a total fertility rate slightly lower than the replacement level—2.0 in 2010. However, U.S. population is expected to continue to increase through 2050 because of immigration.

In sum, there are two population size trends occurring simultaneously that appear to be contradictory: (1) The total number of people on this planet is rising and is expected to continue to increase over the coming decades; and (2) fertility rates are so low in some countries that the countries' populations are likely to decline over the coming years. As we discuss later in this chapter, each of these trends presents a set of problems and challenges.

The Aging of the World's Population

Another demographic trend that presents its own set of challenges is the increasing number and proportion of older individuals in the total population. Between 2010 and 2050, the percentage of older individuals (ages 65 and over) in the world population is expected to double, and will nearly double in the United States (see Figures 12.5, 12.6, and 12.7). The United Nations (2011) projects that, in 2080, more than 2 billion people—one in five people in the world—will be 65 or older. In the United States, the number and share of people age 80 and older is also growing.

The aging of the world's population is due to both declining levels of fertility and to increased longevity. In more than 70 countries—including China, Japan, and all European countries—fertility rates have fallen well below the 2.1 children replacement level (United Nations 2011). Globally, life expectancy increased from 49 in the 1950s to 68 between 2005 and 2010, and is expected to rise to 81 between 2095 and 2100 (United Nations 2011). In the United States, another factor contributing to the aging of the population is the aging of an unusually large generation of Americans known as the **baby boomers**—the generation of Americans born between 1946 and 1964, a period of high birthrates. The first of the baby boomer generation turned 65 in 2011.

Population aging increases pressure on a society's ability to support its elderly members. A commonly used indicator of this pressure is the **elderly support ratio**, calculated as the number of working-age people divided by the number of people 65 or older. Globally, "working age" is considered to be ages 15 to 64; in the United States, "working age" is considered to be ages 20 to 64. As the proportion of older people increases in a population, fewer working-age adults support the elderly population. In 1950, worldwide, there were 12 people of working age (15 to 64) for every person age 65 and older. In 2010, the elderly support ratio was 9 to 1, and by 2050, it is expected to drop to 4 to 1. Germany, Italy, and Japan have an elderly support ratio of 3 to 1—the lowest in the world (Population Reference Bureau 2010). By 2050, Japan will have only one working-age adult for every elderly person; Germany and Italy will each have two.

The decrease in the elderly support ratio raises concerns about whether there will be enough workers to take care of the older population. Population aging raises other concerns as well: How will societies provide housing, medical care, transportation, and other needs of the increasing elderly population? Later in this chapter, we look at the problems and challenges of meeting the needs of a growing elderly population.

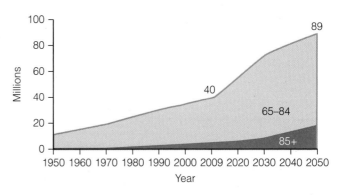

Figure 12.7: U.S. Population Ages 65 and Older, 1950 to 2050
Source: Jacobsen et al. 2011. From Linda A. Jacobsen et al., "America's Aging Population," *Population Bulletin* 66, no. 1. Copyright © 2011 Population Reference Bureau. All rights reserved. Reprinted with permission.

Between 2010 and 2050, the percentage of older individuals (ages 65 and over) in the world population is expected to double, and will nearly double in the United States.

Sociological Theories of Population Growth and Aging

The three main sociological perspectives—structural functionalism, conflict theory, and symbolic interactionism—can be applied to the study of population and aging.

Structural-Functionalist Perspective

Structural functionalism focuses on how changes in one aspect of the social system affect other aspects of society. For example, the **demographic transition theory** of population describes how industrialization has affected population growth. According to this theory, high fertility rates are necessary in traditional agricultural societies to offset high mortality and to ensure continued survival of the population. As a society becomes industrialized and urbanized, improved sanitation, health, and education lead to a decline in mortality. The increased survival rate of infants and children along with the declining economic value of children leads to a decline in fertility rates. About one-third of the world's countries have completed the demographic transition—the progression from a population with short lives and large families to one in which people live

baby boomers The generation of Americans born between 1946 and 1964, a period of high birthrates.

elderly support ratio The ratio of working-age adults (15 to 64) to adults ages 65 and older in a population.

demographic transition theory A theory that attributes population growth patterns to changes in birth rates and death rates associated with the process of industrialization.

longer and have smaller families (Cincotta et al. 2003). Many countries with low fertility rates have entered what is known as a "second demographic transition," in which fertility falls below the two-child replacement level. This second demographic transition has been linked to greater educational and job opportunities for women, increased availability of effective contraception, and the rise of individualism and materialism (Population Reference Bureau 2004).

The growing elderly population is a social change that has led to a number of other social changes, such as increased pressure on the workforce and on federal programs to support the aging population (e.g., Medicare and Social Security). Increased longevity also affects families, who often bear the brunt of elder care. The problems and challenges associated with caring for the aging population are discussed later in this chapter.

The structural-functionalist perspective also focuses attention on the unintended, or latent, consequences of social behavior. Although the intended, or manifest, function of modern contraception is to control and limit childbearing, there have also been far-reaching unintended effects of contraception on the social and economic status of women. The development of modern contraceptives—particularly the birth control pill—has led to fewer births to high school- and college-aged women, increased age at first marriage, and increased participation by women in the workforce. "The advent of the pill allowed women greater freedom in career decisions, by allowing them to invest in higher education and a career with far less risk of an unplanned pregnancy" (Sonfield 2011, p. 9). Another example of an unintended consequence can be found in the findings of several studies that link Social Security programs in various countries and lowered fertility (Kornblau 2009). One reason that people have children is to provide security in old age. Social Security reduces the need to have children to secure care in old age.

Conflict Perspective

The conflict perspective focuses on how wealth and power, or the lack thereof, affect population problems. In 1798, Thomas Malthus predicted that the population would grow faster than the food supply and that masses of people were destined to be poor and hungry. According to Malthusian theory, food shortages would lead to war, disease, and starvation, which would eventually slow population growth. However, conflict theorists argue that food shortages result primarily from inequitable distribution of power and resources (Livernash & Rodenburg 1998).

Conflict theorists also note that population growth results from pervasive poverty and the subordinate position of women in many less developed countries. Poor countries have high infant and child mortality rates. Hence, women in many poor countries feel compelled to have many children to increase the chances that some will survive into adulthood. Their subordinate position prevents many women from limiting their fertility. In many developing countries, a woman must get her husband's consent before she can receive any contraceptive services. Thus, according to conflict theorists, population problems result from continued economic and gender inequality.

In the United States, adults ages 65 to 74 have the highest rate of voting of any age group.

Some conflict theorists view the elderly population as a special interest group that competes with younger populations for scarce resources. Debates about funding programs for the elderly (such as Social Security and Medicare) versus funding for youth programs (such as public schools and child health programs) largely represent conflicting interests of the young versus the old. A growing elderly population means that the elderly have increased power in political issues. In the United States, adults ages 65 to 74 have the highest rate of voting of any age group (File & Crissey 2010).

Symbolic Interactionist Perspective

The symbolic interactionist perspective focuses on how meanings, labels, and definitions learned through interaction affect population problems. For example, many societies are characterized by **pronatalism**—a cultural value that promotes having children.

pronatalism A cultural value that promotes having children.

Throughout history, many religions have worshiped fertility and recognized it as being necessary for the continuation of the human race. In many countries, religions prohibit or discourage birth control, contraceptives, and abortion. Women in pronatalistic societies learn through interaction with others that deliberate control of fertility is socially unacceptable. Women who use contraception in communities in which family planning is not socially accepted face ostracism by their community, disdain from relatives and friends, and even divorce and abandonment by their husbands (Women's Studies Project 2003). However, once some women learn new definitions of fertility control, they become role models and influence the attitudes and behaviors of others in their personal networks (Bongaarts & Watkins 1996).

The symbolic interactionist perspective emphasizes the importance of examining social meanings and definitions associated with aging. "Old age" is largely a social construct; there is no biological marker that indicates when a person is "old." Rather, old age is a matter of social definition. In the United States, and much of the world, a person is considered to be "old" (or a "senior citizen") when they reach 65, as this is the age that company pension plans and Social Security have used to define when a person retires and collects benefits. Due to changes in the Social Security system, the retirement age for receiving full Social Security benefits has been increased to 67, yet we continue to use 65 to define "elderly."

What Do You Think? There are expressions today that suggest that "60 is the new 40" or "70 is the new 50." As life expectancy increases and people redefine 60 as today's 40, do you think that the designation of "senior citizen" or "old" will change from 65 to a higher number? What age do you think of as the beginning of "old age?"

We also learn both positive and negative meanings associated with old age, such as "wise" and "experienced" as well as "frail" and "impaired" (Kornadt & Rothermund 2010). In U.S. culture, negative labels of older people, such as "crone," "old geezer," and "old biddy" are predominant, and reflect ageism—a topic we discuss later in this chapter.

Finally, the symbolic interactionist perspective reminds us that understanding the quality of life among the elderly requires consideration of how the elderly subjectively define their experiences. For example, to assess social isolation among the elderly, researchers often use objective measures such as the frequency of social contacts among the elderly. But it is also important to consider how the elderly subjectively experience their level of social contact, as some older adults have a preference for engaging in solitary activities and experience being alone as enjoyable (Cloutier-Fisher et al. 2011).

Social Problems Related to Population Growth and Aging

Social problems related to population growth include environmental problems; poverty, unemployment, and global insecurity; and poor maternal and infant health. Problems related to the aging of the population include ageism—prejudice and discrimination against older individuals—employment and retirement concerns of older Americans, and the challenge of meeting the various needs of the elderly population, particularly retirement income and health care. Health issues and Medicare are discussed in Chapter 2, and elder abuse is discussed in Chapter 5.

Environmental Problems and Resource Scarcity

According to a survey of faculty at the State University of New York College of Environmental Science and Forestry (2009), overpopulation is the world's top environmental problem, followed closely by climate change and the need to replace fossil fuels with renewable energy sources. As we discuss in Chapter 13, population growth places increased demands on natural resources, such as forests, water, cropland, and oil, and results in increased waste and pollution.

According to the United Nations Environment Programme (UNEP), half of the planet's forests have been cleared for human land use and, in 2025, two-thirds of the world's population will be living in countries with water scarcity or stress, and the world's fisheries will be depleted by the middle of this century (Engelman 2011). The countries that suffer most from shortages of water, farmland, and food are developing countries with the highest population growth rates. However, countries with the largest populations do not necessarily have the largest impact on the environment. This is because the demands that humanity makes on the earth's natural resources—each person's **environmental footprint**—is determined by the patterns of production and consumption in that person's culture. The environmental footprint of an average person in a high-income country is much larger than that of someone in a low-income country. Hence, although population growth is a contributing factor in environmental problems, patterns of production and consumption are at least as important in influencing the effects of population on the environment.

> **What Do You Think?** The growing world population raises questions about how we can provide enough food, water, energy, housing, etc., to meet everyone's needs. Gar Smith (2009) argues that "meeting the 'growing demands of a growing population' is not solving the problem: It's perpetuating it" (p. 15). What does he mean by this statement? Do you agree? Why or why not?

Poverty, Unemployment, and Global Insecurity

Poverty and unemployment are problems that plague countries with high population growth. Less developed, poor countries with high birthrates do not have enough jobs for a rapidly growing population, and land for subsistence farming becomes increasingly scarce as populations grow. In some ways, poverty leads to high fertility, because poor women are less likely to have access to contraception and are more likely to have large families in the hope that some children will survive to adulthood and support them in old age. But high fertility also exacerbates poverty, because families have more children to support and national budgets for education and health care are stretched thin.

A Population Institute report titled *Breeding Insecurity: Global Security Implications of Rapid Population Growth* warns that rapid population growth is a contributing factor to global insecurity, including civil unrest, war, and terrorism (Weiland 2005). Although world population is, overall, aging, some countries in Africa and the Middle East are experiencing a "youth bulge"—a high proportion of 15- to 29-year-olds relative to the adult population. Youth bulges result from high fertility rates and declining infant mortality rates, a common pattern in developing countries today. The combination of a youth bulge with other characteristics of rapidly growing populations, such as resource scarcity, high unemployment rates, poverty, and rapid urbanization, sets the stage for political unrest. "Large groups of unemployed young people, combined with overcrowded cities and lack of access to farmland and water create a population that is angry and frustrated with the status quo and thus is more likely to resort to violence to bring about change" (Weiland 2005, p. 3).

Poor Maternal, Infant, and Child Health

As noted in Chapter 2, maternal deaths (deaths related to pregnancy and childbirth) are the leading cause of mortality for reproductive-age women in the developing world.

environmental footprint The demands that humanity makes on the Earth's natural resources.

Directions: Put a number in the blank that shows how often you have experienced each of the following 10 events. In the following items, "age" means old age.

___ 1. I was told a joke that makes fun of old people.
___ 2. I was given a birthday card that pokes fun at old people.
___ 3. I was ignored or not taken seriously because of my age.
___ 4. I was called an insulting name related to my age.
___ 5. I was patronized or "talked down to" because of my age.
___ 6. I was treated with less dignity and respect because of my age.
___ 7. A doctor or nurse assumed my ailments were caused by my age.
___ 8. Someone assumed I could not hear because of my age.
___ 9. Someone assumed I could not understand because of my age.
___ 10. Someone told me "You're too old for that."

Comparison Data: When 152 U.S. adults over age 60 completed the Ageism Survey, the following percentages indicated that they had experienced the 10 types of ageism in the survey either once, twice, or more:

Item	Experienced once	Experienced twice or more
1. Jokes that poke fun	19	49
2. Birthday card that pokes fun	14	23
3. Ignored	16	19
4. Insulting name	8	9
5. Patronized	16	22
6. Treated with less dignity	12	16
7. Assumed ailment caused by age	22	20
8. Assumed deaf	10	20
9. Assumed could not understand	16	17
10. "You're too old"	17	24

Source: Palmore, Erdman B. 2004. "Research Note: Ageism in Canada and the United States." *Journal of Cross-Cultural* Gerontology 19(1):41–46. Copyright © 2004 by Springer. Reprinted by permission

Having several children at short intervals increases the chances of premature birth, infectious disease, and death for the mother or the baby. Childbearing at young ages (teens) also increases the risks of health problems and death for both women and infants (United Nations Population Division 2009). In developing countries, one in four children is born unwanted, increasing the risk of neglect and abuse. In addition, the more children a woman has, the fewer the parental resources (parental income, time, and maternal nutrition) and social resources (health care and education) available to each child.

Ageism: Prejudice and Discrimination toward the Elderly

Ageism refers to negative stereotyping, prejudice, and discrimination based on a person's or group's perceived chronological age. Ageism is reflected in negative stereotypes of the elderly, such as that they are slow, they don't change their ways, they are grumpy, they are poor drivers, they can't/don't want to learn new things, they are incompetent, and they are physically and/or cognitively impaired (Nelson 2011). Ageism also occurs when older individuals are treated differently because of their age, such as when they are spoken to loudly in simple language, assuming they cannot understand normal speech, or when they are denied employment due to their age. Another form of ageism—**ageism by invisibility**—occurs when older adults are not included in advertising and educational materials. Before reading further, you may want to complete the Ageism Survey in this chapter's Self & Society feature.

Old age is stereotypically viewed as a negative time during which older individuals suffer a loss of identity (retirement from job), loss of respect from society, and increasing dependence on others. Although stereotypes of old people may accurately describe a number of old people, for many older individuals, theses stereotypes do not apply. Contrary to negative views of aging, people in their later years can be productive and fulfilled (see Table 12.1).

ageism Negative stereotyping, prejudice, and discrimination based on a person's or group's perceived chronological age.

ageism by invisibility Occurs when older adults are not included in advertising and educational materials.

TABLE 12.1 Accomplishments of Famous Older Individuals

At 70, fitness guru Jack LaLanne towed 70 boats, carrying a total of 70 people, a mile and a half through Long Beach Harbor while handcuffed and shackled.

At 72, feminist author Betty Friedan published *The Fountain of Age,* where she debunks misconceptions about aging.

At 80, George Burns won an Academy Award for his performance in *The Sunshine Boys.*

At 81, Benjamin Franklin facilitated the compromise that led to the adoption of the U.S. Constitution.

At 82, Winston Churchill wrote *A History of the English-Speaking Peoples.*

At 85, actress Mae West starred in the film *Sextette.*

At 85, Coco Chanel was the head of a fashion design firm.

At 87, Mary Baker Eddy created the newspaper *Christian Science Monitor.*

At 88, actress Betty White became the oldest person to ever host *Saturday Night Live* (as of May 2010).

At 89, Albert Schweitzer headed a hospital in Africa.

At 90, Pablo Picasso was producing drawings and engravings.

At 93, George Bernard Shaw wrote the play *Farfetched Fables.*

At 94, philosopher Bertrand Russell was active in promoting peace in the Middle East.

At 100, Grandma Moses, noted for her rural American landscapes, was painting. She only started painting at age 78, but by the time she died at 101 she had created over 1,500 works of art.

Ageism is embedded in our culture and is much more widely accepted than other "isms" such as racism, sexism, and heterosexism. Nelson (2011) states that "there is no other group like the elderly about which we feel free to openly express stereotypes and even subtle hostility" (p. 40). According to Margaret Gullette (2011), "ageism is to the twenty-first century what sexism, racism, homophobia and ableism were earlier in the

Many older adults remain active. At age 88, Betty White became the oldest person to have ever hosted *Saturday Night Live.*

AP Images/Dana Edelson/NBC

twentieth—entrenched and implicit systems of discrimination, without adequate movements of resistance to oppose them" (p. 15). Ageism is different from the other "isms" in that everyone is vulnerable to ageism if they live long enough.

Ageism is different from the other "isms" in that everyone is vulnerable to ageism if they live long enough.

What Do You Think? Ageism is perpetuated in a variety of ways that we commonly accept as harmless fun. For example, birthday greeting cards for aging adults often communicate the message, "Sorry to hear you are another year older." Nelson (2011) remarks, "think about the outrage that would ensue if there was a section of cards that communicated the message 'sorry to hear you're Black' or 'ha ha ha too bad you're Jewish'—yeah, it wouldn't go over so well. So why does society allow, and even condone, the same message directed against older persons?" (p. 41). We give adult birthday gag gifts that make fun of old people with the theme of "over the hill" and we tell or forward jokes that make fun of the elderly, thinking the jokes are humorous rather than offensive. When people are forgetful, they say they are "having a senior moment" without considering that this statement reflects ageism. After considering these points, do you think you will view such acts of making fun of the elderly as harmless fun, or as examples of ageism?

Another indicator of ageism in our society is the negative view of wrinkles, gray hair, and other physical signs of aging. Millions of Americans purchase products or treatments to make them look younger, spending substantial sums of money and undergoing unnecessary and often risky medical procedures to look younger.

Unlike more traditional societies that honor and respect their elders, ageism is widespread in modern societies. With the invention of the printing press, elders lost their special status as the keepers of a culture's stories and knowledge (Nelson 2011). Ageism also stems partly from fear and anxiety surrounding aging and death. Many people are uncomfortable around the topic of death and don't like to acknowledge death as a natural part of the life cycle. We use euphemisms to avoid talking about death: We say someone "passed away," "has gone to a better place," "is resting in peace," "kicked the bucket," "has departed," and so on. Old people are reminders of our mortality, and as such, take on negative social meanings.

Employment Age Discrimination Most (80 percent of) Americans think they will continue working full- or part-time after they reach retirement age, either because they will want to (44 percent) or because they will have to (36 percent) (Jones 2011). One of the obstacles older workers face in finding and keeping employment is age discrimination. In an experimental study, Lahey (2008) found that younger job seekers were 40 percent more likely to be offered a job interview than older job seekers with similar resumes.

Older workers may be more vulnerable to being "let go" because, although they have seniority on the job, they may also have higher salaries, and businesses that need to cut payroll expenses can save more money by letting higher-salaried personnel go. Prospective employers may view older job applicants as "overqualified" for entry-level positions, less productive than younger workers, and/or more likely to have health problems

How much money do you spend on products that promise to make you looking young?

In this feature, we briefly describe a national research study entitled, *The Elder Care Study: Everyday Realities and Wishes for Change* (Aumann et al. 2010). This research provides both quantitative and qualitative data about the experience of providing care to elderly family members.

Sample and Methods

The 2008 National Study of the Changing Workforce (NSCW) gathered data through telephone interviews with a nationally representative sample of 3,502 employed people using a random-digit dial procedure. The response rate was 54.6 percent. From this initial study, 1,589 caregivers—both those who were currently providing care or who had provided care to someone who had died within the past five years—were asked to participate in a telephone interview about their experiences in providing care to an elderly family member. A subsample of 421 family caregivers agreed to participate in a follow-up interview exploring their experiences caring for an elderly relative or in-law, and of these, 140 were successfully contacted and interviewed.

Selected Quantitative Findings

In response to the question, "Within the past five years, have you provided special attention or care for a relative or in-law 65 years old or older—helping with things that were difficult or impossible for them to do themselves?", nearly half—42 percent—responded yes. Three-quarters of the elderly relatives cared for by participants in the qualitative study were age 75 and older.

The types of care family members provided to elders encompass a wide range of tasks and responsibilities, including direct, in-person care (e.g., preparing meals, performing housework, providing transportation to doctor appointments, bathing, etc.) and indirect care (e.g., shopping, arranging for doctor appointments and other services, managing finances, etc.). Most elders in this study (67 percent) lived in their own homes. Nearly one in four caregivers live with their elderly relative or in-law, either in the caregiver's home (18 percent) or in the elderly person's home (6 percent), and 52 percent live 20 minutes or less from the person for whom they are providing care.

Three quarters (76 percent) of caregivers relied solely on themselves and their families to care for their elderly relative, with no paid assistance. Elder care providers in this study provided elder care for an average of 4.1 years; one in four provided care for 5 years or more. Women (20 percent) and men (22 percent) were equally likely to provide care for elders, although on average, women spent more time than men providing care (9.1 hours a week for women versus 5.7 hours for men). Many of these caregivers are in the "sandwich generation"; 46 percent of women caregivers and 40 percent of men caregivers also have children under the age of 18 at home. The majority of caregivers report not having enough time for their children (71 percent), their spouse/partner (63 percent), and themselves (63 percent).

Selected Qualitative Findings

Family caregivers expressed the wish for more support from workplaces in the form of greater job flexibility and more time off for elder care, especially paid time off without having to use vacation time. Caregivers also expressed wanting more active involvement and help from other family members. But despite the challenges and frustration involved

that could affect not only their productivity, but the cost of employer-based group insurance premiums. Employers may also be concerned about the ability of older workers to learn new skills and adapt to new technology.

Family Caregiving for Our Elders

Many adults have or will provide care and/or financial support for aging spouses, parents, grandparents, and in-laws. Adults who care for their aging parents while also taking care of their own children are referred to as members of the **sandwich generation**—they are "sandwiched" in between taking care of both parents and children.

For thousands of years, caring for older adults has been an important function of the family. For example, Chinese culture embraces the tenet of *filial piety*, which entails respecting, obeying, pleasing, and offering support and care to parents. Due to social, economic, and cultural changes, it has become more difficult for families throughout the world to care for aging parents. This chapter's *Social Problems Research Up Close* feature examines the challenges of elder care in U.S. families.

Countries around the world have used various methods to encourage parental support. The United States and Taiwan offer tax deductions and credits to adult children caring for elderly parents. In China, which has the largest aging population in the world, some parents have taken their adult children to court for failing to support them. Millions of Chinese families have signed a **Family Support Agreement**—a voluntary contract between older parents and adult children that specifies the details of how the adult children will provide parental care (Chou 2011).

sandwich generation A generation of people who care for their aging parents while also taking care of their own children.

Family Support Agreement In China, a voluntary contract between older parents and adult children that specifies the details of how the adult children will provide parental care.

in the caregiving role, many caregivers expressed appreciation for the opportunity to spend time with their elderly relatives and to establish a closer relationship. Interestingly, former caregivers (whose relatives had died) were much more likely to report positive changes in their relationship with their elderly relative during caregiving than were current caregivers. The researchers explained,

It appears that the death of the elder alters caregivers' perspective on the caregiving experience and its impact on their relationship. It is possible that the demands and challenges of family caregiving may negatively impact the caregiver's perception of the relationship with the elder during the caregiving experience. Quite possibly, caregivers do not have enough time or mental resources to reflect on the caregiving experience and the relationship with the care recipient until after the caregiving experience is over. The grieving process can thus be seen as a healing process. (p. 24)

Many caregivers reported deriving satisfaction from helping their elder avoid being placed in a nursing home. One caregiver said,

Knowing that she is not in a nursing home, knowing that there is no one

harming her, no one taking advantage of her. . . . It is awful. I work in that industry, and the stories are horrifying, so knowing that she is safe in her own home—those are the comforts. And knowing that I can do everything I can for her. (p. 22)

Many caregivers also described learning valuable lessons from their caregiving experience, including the importance of planning ahead for one's own aging and elder care. When asked what they hoped for in their own aging, caregivers expressed the desire to (1) not "burden" others, especially their children; (2) not burden themselves or others with unaffordable expenses; and (3) not end up in a nursing home.

Finally, researchers noted that,

An alarming theme that emerged from our interviews is that family caregivers overwhelmingly seem to view aging and receiving elder care as profoundly negative, depressing processes to be avoided if at all possible. People seem to dread the idea of aging and needing care so much they say they would rather be killed in some other way or even commit suicide. (p. 43)

As one caregiver expressed,

I don't even want to think about it. I want to pass in my sleep of old age. It's an ugly time of life—the last few years of suffering. I would rather die in a car wreck than put anyone through what I had to go through taking care of my mother. (p. 41)

The findings of The Elder Care Study point to the need for better models of elder care that involve more support from the workplace, family/friends, and the health care system. The researchers make this point at the conclusion of their report, where they quote a doctor, who, when standing in a hospital next to an elderly person, said the following:

Look at that bed and imagine how many more people are going to be in beds just like this in the coming years. You and I will be in those beds someday! We *have* to make things better than they are now. (p. 44)

Source: Aumann, Kerstin, Ellen Galinsky, Kelly Sakai, Melissa Brown, and James T. Bond. 2010. *The Elder Care Study: Everyday Realities and Wishes for Change*. Families and Work Institute. Available at http://www.familiesandwork.org

What Do You Think? In Singapore, the Maintenance of Parents Act of 1995 makes supporting parents a legal obligation and parents can sue children who fail to support them. And under French law, children are obligated to honor and respect their parents, pay them an allowance, provide or fund a home for them, and to stay informed of their parents' state of health and intervene if there are medical problems. France also has a law that obligates parents to leave their estates to their children. What do you think about such policies? Do you think adult children should be legally responsible for providing support for their aging parents? And should parents be legally obligated to leave their estates to their children?

Retirement Concerns of Older Americans and the Role of Social Security

One of the concerns as we age is financial planning for retirement. A majority (53 percent) of nonretired U.S. adults do not think they will have enough money to live comfortably in retirement, up from about a third in 2002 (Newport 2011). These concerns are not unfounded: The Center for Retirement Research estimates that more than half of U.S. households are at high risk of experiencing a significant drop in living standards at retirement (Morrissey 2011). According to one source, only 57 percent of

Figure 12.8: Framework for Understanding Retirement Planning
Source: Edwards et al. 2011. From Edwards, Kathryn A., Anna Turner, and Alexander Hertel-Fernandez. *A Young Person's Guide To Social Security*. Copyright © 2011 Economic Policy Institute. Reprinted with permission.

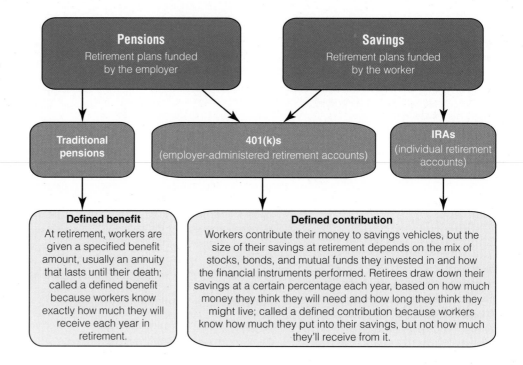

Americans have enough retirement savings to replace one year's worth of salary; only 11 percent have more than four years worth of salary saved in retirement accounts (Greenblatt 2011).

Types of Retirement Plans Retirement plans include (1) traditional pensions, which are "defined benefit" plans, in which retirees receive a specified annual amount until their death; and (2) "defined contribution" plans in which workers contribute money to 401k plans or individual retirement accounts (IRAs), without any guarantee of what their future benefits will be (see Figure 12.8).

Pensions, 401ks, and IRAs provide limited financial security for old age. Fewer private employers are offering guaranteed pensions, and budget shortfalls threaten the pensions and other retirement benefits for government workers. Defined contribution retirement plans (401ks and IRAs) are risky because they involve investments in the market and their value fluctuates. In the recent economic crisis, the stock market took a huge hit, producing losses that decimated the IRAs and 401ks of older Americans. Many workers who were planning to retire could no longer afford to stop working. Others who had recently retired and then lost a lot of money in the market felt compelled to reenter the labor force. Although some older workers want to retire but can't afford to, other older workers want to continue working but are forced to retire due to health problems/disability, job cuts or displacement, or the need to care for parents or spouses (Szinovacz 2011).

The Role of Social Security in Retirement In addition to retirement plans and savings, most workers are eligible to receive Social Security retirement benefits when they reach retirement age. **Social Security,** actually titled "Old Age, Survivors, Disability, and Health Insurance," is a federal insurance program established in 1935 that protects against loss of income due to retirement, disability, or death. More than two-thirds (69 percent) of Social Security benefits paid in 2009 were retirement benefits, 19 percent were disability benefits, and 12 percent were survivor benefits (to dependent spouses and children) (Edwards et al. 2011). The amount a person receives from Social Security is based on how much that person earned during their working history—higher lifetime earnings result in higher benefits. Benefit payments also depend on the age at which a person retires. The minimum age for receiving full benefits was 65 years for many years, but in

Social Security Also called "Old Age, Survivors, Disability, and Health Insurance," a federal program that protects against loss of income due to retirement, disability, or death.

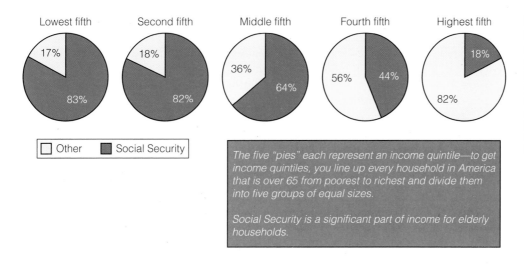

Figure 12.9: **Share of income from Social Security of households 65 or over by income quintile, 2008**
Source: Edwards et al. 2011. From Edwards, Kathryn A., Anna Turner, and Alexander Hertel-Fernandez. *A Young Person's Guide To Social Security.* Copyright © 2011 Economic Policy Institute. Reprinted with permission.

1983, Congress phased in a gradual increase in the full retirement age from 65 to 67. People born in 1960 and later are subject to the new retirement age of 67. Retirees can claim reduced benefits as early as age 62; they receive a larger benefit if they wait until age 70 to claim benefits. In the United States, spouses are entitled to one-half of their partners' benefits regardless of their own work histories and Social Security contributions.

In June 2011, the average monthly Social Security benefit to retired workers was $1,180.80, which totals about $14,000 a year (Social Security Administration 2011). When Social Security was established in 1935, it was not intended to be a person's sole economic support in old age; rather, it was meant to supplement other savings and assets. But Social Security is a major source of family income for most older Americans: For more than half of Americans over age 65, Social Security provides more than half of their income, and without Social Security income, nearly half of all seniors would be living in poverty (see Figure 12.9). Instead, as noted in Chapter 6, poverty rates for U.S. adults ages 65 and older are lower than for any other age group. Because Social Security payments are based on the number of years of paid work and preretirement earnings, women and minorities, who often earn less during their employment years, receive less in retirement benefits.

> Without Social Security income, nearly half of all seniors would be living in poverty.

How Is Social Security Funded? Social Security is funded by workers through a payroll tax called the Federal Insurance Contributions Act (FICA) that comprises 12.4 percent of a worker's wages (6.2 percent is deducted from the worker's paycheck and 6.2 percent is paid by the employer). Self-employed workers pay the entire FICA tax. FICA taxes are paid on wages up to a certain level, or tax cap, which changes each year based on average U.S. wages. In 2011, the tax cap was $106,800, meaning that people whose wages were more than this amount paid FICA tax only on the first $106,800 of their earnings.

Another source of funding for Social Security is a tax on higher-income beneficiaries. For most recipients, Social Security benefits are not taxed, but for recipients who have income from other sources that is above a certain amount ($25,000 for individuals; $32,000 for couples), a portion of Social Security benefits is taxed to help finance Social Security.

Finally, Social Security funds are held in a trust fund and invested in securities guaranteed as to both principal and interest by the federal government. Social Security benefits are paid out of this trust fund, and each year a Board of Trustees issues a report on the financial status of the trust fund.

Is Social Security in Crisis? In 2010, Social Security's income ($781 billion) exceeded its expenditures ($713 billion) (Social Security Trustees 2011). However, a number of factors threaten the long-term ability of Social Security to meet its financial obligations to future retirees, including the retirement of the baby boomers, increasing longevity, the declining elderly support ratio (fewer workers per beneficiary), widening wage inequality (which means that more income is not subject to Social Security taxes), high rates of unemployment, and wage stagnation.

Since 1984, surpluses have been accumulating in the trust fund, creating significant reserves for the baby boomers' retirement. According to the most recent report of the Social Security Trustees (2011), Social Security's total income, including interest earnings on trust fund assets, will be sufficient to cover annual costs until 2023, and trust fund reserves will be exhausted in 2036; thereafter, funding from taxes would cover only about three-quarters of benefits through 2085. The bottom line seems to be that, in the short-term, Social Security is not "broke" and there is no immediate crisis. But long-term changes to the Social Security system will be needed to ensure that the program is able to meet its financial obligations to future retirees. Options for Social Security reform are discussed in the following "Strategies for Action" section of this chapter.

Strategies for Action: Responding to Problems of Population Growth and Population Aging

Whereas some countries are struggling to slow population growth, others are challenged with maintaining or even increasing their populations. In some countries with below-replacement fertility levels, population strategies have focused on *increasing* rather than decreasing the population. For example, in Australia and Japan, the government has paid women a monetary bonus for having babies (Lalasz 2005; Wiseman 2005). Aside from monetary rewards, many countries encourage childbearing by implementing policies designed to help women combine child rearing with employment. For example, as noted in Chapters 5 and 7, many European countries have generous family leave policies and universal child care. Another way to increase population is to increase immigration. Spain, for example, has eased restrictions on immigration as a way to gain population.

What Do You Think? One strategy for encouraging childbearing in European countries with low fertility rates is to provide work–family supports to make it easier for women to combine childbearing with employment. If the United States offered more generous work–family benefits, such as paid parenting leave and government-supported child care, would the U.S. birthrate increase? Would such policies affect the number of children you would want to have?

TABLE 12.2 Total Fertility Rates by Region: 1950–1955 to 2010–2015		
	1950–1955	2010–2015
World	4.95	2.45
More developed	2.81	1.71
United States	3.45	2.08
Less developed (excluding (least developed)	6.01	2.31
Least developed	6.54	4.10

Source: United Nations 2011.

Efforts to Curb Population Growth: Reducing Fertility

To some population experts, stopping population growth is a critical issue. "Zero population growth, which characterized human population for more than 99 percent of its history, must be achieved once again, at least as a long-term average, if the human species is to survive" (McFalls 2007, p. 27). Although worldwide fertility rates have fallen significantly since the 1950s, they are still high in many less developed regions (see Table 12.2). This chapter's *Animals and Society* feature discusses the problem of pet overpopulation, and the importance of controlling the fertility of our dog and cat populations.

More than a third of U.S. households include dogs or cats as pets, also referred to as companion animals (Coate & Knight 2010). Many pets live out their lives in homes where they are loved and well cared for. But each year, about 8 million stray and unwanted cats and dogs are taken to animal shelters across the United States, and about half of these animals end up being euthanized, most commonly through intravenous injection (American Humane Association 2011; Humane Society 2009). In fact, shelter euthanasia is the number one cause of death for both dogs and cats in the United States (American Humane Association 2011).

According to the National Council on Pet Population Study and Policy (2009), the top 10 reasons for taking a companion dog or cat to an animal shelter include moving, cost of pet maintenance, and allergies (see Table 1 within this feature). Regarding cost, many people do not realize that, aside from the initial purchase price of a dog or cat, the cost of maintaining a pet over a 12-year lifetime can cost $22,000 or more for food, treats, vet visits, grooming, toys, medications, training, and other expenses (Greenfield 2011). City dwellers that hire dog walkers can spend an additional $5,000 a year just for this service.

Animal shelters and pet rescue groups have seen a surge in abandoned pets as a result of the increase in foreclosures due to the housing crisis (Brenoff 2011). Although it is a crime in all 50 states to abandon a pet, some families evicted from their homes leave their pets in the yard or in the house to fend for themselves. Animal shelters in college towns see an increase in abandoned pets at the end of the school year when many students move away and can't take their pets with them (Kidd 2009).

Not all abandoned pets are taken to shelters; some are left in the street or are dropped off in parking lots or other public places. Abandoned pets left out in the wild struggle with starvation and disease and pose a threat to both humans and other animals, such as endangered birds. In the wild, unsterilized cats and dogs can multiply quickly. Consider that if an unsterilized cat gives birth to two litters a year, that can result in 400,000 cats over seven years (Shikina 2011.)

The millions of unwanted and feral dogs and cats is part of a larger problem of pet overpopulation. The American Society for the Prevention of Cruelty to Animals (ASPCA) suggests that the best method for reducing pet overpopulation is sterilization and recommends that (1) all companion dogs and cats, except those that are a part of a responsible breeder's program, be spayed or neutered at an early age (2 months); and (2) all communities have accessible and affordable spay/neuter programs (ASPCA 2011). Many vets and animal welfare agencies offer low-cost or free spaying and neutering, and some organizations have mobile clinics that travel to low-income or rural areas where people are not able to transport a pet to a vet.

Despite the problem of pet overpopulation, about a third of pet owners in the United States do not spay or neuter their pet (American Humane Association 2011). Some pet owners want to breed their pets, either for profit or because they think it would be fun to have a litter of puppies or kittens. Others mistakenly believe that their pet won't become accidentally pregnant.

Some states and municipalities have laws to try to curb pet overpopulation. Rhode Island mandates that dogs and cats must be spayed or neutered before being released from a shelter; a Los Angeles ordinance requires pet owners to sterilize their dogs or cats by age 4 months; and some cities, including Austin, Texas, Albuquerque, New Mexico, and West Hollywood, California, completely ban the retail sales of dogs and cats (Humane Society 2010; Shikina 2011). Hawaii is considering a bill that would require pet retailers to sterilize all cats and dogs before selling them (Shikina 2011). And some major pet stores, including PETCO and PetSmart, refuse to sell puppies and kittens at their stores (although many major pet stores allow local animal and shelter groups to display dogs and cats for adoption).

Finally, people wanting to acquire a dog or cat are encouraged to adopt from an animal shelter or nonprofit animal rescue organization. Each year, an estimated 17 million Americans acquire a pet dog or cat, but only about 20 percent adopt their pet from a shelter (American Humane Association 2011). Dogs and cats that are not adopted or rescued from a shelter are, sadly, destroyed.

TABLE 1 Top Ten Reasons for Pet Relinquishment to U.S. Shelters	
DOGS	**CATS**
1. Moving	1. Too many in house
2. Landlord issues	2. Allergies
3. Cost of pet maintenance	3. Moving
4. No time for pet	4. Cost of pet maintenance
5. Inadequate facilities	5. Landlord issues
6. Too many pets in home	6. No homes for littermates
7. Pet illness	7. House soiling
8. Personal problems	8. Personal problems
9. Biting	9. Inadequate facilities
10. No homes for littermates	10. Doesn't get along with other pets

Source: Copyright © 2009 National Council on Pet Population Study and Policy. Reprinted with permission.

Recognizing that men play a crucial role in family planning decisions, family planning programs are making efforts to include men in family planning education and services.

Family Planning and Contraception Since the 1950s, governments and nongovernmental organizations such as the International Planned Parenthood Federation have sought to lower fertility through family planning programs that provide reproductive health services and access to contraceptive information and methods. Worldwide, a slight majority (55 percent) of married women ages 15 to 49 uses some form of modern contraception; in the United States, 73 percent of women use modern contraception. However, in least developed countries, fewer than one in four (23 percent) married women use modern contraception (Population Reference Bureau 2010). Worldwide, about 215 million women who don't want to be pregnant are not using modern contraception; 78 million use traditional, less reliable methods of family planning; and 137 million use no method at all (Ryerson 2011).

Population expert William Ryerson (2011) claims that, in most countries, lack of access to family planning is a very minor reason for not using contraception. In Nigeria, less than one percent of nonusers who don't want to be pregnant cite lack of access as the reason. Other reasons for not using modern contraception include (1) the belief that modern methods of contraception are dangerous; (2) male partners are opposed to using modern contraception; (3) the belief that one's religion opposes the use of contraception; and (4) the belief that God should decide how many children a woman has (Ryerson 2011). Encouraging women to use modern contraception requires providing access to affordable contraceptive methods, but providing access is not enough. Women and men need education to dispel myths about the dangers of using contraception and to understand the health and economic benefits of delayed, spaced, and limited childbearing. Involving men in family planning services is important because, although men play a central role in family planning decisions, they often do not have access to information and services that would empower them to make informed decisions about contraceptive use (Women's Studies Project 2003).

Economic Development Although fertility reduction can be achieved without industrialization, economic development may play an important role in slowing population growth. Families in poor countries often rely on having many children to provide enough labor and income to support the family. Economic development decreases the economic value of children and is also associated with more education for women and greater gender equality.

Economic development tends to result in improved health status of populations. Reductions in infant and child mortality are important for fertility decline, because couples no longer need to have many pregnancies to ensure that some children survive into adulthood. Finally, the more developed a country is, the more likely women are to be exposed to meanings and values that promote fertility control through their interaction in educational settings and through media and information technologies (Bongaarts & Watkins 1996).

> Families in poor countries often rely on having many children to provide enough labor and income to support the family.

The Status of Women: The Importance of Education and Employment Throughout the developing world, the primary status of women is that of wife and mother. Women in developing countries traditionally have not been encouraged to seek education or employment; rather, they are encouraged to marry early and have children.

Improving the status of women by providing educational and occupational opportunities is vital to curbing population growth. Educated women are more likely to marry later, want smaller families, and use contraception. Data from many countries have shown that women with at least a secondary-level education eventually give birth to between one-third and one-half as many children as women with no formal education (Population Reference Bureau 2007).

Better-educated women tend to delay marriage and exercise more control over their reproductive lives, including decisions about childbearing. In addition, "education can result in smaller family size when the education provides access to a job that offers a promising alternative to early marriage and childbearing" (Population Reference Bureau 2004, p. 18). Providing employment opportunities for women is also important to slow population growth, because high levels of female labor force participation and higher wages for women are associated with smaller family size. Primary school enrollment of both young girls and boys is also related to declines in fertility rates. In countries where primary school enrollment is widespread or nearly universal, fertility declines more rapidly because (1) schools help spread attitudes about the benefits of family planning, and (2) universal education increases the cost of having children, because parents sometimes are required to pay school fees for each child and because they lose potential labor that children could provide (Population Reference Bureau 2004). However, providing access to contraception and increasing girls' education are unlikely to slow population growth without changes in male attitudes toward family planning. Indeed, attempts to provide free primary education may increase African men's desire for more children because the costs are decreased (Frost & Dodoo 2009).

Another important component of family planning and reproductive health programs involves changing traditional male attitudes toward women. According to traditional male gender attitudes, (1) a woman's most important role is being a wife and mother, (2) it is a husband's right to have sex with his wife at his demand, and (3) it is a husband's right to refuse to use condoms and to forbid his wife to use any other form of contraception. In parts of Africa, men express their reluctance to use condoms by arguing that "you wouldn't eat a banana with the peel on or candy with the wrapper on" (Frost & Dodoo 2009). A number of programs around the world work with groups of boys and young men to change such traditional male gender attitudes (Schueller 2005).

> Better-educated women tend to delay marriage and exercise more control over their reproductive lives, including decisions about childbearing.

Access to Safe Abortion Abortion is a sensitive and controversial issue that has religious, moral, cultural, legal, political, and health implications (see also Chapter 14). Worldwide, one in five pregnancies ends in abortion, and one in ten pregnancies ends in unsafe abortion, including those performed in unhygienic conditions by unskilled providers (such as traditional or religious healers and herbalists) and those that are self-induced by a woman inserting a foreign object into her uterus, ingesting toxic substances, or inflicting trauma to the abdomen (Mesce & Clifton 2011). Of the estimated 42 million abortions each year, about 47,000 girls and women die. Almost all unsafe abortions take place in developing countries (Mesce & Clifton 2011).

More than one-quarter of the world's population lives in countries where abortion is prohibited or allowed only to save the life of the woman. Where abortion is illegal, it is very often unsafe, and where abortion is legal, and widely accessible through formal health systems, it is highly safe (Barot 2011). Because the abortion rate is similar in regions with legal abortion and in regions with restrictive abortion laws, Barot (2011) notes that "legal restrictions on abortion largely do not affect whether women will get an abortion, but they can have a major impact on whether abortion takes place under safe or unsafe conditions and, therefore, whether it jeopardizes women's health and lives" (p. 25).

Some couples in China are eligible for an exemption from the one-child policy and are encouraged to have a second child.

China's One-Child Policy In 1979, China initiated a national family planning policy that encourages families to have only one child by imposing a monetary fine on couples that have more than one child. The implementation and enforcement of this policy varies from one province to another, and there are a number of exemptions that allow some couples to have two or even more children. Nevertheless, the one-child policy has been effective in reducing population growth in China, which has the highest rate of modern contraceptive use in the world. Of married women ages 15 to 49 in China, 86 percent use modern contraceptives (Population Reference Bureau 2010).

China has been criticized for using extreme measures to enforce its one-child policy, including steep fines, seizure of property, and forced sterilizations and abortions. In addition, because of a traditional preference for male heirs, many Chinese couples have aborted female fetuses with the hope of having a boy in a subsequent pregnancy. This has led to an imbalanced sex ratio, with many more Chinese males than females.

Another problem with the one-child policy is that, as older Chinese retire, there are fewer workers to take their place and to support pensions for the elderly. Largely due to concerns of these imbalances in the population structure, China is considering adopting a two-child policy (Branigan 2011).

Voluntary Childlessness In the United States, as in other countries of the world, the cultural norm is for women and couples to, sooner or later, want to have children. However, a small but growing segment of U.S. women and men does not want children and chooses to be childfree. In the United States, 7 percent of women ages 35 to 44 are voluntarily childless, making voluntary childlessness more common than involuntary childlessness (Hollander 2007). In general, childfree couples are more educated, live in urban areas, are less religious, and do not adhere to traditional gender ideology (Parks 2005). In a study of 121 childless-by-choice women, the top reason women gave for not wanting children is that they simply love their life as it is (Scott 2009). Other reasons included valuing freedom and independence and not wanting to take on the responsibility. Three-quarters of the women said they "had no desire to have a child, no maternal/paternal instinct." Another reason some individuals choose not to have children is concern for overpopulation and a deep caring for the health of the planet. Yet, ironically, voluntarily childless individuals are often criticized as being selfish and individualistic, as well as less well adjusted and less nurturing (Parks 2005). In this chapter's *The Human Side* feature, one man explains why he chose to have a vasectomy and not have children.

Combating Ageism and Age Discrimination in the Workplace

One strategy to combat ageism involves incorporating positive views of aging in educational lessons, beginning in preschool, and in other forms of media. Such lessons should teach about aging as a normal part of life, rather than something to fear or be embarrassed about. Younger people should have opportunities to learn from the wisdom, experience, and life perspective of older individuals.

In 1967, Congress passed the Age Discrimination in Employment Act (ADEA), which was designed to ensure continued employment for people between the ages of 40 and 65. In 1986, the upper age limit was removed, making mandatory retirement illegal in most occupations. Under ADEA, it is illegal to discriminate against people because of their age with respect to hiring, firing, promotion, layoff, compensation, benefits, job assignments, and training. Age discrimination is difficult to prove. Nevertheless, thousands of age discrimination cases are filed with the Equal Employment Opportunity Commission (EEOC) annually.

Matt Leonard lives in California, where he works on climate justice and energy issues with the organizations Greenpeace and Rising Tide North America. In this The Human Side feature, Matt Leonard explains why he has chosen to have a vasectomy—a simple surgical procedure that makes men unable to cause pregnancy.

Courtesy of Matt Leonard

Last year, I . . . had a vasectomy. While it's actually a very common procedure (nearly 500,000 are performed every year in the United States), it raises eyebrows—and a lot of questions.

The first one is always simply: *Why?*

Although this was a very personal decision for me, it was also a choice I made out of larger societal, political, and environmental motivations. I consider the environmental ones paramount. In an economic system that demands infinite growth with finite resources, not doubling my own consumption is one small stone in a big river.

More importantly, I live in the United States, and any child I had would have been raised here and would consume (despite my best efforts) far more resources than I am comfortable accepting. Living even a modest lifestyle in the United States comes as a direct result of the oppression, domination, and deaths of many unseen people, not to mention the exploitation of natural resources at rates that threaten the ability of our planet to sustain life. These facts shouldn't be cause for guilt or shame; instead, they should spur us to organize to confront the systems and institutions that have created these problems. On a personal level, contributing another person to the system that I have spent my adult life fighting is just not something I'm willing to do. . . .

The next question is usually: *But what if you change your mind?*

I view my decision as permanent. As I see it, I already made the decision years ago not to have children, based on sound, rational reasons. If I change my mind in the future, I believe that change would be fundamentally selfish, and I am comfortable committing myself to rational reasons now.

People typically follow up with: *Aren't there other forms of birth control?*

Yes, of course, and most of us here in the United States are lucky to be able to choose the form that is best for our lifestyles, our preferences, and our relationships. A vasectomy fit my needs best.

I guess there's always abstinence, but that's no fun, right? I suppose the rhythm method is an option, but almost everyone knows how [in]effective that is. Condoms are fine and dandy in many situations, but they have their downsides as well, and can seem pointless if you are in a monogamous relationship.

All the other common birth control methods have one aspect in common: They place the onus on women. Not only does our society expect women to deal with the logistics of birth control, but these methods also have severe physiological drawbacks, from roller-coaster hormonal changes to intensifying menstruation cycles to weight and skin changes. Although these methods have come a long way in a few decades, they still burden women and their bodies. Is it any coincidence that in a male-dominated society, the medical establishment has thus far focused on birth control methods that leave the burden solely on women?

For men, vasectomies are simple. There are almost no side effects and no long-term impacts; it's a quick, low-cost, outpatient procedure. Having decided that I want to take an active role in birth control, a vasectomy is fair, easy, and it confronts my privilege on this issue.

What if you decide you want children in the future? people ask.

Many of my friends whom I deeply respect have chosen to have children or will do so in the future. Some people do feel that there is something special and important about having a blood-related child. I just don't share that feeling.

There are thousands of beautiful children all over the world who need parents, and if I ever decide that being a father is something I want in my life, I would be remiss to ignore the existing children needing support and love. For me, adoption is the best option. We need more parents in this world, not more kids.

Finally: *But don't we need the smart, progressive people to reproduce?*

I'm of the nurture-over-nature camp. I think the whole "passing on genes" obsession can sometimes border on eugenics. I'm fairly confident there is no gene that instructs your child to fight for justice, peace, and sustainability. That comes from living those values and instilling them in the communities we are a part of. That's what I want to prioritize in my life—and I feel I can share those things more effectively without a child.

And besides—I've got messed-up teeth, I'm legally blind, bald, and have a history of heart disease. Let Matt Damon pass on his genes instead.

Options for Reforming Social Security

Social Security is the most solvent part of the U.S. government. It is funded through its own separate tax and interest from its trust fund to pay beneficiaries. No other government program or agency is fully funded. Social Security is also a very efficient program. Although it collects taxes from more than 90 percent of the workforce and sends benefits to more than 50 million Americans, Social Security spends less than one cent of every dollar on administration.

Senior citizens generally oppose proposals to cut Social Security benefits.

Nevertheless, as discussed earlier in this chapter, due to a number of factors, unless changes are made in years to come, Social Security won't have enough funds to provide payouts to future retirees. Options for reforming Social Security include strategies for increasing Social Security revenue, cutting benefits, and expanding Social Security benefits. The following outlines these options (Edwards et al. 2011; Morrissey 2011).

Cut Social Security Benefits Some reform proposals call for cuts in Social Security benefits, which would create a significant financial burden on households that depend on Social Security. Aside from simply reducing the amount of benefits paid to recipients, another way to cut benefits is to increase the retirement age. In 1983, the retirement age (the age at which a worker could collect full Social Security retirement benefits) was raised from 65 to 67 to be phased in over 23 years. Some policy makers suggest raising the retirement age even further to 68, 69, or even 70. However, due to the link between higher socioeconomic status and longer life expectancy, raising retirement age imposes the greatest burden on low earners who have lower life expectancies and therefore fewer years to collect Social Security. In addition, older workers already face employment discrimination, and finding or keeping employment is difficult for seniors. Finally, the older workers become, the more likely they will experience health problems, and hence file disability claims.

Increase Social Security Revenue Instead of cutting benefits, many advocates for the elderly suggest raising revenues. One option for increasing Social Security revenue is simply to raise the tax that funds Social Security (the payroll or FICA tax) from its current rate of 12.4 percent (6.2 percent paid by employer; 6.2 percent paid by employee) to a higher rate. To offset gains in life expectancy, Social Security taxes increased 19 times, from 1 percent between 1937 and 1949 to 6.2 percent in 1990 (paid by both employer and employee). As of this writing (August 2011), Social Security taxes have not increased since 1990—the longest period without an increase to the payroll tax. The downside to raising the payroll tax is that it could result in fewer jobs, as employers would have a higher financial burden. Paying a higher payroll tax would also disproportionately burden lower-wage earners.

Another option is to raise or even eliminate the tax cap, so that more earnings are taxed, providing more funds to the Social Security program. This would affect higher-income earners, who currently earn more than the tax cap, but only pay payroll taxes on the first $106,800 (the tax cap in 2011) of their earnings.

Finally, policies that increase employment and wages of all workers would lead to more revenue for Social Security. Because health insurance costs are deducted from taxable income, policies to reduce health care costs would also indirectly increase Social Security funding.

Expand Social Security Benefits In this current economic climate, people need more, not less, economic support. Options for increasing Social Security benefits include raising the minimum benefit amount, offering unemployed parents who are taking care of their children wage credits, and increasing the benefits for the very old (85 years and older). Another proposal involves restoring a student benefit (which existed between 1965 and 1985) so that children of the retired, deceased, or disabled can continue to receive benefits until age 22 if attending college or vocational school.

Not surprisingly, proposals to cut Social Security benefits or raise retirement age for receiving benefits are vehemently opposed by seniors and advocacy groups for older adults. Raising the tax cap would go far to ensure the future financial viability of Social Security, but this option is generally not supported by the wealthy segment of the population who would bear the burden of higher taxes. Until legislators enact changes to prevent the long-term Social Security deficit, there is little chance of seeing Social Security benefits increased.

Understanding Problems of Population Growth and Aging

What can we conclude from our analysis of population growth and aging? First, although fertility rates have declined significantly in recent years and although many countries are experiencing a decline in their fertility rates, world population will continue to grow for several decades. This growth will occur largely in developing regions. Finance columnist Paul Farrell (2009) says that population growth is "the key variable in every economic equation . . . impacting every other major issue facing world economies . . . from peak oil to global warming . . . from foreign policy to nuclear threats . . . from religion to science . . . everything." Given the problems associated with population growth, such as environmental problems and resource depletion, global insecurity, poverty and unemployment, and poor maternal and infant health, most governments recognize the value of controlling population size and support family planning programs. However, efforts to control population must go beyond providing safe, effective, and affordable methods of birth control. Slowing population growth necessitates interventions that change the cultural and structural bases for high fertility rates. Reducing fertility necessitates improving the status of women so that women have more power to control their choices regarding contraception and reproduction and have more life options other than being a wife and mother. Addressing problems associated with population growth also requires the willingness of wealthier countries to commit funds to providing reproductive health care to women, improving the health of populations, and providing universal education for people throughout the world.

As fertility rates decline and life expectancy increases, the United States and other countries are experiencing population aging. Many areas of social life are affected by the aging of the population, and more attention is being given to concerns related to older people. What it means to be old is culturally defined. Unfortunately, many cultural meanings associated with the aged are negative stereotypes. Prejudice toward and discrimination against older people constitutes ageism, but unlike other "isms"—racism, sexism, heterosexism—ageism is more socially acceptable and will affect all of us who reach a certain age.

Although older individuals struggle against ageism, families and governments face challenges of meeting the needs of seniors. Caring for elders has traditionally been an important function of the family, but social, cultural, and economic changes have made it more challenging for families to fulfill this function. Social Security plays an important role in supporting many older retirees, but due to projections of a future shortfall, some reforms to Social Security are necessary. What type of reform measures legislators enact continues to be hotly debated. Until such reform measures are in place, the future of Social Security and the well-being of older Americans is uncertain.

Like other social problems, slowing population growth and meeting the needs of aging populations require political will and leadership. Given all the pressing concerns in the world, control of population may not seem like a priority. But as finance columnist Paul Farrell (2009) warns,

> Population is the core problem that, unless confronted and dealt with, will render all solutions to all other problems irrelevant. Population is the one variable in an economic equation that impacts, aggravates, irritates, and accelerates all other problems.

And if taking care of older populations is not a priority now, those of us who are not yet "old" will suffer the consequences in the future, if, or when, we enter our "golden years."

- **How long did it take for the world's population to reach 1 billion? How long did it take for it to reach 6 billion?**
  It took thousands of years for the world's population to reach 1 billion, and just another 300 years for the population to grow from 1 billion to 6 billion.

- **Where is most of the world's population growth occurring?**
  Most world population growth is in developing countries, primarily in Africa and Asia.

- **Between 2010 and 2050, how is the percentage of older individuals (ages 65 and over) expected to change globally and in the United States?**
  Between 2010 and 2050, the percentage of older individuals (ages 65 and over) in the world population is expected to double, and will nearly double in the United States.

- **What factors are contributing to population aging?**
  Population aging is a function of lowered fertility as well as increased longevity. In the United States, population aging is also occurring because the baby boomers are reaching their senior years.

- **What is the demographic transition?**
  The demographic transition is the progression from a population with short lives and large families to one in which people live longer and have smaller families. About one-third of countries have completed the demographic transition.

- **Many countries are experiencing below-replacement fertility (fewer than 2.1 children born to each woman). Why are some countries concerned about their low fertility?**
  In countries with below-replacement fertility, there are or will be fewer workers to support a growing number of elderly retirees and to maintain a productive economy.

- **What kinds of environmental problems are associated with population growth?**
  Population growth places increased demands on natural resources, such as forests, water, cropland, and oil, and results in increased waste and pollution. Although population growth is a contributing factor in environmental problems, patterns of production and consumption are at least as important in influencing the effects of population on the environment.

- **Why is population growth considered a threat to global security?**
  In developing countries, rapid population growth results in a "youth bulge"—a high proportion of 15- to 29-year-olds relative to the adult population. The combination of a youth bulge with other characteristics of rapidly growing populations, such as resource scarcity, high unemployment rates, poverty, and rapid urbanization, sets the stage for civil unrest, war, and terrorism, because large groups of unemployed young people resort to violence in an attempt to improve their living conditions.

- **How is ageism different from other "isms" (racism, sexism, heterosexism)?**
  Ageism is more widely accepted than other "isms" and, unlike other "isms," everyone is vulnerable to experiencing ageism if they live long enough.

- **How important is Social Security income for senior citizens in the United States?**
  For more than half of Americans over age 65, Social Security provides more than half of their income, and without Social Security income, nearly half of all seniors would be living in poverty.

- **Is Social Security in crisis?**
  In the short-term, Social Security is not "broke" and there is no immediate crisis. But long-term changes to the Social Security system will be needed to ensure that the program is able to meet its financial obligations to future retirees.

- **What is the "sandwich generation?"**
  The "sandwich generation" refers to adults who care for their aging parents while also taking care of their own children—they are "sandwiched" in between taking care of both parents and children.

- **Efforts to curb population growth include what strategies?**
  Efforts to curb population growth include strategies to reduce fertility by providing access to family planning services, involving men in family planning, implementing a one-child policy as in China, and improving the status of women by providing educational and employment opportunities. Achievements in economic development and health are also associated with reductions in fertility.

- **What are three general options for reforming Social Security?**
  Options for reforming Social Security include strategies for increasing Social Security revenue, cutting benefits, and expanding Social Security benefits.

TEST YOURSELF

1. In 2010, world population was 6.9 billion. In 2050, world population is projected to be ___ billion.
   a. 6.5
   b. 6.8
   c. 7.7
   d. 9.3

2. How many countries have achieved below-replacement fertility rates?
   a. None
   b. 5
   c. 28
   d. More than 70

3. What would happen if every country in the world achieved below-replacement fertility rates?
   a. Population growth would stop and world population would remain stable.
   b. World population would immediately begin to decline.
   c. World population would continue to grow for several decades.
   d. World population would decline, but then go up again.

4. For more than half of Americans over age 65, Social Security provides more than half of their income.
   a. True
   b. False

5. Conflict theorists argue that food shortages result primarily from overpopulation of the planet.
   a. True
   b. False

6. Pronatalism is a cultural value that promotes which of the following?
   a. Car ownership
   b. Abstaining from sex until one is married
   c. Having children
   d. Urban living

7. According to the Ageism Survey, jokes and birthday cards that poke fun at the elderly represent a form of ageism.
   a. True
   b. False

8. What is the number one cause of death for dogs and cats in the United States?
   a. Getting hit by a car
   b. Shelter euthanasia
   c. Cancer
   d. Eating something poisonous

9. U.S. women with advanced education are more likely than women with less education to voluntarily choose to have no children.
   a. True
   b. False

10. In 1983, the retirement age (the age at which a worker could collect full Social Security retirement benefits) was raised from 65 to 70 to be phased in over 23 years.
   a. True
   b. False

Answers: 1: d; 2: d; 3: c; 4: a; 5: b; 6: c; 7: a; 8: b; 9: a; 10: b.

KEY TERMS

ageism 385
ageism by invisibility 385
baby boomers 381
demographic transition theory 381
doubling time 378

elderly support ratio 381
environmental footprint 384
Family Support Agreement 388
population momentum 380
pronatalism 382

replacement-level fertility 380
sandwich generation 388
Social Security 390
total fertility rates 379

MEDIA RESOURCES

Turning to Video

 Watch the ABC video, *A Family Affair: Costs of Care* (running time 2;16), available through **CengageBrain.com.** This video describes the financial and emotional costs of taking care of elderly family members. Who has cared for elders in your family? How do you want to be cared for in your elder years?

Online Study Resources

Log in to **www.cengagebrain.com** to access the resources your instructor has assigned. For this book, you can access:

CourseMate

Access chapter-specific learning tools, including learning objectives, practice quizzes, videos, Internet exercises, flash cards, and glossaries, as well as web links, and more in your Sociology CourseMate.

13

Environmental Problems

"The world will no longer be divided by the ideologies of 'left' and 'right,' but by those who accept ecological limits and those who don't."

—Wolfgang Sachs

© Derek Dammann/iStockphoto.com

IN AUGUST 2011, more than 1,200 people, including actress Daryl Hannah, were arrested for participating in peaceful protests at the White House to voice opposition to the proposed Keystone XL oil pipeline, which would carry 900,000 barrels a day of tar sands oil from Canada to the Gulf, crossing Montana, South Dakota, Nebraska, Kansas, Oklahoma, and Texas. **Tar sands** are large, naturally occurring deposits of sand, clay, water, and a dense form of petroleum (that looks like tar). **Tar sands oil** has been referred to as the world's dirtiest oil. Converting tar sands into liquid fuel requires energy and generates high levels of greenhouse gases (that cause global warming and climate change), and also leaves behind large amounts of toxic waste. The mining of tar sands requires large amounts of water and involves destruction of forests and wetlands, which disrupts wildlife habitats. And transporting the tar sands oil through a pipeline involves risk of leakage that could affect the safety of drinking water (Swift et al. 2011). The protest against the proposed Keystone XL pipeline is one of the largest civil disobedience actions in decades. Before her arrest, Daryl Hannah said, "Sometimes it's necessary to sacrifice your freedom for a greater freedom. . . . And we want to be free

The protest against the Keystone XL oil pipeline may be the largest organized action of civil disobedience in decades.

from the horrible death and destruction that fossil fuels cause, and have a clean energy future" (quoted in Martinez 2011). NASA scientist James Hansen said if the Canadian tar sands are fully developed, it is "essentially game over for the climate" (quoted in Elk 2011).

In this chapter, we focus on environmental problems that threaten the lives and well-being of people, plants, and animals all over the world—today and in future generations. After examining how globalization affects environmental problems, we view environmental issues through the lens of structural functionalism, conflict theory, and symbolic interactionism. We then present an overview of major environmental problems, examining their social causes and exploring strategies to reduce or alleviate them.

The Global Context: Globalization and the Environment

Two aspects of globalization that have affected the environment are (1) the permeability of international borders to pollution and environmental problems, and (2) the growth of free trade and transnational corporations.

Permeability of International Borders

Environmental problems such as climate change and destruction of the ozone layer (discussed later in this chapter) extend far beyond their source to affect the entire planet. For example, toxic chemicals (such as polychlorinated biphenyls [PCBs]) from the Southern Hemisphere have been found in the Arctic. In as few as five days, chemicals from the tropics can evaporate from the soil, ride the winds thousands of miles north, condense in the cold air, and fall on the Arctic in the form of toxic snow or rain (French 2000). This phenomenon was discovered in the mid-1980s, when scientists found high levels of PCBs in the breast milk of Inuit women in the Canadian Arctic region.

tar sands Large, naturally occurring deposits of sand, clay, water, and a dense form of petroleum (that looks like tar).

tar sands oil Oil that results from converting tar sands into liquid fuel. It is known as the world's dirtiest oil because producing it requires energy and generates high levels of greenhouse gases (that cause global warming and climate change), and also leaves behind large amounts of toxic waste.

Bryan & Cherry Alexander Photography/Alamy

Toxic chemicals travel thousands of miles from the Southern Hemisphere to the Arctic, where they have been found in the breast milk of Inuit women.

Another environmental problem involving permeability of borders is **bioinvasion:** the intentional or accidental introduction of organisms in regions where they are not native. Bioinvasion is largely a product of the growth of global trade and tourism (Chafe 2005). Invasive species compete with native species for food, start an epidemic, or prey on native species. Nonnative wood-boring insects, which are among the 450 nonnative insects that are established in the United States, cost an estimated $1.7 billion in local government expenses and $830 million in lost residential property values every year (Aukema et al. 2011). Other nonnative insects are red fire ants, which traveled from Paraguay and Brazil on shiploads of lumber to Mobile, Alabama in 1957, and have since spread throughout the southern states, causing damage to gardens and yards, invading the food supplies (seeds, young plants, and insects) of other animals, and harming humans with their painful sting (Hilgenkamp 2005). You might be surprised to learn that the domestic cat is considered among the world's 100 worst invasive species. Native to northeast Africa, cats have spread to every part of the world and are responsible for the decline and extinction of many species of birds (Global Invasive Species Database 2006).

The Growth of Transnational Corporations and Free Trade Agreements

As discussed in Chapter 7, the world's economy is dominated by transnational corporations, many of which have established factories and other operations in developing countries where labor and environmental laws are lax. Transnational corporations have been implicated in environmentally destructive activities—from mining and cutting timber to dumping toxic waste.

The World Trade Organization (WTO) and free trade agreements such as the North American Free Trade Agreement (NAFTA) and the Free Trade Area of the Americas (FTAA) allow transnational corporations to pursue profits, expand markets, use natural resources, and exploit cheap labor in developing countries while weakening the ability of governments to protect natural resources or to implement environmental legislation. Transnational corporations have influenced the world's most powerful nations to institutionalize an international system of governance that values commercialism, corporate rights, and "free" trade over the environment, human rights, worker rights, and human health (Bruno & Karliner 2002).

Under NAFTA's Chapter 11 provisions, corporations can challenge local and state environmental policies, federal-controlled substances regulations, and court rulings if such regulatory measures and government actions negatively affect the corporation's profits. Any country that decides, for example, to ban the export of raw logs as a means of conserving its forests or, as another example, to ban the use of carcinogenic pesticides, can be charged under the WTO by member states on behalf of their corporations for obstructing the free flow of trade and investment. A secret tribunal of trade officials would then decide whether these laws were "trade-restrictive" under the WTO rules and should therefore be struck down. Once the secret tribunal issues its edict, no appeal is possible. The convicted country is obligated to change its laws or face the prospect of perpetual trade sanctions (Clarke 2002, p. 44). For example, in the late 1990s, Ethyl, a U.S. chemical company, used NAFTA rules to challenge Canada's decision to ban the gasoline additive methylcyclopentadienyl manganese tricarbonyl (MMT), which is believed to have harmful effects on human health. Ethyl won the suit, and Canada paid $13 million in damages and legal fees to Ethyl and reversed the ban on MMT (Public Citizen 2005).

bioinvasion The intentional or accidental introduction of plant, animal, insect, and other species in regions where they are not native.

Sociological Theories of Environmental Problems

The three main sociological theories—structural functionalism, conflict theory, and symbolic interactionism—provide insights into social causes of and responses to environmental problems.

Structural-Functionalist Perspective

Structural functionalism focuses on how changes in one aspect of the social system affect other aspects of society. For example, agriculture, forestry, and fishing provide 50 percent of all jobs worldwide and 70 percent of jobs in sub-Saharan Africa, East Asia, and the Pacific (World Resources Institute 2000). As croplands become scarce or degraded, as forests shrink, and as marine life dwindles, millions of people who make their living from these natural resources must find alternative livelihoods. By 2020, an estimated 50 million people globally will be **environmental refugees**—individuals who have migrated because they can no longer secure a livelihood as a result of environmental problems (Zelman 2011). As individuals lose their source of income, so do nations. In one-fourth of the world's nations, crops, timber, and fish contribute more to the nation's economy than industrial goods do (World Resources Institute 2000).

The structural-functionalist perspective raises our awareness of latent dysfunctions—negative consequences of social actions that are unintended and not widely recognized. For example, the more than 840,000 dams worldwide provide water to irrigate farmlands and supply some of the world's electricity. Yet dam building has had unintended negative consequences for the environment, including the loss of wetlands and wildlife habitat, the emission of methane (a gas that contributes to global warming) from rotting vegetation trapped in reservoirs, and the alteration of river flows downstream, which kills plant and animal life (Environmental Defense Fund 2001). Dams have also displaced millions of people from their homes.

The expanding production of biofuels also has unintended consequences. More than one-third of U.S. corn production in 2008 was used to produce ethanol (an alternative fuel discussed later in this chapter), and about half of the vegetable oils produced in the European Union was used for biodiesel fuel. Although biofuels help reduce dependence on fossil fuels, they have caused significant increases in global food prices (World Water Assessment Program 2009). As philosopher Kathleen Moore pointed out, "Sometimes in maximizing the benefits in one place, you create a greater harm somewhere else" (quoted by Jensen 2001, p. 11).

Conflict Perspective

The conflict perspective focuses on how wealth, power, and the pursuit of profit underlie many environmental problems. Wealthy nations, which, per capita, consume more natural resources and generate higher amounts of pollution and waste, exploit less developed nations for raw materials, labor, and as a market to sell goods to (Barbosa 2009).

The capitalistic pursuit of profit encourages making money from industry regardless of the damage done to the environment. McDaniel (2005) noted that "our culture tolerates environmentalism only so long as it has minimal impact on big business. . . . In an economically centered culture, jobs come first, not the health of people or the environment" (pp. 22–23). To maximize sales, manufacturers design products intended to become obsolete. As a result of this **planned obsolescence,** consumers continually throw away used products and purchase replacements. Industry profits at the expense of the environment, which must sustain the constant production and absorb ever-increasing amounts of waste.

Industries also use their power and wealth to influence politicians' environmental and energy policies as well as the public's beliefs about environmental issues. In 2009, oil and gas industries spent $175 million on lobbying—nearly eight times what pro-environmental groups spent—which may explain why climate legislation that year came

environmental refugees Individuals who have migrated because they can no longer secure a livelihood as a result of deforestation, desertification, soil erosion, and other environmental problems.

planned obsolescence The manufacturing of products that are intended to become inoperative or outdated in a fairly short period of time.

The capitalistic pursuit of profit encourages making money from industry regardless of the damage done to the environment.

to a halt (Mackinder 2010). ExxonMobil, the world's largest oil company, has spent millions of dollars on lobbying and has also funded numerous organizations and individuals that have tried to discredit scientific findings that link fossil fuel burning to global climate change (Mooney 2005; Sheppard 2011). Government policies that support ethanol have been linked to political financial contributions by major players in the ethanol industry, such as Archer Daniels Midland, the nation's largest ethanol producer (Food & Water Watch and Network for New Energy Choices 2007). President Obama, who called nuclear power an "important part" of his energy agenda—even after the 2011 Fukushima nuclear power plant disaster, had received generous campaign contributions from Exelon Corporation—the company that operates all 11 of Illinois's nuclear reactors (McCormick 2011).

What Do You Think? About one in five members of Congress own stocks in oil or gas companies. In 2008, 115 members of Congress reported, in total, a minimum of $25.6 million in oil and gas investments (Beckel 2010). Do you think that members of Congress, who make and vote on energy policies, should be allowed to have investments in energy-related industries? Would you feel the same way about members of Congress who invested in renewable energy industries, such as solar and wind power?

The conflict perspective is also concerned with **environmental injustice** (also known as **environmental racism**)—the tendency for marginalized populations and communities to disproportionately experience adversity due to environmental problems. Although environmental pollution and degradation and depletion of natural resources affect us all, some groups are more affected than others. For example, although developing countries have emitted far more greenhouse gases (that cause global warming and climate change), the effects of global climate change are expected to be felt most severely by poor, developing nations (Miranda et al. 2011). In the United States, polluting industries and industrial and waste facilities are often located in minority communities (Bullard 2000). More than half (56 percent) of people living within 1.8 miles of a commercial hazardous waste site are people of color (Bullard et al. 2007). Rates of poverty are also higher among households located near hazardous waste facilities. However, "racial disparities are more prevalent and extensive than socioeconomic disparities, suggesting that race has more to do with the current distribution of the nation's hazardous waste facilities than poverty" (Bullard et al. 2007, p. 60). In North Carolina, hog industries—and the associated environmental and health risks associated with hog waste—tend to be located in communities with large black populations, low voter registration, and low incomes (Edwards & Driscoll 2009).

Symbolic Interactionist Perspective

The symbolic interactionist perspective focuses on how meanings, labels, and definitions learned through interaction and through the media affect environmental problems. Whether an individual recycles, drives a sport-utility vehicle (SUV), or joins an environmental activist group is influenced by the meanings and definitions of these behaviors that the individual learns through interaction with others.

Large corporations and industries commonly use marketing and public relations strategies to construct favorable meanings of their corporation or industry. The term **greenwashing** refers to the way in which environmentally and socially damaging

environmental injustice Also known as environmental racism, the tendency for marginalized populations and communities to disproportionately experience adversity due to environmental problems.

greenwashing The way in which environmentally and socially damaging companies portray their corporate image and products as being "environmentally friendly" or socially responsible.

Go into any large grocery or "big box" store today and take notice of how many products are advertised as "green," "all natural," or "earth-friendly." Marketing companies know that "green" sells, as consumers are becoming more eco-minded. But how valid are the environmental claims made on the labels of the products we buy? Are the claims trustworthy? Or do marketers deliberately mislead consumers regarding the environmental practices of a company or the environmental benefits of a product or service—a practice known as greenwashing? In this *Social Problems Research Up Close* feature, we present a study that attempts to answer these questions.

Sample and Methods

In 2008 and 2009, TerraChoice, an environmental marketing company, sent researchers into leading "big box" retailers in the United States, Canada, the United Kingdom, and Australia with instructions to record every product making an environmental claim. For each product, the researchers recorded product details, claim(s) details, any supporting information, and any explanatory detail or offers of additional information or support. In the United States and Canada, a total of 2,219 products making 4,996 green claims were recorded. These claims were tested against guidelines provided by the U.S. Federal Trade Commission, Competition Bureau of Canada, Australian Competition and Consumer Commission, and the International Organization for Standardization (ISO) 14021 standard for environmental labeling. The researchers wanted to answer the following questions about each green claim they identified: (1) Is the claim truthful? (2) Does the company offer validation for its claim from an independent and trusted third party? (3) Is the claim specific, using

terms that have agreed-upon definitions, not vague ones like "natural" or "nontoxic"? (4) Is the claim relevant to the product it accompanies? (5) Does the claim address the product's principal environmental impact(s) or does it distract consumers from the product's real problems?

Selected Findings and Conclusions

This study found that green claims are most common for products related to children (toys and baby products), cosmetics, and cleaning products. Of the 2,219 North American products surveyed, over 98 percent committed at least one of the following seven "sins" of greenwashing:

1. *Sin of the hidden trade-off*, committed by suggesting a product is "green" based on an unreasonably narrow set of attributes without attention to other important environmental issues. Paper, for example, is not necessarily environmentally preferable just because it comes from a sustainably harvested forest. Other important environmental issues in the papermaking process, including energy, greenhouse gas emissions, and water and air pollution, may be equally or more significant.
2. *Sin of no proof*, committed by an environmental claim that cannot be substantiated by easily accessible supporting information or by a reliable third-party certification. Common examples are facial or toilet tissue products that claim various percentages of postconsumer recycled content without providing any evidence.
3. *Sin of vagueness*, committed by every claim that is so poorly defined or

broad that consumers are likely to misunderstand its real meaning. "All natural" is an example. Arsenic, uranium, mercury, and formaldehyde are all naturally occurring, and poisonous. "All natural" isn't necessarily "green."
4. *Sin of irrelevance*, committed by making an environmental claim that may be truthful but is unimportant or unhelpful for consumers seeking environmentally preferable products. "CFC-free" is a common example, because it is a frequent claim despite the fact that CFCs are banned by law.
5. *Sin of lesser of two evils*, committed by claims that may be true within the product category, but that may distract the consumer from the greater environmental impacts of the category as a whole. Organic cigarettes are an example of this category, as are fuel-efficient sport-utility vehicles.
6. *Sin of fibbing*, the least frequent sin, is committed by making environmental claims that are simply false. The most common examples were products falsely claiming to be Energy Star–certified or Energy Star–registered.
7. *Sin of worshiping false labels* is committed by a product that, through either words or images, gives the impression of third-party endorsement where no such endorsement actually exists.

Although many companies are making meaningful efforts to minimize their impact on the environment, this study alerts consumers to the widespread practice of greenwashing and suggests that there is a need for more accountability and transparency in the marketing of products as "green."

Source: TerraChoice Group Inc. 2009.

companies portray their corporate image, products, and services as being "environmentally friendly" or socially responsible. Greenwashing is commonly used by public relations firms that specialize in damage control for clients whose reputations and profits have been hurt by poor environmental practices. For example, coal is associated with the devastation of communities through the mining practice of mountaintop removal, and the burning of coal is the biggest contributor to pollution that causes global warming. The coal industry has spent enormous sums to convince the public that coal is clean. The "clean coal" campaign has invited widespread criticism from environmentalists: "Saying coal is clean is like talking about healthy cigarettes. There's no such thing as clean coal" (Beinecke 2009).

Although greenwashing involves manipulation of public perception to maximize profits, many corporations make genuine and legitimate efforts to improve their operations, packaging, or overall sense of corporate responsibility toward the environment. For example, in 1990, McDonald's announced that it was phasing out foam packaging and switching to a new, paper-based packaging that is partially degradable. But many environmentalists are not satisfied with what they see as token environmentalism, or as Peter Dykstra of Greenpeace suggested, 5 percent of environmental virtue to mask 95 percent of environmental vice (Hager & Burton 2000).

> **What Do You Think?** Earth Day events have been sponsored by corporations such as Office Depot, Texas Instruments, Raytheon Missile Systems, and Waste Management (a Houston-based company responsible for numerous hazardous waste sites). Some environmentalists, including Earth Day's founder, former senator Gaylord Nelson, consider the participation of corporations in Earth Day evidence of the celebration's success. But other environmentalists accuse corporations of using their financial support of Earth Day as a public relations greenwashing strategy. Do you think that organizers of Earth Day events should accept sponsorship from corporations with poor environmental records? Why or why not?

Corporations also engage in **pinkwashing**—the practice of using the color pink and pink ribbons and other marketing strategies that suggest a company is helping to fight breast cancer, even when the company may be using chemicals linked to cancer. An example of pinkwashing is the cosmetic company Avon's "Kiss Goodbye to Breast Cancer" campaign with a fund-raising lipstick in six shades (Courageous Spirit, Crusade Pink, Faithful Heart, Inspirational Life, Strength, and Triumph). "Those lipsticks may have contained ingredients that disrupt hormone functions (which is in turn linked to breast cancer). The use of hormone disruptors is not uncommon in the cosmetics industry, and is not currently prohibited by U.S. law" (Lubitow & Davis 2011, p. 142).

Environmental Problems: An Overview

Over the past 50 years, humans have altered **ecosystems**—the complex and dynamic relationships between forms of life and the environments they inhabit—more rapidly and extensively than in any other comparable period of time in history (Millennium Ecosystem Assessment 2005). As a result, humans have created environmental problems, including depletion of natural resources; air, land, and water pollution; global warming and climate change; environmental illness; threats to biodiversity; and light pollution. Because many of these environmental problems are related to the ways that humans produce and consume energy, we begin this section with an overview of global energy use.

pinkwashing The practice of using the color pink and pink ribbons to indicate a company is helping to fight breast cancer, even when the company may be using chemicals linked to cancer.

ecosystems The complex and dynamic relationships between forms of life and the environments they inhabit.

Energy Use Worldwide

When Hurricane Irene blew up the East Coast in August 2011, it left millions of people and businesses—from South Carolina to Maine—without power. Connecticut experienced the biggest power outage in its history, and in Virginia, power was out in more than 300 critical services, including hospitals, fire stations, and emergency call centers

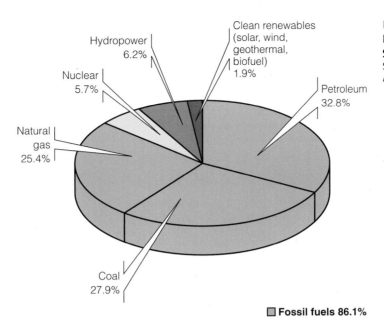

Figure 13.1: World Energy Production by Source, 2007
Source: Energy Information Administration 2010.

Hydropower 6.2%

Clean renewables (solar, wind, geothermal, biofuel) 1.9%

Nuclear 5.7%

Petroleum 32.8%

Natural gas 25.4%

Coal 27.9%

■ Fossil fuels 86.1%

(Kahn 2011). Until we experience a prolonged power outage, most of us take the availability of electricity for granted, and don't think about how dependent we are on energy. Being mindful of environmental problems means seeing the connections between energy use and our daily lives.

> Everything we consume or use—our homes, their contents, our cars and the roads we travel, the clothes we wear, and the food we eat—requires energy to produce and package, to distribute to shops or front doors, to operate, and then to get rid of. We rarely consider where this energy comes from or how much of it we use—or how much we truly need. (Sawin 2004, p. 25)

Most of the world's energy comes from fossil fuels, which include petroleum (or oil), coal, and natural gas (Energy Information Administration 2010) (see Figure 13.1). As you continue reading this chapter, notice that the major environmental problems facing the world today—air, land, and water pollution, destruction of habitats, biodiversity loss, global warming, and environmental illness—are linked to the production and use of fossil fuels.

The next most common source of energy is hydroelectric power (6.2 percent), which involves generating electricity from moving water. As water passes through a dam into a river below, a turbine in the dam produces energy. Although hydroelectric power is nonpolluting and inexpensive and is considered to be a clean and renewable form of energy, it is criticized for affecting natural habitats. For example, dams make certain fish unable to swim upstream to reproduce. Other forms of clean, renewable energy include solar, wind, geothermal (heat from the earth), and biofuel (such as animal dung and wood). In 2010, for the first time, total worldwide capacity of renewable energy, including hydropower and other clean renewables, exceeded that of nuclear energy (Schneider et al. 2011).

Nuclear power, accounting for 5.7 percent of world energy production, is associated with a number of problems related to radioactive nuclear waste—problems that are discussed later in this chapter. As of 2011, there were 437 nuclear power reactors in the world. In the United States, 65 nuclear power plants were powered by a total of 104 nuclear reactors (Clemmitt 2011). Most U.S. nuclear power plants are east of the Mississippi River (see Figure 13.2).

The safety of nuclear power is highly questionable, as equipment failures and natural disasters can result in leaks of harmful radiation (see this chapter's photo essay). The worst nuclear power plant accident in the United States occurred in 1979, when the

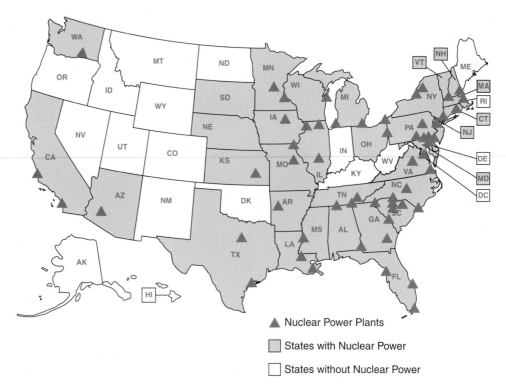

▲ Nuclear Power Plants

☐ States with Nuclear Power

☐ States without Nuclear Power

cooling system at the Three Mile Island reactor failed, causing a partial meltdown. A small amount of radioactive gas was vented from the building to prevent an explosion. Cleanup of the site cost $1 billion and took 14 years (Clemmitt 2011).

Depletion of Natural Resources: Our Growing Environmental Footprint

Humans have used more of the earth's natural resources since 1950 than in the million years preceding 1950 (Lamm 2006). Population growth, combined with consumption patterns, is depleting natural resources such as forests, water, minerals, and fossil fuels.

Water supplies around the world are dwindling, while the demand for water continues to increase because of industrialization, rising living standards, and changing diets that include more food products that require larger amounts of water to produce: milk, eggs, chicken, and beef. Currently, 50 countries are facing moderate or severe water stress; by the year 2030, nearly half the world's population will be living in areas of high water stress (WWF 2008; World Water Assessment Program 2009). With 70 percent of freshwater use going to agriculture, water shortages threaten food production and supply.

The world's forests are also being depleted. The demand for new land, fuel, and raw materials has resulted in **deforestation**—the conversion of forestland to nonforestland. Global forest cover has been reduced by half of what it was 8,000 years ago (Gardner 2005). Every year, about 13 million hectares of the world's forests (an area about the size of Greece) are cut down and converted to land use (UNEP 2008). The major causes of deforestation are the expansion of agricultural land, human settlements, wood harvesting, and road building.

Deforestation displaces people and wild species from their habitats; soil erosion caused by deforestation can cause severe flooding; and, as we explain later in this chapter, deforestation contributes to global warming. Deforestation also contributes to **desertification**—the degradation of semiarid land, which results in the expansion of desert land that is unusable for agriculture. Overgrazing by cattle and other herd animals also contributes to desertification. The problem of desertification is most severe in Africa (Reese 2001). As more land turns into desert, populations can no longer sustain a livelihood on the land, and so they migrate to urban areas or other countries, contributing to social and political instability.

deforestation The conversion of forestland to nonforestland.

desertification The degradation of semiarid land, which results in the expansion of desert land that is unusable for agriculture.

The demands that humanity makes on the earth's natural resources are known as the **environmental footprint.** A person's environmental footprint is determined by the patterns of production and consumption in that person's culture. Environmental footprints are a measure of how much of the earth people require to provide the natural resources they consume, and are often expressed in terms of global hectares per person (1 hectare equals about 2.47 acres). In 2007, the world's population used 1.5 planet Earths to support our consumption. In other words, the amount of natural resources we used in 2007 would take 1.5 years for the Earth to regenerate (Ewing et al. 2010). If we continue current patterns of consumption, we would need the equivalent of two planets to support us by 2030 (WWF 2008).

> If we continue current patterns of consumption, we would need the equivalent of two planets to support us by 2030.

Air Pollution

Transportation vehicles, fuel combustion, industrial processes (such as burning coal and processing minerals from mining), and solid waste disposal have contributed to the growing levels of air pollutants, including carbon monoxide, sulfur dioxide, arsenic, nitrogen dioxide, mercury, dioxins, and lead. Air pollution, which is linked to heart disease, lung cancer, emphysema, chronic bronchitis, and asthma, kills about 3 million people a year (Pimentel et al. 2007). In the United States, about half of the population lives in areas where they are exposed to unhealthy levels of air pollution (ozone or particulate pollution) (American Lung Association 2011).

Indoor Air Pollution When we hear the phrase *air pollution,* we typically think of industrial smokestacks and vehicle exhausts pouring gray streams of chemical matter into the air. But indoor air pollution is also a major problem, especially in poor countries where people commonly cook food and heat their home by burning dung, wood, crop waste, or coal on open fires or stoves without chimneys (World Health Organization 2005). Exposure to this indoor smoke increases risk of pneumonia, chronic respiratory disease, asthma, cataracts, tuberculosis, and lung cancer, and is responsible for up to 1.6 million deaths a year (World Health Organization 2010). Exposure is particularly high among women and children, who spend the most time near the domestic hearth or stove.

Indoor pollution is a serious problem in developing countries. As this woman cooks food for her family, she is exposed to harmful air contaminants from the fumes.

Even in affluent countries, much air pollution is invisible to the eye and exists where we least expect it—in our homes, schools, workplaces, and public buildings. Sources of indoor air pollution include lead dust (from old lead-based paint); secondhand tobacco smoke; by-products of combustion (e.g., carbon monoxide) from stoves, furnaces, fireplaces, heaters, and dryers; and other common household, personal, and commercial products (American Lung Association 2005). Some of the most common indoor pollutants include carpeting (which emits more than a dozen toxic chemicals); mattresses, sofas, and pillows (which emit formaldehyde and fire retardants); pressed wood found in kitchen cabinets and furniture (which emits formaldehyde); and dry-cleaned clothing (which emits perchloroethylene). Air fresheners, deodorizers, and disinfectants emit the pesticide paradichlorobenzene. Potentially harmful organic solvents are present in numerous office supplies, including glue, correction fluid, printing ink, carbonless paper, and felt-tip markers. Many homes today contain a cocktail of toxic chemicals: "Styrene (from plastics), benzene (from plastics and rubber), toluene and

environmental footprint The demand that each person makes on the earth's natural resources, often measured in terms of global hectares (gha) per person.

Photo Essay

Deepwater Horizon Oil Rig Explosion and Fukushima Nuclear Power Plant Accident

The price we pay for energy includes the human, environmental, and financial costs associated with unexpected accidents in energy-production industries. This photo essay looks at two major energy-related disasters: the 2010 British Petroleum (BP) Deepwater Horizon oil rig explosion and the 2011 Fukushima Daiichi nuclear power plant accident. (See Chapter 7 for discussion of the 2010 explosion at the Massey Energy Upper Big Branch coal mine in West Virginia.)

Deepwater Horizon Oil Rig Explosion.

In April 2010, the Deepwater Horizon oil drilling rig, owned by Transocean and leased to BP, exploded over the Macondo oil well, killing 11, injuring 17 oil rig workers, and opening a gusher that released oil and methane gas into the Gulf of Mexico. By the time the oil well was sealed months later, over 4 million barrels of oil had spilled into the Gulf, creating what may be the worst environmental disaster in U.S. history. In the decade prior to the Deepwater Horizon accident, there had been 948 fires and explosions in the Gulf of Mexico oil industry. A presidential commission on the 2010 Gulf oil spill concluded that the disaster was caused by a series of preventable human and engineering mistakes made by BP, Halliburton, and Transocean, and by lack of regulatory oversight by the government (National Commission on the BP Deepwater Horizon Oil Spill and Offshore Drilling 2011).

Although much of the oil in the Deepwater Horizon disaster was recovered by burning and skimming, 156 million gallons were left in the environment. In an attempt to lessen the impact of the oil slick on the ecosystem, BP used 1.84 million gallons of chemical dispersants—detergent-like compounds that break up spilled oil into tiny droplets that would mix with water. But using dispersants may have added new risks to the harm already done. Some of the 57 chemicals found in dispersants are associated with cancer, skin and eye irritation, respiratory problems, kidney problems, and toxicity to aquatic life (Foster 2011). The long-term toll of the oil spill and the use of dispersants on the Gulf water and coastline ecosystems is unknown and could last for decades (NRDC 2011).

Fukushima Nuclear Power Plant Accident.

Following a 9.0 magnitude earthquake and tsunami on March 11, 2011, a series of equipment failures and nuclear meltdowns occurred at the Fukushima Daiichi nuclear power plant in Japan. The accident was assessed as Level 7 on the International Nuclear Event Scale—the maximum scale value that is defined as "[A] major release of radioactive material with widespread health and environmental effects requiring implementation of planned and extended countermeasures" (Jamail 2011). The

AP Photo/Anonymous/US Coast Guard

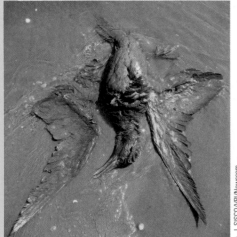

▲ Dead birds, sea turtles, dolphins, and whales were found, covered in oil, along the 650 miles of coastline that were affected by the BP oil spill.

▼ Fishing and tourist industries suffered economic losses in the tens of billions due to the Gulf oil spill in 2010.

CANNOT FISH OR SWIM
HOW THE
BP HELL ARE WE
SUPPOSE TO FEED
OUR KIDS NOW?
Wesley Bland

LEE CELANO/Reuters/Landov

A. J. SISCO/VPI/Newscom

412

◄ The 2010 Deepwater Horizon oil drilling rig explosion is unprecedented in its size, duration, location, and use of toxic chemical dispersants.

▼ Following the Fukushima nuclear power plant accident, there were many protests against nuclear power in Japan, as well as in other countries.

only other nuclear accident rated at Level 7 prior to Fukushima was the 1986 Chernobyl nuclear power plant accident in Ukraine.

Following the accident at the Fukushima nuclear power plant, a mandatory evacuation was issued for people living within 12 miles of the plant; those living outside this radius were advised to stay indoors, close doors and windows, turn off the air conditioner, cover their mouths with masks, and avoid drinking tap water. In the weeks following the accident, the evacuation zone expanded to 18 miles or more, resulting in a total of about 200,000 people being forced to leave their homes.

As of August 2011, the damaged Fukushima nuclear plant was still leaking radiation. An accurate assessment of the severity of the Fukushima disaster is still not available at the time this book goes to press, and the total amount of radioactivity released in the accident may not be known for years. But independent testing found that the amount of radiation released over a period of about five months following the Fukushima nuclear disaster is the equivalent to more than 29 Hiroshima-type atomic bombs and the amount of uranium released is equivalent to 20 Hiroshima bombs (Jamail 2011). The areas surrounding the Fukushima nuclear power plant could remain uninhabitable for decades due to high radiation.

◄ The Fukushima nuclear power plant accident is the most serious nuclear accident since the 1986 Chernobyl nuclear power plant accident.

▼ Following the Fukushima nuclear power plant accident, many Japanese people living near the power plant were evacuated from their homes and advised to wear masks to protect against radiation. These people are standing in line to get into an evacuation center.

xylene, trichloroethylene, dichloromethane, trimethylbenzene, hexanes, phenols, pentanes, and much more outgas from our everyday furnishings, construction materials, and appliances" (Rogers 2002).

Destruction of the Ozone Layer The ozone layer of the earth's atmosphere protects life on earth from the sun's harmful ultraviolet rays. Yet the ozone layer has been weakened by the use of certain chemicals, particularly chlorofluorocarbons (CFCs), used in refrigerators, air conditioners, spray cans, and other applications. The depletion of the ozone layer allows hazardous levels of ultraviolet rays to reach the earth's surface and is linked to increases in skin cancer and cataracts, weakened immune systems, reduced crop yields, damage to ocean ecosystems and reduced fishing yields, and adverse effects on animals. Despite measures that have ended production of CFCs, the ozone is not expected to recover significantly for about another decade. This is because CFCs already in the atmosphere remain for 40 to 100 years.

Acid Rain Air pollutants, such as sulfur dioxide and nitrogen oxide, mix with precipitation to form **acid rain.** Polluted rain, snow, and fog contaminate crops, forests, lakes, and rivers. As a result of the effects of acid rain, all the fish have died in a third of the lakes in New York's Adirondack Mountains (Blatt 2005). Because winds carry pollutants in the air, industrial pollution in the Midwest falls back to earth as acid rain on southeast Canada and the northeast New England states. In China, most of the electricity comes from burning coal, which creates sulfur dioxide pollution and acid rain that falls on one-third of China, damaging lakes, forests, and crops (Woodward 2007). Acid rain also deteriorates the surfaces of buildings and statues. "The Parthenon, Taj Mahal, and Michelangelo's statues are dissolving under the onslaught of the acid pouring out of the skies" (Blatt 2005, p. 161).

Global Warming and Climate Change

Global warming refers to the increasing average global air temperature, caused mainly by the accumulation of various gases (greenhouse gases) that collect in the atmosphere. According to the Intergovernmental Panel on Climate Change (IPCC)—a team of more than 1,000 scientists from 113 countries—"Warming of the climate system is unequivocal, as is now evident from observations of increases in global average air and ocean temperatures, widespread melting of snow and ice, and rising global average sea level" (2007a, p. 5). Average global surface temperatures have increased by about 0.74°C over the past century (between 1906 and 2005) (Intergovernmental Panel on Climate Change 2007a). If current greenhouse gas emissions trends continue, average global temperatures may rise another 2°C by 2035. Although 2°C may not seem significant, the effects of a global temperature increase of 2 degrees are expected to be catastrophic (Global Humanitarian Forum 2009).

Causes of Global Warming The prevailing scientific view is that **greenhouse gases**—primarily carbon dioxide (CO_2), methane, and nitrous oxide—accumulate in the atmosphere and act like the glass in a greenhouse, holding heat from the sun close to the earth. Most scientists believe that global warming has resulted from the marked increase in global atmospheric concentrations of greenhouse gases since industrialization began. Global increases in carbon dioxide concentration are due primarily to the use of fossil fuels.

Deforestation also contributes to increasing levels of carbon dioxide in the atmosphere. Trees and other plant life use carbon dioxide and release oxygen into the air. As forests are cut down or are burned, fewer trees are available to absorb the carbon dioxide. The greenhouse gases methane and nitrous oxide are primarily due to agriculture (Intergovernmental Panel on Climate Change 2007a). Despite the scientific evidence that human activity causes global warming, nearly half (46 percent) of U.S. adults believe that global warming is due more to natural changes in the environment (Newport 2010).

The growth of greenhouse gas emissions is strongest in developing countries, particularly China, which emits more carbon dioxide than any other nation. In 2010, China consumed nearly half of all coal worldwide and surpassed the United States as the world's largest

---

acid rain The mixture of precipitation with air pollutants, such as sulfur dioxide and nitrogen oxide.

global warming The increasing average global air temperature, caused mainly by the accumulation of various gases (greenhouse gases) that collect in the atmosphere.

greenhouse gases Gases (primarily carbon dioxide, methane, and nitrous oxide) that accumulate in the atmosphere and act like the glass in a greenhouse, holding heat from the sun close to the earth.

consumer of energy (BP 2011). With less than 5 percent of the world's population, the United States produces 20 percent of the world's fossil fuel CO_2 emissions (Flavin & Engelman 2009). In North America, carbon emissions per person far exceed that of other regions of the world (see Figure 13.3).

Even if greenhouse gases are stabilized, global air temperature and sea level are expected to continue to rise for hundreds of years. That is because global warming that has already occurred contributes to further warming of the planet—a process known as a *positive feedback loop*. For example, the melting of Siberia's frozen peat bog could release billions of tons of methane, a potent greenhouse gas, into the atmosphere (Pearce 2005). And the melting of ice and snow—another result of global warming—exposes more land and ocean area, which absorbs more heat than ice and snow, further warming the planet.

Even if greenhouse gases are stabilized, global air temperature and sea level are expected to continue to rise for hundreds of years.

Effects of Global Warming and Climate Change Climate change kills an estimated 30,000 people per year, mostly in the developing world (Global Humanitarian Forum 2009). The majority of these deaths are attributed to crop failure leading to malnutrition and water problems such as flooding and drought. The effects of global warming and climate change also include the following (Intergovernmental Panel on Climate Change 2007b; National Assessment Synthesis Team 2000; UNEP 2007):

Melting Ice and Sea-Level Rise. Over the 20th century, global sea levels rose an average of 4.4 to 8.8 inches (Leitzell 2011). Recent forecasts predict that sea-level rise could reach 3 to 6.5 feet over the 21st century (Mulrow & Ochs 2011). The two major factors that are causing a rise in the sea level are (1) thermal expansion caused by the warming of the oceans (water expands as it warms), and (2) the melting of glaciers and the Greenland and Antarctic ice sheets. Scientists say the Arctic Ocean in summer could be ice-free by the end of the century (Leitzell 2011). Rising sea levels pose a threat to 10 percent of the world's population that live in coastal areas, and 13 of the world's 20 largest cities that are located in coastal areas (Mulrow & Ochs 2011). As sea levels rise, some island countries, as well as some barrier islands off the U.S. coast, are likely to disappear, and low-lying coastal areas will become increasingly vulnerable to storm surges and flooding.

Flooding and Spread of Disease. Increased heavy rains and flooding caused by global warming contribute to increases in drownings and increases in the number of people exposed to insect- and water-related diseases, such as malaria and cholera. Flooding, for example, provides fertile breeding grounds for mosquitoes that carry a variety of diseases including encephalitis, dengue fever, yellow fever, West Nile virus, and malaria (Knoell 2007). With the warming of the planet, mosquitoes are now living in areas in which they previously were not found, placing more people at risk of acquiring one of the diseases carried by the insect.

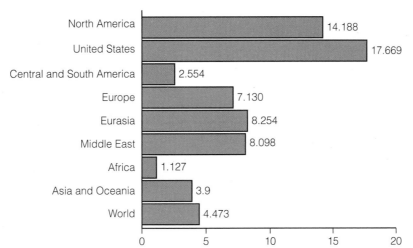

Figure 13.3: Per Capita Carbon Dioxide Emissions,* by Region, 2009
*Unit of measurement = metric tons.
Source: Energy Information Administration 2010.

Shrinking Arctic ice threatens the survival of the polar bear.

Threat of Species Extinction. Scientists have predicted that, in certain areas of the world, global warming will lead to the extinction of up to 43 percent of plant and animal species, representing the potential loss of 56,000 plant species and 3,700 vertebrate species (Malcolm et al. 2006). The U.S. Geological Survey (2007) predicts that, due to the effects of climate change, the entire polar bear population of Alaska may be extinct in the next 43 years.

Extreme Weather: Hurricanes, Droughts, and Heat Waves. Rising temperatures are causing drought in some parts of the world and too much rain in other parts. Globally, the proportion of land surface in extreme drought is predicted to increase from 1 percent to 3 percent (present day) to 30 percent by 2090 (Intergovernmental Panel on Climate Change 2007a). Warmer tropical ocean temperatures can cause more intense hurricanes (Chafe 2006). With rising temperatures, an increase in the number, intensity, and duration of heat waves is expected, with the accompanying adverse health effects (Intergovernmental Panel on Climate Change 2007b). Droughts, as well as floods, can be devastating to crops and food supplies.

Forest Fires. Another effect of global warming is an increase in the number and size of forest fires (Westerling et al. 2006). Warmer temperatures dry out brush and trees, creating ideal conditions for fires to spread. Global warming also means that spring comes earlier, making the fire season longer.

Land Pollution

About 30 percent of the world's surface is land, which provides soil to grow the food we eat. Increasingly, humans are polluting the land with nuclear waste, solid waste, and pesticides. In 2011, 1,287 hazardous waste sites (also called Superfund sites) were on the National Priority List (EPA 2011a).

Nuclear Waste Nuclear waste, resulting from both nuclear weapons production and nuclear reactors or power plants, contains radioactive plutonium, a substance linked to cancer and genetic defects. Radioactive plutonium has a half-life of 24,000 years,

In 2011, Texas experienced its worst drought on record, fueling wildfires that destroyed more than 33,000 acres.

meaning that it takes 24,000 years for the radioactivity to be reduced by half (Mead 1998). Thus, nuclear waste in the environment remains potentially harmful to human and other life for thousands of years.

In the United States, nuclear waste is being stored temporarily in 121 aboveground sites in 39 states. The first planned U.S. repository for nuclear waste was in Yucca Mountain, 100 miles northwest of Las Vegas. However, President Obama has rejected this plan, saying there are too many questions about whether nuclear waste storage at Yucca Mountain would be safe. The question remains about how to safely dispose of nuclear waste.

Nuclear plants have about 52,000 tons of radioactive spent fuel, with about 10,000 tons of that amount sealed in casks (Vedantam 2005b). Because of inadequate oversight and gaps in safety procedures, radioactive spent fuel is missing or unaccounted for at some U.S. nuclear power plants, which raises serious safety concerns (Vedantam 2005a). Accidents at nuclear power plants, such as the 2011 accident at Fukushima, Japan, and the potential for nuclear reactors to be targeted by terrorists add to the actual and potential dangers of nuclear power plants.

Recognizing the hazards of nuclear power plants and their waste, Germany became the first country to order all of its 19 nuclear power plants shut down by 2020 ("Nukes Rebuked" 2000). Belgium is also phasing out nuclear reactors, and Austria, Denmark, Italy, and Iceland have prohibitions against nuclear energy. Nevertheless, in April 2011, there were 437 operating nuclear reactors worldwide, with 64 more under construction (Schneider et al. 2011). The United States had the most operating reactors (104) followed by France (59), Japan (55), and Russia (31).

Solid Waste In 1960, each U.S. citizen generated 2.7 pounds of garbage on average every day. This figure increased to 3.7 pounds in 1980, and to 4.3 pounds in 2009 (see Figure 13.4) (EPA 2010). This figure does not include mining, agricultural, and industrial waste; demolition and construction wastes; junked autos; or obsolete equipment wastes. Just over half of this waste is dumped in landfills; the rest is recycled or composted. The availability of landfill space is limited, however. Some states have passed laws that limit the amount of solid waste that can be disposed of; instead, they require that bottles and cans be returned for a deposit or that lawn clippings be used in a community composting program.

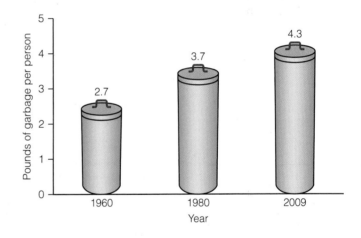

Figure 13.4: Pounds of Garbage per Person, per Day, United States, 1960–2009
Source: EPA 2010.

What Do You Think? Plastic bags, commonly used in grocery and retail stores, are associated with a number of environmental problems. In the United States, most plastic shopping bags, which contain toxic chemicals, end up in landfills where it takes 1,000 years for them to degrade (Cheeseman 2007). Plastic bags also end up in oceans, where marine life can choke or starve after swallowing them. Some countries, such as South Africa, China, Thailand, and Bangladesh, have banned plastic shopping bags. In 2007, San Francisco became the first U.S. city to ban plastic bags from supermarkets and chain pharmacies. Other municipalities, including Westport, Connecticut; Brownsville, Texas; and North Carolina's Outer Banks region have also banned plastic shopping bags. Would you support a ban, or a tax, on plastic bags in your community?

Solid waste includes discarded electrical appliances and electronic equipment, known as **e-waste.** Ever think about where your discarded computer, cell phone, CD player, television, or other electronic product ends up when you replace it with a newer model? Most

e-waste Discarded electrical appliances and electronic equipment.

discarded electronics end up in landfills, incinerators, or hazardous waste exports; only 25 percent are recycled (EPA 2011b). The main concern about dumping e-waste in landfills is that hazardous substances, such as lead, cadmium, barium, mercury, PCBs, and polyvinyl chloride, can leach out of e-waste and contaminate the soil and groundwater.

Pesticides Pesticides are used worldwide for crops and gardens; outdoor mosquito control; the care of lawns, parks, and golf courses; and indoor pest control. Pesticides contaminate food, water, and air and can be absorbed through the skin, swallowed, or inhaled. Many common pesticides are considered potential carcinogens and neurotoxins (Blatt 2005). Even when a pesticide is found to be hazardous and is banned in the United States, other countries from which we import food may continue to use it. In an analysis of 1,398 domestic and 3,655 imported food samples, pesticide residues were detected in 35 percent of the domestic samples and 23 percent of the imported samples (Food and Drug Administration 2010). Pesticides also contaminate our groundwater supplies.

Water Pollution

Our water is being polluted by a number of harmful substances, including pesticides, vehicle exhaust, acid rain, oil spills, sewage, and industrial, military, and agricultural waste. Water pollution is most severe in developing countries, where more than 1 billion people lack access to clean water. In developing nations, more than 80 percent of untreated sewage is dumped directly into rivers, lakes, and seas that are also used for drinking and bathing (World Water Assessment Program 2009).

In the United States, one indicator of water pollution is the thousands of fish advisories issued by the U.S. Environmental Protection Agency (EPA) that warn against the consumption of certain fish caught in local waters because of contamination with pollutants such as mercury and dioxin. The EPA advises women who may become pregnant, pregnant women, nursing mothers, and young children to avoid eating certain fish altogether (swordfish, shark, king mackerel, and tilefish) because of the high levels of mercury (EPA 2004).

Pollutants in drinking water can cause serious health problems and even death. At Camp Lejeune—a Marine Corps base in Onslow County, North Carolina—as many as 1 million people were exposed to water contaminated with trichloroethylene (TCE), an industrial degreasing solvent, and perchloroethylene (PCE), a dry-cleaning agent from 1957 until 1987 (Sinks 2007). Exposure to these chemicals has been linked to a number of health problems, including kidney, liver, and lung damage, as well as cancer, childhood leukemia, and birth defects. In this chapter's *The Human Side* feature, a retired Marine tells the story of his daughter's illness and death that he believes resulted from the contaminated water at Camp Lejeune.

Water pollution also affects the health and survival of fish and other marine life. In the Gulf of Mexico, as well as in the Chesapeake Bay and Lake Erie, there are areas known as "dead zones" that—due to pollution runoff from agricultural uses of fertilizer—have oxygen levels so low they cannot support life (Scavia 2011).

In recent years, there has been increasing public concern about the effects of hydraulic fracturing, or "**fracking**"—a process used in natural gas production that involves injecting at high pressure a mixture of water, sand, and chemicals into deep underground wells to break apart shale rock and release gas. Critics claim that fracking causes methane gas to leak into people's water and sometimes their homes and that toxic fracturing chemicals are spilled into streams near drill sites (Soraghan 2011). A study of 68 private drinking water wells in Pennsylvania and New York found that methane contamination rose sharply with proximity to fracking sites (Holzman 2011).

Chemicals, Carcinogens, and Health Problems

About 3 million tons of toxic chemicals are released into the environment each year (Pimentel et al. 2007). Chemicals in the environment enter our bodies via the food and water we consume, the air we breathe, and the substances with which we come in contact.

fracking Hydraulic fracturing, commonly referred to as "fracking," involves injecting a mixture of water, sand, and chemicals into drilled wells to crack shale rock and release natural gas into the well.

For example, bisphenol A (BPA), a chemical that disrupts endocrine function, is commonly found in food packaging, including cans, plastic wraps, and food storage containers (Betts 2011). During a 2004 World Health Organization convention, 44 hazardous chemicals were found in the bloodstreams of top European Union officials (Schapiro 2004). And in a study of umbilical cord blood of 10 newborns, researchers found an average of 200 industrial chemicals, pesticides, and other pollutants (Environmental Working Group 2005).

The *12th Report on Carcinogens* (U.S. Department of Health and Human Services 2011) lists 240 chemical substances that are "known to be human carcinogens" or "reasonably anticipated to be human carcinogens," meaning that they are linked to cancer. These 246 chemical substances may constitute only a fraction of actual human carcinogens. In the United States, the EPA has required testing on only about 200 of the more than 80,000 chemicals that have been on the market since 1976 (Gearhart 2009).

Most cancer researchers believe that the environment in which we live and work may be a major contributor to the development of cancer—a disease that develops in half of U.S. men and a third of U.S. women (U.S. Department of Health and Human Services 2011). For example, in a review of 152 research studies of environmental pollution and breast cancer, researchers concluded that the evidence supports a link between breast cancer and a number of environmental pollutants (Brody et al. 2007). Many of the chemicals we are exposed to in our daily lives can cause not only cancer but also other health problems, such as infertility, birth defects, and a number of childhood developmental and learning problems (Fisher 1999; Kaplan & Morris 2000; McGinn 2000; Schapiro 2007). Chemicals found in common household, personal, and commercial products can result in a variety of temporary acute symptoms, such as drowsiness, disorientation, headache, dizziness, nausea, fatigue, shortness of breath, cramps, diarrhea, and irritation of the eyes, nose, throat, and lungs. Long-term exposure can affect the nervous system, reproductive system, liver, kidneys, heart, and blood. Fragrances, which are found in perfumes and colognes, shampoos, deodorants, laundry detergents, tampons, air "fresheners," and a host of other consumer products, are known to be respiratory irritants. Some of the 4,000 ingredients that are used in fragrance manufacturing have been linked with cancer, birth defects, neurotoxic effects, and endocrine disruption (Bradshaw 2010).

This mother puts sunscreen on her child to protect against sunburn. But sunscreen, like other personal care products, may contain harmful chemicals. Learn what chemicals are in the personal care products you use at the Environmental Working Group's website Skin Deep at http://www.cosmeticsdatabase.com.

What Do You Think? Some businesses and local governments are voluntarily limiting fragrances in the workplace or banning them altogether to accommodate employees who experience ill effects from them. What do you think about banning fragrances in the workplace or other public places?

Children are more vulnerable than adults to the harmful effects of most pollutants for a number of reasons. For instance, children drink more fluids, eat more food, and inhale more air per unit of body weight than do adults; in addition, crawling and a tendency to put their hands and other things in their mouths provide more opportunities for children to ingest chemical or heavy metal residues.

Multiple Chemical Sensitivity Disorder Multiple chemical sensitivity (MCS), also known as environmental illness, is a condition whereby individuals experience adverse reactions when exposed to low levels of chemicals found in everyday substances (vehicle exhaust, fresh paint, housecleaning products, perfume and other fragrances,

multiple chemical sensitivity Also known as "environmental illness," a condition whereby individuals experience adverse reactions when exposed to low levels of chemicals found in everyday substances.

Jerry Ensminger holds a portrait of his daughter Janey, whose death from leukemia is believed to have been caused by contaminated water at Camp Lejeune.

Retired Marine Master Sergeant Jerry Ensminger is one of about 900 individuals who have filed administrative claims worth $4 billion against Camp Lejeune in Onslow County, North Carolina, for damages that allegedly resulted from drinking and bathing in water contaminated with the dry-cleaning agent

perchloroethylene (PCE) and the industrial degreasing solvent trichloroethylene (TCE) (Hefling 2007).

Here, we present Jerry Ensminger's account of his daughter, Janey, who died in 1985, of leukemia at age 9. Ensminger's story is excerpted from the book *Poisoned Nation* (2007). Janey's illness and death is believed to have been caused by contaminated water at Camp Lejeune that Janey's mother was exposed to during her pregnancy with Janey. If passed, two bills before Congress—the Caring for Camp Lejeune Veterans Act and the Janey Ensminger Act—would provide health care to veterans and their families who have health problems caused by exposure to contaminated water at Camp Lejeune.

My little girl died in my arms and fifteen years later, I found out that the people I had faithfully served for almost 25 years knew she was being poisoned by the water all along. . . . It was 1997 before I finally found out why my daughter died.

Even then, it was just by chance. A local TV station had picked up the story. The evening news was turned on in the living room, and I was carrying my dinner in on a plate from the kitchen. All of a sudden I heard the newscaster saying that the water at Camp Lejeune had been highly contaminated from 1968 to 1985, and that the chemicals it contained had been linked to childhood leukemia. When I heard that, I just dropped my plate right on the floor and began shaking. The next day . . . I started reading everything I could find and making contacts with everyone I knew. There are stages you go through when you lose a child to a catastrophic illness. First you go into shock; then you start wondering why it happened to your child. So, years ago, I checked my family history and her mother's and found there was nothing on either side. But that nagging question of why Janey got leukemia had stayed with me throughout her illness, her death, and

TABLE 13.1 Threatened Species Worldwide, 2011

CATEGORY	NUMBER OF THREATENED* SPECIES IN 2011
Mammals	1,134
Birds	1,240
Amphibians	1,910
Reptiles	664
Fishes (bony)	1,816
Invertebrates (insects, snails, mollusks, etc.)	3,394
Plants	9,098
Other (fungi, lichens, brown algae)	9
Total	16,617

*Threatened species include those classified as critically endangered, endangered, and vulnerable.

Source: IUCN 2011.

synthetic building materials, and numerous other petrochemical-based products). Symptoms of MCS include headache, burning eyes, difficulty breathing, stomach distress or nausea, loss of mental concentration, and dizziness. The onset of MCS is often linked to acute exposure to a high level of chemicals or to chronic long-term exposure. Individuals with MCS often avoid public places and/or wear a protective breathing filter to avoid inhaling the many chemical substances in the environment. Some individuals with MCS build houses made from materials that do not contain the chemicals that are typically found in building materials.

In a national study on the prevalence of multiple chemical sensitivity, 11.2 percent of U.S. adults reported an unusual hypersensitivity to common chemical products such as perfume, fresh paint, and household cleaning products, and 2.5 percent said they had been diagnosed with MCS (Caress & Steinemann 2004). Two-thirds of those with hypersensitivity described their symptoms as either severe or moderately severe. More than a third (39.5 percent) of the sample reported having trouble shopping in public places due to chemical sensitivity.

Threats to Biodiversity

There are an estimated 8.7 million species of life on earth (some scientists believe the number is much higher), 1.3 million of which have been named and catalogued (Zimmer 2011).

for fourteen and a half years after it. And in that moment, that one moment, when I heard the newscast and dropped the plate of food right out of my hands, it all became clear. I suddenly knew why my little girl died.

I signed up with the Marines to serve my country, but I never signed anything that gave them the right to kill my child, to knowingly poison her. . . . I've said that publicly and the Marine Corps has never refuted anything I've said, including that the contamination went back to the 1950s. I'm sure they know that from the geological studies. . . .

The day she died . . . Janey was in a lot of pain, so they suggested that she take morphine. She didn't want to, because she had tried it before and it made her so tired. But this time she just couldn't handle the pain. Janey went through hell for nearly two and a half years, and I went through hell with her. . . . I was there with her every step of the way. Her mother couldn't handle it. Every time Janey went to the hospital,

I was the parent who went with her. Sometimes, she was screaming in my ear, "Daddy, don't let them hurt me." Like when she had the bone marrow transplant and the spinal taps. The last time she went into the hospital was the last day of July, just before her ninth birthday. She didn't come out until September 20, and that was in a casket.

About a week before she died, the head of hematology had come in talking about a new form of therapy. He said it would cause severe burns and ulcers, and they didn't recommend it, but Janey looked at them, blinking to control her tears, and said, "This is my life you are talking about, and I'm not giving up. Let's try it."

The ulcers were all over her mouth, her legs, inside her nose and her vagina. The day she died, she was in such intense pain she could hardly speak, but finally she managed to whisper, "I want to die peacefully." When Janey said that, I started sobbing. She hugged me and said, "Stop it, stop crying." I said, "I can't help it, I love you." "I know you do," she replied. "I

love you too. But, Daddy, I hurt so bad." "Do you want some morphine?" I asked. She was already being given methadone. "Yes, Daddy," she said, "I'm ready."

When the nurse heard that Janey wanted morphine, she knew the time had come. Then, just as they started to give it to her, she said, "Wait. Stop. I want some for my daddy, too."

"This is a very powerful pain medicine. I can't give it to your daddy," the nurse said.

"But my daddy hurts, too," Janey answered. You see, I always took a little of whatever she took to show her I was with her. The morphine killed her. It's a respiratory depressant. . . .

The organization I served faithfully for 24 and a half years knew about this all along, and they never said anything, well, shame on them.

Source: From Hefling 2007; Sinks 2007; Schwartz-Nobel 2007.

This enormous diversity of life, known as **biodiversity,** provides food, medicines, fibers, and fuel; purifies air and freshwater; pollinates crops and vegetation; and makes soils fertile.

One species of life on earth goes extinct every three hours (Leahy 2009). As shown in Table 13.1, 16,617 species worldwide are threatened with extinction. Most extinctions today result from habitat loss caused by deforestation and urban sprawl, overharvesting, global warming, and bioinvasion—environmental problems caused by human activity (Cincotta & Engelman 2000; Leahy 2009).

> One species of life on earth goes extinct every three hours.

Light Pollution

Light pollution refers to artificial lighting that is annoying, unnecessary, and/or harmful to life forms on earth. The United States, like much of the rest of the world, has become increasingly "lit up" with artificial light. Almost all people in developed societies use artificial light, reducing the natural period of darkness at night. Yet, "darkness is as essential to our biological welfare, to our internal clockwork, as light itself" (Klinkenborg 2008, p. 109). Some research suggests that exposure to artificial light at night contributes to sleep disorders, depression, and other mood disorders. Other studies have found a link between exposure to artificial light at night and breast cancer (Chepesiuk 2009).

Light pollution also has adverse effects on the migration, feeding, and reproductive patterns of many animal species (Chepesiuk 2009; Klinkenborg 2008). For example, frogs have been found to inhibit their mating calls and bats alter their feeding behavior when they are exposed to artificial light at night. Sea turtles, which have a natural predisposition to nest on dark beaches, find fewer of them to nest on. When sea turtle eggs hatch, the babies instinctually go toward the brighter, reflective sea horizon, but are confused by artificial lighting near the beach and may wander off toward the artificially lit beach development.

biodiversity The diversity of living organisms on earth.

light pollution Artificial lighting that is annoying, unnecessary, and/or harmful to life forms on earth.

Social Causes of Environmental Problems

Various structural and cultural factors have contributed to environmental problems. These include population growth, industrialization and economic development, and cultural values and attitudes such as individualism, consumerism, and militarism.

Population Growth

The world's population is growing; in 2011 world population reached the 7 billion mark. Population growth places increased demands on natural resources and results in increased waste. As Hunter (2001) explained:

> Global population size is inherently connected to land, air, and water environments because each and every individual uses environmental resources and contributes to environmental pollution. While the scale of resource use and the level of wastes produced vary across individuals and across cultural contexts, the fact remains that land, water, and air are necessary for human survival. (p. 12)

However, population growth itself is not as critical as the ways in which populations produce, distribute, and consume goods and services. As shown in Figure 13.5, regions with the highest populations have the lowest environmental footprint.

Industrialization and Economic Development

Many of the environmental problems confronting the world are associated with industrialization and economic development. Industrialized countries, for example, consume more energy and natural resources and contribute more pollution to the environment than poor countries.

The relationship between level of economic development and environmental pollution is curvilinear rather than linear. For example, industrial emissions are minimal in regions with low levels of economic development and are high in the middle-development range as developing countries move through the early stages of industrialization. However, at more advanced stages of industrialization, industrial emissions ease because heavy-polluting manufacturing industries decline, "cleaner" service industries increase, and because rising incomes are associated with a greater demand for environmental quality

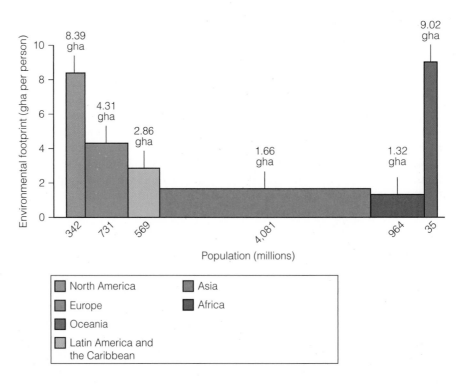

Figure 13.5: **Environmental Footprint and Population by Region, 2007**
Source: Ewing et al. 2010.

and cleaner technologies. However, a positive linear correlation has been demonstrated between per capita income and national carbon dioxide emissions (Hunter 2001).

In less developed countries, environmental problems are largely the result of poverty and the priority of economic survival over environmental concerns. Vajpeyi (1995) explained:

> Policymakers in the Third World are often in conflict with the ever-increasing demands to satisfy basic human needs—clean air, water, adequate food, shelter, education—and to safeguard the environmental quality. Given the scarce economic and technical resources at their disposal, most of these policymakers have ignored long-range environmental concerns and opted for short-range economic and political gains. (p. 24)

In considering the effects of economic development on the environment, note that the ways we measure economic development and the "health" of economies also influence environmental outcomes. Two primary measures of economic development and the health of economies are gross domestic product (GDP) and consumer spending. But these measures overlook the social and environmental costs of the production and consumption of goods and services. Until definitions and measurements of "economic development" and "economic health" reflect these costs, the pursuit of economic development will continue to contribute to environmental problems.

Cultural Values and Attitudes

Cultural values and attitudes that contribute to environmental problems include individualism, consumerism, and militarism.

Individualism Individualism, which is a characteristic of U.S. culture, puts individual interests over collective welfare. Even though recycling is good for our collective environment, many individuals do not recycle because of the personal inconvenience involved in washing and sorting recyclable items. Similarly, individuals often indulge in countless behaviors that provide enjoyment and convenience for themselves at the expense of the environment: long showers, recreational boating, frequent meat eating, the use of air conditioning, and driving large, gas-guzzling SUVs, to name just a few.

Consumerism Consumerism—the belief that personal happiness depends on the purchasing of material possessions—also encourages individuals to continually purchase new items and throw away old ones. The media bombard us daily with advertisements that tell us life will be better if we purchase a particular product. Consumerism contributes to pollution and environmental degradation by supporting polluting and resource-depleting industries and by contributing to waste.

Militarism The cultural value of militarism also contributes to environmental degradation (see also Chapter 15). "It is generally agreed that the number one polluter in the United States is the American military. It is responsible each year for the generation of more than one-third of the nation's toxic waste . . . an amount greater than the five largest international chemical companies combined" (Blatt 2005, p. 25). Toxic substances from military vehicles, weapons materials, and munitions pollute the air, land, and groundwater in and around military bases and training areas. The Pentagon has asked Congress to loosen environmental laws for the military, and the EPA is forbidden to investigate or sue the military (Blatt 2005; Janofsky 2005).

Strategies for Action: Responding to Environmental Problems

Responses to environmental problems include environmental activism, environmental education, the use of green energy, modifications in consumer products and behavior, efforts to slow population growth, and government regulations and legislation. Sustainable economic development, international cooperation and assistance, and institutions of higher education also play important roles in alleviating environmental problems.

The U.S. environmental movement may be the largest single social movement in the United States.

Environmental Activism

With more than 6,500 national and 20,000 local environmental organizations with a combined membership of between 20 million and 30 million, the U.S. environmental movement may be the largest single social movement in the United States (Brulle 2009). A Gallup survey found that one in five (20 percent) U.S. adults reports being an active participant in the environmental movement (Gallup Organization 2011) (see Table 13.2). And more than one in four (27.3 percent) college freshmen believe that getting involved in programs to clean up the environment is an essential or very important objective (Pryor et al. 2010).

> **What Do You Think?** Gallup Poll results indicate that women are more likely than men to worry about the environment, to be active in or sympathetic toward the environmental movement, and to give precedence to the environment over economic and energy concerns. Why do think this gender difference exists?

TABLE 13.2 Involvement in the Environmental Movement*

INVOLVEMENT	PERCENTAGE OF U.S. ADULTS
Active participant	20
Sympathetic but not active	42
Neutral	27
Unsympathetic	9
No opinion	2

*In a 2011 Gallup survey, a national sample of U.S. adults was asked: "Do you think of yourself as an active participant in the environmental movement, sympathetic towards the movement, but not active, neutral, or unsympathetic towards the movement?"

Source: Gallup Organization 2011.

Environmental organizations exert pressure on government and private industry to initiate or intensify actions related to environmental protection. Environmentalist groups also design and implement their own projects and disseminate information to the public about environmental issues. Environmental organizations use the Internet and e-mail to send e-mail action alerts to members informing them when Congress and other decision makers threaten the health of the environment. These members can then send e-mails and faxes to Congress, the president, and business leaders, urging them to support policies that protect the environment.

Religious Environmentalism From a religious perspective, environmental degradation can be viewed as sacrilegious, sinful, and an offense against God (Gottlieb 2003a). "The world's dominant religions—as well as many people who identify with the 'spiritual' rather than with established faiths—have come to see that the environmental crisis involves much more than assaults on human health, leisure, or convenience. Rather, humanity's war on nature is at the same time a deep affront to one of the essentially divine aspects of existence" (Gottlieb 2003b, p. 489). This view has compelled religious groups to take an active role in environmental activism.

For example, the National Association of Evangelicals, an umbrella group of 51 church denominations, adopted a platform called For the Health of the Nation: An Evangelical Call to Civic Responsibility. This platform, which has been signed by nearly 100 evangelical leaders, calls on the government to "protect its citizens from the effects of environmental degradation" (Goodstein 2005). Larry Schweiger, president of the National Wildlife Federation, welcomes evangelicals as allies and explains that conservative lawmakers who might not pay attention to what environmental groups say may be more likely to pay attention to what the faith community is saying.

Radical Environmentalism The **radical environmental movement** is a grassroots movement of individuals and groups that employs unconventional and often illegal means of protecting wildlife or the environment. Radical environmentalists believe in what is known as **deep ecology**: the view that maintaining the earth's natural systems should take precedence over human needs, that nature has a value independent of human existence,

radical environmental movement A grassroots movement of individuals and groups that employs unconventional and often illegal means of protecting wildlife or the environment.

deep ecology The view that maintaining the earth's natural systems should take precedence over human needs, that nature has a value independent of human existence, and that humans have no right to dominate the earth and its living inhabitants.

This Greenpeace activist is climbing the 630-foot chimney at Kingsnorth coal power plant.

and that humans have no right to dominate the earth and its living inhabitants (Brulle 2009). The best-known radical environmental groups are the Earth Liberation Front (ELF) and the Animal Liberation Front (ALF), which are international underground movements consisting of autonomous individuals and small groups who engage in "direct action" to (1) inflict economic damage on those profiting from the destruction and exploitation of the natural environment, (2) save animals from places of abuse (e.g., laboratories, factory farms, and fur farms), and (3) reveal information and educate the public on atrocities committed against the earth and all the species that populate it.

In 2007, six Greenpeace environmental activists climbed a 630-foot chimney at Kingsnorth coal power plant in England, intending to shut down the plant by occupying the chimney. They planned to write the words, "Gordon, bin it" on the chimney to pressure Prime Minister Gordon Brown to stop the building of new coal power plants, but after writing "Gordon," the six activists were served with an injunction and came down from the chimney. The "Kingsnorth Six," as they are called, were criminally charged for property damage; it cost 35,000 euros (equal to US $53,000 at the time) to remove the graffiti. Jurors in the case found the Kingsnorth Six "not guilty," accepting the defense arguments that the six activists had a "lawful excuse" to damage property at the Kingsnorth power station to prevent even greater damage caused by global warming (McCarthy 2008). The Criminal Damage Act of 1971 allows individuals to damage property to prevent even greater damage—such as breaking down the door of a burning house to put out a fire. James Hansen, a top NASA climate scientist, testified for the defense, and told the jury that carbon dioxide emissions from the Kingsnorth power plant would contribute to climate change.

The Kingsnorth Six Greenpeace activists were acquitted of criminal charges of property damage as jurors agreed that the defendants had a "lawful excuse" for their actions. But other radical environmentalists have been prosecuted as terrorists. **Ecoterrorism** is defined as any crime intended to protect wildlife or the environment that is violent, puts human life at risk, or results in damages of $10,000 or more (Denson 2000). Many environmentalists question whether "terrorist" is an appropriate label and argue that the real terrorists are corporations that plunder the earth.

ecoterrorism Any crime intended to protect wildlife or the environment that is violent, puts human life at risk, or results in damages of $10,000 or more.

The Role of Corporations in the Environmental Movement Corporations are major contributors to environmental problems and often fight against environmental efforts that threaten their profits. However, some corporations are joining the environmental movement for a variety of reasons, including pressure from consumers and environmental groups, the desire to improve their public image, genuine concern for the environment, and/or concern for maximizing current or future profits.

In 1994, out of concern for public and environmental health, Ray Anderson, founder and chairman of Interface carpet company, set a goal of being a sustainable company by 2020—"a company that will grow by cleaning up the world, not by polluting or degrading it" (McDaniel 2005, p. 33). Anderson envisioned recycling all the materials used, not releasing any toxins into the environment, and using solar energy to power all production. The company has made significant progress toward these goals, reducing use of fossil fuels by 45 percent, and reducing water and landfill use by as much as 80 percent.

Rather than hope that industry voluntarily engages in eco-friendly practices, corporate attorney Robert Hinkley suggested that corporate law be changed to mandate socially responsible behavior. Hinkley explained that corporations pursue profit at the expense of the public good, including the environment, because corporate executives are bound by corporate law to try to make a profit for shareholders (Cooper 2004). Hinkley suggested that corporate law should include a Code for Corporate Citizenship that would say the following: "The duty of directors henceforth shall be to make money for shareholders but not at the expense of the environment, human rights, public health and safety, dignity of employees, and the welfare of the communities in which the company operates" (quoted by Cooper 2004, p. 6).

Environmental Education

One goal of environmental organizations and activists is to educate the public about environmental issues and the seriousness of environmental problems. Being informed about environmental issues is important because people who have higher levels of environmental knowledge tend to engage in higher levels of pro-environment behavior. For example, environmentally knowledgeable people are more likely to save energy in the home, recycle, conserve water, purchase environmentally safe products, avoid using chemicals in yard care, and donate funds to conservation (Coyle 2005).

> Being informed about environmental issues is important because people who have higher levels of environmental knowledge tend to engage in higher levels of pro-environment behavior.

A main source of information about environmental issues for most Americans is the media. However, because corporations and wealthy individuals with corporate ties own the media, unbiased information about environmental impacts of corporate activities may not readily be found in mainstream media channels. Indeed, the public must consider the source in interpreting information about environmental issues. Propaganda by corporations sometimes comes packaged as "environmental education." Hager and Burton (2000) explained: "Production of materials for schools is a growth area for public relations companies around the world. Corporate interests realize the value of getting their spin into the classrooms of future consumers and voters" (p. 107).

Wind energy is harnessed by turbines such as those pictured in this photo of a wind farm in Altamont Pass, California.

"Green" Energy

Increasing the use of **green energy**—energy that is renewable and nonpolluting—can help alleviate environmental problems associated with fossil fuels. Also known as clean energy, green energy sources include solar power, wind power, biofuel, and hydrogen.

> **What Do You Think?** The World Bank defines "clean energy" as energy that does not produce carbon dioxide when generated. Under this definition, nuclear energy is considered "clean energy." When Obama, in 2011, announced he wanted 80 percent of the nation's electricity to come from clean energy sources by 2035, his definition of "clean" included nuclear power and even natural gas and "clean coal." Do you think that nuclear power, natural gas, and coal is clean energy?"

Solar and Wind Energy Solar power involves converting sunlight to electricity through the use of photovoltaic cells. More than half of solar photovoltaic cells are installed in Germany; the United States has only 6 percent (Russell 2011). Other forms of solar power include the use of solar thermal collectors, which capture the sun's warmth to heat building space and water, and "concentrating solar power plants," which use the sun's heat to make steam to turn electricity-producing turbines.

Wind turbines, which turn wind energy into electricity, are operating in 82 countries. The United States is the world's leading generator of wind energy (Kitasei 2011). One disadvantage of wind power is that wind turbines have been known to result in bird mortality. However, this problem has been mitigated in recent years through the use of painted blades, slower rotational speeds, and careful placement of wind turbines.

Biofuel Biofuels are fuels derived from agricultural crops. Two types of biofuels are ethanol and biodiesel.

Ethanol is an alcohol-based fuel that is produced by fermenting and distilling corn or sugar. Ethanol is blended with gasoline to create E85 (85 percent ethanol and 15 percent gasoline). Vehicles that run on E85, called flexible fuel vehicles, have been used by the

green energy Also known as clean energy, energy that is nonpolluting and/or renewable, such as solar power, wind power, biofuel, and hydrogen.

government and in private fleets for years and have just recently become available to consumers. Most ethanol is produced in the United States and Brazil (Shrank 2011).

A problem associated with ethanol fuel is that increased demand for corn, which is used to make most ethanol in the United States, has driven up the price of corn, resulting in higher food prices (many processed food items contain corn, and animal feed is largely corn). And as corn prices rise, so too do those of rice and wheat because the crops compete for land. Rising food prices threaten the survival of the world's poorest 2 billion people who depend on grain to survive. The grain it takes to fill a 25-gallon tank with ethanol would feed one person for an entire year (Brown 2007).

Increased corn and/or sugar cane production to meet the demand for ethanol also has adverse environmental effects, including increased use and runoff of fertilizers, pesticides, and herbicides; depletion of water resources; and soil erosion. In addition, tropical forests are being clear-cut to make room for "energy crops," leaving less land for conservation and wildlife (Price 2006); biofuel refineries commonly run on coal and natural gas (which emit greenhouse gases); farm equipment and fertilizer production require fossil fuels; and the use of ethanol involves emissions of several pollutants. Finally, even if 100 percent of the U.S. corn crop were used to produce ethanol, it would only displace less than 15 percent of U.S. gasoline use (Food & Water Watch and Network for New Energy Choices 2007).

Biodiesel fuel is a cleaner-burning diesel fuel made from vegetable oils and/or animal fats, including recycled cooking oil. Some individuals who make their own biodiesel fuel obtain used cooking oil from restaurants at no charge.

Hydrogen Power Hydrogen, the most plentiful element on earth, is a clean burning fuel that can be used for electricity production, heating, cooling, and transportation. Many see a movement to a hydrogen economy as a long-term solution to the environmental and political problems associated with fossil fuels. Further research is needed, however, to develop nonpolluting and cost-effective ways to extract and transport hydrogen.

Carbon Capture and Storage Coal-fired power plants emit more carbon dioxide than any other source. One proposal to reduce CO_2 emissions is carbon capture and storage (CCS)—a process of removing CO_2 from the smokestacks of coal-burning plants and storing it deep underground. The technology required for this process, which is still in the development stage, is expensive and requires large inputs of energy. The development of carbon capture and storage technology also promotes continued use of coal, and diverts or reduces investments in renewable energy such as solar, wind, and geothermal energy. Finally, scientists are concerned that stored carbon dioxide could leak out into the environment and cause sudden and drastic climate change (Miller & Spoolman 2009).

Modifications in Consumer Behavior

In the United States and other industrialized countries, many consumers are making "green" choices in their behavior and purchases that reflect concern for the environment. In some cases, these choices carry a price tag, such as paying more for organically grown food or for clothing made from organic cotton. Consumers are also motivated to make green purchases that save money. For example, after gas prices topped $4 a gallon in 2008, sales of gas-guzzling SUVs dropped, and sales of more fuel-efficient cars, such as hybrids, increased. Consumers often consider their utility bill when they choose energy-efficient appliances and electrical equipment. Although some eco-minded individuals choose "green" products and services, others choose to reduce their overall consumption and "buy nothing" rather than "buy green." For example, many consumers are choosing not to buy bottled water and to drink tap water instead. The switch from bottled to tap is partly fueled by the need to cut down on unnecessary spending in hard economic times, but environmental concerns are also a factor. The production and transportation of bottled water uses fossil fuels, and the disposal of plastic water bottles adds to our already overburdened landfills.

Although the average size of new housing in the United States has increased considerably, some homeowners are choosing to downsize their housing. For some, the driving force behind housing downsizing is economic, but others are moving into smaller dwellings out of concern for the environment.

Table 13.3 presents tips for how consumers can reduce the amount of carbon dioxide each of us produces.

Tiny houses have less impact on the environment.

Green Building The U.S. Green Building Council developed green building standards known as Leadership in Energy and Environmental Design (LEED). These standards consist of 69 criteria to be met by builders in six areas, including energy use and emissions, water use, materials and resource use, and sustainability of the building site. LEED buildings include the Pentagon Athletic Center, the Detroit Lions' football training facility, and the David L. Lawrence Convention Center in Pittsburgh.

Slow Population Growth

As discussed in Chapter 12, slowing population growth is an important component of efforts to protect the environment. A recent study concluded that providing the 200 million women worldwide who want access to contraception but currently don't have it would reduce projected world population in 2050 by a half billion people, and would prevent the emission of at least 34 gigatons of carbon dioxide (a gigaton equals one billion tons) (Wire 2009).

Although Americans who are concerned about the environment may think about how their home energy use, travel, food choices, and other lifestyle behaviors affect the environment, they rarely consider the environmental impact of their reproductive choices. Researchers at Oregon State University estimate that each child born in the United States, and the children that child has as an adult, adds 9,441 metric tons of carbon dioxide to the environment—the equivalent of burning 972,160 gallons of gas (Murtaugh & Schlax 2009).

TABLE 13.3 Top Ten Things You Can Do to Fight Global Warming

National Geographic's Green Guide (www.thegreenguide.com) lists the following tips for consumers to reduce the amount of carbon dioxide they produce each year. All of the following CO_2 reductions listed are on an annual basis:

1. Replace five incandescent light bulbs in your home with compact fluorescent bulbs (CFLs): Swapping those 75-watt incandescent bulbs with 19-watt CFLs can cut 275 pounds of CO_2.

2. Instead of short-haul flights of 500 miles or so, take the train and bypass 310 pounds of CO_2.

3. Sure it may be hot, but get a fan, set your thermostat to 75°F and blow away 363 pounds of CO_2.

4. Replace refrigerators more than 10 years old with today's more energy-efficient Energy Star models and save more than 500 pounds of CO_2.

5. Shave your eight-minute shower to five minutes for a savings of 513 pounds.

6. Caulk, weatherize, and insulate your home. If you rely on natural gas heating, you'll stop 639 pounds of CO_2 from entering the atmosphere (472 pounds for electric heating). And this summer, you'll save 226 pounds from air conditioner use.

7. Whenever possible, dry your clothes on a line outside or a rack indoors. If you air-dry half your loads, you'll dispense with 723 pounds of CO_2.

8. Trim down on the red meat. Because it takes more fossil fuels to produce red meat than fish, eggs, and poultry, switching to these foods will slim your CO_2 emissions by 950 pounds.

9. Leave the car at home and take public transportation to work. Taking the average U.S. commute of 12 miles by light-rail will leave you 1,366 pounds of CO_2 lighter than driving. The standard, diesel-powered city bus can save 804 pounds, and heavy rail subway users save 288.

10. Finally, support the creation of wind, solar, and other renewable energy facilities by choosing green power if offered by your utility. To find a green power program in your state, call your local utility or visit the U.S. Department of Energy's Green Power Network page at http://apps.3eere.energy.gov/greenpower/

Source: National Geographic 2007. From National Geographic's Green Guide. Copyright © 2007 National Geographic Society. All Rights reserved. Reprinted with permission.

TABLE 13.4 Carbon Reduction Figures for Various Lifestyle Behaviors*

BEHAVIOR	AMOUNT OF CO_2 NOT RELEASED INTO THE ENVIRONMENT
Replace windows with energy-efficient windows	12 metric tons
Recycle newspapers, magazines, glass, plastic, aluminum, and steel cans	17 metric tons
Replace ten 75-watt incandescent bulbs with 25-watt energy-efficient lights	36 metric tons
Increase auto gas mileage from 20 mpg to 30 mpg	148 metric tons
Reduce number of children by one	9,441 metric tons

*Calculated over an 80-year period.

Source: Murtaugh and Schlax 2009.

As shown in Table 13.4, reducing the number of children a woman has by one saves far more carbon dioxide from being released in the environment than do many other environmentally friendly lifestyle choices.

Government Policies, Programs, and Regulations

Government policies and regulations can play an important role in protecting and restoring the environment. Before reading further, assess your attitudes toward government interventions to reduce global warming in this chapter's *Self and Society* feature.

Cap and Trade Programs Cap and trade programs are a free-market approach used to control pollution by providing economic incentives to power plants and other industries for achieving reductions in the emissions of pollutants. In a cap and trade system, a limit is set on the amount of carbon dioxide that can be released into the air. Polluters buy credits allowing them to emit a limited amount of carbon dioxide. They can sell leftover credits to other polluters, creating a monetary incentive to reduce emissions. Twenty-three U.S. states have joined regional agreements to lower carbon emissions through the cap and trade system (Pew Center on the States 2009). In 2010, California adopted the nation's most stringent rules to curb greenhouse gas emissions; the rules reward industries that cut emissions by allowing them to sell carbon credits to other industries. Critics of the cap and trade approach argue that it fails to achieve the lowest possible emissions because it does not require all plants to use the best available technology to reduce emissions. By allowing some plants to have higher emissions, it also exposes populations living near these high-emissions plants to excessive air pollution.

In 2009, the U.S. House passed the American Clean Energy and Security Act, which sought to establish a federal cap and trade system, but the bill stalled in the Senate. Hopes for passing cap and trade legislation were diminished when dozens of Republicans who are skeptical about global warming were elected to Congress in the November 2010 mid-term elections.

Policies and Regulations on Energy In 2004, more than 20 countries committed to specific targets for the renewable share of total energy use (UNEP 2007). A number of states have set goals of producing a minimum percentage of electricity from wind power, solar power, or other renewable sources (Prah 2007). In addition, more than 70 mayors and other local leaders from around the world signed the Urban Environmental Accords, pledging to obtain 10 percent of energy from renewable resources by 2012, and to reduce greenhouse gases 25 percent by 2030 (Stoll 2005). Economic stimulus packages in the United States and elsewhere are targeting renewable energy, with the promise of investments in renewable energy as well as jobs. The American Recovery and Reinvestment Act of 2009 included more than $70 billion in spending and tax credits for clean energy and transportation programs (Makower et al. 2009).

Taxes Some environmentalists propose that governments use taxes to discourage environmentally damaging practices and products (Brown & Mitchell 1998). In the 1990s, a number of European governments increased taxes on environmentally harmful activities and products (such as gasoline, diesel, and motor vehicles) and decreased taxes on income and labor (Renner 2004). As a result of high gasoline taxes in Europe, gas there costs as much as $8 a gallon, which has increased consumer demand for small, fuel-efficient cars. Raising gasoline taxes in the United States is highly unpopular with voters and consumers. On the other hand, tax incentives and credits are used for renewable energy, hybrid and electric cars, and energy efficiency.

Answer each of the following questions:

1. Some people believe that the United States government should limit the amount of greenhouse gases thought to cause global warming that U.S. businesses can produce. Other people believe that the government should not limit the amount of greenhouse gases that U.S. businesses put out. What do you think?

2. Do you favor or oppose each of the following as a way for the federal government to try to reduce future global warming?
   a. Increased taxes on electricity so people use less of it
   b. Increased taxes on gasoline so people either drive less, or buy cars that use less gas
   c. Tax breaks for companies to produce more electricity from water, wind, and solar power

3. For each of the following, do you think the government should require by law, encourage with tax breaks but not require, or stay out of entirely?
   a. Building cars that use less gasoline
   b. Building air conditioners, refrigerators, and other appliances that use less electricity
   c. Building new homes and offices that use less energy for heating and cooling

4. Do you think that U.S. actions to reduce global warming in the future would hurt the U.S. economy, help the economy, or have no effect on the U.S. economy?

5. Do you think the United States should take action on global warming only if other major industrial countries such as China and India agree to do equally effective things, that the United States should take action even if these other countries do less, or that the United States should not take action on this at all?

6. Do you think most U.S. business leaders *want* the federal government to do things to stop global warming, or do you think most U.S. business leaders *do not want* the federal government to do things to stop global warming?

7. How important is the issue of global warming to you personally—extremely important, very important, somewhat important, not too important, or not at all important?

Comparison Data

Telephone interviews with a sample of 1,000 U.S. adults from across the nation found that, overall, Americans want government to be involved in reducing global warming:

1. Government should limit greenhouse gases from U.S. businesses (76%)
   Government should not limit greenhouse gases from U.S. businesses (20%)
   Don't know (3 percent)

2. To reduce future global warming, government should . . .

	Favor	Oppose
a. Increase taxes on electricity	22%	78%
b. Increase taxes on gasoline	28%	71%
c. Give companies tax breaks to produce more electricity from water, wind, and solar power	84%	15%

3. Should the government require by law, encourage with tax breaks but not require, or stay out of entirely?

	Require	Encourage	Stay out of
a. Building cars that use less gasoline	31%	50%	19%
b. Building air conditioners, refrigerators, and other appliances that use less electricity	29%	51%	20%
c. Building new homes and offices that use less energy for heating and cooling	24%	56%	20%

4. U.S. actions to reduce global warming in the future would:
   Hurt U.S. economy (20%)
   Help U.S. economy (56%)
   Not affect economy (23%)

5. The United States should take action on global warming:
   Only if other countries do (14%)
   Even if other countries do less (68%)
   Not take action at all (18%)

6. Most U.S. business leaders _____ the government to do things to stop global warming.
   do want (25%)   do not want (72%)   don't know (3%)

7. How important is the issue of global warming to you personally?
   Extremely or very important (46%)
   Somewhat important (30%)
   Not too important (12%)
   Not at all important (12%)

Source: Adapted from Jon Krosnick. 2010. *Global Warming Poll*. Stanford University. Available at http://woods.stanford.edu. Used by permission.

Fuel Efficiency Standards The Energy Independence and Security Act of 2007, which requires a 40 percent increase in fuel economy to 35 miles per gallon by 2020, lacked details instructing automakers how to comply. In 2009, President Obama signed an order requiring the Department of Transportation (DOT) to enforce the new fuel-economy standards. In 2011, the first-ever national fuel efficiency standards for trucks and buses were finalized, mandating that heavy-duty vehicles made from 2014 to 2018 have higher fuel efficiency and lower emissions. When fully implemented, these new standards are projected to save as much oil as the United States imports from Iraq (Environmental Defense Fund 2011).

Policies on Chemical Safety In 2003, the European Union drafted legislation known as Registration, Evaluation, and Authorization and Restriction of Chemicals (REACH) that requires chemical companies to conduct safety and environmental tests to prove that the chemicals they are producing are safe. If they cannot prove that a chemical is safe, it will be banned from the market (Rifkin 2004). The European Union has become a world leader in environmental stewardship by placing the "precautionary principle" at the center of EU regulatory policy. The precautionary principle requires industry to prove that their products are safe. In contrast, in the United States, chemicals are assumed to be safe unless proven otherwise, and the burden is put on the consumer, the public, or the government to prove that a chemical causes harm.

> **What Do You Think?** In general, the European Union seems to be more concerned than the United States about health effects of chemicals—as evidenced by their stricter controls and bans on chemicals. If the United States had a national health insurance plan similar to that of European countries, in which the federal government paid for health care, do you think the U.S. government would enact tougher controls on hazardous chemicals and other environmental issues?

International Cooperation and Assistance

Global environmental concerns call for global solutions forged through international cooperation and assistance. For example, the 1987 Montreal Protocol on Substances That Deplete the Ozone Layer forged an agreement made by 70 nations to curb the production of CFCs (which contribute to ozone depletion and global warming).

In 1997, delegates from 160 nations met in Kyoto, Japan, and forged the **Kyoto Protocol**—the first international agreement to place legally binding limits on greenhouse gas emissions from developed countries. The United States, the world's largest producer of greenhouse gas emissions, rejected the Kyoto Protocol in 2001. As of August 2011, 191 countries had signed and ratified the Kyoto Protocol; the United States had not ratified the Kyoto Protocol.

Avoiding dangerous climate change will require rich countries to cut carbon emissions at least 80 percent by the end of the 21st century, with cuts of 30 percent by 2020—significantly more than the cuts required under the Kyoto Protocol (United Nations Development Programme 2007). In 2009, world leaders met at a climate summit meeting in Copenhagen, where the United States, China, and dozens of other countries signed on to the Copenhagen Accord—a voluntary agreement to curb climate change by cutting greenhouse gas emissions. But the Copenhagen Accord has no provisions for enforcement of the agreement, and even if each country actually meets its pledge, the pledges are "drastically inadequate" to meet the cuts that climate scientists say are needed (Brecher 2011). Some countries do not have the technical or economic resources to implement the requirements of environmental treaties. Wealthy industrialized countries can help less developed countries address environmental concerns through economic aid. Because industrialized countries have more economic and technological resources, they bear the primary responsibility for leading the nations of the world toward environmental cooperation.

Sustainable Economic Development

Achieving global cooperation on environmental issues is difficult, in part, because developed countries (primarily in the Northern Hemisphere) have different economic agendas from those of developing countries (primarily in the Southern Hemisphere). The northern agenda emphasizes preserving wealth and affluent lifestyles, whereas the southern agenda focuses on overcoming mass poverty and achieving a higher quality of life. Southern countries are concerned that northern industrialized countries—having

Kyoto Protocol The first international agreement to place legally binding limits on greenhouse gas emissions from developed countries.

already achieved economic wealth—will impose international environmental policies that restrict the economic growth of developing countries.

As discussed in Chapter 6, development involves more than economic growth and the alleviation of poverty. The human development approach views the well-being of populations in terms of not only their income, but their access to education, and their ability to lead long, healthy lives in societies that respect and value everyone. The 2010 *Human Development Report* states that "the main threat to maintaining progress in human development comes from the increasingly evident unsustainability of production and consumption" (UNDP 2010, p. 81). The long-term environmental, social, and economic health of societies requires **sustainable development**—development that enables human populations to have fulfilling lives without degrading the planet. "The aim here is for those alive today to meet their own needs without making it impossible for future generations to meet theirs. . . . This in turn calls for an economic structure within which we consume only as much as the natural environment can produce, and make only as much waste as it can absorb" (McMichael et al. 2000, p. 1067).

Sustainable development requires the use of clean, renewable energy. Renewable energy projects in developing countries have demonstrated that providing affordable access to green energy helps to alleviate poverty by providing energy for creating business and jobs and by providing power for refrigerating medicine, sterilizing medical equipment, and supplying fresh water and sewer services needed to reduce infectious disease and improve health (Flavin & Aeck 2005).

The Role of Institutions of Higher Education

Colleges and universities can play an important role in efforts to protect the environment by encouraging use of bicycles on campus, using hybrid and electric vehicles, establishing recycling programs, using local and renewable building materials for new buildings, involving students in organic gardening to provide food for the campus, using clean energy, and incorporating environmental education into the curricula. A growing number of colleges and universities are establishing **Green Revolving Funds** (GRFs), which are funds dedicated to financing cost-saving energy-efficiency upgrades and other projects that decrease resource use and minimize environmental impact (Sustainable Endowments Institute 2011). The resulting savings in operating expenses are returned to the fund and then reinvested in additional projects. In some colleges and universities, students play a role in deciding how Green Revolving Funds are spent and in implementing the various energy upgrades and projects. Nevertheless, David Newport, director of the environmental center at the University of Colorado at Boulder, believes that institutions of higher education are not doing enough to promote sustainability. "We're supposed to be on the leading edge, and we're behind the curve. . . . There are, what 4,500 colleges in the United States, and how many of them are really doing something? Less than 100 or 200?" (quoted by Carlson 2006, p. A10).

Understanding Environmental Problems

Environmental problems are linked to corporate globalization, rapid and dramatic population growth, expanding world industrialization, patterns of excessive consumption, and reliance on fossil fuels for energy. The Global Footprint Network (2010) offers the following analysis of environmental problems:

> Climate change is not the problem. Water shortages, overgrazing, erosion, desertification and the rapid extinction of species are not the problem. Deforestation, reduced cropland productivity and the collapse of fisheries are not the problem. Each of these crises, though alarming, is a symptom of a single, over-riding issue. Humanity is simply demanding more than the earth can provide.

Whether we understand environmental problems as resulting from a complex set of causes, or from one, simple underlying cause such as consuming more than the earth can provide, we cannot afford to ignore the growing evidence of the irreversible effects

sustainable development
Occurs when human populations can have fulfilling lives without degrading the planet.

**Green Revolving Funds
(GRFs)** College and university funds that are dedicated to financing cost-saving energy-efficiency upgrades and other projects that decrease resource use and minimize environmental impacts.

of global warming and loss of biodiversity, and the adverse health effects of toxic waste and other forms of pollution.

Many Americans believe in a "technological fix" for the environment—that science and technology will solve environmental problems. Paradoxically, the same environmental problems that have been caused by technological progress may be solved by technological innovations designed to clean up pollution, preserve natural resources and habitats, and provide clean forms of energy. But leaders of government and industry must have the will to finance, develop, and use technologies that do not pollute or deplete the environment. When asked how companies can produce products without polluting the environment, Robert Hinkley suggested that, first, it must become a goal to do so:

> I don't have the technological answers for how it can be done, but neither did President John F. Kennedy when he announced a national goal to land a man on the moon by the end of the 1960s. The point is that, to eliminate pollution, we first have to make it our goal. Once we've done that, we will devote the resources necessary to make it happen. We will develop technologies that we never thought possible. But if we don't make it our goal, then we will never devote the resources, never develop the technology, and never solve the problem. (quoted by Cooper 2004, p. 11)

But the direction of technical innovation is largely in the hands of big corporations that place profits over environmental protection. Unless the global community challenges the power of transnational corporations to pursue profits at the expense of environmental and human health, corporate behavior will continue to take a heavy toll on the health of the planet and its inhabitants. Because oil has been implicated in political and military conflicts involving the Middle East (see Chapter 15), such conflicts are likely to continue as long as oil plays the lead role in providing the world's energy.

Global cooperation is also vital to resolving environmental concerns but is difficult to achieve because rich and poor countries have different economic development agendas: Developing poor countries struggle to survive and provide for the basic needs of their citizens; developed wealthy countries struggle to maintain their wealth and relatively high standard of living. Can both agendas be achieved without further pollution and destruction of the environment? Is sustainable economic development an attainable goal? With mounting concern about climate change, the health impacts of air pollution, rising oil prices, and the need to ensure energy access to all, governments worldwide have strengthened their commitment to sustainable, renewable energy policies and projects (UNEP 2007).

In the United States, there is significant opposition to governmental regulations designed to address environmental problems. Opponents of regulation argue that rules and restrictions are harmful to the economy and destroy jobs. Yet, an analysis of the costs and benefits of the Clean Air Act Amendments of 1990 found that value of the benefits ($1.3 trillion) was 25 times the cost ($53 billion). In 2010 alone, the Clean Air Act Amendments of 1990 saved an estimated 160,000 lives (Shapiro & Irons 2011). Lax regulation was a contributing factor in the BP Deepwater Horizon oil spill of 2010—the largest oil spill in U.S. history, which had devastating effects on jobs and the economy in the Gulf states. The claim that environmental regulation hurts jobs is not supported by the evidence, and the costs of regulations are far outweighed by the benefits to health, safety, and well-being (Shapiro & Irons 2011). Protecting the environment does not mean that jobs must be sacrificed in the process. Indeed, transitioning to a low-carbon economy can simultaneously protect the environment, create national energy independence, lower energy prices, *and* create jobs (Brecher 2011).

Our collective response to the precarious state of the environment constitutes a test that we cannot afford to fail. As environmentalist Bill McKibben (2008) notes,

> The next few years are a kind of final exam for the human species. Does that big brain really work or not? It gave us the power to build coal-fired power plants and SUVs and thereby destabilize the working of the earth. But does it give us the power to back away from those sources of power, to build a world that isn't bent on destruction? Can we think, and feel, our way out of this, or are we simply doomed to keep acting out the same set of desires for MORE that got us into this fix?

The 2004 Nobel Peace Prize was awarded to Wangari Maathai for leading a grassroots environmental campaign called the Green Belt Movement, which is responsible for planting 30 million trees across Kenya. Maathai is the first person to be awarded the Nobel Peace Prize for environmental work.

© Micheline Pelletier/Corbis

In 2004, the Nobel Peace Prize was given to Wangari Maathai for leading a grassroots environmental campaign to plant 30 million trees across Kenya. This was the first time ever that the Nobel Peace Prize was awarded to someone for accomplishments in restoring the environment. In her acceptance speech, Maathai explained, "A degraded environment leads to a scramble for scarce resources and may culminate in poverty and even conflict" (quoted by Little 2005, p. 2). With ongoing conflict around the globe, it is time for world leaders to recognize the importance of a healthy environment for world peace and to prioritize environmental protection in their political agendas.

CHAPTER REVIEW

• How do free trade agreements pose a threat to environmental protection?
Free trade agreements such as NAFTA and the FTAA provide transnational corporations with privileges to pursue profits, expand markets, use natural resources, and exploit cheap labor in developing countries while weakening the ability of governments to protect natural resources or to implement environmental legislation.

• What are environmental refugees?
Environmental refugees are individuals who have migrated because they can no longer secure a livelihood as a result of deforestation, desertification, soil erosion, and other environmental problems.

• What is greenwashing?
Greenwashing refers to the ways in which environmentally and socially damaging companies portray their corporate image and products as being "environmentally friendly" or socially responsible.

• Where does most of the world's energy come from?
Most of the world's energy comes from fossil fuels, which include oil, coal, and natural gas. This is significant because many of the serious environmental problems in the world today, including global warming and climate change, biodiversity loss, and pollution, stem from the use of fossil fuels.

• What are the major causes and effects of deforestation?
The major causes of deforestation are the expansion of agricultural land, human settlements, wood harvesting, and road building. Deforestation displaces people and wild species from their habitats, contributes to global warming, and contributes to desertification, which results in the expansion of desert land that is unusable for agriculture. Soil erosion caused by deforestation can cause severe flooding.

• What are the effects of air pollution on human health?
Air pollution, which is linked to heart disease, lung cancer, and respiratory ailments, such as emphysema, chronic bronchitis, and asthma kills about 3 million people a year.

• What are some examples of common household, personal, and commercial products that contribute to indoor pollution?
Some common indoor air pollutants include carpeting, mattresses, drain cleaners, oven cleaners, spot removers,

shoe polish, dry-cleaned clothes, paints, varnishes, furniture polish, potpourri, mothballs, fabric softener, caulking compounds, air fresheners, deodorizers, disinfectants, glue, correction fluid, printing ink, carbonless paper, and felt-tip markers.

- **What is the primary cause of global warming?**
The prevailing view on what causes global warming is that greenhouse gases—primarily carbon dioxide, methane, and nitrous oxide—accumulate in the atmosphere and act like the glass in a greenhouse, holding heat from the sun close to the earth. The primary greenhouse gas is carbon dioxide, which is released into the atmosphere by burning fossil fuels.

- **How does global warming contribute to further global warming?**
As global warming melts ice and snow, it exposes more land and ocean area, which absorbs more heat than ice and snow, further warming the planet. The melting of Siberia's frozen peat bog—a result of global warming—could release billions of tons of methane, a potent greenhouse gas, into the atmosphere and cause further global warming. This process, whereby the effects of global warming cause further global warming, is known as a positive feedback loop.

- **What is the relationship between level of economic development and environmental pollution?**
There is a curvilinear relationship between level of economic development and environmental pollution. In regions with low levels of economic development, industrial emissions are minimal, but emissions rise in countries that are in the middle economic development range as they move through the early stages of industrialization. However, at more advanced stages of industrialization, industrial emissions ease because heavy-polluting manufacturing industries decline, "cleaner" service industries increase, and because rising incomes are associated with a greater demand for environmental quality and cleaner technologies.

- **What does the term "environmental injustice" refer to?**
Environmental injustice, also called environmental racism, refers to the tendency for marginalized populations and communities to disproportionately experience adversity due to environmental problems. For example, in the United States, polluting industries, industrial and waste facilities, and transportation arteries (which generate vehicle emissions pollution) are often located in minority communities.

- **What are some of the concerns about nuclear energy?**
Nuclear waste contains radioactive plutonium, a substance linked to cancer and genetic defects. Nuclear waste in the environment remains potentially harmful to human and other life for thousands of years, and disposing of nuclear waste is problematic. Accidents at nuclear power plants, such as the 2011 Fukushima disaster, and the potential for nuclear reactors to be targeted by terrorists add to the actual and potential dangers of nuclear power plants.

- **How much garbage did each person in the United States generate in 1960? 1980? 2009?**
In 1960, each U.S. citizen generated 2.7 pounds of garbage on average every day. This figure increased to 3.7 pounds in 1980, and to 4.3 pounds in 2009.

- **Why are women who may become pregnant, pregnant women, nursing mothers, and young children advised against eating certain types of fish?**
The U.S. Environmental Protection Agency advises women who may become pregnant, pregnant women, nursing mothers, and young children to avoid eating certain fish altogether (swordfish, shark, king mackerel, and tilefish) because of the high levels of mercury.

- **How often does a species of life on earth become extinct?**
One species of life on earth goes extinct every three hours.

- **What social and cultural factors contribute to environmental problems?**
Social and cultural factors that contribute to environmental problems include population growth, industrialization and economic development, and cultural values and attitudes such as individualism, consumerism, and militarism.

- **What are some of the strategies for alleviating environmental problems?**
Strategies for alleviating environmental problems include efforts to lower fertility rates and slow population growth, environmental activism, environmental education, the use of "green" energy, modifications in consumer products and behavior, and government regulations and legislation. Sustainable economic development and international cooperation and assistance also play important roles in alleviating environmental problems.

- **According to 2004 Nobel Peace Prize winner Wangari Maathai, why is environmental protection important for national and international security?**
In her acceptance speech for the 2004 Nobel Peace Prize, Wangari Maathai explained that "a degraded environment leads to a scramble for scarce resources and may culminate in poverty and even conflict."

TEST YOURSELF

1. Tar sands oil is known as the world's _____ oil.
   a. most expensive
   b. least expensive
   c. cleanest
   d. dirtiest

2. Red fire ants are an example of:
   a. A bioinvasion
   b. An extinct species
   c. A threatened species
   d. An alternative fuel

3. Which innovation has resulted in increased food prices?
   a. Solar power
   b. Bottled water
   c. Wind turbines
   d. Biofuels

4. If greenhouse gases were to be stabilized today, global air temperature and sea level would be expected to
   a. Remain at their current level
   b. Decrease immediately
   c. Begin to decrease within 20 years
   d. Continue to rise for hundreds of years

5. The United States has more operating nuclear reactors than any other country.
   a. True
   b. False

6. Most solid waste in the United States is recycled.
   a. True
   b. False

7. In spring 2007, which U.S. city became the first to ban plastic bags from supermarkets and chain pharmacies?
   a. Miami
   b. San Francisco
   c. Atlanta
   d. Portland

8. In the United States, the EPA has required testing on all of the more than 80,000 chemicals that have been on the market since 1976.
   a. True
   b. False

9. Which of the following is the number one polluter in the United States?
   a. The military
   b. Dow Chemical
   c. Archer Daniels Midland
   d. ExxonMobil

10. A growing number of _____ are establishing Green Revolving Funds (GRFs), which are funds dedicated to financing cost-saving energy-efficiency upgrades and other projects that decrease resource use and minimize environmental impact.
   a. State governments
   b. Colleges and universities
   c. Manufacturing industries
   d. Countries

Answers: 1: d; 2: a; 3: d; 4: d; 5: a; 6: b; 7: b; 8: b; 9: a; 10: b.

KEY TERMS

acid rain 414
biodiversity 421
bioinvasion 404
deep ecology 424
deforestation 410
desertification 410
ecosystems 408
ecoterrorism 425
environmental footprint 411
environmental injustice 406

environmental racism 406
environmental refugees 405
e-waste 417
fracking 418
global warming 414
green energy 427
greenhouse gases 414
Green Revolving Funds (GRFs) 433
greenwashing 406
Kyoto Protocol 432

light pollution 421
multiple chemical sensitivity 419
pinkwashing 408
planned obsolescence 405
radical environmental movement 424
sustainable development 433
tar sands 403
tar sands oil 403

MEDIA RESOURCES

Turning to Video

▶❙❙ Watch the BBC video *Powered by Coal* (running time 11:12), available through **CengageBrain.com.** This video describes the environmental and human health problems associated with the use of coal for energy. Would you want to live near a coal power plant?

Online Study Resources

Log in to **www.cengagebrain.com** to access the resources your instructor has assigned. For this book, you can access:

CourseMate

Access chapter-specific learning tools including learning objectives, practice quizzes, videos, Internet exercises, flash cards, and glossaries, as well as web links, and more in your Sociology CourseMate.

Dan McCoy/Rainbow/Getty Images

14

Science and Technology

"We have arranged things so that almost no one understands science and technology. This is a prescription for disaster. We might get away with it for a while, but sooner or later this combustible mixture of ignorance and power is going to blow up in our faces."

—Carl Sagan, astronomer and astrobiologist

CHARLES OKEKE WAS only 30 years old when a blood clot destroyed his heart. He was one of the lucky ones though. In 1996, he had a heart transplant and, for 13 years, the computer consultant lived an idyllic life (Ashton 2010). But in 2008, once again, his heart failed. It was then that Mr. Okeke was outfitted with a "Total Artificial Heart" powered by a 400-pound machine nicknamed "Big Blue." Living in the hospital tethered to a machine for almost two years, the father of three states, "for the longest time I could not physically put my hand to my chest because it felt so weird" (p. 1). In 2010, Charles Okeke became the first person in history to receive the appropriately named "Freedom Driver," a 13-pound portable "Big Blue" replacement, carried in a backpack (SynCardia 2011). In 2011, Charles Okeke received a dual heart and kidney transplant. He is back at home, doing well, and glad to be sleeping in his own bed.

Courtesy of the Mayo Clinic

Pictured, Charles Okeke having dinner with his wife, Natalie, in a dining room at the Mayo Clinic Hospital. Before receiving his "mini" artificial heart, Okeke had to rely on this 400-pound artificial heart to keep him alive. He remained attached to it 24/7, and was unable to leave the hospital for over a year.

Many of the technologies available today seem futuristic. But such technologies—from virtual reality, cloning, and teleportation to artificial hearts—are no longer just the stuff of popular science fiction movies. Virtual reality is now used to train workers in occupations as diverse as medicine, engineering, and professional football. The ability to genetically replicate embryos has sparked worldwide debate over the ethics of reproduction, California Institute of Technology scientists have transported a ray of light from one location to another, and Charles Okeke lived for almost two years with an artificial heart. Just as the telephone, the automobile, the television, and countless other technological innovations have forever altered social life, so will technologies that are more recent.

Science and technology go hand in hand. **Science** is the process of discovering, explaining, and predicting natural or social phenomena. A scientific approach to understanding acquired immunodeficiency syndrome (AIDS), for example, might include investigating the molecular structure of the virus, the means by which it is transmitted, and public attitudes about AIDS. **Technology,** as a form of human cultural activity that applies the principles of science and mechanics to the solution of problems, is intended to accomplish a specific task—in this case, the development of an AIDS vaccine.

Societies differ in their level of technological sophistication and development. In agricultural societies, which emphasize the production of raw materials, the use of tools to accomplish tasks previously done by hand, or **mechanization,** dominates. As societies move toward industrialization and become more concerned with the mass production of goods, automation prevails. **Automation** involves the use of self-operating machines, as in an automated factory in which autonomous robots assemble automobiles. Finally, as a society moves toward postindustrialization, it emphasizes service and information professions (Bell 1973). At this stage, technology shifts toward **cybernation,** whereby machines control machines—making production decisions, programming robots, and monitoring assembly performance.

What are the effects of science and technology on humans and their social world? How do science and technology help to remedy social problems, and how do they contribute to social problems? Is technology, as Postman (from 1992) suggested, both a friend and a foe to humankind? We address each of these questions in this chapter.

science The process of discovering, explaining, and predicting natural or social phenomena.

technology Activities that apply the principles of science and mechanics to the solutions of a specific problem.

mechanization Dominant in an agricultural society, the use of tools to accomplish tasks previously done by hand.

automation Dominant in an industrial society, the replacement of human labor with machinery and equipment that are self-operating.

cybernation Dominant in postindustrial societies, the use of machines to control other machines.

The Global Context:
The Technological Revolution

Less than 50 years ago, traveling across state lines was an arduous task, a long-distance phone call was a memorable event, and mail carriers brought belated news of friends and relatives from far away. Today, travelers journey between continents in a matter of hours, and for many, e-mail, faxes, instant messaging, texting, and electronic fund transfers have replaced previously conventional means of communication.

The world is a much smaller place than it used to be, and it will become even smaller as the technological revolution continues. In 2011, the Internet had 2.1 billion users in more than 200 countries with 240 million users in the United States (Internet Statistics 2011a). Of all Internet users, the highest proportion come from Asia (44.0 percent), followed by Europe (22.7 percent), North America (13.0 percent), Latin America and the Caribbean (10.3 percent), Africa (5.7 percent), the Middle East (3.3 percent), and Oceania/Australia (1.0 percent) (Internet Statistics 2011a).

Although the **penetration rate,** i.e., the percentage of people who have access to and use the Internet in a particular area, is higher in industrialized countries, there is some movement toward the Internet becoming a truly global medium as Africans, Middle Easterners, and Latin Americans increasingly "get online." For example, although Internet use in the United States grew 152 percent between 2000 and 2010, the number of Internet users in Nigeria increased by 21,891 percent during the same time period (Internet Statistics 2011b).

The movement toward globalization of technology is, of course, not limited to the use and expansion of the Internet. The world robot market and the U.S. share of it continues to expand, Microsoft's Internet platform and support products are sold all over the world, scientists collect skin and blood samples from remote islanders for genetic research, a global treaty regulating trade of genetically altered products has been signed by more than 100 nations, and Intel's central processing units (CPUs) power an estimated 80 percent of the world's personal computers (PCs) (Randewich 2011).

To achieve such scientific and technological innovations, sometimes called research and development (R&D), countries need material and economic resources. *Research* entails the pursuit of knowledge; *development* refers to the production of materials, systems, processes, or devices directed to the solution of practical problems. According to the National Science Foundation (NFS), the United States spends over $398 billion a year on research and development, 40 percent of the world's R&D expenditures (Galama & Hosek 2008; NSF 2010). As in most other countries, U.S. funding sources are primarily from private industry, 67 percent of the total, followed by the federal government and nonprofit organizations such as research institutes at colleges and universities (NSF 2010).

The United States leads the world in science and technology, although there is some evidence that we are falling behind (Price 2008; Dutta & Mia 2011). A report by the Information Technology and Innovation Foundation (ITIF) concludes that, in 2011, the United States, when compared to 44 other countries and regions (ITIF 2011),

- Ranked 4th overall in global competitiveness behind Singapore, Finland, and Sweden.
- Ranked 6th in the rate of science and technology researchers but 39th in the increase of science and technology researchers between 1999 and 2008.
- Ranked 14th in the number and quality of science and technology publications but 36th in the increase in the number and quality of science and technology publications between 1996 and 2009.
- Ranked 5th in corporate investments in information technology but 21st in the increase in corporate investment in information technology between 1999 and 2008.
- Ranked 11th in broadband telecommunications, but 21st in the increase in broadband telecommunications between 2002 and 2009.

Interestingly, if U.S. states were treated as countries in terms of global competitiveness, Massachusetts would be the most competitive country in the world, followed by California and Connecticut.

penetration rate The percentage of people who have access to and use the Internet in a particular area.

The decline of U.S. supremacy in science and technology is likely to be the result of several interacting forces (Lemonick 2006; ITIF 2009; Price 2008; World Bank 2009). First, the federal government has been scaling back its investment in research and development in response to fiscal deficits. Second, corporations, the largest contributors to research and development, have begun to focus on short-term products and higher profits as pressure from stockholders mounts. Third, developing countries, most notably China and India, are expanding their scientific and technological capabilities at a faster rate than the United States. Although the United States is ranked 27th in its change score for business research and development investments between 1999 and 2008, China is ranked second (ITIF 2011).

Fourth, there has been a drop in science and math education in U.S. schools, both in terms of quality and quantity. The United States still grants the highest proportion of science and engineering PhD degrees in the world in recent years, but the number has declined, whereas other countries' rates have increased. In response to such concerns, in a 2009 letter to President Obama, the chair of the National Science Board recommended that:

> . . . the new Administration . . . advance **STEM** (science, technology, engineering, and mathematics) education for all American students, to nurture innovation, and to ensure the long-term economic prosperity of the Nation. The urgency of this task is underscored by the need to ensure that the United States continues to excel in science and technology . . . and [to] guarantee that all American students are provided the educational resources and tools needed to participate fully in the science and technology–based economy of the 21st century. (Kalil 2009, p. 1)

Finally, Mooney and Kirshenbaum (2009) document "unscientific America"—the tremendous disconnect between the citizenry, media, politicians, religious leaders, education, and the entertainment industry (e.g., *CSI, Grey's Anatomy*) on the one hand, and science and scientists on the other. Post-World War II America, in part because of the Cold War, invested in R&D, leading to such scientific and technological advances as the space program, the development of the Internet, and the decoding of the genome. Further, a 2010 survey indicates that Americans, when asked to rank order 16 elements of

> . . . the United States still grants the highest proportion of science and engineering PhD degrees in the world in recent years, but the number has declined, whereas other countries' rates have increased.

STEM An acronym for science, technology, engineering and mathematics.

Forty-four percent of all Internet users live in Asia. That translates to nearly 1 billion people, nearly half of whom live in China. The sheer number of people online in China is a function of China's population. However, China's penetration rate is only 35 percent compared to, for example, Japan that has 78 percent of its population online.

© baobao ou/Getty Images

Answer each of the following questions. Calculate the number of questions you answered correctly and then compare your score to a sample of 1,005 U.S. adults who were asked exactly the same questions during a phone survey in 2009.

1. Which over-the-counter drug do doctors recommend that people take to help prevent heart attacks?
   a. Antacids
   b. Cortisone
   c. Aspirin
   d. Allergy medications

2. According to most astronomers, which of the following is no longer considered a planet?
   a. Neptune
   b. Pluto
   c. Saturn
   d. Mercury

3. Which of the following may cause a tsunami?
   a. A very warm ocean current
   b. A large school of fish
   c. A melting glacier
   d. An earthquake under the ocean

4. The global positioning system, or GPS, relies on which of these to work?
   a. Satellites
   b. Stars
   c. Magnets
   d. Lasers

5. What gas do most scientists believe causes temperatures in the atmosphere to rise?
   a. Hydrogen
   b. Helium
   c. Carbon dioxide
   d. Radon

6. How are stem cells different from other cells?
   a. They can develop into many different types of cells
   b. They are found only in bone marrow
   c. They are found only in plants

7. What have scientists recently discovered on Mars?
   a. Platinum
   b. Plants
   c. Mold
   d. Water

8. The continents on which we live have been moving their location for millions of years and will continue to move in the future.
   a. True
   b. False

9. Antibiotics will kill viruses as well as bacteria.
   a. True
   b. False

10. Electrons are smaller than atoms.
    a. True
    b. False

11. Lasers work by focusing sound waves.
    a. True
    b. False

12. All radioactivity is man-made.
    a. True
    b. False

Answers: 1: c; 2: b; 3: d; 4: a; 5: c; 6: a; 7: d; 8: a; 9: b; 10: a; 11: b; 12: b.

Source: Pew Research Center, 2009a. "Public Praises Science; Scientists Fault Public, Media." July 9, Available at. http://pewresearch.org/sciencequiz

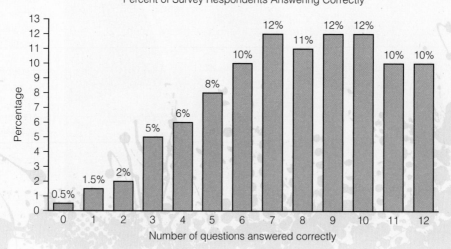

Percent of Survey Respondents Answering Correctly

American life, assign science and technology the highest ranking (Harris Poll 2010). Yet, despite these significant contributions and primarily positive attitudes toward science and scientists, most Americans know very little about science (see this chapter's *Self and Society*), and there are divergent patterns of thinking in reference to some of today's most important issues (see Table 14.1). For example, 32 percent of the American public, compared to 87 percent of American scientists, believe that humans evolved from other living things due to a natural process (Pew 2009a).

What Do You Think? Scientific discoveries and technological developments require the support of a country's citizens and political leaders. For example, although abortion has been technically possible for years, millions of the world's citizens live in countries where abortion is either prohibited or permitted only when the life of the mother is in danger. Can you name other scientific discoveries or technological developments that are technically possible, but likely to be rejected by large segments of the population?

Postmodernism and the Technological Fix

Many Americans believe that social problems can be resolved through a **technological fix** (Weinberg 1966) rather than through social engineering. For example, a social engineer might approach the problem of water shortages by persuading people to change their lifestyle: use less water, take shorter showers, and wear clothes more than once before washing. A technologist would avoid the challenge of changing people's habits and motivations and instead concentrate on the development of new technologies that would increase the water supply.

Social problems can be tackled through both social engineering and a technological fix. In recent years, for example, social engineering efforts to reduce drunk driving have included imposing stiffer penalties for drunk driving and disseminating public service announcements, such as "Friends don't let friends drive drunk." An example of a technological fix for the same problem is the development of car airbags, which reduce injuries and deaths resulting from car accidents.

Not all individuals, however, agree that science and technology are good for society. **Postmodernism,** an emerging worldview holds that rational thinking and the scientific perspective have fallen short in providing the "truths" they were once presumed to hold. During the industrial era, science, rationality, and technological innovations were thought to hold the promises of a better, safer, and more humane world. Today, postmodernists question the validity of the scientific enterprise, often pointing to the unforeseen and unwanted consequences of resulting technologies. Automobiles, for example, began to be mass-produced in the 1930s, in response to consumer demands. But the proliferation of automobiles has also led to increased air pollution and the deterioration of cities as suburbs developed and, today, traffic fatalities are the number one cause of accident-related deaths.

TABLE 14.1 General Public and Scientists Views on Selected Science and Technology Issues, 2009

	PUBLIC (%)	SCIENTISTS (%)
Think that humans, other living things have evolved due to natural processes	32	87
Think that earth is getting warmer because of human activity	49	84
Favor use of animals in scientific research	52	93
Favor federal funding for embryonic stem cell research	58	93
Favor building more nuclear power plants	51	70
Say that all parents should be required to vaccinate their children	69	82

Source: Pew 2009a. From "Public Praises Science; Scientists Fault Public, Media: Scientific Achievements Less Prominent than a Decade Ago", July 9, 2009, Pew Research Center For the People & the Press, a project of the Pew Research Center. Reprinted with permission.

technological fix The use of scientific principles and technology to solve social problems.

postmodernism A worldview that questions the validity of rational thinking and the scientific enterprise.

Sociological Theories of Science and Technology

Each of the three major sociological frameworks helps us to better understand the nature of science and technology in society.

The world was made a smaller place by the Pony Express in the late 1800s. Today, the iPhone, combining a number of technological feats, makes the world even smaller.

Structural-Functionalist Perspective

Structural functionalists view science and technology as emerging in response to societal needs—that "science was born indicates that society needed it" (Durkheim 1973/1925). As societies become more complex and heterogeneous, finding a common and agreed-on knowledge base becomes more difficult. Science fulfills the need for an assumed objective measure of "truth" and provides a basis for making intelligent and rational decisions. In this regard, science and the resulting technologies are functional for society.

If society changes too rapidly as a result of science and technology, however, problems may emerge. When the material part of culture (i.e., its physical elements) changes at a faster rate than the nonmaterial part (i.e., its beliefs and values), a **cultural lag** may develop (Ogburn 1957). For example, the typewriter, the conveyor belt, and the computer expanded opportunities for women to work outside the home. With the potential for economic independence, women were able to remain single or to leave unsatisfactory relationships and/or establish careers. But although new technologies have created new opportunities for women, beliefs about women's roles, expectations of female behavior, and values concerning equality, marriage, and divorce have lagged behind.

Robert Merton (1973), a structural functionalist and founder of the subdiscipline sociology of science, also argued that scientific discoveries or technological innovations may be dysfunctional for society and may create instability in the social system. For example, the development of time-saving machines increases production, but it also displaces workers and contributes to higher rates of employee alienation. Defective technology can have disastrous effects on society. In 2007, approximately 17,000 passengers were grounded at the Los Angeles airport because of a network card that sent incorrect information to the United States Customs and Border Protection network. For nearly eight hours, no one could leave or enter the United States (Associated Press 2007).

Conflict Perspective

Conflict theorists, in general, argue that science and technology benefit a select few. For some conflict theorists, technological advances occur primarily as a response to capitalist needs for increased efficiency and productivity and thus are motivated by profit. As McDermott (1993) noted, most decisions to increase technology are made by "the immediate practitioners of technology, their managerial cronies, and for the profits accruing to their corporations" (p. 93). In the United States, private industry spends more money on research and development than the federal government does. The Dalkon Shield (IUD)

cultural lag A condition in which the material part of culture changes at a faster rate than the nonmaterial part.

and silicone breast implants are examples of technological advances that promised millions of dollars in profits for their developers. However, the rush to market took precedence over thorough testing of the products' safety. Subsequent lawsuits filed by consumers who argued that both products had compromised the physical well-being of women resulted in large damage awards for the plaintiffs.

Science and technology also further the interests of dominant groups to the detriment of others. The need for scientific research on AIDS was evident in the early 1980s, but the required large-scale funding was not made available so long as the virus was thought to be specific to homosexuals and intravenous drug users. Only when the virus became a threat to mainstream Americans were millions of dollars allocated to AIDS research. Hence, conflict theorists argue that granting agencies act as gatekeepers to scientific discoveries and technological innovations. These agencies are influenced by powerful interest groups and the marketability of the product rather than by the needs of society.

When the dominant group feels threatened, it may use technology as a means of social control. For example, the use of the Internet is growing dramatically in China, the world's largest Internet market. Censorship has been consolidated under the State Internet Information Office, also known as the "great firewall of China" (Chen 2011). A study by Harvard Law School researchers indicates that, of the 204,000 websites accessed, nearly 20,000 were inaccessible. Top Google search results for such words as "Tibet," "equality," "Taiwan," and "democracy China" were consistently blocked (Associated Press 2010). China is not alone, however. A study by OpenNet found that Saudi Arabia, Burma, Iran, Vietnam, Uzbekistan, Afghanistan, and Turkmenistan also have pervasive censorship of the Internet (OpenNet 2011).

Finally, conflict theorists as well as feminists argue that technology is an extension of the patriarchal nature of society that promotes the interests of men and ignores the needs and interests of women. As in other aspects of life, women play a subordinate role in reference to technology in terms of both its creation and its use. For example, washing machines, although time-saving devices, disrupted the communal telling of stories and the resulting friendships among women who gathered together to do their chores. Bush (1993) observed that, in a "society characterized by a sex-role division of labor, any tool or technique . . . will have dramatically different effects on men than on women" (p. 204).

Symbolic Interactionist Perspective

Knowledge is relative. It changes over time, over circumstances, and between societies. We no longer believe that the world is flat or that the Earth is the center of the universe, but such beliefs once determined behavior because individuals responded to what they thought to be true. The scientific process is a social process in that "truths"—socially constructed truths—result from the interactions between scientists, researchers, and the lay public.

Kuhn (1973) argued that the process of scientific discovery begins with assumptions about a particular phenomenon (e.g., the world is flat). Because unanswered questions always remain about a topic (e.g., why don't the oceans drain?), science works to fill these gaps. When new information suggests that the initial assumptions were incorrect (e.g., the world is not flat), a new set of assumptions or framework emerges to replace the old one (e.g., the world is round). It then becomes the dominant belief or paradigm.

Symbolic interactionists emphasize the importance of this process and the effect that social forces have on it. Lynch et al. (2008) describe the media's contribution in framing societal beliefs about racial discrimination, racism, and genetic determinism. Social forces also affect technological innovations, and their success depends, in part, on the social meaning assigned to any particular product. As social constructionists argue, individuals socially construct reality as they interpret the social world around them, including the meaning assigned to various technologies. If claims makers can successfully define a product as impractical, cumbersome, inefficient, or immoral, the product is unlikely to gain public acceptance. Such is the case with RU-486, an oral contraceptive known as the "abortion pill" that is widely used in France, Great Britain, and China, but availability of which, although legal in the United States, is opposed by a majority of Americans (Gallup 2000; Gottlieb 2000). In 2009, over objections from the Vatican, RU-486 was approved for use in Italy (Rizzo 2009).

Cyberstalking, pornography on the Internet, and identity theft are crimes that were unheard of before the computer revolution and the enormous growth of the Internet. One such "high-tech" crime, computer hacking, ranges from childish pranks to deadly viruses that shut down corporations. In this classic study, Jordan and Taylor (1998) enter the world of hackers, analyzing the nature of this illegal activity, hackers' motivations, and the social construction of the "hacking community."

Sample and Methods

Jordan and Taylor (1998) researched computer hackers and the hacking community through 80 semi-structured interviews, 200 questionnaires, and an examination of existing data on the topic. As is often the case in crime, illicit drug use, and other similarly difficult research areas, a random sample of hackers was not possible. Snowball sampling is often the preferred method in these cases; that is, one respondent refers the researcher to another respondent, who then refers the researcher to another respondent, and so forth. Through their analysis, the investigators provide insight into this increasingly costly social problem and the symbolic interactionist notion of "social construction"—in this case, of an online community.

Findings and Conclusions

Computer hacking, or "unauthorized computer intrusion," is an increasingly serious problem, particularly in a society dominated by information technologies. Unlawful entry into computer networks or databases can be achieved by several means, including (1) guessing someone's password, (2) tricking a computer about the identity of another computer (called "IP spoofing"), or (3) "social engineering," a slang term referring to getting important access information by stealing documents, looking over someone's shoulder, going through their garbage, and so on.

Hacking carries with it certain norms and values, because, according to Jordan and Taylor (1998), the hacking community can be thought of as a culture within a culture. The two researchers identified six elements of this socially constructed community:

- *Technology*. The core of the hacking community is the technology that allows it to occur. As one professor who was interviewed stated, the young today have "lived with computers virtually from the cradle, and therefore have no trace of fear, not even a trace of reverence."
- *Secrecy*. The hacking community must, on the one hand, commit secret acts because their "hacks" are illegal. On the other hand, much of the motivation for hacking requires publicity to achieve the notoriety often sought. In addition, hacking is often a group activity that bonds members together. As one hacker stated, hacking "can give you a real kick some time. But it can give you a lot more satisfaction and recognition if you share your experiences with others."
- *Anonymity*. Whereas secrecy refers to the hacking act, anonymity refers to the importance of the hacker's identity remaining unknown. Thus, for example, hackers and hacking groups take on names such as Legion of Doom, the Inner Circle I, Mercury, and Kaos, Inc.
- *Membership fluidity*. Membership is fluid rather than static, often characterized by high turnover rates, in part, as a response to law enforcement pressures. Unlike more structured organizations, there are no formal rules or regulations.
- *Male dominance*. Hacking is defined as a male activity; consequently, there are few female hackers. Jordan and Taylor (1998) also note, after recounting an incident of sexual harassment, that "the collective identity hackers share and construct . . . is in part misogynist" (p. 768).
- *Motivation*. Contributing to the articulation of the hacking communities' boundaries are the agreed-upon definitions of acceptable hacking motivations, including (1) addiction to computers, (2) curiosity, (3) excitement, (4) power, (5) acceptance and recognition, and (6) community service through the identification of security risks.

Finally, Jordan and Taylor (1998, p. 770) note that hackers also maintain group boundaries by distinguishing between their community and other social groups, including "an antagonistic bond to the computer security industry (CSI)." Ironically, hackers admit a desire to be hired by the CSI, which would not only legitimize their activities but also give them a steady income.

Jordan and Taylor conclude that the general fear of computers and of those who understand them underlies the common, although inaccurate, portrayal of hackers as pathological, obsessed computer "geeks." When journalist Jon Littman asked hacker Kevin Mitnick if he was being demonized because of increased dependence on and fear of information technologies, Mitnick replied, "Yeah. . . . That's why they're instilling fear of the unknown. That's why they're scared of me. Not because of what I've done, but because I have the capability to wreak havoc" (Jordan & Taylor 1998, p. 776).

Not only are technological innovations subject to social meaning, but who becomes involved in what aspects of science and technology is also socially defined. Men, for example, far outnumber women in earning computer science degrees, as many as ten to one at some schools. Further, although women make up 47.3 percent of the general workforce, they make up just 27 percent of computer scientists and 20 percent of computer programmers (U.S. Bureau of the Census 2011a). Societal definitions of men as being rational, mathematical, and scientifically minded and as having greater mechanical aptitude than women are, in part, responsible for these differences. This chapter's *Social Problems Research Up Close* feature highlights one of the consequences of the masculinization of technology, as well as the ways in which computer hacker identities and communities are socially constructed.

Technology and the Transformation of Society

A number of modern technologies are considerably more sophisticated than technological innovations of the past. Nevertheless, older technologies have influenced the nature of work as profoundly as the most mind-boggling modern inventions. Postman (1992) described how the clock—a relatively simple innovation that is taken for granted in today's world—profoundly influenced the workplace and with it the larger economic institution:

> The clock had its origin in the Benedictine monasteries of the twelfth and thirteenth centuries. The impetus behind the invention was to provide a more or less precise regularity to the routines of the monasteries, which required, among other things, seven periods of devotion during the course of the day. The bells of the monastery were to be rung to signal the canonical hours; the mechanical clock was the technology that could provide precision to these rituals of devotion. . . . What the monks did not foresee was that the clock is a means not merely of keeping track of the hours but also of synchronizing and controlling the actions of men. And thus, by the middle of the fourteenth century, the clock had moved outside the walls of the monastery, and brought a new and precise regularity to the life of the workman and the merchant. . . . In short, without the clock, capitalism would have been quite impossible. The paradox . . . is that the clock was invented by men who wanted to devote themselves more rigorously to God; it ended as the technology of greatest use to men who wished to devote themselves to the accumulation of money. (pp. 14–15)

Today, technology continues to have far-reaching effects not only on the economy but also on every aspect of social life. In the following section, we discuss societal transformations resulting from various modern technologies including workplace technology, computers, the Internet, and science and biotechnology.

> . . . technology continues to have far-reaching effects not only on the economy but also on every aspect of social life.

Technology and the Workplace

All workplaces—from government offices to factories and from supermarkets to real estate agencies—have felt the impact of technology. Some technology lessens the need for supervisors and makes control by employers easier. For example, employees in the Department of Design and Construction in New York City must scan their hands each time they enter or leave the workplace. The use of identifying characteristics such as hands, fingers, and eyes is part of a technology called *biometrics*. Union leaders "called the use of biometrics degrading, intrusive and unnecessary and said that experimenting with the technology could set the stage for a wider use of biometrics to keep tabs on all elements of the workday" (Chan 2007, p. 1).

Technology can also make workers more accountable by gathering information about their performance. Further, through time-saving devices such as personal digital assistants (PDAs) and battery-powered store-shelf labels, technology can enhance workers' efficiency. New medical software marketed by Wal-Mart at a cost of $25,000 a year for one physician and $10,000 a year for each additional physician in a practice will not only save time and save money on costly recordkeeping but is also likely to improve patient care (Lohr 2009).

However, technology can also contribute to worker error. In a recent study of a popular hospital computer system, researchers found several ways that the computerized drug-ordering program endangered the health of patients. For example, the software program warned a doctor of a patient's drug allergy only *after* the drug was ordered and, rather than showing the usual dose of a particular drug, the program showed the dosage available in the hospital pharmacy (DeNoon 2005).

Technology is also changing the location of work (see Chapter 7). In 2011, American Telephone and Telegraph (AT&T) conducted a national survey of small business owners, and 40 percent reported that mobile devices allowed their employees to work away

Chris Willson/Alamy

from the office (AT&T 2011). The number of telecommuters, however, is difficult to estimate. If telecommuters are defined as anyone who works "primarily from the home," the number may be as low as 2.9 million. If telecommuters are defined as anyone who "works from home at least once a month," then the estimate may be as high as 33.7 million (Meinert 2011).

Telecommuting has increased as the result of several interacting social forces over time. First, a troubled economy and escalating gas prices make telecommuting a rational response to increase costs. Second, high-speed Internet and the increase in wireless access make telecommuting easier and more efficient. Finally, telecommuting is growing as an employee option as employers increasingly embrace the importance of balancing work and family obligations (World at Work 2009; Meinert 2011).

Telepresencing, a much more technologically sophisticated version of teleconferencing, allows life-sized participants in the virtual presence of one another to realistically communicate through broadcast quality sound and images (Houlahan 2006; Sharkey 2009). The telepresence industry includes such giants as Microsoft and Cisco Systems, both of which have invested millions of dollars into the R&D of this new technology.

Robots are no longer just the lead characters in movies, video games, and comic books. Pictured here, a robot greets visitors at Tokyo International Airport. Robots are also used at the airport as porters to carry heavy luggage.

> **What Do You Think?** Telepresencing is no longer confined to the boardroom. Segway-like robots that allow real-time face-to-face conversations for anyone that can afford the $3,000-plus price tag are now available. Telepresence robots represent you anywhere from work conferences to meetings with the boss (Ganapati 2010). Using a combination of sensors, cameras, speakers, and microphones, telepresence robots move smoothly and quietly, and are easy to drive—all you need is a computer and a browser. How would you feel if you signed up for a face-to-face class and your professor turned out to be a telepresence robot?

Robotic technology has also revolutionized work. Although the economic downturn of 2009 and 2010 led to a reduction in sales of robotic equipment, 2011 sales are promising (IFR 2011). Ninety percent of robots work in factories, and more than half of these are used in heavy industry, such as automobile manufacturing. Robots are most commonly used for materials handling and spot welding, although there has recently been a trend toward integrating them more fully into the manufacturing process (Pethokoukis 2004). An employer's decision to use robotics depends on direct costs (e.g., initial investment) and indirect costs (e.g., unemployment compensation), the feasibility and availability of robots to perform the desired tasks, and the resulting increased rate of productivity. Use of robotics may also depend on whether labor unions resist the replacement of workers by machines.

Technology has also changed the nature of work. Federal Express not only created a FedEx intranet for its employees, they allow customers to enter their package-tracking database, saving the company millions of dollars a year. The used car industry has been revolutionized by the disabler—a remote device wired to a car's ignition system. When a customer fails to make a car payment on time, the device prevents the car from starting. Because the device has made it less likely that borrowers will default on their loans, there is some evidence that dealerships have become more willing to qualify low-income, low-credit customers (Welsh 2009).

telepresencing A sophisticated technology that allows life-sized participants in the virtual presence of one another to realistically communicate through broadcast quality sound and images.

Automation means that machines can now perform the labor originally provided by human workers, such as robots performing tasks on automobile assembly lines.

© Adam Lubroth/Riser/Getty Images

The Computer Revolution

Early computers were much larger than the small machines we have today and were thought to have only esoteric uses among members of the scientific and military communities. In 1951, only about a half dozen computers existed (Ceruzzi 1993). The development of the silicon chip and sophisticated microelectronic technology allowed tens of thousands of components to be imprinted on a single chip smaller than a dime. The silicon chip led to the development of laptop computers, cellular phones, digital cameras, the iPad, and portable DVDs. The silicon chip also made computers affordable. Although the first PC was developed only 20 years ago, today 77 percent of adult Americans use a computer, at least occasionally, compared to 65 percent of adult Americans in 2000 (U.S. Bureau of the Census 2011a) (see Table 14.2).

Americans are more likely to use computers at home rather than at work (U.S. Bureau of the Census 2011a). As with computer use in general, computer use in these two locations is associated with demographic variables. With the exception of age (computer use at home is highest for 15- to 24-year-olds, whereas computer use at work is highest for 35- to 44-year-olds), computer use at home and at work follows the same pattern of overall computer use—the more educated you are and the higher your income, the higher the probability that you use a computer at home or at work (U.S. Bureau of the Census 2009; 2011a). The most common computer activities include accessing the Internet, sending e-mail, using a search engine, getting the news or e-mail, followed by word processing, working with spreadsheets or databases, and accessing or updating calendars or schedules.

More than 76.6 percent of American households own computers (BLS 2010). Desktop ownership has decreased since 2006, and the proportion of Americans who own laptop computers has increased from 30 percent to 52 percent between 2006 and 2010 (Smith 2010). Globally, Israel has the highest rate of computer ownership (122 computers for every 100 people), and Honduras has one of the lowest, with just 2.5 computers for every 100 people (*The Economist* 2008; U.S. Bureau of the Census 2011a).

Not surprisingly, computer education has also mushroomed in the last two decades. In 1980, 11,154 U.S. college students earned a bachelor's degree in computer and information sciences; by 2006, that number had increased to 38,476 (U.S. Bureau of the Census 2011a). Universities increasingly require their students to have laptop computers, provide wireless campus corridors, and spend millions of dollars on hardware and software.

TABLE 14.2 Computer and Internet Use by Select Characteristics, 2000 and 2010*

CHARACTERISTIC	ADULT COMPUTER USERS (%)		ADULT INTERNET USERS (%)	
	2000	2010	2000	2010
Total Adults	65	77	53	79
Age				
18 to 29 years old	82	89	72	95
30 to 49 years old	76	86	62	87
50 to 64 years old	61	78	48	78
65 years old and over	21	42	15	42
Sex				
Female	64	76	51	79
Male	66	78	56	79
Race/Ethnicity				
White, non-Hispanic	66	79	55	80
Black, non-Hispanic	59	72	42	71
English-speaking Hispanic	64	74	48	82
Educational				
Less than high school	28	47	19	52
High school graduate	56	67	41	67
Some college	80	89	69	90
College graduate or higher	88	94	79	96
Annual Household Income				
Less than $30,000	48	58	35	63
$30,000 to $49,999	74	82	61	84
$50,000 to $74,999	85	89	74	89
$75,000 or more	90	96	81	95

*Adults who use a computer or the Internet at home, work, school, or anywhere else at least on an occasional basis.

Source: U.S. Bureau of the Census 2011a.

Computers are big business, and the United States is one of the most successful producers of computer technology in the world, boasting several of the top companies—Dell, Acer, Hewlett-Packard, Toshiba, and Apple. Retail sales of computers exceed $20.3 billion annually, with Americans spending more than $7.6 billion on software alone (U.S. Bureau of the Census 2009). Spending on computers, although recently slowing due to the struggling economy (Vance 2009a), is beginning to show signs of recovery (U.S. Bureau of the Census 2011b). For example, Apple posted record revenues in 2010, on the heels of releasing several upgraded portables including the iPad, the iPhone 4, and the MacBook Pro (Michaels & Snell 2010). Other computer companies, including Dell and IBM, also saw double-digit increases in 2011 (Kopytoff 2011).

Computer software is, as noted, also big business and, in some cases, too big. In 1999, a federal judge found that Microsoft Corporation was in violation of antitrust laws, laws that prohibit unreasonable restraint of trade (*U.S. v. Microsoft Corporation 1999*). At issue were Microsoft's Windows operating system and the vast array of Windows-based applications (e.g., spreadsheets, word processors, tax software)—applications that *only* work with Windows. As a result of the appellate process and other delays, Microsoft's compliance with the final judgment remains incomplete and continues to be monitored by the courts (U.S. Department of Justice 2011).

Novell, alleging its word-processing application WordPerfect was unfairly "squeezed out" of the market, also sued Microsoft. The case was dismissed in 2004, but overturned at the appellate level, leading to the present litigation between the two rivals (Rigby & Chang 2011). Further, Microsoft is a defendant in a patent violation case along with Apple, Inc., LG Electronics, and several other defendants. The legal issue concerns whether or not the accused violated a patent relating to touch-pad technology (Magee 2009). Finally, Google has filed suit against the U.S. Department of the Interior, alleging that Microsoft's e-mail application Outlook was the only one considered by the government, thereby unduly restricting competition in violation of federal law that requires "open and competitive procedures in soliciting contracts" (Sarno 2010, p.1).

Information and Communication Technology and the Internet

Information and communication technology, or ICT, refers to any technology that carries information. Most information technologies were developed within a 100-year span: taking pictures and telegraphy (1830s), rotary power printing (1840s), the typewriter (1860s), transatlantic cable (1866), the telephone (1876), motion pictures (1894), wireless telegraphy (1895), magnetic tape recording (1899), radio (1906), and television (1923) (Beniger 1993). The concept of an "information society" dates back to the 1950s, when an economist identified a work sector he called "the production and distribution of knowledge." In 1958, 31 percent of the labor force was employed in this sector—today, more than 50 percent is. When this figure is combined with those in service occupations, more than 75 percent of the labor force is involved in the information society.

The **Internet** is an international information infrastructure—a network of networks—available through universities, research institutes, government agencies, libraries, and businesses. In 2010, 79 percent of all Americans used the Internet from some location compared to 53 percent in 2000 (see Table 14.2). U.S. users are equally likely to be male or female but are more likely to be under 50 years of age, to have a college degree, and to have an annual household income of $75,000 or more (U.S. Bureau of the Census 2011a). The most active Internet users are connected to broadband—that is, services that provide high-speed (DSL and cable) access rather than dial-up service—and are high-income young males (McGann 2005; U.S. Bureau of the Census 2011a). About 55 percent of all Internet users have used wireless connections at home, at the office, or at some other location to log on to the Internet (Pew 2009b).

The Internet, or the "Web" as it is most commonly known, has evolved to what is now called **Web 2.0**—a platform for millions of users to express themselves online in the common areas of cyberspace (Grossman 2006; Pew 2007). The development of Web 2.0 is a story about

> community and collaboration on a scale never seen before. It's about the cosmic compendium of knowledge that is Wikipedia and the million-channel people's network of YouTube and the online metropolis MySpace. It's about the many wresting power from the few and helping one another for nothing and how that will not only change the world, but also change the way the world changes. (Grossman 2006, p. 1)

Wireless access to the Internet is also altering Internet use. The growth of mobile Internet capabilities has led to "always-present" connectivity as users working on laptops and BlackBerrys, iPhones, and iPads access the Internet from coffee shops, classrooms, and shopping malls (Horrigan 2009). Although varying significantly in motivation for use, attitudes toward the Internet, and demographic characteristics, 39 percent of Americans have wireless access to the Internet. Figure 14.1 displays Internet activities of Americans by generational category (Pew 2010).

E-Commerce **E-commerce** is the buying and selling of goods and services over the Internet. Despite a slowdown in the economy, or perhaps because of it, online business-to-customer sales increased 2.1 percent between 2008 and 2009, and now represent 47 percent of all retail sales in the United States (U.S. Bureau of the Census 2011c). The largest category of business-to-customer online sales are for drugs, health aids, and beauty aids (29.0 percent), followed by miscellaneous items (e.g., jewelry, collectibles) (13.4 percent), clothing, clothing accessories, and footwear (10.6 percent), and computer hardware (10.2 percent) (U.S. Bureau of the Census 2011a). Most (about 35 percent) business-to-customer online sales are initiated through online searches by two distinct groups of shoppers—bargain hunters looking for the best price and convenience shoppers who are more interested in saving time than money.

> The growth of mobile Internet capabilities has led to "always-present" connectivity as users working on laptops and BlackBerrys, iPhones, and iPads access the Internet from coffee shops, classrooms, and shopping malls.

What Do You Think? The advent and proliferation of Wi-Fi (i.e., wireless access to the Internet) has facilitated a variety of Internet software and hardware innovations such as smartphones, the iPad 2, netbooks, and thousands of downloadable applications ("apps"). It has even led to institutional shifts; for example, e-commerce now includes m-commerce or the ability to make financial transactions (e.g., mobile banking) from a mobile device such as a cell phone. Do you think that the advantages of 24/7 access to the Internet outweigh the disadvantages that accompany "always-present" technologies?

Internet An international information infrastructure available through universities, research institutes, government agencies, libraries, and businesses.

Web 2.0 A platform for millions of users to express themselves online in the common areas of cyberspace.

e-commerce The buying and selling of goods and services over the Internet.

Figure 14.1:
Online Activities by Generation, 2010
Source: Pew 2010. From "Generations 2010", December 16, 2010, Pew Research Center For the People & the Press, a project of the Pew Research Center. Reprinted with permission.

Key:

90–100%	40–49%
80–89%	30–39%
70–79%	20–29%
60–69%	10–19%
50–59%	0–9%

Millennials Ages 18-33	Gen X Ages 34-45	Younger Boomers Ages 46-55	Older Boomers Ages 56-64	Silent Generation Ages 65-73	G.I. Generation Age 74+
Email	Email	Email	Email	Email	Email
Search	Search	Search	Search	Search	Search
Health info	Health info	Health info	Health info	Health info	Health info
Social network sites	Get news	Get news	Get news	Get news	Buy a product
Watch video	Govt website	Govt website	Govt website	Travel reservations	Get news
Get news	Travel reservations	Travel reservations	Buy a product	Buy a product	Travel reservations
Buy a product	Watch video	Buy a product	Travel reservations	Govt website	Govt website
IM	Buy a product	Watch video	Bank online	Watch video	Bank online
Listen to music	Social network sites	Bank online	Watch video	Financial info	Financial info
Travel reservations	Bank online	Social network sites	Social network sites	Bank online	Religious info
Online classifieds	Online classifieds	Online classifieds	Online classifieds	Rate things	Watch video
Bank online	Listen to music	Listen to music	Financial info	Social network sites	Play games
Govt website	IM	Financial info	Rate things	Online classifieds	Online classifieds
Play games	Play games	IM	Listen to music	IM	Social network sites
Read blogs	Financial info	Religious info	Religious info	Religious info	Rate things
Financial info	Religious info	Rate things	IM	Play games	Read blogs
Rate things	Read blogs	Read blogs	Play games	Listen to music	Donate to charity
Religious info	Rate things	Play games	Read blogs	Read blogs	Listen to music
Online auction	Online auction	Online auction	Online auction	Donate to charity	Podcasts
Podcasts	Donate to charity	Donate to charity	Donate to charity	Online auction	Online auction
Donate to charity	Podcasts	Podcasts	Podcasts	Podcasts	Blog
Blog	Blog	Blog	Blog	Blog	IM
Virtual worlds	Virtual worlds	Virtual worlds	Virtual worlds	Virtual worlds	Virtual worlds

Shopping online at retail websites is only one component of e-commerce. In 2008, nearly half of online users reported visiting online classified sites such as Craigslist and, globally, 30 percent or more of Internet users do their banking on the Internet (Jones 2009; Budde 2009). Additionally, online consumers often use the Internet to research products and to chat with others about purchasing decisions. In a national survey, 56 percent of music buyers, 39 percent of cell phone purchasers, and 49 percent of prospective real estate investors, used the Internet to do product research. Less than 12 percent of respondents in each of the three product categories reported that "online information had a major impact on their purchasing decision" (Horrigan 2008, p. 6).

Health and Digital Medicine The Internet acts as the third most likely source of health information preceded only by health professionals, and friends or family members (Fox 2009; Pew 2011a) (also see Chapter 2). Most online searches result in useful information that affects health care decisions, including decisions about seeing a doctor, how to deal with a specific disease, and diet and exercise information. Peer-to-peer medical help is also important, as 23 percent of chronically ill Internet users report going online to meet others with similar health problems (Pew 2011b).

Computerized health records are part of the Obama administration's health care reform package and have bipartisan support despite rancor over health care reform in general (Lohr 2011). Besides providing medical information, there is considerable evidence that online medical records help improve medical care:

> A paper record is a passive, historical document. An electronic health record can be a vibrant tool that reminds and advises doctors. It can hold information on a patient's visits, treatments and conditions, going back years, even decades. It can be summoned with a mouse click, not hidden in a file drawer in a remote location and thus useless in medical emergencies. (Lohr 2008, p. 1)

Digital patient records are thus the first step in creating **learning health systems** whereby physicians, looking across patient populations, can identify successful treatments or detect harmful interactions (Lohr 2011).

There is also evidence that technology can help mediate the soaring cost of health care and save as much as $80 billion a year in the United States alone (Atkinson & Castro 2008). Much of the savings comes from "increases in efficiency, such as shorter hospital stays because of better coordination, better productivity for nurses, and more efficient drug utilization" (p. 27). Other important sources of cost effectiveness include electronic claims processing and reducing medical errors through more effective diagnostic and treatment interventions.

Further, doctors can provide online services for patients. Presently, several websites provide 24/7 answers to health-related questions (e.g., healthcaremagic.com) at a cost significantly lower than an office visit. There is also a trend toward patients and physicians meeting face-to-face online via video conferencing, satellite hookups, and Skype-type applications (Freudenheim 2010).

The Search for Knowledge and Information The Internet, perhaps more than any other technology, is the foundation of the information society. Whether reading an online book, mapping directions, visiting the Louvre, or accessing Wikipedia, the Internet provides millions of surfers with instant answers to questions previously requiring a trip to the library. As a matter of fact, after e-mail, searching for information is the second most common online activity of members of all generations (see Figure 14.1). Even YouTube, at first a source of entertainment, has become a reference tool and is now the number-two search engine, edging out Yahoo (Helft 2009).

There is concern, however, that the very way in which the "Google generation" reads, thinks, and approaches problems has been altered by the new technology. Scholars at University College London, after examining computer logs of visitors to online research sites, found that visitors jumped from one site to the next, rarely returned to previous sites visited, and infrequently read more than a couple of pages of an article before jumping to another site. They concluded that:

> It is clear that users are not reading online in the traditional sense; indeed, there are signs that new forms of "reading" are emerging as users "power browse" horizontally through titles, contents pages, and abstracts going for quick wins. It almost seems that they go online to avoid reading in the traditional sense. (UCL 2008, p. 10)

Finally, with the lightening growth of the Internet, there are concerns that information is often outdated, difficult to find, and limited in scope (Mateescu 2010). Some

learning health systems The result of electronic records whereby physicians can look across patient populations and identify successful treatments or detect harmful interactions.

are already looking ahead to a time when search engines look for information not syntactically (i.e., based on *combinations* of words and phrases) but semantically (i.e., based on the *meaning* of words and phrases). The **Semantic Web,** sometimes referred to as Web 3.0, entails not only pages of information but pages that describe the interrelationship between the pages of information resulting in "smart media" (Semantic Media 2011).

Games and Entertainment Over half of all Americans play video games, although less than a quarter play video games online; only 9 percent play massive multiplayer online games (MMOG) such as World of Warcraft, and less than 2 percent have visited a virtual world such as Second Life (Lenhart 2008). Video games are big business, a $25.1 billion industry in 2010 (ESA 2011). For example, Angry Birds, a video game for Apple's iPad, has "40 million active users, 75 million paid and ad-supported downloads, and 2 million plush dolls sold" (Marin 2011, p. 1).

Although traditionally "gamers" were teenage boys (see this chapter's Social Problems Research Up Close), the genre is spreading to new audiences—women, adults, and older Americans (ESA 2011). Despite changing demographics, video game characters remain overwhelmingly male and, when female, are often hyper-sexualized (e.g., wearing provocative clothes and with exaggerated breast size) victims or prizes. As in other media, character portrayals impact beliefs and attitudes. Researchers Behm-Morawitz and Mastro (2009) report that undergraduates playing the video game *Tomb Raider* with the hyper-sexualized character Lara Croft were significantly more likely to have "less favorable attitudes toward women's cognitive capabilities" than undergraduates who were not exposed to the hyper-sexualized character (p. 820).

Image Courtesy of The Advertising Archives

Video games, a billion-dollar industry, most often have male lead characters. When females are present, they are often portrayed as either victim or prize, and are hyper-sexualized.

Finally, competition for dollars extends beyond video games. YouTube, in competition with Hulu, has partnered with MGM, CBS, Sony, and Lionsgate to show full-length movies and television shows (Stelter & Helft 2009). In fact, in a 2011 survey of 2,309 Americans, over half of the respondents with cable reported they would give up cable in favor of watching movies and shows on the Internet if conditions were right (Harris Interactive 2011).

Politics and e-Government Technology is changing the world of politics. In 2010, approximately 73 percent of U.S. adult Internet users went online to find news or information about the 2010 midterm elections, or to send or receive political messages through e-mail, instant messaging, Twitter, and the like (Smith 2011).

When Iranian President-elect Mahmoud Ahmadinejad banned foreign media from the streets of Tehran, protestors contesting the election used Twitter, social networking sites, Flickr, and YouTube to stay in contact with one another and to transmit messages and images around the world. More recently, Facebook played an essential role in several of the Middle East uprisings (Preston 2011). Said Elliot Schrage, vice president for global communications, public policy, and marketing of Facebook, "We've witnessed brave people of all ages coming together to effect a profound change in their country. Certainly, technology was a vital tool in their efforts but we believe their bravery and determination mattered most" (quoted in Preston 2011, p. 1).

The United States is just one of over 100 countries worldwide that hosts a government website. In 2010, the United Nations ranked e-government sites on a variety of criteria including information delivery, ease of obtaining information, delivery of public services, and citizen–government interaction. The overall ranking, as indicated by the e-government development index, reveals that the Republic of Korea is ranked first in the world followed by the United States, Canada, the United Kingdom, and the Netherlands. In general, developed countries' e-government capabilities are greater

Semantic Web Sometimes called Web 3.0, a version of the Internet in which pages not only contain information but also describe the interrelationship between pages; sometimes called smart media.

than those in developing countries, with the majority of countries ranked in the top 20 being high-income nations (United Nations 2010).

Social Networking and Blogging Social network sites (e.g., Facebook, Twitter) and blogs comprise a sector of the Internet called **membership communities.** Membership communities have changed in recent years in three substantively significant ways. First, the *number of people* who visit membership communities has increased. In 2010, Facebook topped 500 million members around the world (Wortham 2010). Second, the *amount of time* members spend at a membership community site has grown dramatically. Globally, one in every 11 minutes online can be accounted for by time on a social network or a blogging site (Nielsen 2009a).

Finally, *who joins* membership communities is changing. Although it is true that adolescents and young adults remain more likely to use social networking sites than older adults, between December 2008 and May 2010, people over the age of 45 more than doubled their participation (Wayne 2010). Similarly, Twitter, the microblogging site that asks "What are you doing?" had nearly 3 million visitors between the ages of 35 and 49 in 2008—42 percent of the site's total audience (Nielsen 2009b).

What Do You Think? Professor Laurence Thomas of Syracuse University walked out of his classroom after a student, in the front row of a large lecture hall, was seen sending a text (Jaschik 2008). Dr. Thomas's behavior has been both praised and criticized. What do you think? What should a professor do if students are not paying attention?

Science and Biotechnology

Although recent computer innovations and the establishment of the Internet have led to significant cultural and structural changes, science and its resulting biotechnologies have produced not only dramatic changes but also hotly contested issues with public policy implications. In this section, we look at some of the issues raised by developments in genetics, food and biotechnology, and reproductive technologies.

Genetics Molecular biology has led to a greater understanding of the genetic material found in all cells—DNA (deoxyribonucleic acid)—and with it the ability for **genetic screening.**

> If you could uncoil a strip of DNA, it would reach 6 feet in length, a code written in words of four chemical letters: A, T, G, and C. Fold it back up, and it shrinks to trillionths of an inch, small enough to fit in any one of our 100 trillion cells, carrying the recipe for how to create human beings from scratch. (Gibbs 2003, p. 42)

Currently, researchers are trying to complete genetic maps that will link DNA to particular traits. There is some evidence that personality characteristics are inherited; other evidence links certain conditions previously thought of as psychological in nature (e.g., addiction, anorexia, and autism) as, at least in part, genetically induced (Harmon 2006). Already, specific strands of DNA have been identified as carrying physical traits such as eye color and height, as well as such diseases as sickle-cell disease, breast cancer, cystic fibrosis, prostate cancer, depression, and Alzheimer's (ORNL 2011).

The U.S. Human Genome Project (HGP), a 13-year effort to decode human DNA, is now complete. Conclusion of the project is transforming medicine:

> All diseases have a genetic component whether inherited or resulting from the body's response to environmental stresses like viruses or toxins. The successes of the HGP have . . . enabled researchers to pinpoint errors in genes—the smallest units of heredity—that cause or contribute to disease. The ultimate goal is to use this information to develop new ways to treat, cure, or even prevent the thousands of diseases that afflict humankind. (Human Genome Project 2007, p. 1)

membership communities Internet sites where participation requires membership and members regularly communicate with one another for personal and/or professional reasons.

genetic screening The use of genetic maps to detect predispositions to human traits or disease(s).

The hope is that, if a defective or missing gene can be identified, possibly a healthy duplicate can be acquired and transplanted into the affected cell. This is known as **gene therapy.** Alternatively, viruses have their own genes that can be targeted for removal. Experiments are now under way to accomplish these biotechnological feats.

Food and Biotechnology **Genetic engineering** is the ability to manipulate the genes of an organism in such a way that the natural outcome is altered. Genetically modified (GM) food, also known as genetically engineered food, and genetically modified organisms involve this process of DNA recombination—scientists transferring genes from one plant into the genetic code of another plant.

In the United States, genetically modified organisms (GMOs) are in an estimated 80 percent of all packaged food sold in the United States and Canada (Spaeth 2011). Yet, a national survey of U.S. adults found that less than half (41 percent) were aware that foods containing GM ingredients are currently sold in stores, and although most Americans are likely to consume foods with GM ingredients every day, less than one-third of the survey sample (26 percent) said they had consumed food containing GM ingredients (Pew 2006).

Biotechnology companies and other supporters of GM foods commonly cite the alleviation of hunger and malnutrition as a main benefit, claiming that this technology can enable farmers to produce crops with higher yields. Critics of GM foods argue that the world already produces enough food for all people to have a healthy diet. According to the World Hunger Education Service (2011), if all the food produced worldwide were distributed equally, every person would be able to consume 2,720 calories a day. Biotechnology, critics argue, will not alter the fundamental causes of hunger, which are poverty and lack of access to food and to land on which to grow food.

> Human health concerns [of genetically modified foods] include possible toxicity, carcinogenicity, food intolerance, antibiotic resistance buildup, decreased nutritional value, and food allergens. . . .

Biotechnology companies claim that GM foods approved by the Food and Drug Administration are safe for human consumption, and they even cite potential health benefits such as the use of genetic modification to remove harmful allergens from foods or to improve nutritional benefits (Kaplan 2009; ORNL 2011). But critics claim that research on the effects of GM crops and foods on human health is inadequate, especially concerning long-term effects. Human health concerns include possible toxicity, carcinogenicity, food intolerance, antibiotic resistance buildup, decreased nutritional value, and food allergens in GM foods.

Biotechnology skeptics are also concerned about the environmental effects of GM crops. Biotechnology companies claim that crops that are genetically designed to repel insects negate the need for chemical (pesticide) control and thus reduce pesticide poisoning of land, water, animals, foods, and farmworkers. However, critics are concerned that insect populations can build up resistance to GM plants with insect-repelling traits, which would necessitate increased rather than decreased use of pesticides. Similarly, the use of Roundup, a top-selling weed killer, has lead to the development of "superweeds" requiring the use of more rather than less toxic herbicides (Neuman & Pollack 2010).

GM seed contamination is also of concern. First, organic farmers fear that cross-pollination will contaminate neighboring crops, reducing demand and thus sales for organically grown produce (Pollack 2011). Second, to maintain control over their products, some biotechnology companies have developed "terminator" seeds, which cause the plant to produce sterile seeds. For example, Monsanto initially agreed not to market its terminator technology but later adopted a positive stance on genetic seed sterilization, suggesting that the commercialization of terminator technology might occur in the future (ETC Group 2003). If the seed sterility trait in terminator crops inadvertently contaminates traditional crops and plant life, the ramifications would be devastating to life on earth.

gene therapy The transplantation of a healthy gene to replace a defective or missing gene.

genetic engineering The manipulation of an organism's genes in such a way that the natural outcome is altered.

The first genetically engineered crop was introduced for commercial production in 1996. Today, more than 200 million acres are devoted to these crops, with the United States being the largest producer in the world.

Biotechnology critics also raise concerns about insufficient safeguards and regulatory mechanisms. In 2000, worldwide concern about the safety of GM crops resulted in 160 nations signing the landmark Biosafety Protocol, which requires producers of a GM food to demonstrate that it is safe before it is widely used. The Biosafety Protocol also allows countries to ban the importation of GM crops based on suspected health, ecological, or social risks. In 2011, officials from 15 countries and the European Union signed a supplementary agreement that addressed liability and regress from damages resulting from living modified organisms (UNEP 2011).

Reproductive Technologies The evolution of "reproductive science" has been furthered by scientific developments in biology, medicine, and agriculture. At the same time, however, its development has been hindered by the stigma associated with sexuality and reproduction, its link with unpopular social movements (e.g., contraception), and the feeling that such innovations challenge the natural order (Clarke 1990). Nevertheless, new reproductive technologies have been and continue to be developed.

In **in vitro fertilization (IVF),** an egg and a sperm are united in an artificial setting, such as a laboratory dish or test tube. Although the first successful attempt at IVF occurred in 1944, the first test-tube baby, Louise Brown, was not born until 1978. Perhaps the most famous test-tube babies are the Suleman octuplets born to an unemployed, single California mother who already had six children. The case of "Octomom" made national headlines and led to a debate over fertility clinics and the ethics of IVF. Presently, no laws restrict the number of embryos a woman may receive, although medical guidelines suggest that physicians take into consideration the age, environment, and mental and physical condition of the mother-to-be (Archibold 2009, p. 1).

Other concerns surround the disposal of the nearly 400,000 frozen embryos in U.S. fertility clinics. A survey of patients at nine fertility clinics reveals that, among patients who did not want any more children (i.e., had surplus embryos), 66 percent were willing to donate their embryos to research and 20 percent said they would likely keep the embryos indefinitely (Grady 2008). Further criticisms of IVF are often based on traditional definitions of the family and the legal complications created when a child can have as many as five potential parental ties—egg donor, sperm donor, surrogate mother, and the

in vitro fertilization (IVF) The union of an egg and a sperm in an artificial setting such as a laboratory dish.

two people who raise the child (depending on the situation, IVF may not involve donors and/or a surrogate). Litigation over who the "real" parents are has already occurred. Over 50,000 children a year are born as a result of IVF technology.

Perhaps more than any other biotechnology, abortion epitomizes the potentially explosive consequences of new technologies (also see Chapter 13). **Abortion** is the removal of an embryo or fetus from a woman's uterus before it can survive on its own. Globally, according to Singh et al. (2009):

- Of the estimated 208 million pregnancies in 2008, 40 percent are unintended; 33 million (16 percent) resulted in unintended births; and 41 million ended in abortions (20 percent).
- The lowest abortion rate in the world is in western Europe (12 abortions per 1,000 women between the ages of 15 and 44); the rate is 31 in Latin America, 29 in Africa and Asia, and 21 in North America.
- The rate of abortions is declining worldwide; it is declining faster in developed countries where abortions are generally safe and legal than in developing countries where more than half of abortions are unsafe and illegal.
- Five million women are hospitalized each year, and another 70,000 die from complications associated with unsafe abortions.
- Contraception use has increased in many parts of the world.
- Twenty-two countries or areas within countries have changed their abortion laws since 1997: Nineteen cases liberalized their laws and three made them more restrictive.

Abortion laws vary dramatically around the world. The majority of women live in countries where abortion is allowed at least under some circumstances (Singh et. al 2009). However, 6 percent of women live in countries where abortion is banned under all circumstances. Nicaraguan law, for example, prohibits abortion even in the case of rape, incest, or deformity of the fetus. According to Amnesty International, the new Nicaraguan law has led to an increase in maternal deaths and hospital admissions associated with birth-related complications (Busari 2009). In Nicaragua, women and girls who seek an abortion and the doctors and nurses who provide abortion services receive prison sentences.

In the United States, since the U.S. Supreme Court's ruling in *Roe v. Wade* in 1973, abortion has been legal. However, recent Supreme Court decisions have limited the scope of the *Roe v. Wade* decision. In *Planned Parenthood of Southeastern Pennsylvania v. Casey* (1992), the court ruled that a state may restrict the conditions under which an abortion is granted, such as requiring a 24-hour waiting period or parental consent for minors. In 2005, the Supreme Court considered "whether laws requiring parental notification before a minor can get an abortion must make an explicit exception when the minor's health is at stake" (Barbash 2005, p. 1). In 2006, the Supreme Court unanimously affirmed the need for a medical emergency exception for teenagers seeking an abortion (Greenhouse 2006).

Recently, Republicans in Congress have begun seeking restrictions on a woman's legal right to an abortion. For example, a bill in the U.S. House of Representatives, if passed, would eliminate federal funding for any women's health care clinic that offers abortions. Further, a second bill would eliminate tax breaks for employers who provide health care policies for their employees that cover abortion services (Steinhauer 2011).

Additional challenges to *Roe v. Wade* are being fought at the state level. Historically, abortions are banned when the fetus is considered viable, usually around 22 to 26 weeks from conception. Nebraska has recently passed a law that bans abortions after 20 weeks "unless there is imminent danger to the woman's life or physical health" (Eckholm 2011, p. 2) and several other states are considering similar legislation. Additionally, a new Oklahoma law requires that a woman *have* and *view* an ultrasound as a doctor or technician describes, in detail, the anatomical characteristics of the fetus. There are no exceptions for rape and incest victims (McKinley 2010).

abortion The intentional termination of a pregnancy.

Many of the state restrictions on abortion are a result of the pro-life movement's success in redefining the abortion issue as one concerning *fetal rights* rather than *women's rights* (Greenhouse 2007). The Unborn Victims of Violence Act of 2004 protects all children in utero, regardless of the stage of development. Whereas the original intent of the act was to protect the fetus from violent crimes, observers note that, if a fetus is defined as a human being with the same rights as women and men in law, then it may in effect overturn *Roe v. Wade*. Presently antiabortion activists are raising money and pressuring state legislatures to get "personhood" measures on local and state ballots (Abcarian 2009).

Most recent debates concern intact dilation and extraction (D & E) abortions, which often take place in the second trimester of pregnancy. Opponents refer to such abortions as **partial birth abortions** because the limbs and the torso are typically delivered before the fetus has expired. However, former National Organization for Women president Kim Gandy states, "Try as you might, you won't find the term 'partial birth abortion' in any medical dictionary. That's because it doesn't exist in the medical world—it's a fabrication of the anti-choice machine" (U.S. Newswire 2003, p. 1). D & E abortions are performed because the fetus has a serious defect, the woman's health is jeopardized by the pregnancy, or both. In 2003, a federal ban on partial birth abortions was signed into law (White House 2003). Several constitutional challenges to the ban have occurred and, in 2004, a federal judge ruled that the ban was unconstitutional because it imposes an "undue burden on a woman's right to choose an abortion" (Willing 2005). However, in a major victory for the Bush administration, in 2007, the U.S. Supreme Court upheld the Partial-Birth Abortion Ban Act in a 5–4 decision (Greenhouse 2007). The significance of the case lies in the fact that it is the first time the U.S. Supreme Court has upheld a ban on any type of abortion procedure. Thirty-one states have also enacted partial birth abortion bans (Guttmacher Institute 2011).

Feminists, including U.S. Supreme Court Justice Ruth Bader Ginsburg, strongly oppose the ban, arguing that it is just one step closer to making all abortions illegal. They are also quick to note that the ban was not supported by "the American Medical Association, the American College of Obstetricians and Gynecologists, the American Medical Women's Association, the American Nurses Association, or the American Public Health Association" (U.S. Newswire 2003, p. 1). Says Eleanor Smeal, president of the Feminist Majority Foundation, "[I]n upholding the Bush administration's abortion ban . . . the U.S. Supreme Court showed its true colors: that it does not care about the health, well-being, and safety of American women. . ." (Smeal 2007). Further, Justice Ginsburg, writing a strongly worded dissent, states, "this way of thinking reflects ancient notions of women's place in the family and under the Constitution—ideas that have long been discredited" (quoted in Greenhouse 2007, p. 1).

Abortion is a complex issue for everyone, but especially for women, whose lives are most affected by pregnancy and childbearing. Women who have abortions are disproportionately poor, unmarried minority women who say that they intend to have children

> Abortion is . . . a complex issue for societies, which must respond to the pressures of conflicting attitudes toward abortion and the reality of high rates of unintended and unwanted pregnancy.

partial birth abortions Also called an intact dilation and extraction (D & E) abortion, the procedure may entail delivering the limbs and the torso of the fetus before it has expired.

TABLE 14.3 Attitudes toward Abortion by Political Affiliation and Religion, 2011

	LEGAL IN ALL/MOST CASES %	ILLEGAL IN ALL/MOST CASES %	DON'T KNOW %
Total	54	42	4
Republican	34	64	2
Democrat	65	31	4
Independent	58	38	4
Protestant	47	49	4
White evangelical	34	64	2
White mainline	60	37	3
Catholic	52	45	3
White Catholic	54	44	2
Unaffiliated	71	26	2

SOURCE: Pew 2011c. From "Fewer Are Angry at Government, But Discontent Remains High Republicans, Tea Party Supporters More Mellow", March 3, 2011, Pew Research Center For the People & the Press, a project of the Pew Research Center. Reprinted with permission.

in the future. Abortion is also a complex issue for societies, which must respond to the pressures of conflicting attitudes toward abortion and the reality of high rates of unintended and unwanted pregnancy. The debate over abortion has also complicated passage of federal health care reform as conservatives, citing a 30-year ban on using taxpayer's money to pay for elective abortions, battle the Obama administration and abortion rights supporters (Kirkpatrick 2009).

Attitudes toward abortion tend to be polarized between two opposing groups of abortion activists—pro-choice and pro-life. As recently as 2009, public opinion surveys indicated a fairly equal split between the two positions. However, in a 2011 poll, the majority of Americans responded that "Abortions should be legal in all or most cases" (see Table 14.3) (Pew 2011c). Advocates of the pro-choice movement hold that freedom of choice is a central human value, that procreation choices must be free of government interference, and that because the woman must bear the burden of moral choices, she should have the right to make such decisions. Alternatively, pro-lifers hold that the unborn fetus has a right to live and be protected, that abortion is immoral, and that alternative means of resolving an unwanted pregnancy should be found.

Cloning, Therapeutic Cloning, and Stem Cells In July 1996, scientist Ian Wilmut of Scotland successfully cloned an adult sheep named Dolly. To date, cattle, goats, mice, pigs, cats, rabbits, and horses have also been cloned. This technological breakthrough caused worldwide concern about the possibility of human cloning, leading the United Nations to adopt a declaration that called for governments to ban all forms of cloning that are at odds with human dignity and the preservation of human life (Lynch 2005). However, in 2009, U.N. General Assembly members could not agree on a draft proposal and gave up its efforts to develop a worldwide cloning ban (Associated Press 2009).

One argument in favor of developing human cloning technology is its medical value; it may potentially allow everyone to have "their own reserve of therapeutic cells that would increase their chance of being cured of various diseases, such as cancer, degenerative disorders, and viral or inflammatory diseases" (Kahn 1997, p. 54). Human cloning could also provide an alternative reproductive route for couples who are infertile and for those in which one partner is at risk for transmitting a genetic disease.

Arguments against cloning are largely based on moral and ethical considerations. Critics of human cloning suggest that, whether used for medical therapeutic purposes or as a means of reproduction, human cloning is a threat to human dignity (Human Cloning Prohibition Act of 2007). For example, cloned humans would be deprived of their individuality, and as Kahn (1997, p. 119) pointed out, "creating human life for the sole purpose of preparing therapeutic material would clearly not be for the dignity of the life created."

therapeutic cloning Use of stem cells to produce body cells that can be used to grow needed organs or tissues; regenerative cloning.

stem cells Undifferentiated cells that can produce any type of cell in the human body.

Therapeutic cloning uses stem cells from human embryos. **Stem cells** can produce any type of cell in the human body and thus can be "modeled into replacement parts for people suffering from spinal cord injuries or regenerative diseases, including Parkinson's and diabetes" (Eilperin & Weiss 2003, p. A6). In 2009, the U.S. Food and Drug Administration approved the first trials of embryonic stem cell therapy for paralyzed patients with spinal cord injuries (Park 2009, p. 2). This chapter's *The Human Side* on page 464 recounts the use of stem cells to help a young boy recover from a life-threatening disease.

Today, the use of embryonic stem cells is being replaced by induced pluripotent stem cells (IPSCs). Because of IPSCs, it is "now possible for researchers to churn out unlimited quantities of a patient's stem cells, which can then be turned into any of the cells that the body might need to repair or replace" (Park 2009, p. 3). For example, using IPSCs, researchers generated nerve cells from skin cells of individuals both with and without Rett syndrome, an autism spectrum disorder. They discovered that the created neurons from individuals with Rett syndrome, when compared to those created from individuals without Rett syndrome, were less likely to communicate with neighboring neurons, were smaller in size, and made fewer physical connections, i.e., synapses. Thus, through the use of IPSCs, researchers have a better understanding of how and why the nervous systems of patients with Rett syndrome are affected (NIH 2010).

In 2009, President Obama lifted the ban on federal funding of embryonic stem cell research and directed the National Institute of Health (NIH) to develop research support guidelines. Abortion opponents criticized the executive order for publicly financing research that destroys embryos, something conservatives argue is morally wrong. The resulting legal challenge to the public funding of embryonic stem cell research is yet undecided (Kaiser 2011).

The research community in turn was critical of the administration's decision to limit federal financing to research on surplus embryos from fertility clinics (Harris 2009). The new guidelines allow for the expansion of stem cell lines previously restricted, but prohibit federal funding of stem cell lines created solely for the purpose of research or created through therapeutic cloning (Rovner & Gold 2009; Harris 2009). Despite government restrictions on therapeutic cloning, a recent survey indicates that more than two-thirds of respondents approve of the use of therapeutic cloning to cure cancer or treat cardiovascular disease (Evans & Kelly 2011).

Despite what appears to be a universal race to the future and the indisputable benefits of scientific discoveries such as the workings of DNA and the technology of IVF and stem cells, some people are concerned about the duality of science and technology. Science and the resulting technological innovations are often life assisting and life giving; they are also potentially destructive and life threatening. The same scientific knowledge that led to the discovery of nuclear fission, for example, led to the development of both nuclear power plants and the potential for nuclear destruction. Thus, we now turn our attention to the problems associated with science and technology.

Societal Consequences of Science and Technology

Scientific discoveries and technological innovations have implications for all social actors and social groups. As such, they also have consequences for society as a whole. Figure 14.2 displays the public's relative assessment of the costs and benefits of scientific research. Although most Americans believe that the benefits of scientific research outweigh the potential for harmful results, there is no denying that science and the resulting technologies have some negative consequences.

Social Relationships, Social Networking, and Social Interaction

Technology affects social relationships and the nature of social interaction. The development of telephones has led to fewer visits with friends and relatives; with the advent of DVRs, cable television, and video streaming, the number of places where social life occurs (e.g., movie theaters) has declined. Even the nature of dating has changed as computer networks facilitate instant messaging, cyberdates, and "private" chat rooms. As technology increases, social relationships and human interaction are transformed.

Technology also makes it easier for individuals to live in a cocoon—to be self-sufficient in terms of finances (e.g., Quicken), entertainment (e.g., Hulu), work (e.g., telecommuting),

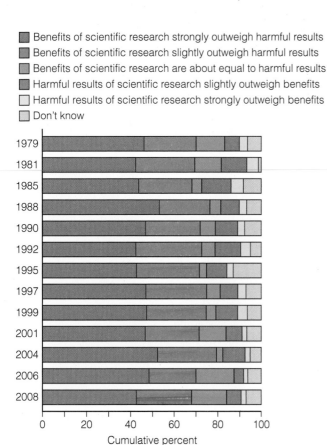

Benefits of scientific research strongly outweigh harmful results
Benefits of scientific research slightly outweigh harmful results
Benefits of scientific research are about equal to harmful results
Harmful results of scientific research slightly outweigh benefits
Harmful results of scientific research strongly outweigh benefits
Don't know

1979
1981
1985
1988
1990
1992
1995
1997
1999
2001
2004
2006
2008

0   20   40   60   80   100
Cumulative percent

NOTE: Includes all years for which data was collected.

Figure 14.2: Public Assessment of Scientific Research, 1979–2008
Source: NSF 2010.

news (e.g., Twitter), recreation (e.g., Wii), shopping (e.g., Amazon), communication (e.g., e-mail, texting), family conferences (e.g., Skype), and many other aspects of social life. For example, Facebook, the fastest-growing social networking site on the Internet, adds nearly a million users a day. The popularity of social networking sites has led to fears over the privacy and security of information posted and has led some to abandon membership communities (Della Cava 2010).

Although technology can bring people together, it can also isolate them from each other (Klotz 2004). For example, children who use a home computer "spend much less time on sports and outdoor activities than non-computer users" (Attewell et al. 2003, p. 277). A study of more than 1,500 U.S. Internet users between the ages of 18 and 64 found that, for every hour a respondent was on the Internet, there was a corresponding 23.5-minute reduction in face-to-face interaction with family members (Nie et al. 2004). Further, there is evidence that members of social networking sites are less likely to socialize with their neighbors or to rely on them for care and assistance (Pew 2009c).

Loss of Privacy and Security

Schools, employers, and the government are increasingly using technology to monitor individuals' performance and behavior. A 2005 study reports that 36 percent of companies use "keystroke monitoring so they can read what people type as well as track how much time they spend at the computer" and "55 percent retain and review e-mail messages" (MacMillan 2005, p. 1). In 2010, the Wisconsin Supreme Court held that personal e-mails sent and received (from and to a public school teacher) on state-owned computers (in a public school district) "are not part of government's business" and are, therefore, not part of the public record (Supreme Court of Wisconsin 2010).

In 2010, identity theft was the number-one complaint filed with the Federal Trade Commission for the 11th year in a row (FTC 2010; see also Chapter 4). Concerns over online privacy are so great that, in 2011, officials from the Federal Trade Commission (FTC), the Federal Communications Commission (FCC), and the National Telecommunications and Information Administration (NTIA) testified before a House Subcommittee on issues related to online privacy (Heitmann 2011).

Through computers, individuals can obtain access to someone's phone bills, tax returns, medical reports, credit histories, bank account balances, and driving records. Widman (2011) identifies several notable security breaches:

- The U.S. Department of Veteran Affairs sent a hard drive containing unencrypted information for repair. When the disc could not be fixed, it was recycled—with the personal information from over 75 million veterans on it.
- A laptop containing the records (e.g., names, Social Security numbers) of over 1,000,000 people was stolen from the vehicle of an Oklahoma Department of Human Services's employee who had left the computer in his car.
- The computer network of TJX, the parent company of T.J. Maxx and Marshalls, had its computer system hacked, compromising the credit card records of tens of millions of customers.
- A credit card processing company was responsible for as many as 40 million MasterCard holders being at risk of having their data stolen. At a minimum, 200,000 did.

Although just inconvenient for some, unauthorized disclosure of, for example, medical records, is potentially devastating for others. If a person's medical records indicate that he or she is human immunodeficiency virus (HIV)-positive, that person could be in danger of losing his or her job or health benefits. If DNA testing of hair, blood, or skin samples reveals a condition that could make the person a liability in an insurer's or employer's opinion, the individual could be denied insurance benefits, medical care, or even employment. In response to such fears, the **Genetic Information Nondiscrimination Act of 2008** (GINA) was passed. GINA is a federal law that prohibits discrimination in health coverage or employment based on genetic information (Department of Health and Human Services 2009).

Technology has created threats not only to the privacy of individuals but also to the security of entire nations. Computers can be used (or misused) in terrorism and warfare to cripple the infrastructure of a society and to tamper with military information and communication operations (see Chapter 15). In 2009, cyber attacks successfully stalled or slowed down 27 American and South Korean government and commercial websites, including websites of the Treasury Department, the New York Stock Exchange, the Secret Service, and the White House (Sang-Hun & Markoff 2009).

Unemployment, Immigration, and Outsourcing

Some technologies replace human workers—robots replace factory workers, word processors displace secretaries and typists, and computer-assisted diagnostics reduce the need for automobile mechanics. In 2009, Microsoft introduced "Laura," a personal assistant that resides on your desktop, senses moods and personality nuances, and follows directions (Vance 2009b).

Unemployment rates can also increase when companies **outsource** (sometimes called off-shore) jobs to lower-wage countries:

> The globalization of work tends to start from the bottom up. The first jobs to be moved abroad are typically simple assembly tasks, followed by manufacturing, and later skilled work like computer programming. At the end of this progression is the work done by scientists and engineers in research and development laboratories. (Lohr 2006, p. 1)

For example, in the United States, "offshore outsourcing in information technology, finance and other back office functions such as human resources has nixed 1.1 million jobs since 2008 and will result in another 1.3 million positions lost by 2014 . . . " (Dignan 2010). It should be noted, however, that economic trends impact receiving countries as well, in other words, jobs lost in one country impact outsourcing in another. In 2009, an estimated 1.5 million jobs, the majority in financial services, textiles, information technology, exports, and automobile manufacturing, were lost in India due to the global economic crisis (Joseph 2010).

Finally, there is some concern about the number of immigrant employees that are in the United States on H-1B visas. H-1B visas permit employers to temporarily hire foreign workers in certain specialty occupations including high-tech industries. The need for immigrant high-tech employees is a consequence of the lack of American counterparts in STEM occupations, that is, science, technology, engineering and mathematics (Price 2008). Currently, the number of H-1B visas issued each year by the federal government is limited to 65,000 (U.S. Citizenship and Immigration Services 2011). However, critics of

Genetic Information Nondiscrimination Act of 2008 A federal law that prohibits discrimination in health coverage or employment based on genetic information.

outsource A practice in which a business subcontracts with a third party, often in low-wage countries such as China and India, for services.

Tommy Bennett's big brown eyes and sweet demeanor make it that much harder to accept his plight. Just 3 years old, he blithely endures the constant barrage of drugs, needles, and tests as though he instinctively knows that they are destined to cure him. The development of new technologies has produced new forms of work and new demands for highly skilled workers in certain segments of the labor market.

Born with a rare, degenerative disease called Sanfilippo syndrome, Tommy lacks a critical enzyme needed for proper organ and brain development. Without enzymes, Tommy will die by adolescence. With the enzymes, Tommy's brain may unlock the potential to allow him to talk, dress, and care for himself. Such skills have eluded his two affected siblings, 4-year-old Hunter and 6-year-old Ciara. Ciara had just been diagnosed with Sanfilippo syndrome when their mom became pregnant with Tommy.

Since that time, they have searched desperately for someone willing to take a chance on helping Ciara, Hunter, and Tommy. The Bennetts found hope at Duke University Medical Center, the only program in the country willing to apply the benefits of stem cells—derived from newborn babies' umbilical cords—to treat this disease.

Proof of a Sanfilippo cure remains elusive, and Tommy is only the sixth Sanfilippo patient ever to have received a stem cell transplant. Yet if the transplant is to help, Tommy is a good candidate. He is young enough that the disease has only just begun to wreak havoc on his brain and organs. His siblings have progressed too far to be helped. Still, the sting of disappointment was palpable when doctors deemed Tommy the only viable candidate.

Thankful as they are for the opportunity, the Bennetts have embarked on a costly gamble—financially, emotionally, and physically. The Bennetts uprooted their kids and moved 900 miles away from family and friends to undergo a series of grueling tests before transplant could begin.

Then came the real test of endurance. Confined to the unit for four straight weeks, Tommy's small body was ravaged by toxic doses of chemotherapy designed to wipe out his immune system and make way for a new one that might provide the crucial enzyme.

Alicia took on hospital duty, caring for Tommy night and day, and catching a few winks of sleep as time permitted on a pull-down cot. John assumed full-time care for Ciara and Hunter at a rented apartment nearby, no easy task given Ciara's penchant for 3 a.m. awakenings. The process is clearly daunting, yet the transplant itself is deceptively simple. It takes just fifteen minutes for a bag of red liquid to drip intravenously into a child's bloodstream. Nurses literally squeeze every last drop from the bag, lest they lose a single stem cell that floats amidst a billion blood and supporting cells.

Every parent knows that stem cells hold the key to their child's survival. If they grow, the child has a fighting chance to live. If they do not, the child has probably exhausted his or her last resort at a cure.

Then comes the waiting, and the familiar refrain, "Grow, Cells, Grow." The words resonate within the halls, grace the walls of every room, and are sprinkled throughout cards of love and hope. Parents recite them like a battle cry designed to incite soldiers in action.

Indeed, stem cells are like tiny soldiers who descend upon bone marrow and rescue it from near-certain demise. So powerful are stem cells that it takes only ten to a hundred of them to restore a child's entire blood-forming and immune system—in Tommy's case providing the missing enzymes. Moreover, they know exactly where to go and what function to perform.

Yet such remarkable power is not without its drawbacks. Stem cells can attack the last remnants of the child's immune system, a complication called graft-versus-host disease. Stem cells take time to grow and mature, leaving the child's developing immune system vulnerable to minor infections that could prove deadly.

Children also suffer mightily from the dangerously high levels of chemotherapy needed to wipe out their immune system: Often, their mucous linings literally slough off from within, causing severe diarrhea and vomiting. Nausea, painful sores, fatigue, and stomach pains also plague the children as the chemo exerts its effects.

Luckily, Tommy endured far less of the usual symptoms of his transplant, but only time will tell if the new cells have become his own. A year must pass before his new immune system will be running at full force. A lot can happen in that time, but hope, prayer, and a will to overcome will be on their side.

Sadly, Tommy Bennett died on November 24, 2003.

Source: Reprinted from *Phi Kappa Phi Forum*, Vol. 83, No. 1 (winter 2003). Copyright © by Becky Levine. By permission of the publishers.

the H-1B visa limitations are pressuring the federal government to increase the number of visas available annually.

The H-1B visa program is not, however, without problems for the immigrant employee. Worker's visas are temporary—valid for a maximum of six years—at which time the visa holder must leave the United States unless permanent residency has been granted. Employees with H-1B visas are often paid less than American employees are, and cannot voluntarily quit their jobs for fear of deportation (Price 2008; U.S. Citizenship and Immigration Services 2011).

The Digital Divide

One of the most significant social problems associated with science and technology is the increased division between the classes. In a now oft-quoted statement, Welter (1997) notes:

It is a fundamental truth that people who ultimately gain access to, and who can manipulate, the prevalent technology are enfranchised and flourish. Those individuals (or cultures) that are denied access to the new technologies, or cannot master and pass them on to the largest number of their offspring, suffer and perish. (p. 2)

The fear that technology will produce a "virtual elite" is not uncommon. Several theorists hypothesize that, as technology displaces workers—most notably the unskilled and uneducated—certain classes of people will be irreparably disadvantaged—the poor, minorities, and women. There is even concern that biotechnologies will lead to a "genetic stratification," whereby genetic screening, gene therapy, and other types of genetic enhancements are available only to the rich.

Globally, the digital divide reflects the economic and social conditions of a country. In general, wealthier and more educated countries have more technology than poorer countries with a less educated populace, and the gap is growing (see Figure 14.3) (Wakefield 2010; White et al. 2011). For example, 90.6 percent of the population of Iceland is online compared to less than one percent of the population in Myanmar, Bangladesh, Ethiopia, Congo, and Cambodia.

Similarly, in the United States, the wealthier the family, the more likely the family is to have broadband Internet access. Of households with annual incomes of more than $75,000, 87 percent have broadband access. However, only 45 percent of households with income levels below $30,000 a year have broadband access (U.S. Bureau of the Census 2011c). Computer use and Internet use follow a similar pattern (see Table 14.2).

Income, however, cannot explain all of the variation in broadband home access by race and ethnicity (U.S. Department of Commerce 2010). Some of the differences in Internet use and broadband access are a function of housing patterns. Inner-city neighborhoods are disproportionately populated by racial and ethnic minorities and are simply less likely to be "wired," that is, to have the telecommunications hardware

> Several theorists hypothesize that, as technology displaces workers—most notably the unskilled and uneducated—certain classes of people will be irreparably disadvantaged—the poor, minorities, and women.

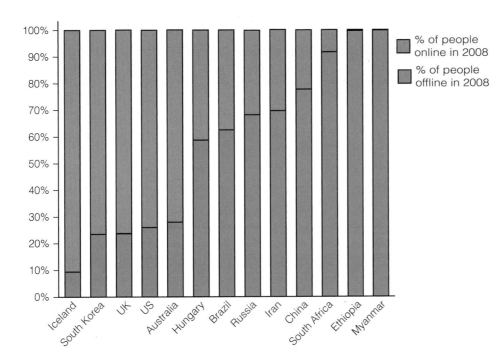

Figure 14.3: People On- and Offline by Country, 2008
Source: Wakefield 2010.

necessary for access to online services. In fact, cable and telephone companies are less likely to lay fiber optic cables in these areas—a practice called "information apartheid" or "electronic redlining."

Racial and ethnic minorities' lack of access to computers and the Internet, although signaling a type of digital divide, may be less common than what researchers are now calling the **participation gap**. Black and Latino children, for example, are more likely to access the Internet via mobile devices, use Twitter, and use the Internet to play games, participate in social networking, and watch video games than their advantaged, white counterparts (McCollum 2011). Further, some research indicates that having a home computer, although increasing computer skills, actually lowers academic achievement rather than increasing it, as one might expect (Malamud & Pop-Eleches 2010).

There are few gender disparities in computer use and access in developed countries such as the United States and Japan. However, in developing counties, women play a subordinate role in information communication technologies (ICT), which affects their employability:

> The perception of women being passive consumers of ICT rather than producers extends to their work-related use as well, where one continues to see a feminization of lower level ICT jobs. . . . Women continue to be concentrated in tedious, repetitive tasks as when they were during the first wave of industrialization, in manufacturing sectors such as textiles, clothing, and electronics. The lower skilled ICT jobs that women typically find themselves in are word-processing and data entry. (Thas et al. 2007, p. 10)

Concern over accessibility to broadband connectivity has led to a debate over net neutrality. **Net neutrality** advocates hold that Internet users should be able to visit any website and access any content without Internet service providers (ISP) (e.g., cable or telephone companies) acting as gatekeepers by controlling, for example, the speed of downloads. Why would an ISP do that? Hypothetically, if Internet service provider company X signs an agreement with search engine Y, then it's in the best interest of Internet service provider X to slow down all other search engines' performances so that you will switch to search engine Y. Internet service providers argue that Internet users, be they individuals or corporations, who use more than their "fair share" of the Internet should pay more. Why should you pay the same monthly fee as your neighbor who nightly downloads full-length movie files? Others fear any government regulation of the Internet and/or prefer a strictly market model.

In response to the debate, the Internet Freedom Preservation Act of 2009 has been introduced into Congress. The introduction to the bill states that as "the Nation becomes more reliant upon such Internet technologies and services, unfettered access to the Internet to offer, access, and utilize content, services, and applications is vital" (Internet Freedom Preservation Act 2009, p. 1). The proposed legislation, which remains in committee (Library of Congress 2011), requires that Internet service providers "not block, interfere with, discriminate against, impair, or degrade the ability of any person to use an Internet access service" (p. 1).

Mental and Physical Health

Youth between the ages of 8 and 18 spend an average of 7 hours and 38 minutes a day consuming some type of media, and because more than one medium can be consumed at a time, the actual total exposure to media per day is 10 hours and 45 minutes (Kaiser Family Foundation 2010). Media consumption of all types, except for reading, have increased over the last decade, and heavier media consumption is associated with lower levels of reported personal contentment, boredom, having fewer friends, and not being happy at school (see Figure 14.4).

participation gap The tendency for racial and ethnic minorities to participate in information and communication technologies (e.g., using smartphones to access the Internet rather than a computer) that place them in a disadvantaged position (e.g., difficult to research a term paper on a smartphone).

Net neutrality A principle that holds that Internet users should be able to visit any website and access any content without Internet service provider interference.

Among all 8- to 18-year-olds, percent of heavy, moderate, and light media users who say they get mostly:[†]			
	Heavy Users	**Moderate Users**	**Light Users**
Good grades [A's and B's]	51%[a]	65%[b]	66%[b]
Fair/poor grades [C's or below]	47%[a]	31%[b]	23%[c]
Among all 8- to 18-year-olds, percent of heavy, moderate, and light media users who say they:[††]			
Have a lot of friends	93%	91%	91%
Get along well with their parents	84%[a]	90%[b]	90%[ab]
Have been happy at school this year	72%[a]	81%[b]	82%[b]
Are often bored	60%[a]	53%[b]	48%[b]
Get into trouble a lot	33%[a]	21%[b]	16%[b]
Are often sad or unhappy	32%[a]	23%[b]	22%[b]

Note: Statistical significance should be read across rows.
[†]Students whose schools don't use grades are not shown.
[††]Percent who say each statement is "a lot" or "somewhat" like them.

Figure 14.4: **Impact of Media Consumption on 8- to 18-year-olds, 2010**
Source: Kaiser Family Foundation 2010. From "Report: Generation M2: Media in the Lives of 8- to 18-Year-Olds", (#8010), The Henry J. Kaiser Family Foundation, January 2010. This information was reprinted with permission from the Henry J. Kaiser Family Foundation. The Kaiser Family Foundation, a leader in health policy analysis, health journalism and communication, is dedicated to filling the need for trusted, independent information on the biggest health issues facing our nation and its people. The Foundation is a non-profit private operating foundation, based in Menlo Park, California.

Technology changes what we do, and when and how we do it. For example, a study on the relationship between sleep and technology use found that 95 percent of Americans use some kind of technology the hour before going to sleep (National Sleep Foundation 2011). Watching television, surfing on the net, texting, or playing video games are disruptive to sleep patterns and may be responsible for the millions of Americans who suffer from insomnia, restlessness, and general "sleepiness" during the week.

It's not just behavior patterns that are altered by technology use. Research indicates that technology can be dangerous. Drivers using cell phones are four times more likely to cause an accident as those who do not use cell phones and as likely to cause a crash as a legally drunk driver (Richtel 2009). Cell phones emit low levels of radioactivity associated with, according to some researchers, increased risks of brain cancer (Davis 2010).

Further, the multi-tasking that is associated with technology is linked to distraction, a false sense of urgency, and the inability to focus. Stanford researchers constructed experiments to test the reactions of media multi-taskers compared to non-media multi-taskers. Over various trials, media multi-taskers were unable to ignore irrelevant information, had poorer memories when asked to remember a sequence of alphabetical letters, and were unable to perform primary tasks if distracted (Ophir et al. 2009).

Some technologies have unknown risks. Biotechnology, for example, has promised and, to some extent, has delivered everything from lifesaving drugs to hardier pest-free tomatoes. Limbs are being replaced by bionic devices controlled by the recipient's thoughts (Brown 2006), and micro-sized telescopes are being implanted in the eyes of people with damaged retinas (Eisenberg 2009). However, biotechnologies have also created **technology-induced diseases.** According to the American Academy of Environmental Medicine, genetically modified foods, have "more than a casual association" with "adverse health effects" (AAEM 2009, p. 1).

Technological innovations are, for many, a cause of anguish, stress, and fear, particularly when the technological changes are far-reaching. About one-third of Americans believe that "science is going too far and hurting society," and 31 percent agree with the statement that "technology is making life too complicated for me" (Pew 2007). Moreover, nearly 60 percent of workers report being "technophobes," fearful of technology (Smith 2008), and the publication of a recent handbook on Internet addiction attests to the reality of the problem (Young & de Abreu 2011).

technology-induced diseases Diseases that result from the use of technological devices, products, and/or chemicals.

Malicious Use of the Internet

Some Internet users access the Internet for malicious purposes including but not limited to cybercrime and prostitution (see Chapter 4), hacking (see this chapter's Social Problems Research Up Close), piracy, electronic aggression (e.g., cyber-bulling), and "questionable content" sites. **Internet piracy** entails illegally downloading or distributing copyrighted material (e.g., music, games, or software). Court cases indicate that trade organizations (e.g., the Recording Industry Association of American) are pursuing criminal cases against violators, and courts are imposing strict penalties. In 2009, a student at Boston University was assessed $675,000 in damages for sharing 30 songs (*The Economist* 2009).

Malware is a general term that includes any spyware, crimeware, worms, viruses, and adware that is installed on owners' computers without their knowledge. Malware costs billions of dollars a year—$13 billion worldwide in 2006 (Thas et al. 2007). There is some evidence that, despite the best efforts of computer security experts, the proliferation of malicious software exceeds our knowledge or ability to stop it (Markoff 2008). Of late, malware has been discovered on factory computers and computerized industrial equipment (Richmond 2010).

Electronic aggression is defined as any kind of aggression that takes place with the use of technology (David-Ferdon & Hertz 2009). For example, **cyber-bullying** refers to the use of electronic devices (e.g., websites, e-mail, instant messaging, or text messaging) to send or post negative or hurtful messages or images about an individual or a group (Kharfen 2006). Estimates of the frequency of involvement in cyber-bullying either as victim, perpetrator, or both range dramatically, although electronic aggression researchers generally agree that texting is the most common means of cyber-bullying (David-Ferdon & Hertz 2009). Because cyber-bullying is capable of reaching wider audiences and thus doing more harm, many states and school districts have begun creating cyber-bullying disciplinary policies.

In 2008, a federal jury convicted Lori Drew of posing as a 13-year-old boy on a fraudulent MySpace account to get information from a 13-year-old friend of her daughter's, Megan Meier, who had allegedly spread lies about her. As "Josh Evans," Ms. Drew flirted with and then "broke up" with Megan with an e-mail that read "The world would be a better place without you" (McCarthy & Michels 2009, p. 3). Despite the fact that Megan Meier hung herself after receiving the e-mail, the conviction was later dismissed.

Finally, Optenet (2008), a global Internet security vendor, analyzed the content of nearly three million randomly selected uniform resource locators (URLs) from around the world and classified them by content. The largest single category of Internet content was pornography (36.2 percent), which had over three times the volume of the next largest category, e-commerce/shopping (10.8 percent). All "questionable content" sites measured by the researchers increased between 2006 and 2007, including websites containing content related to the *promotion* of anorexia and bulimia (+470.0 percent), violence (+125.6 percent), racism (+70.3 percent), drug use (+62.3 percent), and child pornography (+18.0 percent). However, pornography websites in general, as a portion of the total number of websites, decreased 8.9 percent between 2006 and 2007.

Internet piracy Illegally downloading or distributing copyrighted material (e.g., music, games, software).

malware A general term that includes any spyware, viruses, and adware that is installed on an owner's computer without their knowledge.

cyber-bullying The use of electronic devices (e.g., websites, e-mail, instant messaging, text messaging) to send or post negative or hurtful messages or images about an individual or a group.

The Challenge to Traditional Values and Beliefs

Technological innovations and scientific discoveries often challenge traditionally held values and beliefs, in part because they enable people to achieve goals that were previously unobtainable. Before recent advances in reproductive technology, for example, women could not conceive and give birth after menopause. Technology that allows post-menopausal women to give birth challenges societal beliefs about childbearing and the role of older women. The techniques of egg retrieval, in vitro fertilization, and gamete intrafallopian transfer make it possible for two different women to each make a biological contribution to the creation of a new life. Such technology requires society to re-examine its beliefs about what a family is and what a mother is. Should family be defined by custom, law, or the intentions of the parties involved?

Medical technologies that sustain life lead us to rethink the issue of when life should end. The increasing use of computers throughout society challenges the traditional value of privacy. New weapons systems make questionable the traditional idea of war as something that can be survived and even won. And cloning causes us to wonder about our traditional notions of family, parenthood, and individuality. Toffler (1970) coined the term **future shock** to describe the confusion resulting from rapid scientific and technological changes that unravel our traditional values and beliefs.

What Do You Think? Perhaps no technologies are as futuristic and as controversial as nanotechnologies. Nanotechnologies are a classification of different technologies, all of which have one thing in common—they are small, very, very small. A nanometer is one-billionth of a meter. To put this in perspective, a piece of paper is 100,000 nanometers thick (NNI 2009). That nanotechnologies will change our lives and the social world we live in is indisputable. Would you wear a T-shirt that, through nanotechnology, converts the energy your movements generate to electricity that can then power your laptop? Would you be comfortable with paint that turns colors because you told it to? How about nano-sized medical robots that are injected into your blood stream to look for and fight off diseases? Do any of these innovations, all presently proposed by nanotechnology researchers, make you feel a little uneasy? Why or why not?

Strategies for Action: Controlling Science and Technology

As technology increases, so does the need for social responsibility. Nuclear power, genetic engineering, cloning, and computer surveillance all increase the need for social responsibility: "Technological change has the effect of enhancing the importance of public decision making in society, because technology is continually creating new possibilities for social action as well as new problems that have to be dealt with" (Mesthene 1993, p. 85). In the following sections, we address various aspects of the public debate, including science, ethics, the law, the role of corporate America, and government policy.

Science, Ethics, and the Law

Science and its resulting technologies alter the culture of society through the challenging of traditional values. Public debate and ethical controversies, however, have led to structural alterations in society as the legal system responds to calls for action. For example, several states now have what are called genetic exception laws. **Genetic exception laws** require that genetic information be handled separately from other medical information,

future shock The state of confusion resulting from rapid scientific and technological changes that unravel our traditional values and beliefs.

genetic exception laws Laws that require that genetic information be handled separately from other medical information.

Annually, "tens of millions of animals are used for biomedical research, chemical testing, and training" (Bowman 2011, p. 1). An estimated 1 million are dogs and cats, monkeys and apes, hamsters and guinea pigs, rabbits, pigs, and sheep, and other farm animals. All else, including birds, fish, rats and mice, frogs and lizards, number somewhere between 80 to 100 million. Each has or will be subject to **vivisection,** the practice of cutting into or otherwise harming living, nonhuman animals for the purpose of scientific research.

The use of animals for research purposes is not a new phenomenon, but the century-old debate over the morality of using animals in research has recently picked up momentum as animal rights groups such as PETA (People for the Ethical Treatment of Animals) and the Animal Liberation Front make inroads into the American collective consciousness. Additionally, there is evidence that pets are assuming an increasingly important role in our emotional lives (Power 2008). In a national survey of pet owners, 57 percent of respondents stated that if they were stranded on a desert island and could have only one companion, they would choose their pet (Cohen 2002).

The issues surrounding the vivisection of nonhuman animals are not easily resolved. Should humans' rights trump those of nonhuman animals? Do nonhuman animals, as human animals, have a right to a clean, safe, and pain-free environment? Should nonhuman animals die in the hopes of gathering data that might prolong or save human life?

Although a recent public opinion poll found that 59 percent of Americans believe that "medical testing on animals" is "morally acceptable" (Saad 2010), the number of Americans supporting nonhuman animal–based research has been steadily declining (Bowman 2011).

The arguments both for (at one end of the continuum) and against (at the other end of the continuum) the use of animals in research generally fall into three camps—yes, no, and sometimes. Advocates argue that any distinction that differentiates dogs and chimpanzees, for example, from birds and fish, is an artificial one. The only substantively significant distinction to be made is between human animals and nonhuman animals.

That said, if your daughter was ill and you had the choice of giving her drug A (which killed all of the dogs and cats it was tested on), drug B (which killed some of the dogs and cats it was tested on), or drug C (which killed none of the dogs and cats it was tested on), which would you choose? Obviously, the rational answer is drug C, proving that, or so many scientists would argue, using animals in research provides us with important information in making life and death decisions. Test on animals or your child dies—your choice. In a recent survey of biomedical scientists, 90 percent responded that the use of animals in research is "essential" (*Nature News* 2011).

Finally, the Food and Drug Administration *requires* that drugs and procedures be tested on nonhuman animals before they can be consumed by or used on human animals. And to opponents who are quick to point out that the human and nonhuman animals are too different to result in any meaningful findings, Holland (2010) responds:

> Pigs are clearly not human, but heart valves in both species are nearly interchangeable. Some viruses that attack rodents, cats, and primates can do the same damage to humans. And diseases in some species have highly similar "first cousins" that affect people. The key issue for researchers is that the biological mechanisms for many human and animal viruses are highly similar, if not the same. So understanding how to stop one form of virus

leading to what is sometimes called patient shadow files (Legay 2001). The logic of such laws rests with the potentially devastating effects of genetic information being revealed to insurance companies, other family members, employers, and the like. Examples of protected tests include the BRCA1 and BRAC2 tests for breast cancer and carrier screening for such conditions as sickle cell anemia and cystic fibrosis (Genetics and Public Policy Center 2010).

Are such regulations necessary? In a society characterized by rapid technological and thus social change—a society in which custody of frozen embryos is part of the divorce agreement—many would say yes. Cloning, for example, is one of the most hotly debated technologies in recent years. Bioethicists and the public vehemently debate the various costs and benefits of this scientific technique. Despite such controversy, however, the chairman of the National Bioethics Advisory Commission warned nearly 15 years ago that human cloning will be "very difficult to stop" (McFarling 1998). At

vivisection The practice of cutting into or otherwise harming living, non-human animals for the purpose of scientific research.

in humans can provide the key to doing the same for animals and vice versa.

Hanson (2010), however, convincingly argues that such justifications are an ethical "bait and switch." Few of us could even bear to look at pictures of "monkeys, with their electrode-implanted brains and bolted heads, being put through their paces in a desperate attempt to get a life sustaining sip of water" (p. 3). Yet, researchers argue that such testing techniques may lead to finding the cause or cure for Alzheimer's disease. Hanson, a renowned neuroscientist and PETA member, continues noting that "these experiments . . . are so thoroughly unrelated to the neuropathology of Alzheimer's . . . that in more than 28 years of research in the neuroscience of the disease, I have never come across a single reference to them in any scientific literature on neurodegenerative diseases" (p. 3).

Moderates acknowledge that the use of animals in scientific research may be necessary but that the issue is multifaceted. Henry and Pulcino's (2009) study of the characteristics that influence attitudes toward the use of animals in research is a case in point. In a pen and pencil experiment, the researchers varied three independent variables: (1) type of animal used (chimpanzees, dogs, or mice), (2) level of harm to the animal (no harm, serious injury, death), and (3) severity of the disease being studied (eczema, rheumatoid arthritis, cancer). In addition to individual differences (for example, females were less tolerant of the use of animals in research than males), all of the independent variables were statistically

significant in predicting what a given subject would recommend be done. Subjects were more likely to be in favor of using animals for testing if mice were being used, there was no harm, and the disease under investigation was cancer. Although the results are not particularly surprising, the lack of significant interactions is. Respondents considered each of three research variables independently of one another rather than simultaneously.

There are laws that protect the use of animals in scientific research. The Laboratory Animal Welfare Act was passed in 1966, after an article appeared in *Life Magazine* entitled, "Concentration Camps for Lost and Stolen Pets." A later amendment required that oversight "care and use" committees be established at research institutions but most often are staffed by animal use advocates (Hanson 2010). Further, in 2009, Japan, Canada, the United States, and the European Union signed an international agreement that would reduce the number of animals used in product safety testing (NIEHS 2009). The new agreement, however, has little impact on the use of animals in biotechnology research.

There are also norms within the culture of "doing science," some of which have become formalized: minimize pain and suffering, use as few animals as possible, hire caretakers (e.g., veterinarians), look for alternative means of accomplishing the same end, and, given a choice, use lower-level animals rather than higher-level animals (Research Animal Resources 2003). The problem is that there is overwhelming evidence that

all animals suffer, and even "lower-level" animals, such as mice, grimace when experiencing pain and laugh when being tickled, in other words, they have emotions (Flecknell 2010; Ferdowsian 2010).

Ironically, the same thing that led to the use of animals in research, the pursuit of scientific and technological know-how, may lead to ending the use of animals in scientific research. For example, a "surrogate in-vitro human immune system . . . has been developed [in a test tube] to help predict an individual's immune response system" (Ferdowsian 2010). Further, the ICCVAM (Interagency Coordinating Committee on the Validation of Alternative Methods), established in 2000 under the National Institute of Health, is designed to

promote the development, validation, and regulatory acceptance of new, revised, and alternative regulatory safety testing methods. Emphasis is on alternative methods that will reduce, refine (less pain and distress), and replace the use of animals in testing while maintaining and promoting scientific quality and the protection of human health, animal health, and the environment. (ICCVAM 2011)

The issue is complex and there are no easy answers. But asking the right questions is a step in the right direction. Perhaps as philosopher Jeremy Bentham said over 200 years ago, the right question is not, "Can they speak?" nor "Can they reason?" but "Do they suffer?'

present, 15 states have laws pertaining to human cloning, with some prohibiting cloning for reproductive purposes, some prohibiting therapeutic cloning, and still others prohibiting both (NCSL 2008).

Should the choices that we make as a society be dependent on what we can do or what we should do? Whereas scientists and the agencies and corporations who fund them often determine what we *can* do, who should determine what we *should* do (see this chapter's Animal and Society feature)? Although such decisions are likely to have a strong legal component—that is, they must be consistent with the rule of law and the constitutional right of scientific inquiry—legality or the lack thereof often fails to answer the question, "What should be done?" *Roe v. Wade* (1973) did little to quash the public debate over abortion and, more specifically, the question of when life begins. Thus, it is likely that the issues surrounding the most controversial of technologies will continue throughout the 21st century with no easy answers.

Technology and Corporate America

As philosopher Jean-Francois Lyotard noted, knowledge is increasingly produced to be sold. The development of genetically altered crops, the commodification of women as egg donors, and the harvesting of regenerated organ tissues are all examples of potentially market-driven technologies. Like the corporate pursuit of computer technology, profit-motivated biotechnology creates several concerns.

First is the concern that only the rich will have access to lifesaving technologies such as genetic screening and cloned organs. Such fears are justified. Companies with obscure names such as Progenitor International Research, Millennium Pharmaceuticals, Darwin Molecular, and Myriad Genetics have been patenting human life. Myriad Genetics has nine patents on the breast and ovarian cancer genes (Bollier 2009). However, because of the resulting **gene monopolies** and the associated astronomical patient costs for genetic screening and treatment, in 2009, the American Civil Liberties Union (ACLU) and several plaintiffs filed a lawsuit against Myriad Genetics, alleging that the patents are "invalid and unconstitutional" (Genomics Law Report 2011). In 2011, Myriad Genetics appealed an unfavorable lower court decision and the case now rests in the hands of the U.S. Court of Appeals.

The commercialization of technology causes several other concerns, including issues of quality control and the tendency for discoveries to remain closely guarded secrets rather than collaborative efforts (Rabino 1998; Lemonick & Thompson 1999; Mayer 2002; Crichton 2007). In addition, industry involvement has made government control more difficult because researchers depend less and less on federal funding. More than 67 percent of research and development in the United States is supported by private industry using their own company funds (NSF 2010).

Runaway Science and Government Policy

Science and technology raise many public policy issues. Policy decisions, for example, address concerns about the safety of nuclear power plants, the privacy of electronic mail, the hazards of chemical warfare, and the ethics of cloning. In creating science and technology, have we created a monster that has begun to control us rather than the reverse? What controls, if any, should be placed on science and technology? And are such controls consistent with existing law? Consider the use of the file-sharing network BitTorrent to download music and movie files (the question of intellectual property rights and copyright infringement); a Utah law limiting children's access to material on the Internet (free speech issues); and Acxiom, the "cookie"-collecting company that helps corporations customize advertising on websites by tracing clicks and keystrokes (Fourth Amendment privacy issues) (ACLU 2005; Kaplan 2000; Schaefer 2001; *The Economist* 2009; Clifford 2009).

> Policy decisions . . . address concerns about the safety of nuclear power plants, the privacy of electronic mail, the hazards of chemical warfare, and the ethics of cloning.

The government, often through Congress, regulatory agencies, or departments, prohibits the use of some technologies (e.g., assisted-suicide devices) and requires others (e.g., seat belts). For example, in 2011, a bill was introduced in the U.S. House of Representatives that (1) supports the use of embryonic stem cells, including human embryonic stems cells, (2) defines the types of human embryonic stem cells eligible for use in research (e.g., donated from IVF clinics), (3) mandates that the Department of Health and Human Services maintain, review, and update guidelines in support of human stem cell research, and (4) prohibits public funds be used for human cloning (Stem Cell Research Advancement Act 2011).

Through financial support or the lack thereof, the government also promotes or discourages certain technologies. To help ameliorate the "digital divide," the 2009 economic stimulus package (see Chapter 7) provided $7.2 billion to expand broadband Internet services to underserved populations, primarily the 40 million poor and/or people living in rural areas who do not have broadband access (Cauley 2009).

gene monopolies Exclusive control over a particular gene as a result of government patents.

Further, in his 2011 State of the Union address, the president spoke of the importance of science and technological innovation:

> With more research and incentives, we can break our dependence on oil with biofuels, and become the first country to have a million electric vehicles on the road by 2015. . . . Within 25 years, our goal is to give 80 percent of Americans access to high-speed rail. . . . And to help pay for it, I'm asking Congress to eliminate the billions in taxpayer dollars we currently give to oil companies. . . . Within the next five years, we'll make it possible for businesses to deploy the next generation of high-speed wireless coverage to 98 percent of all Americans . . . connecting every part of America to the digital age.

The federal government has also instituted several initiatives dealing with technology-related crime (see Chapter 4). Of late, the issue of online pornography has come to the forefront. In 2011, after a two-year investigation, Jason Farineau was convicted of distributing child pornography over the Internet via file-sharing software, and possession of child pornography. Farineau, 29, was sentenced to over 17 years in prison (U.S. Attorney's Office 2011). Because of the ease of committing such offenses, the Federal Bureau of Investigation (FBI) is pushing Internet service providers (ISPs) to maintain records of "origin and destination information" of their customers, in other words, to keep logs of the websites they visit for up to two years (McCullagh 2010).

The FBI's "cyber-mission" includes but is not limited to fighting online child pornography. It also includes (1) preventing serious computer security breaches, (2) identifying and terminating the activity of online sexual predators, (3) preserving national security and the integrity of intellectual property, and (4) dismantling transnational organized crime (FBI 2011). Chapter 4 contains an in-depth discussion of transnational crime.

Finally, the government has several science and technology boards and initiatives, including the National Science and Technology Council, the Office of Science and Technology Policy, the President's Council of Advisors on Science and Technology, and the U.S. National Nanotechnology Initiative. These agencies advise the president on matters of science and technology, including research and development, implementation, national policy, and coordination of different initiatives.

In 1994, Luis Gonzalez-Bunster, just one week out of high school, was seriously injured in a car accident. Paralyzed from the chest down, he remains in a wheelchair but is more hopeful than ever that stem cell therapy will restore his spinal cord function. In 2009, Luis rode a 28-day, 500-mile marathon using his hand-powered bicycle to raise awareness about stem cell research (Sterns 2009).

What Do You Think? On July 21, 1969, the United States became the first country to put a person on the moon. Commander Neil Armstrong stepped onto the Sea of Tranquility region of the lunar surface and, as he put his left foot down declared, "That's one small step for man, one giant leap for mankind." Just a little over 40 years later, in 2011, the space program ended with Endeavour's final mission. Do you think the U.S. space program should continue to be funded?

Understanding Science and Technology

What are we to understand about science and technology from this chapter? As structural functionalists argue, science and technology evolve as a social process and are a natural part of the evolution of society. As society's needs change, scientific discoveries and technological innovations emerge to meet these needs, thereby serving the functions of the whole. Consistent with conflict theory, however, science and technology also meet the needs of select groups and are characterized by political components. As Winner (1993) noted, the structure of science and technology conveys political messages, including "power is centralized," "there are barriers between social classes," "the world is hierarchically structured," and "the good things are distributed unequally" (p. 288).

The scientific discoveries and technological innovations that society embraces as truth itself are socially determined. Research indicates that science and the resulting technologies have both negative and positive consequences—a **technological dualism.**

technological dualism The tendency for technology to have both positive and negative consequences.

Technology saves lives, time, and money; it also leads to death, unemployment, alienation, and estrangement. Weighing the costs and benefits of technology poses ethical dilemmas, as does science itself. Ethics, however, "is not only concerned with individual choices and acts. It is also and, perhaps, above all concerned with the cultural shifts and trends of which acts are but the symptoms" (McCormick & Richard 1994, p. 16).

Thus, society makes a choice by the very direction it follows. These choices should be made on the basis of guiding principles that are both fair and just, such as those listed here (Goodman 1993; Winner 1993; Eibert 1998; Buchanan et al. 2000; Murphie & Potts 2003):

1. Science and technology should be prudent. Adequate testing, safeguards, and impact studies are essential. Impact assessment should include an evaluation of the social, political, environmental, and economic factors.

2. No technology should be developed unless all groups, and particularly those who will be most affected by the technology, have at least some representation "at a very early stage in defining what that technology will be" (Winner 1993, p. 291). Traditionally, the structure of the scientific process and the development of technologies have been centralized (i.e., decisions have been made by a few scientists and engineers); decentralization of the process would increase representation.

3. Means should not exist without ends. Each new innovation should be directed to fulfilling a societal need rather than the more typical pattern in which a technology is developed first (e.g., high-definition television) and then a market is created (e.g., "You'll never watch a regular TV again!"). Indeed, from the space program to research on artificial intelligence, the vested interests of scientists and engineers, whose discoveries and innovations build careers, should be tempered by the demands of society.

What the 21st century will hold, as the technological transformation continues, may be beyond the imagination of most of society's members. Technology empowers; it increases efficiency and productivity, extends life, controls the environment, and expands individual capabilities. According to a National Intelligence Council report, "Life in 2015 will be revolutionized by the growing effort of multidisciplinary technology across all dimensions of life: social, economic, political, and personal" (NIC 2003, p. 1).

As we proceed further into the first computational millennium, one of the great concerns of civilization will be the attempt to reorder society, culture, and government in a manner that exploits the digital bonanza yet prevents it from running roughshod over the checks and balances so delicately constructed in those simpler pre-computer years.

CHAPTER REVIEW

• **What are the three types of technology?**
The three types of technology, escalating in sophistication, are mechanization, automation, and cybernation. Mechanization is the use of tools to accomplish tasks previously done by hand. Automation involves the use of self-operating machines, and cybernation is the use of machines to control machines.

• **What are some of the reasons the United States may be "losing its edge" in scientific and technological innovations?**
The decline of U.S. supremacy in science and technology is likely to be the result of four interacting social forces. First, the federal government has been scaling back its investment in research and development. Second, corporations have begun to focus on short-term products and higher profits, and third, there has been a drop in science and math education in U.S. schools in terms of both quality and quantity. Finally, as documented in the book *Unscientific American,* there is a disconnect between American society and the principles of science.

• **What are some Internet global trends?**
In 2011, the Internet had 2.1 billion users in more than 200 countries with 240 million users in the United States. Of all Internet users, the highest proportion come from Asia (44.0 percent), and the lowest from Oceania/Australia (1.0 percent) (Internet Statistics 2011a). Penetration rates vary but, in general, are higher in wealthier rather than poorer nations.

• **According to Kuhn, what is the scientific process?**
Kuhn describes the process of scientific discovery as occurring in three steps. First are assumptions about a particular phenomenon. Next, because unanswered questions always remain about a topic, science works to start filling in the gaps. Then, when new information suggests that the initial assumptions were incorrect, a new set of assumptions or framework emerges to replace the old one. It then becomes the dominant belief or paradigm until it is questioned and the process repeats.

- **What is meant by the computer revolution?**
  The silicon chip made computers affordable. Today, 89 percent of Americans occasionally use a computer from some location, and 79 percent occasionally use the Internet from some location. The comparable statistics in 1995 for computer use and Internet use respectively are 54 percent and 14 percent.

- **What is the Human Genome Project?**
  The U.S. Human Genome Project is an effort to decode human DNA. The 13-year-old project is now complete, allowing scientists to "transform medicine" through early diagnosis and treatment as well as possibly preventing disease through gene therapy. Gene therapy entails identifying a defective or missing gene and then replacing it with a healthy duplicate that is transplanted to the affected area.

- **What is the legal status of abortion in the United States?**
  In the United States, since the U.S. Supreme Court's ruling in *Roe v. Wade* in 1973, abortion has been legal. However, recent Supreme Court and state court decisions have limited the scope of *Roe v. Wade*.

- **How are some of the problems of the Industrial Revolution similar to the problems of the technological revolution?**
  The most obvious example is in unemployment. Just as the Industrial Revolution replaced many jobs with technological innovations, so too has the technological revolution.

Furthermore, research indicates that many of the jobs created by the Industrial Revolution, such as working on a factory assembly line, were characterized by high rates of alienation. Rising rates of alienation are also a consequence of increased estrangement as high-tech employees work in "white-collar factories."

- **What is meant by outsourcing and why is it important?**
  Outsourcing is the practice of a business subcontracting with a third party, often in low-wage countries such as China and India, for services. The problem with outsourcing is that it tends to lead to higher rates of unemployment in the export countries.

- **What is the digital divide?**
  The digital divide is the tendency for technology to be most accessible to the wealthiest and most educated. For example, some fear that there will be "genetic stratification," whereby the benefits of genetic screening, gene therapy, and other genetic enhancements will be available to only the richest segments of society.

- **What is meant by the commercialization of technology?**
  The commercialization of technology refers to profit-motivated technological innovations. Whether it be the isolation of a particular gene, genetically modified organisms, or the regeneration of organ tissues, where there is a possibility for profit, private enterprise will be there.

TEST YOURSELF

1. Which of the following technologies is associated with industrialization?
   a. Mechanization
   b. Cybernation
   c. Hibernation
   d. Automation
2. Analysis of the data over time suggests that the science and technology gap between poorer and richer nations is getting greater rather than shrinking.
   a. True
   b. False
3. The U.S. government, as part of the technological revolution, spends more money on research and development than educational institutions and corporations combined.
   a. True
   b. False
4. Which theory argues that technology is often used as a means of social control?
   a. Structural functionalism
   b. Social disorganization
   c. Conflict theory
   d. Symbolic interactionism
5. The majority of women live in countries where abortions are banned under all circumstances.
   a. True
   b. False
6. The ability to manipulate the genes of an organism to alter the natural outcome is called

   a. gene therapy
   b. gene splicing
   c. genetic engineering
   d. genetic screening
7. Genetically modified foods have been documented as harmless to humans by the Food and Drug Administration.
   a. True
   b. False
8. In 2007, the U.S. Supreme Court upheld the Partial Birth Abortion Ban in a 5-to-4 decision.
   a. True
   b. False
9. The practice of outsourcing entails
   a. creating high-tech jobs in the United States for immigrants
   b. hiring temporary workers to cover for employees who are absent
   c. allowing company workers to work from home
   d. subcontracting jobs often to workers in low-wage countries
10. The Genetic Information Nondiscrimination Act (GINA)
   a. makes human cloning illegal in the United States
   b. establishes a criminal penalty for human cloning in the United States
   c. is a federal law that prohibits discrimination in health coverage or employment based on genetic information
   d. all of the above

Answers: 1: d; 2: a; 3: b; 4: c; 5: b; 6: c; 7: b; 8: a; 9: d; 10: c.

KEY TERMS

MEDIA RESOURCES

Turning to Video

▶❚❚ Watch the BBC video *The Family and Technology* (running time 2:11) available through **CengageBrain .com**. This video examines the impact of new technologies on family functioning. How have technologies such as the cell phone, the Internet, and video games impacted your family?

Online Study Resources

Log in to **www.cengagebrain.com** to access the resources your instructor has assigned. For this book, you can access:

CourseMate

Access chapter-specific learning tools, including learning objectives, practice quizzes, videos, Internet exercises, flash cards, and glossaries, as well as web links, and more in your Sociology CourseMate.

ROMEO GACAD/AFP/Getty Images

Conflict, War, and Terrorism

"Every gun that is made, every warship launched, every rocket fired, signifies in the final sense a theft from those who hunger and are not fed, those who are cold and not clothed."

—General Dwight D. Eisenhower, former U.S. president and military leader

In this photo taken Thursday, July 29, 2010, Staff Sgt. Melinda Miller hugs Gina after a workout on an obstacle course at Peterson Air Force Base in Colorado Springs, Colo. Gina was a playful 2-year-old German shepherd when she went to Iraq as a highly trained bomb-sniffing dog with the U.S. military, but months of door-to-door searches and noisy explosions left her cowering and fearful. After she came home to Peterson Air Force Base in June 2009, a military veterinarian diagnosed her with post-traumatic stress disorder.

THINGS WERE STARTING to turn around for the 28-year-old Marine vet Clay Hunt. After two tours of duty, one in Iraq and one in Afghanistan, he suffered from PTSD (post-traumatic stress disorder) and the guilt of surviving what his buddies hadn't but he, unlike so many other vets, had done everything right (Hefling 2011; Wise 2011). He moved closer to his family, took his medications and sought counseling, and volunteered in both Haiti and Chile after the earthquakes. He even made public service announcements warning other vets of the dangers of PTSD—reaching out to them to get the help he had. Nonetheless, when Clay got a construction job, bought a truck, and started dating again, his friends and relatives were relieved. After all, he had made plans for a big reunion with his fellow marines the following weekend. But Clay never made it to the reunion. On March 31, 2011, this young man with movie-star good looks and a friendly, easy manner bolted himself inside his Houston apartment and shot himself. Ironically, the Purple Heart recipient had a tattoo on his arm from J.R.R. Tolkien's *Lord of the Rings:* "Not all those who wander are lost."

War is one of the great paradoxes of human history. It both protects and annihilates. It creates and defends nations but may also destroy them. **War,** the most violent form of conflict, refers to organized armed violence aimed at a social group in pursuit of an objective. Wars have existed throughout human history and continue in the contemporary world. Whether war is just or unjust, defensive or offensive, it involves the most horrendous atrocities known to humankind. This is especially true in the 21st century, when nearly all wars are fought in populated areas rather than on remote battlefields, having deadly consequences for civilians. Thus, war is not only a social problem in and of itself, but it also contributes to a host of other social problems—death, disease, and disability, crime and immorality, psychological terror, loss of economic resources, and environmental devastation. In this chapter, we discuss each of these issues within the context of conflict, war, and terrorism, the most threatening of all social problems.

The Global Context: Conflict in a Changing World

As societies have evolved and changed throughout history, the nature of war has also changed. Before industrialization and the sophisticated technology that resulted, war occurred primarily between neighboring groups on a relatively small scale. In the modern world, war can be waged between nations that are separated by thousands of miles as well as between neighboring nations. Increasingly, war is a phenomenon internal to states, involving fighting between the government and rebel groups or among rival contenders for state power. Indeed, Figure 15.1 documents that wars between states, that is, interstate wars, recently made up the smallest percentage of armed conflicts. In the following sections, we examine how war has changed our social world and how our changing social

war Organized armed violence aimed at a social group in pursuit of an objective.

world has affected the nature of war in the industrial and postindustrial information age.

War and Social Change

The very act that now threatens modern civilization—war—is largely responsible for creating the advanced civilization in which we live. Before large political states existed, people lived in small groups and villages. War broke the barriers of autonomy between local groups and permitted small villages to be incorporated into larger political units known as chiefdoms. Centuries of warfare between chiefdoms culminated in the development of the state. The **state** is "an apparatus of power, a set of institutions—the central government, the armed forces, the regulatory and police agencies—whose most important functions involve the use of force, the control of territory, and the maintenance of internal order" (Porter 1994, pp. 5–6). The creation of the state in turn led to other profound social and cultural changes:

> And once the state emerged, the gates were flung open to enormous cultural advances, advances undreamed of during—and impossible under—a regimen of small autonomous villages. . . . Only in large political units, far removed in structure from the small autonomous communities from which they sprang, was it possible for great advances to be made in the arts and sciences, in economics and technology, and indeed in every field of culture central to the great industrial civilizations of the world. (Carneiro 1994, pp. 14–15)

Industrialization and technology could not have developed in the small social groups that existed before military action consolidated them into larger states. Thus, war contributed indirectly to the industrialization and technological sophistication that characterize the modern world. Industrialization, in turn, has had two major influences on war. Cohen (1986) calculated the number of wars fought per decade in industrial and preindustrial nations and concluded that "as societies become more industrialized, their proneness to warfare decreases" (p. 265). Thus, for example, in 2010, there were 15 major armed conflicts, the majority of which were in less developed countries in Africa and Asia (SIPRI 2011a).

Although industrialization may decrease a society's propensity to war, it also increases the potential destruction of war. With industrialization, military technology became more sophisticated and more lethal. Rifles and cannons replaced the clubs, arrows, and swords used in more primitive warfare and, in turn, were replaced by tanks, bombers, and nuclear warheads. Today, the use of new technologies such as high-performance sensors, information processors, directed energy technologies, precision-guided munitions, and computer worms and viruses, has changed the very nature of conflict, war, and terrorism. According to one defense analyst (Billitteri 2010, p. 1):

> Unmanned "drone" aircraft controlled from remote video consoles are being used in increasing numbers by the U.S. military in Afghanistan and by the CIA in Pakistan and other places outside of recognized war zones. . . . The U.S. military now possesses some 7,000 drones, and more than 40 nations, including Iran and China, have drone technology. Unmanned aircraft are being used for everything from border control and environmental monitoring to drug interdiction and building inspections.

■ Total Conflict ■ Internal Conflict ■ Interstate Conflict

Figure 15.1: Global Trends in Violent Conflict, 1946–2007
Source: Hewitt et al. 2010, p. 19. From Hewitt, J. Joseph (2010). "Trends in Global Conflict, 1946–2007", in J. Joseph Hewitt, Jonathan Wikenfeld, and Ted Robert Gurr (eds.). *Peace and Conflict 2010: Executive Summary*. College Park (Md.) Center for International Development and Conflict Management, University of Maryland, p. 19. Copyright © 2010 University of Maryland. Reprinted with permission.

. . . [T]he use of new technologies such as high-performance sensors, information processors, directed energy technologies, precision-guided munitions, and computer worms and viruses, has changed the very nature of conflict, war, and terrorism.

state The organization of the central government and government agencies such as the military, police, and regulatory agencies.

A soldier prepares an unmanned Predator drone for a mission. In the 20th century, industrialization spurred technological innovations that transformed warfare more rapidly than at any other time in human history.

In the postindustrial information age, computer technology has not only revolutionized the nature of warfare, it has made societies more vulnerable to external attacks. For example, in 2009, North Korea attacked computer systems in the United States and in South Korea, resulting in brief disruptions to government and major commercial networks (Sang-Hun & Markoff 2009). As a consequence of such attacks, in 2011, the Pentagon announced that major cyberattacks would be considered acts of war; one official was quoted as saying "If you shut down our power grid, maybe we will put a missile down one of your smokestacks" (*Defense News* 2011).

The Economics of Military Spending

The increasing sophistication of military technology has commanded a large share of resources, totaling $1.63 trillion worldwide in 2010, or about 2.6 percent of the total global domestic product (Perlo-Freeman et al. 2011). Global military spending has been increasing since 1998, with dramatic increases between 2002 and 2008, as a consequence of expenditures for U.S.-led operations after September 11. "Over the period 2001–10, US military spending increased by 81 percent, compared to 32 percent in the rest of the world" (Perlo-Freeman et al. 2011, p. 157). U.S. funding for war since the 9/11 attacks surpassed $1 trillion in 2011 (Belasco 2011). Table 15.1 shows the government's estimates of the costs of the wars in Iraq and Afghanistan, as well as for Operation Noble Eagle, which deploys reservists for homeland defense.

The material costs of war often continue long after the fighting stops. On August 19, 2010, the last U.S. combat brigade withdrew from Iraq, formally ending Operation Iraqi Freedom (Sanok & Freier 2010), the military's name for the invasion and occupation of the country. A little over a year later, President Obama announced that the remaining troops in Iraq would be coming home (Horsley & McEvers 2011). Nonetheless, some analysts estimate that the long-term costs of the Iraq War (e.g., cost of military trainers, lifetime care for the wounded, equipment replacement, interest on debt payments, and the impact on oil prices) may be as high as three to four trillion dollars (Herszenhorn 2008; Stiglitz & Bilmes 2010). In a recent survey about military spending, 22 percent of Americans responded "too little" was being spent, 35 percent responded that the amount was "about right," and 39 percent said that "too much" was being spent (Newport 2011a).

TABLE 15.1 Annual Costs of Wars in Iraq and Afghanistan, and Enhanced Security, 2001–2010 (Estimated, in Billions)

OPERATION	2001 AND 2002	2003	2004	2005	2006	2007	2008	2009	2010	2001–2010	2012 REQUEST 2011	2012
Iraq	0	53	75.9	85.5	101.6	131.2	142.1	95.5	71.3	756.2	49.3	17.7
Afghanistan	20.8	14.7	14.5	20	19	36.9	42.1	59.5	93.8	324.4	118.6	113
Domestic security	13	8	3.7	2.1	0.8	0.5	0.1	0.1	0.1	28.5	0.1	0.1
Unallocated	0	5.5	0	0	0	0	0	0	0	5.5	0	0
TOTAL*	33.8	81.1	94.1	107.6	121.4	171	183.3	155.1	165.3	1283.3	168.1	131.7

*Totals may not add due to rounding.

Source: Belasco 2011.

According to official sources, in 2011 alone, the United States will spend over $118 billion in Afghanistan "for military operations, base security, reconstruction, foreign aid, embassy costs, and veteran's health care" (Dwyer 2011). In June 2011, President Obama ordered the withdrawal of 10,000 U.S. troops in Afghanistan by the end of the year, followed by another 23,000 by the summer of 2012. This will leave 70,000 U.S. troops in Afghanistan until 2014, the administration's target date for complete withdrawal of troops (Sciutto 2011).

The **Cold War,** the state of political tension, economic competition, and military rivalry that existed between the United States and the former Soviet Union for nearly 50 years, provided justification for large expenditures for military preparedness. However, the end of the Cold War, along with the rising national debt, resulted in cutbacks in the U.S. military budget in the 1990s.

Today, military spending has nearly returned to the levels during the Cold War. In 2010, the United States accounted for 43 percent of the world's military spending, the largest single percentage of any nation. The next highest military spenders are China, United Kingdom, France, and Russia, which, along with the United States, account for more than 61 percent of the world's military expenditures (SIPRI 2011b). U.S. national defense outlays, amounting to $698 billion in 2010, include expenditures for salaries of military personnel, research and development, and weapons, but do not include expenditures for overseas wars or veterans benefits (SIPRI 2011b).

What Do You Think? A recent report by the Military Leadership Diversity Commission (2011) recommends, among other things, that the Department of Defense and the military services eliminate "combat exclusion policies" for women. Do you think that women should be allowed to serve in full combat positions in the armed forces?

The U.S. government not only spends money on its own military and defense but also sells military equipment to other countries, either directly or by helping U.S. companies sell weapons abroad. Although the purchasing countries may use these weapons to defend themselves from hostile attack, foreign military sales may pose a threat to the United States by arming potential antagonists. For example, the United States, the world's leading arms-exporting nation, supplied weapons to Iraq to use against Iran during the war between 1980 and 1987. These same weapons were then used against Americans in the Gulf War and Iraq War (Silverstein 2007). Similarly, after the Soviet invasion of Afghanistan in 1979, the United States funded Afghan rebel groups. Years after the Soviets left Afghanistan, rebels continued to fight for control of the country. Using weapons supplied by the United States, the Taliban took over much of Afghanistan and sheltered al Qaeda and Osama bin Laden—also a former recipient of U.S. support—as they planned the attacks on September 11 (Rashid 2000; Bergen 2002).

The United States regularly transfers arms to countries in active conflict. For example, in 2006 and 2007, the United States sold about $9.8 billion in arms to allies for use in war zones in Pakistan, Iraq, Israel, Afghanistan, and Colombia (Berrigan 2009). A 2005 report titled *U.S. Weapons at War: Promoting Freedom or Fueling Conflict?* concluded that, far "from serving as a force for security and stability, U.S. weapons sales frequently serve to empower unstable, undemocratic regimes to the detriment of U.S. and global security" (Berrigan & Hartung 2005). For example, during the 2011 so-called "Arab Spring"—a series of popular uprisings against authoritarian regimes in the Middle East—reports surfaced that authorities in Egypt and Bahrain used teargas and weapons obtained through authorized arms agreements with the United States against unarmed civilians (Braun 2011; Wali & Sami 2011).

Cold War The state of military tension and political rivalry that existed between the United States and the former Soviet Union from the 1950s through the late 1980s.

Sociological Theories of War

Sociological perspectives can help us understand various aspects of war. In this section, we describe how structural functionalism, conflict theory, and symbolic interactionism can be applied to the study of war.

Structural-Functionalist Perspective

Structural functionalism focuses on the functions that war serves and suggests that war would not exist unless it had positive outcomes for society. We have already noted that war has served to consolidate small autonomous social groups into larger political states. An estimated 600,000 autonomous political units existed in the world at about 1000 B.C. Today, that number has dwindled to fewer than 200.

Another major function of war is that it produces social cohesion and unity among societal members by giving them a "common cause" and a common enemy. For example, in 2005, *Newsweek* began a new feature about everyday American heroes called "Red, White, and Proud." Unless a war is extremely unpopular, military conflict also promotes economic and political cooperation. Internal domestic conflicts between political parties, minority groups, and special interest groups often dissolve as they unite to fight the common enemy. During World War II, U.S. citizens worked together as a nation to defeat Germany and Japan.

In the short term, war may also increase employment and stimulate the economy. The increased production needed to fight World War II helped pull the United States out of the Great Depression. The investments in the manufacturing sector during World War II also had a long-term impact on the U.S. economy. Hooks and Bloomquist (1992) studied the effect of the war on the U.S. economy between 1947 and 1972, and concluded that the U.S. government "directed, and in large measure, paid for a 65 percent expansion of the total investment in plant and equipment" (p. 304). War can also have the opposite effect, however. In a 2005 restructuring of the military, the Pentagon, seeking a "meaner, leaner fighting machine," recommended shutting down or reconfiguring nearly 180 military installations, "ranging from tiny Army reserve centers to sprawling Air Force bases that have been the economic anchors of their communities for generations" (Schmitt 2005), at a cost of thousands of civilian jobs.

Wars also function to inspire scientific and technological developments that are useful to civilians. For example, innovations in battlefield surgery during World War II and the Korean War resulted in instruments and procedures that later became common practice in civilian hospital emergency wards (Zoroya 2006). Research on laser-based defense systems led to laser surgery, research in nuclear fission facilitated the development of nuclear power, and the Internet evolved from an U.S. Department of Defense research project. In the U.S. airline industry, which owes much of its technology to the development of air power by the U.S. Department of Defense, the distinction between military and civilian technology is important because different government agencies regulate its exports. The U.S. Department of Commerce regulates the export of parts produced for use on commercial airlines whereas the Department of State imposes stricter controls on parts produced for military aircraft to prevent sales to countries at odds with U.S. foreign policy objectives (Millman 2008). Today, **dual-use technologies,** a term referring to defense-funded innovations that also have commercial and civilian applications, are quite common. For example, "almost all information technology is dual use. We both use the same operating systems, the same networking protocols, the same applications, and even the same security software" (Schneier 2008, p. 1).

As structural functionalists argue, a major function of war is that it produces unity among societal members. War provides a common cause and a common identity. Societal members feel a sense of cohesion, and they work together to defeat the enemy.

Wars . . . function to inspire scientific and technological developments. . . . Research on laser-based defense systems led to laser surgery, research in nuclear fission facilitated the development of nuclear power, and the Internet evolved from an U.S. Department of Defense research project.

dual-use technologies
Defense-funded technological innovations with commercial and civilian use.

War also serves to encourage social reform. After a major war, members of society have a sense of shared sacrifice and a desire to heal wounds and rebuild normal patterns of life. They put political pressure on the state to care for war victims, improve social and political conditions, and reward those who have sacrificed lives, family members, and property in battle. As Porter (1994) explained, "Since . . . the lower economic strata usually contribute more of their blood in battle than the wealthier classes, war often gives impetus to social welfare reforms" (p. 19).

Finally, the U.S. military has historically provided an alternative for the advancement of poor or disadvantaged groups who otherwise face discrimination or limited opportunities in the formal economy. The military's specialized training, tuition assistance programs for a college education, and preferential hiring practices improve the prospects of veterans to find a decent job or career after their service (Military 2007).

Conflict Perspective

Conflict theorists emphasize that the roots of war are often antagonisms that emerge whenever two or more ethnic groups (e.g., Bosnians and Serbs), countries (United States and Vietnam), or regions within countries (the U.S. North and South) struggle for control of resources or have different political, economic, or religious ideologies. In addition, conflict theory suggests that war benefits the corporate, military, and political elites. Corporate elites benefit because war often results in the victor taking control of the raw materials of the losing nations, thereby creating a bigger supply of raw materials for its own industries. Indeed, many corporations profit from defense spending. Under the Pentagon's bid-and-proposal program, for example, corporations can charge the cost of preparing proposals for new weapons as overhead on their Department of Defense contracts. Also, Pentagon contracts often guarantee a profit to the developing corporations. Even if the project's cost exceeds initial estimates, called a cost overrun, the corporation still receives the agreed-on profit. In the late 1950s, President Dwight D. Eisenhower referred to this close association between the military and the defense industry as the **military-industrial complex.**

Contemporary examples of the military-industrial complex include the former Bush administration's direct link to defense spending. Even as the U.S. government was deciding on the war in Iraq, "many former Republican officials and political associates of the Bush administration [were] associated with the Carlyle Group, an equity investment firm with billions of dollars in military and aerospace assets" (Knickerbocker 2002, p. 2). Further, news media have come to rely heavily on retired military professionals with ties to the defense establishment and military contractors for interpretation of the Iraq and Afghanistan wars. This close "intersection of network news and wartime commerce" blurs the line between security policy and private commercial interests (Barstow 2008, p. 1).

The military elite benefit because war and the preparations for it provide prestige and employment for military officials. For example, Military Professional Resources Inc. (MPRI), an organization staffed by former military, defense, law enforcement, and other professionals, operates in more than 40 countries with U.S. and other government contracts involving military and police training, democracy and governance support, disaster management, and other operations in "post-conflict and transitional environments" (MPRI 2011). MPRI is one of four U.S. firms that were under a $300 million contract from the Department of Defense to produce public service announcements, news, and entertainment for the Iraqi media. The three-year project, from 2009 through 2011, supported Iraqi and U.S. political goals by promoting reconciliation and nonsectarian nationalism, as well as pro-military, pro-police, and pro-U.S. attitudes (DeYoung & Pincus 2008; Cary 2010).

According to some estimates, private contractors such as MPRI and many others contributed more than 180,000 civilians to the occupation of Iraq, about 20,000 more than the U.S. military and government employees deployed in country (Miller 2007). Private security companies—for-profit organizations contracted by the U.S. government to perform security functions that the military formerly provided—account for a significant portion of the U.S. forces. In fact, according to estimates from Congressional Research

military-industrial complex A term first used by Dwight D. Eisenhower to connote the close association between the military and defense industries.

Service, as of March 2011, the Department of Defense "had more contractor personnel in Afghanistan and Iraq (155,000) than uniformed personnel (145,000)" (Schwartz & Swain 2011, p. 2). Between 2005 and 2010, the Department of Defense (DOD) obligated about $33.9 billion for civilian contracts to support its operations in Afghanistan, or about 16 percent of its total obligations for the mission during that time. In Iraq, DOD obligated about $15.4 billion on civilian contracts, or 20 percent of its obligations in the country from 2005 to 2010 (Schwarz & Swain 2011).

The North Carolina-based firm Blackwater Worldwide—which changed its name to Xe, pronounced "z," in 2009—received national attention in 2004, when four of its employees in Iraq were killed by a Sunni mob in Fallujah, where their charred corpses were hung along public streets. In September 2007, Blackwater personnel guarding a U.S. diplomatic convoy opened fire at a traffic circle in Baghdad, killing 17 and wounding 24 Iraqi civilians. Company officials initially claimed that their contractors responded proportionately to a nearby attack. In response, and after many years of lodging complaints about alleged indiscriminate firings by private security contractors, the Iraqi government revoked Blackwater's license to operate in Iraq (Tavernise 2007). An FBI investigation of the incident found that most of the killings were in violation of the rules for use of deadly force (Johnston & Broder 2007). The U.S. Department of Justice charged five former Blackwater employees with voluntary manslaughter in January 2009, but the case was dismissed on procedural grounds in December 2009 (CNN 2009).

War also benefits the political elite by giving government officials more power. Porter (1994) observed that "throughout modern history, war has been the level by which . . . governments have imposed increasingly larger tax burdens on increasingly broader segments of society, thus enabling ever-higher levels of spending to be sustained, even in peacetime" (p. 14). Political leaders who lead their country to a military victory also benefit from the prestige and hero status conferred on them.

Finally, feminists and many other analysts often note the overwhelming association between war and gender. By and large, active combat has historically been carried out by men. Nature-based arguments about gender—i.e., that men are innately aggressive or violent and women inherently peaceful—are not generally supported by social science research and do not adequately explain why men are more likely to kill than women. Feminists emphasize the social construction of aggressive masculine identities and their manipulation by elites as important reasons for the association between masculinity and militarized violence (Alexander & Hawkesworth 2008).

The recent entry of women into the U.S. armed forces is changing how women's roles in combat are perceived and even how the military conducts operations in war. The wars in Afghanistan and Iraq "are the first in which tens of thousands of American military women have lived, worked and fought with men for prolonged periods" (Myers 2009, p. 1). Although barred from most combat units, women have served nonetheless in most of the same capacities as male soldiers: in armed patrols, as gunners in vehicles, in bomb disposal units, and as officers leading male troops into combat (Alvarez 2009a). The military houses women in separate quarters, has separate showers and bathrooms for women, allows married couples to live together, and makes contraceptives available to soldiers (Myers 2009). Because women are already serving in combat roles (officially and unofficially) without the negative impact on discipline that opponents have feared, many analysts predict that the military's restrictions on women in combat will eventually be lifted (Alvarez 2009a).

What Do You Think? Since President Clinton signed a bill instituting a "don't ask, don't tell" (DADT) policy in 1993, thousands of gay and lesbian military personnel have been discharged from the military (Stolberg 2010). In September 2011, the repeal of DADT went into effect. Do you think that the full inclusion of gays and lesbians into the military will disrupt unit cohesion?

Although some feminists view women's participation in the military as a matter of equal rights, others object because they see war as an extension of patriarchy and the subordination of women in male-dominated societies. Ironically, because protection of women is perceived as a feature of masculine identity, feminists also point out that war and other conflicts are often justified using "the language of feminism" (Viner 2002). For example, former President Bush used respect for women's rights and protection of women subjugated under the Taliban as a partial justification for the attack on Afghanistan in 2001 (Viner 2002).

Symbolic Interactionist Perspective

The symbolic interactionist perspective focuses on how meanings and definitions influence attitudes and behaviors regarding conflict and war. The development of attitudes and behaviors that support war begins in childhood. American children learn to glorify and celebrate the Revolutionary War, which created our nation. Movies romanticize war, children play war games with toy weapons, and various video and computer games glorify heroes conquering villains.

CHRIS KLEPONIS/AFP/Getty Images

Symbols of patriotism abounded after the death of Osama bin Laden was announced on May 2, 2011. Here, revelers celebrate the death of the al Qaeda leader nearly 10 years after the bombing of the World Trade Center. The majority of Americans believe that bin Laden's death makes America a safer place (Saad 2011).

Symbolic interactionism helps to explain how military recruits and civilians develop a mind-set for war by defining war and its consequences as acceptable and necessary. The word *war* has achieved a positive connotation through its use in various popular public policies—the war on drugs, the war on poverty, and the war on crime. Positive labels and favorable definitions of military personnel facilitate military recruitment and public support of armed forces. In 2005, the Army National Guard launched a $38 million marketing campaign targeting young men and women with advertisements, showing "troops with weapons drawn, helicopters streaking and tanks rolling," all "in an attempt to remind people what the Guard has been about since Colonial Days: fighting wars and protecting the homeland." The new slogan? "The most important weapon in the war on terrorism. You" (Davenport 2005, p. 1).

Many government and military officials convince the masses that the way to ensure world peace is to be prepared for war. Patriotism is a popular sentiment in American society. For example, 75 percent of Americans say they display a U.S. flag at home, at the office, on their car, or on their clothing (Pew Research Center 2011a) (see Table 15.2).

Governments may use propaganda and appeals to patriotism to generate support for war efforts and to motivate individuals to join armed forces. Salladay (2003), for example, notes that those in favor of the war in Iraq have commandeered the language of patriotism, making it difficult but necessary for peace activists to use the same symbols or phrases. In their study of the U.S. peace movement, Woehrle et al. (2008) observed that the government and supporters of U.S. wars often framed the issue as "supporting our boys" or "supporting the troops." This made public discussion about whether a particular war is effective or justifiable appear to be a betrayal of soldiers. By analyzing public statements from leading peace movement groups that opposed the first Gulf War (1991) and the Iraq War, the researchers documented how opponents of these wars developed a counter-narrative that "peace is patriotic" and reframed the issue around how war itself endangered the troops, how government policies failed to provide for the welfare of troops, and how the well-being of civilians was negatively affected by war.

> Governments may use propaganda and appeals to patriotism to generate support for war efforts and to motivate individuals to join armed forces.

TABLE 15.2 Who Flies the Flag?

	DISPLAY FLAG AT HOME, OFFICE, ON CAR, OR ON CLOTHING?*	
	YES	NO
Total	75	25
White	77	23
African American	68	32
Republican	87	13
Democrat	67	33
Independent	76	24
18 to 29 years old	70	30
30 to 49 years old	75	25
50 to 64 years old	81	19
65 years old and older	76	23
College graduate	70	29
Some college	75	25
High school graduate or less	78	21

*Numbers may not sum to 100 because of refusals or responses of "don't know." Poll conducted March 30 to April 3, 2011.

Source: Pew Research Center 2011a. From "75%-A Nation of Flag Wavers", March 0-April 33, 2011, Pew Research Center For the People & the Press, a project of the Pew Research Center. Reprinted with permission. - A Nation of Flag Wavers

To legitimize war, the act of killing in war is not regarded as "murder." Deaths that result from war are referred to as casualties. Bombing military and civilian targets appears more acceptable when nuclear missiles are "peacekeepers" that are equipped with multiple "peace heads." Killing the enemy is more acceptable when derogatory and dehumanizing labels such as Gook, Jap, Chink, Kraut, and Haji convey the attitude that the enemy is less than human.

Such labels are socially constructed as images, often through the media, and are presented to the public. Social constructionists, like symbolic interactionists in general, emphasize the social aspects of "knowing." Thus, Li and Izard (2003) used content analysis to analyze newspaper and television coverage of the World Trade Center and Pentagon attacks on September 11. The researchers examined the first eight hours of coverage of the attacks presented on CNN, ABC, CBS, NBC, and Fox, as well as in eight major U.S. newspapers (including the *Los Angeles Times,* the *New York Times,* and the *Washington Post*). Results of the analysis indicated that newspaper articles tended to have a "human interest" emphasis, whereas television coverage was more often "guiding and consoling." Other results suggested that both media relied most heavily on government sources, that newspapers and the networks were equally factual, and that networks were more homogeneous in their presentation than newspapers. One indication of the importance of the media lies in former President George W. Bush's creation of the Office of Global Communications—"a huge production company, issuing daily scripts on the Iraq War to U.S. spokesmen around the world, auditioning generals to give media briefings, and booking administration stars on foreign news shows" (Kemper 2003, p. 1).

Causes of War

The causes of war are numerous and complex. Most wars involve more than one cause. The immediate cause of a war may be a border dispute, for example, but religious tensions that have existed between the two combatant countries for decades may also contribute to the war. The following section reviews various causes of war.

Conflict over Land and Other Natural Resources

Nations often go to war in an attempt to acquire or maintain control over natural resources, such as land, water, and oil. Michael Klare, author of *Resource Wars: The New Landscape of Global Conflict* (2001), predicted that wars will increasingly be fought over resources as supplies of the most needed resources diminish. Disputed borders have been common motives for war. Conflicts are most likely to arise when borders are physically easy to cross and are not clearly delineated by natural boundaries, such as major rivers, oceans, or mountain ranges.

Water is another valuable resource that has led to wars. Unlike other resources, water is universally required for survival. At various times, the empires of Egypt, Mesopotamia, India, and China all went to war over irrigation rights. In 1998, five years after Eritrea gained independence from Ethiopia, forces clashed over control of the port city Assab and with it, access to the Red Sea.

Not only do the oil-rich countries in the Middle East present a tempting target in themselves, but war in the region can also threaten other nations that are dependent on Middle Eastern oil. Thus, when Iraq seized Kuwait and threatened the supply of oil from the Persian Gulf, the United States and many other nations reacted militarily in the Gulf War. In a document prepared for the Center for Strategic and International Studies, Starr and Stoll (1989) warned that soon

> water, not oil, will be the dominant resource issue of the Middle East. According to World Watch Institute, "despite modern technology and feats of engineering, a secure water future for much of the world remains elusive." The prognosis for Egypt, Jordan, Israel, the West Bank, the Gaza Strip, Syria, and Iraq is especially alarming. If present consumption patterns continue, emerging water shortages, combined with a deterioration in water quality, will lead to more competition and conflict. (p. 1)

Despite such predictions, tensions in the Middle East have erupted into fighting repeatedly in recent years—but not over water. In July 2006, Israel and Lebanon fought a border war that killed a thousand people and displaced a million Lebanese. Civil war erupted in the Palestinian territories between rival parties when Hamas seized control of the Gaza Strip in response to Fatah's refusal to hand over the government after Hamas won legislative elections (BBC 2007a). Further, in the summer of 2007, hundreds were killed when Lebanese security forces attacked Palestinian militants based in Lebanon's refugee camps (BBC 2007b). In December 2008, the Israeli military launched air strikes against Gaza to prepare the way for its January 2009 ground invasion that lasted 22 days. Israeli authorities claimed the campaign was intended to stop Hamas's rocket attacks against civilians in southern Israel (BBC 2009). Under pressure from the United States, Israel and the Palestinian National Authority began direct talks about a "two-state" solution to the conflict in September 2010. By late 2011, the talks had stalled without significant progress toward an agreement. The "Quartet" (i.e., the European Union, the United States, Russia, and the United Nations) has urged that talks resume, and that a settlement be reached by the end of 2012 (BBC 2011).

Conflict over Values and Ideologies

Many countries initiate war not over resources but over beliefs. World War II was largely a war over differing political ideologies: democracy versus fascism. The Cold War involved the clash of opposing economic ideologies: capitalism versus communism. Conflicts over values or ideologies are not easily resolved. They are less likely to end in compromise or negotiation because they are fueled by people's convictions. For example, when a representative sample of American Jews was asked, "Do you agree or disagree with the following statement? 'The goal of Arabs is not the return of occupied territories but rather the destruction of Israel,'" 75 percent agreed, 20 percent disagreed, and 5 percent were unsure (American Jewish Committee 2010).

If ideological differences can contribute to war, do ideological similarities discourage war? The answer seems to be yes; in general, countries with similar ideologies are less likely to engage in war with each other than countries with differing ideological values (Dixon 1994). Democratic nations are particularly disinclined to wage war against one another (Brown et al. 1996; Rasler & Thompson 2005).

Racial, Ethnic, and Religious Hostilities

Racial, ethnic, and religious groups vary in their cultural beliefs, values, and traditions. Thus, conflicts between racial, ethnic, and religious groups often stem from conflicting values and ideologies. Such hostilities are also fueled by competition over land and other scarce natural and economic resources. Gioseffi (1993) noted that "experts agree that the depleted world economy, wasted on war efforts, is in great measure the reason for renewed ethnic and religious strife. 'Haves' fight with 'have-nots' for the smaller piece of the pie that must go around" (p. xviii). Racial, ethnic, and religious hostilities sometimes are perpetuated by a wealthy minority to divert attention away from their exploitations and to maintain their own position of power. Such **constructivist explanations** of ethnic conflict—those that emphasize the role of leaders of ethnic groups in stirring up inter-communal hostility—differ sharply from **primordial explanations,** or those that

constructivist explanations Those explanations that emphasize the role of leaders of ethnic groups in stirring up hatred toward others external to one's group.

primordial explanations Those explanations that emphasize the existence of "ancient hatreds" rooted in deep psychological or cultural differences between ethnic groups, often involving a history of grievance and victimization, real or imagined, by the enemy group.

emphasize the existence of "ancient hatreds" rooted in deep psychological or cultural differences between ethnic groups.

As described by Paul (1998), sociologist Daniel Chirot argued that the recent worldwide increase in ethnic hostilities is a consequence of "retribalization," that is, the tendency for groups, lost in a globalized culture, to seek solace in the "extended family of an ethnic group" (p. 56). Chirot identified five levels of ethnic conflict: (1) multiethnic societies without serious conflict (e.g., Switzerland), (2) multiethnic societies with controlled conflict (e.g., United States and Canada), (3) societies with ethnic conflict that has been resolved (e.g., South Africa), (4) societies with serious ethnic conflict leading to warfare (e.g., Sri Lanka), and (5) societies with genocidal ethnic conflict, including "ethnic cleansing" (e.g., Darfur).

Religious differences as a source of conflict have recently come to the forefront. An Islamic jihad, or holy war, has been blamed for the September 11 attacks on the World Trade Center and Pentagon as well as for bombings in Kashmir, Sudan, the Philippines, Indonesia, Kenya, Tanzania, Saudi Arabia, Spain, and Great Britain. Some claim that Islamic beliefs in and of themselves have led to recent conflicts (Feder 2003). Others contend that religious fanatics, not the religion itself, are responsible for violent confrontations. Wars over differing religious beliefs have led to some of the worst episodes of bloodshed in history, in part, because some religions lend themselves to martyrdom—the idea that dying for one's beliefs leads to eternal salvation. For example, Islamic leader Osama bin Laden claimed that unjust U.S.-Middle East policies are responsible for "dividing the whole world into two sides—the side of believers and the side of infidels" (Williams 2003, p. 18).

> Wars over differing religious beliefs have led to some of the worst episodes of bloodshed in history, in part, because some religions lend themselves to martyrdom— the idea that dying for one's beliefs leads to eternal salvation.

What Do You Think? The release of years of documents on the wars in Iraq and Afghanistan by Wikileaks reveal the deaths of thousands of Iraqi and Afghani civilians by coalition forces (Leigh 2010; Davies & Leigh 2010). In one case, five American soldiers were charged with the murders of at least three civilians in incidents near Kandahar, Afghanistan. One of the four soldiers admitted participating in the murders (Mogelson 2011). Although the remaining defendants deny the accusations, several witnesses and the physical evidence appear to support the charges. Sociologist Stjepan Mestrovic, a war crimes expert, after reading several hundred pages of background information, stated that, "In a dysfunctional unit, we cannot predict who will be the deviant — but we can predict deviance." What do you think he meant by this statement?

Defense against Hostile Attacks

The threat or fear of being attacked may cause the leaders of a country to declare war on the nation that poses the threat. This is an example of what experts in international relations refer to as the **security dilemma**: "actions to increase one's security may only decrease the security of others and lead them to respond in ways that decrease one's own security" (Levy 2001, p. 7). Such situations may lead to war inadvertently. The threat may come from a foreign country or from a group within the country. After Germany invaded Poland in 1939, Britain and France declared war on Germany out of fear that they would be Germany's next victims. Germany attacked Russia in World War I, in part out of fear that Russia had entered the arms race and would use its weapons against Germany. Japan bombed Pearl Harbor hoping to avoid a later confrontation with the U.S. Pacific fleet, which posed a threat to the Japanese military.

In 2001, a U.S.-led coalition bombed Afghanistan in response to the September 11 terrorist attacks. Moreover, in March 2003, the United States, Great Britain, and a loosely

security dilemma A characteristic of the international state system that gives rise to unstable relations between states; as State A secures its borders and interests, its behavior may decrease the security of other states and cause them to engage in behavior that decreases A's security.

coupled "coalition of the willing" invaded Iraq in response to perceived threats of weapons of mass destruction and the reported failure of Saddam Hussein to cooperate with United Nations' weapons inspectors. Yet, in 2005, a presidential commission concluded that the attack on Iraq was based on faulty intelligence and that, in fact, "America's spy agencies were 'dead wrong' in most of their judgments about Iraq's weapons of mass destruction" (Shrader 2005, p. 1). As a result, by 2007, many Americans, more than 60 percent, favored a partial or complete withdrawal from Iraq (CNN/Opinion Research Corporation Poll 2007). Many believe that the public's desire for withdrawal from Iraq was a key factor in the outcome of the 2008 U.S. presidential election. The Obama administration withdrew the last remaining combat brigade from Iraq in August 2010.

Revolutions and Civil Wars

Revolutions and civil wars involve citizens warring against their own government and often result in significant political, economic, and social change. The difference between a revolution and a civil war is not always easy to determine. Scholars generally agree that revolutions involve sweeping changes that fundamentally alter the distribution of power in society (Skocpol 1994). The American Revolution resulted from colonists revolting against British control. Eventually, they succeeded and established a republic where none existed before. The Russian Revolution involved a revolt against a corrupt, autocratic, and out-of-touch ruler, Czar Nicholas II. Among other changes, the revolution led to wide-scale seizure of land by peasants who formerly were economically dependent on large landowners. More recently, after a long history of civil war and semi-autonomous rule, and as part of a 2005 peace deal with northern Sudan, southern Sudan held a referendum and declared independence from northern Sudan. The Republic of South Sudan became the newest member state of the United Nations on July 14, 2011 (Worsnip 2011).

Civil wars may result in a different government or a new set of leaders but do not necessarily lead to such large-scale social change. Because the distinction between a revolution and a civil war depends upon the outcome of the struggle, it may take many years after the fighting before observers agree on how to classify it. Revolutions and civil wars are more likely to occur when a government is weak or divided, when it is not responsive to the concerns and demands of its citizens, and when strong leaders are willing to mount opposition to the government (Barkan & Snowden 2001; Renner 2000).

One of the world's longest-running civil wars came to an end in May 2009. Since 1983, the government of Sri Lanka fought an insurgency led by the Liberation Tigers of Tamil Eelam (LTTE). Also known as the Tamil Tigers, the LTTE were separatist militants who sought to carve an independent state out of the northern and eastern portions of this island country. The war resulted in more than 68,000 deaths (Gardner 2007). The Sri Lankan Army defeated the last remnants of the LTTE and killed their leader in May 2009 (Buncombe 2009). Like many civil wars, the war in Sri Lanka was also a struggle between a majority community (in this case, Sinhalese Buddhists) and a relatively poor and disadvantaged minority community (Hindu Tamils). Despite the end of the war, a political settlement addressing the minority community's grievances and the question of power-sharing has not been achieved. Civil wars have also erupted in newly independent republics created by the collapse of communism in eastern Europe, as well as in Rwanda, Sierra Leone, Chile, Uganda, Liberia, and Sudan. More recently, ranging from civil unrest to civil wars and revolutionary change, the "Arab Spring" of 2011 was characterized as a whole by pro-democracy rebellions in the Middle East.

Nationalism

Some countries engage in war in an effort to maintain or restore their national pride. For example, Scheff (1994) argued that "Hitler's rise to power was laid by the treatment Germany received at the end of World War I at the hands of the victors" (p. 121). Excluded from the League of Nations, punished by the Treaty of Versailles, and ostracized by the world community, Germany turned to nationalism as a reaction to material and symbolic exclusion.

In the late 1970s, Iranian militants seized the U.S. embassy in Tehran and held its occupants hostage for more than one year. President Carter's attempt to use military forces to free the hostages was not successful. That failure intensified doubts about America's ability to use military power effectively to achieve its goals. The hostages in Iran were eventually released after President Reagan took office, but doubts about the strength and effectiveness of the U.S. military still called into question America's status as a world power. Subsequently, U.S. military forces invaded the small island of Grenada because the government of Grenada was building an airfield large enough to accommodate major military armaments. U.S. officials feared that this airfield would be used by countries in hostile attacks on the United States. From one point of view, the large-scale "successful" attack on Grenada functioned to restore faith in the power and effectiveness of the U.S. military.

What Do You Think? The United States has widely used targeted killings against "enemy combatants" in the war on terror, most notably Osama bin Laden but more widely in drone attacks against suspected terrorists in Pakistan, Afghanistan, and Yemen. Some say targeted killing during wartime is legally justifiable under international law and generally considered an act of self-defense (Blum & Heymann 2010). Critics of the policy counter that extrajudicial executions are illegal, even in wartime. They equate targeted killings with assassination, or murder for political reasons, and argue it is a serious violation of U.S. law and international human rights norms (Cohn 2011). What do you think?

Terrorism

Terrorism is the premeditated use, or threatened use, of violence against civilians by an individual or group to gain a political or social objective (Barkan & Snowden 2001; Brauer 2003; Goodwin 2006). Terrorism may be used to publicize a cause, promote an ideology, achieve religious freedom, attain the release of a political prisoner, or rebel against a government. Terrorists use a variety of tactics, including assassinations, skyjackings, suicide bombings, armed attacks, kidnapping and hostage taking, threats, and various forms of bombings. Through such tactics, terrorists struggle to induce fear within a population, create pressure to change policies, or undermine the authority of a government they consider objectionable. Most analysts agree that, unlike war—where a clear winner is more likely—terrorism is unlikely to be completely defeated.

> There can be no final victory in the fight against terrorism, for terrorism (rather than full-scale war) is the contemporary manifestation of conflict, and conflict will not disappear from earth as far as one can look ahead and human nature has not undergone a basic change. But it will be in our power to make life for terrorists and potential terrorists much more difficult. (Laqueur 2006, p. 173)

Types of Terrorism

Terrorism can be either transnational or domestic. **Transnational terrorism** occurs when a terrorist act in one country involves victims, targets, institutions, governments, or citizens of another country. The 1988 bombing of Pan Am Flight 103 over Lockerbie, Scotland, exemplifies transnational terrorism. The incident took the lives of 270 people, including 35 Syracuse University undergraduates returning from an overseas studies program in London. After a 10-year investigation, Abdel Basset Ali al-Megrahi, a Libyan intelligence agent (CNN 2001), was sentenced to life imprisonment in Scotland for his role in preparing the bomb that brought down the plane. In 2003, the Libyan government agreed to pay $2.7 billion in compensation to the victims' families (Smith 2004). After

terrorism The premeditated use or threatened use of violence by an individual or group to gain a political objective.

transnational terrorism Terrorism that occurs when a terrorist act in one country involves victims, targets, institutions, governments, or citizens of another country.

a diagnosis of terminal cancer, the Scottish government released Megrahi "on compassionate grounds" to return to Libya where he received a hero's welcome organized by the Libyan government (Cowell & Sulzberger 2009).

The 2001 attacks on the World Trade Center, the Pentagon, and Flight 93—the most devastating in U.S. history—are also the deadliest examples of transnational terrorism. Al Qaeda, a global alliance of militant Sunni Islamic groups advocating jihad ("holy war") against the West, was also responsible for attacks on U.S. embassies in Kenya and Tanzania (1998) and the bombing of a naval ship, the USS *Cole,* moored in Aden Harbor, Yemen (2000). Al Qaeda has since been linked to deadly bombings in Bali, Indonesia (2002), Madrid (2004), and London (2005). After nearly 20 years of effort to track him down, President Obama announced on May 2, 2011, that Osama bin Laden, the leader of al Qaeda, was dead, killed by Navy Seals and CIA operatives during a raid on a private residential compound in Abbottabad, Pakistan, about 30 miles northeast of Islamabad (Baker et al. 2011). Secretary of Defense Leon Panetta later announced that the U.S. government was "within reach" of defeating al Qaeda (Burns 2011). In a poll taken the day after bin Laden's death was announced, 54 percent of respondents said they thought bin Laden's death would make the United States safer from terrorism, while 28 percent thought the United States would be less safe (Saad 2011).

Many groups other than al Qaeda use terrorism to further their own social and political goals. In fact, the U.S. Department of State identifies 48 "foreign terrorist organizations . . . [that] threaten the security of U.S. nationals or the national security (national defense, foreign relations, or the economic interests) of the United States" (Office of the Coordinator for Counterterrorism 2011). One of the militant groups on this list includes Lashkar-e-Taiba, based in Pakistan. In November 2008, 10 members of Lashkar-e-Taiba laid siege to India's largest city, Mumbai, for 3 days. Armed with assault rifles, pistols, grenades, and at least one bomb, the gunmen stormed Mumbai, opening fire on civilians in public places, including two luxury hotels, a hospital, a railway station, a popular restaurant, a college, a cinema, and a residential compound. At least 166 people were killed during the attacks, and over 300 were wounded (Sengupta 2009).

Domestic terrorism, sometimes called insurgent terrorism (Barkan & Snowden 2001), is exemplified by the 1995 truck bombing of a nine-story federal office building in Oklahoma City, resulting in 168 deaths and the injury of more than 200 people. Gulf War veteran Timothy McVeigh and Terry Nichols were convicted of the crime. McVeigh is reported to have been a member of a paramilitary group that opposes the U.S. government. In 1997, McVeigh was sentenced to death for his actions, and he was executed in 2001 (Barnes 2004). More recently, in 2010, Major Nidal Malik Hasam, an Arab American and a U.S. army psychiatrist, was accused of killing 13 people in a Fort Hood, Texas rampage.

The 2004 bombing of a Russian school by Chechen militants, which killed 324 people, nearly half of them children, is also an act of domestic terrorism, as Chechen rebels continue to fight for an independent state. Since 1968, the Basque separatist group ETA has used bombings and assassinations to fight for the political independence of the Basque population from Spain. In January 2011, weakened by hundreds of arrests in recent years, the group declared a "permanent and general" cease-fire (Tremlett 2011).

JEWEL SAMAD/AFP/Getty Images

Admitted terrorist and U.S. citizen Faisal Shahzad is seen here on a television monitor as federal and state officials brief the media on the 2010 attempted car bombing in New York City's Time Square. Shahzad, who described himself as a "proud terrorist," was sentenced /to life in prison and instructed by the judge to ". . . spend some of the time in prison thinking carefully about whether the Koran wants you to kill lots of people" (Wilson 2010, p. 1).

domestic terrorism Domestic terrorism, sometimes called insurgent terrorism, occurs when the terrorist act involves victims, targets, institutions, governments, or citizens from one country.

Patterns of Global Terrorism

A report by the National Counterterrorism Center (2011) described patterns of terrorism around the world (see Figure 15.2). In 2010:

- There were approximately 11,500 domestic and international terrorist attacks recorded in 72 countries around the world.
- Over 13,200 people lost their lives as a result of these attacks.
- More than 75 percent of the world's terrorist attacks and resulting deaths occurred in the Near East or South Asia.
- Attacks in Afghanistan and Iraq were higher in 2010 than in 2009.
- About 42 percent of those killed lived in South Asia, and 30 percent lived in the Middle East.

In 2010, a random sample of U.S. adults was asked, "How worried are you that you or someone in your family will become a victim of terrorism?" Of respondents, 42 percent said that they were "very worried" or "somewhat worried" (Saad 2010), a sharp drop from 59 percent shortly after the September 11 attacks. In other polls, Americans report that they are more concerned about international terrorism than any other international security matter, including the wars in Afghanistan and Iraq, Iran's and North Korea's nuclear capabilities, and the Israeli-Palestinian conflict (Morales 2009).

> . . . Americans report that they are more concerned about international terrorism than any other international security matter, including the wars in Afghanistan and Iraq, Iran's and North Korea's nuclear capabilities, and the Israeli-Palestinian conflict.

The Roots of Terrorism

In 2003, a panel of terrorist experts came together in Oslo, Norway, to address the causes of terrorism (Bjorgo 2003). Although not an exhaustive list, several causes emerged from the conference:

- A failed or weak state, which is unable to control terrorist operations
- Rapid modernization, when, for example, a country's sudden wealth leads to rapid social change
- Extreme ideologies—religious or secular
- A history of political violence, civil wars, and revolutions
- Repression by a foreign occupation (i.e., invaders to the inhabitants)
- Large-scale racial or ethnic discrimination
- The presence of a charismatic leader

Figure 15.2: Comparison of Terrorist Attacks and Deaths, Iraq Only and Worldwide (excluding Iraq), 2006–2010
Source: National Counterterrorism Center 2011.

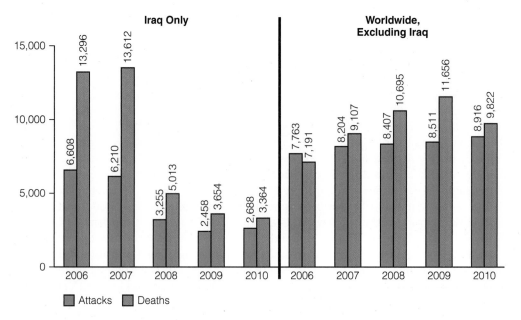

Note that Iraq has several of the characteristics listed here, including rapid modernization (e.g., oil reserves), extreme ideologies (e.g., Islamic fundamentalism), a history of violence (e.g., invasion of Kuwait), large-scale ethnic discrimination (e.g., persecution of Kurdish minority), and a weak state that is unable to control terrorist operations (e.g., the newly elected Iraqi government).

The causes of terrorism listed here, however, are macro in nature. What about social-psychological variables? How do individuals choose to join terrorist organizations or use terrorist tactics? Borum (2003) suggested a four-stage micro-level process. The decision to commit a terrorist act begins with an individual's assessment that *something is not right* (e.g., government-imposed restrictions). Next, individuals *define the situation as unfair* in that the "not right" condition does not apply to everyone (e.g., government-imposed restrictions are imposed on some but not on others). Individuals then begin to *blame specific others for the injustice* (e.g., government leaders) and, finally, to *redefine those who are responsible for the injustice as bad or evil* (e.g., a fascist regime). This process of "ideological development," as portrayed in Figure 15.3, often leads to stereotyping and dehumanizing the enemy, which then facilitates violence. Although the process was developed as a heuristic device, Borum noted that "understanding the mind-set" of a terrorist can help in the fight against terrorism.

America's Response to Terrorism

A government can use both defensive and offensive strategies to fight terrorism (see this chapter's *Animals and Society* feature). Defensive strategies include using metal detectors and X-ray machines at airports and strengthening security at potential targets, such as embassies and military command posts. The Department of Homeland Security (DHS) coordinates such defensive tactics for the U.S. government. DHS was created in 2002, from 22 domestic agencies (e.g., the U.S. Coast Guard, the Immigration and Naturalization Service, and the Secret Service) and, as of 2011, has more than 230,000 employees. With a budget of $56.3 billion in 2011, the mission of DHS is as follows: "We will prevent and deter terrorist attacks and protect against and respond to threats and hazards to the nation. We will ensure safe and secure borders, welcome lawful immigrants and visitors, and promote the free-flow of commerce" (U.S. Department of Homeland Security 2011). The government's capacity to investigate and counteract terrorism, domestically and internationally, is much larger than the Department of Homeland Security. A *Washington Post* investigation of the U.S. counterterrorism industry concluded that nearly 1,300 government organizations and 2,000 private companies are involved in intelligence and counterterrorism (Priest & Arkin 2010). The industry publishes 50,000 intelligence reports each year, and over 854,000 of their employees hold top-secret security clearances. The *Post* report concluded that "The government has built a national security and intelligence system so big, so complex, and so hard to manage, no one really knows if it's fulfilling its most important purpose: keeping its citizens safe" (*Top Secret America* 2010, p. 1).

Offensive strategies include retaliatory raids, such as the U.S. bombing of terrorist facilities in Afghanistan, group infiltration, and preemptive strikes. New legislation facilitates such offensive tactics. In October 2001, the USA PATRIOT Act (Uniting

It's Not Right **It's Not Fair** **It's Your Fault** **You're Evil**

Social and Economic Deprivation → Inequality and Resentment → Blame/Attribution → Generalizing/Stereotyping / Dehumanizing/Demonizing the Enemy (Cause)

Context **Comparison** **Attribution** **Reaction**

Figure 15.3: The Process of Ideological Development
Source: Borum 2003.

There are monuments to them all over the world but none quite as spectacular as the Animals in War Memorial in England. Opened in 2004, the memorial honors the millions of animals that valiantly "suffered and served":

. . . to the mules that were silenced for the Burmese jungle by having their vocal cords severed; the donkeys that collapsed under panniers of ammunition; and the dogs that ripped their paws raw digging for survivors, or had half their faces blown off searching for mines but carried on to find more—they are all remembered; the camels and canaries; the elephants and oxen; the messenger pigeons that flew home on one wing; and even the glow worms, by whose gentle light the soldiers read their maps in the first world war." (p. 1)

Analogous to the Victoria Cross, the most prestigious military award in England, the Dickin Medal honors animals that have shown "conspicuous gallantry or devotion to duty while serving in military conflict" (Treo 2010, p. 1). In 2010, after 62 previous winners—including 32 WW II messenger pigeons, 26 dogs, 3 horses, and a cat—an eight-year-old black Lab named Treo was honored. Treo saved countless military and civilian lives in Afghanistan by locating Taliban-hidden improvised explosive devices (IEDs), many of them in what are called daisy chains—a series of IEDs connected together.

Since 9/11, animals have increasingly been used for guarding U.S. borders and protecting against terrorism. U.S. military forces commonly use dogs because of their keen sense of smell, loyalty, and intelligence. Worldwide, there are presently over 1,300 working K-9 teams, the majority of which are involved in bomb and drug detection (Animals at Arms 2010). However, when the military speaks of "chem dogs," they are referring to dogs that locate weapons of mass destruction. These dogs have such a sophisticated sense of smell that they can detect chemical weapons, or components of chemical weapons, while they are still concealed and before they are released into the atmosphere. Labrador retrievers, Belgian Malinois, and German shepherds, among others, are trained at schools such as the Canine Enforcement Training Center, where each year about 100 certified chemical detector dogs graduate, many of which will become employees of the U.S. Customs and Border Protection (Langley n.d.)

The U.S. Navy, however, is responsible for training two unlikely additions to the military

U.S. Air Force photo/Tech. Sgt. Bennie J. Davis III

Here, Brady Rusk, 12, hugs Eli, his older brother's military working dog, at an early retirement and adoption ceremony at Lackland Air Force Base, Texas. Private first class Colton Rusk's family adopted the black Lab after their son was killed by a sniper in Afghanistan.

lineup—bottlenose dolphins and California sea lions. The Navy's Marine Mammal Program, relatively unknown until the 1990s, trains nearly 80 dolphins and 40 sea lions for a variety of missions (Oliver 2011; Larsen 2011). Both animals have skills that set them apart from other mammals and from each other. When given a cue by a handler, dolphins use *echolocation* or biosonar to search and find a particular target. The dolphin then uses a series of clicks that bounce off the target and echo back to the dolphin's jawbone, which then transmits the information to the brain, creating a "mental image of the object" (Animals at Arms 2010; Simon 2003; U.S. Navy 2010). The dolphin then returns to the handler and through a series of noises and movements conveys information about the target to him or her.

and Strengthening America by Providing Appropriate Tools Required to Intercept and Obstruct Terrorism) was signed into law. The act increases police powers both domestically and abroad. Advocates of the PATRIOT Act argue that, during war, some restrictions of civil liberties are necessary. Moreover, the legislation is not "a substantive shift in policy but a mere revitalization of already established precedents" (Smith 2003, p. 25). Critics hold that the act poses a danger to civil liberties. For example, the original act provides for the *indefinite* detention of immigrants if the immigrant group is defined as a "danger to national security" (Romero 2003). In 2007, Congress revised the act to address some of these concerns, particularly to limit the government's authority to conduct wiretaps. Unlike the earlier PATRIOT Act, which had "sunset" clauses mandating review by Congress before renewal, most of the provisions in the new act are now permanent features of the law.

Among the most controversial U.S. policies in the war on terrorism is the indefinite detention of "enemy combatants" at a military prison and interrogation camp in Guantanamo, Cuba. This is the primary detention center for the Taliban or their allies

Because of this ability, dolphins are also used to detect mines in harbors and other waterways, thereby protecting U.S. citizens and military forces from any number of threats (Larsen 2011; Oliver 2011; NPR 2009). After locating a mine, dolphins are trained to attach a line to the mine, which is then attached to a buoy that floats to the top of the water, thereby informing military personnel where the mine is located. Dolphins not only detect mines that are tethered off the bottom of the ocean, they can also locate mines buried deep under the ocean's floor (Animals at Arms 2010, p. 1).

Dolphins also protect ships, submarines, harbors, nuclear power plants, and U.S. borders against any kind of waterborne threat. After the attack on the USS *Cole,* there were fears that enemy swimmers might try to attach explosive devices to other U.S. ships. In response to the attack, the U.S. Navy created the anti-swimmer dolphin system (Frey 2003).

Anti-swimmer dolphins are able to detect and mark combat swimmers with "an illuminated floating beacon" and, if necessary, use their rock-hard bottlenoses to delay the combat swimmer's movements until U.S. military personnel arrive (Larsen 2011; German 2011, p. 1). Dolphins, who comfortably work in a variety of environments, have served "tours of duty" in Vietnam, the Gulf War, Bahrain, and Iraq, and can be deployed anywhere in the world within a 72-hour period (Frey 2003; Larsen 2011). Recently, sea mammals have been used as sentries around Trident submarines in Kings Bay, Georgia, and Bangor, Washington (German 2011).

Sea lions complete tasks similar to those completed by dolphins but do so in a slightly different manner because of their unique skills. Sea lions can see in the dark as well as underwater, and have underwater directional hearing. Unlike the sheer force that dolphins use to intercept combat swimmers, sea lions use a little finesse (Leinwand 2003, p. 1):

The sea lions are trained to detect swimmers or divers approaching military ships or piers. The animals carry a clamp in their mouths. They approach the swimmer quietly from behind and attach the clamp, which is connected to a rope, to the swimmer's leg. With the person restrained, sailors aboard ships can pull the swimmer out of the water. . . . Navy officials say the sea lions, part of the Shallow Water Intruder Detection System program, are so well-trained that the clamp is on the swimmer before he is aware of it.

Sea lions—and to a lesser extent, dolphins—can also be used to retrieve lost materials but, unlike dolphins, sea lions can go ashore. Since 1962, it is estimated that sea lions in the Marine Mammal Program have "recovered millions of dollars of U.S. Naval torpedoes and instrumentation dropped on the sea floor" (Larsen 2011, p. 1). Sea lions are also useful in retrieving expensive military gear dropped during practice exercises. The sea lion locates the item by listening for the acoustic beacon inside the dropped item and then dives for it, holding a metal plate in its teeth. The metal plate is then attached to the dropped equipment and hauled to the water's surface (German 2011).

Dolphins and sea lions are clearly the first choice for the U.S. Navy, but recently "man's best friend" edged out the two sea mammals. In 2011, Operation Neptune Spear, the mission to kill Bin Laden, used 79 U.S. Navy SEALS (of the human variety) and a dog to enter Abbottabad, Pakistan. However, only 24 commandos slid down the ropes from the hovering Black Hawk helicopter into Bin Laden's compound, and only one had a dog strapped to his back (Frankel 2011a; Jeon 2011). Although details about the dog are as top secret as the identity of the Navy SEALS, it is rumored that, soon after the mission, a Belgian Malinois named Cairo, a member of the Navy SEALS, had a closed-door session with President Obama and the rest of the team (Frankel 2011b).

Like their fellow dolphin and sea lion warriors, dogs form strong relationships with their animal handlers and vice versa. When Private First Class Colton Rusk, a 20-year-old marine, was killed by a sniper, his black Labrador retriever, Eli, crawled on top of him to protect him from further harm. In the obituary, Eli was listed first as one of the surviving members of the grieving family, and the three-year old Lab, after retiring early from the military, was adopted by the Rusks. When Eli arrived at his new home, said mother Kathy Rusk, he ran to Colton's bed, sniffed around, and jumped right in (Bumiller 2011; Roughton 2011).

No machine, no technology, no sophisticated military equipment can do what an animal does or feel what an animal feels.

captured in Afghanistan, as well as suspected terrorists, including al Qaeda members from other regions. Since 2002, as many as 779 detainees have been held at "Gitmo" (Leigh et al. 2011). The Bush administration argued that, because these detainees were not members of a state's army, they were not covered by the **Geneva Conventions,** the principal international treaties governing the laws of war and in particular the treatment of prisoners of war and civilians during wartime. In 2006, the U.S. Supreme Court rejected this argument, ruling that the detainees were subject to minimal protections under the Conventions. Furthermore, in 2008, the Court ruled that the constitution guarantees the right of detainees to challenge their detention in a federal court (Greenburg & de Vogue 2008).

On his second day in office, President Obama ordered the Pentagon to close Guantanamo Bay and all other detention facilities by January 2010. Concerns over security have led many to oppose bringing the detainees to the U.S. mainland for trial and possible incarceration. Nonetheless, in November 2009, the Obama administration announced that Khalid Shaikh Mohammed, the reputed mastermind of the September 11 attacks, would

Geneva Conventions A set of international treaties that govern the behavior of states during wartime, including the treatment of prisoners of war.

Demonstrators stage a mock waterboarding at a protest in front of the White House in March 2008. Such "enhanced interrogation techniques" were used repeatedly by the CIA against suspected terrorists since 9/11. Waterboarding sparked a public debate in the United States and around the world about what constitutes torture and whether the goal of getting important information from potentially dangerous detainees justifies the interrogation methods used.

guerrilla warfare Warfare in which organized groups oppose domestic or foreign governments and their military forces; often involves small groups of individuals who use camouflage and underground tunnels to hide until they are ready to execute a surprise attack.

weapons of mass destruction (WMD) Chemical, biological, and nuclear weapons that have the capacity to kill large numbers of people indiscriminately.

stand trial in a federal courtroom in Manhattan, a few blocks away from the site of the World Trade Center bombings (Savage 2009). In December 2009, the administration announced plans to convert an Illinois state prison to a federal detention center that would house dozens of Guantanamo terrorism suspects (Slevin 2009). In 2011, Congress prohibited the transfer of Guantanamo detainees to U.S. soil. This thwarted the Obama administration's plans to put Khalid Shaikh Mohammed and the other detainees on trial in U.S. courts (Ryan & Khan 2011). As of April 2011, 172 men are imprisoned indefinitely in Guantanamo (Savage et al. 2011).

The treatment of detainees at Guantanamo and at secret detention centers (so-called "black sites") around the world has sparked public debate about interrogation techniques used on suspected terrorists. According to documents released by the Department of Justice in 2009, between 2002 and 2005, interrogators with the CIA and U.S. military intelligence used sleep deprivation, extended periods of standing, prolonged exposure to cold and noise, sexual degradation, and "waterboarding" (i.e., simulated drowning), among many other aggressive techniques (Mazzetti & Shane 2009a). The Bush administration and other supporters argued that such "enhanced interrogation techniques" were necessary to extract useful information that protected society from future terrorist attacks. Opponents, including many officials and security experts, claimed that these practices: (1) did not elicit useful information, (2) were counterproductive politically, (3) were banned by international treaties to which the United States is obligated, and (4) constituted torture. Congress and the White House have since banned these practices. In August 2009, the U.S. Attorney General Eric Holder appointed a federal prosecutor to investigate alleged abuses by the CIA and to consider whether a full criminal investigation was warranted (Mazzetti & Shane 2009b). The probe ended in 2011 without widespread criminal investigations.

Public opinion on this topic is mixed. In response to a question about whether "torture to gain important information from suspected terrorists is justified," 48 percent of American adults said that they believed torture is "often" or "sometimes" justified, whereas 47 percent believed that torture is "rarely" or "never" justified (Pew Research Center for the People & the Press 2009).

Republicans were twice as likely as Democrats to say that torture was "sometimes" justified (49 percent versus 24 percent), and Democrats were nearly three times more likely than Republicans to say that torture was "never" justified (38 percent versus 14 percent).

Combating terrorism is difficult, and recent trends will make it increasingly problematic (Zakaria 2000; Strobel et al. 2001). First, hackers who illegally gain access to classified information can easily acquire data stored on computers. For example, interlopers obtained the fueling and docking schedules of the USS *Cole*. Second, the Internet permits groups with similar interests, once separated by geography, to share plans, fund-raising efforts, recruitment strategies, and other coordinated efforts. Worldwide, thousands of terrorists keep in touch through e-mail, and virtually all terrorist groups maintain websites for recruitment, fund-raising, internal communication, and propagandizing (Weimann 2006). Third, globalization contributes to terrorism by providing international markets where the tools of terrorism—explosives, guns, electronic equipment, and the like—can be purchased. Finally, fighting terrorism under guerrilla warfare-like conditions is increasingly a concern. Unlike terrorist activity, which targets civilians and may be committed by lone individuals, **guerrilla warfare** is committed by organized groups opposing a domestic or foreign government and its military forces.

The possibility of terrorists using weapons of mass destruction is the most frightening scenario of all and, as stated earlier, the motivation for the 2003 war with Iraq. **Weapons of mass destruction (WMD)** include chemical, biological, and nuclear weapons. Anthrax, for example, although usually associated with diseases in animals, is a highly deadly disease in humans and, although preventable by vaccine, has a "lethal lag time." In a hypothetical city of 100,000 people, delaying a vaccination program one day would result in 5,000 deaths; a delay of six days would result in 35,000 deaths. In 2001, trace amounts of anthrax were found in several letters sent to media and political figures, resulting in five deaths and the inspection and closure of several postal facilities (Baliunas 2004). Despite widespread speculation that al Qaeda or Saddam Hussein was responsible for the attacks, investigators soon began to suspect that the source was domestic. In 2008, shortly after the FBI informed Bruce Ivins, a microbiologist at a U.S. Army laboratory in Fort Detrick, Maryland, that they intended to charge him with the crime, the scientist committed suicide (Associated Press 2008).

Other examples of the use of WMD exist. On at least eight occasions, Japanese terrorists dispersed aerosols of anthrax and botulism in Tokyo (Inglesby et al. 1999), and in 2000, a religious cult, hoping to disrupt elections in an Oregon county, "contaminated local salad bars with salmonella, infecting hundreds" (Garrett 2001, p. 76).

In 2004, the poison ricin was detected on a mail-opening machine in Senate majority leader Bill Frist's Washington, DC, office (Associated Press 2005). Furthermore, in 2006, the hit TV series *24* featured a fictional story about a Russian separatist group that stole chemical weapons from the U.S. military and planted them in a gas distribution facility in downtown Los Angeles, threatening to kill thousands unless special agent Jack Bauer (played by Kiefer Sutherland) could catch the terrorists in time. The show builds on real fears among experts, who estimate that Russia possesses about 40,000 metric tons of chemical weapons, the world's largest stockpile by far. The United States, Canada, and the European Union have pledged nearly $2 billion to support Russia's program to secure and destroy all chemical weapons by 2012 (Walker & Tucker 2006). As part of a $7 billion program, Russia has built six facilities to carry out the destruction plan (RIA Novosti 2010). Table 15.3 lists the U.S. government's military goals in combating WMD, including stopping their proliferation, securing existing stockpiles, and working with allies to prevent the use of WMD by potential adversaries (Moroney et al. 2009).

Despite the understandable concerns that Americans have about the potential of terrorist attacks, it is important to remember that death by terrorism is an extraordinarily rare occurrence worldwide, and especially in American society. There has been no major terrorist event in the United States since the attacks of 2001, and as one analyst observed, except for 2001, more Americans have died struck by lightning than by international terrorists (Mueller 2006, p. 13). Nonetheless, fears about terrorist attacks continue to be widespread. For example, in a poll taken the day after the announcement of Osama bin Laden's death during a U.S. raid on his compound, 62 percent said they thought "an act of terrorism is either 'very' or 'somewhat likely' to occur in the U.S. in the next few weeks" (Saad 2011, p. 1).

Joao Silva/The New York Times/Redux Pictures

An Iraqi insurgent takes aim on U.S. positions in Najaf during battle in January 2006. Although the U.S. Army has superior weapons and training, insurgent forces rely on deep knowledge of urban terrain and count on support from the local population.

TABLE 15.3 U.S. Military's Goals in Combating Weapons of Mass Destruction (WMD)
OVERALL COMBATING WMD DESIRED END STATES
The U.S. Armed Forces, in concert with other elements of U.S. national power, deter WMD use.
The U.S. Armed Forces are prepared to defeat an adversary threatening to use WMD and prepared to deter follow-up use.
Existing worldwide WMD are secure, and the U.S. Armed Forces contribute, as appropriate, to secure, reduce, reverse, or eliminate them.
Current or potential adversaries are dissuaded from producing WMD.
Current or potential adversaries' WMD are detected and characterized, and elimination is sought.
Proliferation of WMD and related materials to current and/or potential adversaries is dissuaded, prevented, defeated, or reversed.
If WMD are used against the United States or its interests, the U.S. Armed Forces are capable of minimizing the effects in order to continue operations in a WMD environment and assist U.S. civil authorities, allies, and partners.
The U.S. Armed Forces assist in attributing the source of an attack, respond decisively, and/or deter future attacks.
Allies, partners, and U.S. civilian agencies are capable partners in combating WMD.

Source: Moroney et al. 2009. From Moroney, Jennifer D.P. et al. "Building Partner Capacity to combat Weapons of Mass Destruction." Copyright © 2009 Rand National Defense Institute. Reprinted with permission.

Social Problems Associated with Conflict, War, and Terrorism

Social problems associated with conflict, war, and terrorism include death and disability; rape, forced prostitution, and displacement of women and children; social-psychological costs; diversion of economic resources; and destruction of the environment.

Death and Disability

Many American lives have been lost in wars, including 53,000 in World War I, 292,000 in World War II, 34,000 in Korea, and 47,000 in Vietnam (Leland & Oboroceanu 2010). In Iraq, between March 2003 and July 2011, over 32,000 U.S. troops were wounded and nearly 4,500 were killed (Iraq Coalition Casualty Count 2011). Many civilians and enemy combatants also die or are injured in war. Despite thousands of Iraqi deaths, many Americans are unaware of this tremendous loss of life. In a program the Pentagon developed, American reporters were "embedded" into U.S. military units to provide "journalists with a detailed understanding of military culture and life on the frontlines" (Lindner 2009, p. 21). One of the by-products of this program, however, was that 90 percent of the stories by embedded journalists were written from the perspective of the American soldier, focusing "on the horrors facing the troops, rather than upon the thousands of Iraqis who died" (p. 45).

The impact of war and terrorism extends far beyond those who are killed. Many of those who survive war incur disabling injuries or contract diseases. For example, in South Sudan alone, over 4,200 people were killed or wounded by land mines following the end of civil war with northern Sudan in January 2005 (UNAMO 2011). In 1997, the Mine Ban Treaty, which requires that governments destroy stockpiles within 4 years and clear land mine fields within 10 years, became international law. To date, 156 countries have signed the agreement; 39 countries remain, including China, India, Israel, Russia, and the United States (ICBL 2011). War-related deaths and disabilities also deplete the labor force, create orphans and single-parent families, and burden taxpayers who must pay for the care of orphans and disabled war veterans (see Chapter 2 for a discussion of military health care).

The killing of unarmed civilians is also likely to undermine the credibility of armed forces and make their goals more difficult to defend. In Iraq, for example, the events in Haditha, a city in western Iraq, became international news, outraged Iraqis, and led to intense condemnation of the U.S. mission. In November 2005, after a roadside bomb killed one marine and wounded two others, "Marines shot five Iraqis standing by a car and went house to house looking for insurgents, using grenades and machine guns to clear houses" (Watkins 2007, p. 1). Twenty-four Iraqis were killed, many of them women and children, "shot in the chest and head from close range" (McGirk 2006, p. 3). After *Time* magazine broke the story in March 2006 (McGirk 2006), the military investigated and reversed its claim that the civilians died as a result of the roadside bomb. Eight marines were accused of wrongdoing during the incident; charges against six of them were dismissed, and one was cleared of all charges (Reuters 2008; Puckett & Faraj 2009). The remaining defendant has pled not guilty and awaits trial while on active duty at Camp Pendleton, California (Walker 2011).

What Do You Think? When members of the U.S. military are killed, their remains are flown to Dover Air Force Base in Delaware to be returned to their loved ones. In 1991, during the First Gulf War, the U.S. government banned media coverage of coffins returning home. On February 26, 2009, then Secretary of Defense Robert Gates overturned the 18-year ban and announced that the Pentagon would consult families "on an individual basis" about their wishes, declaring "We ought not presume to make that decision in their place" (quoted in Tyson 2009, p. 2). Do you think the ban should have been lifted? Is there a public interest in allowing media coverage even if families disagree?

The remains of U.S. service members are flown to Dover Air Force Base in Delaware for transfer to loved ones. Between 1991 and 2009, the U.S. public was not permitted to see images of this procedure because of a government ban. In a reversal of that policy, then Defense Secretary Gates announced in February 2009 that the Pentagon would consult with individual families about their wishes before allowing access to the media.

Lastly, individuals who participate in experiments for military research may also suffer physical harm. U.S. representative Edward Markey of Massachusetts identified 31 experiments dating back to 1945, in which U.S. citizens were subjected to harm from participation in military experiments. Markey charged that many of the experiments used human subjects whom were captive audiences or populations considered "expendable," such as elderly individuals, prisoners, and hospital patients. Eda Charlton of New York was injected with plutonium in 1945. She and 17 other patients did not learn of their poisoning until 30 years later. Her son, Fred Shultz, said of his deceased mother:

> I was over there fighting the Germans who were conducting these horrific medical experiments . . . at the same time my own country was conducting them on my own mother. (Miller 1993, p. 17)

Rape, Forced Prostitution, and Displacement of Women and Children

Half a century ago, the Geneva Convention prohibited rape and forced prostitution in war. Nevertheless, both continue to occur in modern conflicts.

Before and during World War II, the Japanese military forced 100,000 to 200,000 women and teenage girls into prostitution as military "comfort women." These women were forced to have sex with dozens of soldiers every day in "comfort stations." Many of the women died as a result of untreated sexually transmitted diseases, harsh punishment, or indiscriminate acts of torture.

Since 1998, Congolese government forces have fought Ugandan and Rwandan rebels. Women have paid a high price for this civil war, in which gang rape is "so violent, so systematic, so common . . . that thousands of women are suffering from vaginal fistula, leaving them unable to control bodily functions and enduring ostracism and the threat of debilitating health problems" (Wax 2003, p. 1). Though much less common than violence against women, aid workers also see increasing incidents of rape and sexual violence against men as "yet another way for armed groups to humiliate and demoralize Congolese communities into submission" (Gettleman 2009, p. 1). United Nations officials called the situation in Congo "the worst sexual violence in the world" (Gettleman 2008, p. 1).

Feminist analyses of wartime rape emphasize that the practice reflects not only a military strategy but also ethnic and gender dominance. For example, Refugees International, a humanitarian aid group, reports that rape is "a systematic weapon of ethnic cleansing" against Darfuris and is "linked to the destruction of their communities" (Boustany 2007, p. 9).

© Georges Gobet/AFP/Getty Images

A child soldier in Liberia points his gun at a cameraman while carting a teddy bear on his back. Although reliable figures are hard to obtain, the UN estimates that about 300,000 child soldiers are fighting in wars worldwide.

Under Darfur's traditional law, prosecution of rapists is nearly impossible: Four male witnesses are required to accuse a rapist in court and single women risk severe corporal punishment for having sex outside of marriage.

War and terrorism also force women and children to flee to other countries or other regions of their homeland. For example, since 1990, at least 17 million children have been forced to leave their homeland because of armed conflict (Save the Children 2005). Refugee women and female children are particularly vulnerable to sexual abuse and exploitation by locals, members of security forces, border guards, or other refugees. In refugee camps, women and children may also be subjected to sexual violation. A 2003 report by Save the Children examined the treatment of women and children in 40 conflict zones. The use of child soldiers was reported in 70 percent of the zones, and trafficking of women and girls was reported in 85 percent of the zones. Wars are particularly dangerous for the very young—"Nine out of ten countries with the highest under-5 mortality rates are experiencing, or emerging from, armed conflict." These include Sierra Leone, Angola, Afghanistan, Niger, Liberia, Somalia, Mali, Chad, and the Democratic Republic of the Congo (Save the Children 2007, p. 13). Despite the *Convention on the Rights of Children* (UNICEF 2011), which was enacted to protect children from the effects of armed conflict, it is estimated that, worldwide, 40 million children of primary school age—nearly one in three of all children in war zones—are prevented from attending school because of armed conflict (Save the Children 2009).

Social-Psychological Costs

Terrorism, war, and living under the threat of war disrupt social-psychological well-being and family functioning. For example, Myers-Brown et al. (2000) report that Yugoslavian children suffered from depression, anxiety, and fear as a response to conflicts in that region, emotional responses not unlike those Americans experienced after the events of 9/11 (NASP 2003). More recently, as a result of the war, there is evidence that many Iraqi children suffer from everything from "nightmares and bedwetting to withdrawal, muteness, panic attacks and violence towards other children, sometimes even to their own parents" (Howard 2007, p. 1). Further, a study of children in postwar Sierra Leone found that over 70 percent of boys and girls whose parents had been killed were at "serious risk" of suicide (Morgan & Behrendt 2009).

Guerrilla warfare is particularly costly in terms of its psychological toll on soldiers. In Iraq, soldiers were repeatedly traumatized as "guerrilla insurgents attack[ed] with impunity," and death was as likely to come from "hand grenades thrown by children, [as] earth-rattling bombs in suicide trucks, or snipers hidden in bombed-out buildings" (Waters 2005, p. 1). In 2008, the U.S. Army estimated that for every 100,000 soldiers deployed to Iraq, 20.2 committed suicide compared to 12 suicides per 100,000 soldiers who were not deployed to Iraq (Alvarez 2009b; U.S. Army Medical Command 2007). This was the first time since the Vietnam War that the suicide rate for U.S. soldiers surpassed that for civilians. The suicide rate increased in 2009 and again in 2010. In both years, when reservists are included in the totals with active duty personnel, more military members committed suicide than died in combat in Iraq and Afghanistan (Donnelly 2011).

The suicide rate increased in 2009 and again in 2010. In both years, when reservists are included in the totals with active duty personnel, more military members committed suicide than died in combat in Iraq and Afghanistan.

What Do You Think? In 2009, Major Nidal Malik Hasan, an army psychiatrist, allegedly shot and killed 13 people, wounding 29 others at a military base in Fort Hood, Texas (Johnson & Shane 2009). Haunted by patients' stories and tired of being harassed because he was Muslim, Hasan—an American citizen—had recently hired an attorney to help him get out of the army (Friedman & Esposito 2009). Clearly, Major Hasan's behavior is a consequence of a complex interplay of psychological and sociological factors but it has led some to ask, should a person be able to leave the military before their term of service has ended? Why or why not?

Military personnel who engage in combat and civilians who are victimized by war may experience a form of psychological distress known as **post-traumatic stress disorder (PTSD),** a clinical term referring to a set of symptoms that can result from any traumatic experience, including crime victimization, rape, or war. Symptoms of PTSD include sleep disturbances, recurring nightmares, flashbacks, and poor concentration (NCPSD 2007). For example, Canadian Lieutenant General Roméo Dallaire, head of the United Nations peacekeeping mission in Rwanda, witnessed horrific acts of genocide. For years after his return, he continued to have images of "being in a valley at sunset, waist deep in bodies, covered in blood" (quoted in Rosenberg 2000, p. 14). PTSD is also associated with other personal problems, such as alcoholism, family violence, divorce, and suicide.

Estimates of PTSD vary widely, although they are consistently higher among combat versus noncombat veterans. In a telephone study of 1,965 military personnel who had been deployed to Iraq and Afghanistan, 14 percent reported current symptoms of PTSD, 14 percent reported current symptoms of major depression, and 9 percent reported symptoms consistent with both conditions (Tanielian et al. 2008). If these estimates are correct, the researchers calculate that, as of April 2008, 303,000 Iraq and Afghanistan veterans were suffering from PTSD or major depression (see this chapter's *The Human Side*). According to one PTSD survivor (Montalván 2011, p. 50):

> Post traumatic stress disorder (PTSD) isn't something you just get over. You don't go back to being who you were. It's more like a snow globe. War shakes you up, and suddenly all those pieces of your life—muscles, bones, thoughts, beliefs, relationships, even your dreams—are floating in the air out of your grip. They'll come down. I'm here to tell you that, with hard work, you'll recover. But they'll never come down where they once were. You're a changed person after combat. Not better or worse, just different.

Figure 15.4 describes the types of traumas to which Afghanistan and Iraq war veterans were exposed.

However, the rate of PTSD among soldiers is difficult to measure for several reasons. First, there is a lag, often of several years, between the time of exposure to trauma and the manifestation of symptoms. Second, soldiers are generally reluctant to report symptoms or to seek help (Wein 2009). When they do seek help, army officials say the Department of Veterans Affairs (VA) is slow to respond (Dao 2009). As of December 2010, despite a major initiative led by the Secretary of Veterans Affairs, 250,000 backlogged PTSD claims (about one-third of all PTSD claims) had not been addressed within the required 125-day response time (Rizzo 2010).

Diversion of Economic Resources

As discussed earlier, maintaining the military and engaging in warfare require enormous financial capital and human support. In 2010, worldwide military expenditures totaled more than $1.63 trillion (Perlo-Freeman et al. 2011). Money that is spent for military purposes could be allocated to social programs. The decision to spend $567 million, equal to the operating cost of the Smithsonian Institution (Center for Defense Information 2003), for one Trident II D-5 missile is a political choice. Similarly, allocating $2.3 billion for a "Virginia"

post-traumatic stress disorder (PTSD) A set of symptoms that may result from any traumatic experience, including crime victimization, war, natural disasters, or abuses.

In January 2006, Olara Otunnu, a former United Nation's advocate for child victims of war, described Uganda as "the worst place in the world to be a child" (Large 2006). For 20 years, starting in 1986, northern Uganda has experienced a guerrilla war that has displaced nearly two million people and caused widespread poverty. For much of the war, the two main combatants were the government of Uganda and the Lord's Resistance Army (LRA), a rebel group led by Joseph Kony, a ruthless and unstable commander who fought to restore his ethnic group to power. Tens of thousands of Ugandan children were forcibly conscripted into the government's army or abducted by the LRA, where they were sexually abused and forced into combat. Many thousands more were killed or wounded.

When the fighting subsided in 2006, more than 800,000 Ugandans were living in government-run camps (Large 2006). Dependent on international aid groups such as the United Nations World Food Programme, UNICEF, and Médecins Sans Frontières for food and medical care, other international groups intervened in the most war-torn areas to develop additional support projects. Among the most important of these projects were those designed to reintegrate child soldiers into the local communities.

Annan et al. (2008), in conducting the Survey of War Affected Youth (SWAY), wanted to provide information to the United Nations and other international agencies to help them develop and improve youth-related social services. Focusing, in part, on the impact of the war on young women and girls, SWAY provides reliable data about this population through a systematic random sample of households in the most affected areas.

Sample and Methods

Between October 2006 and August 2007, the SWAY team employed two methods of data collection. First, they administered a large-scale quantitative survey to young women and girls in two districts of Uganda. Relying on lists used by the World Food Programme for food distribution, 1,100 households were randomly selected. Household members were interviewed and asked to identify all youth living in the household in 1996. From this retrospective sample, Annan and colleagues identified 857 female youth between the ages of 14 and 35 in 2006 (who were between the ages of 4 and 25 in 1996, as the worst violence of the war began). The researchers then identified, located, and interviewed 619 respondents from this sample. Second, Annan et al. (2008) drew a nonrandom subsample of 30 girls from these respondents and conducted in-depth, qualitative interviews with them.

The data the researchers collected focused on the women's and girls' experiences during and after the war, including their experiences with abduction, as "forced wives" assigned to LRA commanders, and with their families, communities, and nongovernmental organizations (NGO) upon returning home. In addition, the SWAY team collected data about the well-being of women and girls in the study, including their recent economic activity, education, physical and mental health, and community participation, among many other variables.

Findings and Conclusions

The data that SWAY researchers collected detailed the extent of violence and identified some important and surprising results. Because Annan and her colleagues had randomly sampled the population, they could use their data to estimate the total amount of violence that young women and girls experienced. They concluded that about 66,000 (20 percent) of all women and girls between the ages of 14 and 30 in 2006 had been abducted by the LRA during the previous 10 years. The researchers report that, "The average age for females at first abduction is approximately 16 years of age, with the majority of females experiencing abduction between the ages of 10 to 18 years of age" (p. 46). Forty percent of abductees were held in captivity for less than 14 days, 25 percent between two weeks and one year, and 19 percent for over a year. Five percent never returned home and are presumed dead. Twenty-five percent were given to LRA members as "forced wives" and, of that number, half gave birth to at least one child as a result of their captivity.

Many of the abducted young women and girls witnessed horrendous acts of violence during their captivity, at rates well above what nonabductees experienced. They witnessed (1) the beating and torture of others (83 percent); (2) the violent death

attack submarine while our schools continue to deteriorate is also a political choice. Between 2001 and 2011, taxpayers paid $1.28 trillion dollars for the cost of the Iraq and Afghanistan Wars, the equivalent of providing (1) one year of low-income health care for 259 million people, or (2) a year's salary for over 19 million elementary school teachers, or (3) one year of Head Start for 165 million children, or (4) 285 million households with renewable electricity (solar photovoltaic), or (5) one year of scholarships for 159 million university students (National Priorities Project 2011). Proposed expenditures for the 2012 fiscal year include more money for national defense than for justice, transportation, veterans' benefits, and natural resources and the environment combined (Office of Management and Budget 2011).

Destruction of the Environment

The environmental damage that occurs during war devastates human populations long after war ends. As the casings for land mines erode, poisonous substances—often carcinogenic—leak into the ground (UNUSA 2004). In 1991, during the Gulf War, Iraqi

of a family member or friend (53 percent); (3) occupied houses being set on fire (42 percent); (4) multiple, simultaneous killings (38 percent); and (5) injuries from combat or a land mine (24 percent).

The abductees were also forced to commit a variety of brutal acts. Seventeen percent said they were forced to kill an opposing soldier; 21 percent were forced to kill an unknown civilian; and 5 percent were forced to kill a family member or friend. In addition, 21 percent were forced to beat or cut a civilian, and 25 percent were made to step on or otherwise abuse a corpse. In each of these categories, less than 1 percent of nonabducted girls and youth committed such acts.

The LRA commanders distributed the young women and girls as "forced wives," partly on the basis of seniority. Half of the senior commanders took five or more wives, whereas low-ranking commanders took, on the average, two wives. When many young women and girls were abducted at the same time, they were distributed randomly among the rebels, as this account from a woman forcibly married after her abduction at 17 illustrates:

> After reaching the LRA camp, I was grouped with other females who were recently abducted. [The commander] gave the orders to his escorts to distribute the wives. Clothes were placed in bags and put into a pile. The men who wanted wives stood nearby and watched as girls were told to pick a bag of clothing. Whoever owned the clothing then became the man to that girl. The clothes I picked belonged to a 35-year-old fighter. (p. 40)

Other forced wives reported that commanders chose wives on the basis of their appearance and their educational level. As one abductee remembered: "The prettiest, more educated girls were the first to be chosen and commanders always chose before fighters. . . . [Commanders] preferred girls with education because they needed them to write down numbers when radio codes were coming in" (p. 43).

Most of the young women and girls (68 percent) who were held for more than two weeks escaped on their own. Of the remainder, the LRA captors released 27 percent and the Ugandan Army rescued 5 percent. Forced wives were more carefully guarded by their captors than other detainees—83 percent escaped, 10 percent were rescued by the Ugandan military, and only 7 percent were released by the LRA.

Surprisingly, only a small percentage of female abductees reported severe psychological distress. Forced wives exhibited a slightly higher rate of distress than other abductees. Problems were most pronounced when the captives returned home to their families, but generally lessened over time. In fact, 83 percent of abductees reported positive relationships with their families and communities. According to Annan et al. (2008), "resilience and acceptance rather than rejection or trauma" was the norm in most families. Despite the violence they were forced to commit in captivity, former abductees were no more likely to commit violence, engage in fights, or report hostility after their release than nonabductees.

The SWAY team used this and other data they created to help aid agencies identify subgroups within the population of abductees who were especially in need of assistance as they returned home. One of these groups was forced wives who gave birth during their abduction: "Girls who return from captivity with children are three times less likely to return to school than those who do not conceive children in captivity and 10 times less likely to return to school than girls who were never abducted" (p. 43). Also, because about one-third of female abductees returned home at the age of 18 or older, the researchers advised that age-based social service programs that targeted children under 18 were missing a significant percentage of the population in need.

Finally, NGO and government aid programs used categories like "abducted" and "forced wives" to determine categories of people requiring the most assistance. Based on the data, the research team concluded that these were "crude and poor predictors of vulnerability" among young women and girls affected by the war. In fact, regardless of whether they had been abducted or not, female youth who came from poor households, were unemployed, estranged from their families, suffered injuries or illness, and exhibited the highest levels of stress were ironically the least likely to receive assistance. By using rigorous methods of sampling and data collection, the *Survey of War Affected Youth* helped support groups revise their category-based approach in targeting the population to one based on "real needs" among those who suffered most from the devastation brought on by the war.

Source: Annan et al. 2008.

troops set 650 oil wells on fire, releasing oil, which covers the surface of the Kuwaiti desert and continues to seep into the ground, threatening underground water supplies. The smoke from the fires that hung over the Gulf region for eight months contained soot, sulfur dioxide, and nitrogen oxides—the major components of acid rain—and a variety of toxic and potentially carcinogenic chemicals and heavy metals. The U.S. Environmental Protection Agency estimates that, in March 1991, about 10 times as much air pollution was being emitted in Kuwait as by all U.S. industrial and power-generating plants combined (Renner 1993; Funke 1994; Environmental Media Services 2002).

Combatants often intentionally exploit natural resources to fuel their efforts. The local elephant population was heavily depleted during the civil war in southern Angola. The rebel group UNITA killed the animals to trade ivory for money to buy weapons. In addition, many elephants were killed or fatally crippled by land mines planted by the guerrillas, though scientists report that, remarkably, the animals seemed to have learned to avoid mined areas (Marshall 2007). Others have been fitted with prosthetic devices

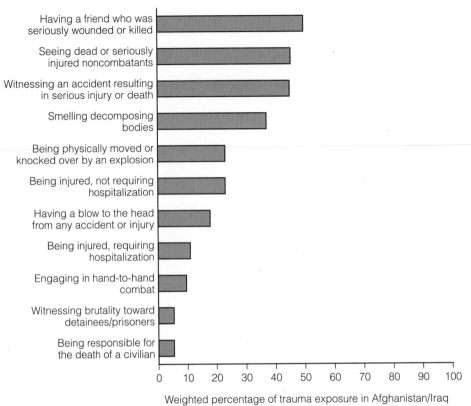

Figure 15.4: Trauma Exposures Reported by Afghanistan/Iraq Service Members (N = 1,965)

Weighted percentage of trauma exposure in Afghanistan/Iraq

The sacrifices associated with military service often impose enormous psychological burdens on soldiers and their families.

nuclear winter The predicted result of a thermonuclear war whereby thick clouds of radioactive dust and particles would block out vital sunlight, lower temperature in the northern hemisphere, and lead to the death of most living things on earth.

for lost limbs. Between 1992 and 1997, UNITA also earned $3.7 billion to support its fighting from the sale of "conflict" or "blood" diamonds (GreenKarat 2007). Depending upon the location of diamonds, mining can be highly destructive to riverbed ecosystems or to areas surrounding open-pit mines.

The ultimate environmental catastrophe facing the planet is thermonuclear war. Aside from the immediate human casualties, poisoned air, poisoned crops, and radioactive rain, many scientists agree that the dust storms and concentrations of particles created by a massive exchange of nuclear weapons would block vital sunlight and lower temperatures in the northern hemisphere, creating a **nuclear winter.** In the event of large-scale nuclear war, most living things on earth would die. For example, a nuclear blast and the resulting blast wave create overpressure—the amount of pressure in excess of ordinary atmospheric levels as measured by pounds per inch (psi). As described in a report sponsored by the U.S. Air Force:

- At 20 psi of overpressure, even reinforced-concrete buildings are destroyed.
- Ten psi will collapse most factories and commercial buildings, as well as wood-frame and brick houses.
- Five psi flattens most houses and lightly constructed commercial and industrial structures.
- Three psi suffices to blow away the walls of steel-frame buildings.
- Even 1 psi will produce flying glass and debris sufficient to injure large numbers of people. (Ochmanek & Schwartz 2008, p. 6)

The fear of nuclear war has greatly contributed to the military and arms buildup, which, ironically, also causes environmental destruction even in times of peace. For example, in practicing military maneuvers, the armed forces demolish natural vegetation, disturb wildlife habitats, erode soil, silt up streams, and cause flooding. Bombs exploded during peacetime leak radiation into the atmosphere and groundwater. From 1945 to 1990, 1,908 bombs were tested—that is, exploded—at more than

Luis Montalván, a 17-year veteran and former captain in the U. S. Army, had courage to spare. He had won a Purple Heart, two Bronze Stars, and a Combat Action Badge, and yet his toughest challenge was yet to come. Returning from Iraq with PTSD and traumatic brain injury, he was unable to function and turned to the bottle for help, isolating himself from the world. But then he met Tuesday, a beautiful, sensitive golden retriever who had been trained as a service dog but who had suffered his own losses. In this excerpt, Luis and Tuesday go on a date with a beautiful woman (pp.179–181).

I let my date step up first, like the traditional gentleman my mother had raised, then stepped into the small entryway with Tuesday.

"No dogs," the bus driver barked.

"Oh, this is my service dog," I said with a smile, expecting her to let me pass.

"I said no dogs, sir."

"But this is my service dog."

She looked Tuesday over, her lips pursed. "That's not a service dog."

"What?"

"I said that's not a service dog, sir."

"Yes, Tuesday is my service dog. See his vest. See my cane."

"Service dog don't wear a vest like that. Service dog has a big handle you hold onto."

"That's a guide dog for the blind," I said, trying to hold myself together. "This is a service dog for the disabled."

"Sir, I know a service dog when I see a service dog, and that ain't no service dog."

I pulled out my cell phone. "Then call the cops," I said angrily, "because I am not getting off this bus."

I was sweating. Big time. It was winter in New York, probably 30 degrees outside, but I could feel the sweat dripping down the back of my neck. I was trying to impress a beautiful, intelligent woman, the first woman I had talked to in a year, and I couldn't even get on a city bus. I mean, it was bad enough having to bring Tuesday with me. I love him but it doesn't exactly say "boyfriend material" when a man has to bring a golden retriever on the first date just to keep it together.

I looked the bus driver straight in the eye. I held out my phone. There was nothing else to do. I couldn't look at my date. I couldn't even look in that direction, because I knew every passenger on the bus was staring at me, and that thought made my PTSD-addled brain reel.

"Please," I said quietly. "I'm on a date. Please let me on."

"No sir," she said loudly, trying to embarrass me.

"Then I'm calling the police," I said angrily, "because you are violating my rights. I hope you are ready to explain to your boss why you wouldn't let a disabled person on your bus."

She gave me a nasty look, waiting for thirty seconds to see if I would back down, then let me pass with a grunt. I felt like throwing up and I was probably shaking, but I had won. I had made it onto a city bus.

Hold it together, Luis, I told myself, as I took a seat beside my date and Tuesday settled between my knees. *Hold it together.*

"Are you all right?"

I took a deep breath and petted the back of Tuesday's head. "I'm okay," I said." That happens sometimes. Right, Tuesday? Right, good boy?" I talk to Tuesday when I'm nervous, even in the middle of conversations.

"I'm sorry."

"Don't be," I said. I looked at her. Smart, beautiful, understanding. She smiled, patted me on the arm, and . . .

"That ain't no service dog."

I looked up. It was the bus driver. She was talking to a woman in the first seat, presumably a friend, but she was intentionally talking loud enough for the whole bus to hear.

Keep it together, Luis. "I think you'll like this restaurant. . . ."

"I've been driving this bus a long time," the bus driver continued, clearly trying to embarrass me. "I know service dogs."

My mind was crumbling. "I think you'll um, I think you'll like . . ."

"Service dog's got a handle."

She was like the voice of PTSD always playing inside my head, always bringing up betrayals.

"Ain't no service dog. I know a service dog."

She was harassing me and wouldn't stop.

"He thinks I don't know a service dog. I know a service dog."

"I'm not deaf," I said in a raised voice. "That's not my disability."

Some of the other passengers laughed. Tuesday turned and nuzzled me with his snout. I grabbed him around the neck, and he leaned into my chest. I could tell by Tuesday's reaction more than anything that I had been shouting. This bus driver was pushing me, harassing me, trying to make me snap.

"Sorry about the dog," she said sarcastically to the people at the next stop. "Man *says* it's a service dog."

I went inside myself. I held Tuesday and tried to beat down the anger. I could feel a migraine coming, but I pushed it away. *Just a few hours,* I thought. *A few hours and it will be over.*

Source: Montalván 2011.

---

35 sites around the world. Although underground testing has reduced radiation, some radioactive material still escapes into the atmosphere and is suspected of seeping into groundwater.

Finally, although arms control and disarmament treaties of the last decade have called for the disposal of huge stockpiles of weapons, no completely safe means of disposing of weapons and ammunition exist. Many activist groups have called for placing weapons in storage until safe disposal methods are found. Unfortunately, the longer the weapons are stored, the more they deteriorate, increasing the likelihood of dangerous leakage. In 2003, a federal judge gave permission, despite objections by

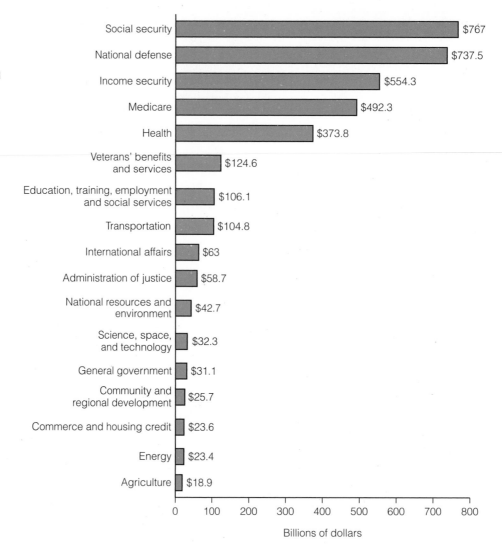

Figure 15.5: Selected Federal U.S. Outlays by Function for 2012 (estimated)
Source: Office of Management and Budget 2011.

Function	Billions of dollars
Social security	$767
National defense	$737.5
Income security	$554.3
Medicare	$492.3
Health	$373.8
Veterans' benefits and services	$124.6
Education, training, employment and social services	$106.1
Transportation	$104.8
International affairs	$63
Administration of justice	$58.7
National resources and environment	$42.7
Science, space, and technology	$32.3
General government	$31.1
Community and regional development	$25.7
Commerce and housing credit	$23.6
Energy	$23.4
Agriculture	$18.9

environmentalists, to incinerate 2,000 tons of nerve agents and mustard gas left from the Cold War era. Although the army said that it is safe to dispose of the weapons, they issued protective gear in case of an "accident" to the nearly 20,000 residents who lived nearby (CNN 2003a).

Strategies for Action: In Search of Global Peace

Various strategies and policies are aimed at creating and maintaining global peace. These include the redistribution of economic resources, the creation of a world government, peacekeeping activities of the United Nations, mediation and arbitration, and arms control.

Redistribution of Economic Resources

Inequality in economic resources contributes to conflict and war because the increasing disparity in wealth and resources between rich and poor nations fuels hostilities and resentment. Therefore, any measures that result in a more equal distribution of economic resources are likely to prevent conflict. John J. Shanahan (1995), retired U.S. Navy vice admiral and former director of the Center for Defense Information, suggested that wealthy nations can help reduce the social and economic roots of conflict by providing economic assistance to poorer countries. Nevertheless, U.S. military expenditures for national defense far outweigh U.S. economic assistance to foreign countries. For instance, the Obama administration's 2012 budget request included $737.5 billion for national security, but just $63 billion for International Affairs (see Table 15.4).

As discussed in Chapter 12, strategies that reduce population growth are likely to result in higher levels of economic well-being. Funke (1994) explained that "rapidly increasing populations in poorer countries will lead to environmental overload and resource depletion in the next century, which will most likely result in political upheaval and violence as well as mass starvation" (p. 326). Although achieving worldwide economic well-being is important for minimizing global conflict, it is important that economic development does not occur at the expense of the environment.

Finally, former United Nations Secretary General Kofi Annan, in an address to the United Nations, observed that it is not poverty per se that leads to conflict but rather the "inequality among domestic social groups" (Deen 2000). Referencing a research report completed by the Tokyo-based United Nations University, Annan argued that "inequality . . . based on ethnicity, religion, national identity, or economic class . . . tends to be reflected in unequal access to political power that too often forecloses paths to peaceful change" (Deen 2000).

TABLE 15.4 United Nations Peacekeeping Operations: Summary Data, 2011	
Military personnel and civilian police serving in peacekeeping operations	98,761
Countries contributing military personnel and civilian police	114
International civilian personnel	5,727
Local civilian personnel	13,655
UN volunteers	2,249
Total number of fatalities in peacekeeping operations since 1948	2,910
Approved budgets for July 1, 2010, to June 30, 2011	$7.83 billion
Estimated total cost of operations from 1948 to June 30, 2010	$69 billion

Source: United Nations 2011.

The United Nations

Founded in 1945 after the devastation of World War II, the United Nations (UN) today includes 193 member states and is the principal organ of world governance. In its early years, the UN's main mission was the elimination of war from society. In fact, the UN charter begins, "We the people of the United Nations—Determined to save succeeding generations from the scourge of war." During the past 65 years, the UN has developed major institutions and initiatives in support of international law, economic development, human rights, education, health, and other forms of social progress. The Security Council is the most powerful branch of the United Nations. Comprised of 15 member states, it has the power to impose economic sanctions against states that violate international law. It can also use force, when necessary, to restore international peace and security.

The United Nations has engaged in 64 peacekeeping operations since 1948 (see Table 15.4) (United Nations 2011).

> United Nations peacekeepers—military personnel in their distinctive blue helmets or blue berets, civilian police and a range of other civilians—help implement peace agreements, monitor cease fires, create buffer zones, or support complex military and civilian functions essential to maintain peace and begin reconstruction and institution building in societies devastated by war. (United Nations 2003, p. 1)

In 2011, the UN was involved in overseeing 15 multinational peacekeeping forces in Afghanistan, Bosnia, East Timor, Sudan, Cyprus, Haiti, Liberia, and Kosovo (United Nations 2011).

In the last few years, the UN has come under heavy criticism. First, in recent missions, developing nations have supplied more than 75 percent of the troops while developed countries—United States, Japan, and Europe—have contributed 85 percent of the finances. As one UN official commented, "You can't have a situation where some nations contribute blood and others only money" (quoted by Vesely 2001, p. 8). Second, a review of UN peacekeeping operations noted several failed missions, including an intervention

Oil smoke from the 650 burning oil wells left in the wake of the Gulf War contains soot, sulfur dioxide, and nitrogen oxides, the major components of acid rain, along with a variety of toxic and potentially carcinogenic chemicals and heavy metals.

in Somalia in which 44 U.S. marines were killed (Lamont 2001). Third, as typified by the debate over the disarming of Iraq, the UN cannot take sides but must wait for a consensus of its members that, if not forthcoming, undermines the strength of the organization (Goure 2003). Even if a consensus emerges, without a standing army, the UN relies on troop and equipment contributions from member states, and there can be significant delays and logistical problems in assembling a force for intervention. The consequences of delays can be staggering. For example, under the Genocide Convention of 1948, the UN is obligated to prevent instances of genocide, defined as "acts committed with the intent to destroy, in whole or in part, a national, racial, ethnical or religious group" (United Nations 1948). In 2000, the UN Security Council formally acknowledged its failure to prevent the 1994 genocide in Rwanda (BBC 2000). After the death of 10 Belgian soldiers in the days leading up to the genocide, and without a consensus for action among the members, the Security Council ignored warnings from the mission's commander about impending disaster and withdrew its 2,500 peacekeepers.

Finally, the concept of the UN is that its members represent individual nations, not a region or the world. And because nations tend to act in their own best economic and security interests, UN actions performed in the name of world peace may be motivated by nations acting in their own interests.

As a result of such criticisms, outgoing UN Secretary General Kofi Annan called on the member states of the UN to approve the most far-reaching changes in the 60-year history of the organization (Lederer 2005). One of the most controversial recommendations concerns the composition of the Security Council, the most important decision-making body of the organization. Annan's recommendation that the 15 members of the Security Council—a body dominated by the United States, Great Britain, France, Russia, and China—be changed to include a more representative number of nations could, if approved, shift the global balance of power. Ban Ki-moon, former foreign minister of South Korea, was elected as the eighth Secretary General in October 2006 (MacAskill et al. 2006), after campaigning for the position on a platform that included support for the UN reforms. So far, the United Nations has not been able to reach agreement about whether and how to expand the membership of the Security Council (Macfarquhar 2010).

What Do You Think? On March 17, 2011, after a month long civil war between opposition forces and forces loyal to Libyan head of state Muammar Gaddafi, the United Nations Security Council passed a resolution authorizing international military intervention in Libya. The resolution empowered UN and NATO forces "to take all necessary measures" short of an occupying force in order "to protect civilians . . . under threat of attack" (United Nations Security Council 2011, p. 3). Near the end of 2011, opposition forces occupied the capital of Libya, Tripoli, and Gaddafi was killed (Gamel & Keath 2011). Given the number of other civil wars in countries around the world, including Uganda, Somalia, Yemen, and Cote d'Ivoire, on what basis should the UN make decisions about intervention? What do you think?

Mediation and Arbitration

Most conflicts are resolved through nonviolent means. Mediation and arbitration are just two of the nonviolent strategies used to resolve conflicts and to stop or prevent war. In mediation, a neutral third party intervenes and facilitates negotiation between representatives or leaders of conflicting groups. Good mediators do not impose solutions but rather help disputing parties generate options for resolving the conflict (Conflict Research Consortium 2003). Ideally, a mediated resolution to a conflict meets at least some of the concerns and interests of each party to the conflict. In other words, mediation attempts to find "win-win" solutions in which each side is satisfied with the solution.

Although mediation is used to resolve conflict between individuals, it is also a valuable tool for resolving international conflicts. For example, former U.S. Senator George Mitchell successfully mediated talks between parties to the conflict in Northern Ireland in 1998. The resulting political agreement continues to hold today. Also, in May 2008, the government of Qatar mediated talks between Lebanon's political parties that resulted in an agreement that averted civil war after an 18-month political crisis. Using mediation as a means of resolving international conflict is often difficult, given the complexity of the issues. However, research by Bercovitch (2003) shows that there are more mediators available and more instances of mediation in international affairs than ever before. For example, Bercovitch identified more than 250 separate mediation attempts during the Balkan Wars of the early and mid-1990s.

Arbitration also involves a neutral third party who listens to evidence and arguments presented by conflicting groups. Unlike mediation, however, the neutral third party in arbitration arrives at a decision that the two conflicting parties agree in advance to accept. For instance, the Permanent Court of Arbitration—an intergovernmental organization based in The Hague since 1899—arbitrates disputes about territory, treaty compliance, human rights, commerce, and investment among any of its 107 member states who signed and ratified either of its two founding legal conventions. Recent cases include a dispute between France, Britain, and Northern Ireland about the Eurotunnel, a dispute between Pakistan and India about the construction of a hydroelectric power project, and boundary disputes between Eritrea and Ethiopia, as well as the government of Sudan and the Sudan People's Liberation Army (Permanent Court of Arbitration 2011).

What Do You Think? Preventing Iran from acquiring nuclear weapons has been a central foreign policy goal for the past several U.S. presidential administrations. The second Bush administration refused to negotiate with Iran on any issue unless Iran first suspended its uranium enrichment program. Following on campaign promises, the Obama administration pursued a policy of direct negotiations with Iran without any preconditions. The latest round of talks, held in January 2011, was inconclusive, and there are currently no plans for another round (Erlanger 2011). Does setting preconditions make it more or less likely that the United States will change Iran's policies? What do you think?

Arms Control and Disarmament

In the 1960s, the United States and the Soviet Union led the world in a nuclear arms race, with each competing to build a larger and more destructive arsenal of nuclear weapons than its adversary. If either superpower were to initiate a full-scale war, the retaliatory powers of the other nation would result in the destruction of both nations as well as much of the planet. Thus, the principle of **mutually assured destruction (MAD)** that developed from nuclear weapons capabilities transformed war from a win-lose proposition to a lose-lose scenario. If both sides would lose in a war, the theory suggested, neither side would initiate war. At its peak year in 1966, the U.S. stockpile of nuclear weapons included more than 32,000 warheads and bombs (see this chapter's *Self and Society* feature).

As their arsenals continued to grow at an astronomical cost, both sides recognized the necessity for nuclear arms control, including the reduction of defense spending, weapons production and deployment, and armed forces. Throughout the Cold War and even today, much of the behavior of the United States and the Soviet Union has been governed by major arms control initiatives. These initiatives include:

arbitration Dispute settlement in which a neutral third party listens to evidence and arguments presented by conflicting groups and arrives at a decision that the parties have agreed in advance to accept.

mutually assured destruction (MAD) A Cold War doctrine referring to the capacity of two nuclear states to destroy each other, thus reducing the risk that either state will initiate war.

- The Limited Test Ban Treaty that prohibited testing of nuclear weapons in the atmosphere, underwater, and in outer space
- The Strategic Arms Limitation Treaties (SALT I and II) that limited the development of nuclear missiles and defensive antiballistic missiles
- The Strategic Arms Reduction Treaties (START I and II) that significantly reduced the number of nuclear missiles, warheads, and bombs
- The Strategic Offensive Reduction Treaty (SORT) that requires that the United States and the Russian Federation each reduce their number of strategic nuclear warheads to between 1,700 and 2,200 by 2012
- The New Strategic Arms Reduction Treaty (new START) that further reduces the number of deployed strategic nuclear warheads to 1,550 for each country and significantly reduces the number of strategic nuclear missile launchers allowed (Atomic Archive 2011)

With the end of the Cold War came the growing realization that, even as Russia and the United States greatly reduced their arsenals, other countries were poised to acquire nuclear weapons or expand their existing arsenals. Thus, the focus on arms control shifted toward **nuclear nonproliferation,** i.e., the prevention of the spread of nuclear technology to nonnuclear states.

Nuclear Nonproliferation Treaty The Nuclear Nonproliferation Treaty (NPT) was signed in 1970, and was the first treaty governing the spread of nuclear weapons technology from the original nuclear weapons states (i.e., the United States, the Soviet Union, the United Kingdom, France, and China) to nonnuclear countries. The NPT was renewed in 2000, and is subject to review every five years. Currently, 189 countries have adopted it. The NPT holds that countries without nuclear weapons will not try to get them; in exchange, the countries with nuclear weapons agree that they will not provide nuclear weapons to countries that do not have them. Signatory states without nuclear weapons also agree to allow the International Atomic Energy Agency to verify compliance with the treaty through on-site inspections (Atomic Archive 2011). Only India, Israel, and Pakistan have not signed the agreement, although each of these states is known to possess a nuclear arsenal. Further, many experts suspect that Iran and Syria—both signatories to the NPT—are developing nuclear weapons programs. Both countries claim that their nuclear reactors are for peaceful purposes (e.g., domestic power consumption), not to develop a nuclear weapons arsenal. However, a report by the International Atomic Energy Agency tentatively concluded that Iran has "sufficient information to be able to design and produce a workable" atomic bomb (Broad & Sanger 2009, p. 1).

In January 2003, North Korea, under suspicion of secretly producing nuclear weapons, announced that it was withdrawing from the treaty, effective immediately (Arms Control Association 2003). In July 2005, after "more than a year of stalemate, North Korea agreed . . . to return to disarmament talks . . . and pledged to discuss eliminating its nuclear-weapons program" (Brinkley & Sanger 2005, p. 1). However, in May 2009, shortly after testing a long-range missile over Japan, North Korea detonated a nuclear device underground, heightening fears that it would launch a nuclear attack against South Korea, China, or Japan (NTI 2011).

Even if military superpowers honor agreements to limit arms, the availability of black-market nuclear weapons and materials presents a threat to global security. One of the most successful nuclear weapons brokers was Pakistan's Abdul Qadeer Khan. Khan, the "father" of Pakistan's nuclear weapons program, sold the technology and equipment required to make nuclear weapons to rogue states such as Iran and North Korea (Powell & McGirk 2005).

In 2003, the United States accused Iran of operating a covert nuclear weapons program. Although Iranian officials responded that their nuclear program was solely for the purpose of generating electricity, concerns escalated with Iran's successful test of a 1,200-mile-range missile in 2009 (Dareini 2009). Hostile public statements about Israel by Iran's president Mahmoud Ahmadinejad deepen already grave concerns about Iran's possible acquisition of nuclear weapons. Analysts fear the development of nuclear weapons by Iran would spur a nuclear competition with Israel, the only other state in the Middle

nuclear nonproliferation
Efforts to prevent the spread of nuclear weapons, or the materials and technology necessary for the production of nuclear weapons.

1. What war saw the first use of a nuclear weapon?
   a. World War I
   b. World War II
   c. The Cold War
2. Nuclear fission involves which process?
   a. Forcing atoms to vibrate at incredibly high speeds
   b. Merging two smaller atoms into one larger atom
   c. Using a neutron to split the nucleus of an atom
3. If two atoms of hydrogen are brought together in nuclear fusion, what do they form?
   a. A helium atom
   b. A hydrogen molecule
   c. Mega-hydrogen
4. What kind of bomb was "Little Boy"?
   a. Gun-triggered fission bomb
   b. Implosion-triggered fission bomb
   c. Teller-Ulam fusion bomb
5. What was the name of the secret U.S. atomic bomb program?
   a. Project U-235
   b. The Oppenheimer Project
   c. The Manhattan Project
6. What is the center of a bomb blast called?
   a. Fallout zone
   b. Hypocenter
   c. Blast radius
7. What is radioactive fallout?
   a. Pressure from the shock wave created by the blast
   b. Clouds of fine radioactive dust particles and debris
   c. A wave of intense heat from the explosion
8. Which of the following Japanese cities did not sustain an atomic blast at the close of World War II?
   a. Hiroshima
   b. Nagasaki
   c. Tokyo
9. What is nuclear winter?
   a. A hypothetical scenario in which nuclear warfare causes clouds of dust to block out the sun
   b. A theoretical situation in which the mass use of nuclear weapons leads to widespread infertility
   c. Another name for the Cold War
10. How high do temperatures at the hypocenter of an atomic blast reach?
    a. Up to 300 million degrees Fahrenheit (167 million degrees Celsius)
    b. Up to 400 million degrees Fahrenheit (223 million degrees Celsius)
    c. Up to 500 million degrees Fahrenheit (300 million degrees Celsius)

Answers: 1. b; 2. c; 3. a; 4. a; 5. c; 6. b; 7. b; 8. c; 9. a; 10. c.

Scoring: Give yourself one point for every correct answer and then assess how well you did using the following scale.

Number of Correct Answers

10	The Pentagon needs you!
9	Congratulations, professor.
8	You're the bomb!
7	Better than most . . .
6	Take a physics course!
5	Time to turn in your lab coat.
4 or below	Don't quit your day job.

Source: Adapted with permission from the Discovery Channel 2011, "Curiosity Challenge: Wild World, Nuclear Bomb Quiz" *Physics, Concepts and Definitions*. Available at http://curiosity.discovery.com

East to possess nuclear weapons, and provoke other states in the region to acquire them (Salem 2007). Nonetheless, analysts disagree about whether there is conclusive evidence that Iran, despite the capability, is designing a nuclear warhead (Shuster 2009).

Many observers consider South Asia the world's most dangerous nuclear rivalry. India and Pakistan are the only two nuclear powers that share a border and have repeatedly fired upon each other's armies while in possession of nuclear weapons (Stimson Center 2007). India first detonated nuclear weapons in 1974. Pakistan detonated six weapons in 1998, a few weeks after India's second round of nuclear tests. Although precise figures are hard to come by, experts estimate that India has about 50 nuclear bombs, and Pakistan has about 60 nuclear bombs (Carnegie Endowment for International Peace 2009). Pakistan, however, is aggressively expanding its nuclear weapons program (Shanker & Sanger 2009). Many fear that a conventional military confrontation between these two countries may someday escalate to an exchange of nuclear weapons. In addition, widespread social unrest in Pakistan coupled with clashes in 2009 between Taliban forces and Pakistan's army 60 miles from the capital, Islamabad, have heightened fears about the security of Pakistan's nuclear arsenal and its vulnerability to seizure by extremist forces (Kerr & Nikitin 2009).

> Even if military superpowers honor agreements to limit arms, the availability of black-market nuclear weapons and materials presents a threat to global security.

AP Photo/Bullit Marquez

As states that want to obtain nuclear weapons are quick to point out, nuclear states that advocate for nonproliferation possess well over 25,000 weapons, a huge reduction in the world's arsenal from Cold War days but still a massive potential threat to the earth. "Do as I say not as I do" is a weak bargaining position. Recognizing this situation and with concern about the possibility of new arms races, many high-level experts have begun to advocate a more comprehensive approach to banning nuclear weapons—with the United States leading the nuclear powers toward complete nuclear disarmament. In January 2007, three former U.S. Secretaries of State (George Schultz, William Perry, and Henry Kissinger) and former chairman of the Senate Armed Services Committee Sam Nunn published an editorial in the *Wall Street Journal* advocating that the United States "take the world to the next stage" of disarmament to "a world free of nuclear weapons" (Schultz et al. 2007). As a start, their proposal advocated the elimination of all short-range nuclear missiles, a complete halt in the production of weapons-grade uranium, and continued reduction of nuclear forces. President Obama visited Russia in July 2009, to sign the New Strategic Arms Reduction Treaty that reduces the strategic nuclear arsenals of the United States and the Russian Federation by at least 25 percent. The Senate ratified the treaty in December 2010 (Sheridan & Branigin 2010). This was reportedly "a first step in a broader effort intended to reduce the threat of such weapons drastically and to prevent their further spread to unstable regions" (Levy & Baker 2009).

Although the United States is the world's top arms exporter, it is also the world's leader in the destruction of conventional weapons. In 2010, the Office of Weapons Removal and Abatement contributed over $160 million to 43 countries to destroy conventional weapons. Since 2001, the United States has helped destroy 90,000 tons of ammunition and 1.5 million weapons (U.S. Department of State 2011).

The Problem of Small Arms

Although the devastation caused by even one nuclear war could affect millions, the easy availability of conventional weapons fuels many active wars around the world. Small arms and light weapons include handguns, submachine guns and automatic weapons, grenades, mortars, land mines, and light missiles. The Small Arms Survey estimated that, in 2006, 639 million firearms were in circulation, with about 59 percent of them being owned legally (Geneva Graduate Institute for International Studies 2006). Civilians purchased 80 percent of these weapons and were the victims of 90 percent of the casualties caused by them. Half a million people die each year as a result of small arms use, 200,000 in homicides and suicides, and the rest during wars and other armed conflicts.

Unlike control of weapons of mass destruction such as chemical and biological weapons, controlling the flow of small arms—especially firearms—is not easy because they have many legitimate uses by the military, by law enforcement officials, and for recreational or sporting activities (Schroeder 2007). Small arms are easy to afford, to use, to conceal, and to transport illegally. The small arms trade is also lucrative. According to official records, the top five exporters of small arms in 2008 were the United States ($715 million), Italy ($562 million), Germany ($472 million), Brazil ($273 million), and Switzerland ($211 million) (Small Arms Survey 2011).

The U.S. Department of State's Office of Weapons Removal and Abatement (WRA) administers a program that has supported the destruction of over 1.3 million small arms and light weapons and 50,000 tons of ammunition in 33 countries since its founding in 2001 (Office of Weapons Removal and Abatement 2010). The availability of these weapons fuels terrorist groups and undermines efforts to promote peace after wars have formally concluded. "If not expeditiously destroyed or secured, stocks of arms and ammunition left over after the cessation of hostilities frequently recirculate into neighboring regions, exacerbating conflict and crime" (U.S. Department of State 2007, p. 1).

Understanding Conflict, War, and Terrorism

As we come to the close of this chapter, how might we have an informed understanding of conflict, war, and terrorism? Each of the three theoretical positions discussed in this chapter reflects the realities of global conflict. As structural functionalists argue, war offers societal benefits—social cohesion, economic prosperity, scientific and technological developments, and social change. Furthermore, as conflict theorists contend, wars often occur for economic reasons because corporate elites and political leaders benefit from the spoils of war—land and water resources and raw materials. The symbolic interactionist perspective emphasizes the role that meanings, labels, and definitions play in creating conflict and contributing to acts of war.

The September 11 attacks on the World Trade Center and the Pentagon and the aftermath—the battle against terrorism, the wars in Iraq and Afghanistan—changed the world Americans live in. For some theorists, these events were inevitable. Political scientist Samuel P. Huntington argued that such conflict represents a **clash of civilizations.** In *The Clash of Civilizations and the Remaking of World Order* (1996), Huntington argued that in the new world order,

> the most pervasive, important and dangerous conflicts will not be between social classes, rich and poor, or economically defined groups, but between people belonging to different cultural entities . . . the most dangerous cultural conflicts are those along the fault lines between civilizations . . . the line separating peoples of Western Christianity, on the one hand, from Muslim and Orthodox peoples on the other. (p. 28)

Although not without critics, the hypothesis of a clash of civilizations has some support. In an interview of almost 10,000 people from nine Muslim states representing half of all Muslims worldwide, only 22 percent had favorable opinions toward the United States (CNN 2003b). Even more significantly, 67 percent saw the September 11 attacks as "morally justified," and the majority of respondents found the United States to be overly materialistic and secular and having a corrupting influence on other nations. Moreover, according to a poll of world public opinion taken every year since 2002, positive attitudes about the United States are consistently lowest in predominantly Muslim countries (Pew Research Center 2011b). Although favorable views of the United States increased dramatically in a small number of Muslim countries after the election of President Obama, the 2011 poll found that only small minorities continued to see the United States favorably in most predominantly Muslim counties, for example, 10 percent in Turkey, 11 percent in Pakistan, 13 percent in Jordan, 18 percent in Palestinian territories, and 20 percent in Egypt.

Conversely, according to a recent poll, 38 percent of Americans have unfavorable views of Islam and 35 percent of the respondents believe that, compared to other religions, Islam is more likely to encourage violence (Pew Research Center for the People & the Press 2010). Moreover, a recent Gallup Poll indicates that barely half (53 percent) of Americans believe that Muslims living in the United States are supportive of the United States, 36 percent believe that Muslims living in the United States are "too extreme in their religious beliefs," and 28 percent believe that U.S. Muslims are "sympathetic to the al Qaeda terrorist organization" (Newport 2011b, p. 1).

Ultimately, we are all members of one community—earth—and have a vested interest in staying alive and protecting the resources of our environment for our own and future generations. But, as we have seen, conflict between groups is a feature of social life that is not likely to disappear. What is at stake—human lives and the ability of our planet to sustain life—merits serious attention. World leaders have traditionally followed the advice of philosopher Carl von Clausewitz: "If you want peace, prepare for war." Thus,

clash of civilizations A hypothesis that the primary source of conflict in the 21st century has shifted away from social class and economic issues and toward conflict between religious and cultural groups, especially those between large-scale civilizations such as the peoples of Western Christianity and Muslim and Orthodox peoples.

nations have sought to protect themselves by maintaining large military forces and massive weapons systems. These strategies are associated with serious costs, particularly in hard economic times. In diverting resources away from other social concerns, defense spending undermines a society's ability to improve the overall security and well-being of its citizens. Conversely, defense-spending cutbacks, although unlikely in the present climate, could potentially free up resources for other social agendas, including lowering taxes, reducing the national debt, addressing environmental concerns, eradicating hunger and poverty, improving health care, upgrading educational services, and improving housing and transportation. Therein lies the promise of a "peace dividend." The hope is that future dialogue on the problems of war and terrorism will redefine national and international security to encompass social, economic, and environmental well-being.

CHAPTER REVIEW

- **What is the relationship between war and industrialization?**
  War indirectly affects industrialization and technological sophistication because military research and development advances civilian-used technologies. Industrialization, in turn, has had two major influences on war: The more industrialized a country is, the lower the rate of conflict, and if conflict occurs, the higher the rate of destruction.

- **What are the latest trends in armed conflicts?**
  Since World War II, wars between two or more states make up the smallest percentage of armed conflicts. In the contemporary era, the majority of armed conflicts have occurred between groups in a single state, who compete for the power to control the resources of the state or to break away and form their own state.

- **In general, how do feminists view war?**
  Feminists are quick to note that wars are part of the patriarchy of society. Although women and children may be used to justify a conflict (e.g., improving women's lives by removing the repressive Taliban in Afghanistan), the basic principles of male dominance and control are realized through war. Feminists also emphasize the social construction of aggressive masculine identities and their manipulation by elites as important reasons for the association between masculinity and militarized violence.

- **What are some of the causes of war?**
  The causes of war are numerous and complex. Most wars involve more than one cause. Some of the causes of war are conflict over land and natural resources; values or ideologies; racial, ethnic, and religious hostilities; defense against hostile attacks; revolution; and nationalism.

- **What is terrorism, and what are the different types of terrorism?**
  Terrorism is the premeditated use, or threatened use, of violence by an individual or group to gain a political or social objective. Terrorism can be either transnational or domestic. Transnational terrorism occurs when a terrorist act in one country involves victims, targets, institutions, governments, or citizens of another country. Domestic terrorism involves only one nation, such as the 1995 truck bombing of a nine-story federal office building in Oklahoma City.

- **What are some of the macro-level "roots" of terrorism?**
  Although not an exhaustive list, some of the macro-level "roots" of terrorism include (1) a failed or weak state, (2) rapid modernization, (3) extreme ideologies, (4) a history of violence, (5) repression by a foreign occupation, (6) large-scale racial or ethnic discrimination, and (7) the presence of a charismatic leader.

- **How has the United States responded to the threat of terrorism?**
  The United States has used both defensive and offensive strategies to fight terrorism. Defensive strategies include using metal detectors and X-ray machines at airports and strengthening security at potential targets, such as embassies and military command posts. The Department of Homeland Security coordinates such defensive tactics. Offensive strategies include retaliatory raids such as the U.S. bombing of terrorist facilities in Afghanistan, group infiltration, and preemptive strikes.

- **What is meant by "diversion of economic resources"?**
  Worldwide, the billions of dollars used on defense could be channeled into social programs dealing with, for example, education, health, and poverty. Thus, defense monies are economic resources diverted from other needy projects.

- **What are some of the criticisms of the United Nations?**
  First, in recent missions, developing nations have supplied more than 75 percent of the troops. Second, several recent UN peacekeeping operations have failed. Third, the UN cannot take sides but must wait for a consensus of its members that, if not forthcoming, undermines the strength of the organization. Fourth, the concept of the UN is that its members represent individual nations, not a region or the world. Because nations tend to act in their own best economic and security interests, UN actions performed in the name of world peace may be motivated by nations acting in their own interests. Finally, the Security Council limits power to a small number of states.

- **What problems do small arms pose?**
  Even after a conflict ends, these weapons circulate in society, making crime worse or falling into the hands of terrorists. Trade in small arms is legal because they have many legitimate uses—for example, by the military, police, and hunters. Because they are small and simple to handle, these weapons are easily concealed and transported, making it difficult to control them.

1. War between states is still the most common form of warfare.
   a. True
   b. False
2. The rise of the modern state is most directly a result of
   a. industrialization and the creation of national markets
   b. innovations in communications technology
   c. the development of armies to control territory
   d. the development of police to control a population
3. Structural-functionalist explanations about war emphasize
   a. that war is a biological necessity
   b. that despite its destructive power, war persists because it fulfills social needs
   c. that war is an anachronism that will eventually disappear
   d. that war is necessary because it benefits political and military elites
4. On the whole, conflicts over values and ideologies are more difficult to resolve than those over material resources.
   a. True
   b. False
5. All wars are a result of unequal distribution of wealth.
   a. True
   b. False
6. Next to defense spending, transfers to foreign governments are the most expensive item in the U.S. government budget.
   a. True
   b. False

7. Which of the following factors is a likely cause of revolutions or civil wars?
   a. A weak or failed state
   b. An authoritarian government that ignores major demands from citizens
   c. The availability of strong opposition leaders
   d. All of the above
8. Primordial explanations of ethnic conflict suggest that
   a. ethnic leaders instigate hostilities to serve their own interests
   b. people become hostile when they blame their frustration with economic hardship on competing ethnic groups
   c. ancient hatreds compel ethnic groups to continue fighting
   d. None of the above
9. The consequences of war and the military on the environment
   a. are prevalent only during wartime
   b. persist in peacetime or for many years after a war is over
   c. do not significantly impact the environment
   d. have been mostly reduced by technological innovations
10. Advocates of nuclear nonproliferation seek to
    a. ban all nuclear weapons
    b. prevent construction of nuclear power plants
    c. prevent new states from acquiring nuclear weapons
    d. None of the above

Answers: 1: b; 2: c; 3: b; 4: a; 5: b; 6: b; 7: d; 8: c; 9: b; 10: c.

arbitration 509
clash of civilizations 513
Cold War 481
constructivist explanations 487
domestic terrorism 491
dual-use technologies 482
Geneva Conventions 495
guerrilla warfare 496

military-industrial complex 483
mutually assured destruction (MAD) 509
nuclear nonproliferation 510
nuclear winter 504
post-traumatic stress disorder (PTSD) 501
primordial explanations 487

security dilemma 488
state 479
terrorism 490
transnational terrorism 490
war 478
weapons of mass destruction (WMD) 496

Turning to Video

▶❙❙ Although a peace agreement between the Sudan government and the opposition party was signed in 2011, the impact of the civil war on the children of Darfur is ongoing. After watching the ABC video *Children of War* (running time 2:21), describe in as much detail as possible what the children's pictures tell you about their experiences.

Online Study Resources

Log in to **www.cengagebrain.com** to access the resources your instructor has assigned. For this book, you can access:

CourseMate

Access chapter-specific learning tools, including learning objectives, practice quizzes, videos, Internet exercises, flash cards, and glossaries, as well as web links, and more in your Sociology CourseMate.

Appendix
Methods of Data Analysis

Description, Correlation, Causation, Reliability and Validity, and Ethical Guidelines in Social Problems Research

There are three levels of data analysis: description, correlation, and causation. Data analysis also involves assessing reliability and validity.

Description

Qualitative research involves verbal descriptions of social phenomena. Having a homeless and single pregnant teenager describe her situation is an example of qualitative research.

Quantitative research often involves numerical descriptions of social phenomena. Quantitative descriptive analysis may involve computing the following: (1) means (averages), (2) frequencies, (3) mode (the most frequently occurring observation in the data), (4) median (the middle point in the data; half of the data points are above the median and half are below), and (5) range (the highest and lowest values in a set of data).

Correlation

Researchers are often interested in the relationship between variables. *Correlation* refers to a relationship between or among two or more variables. The following are examples of correlational research questions: What is the relationship between poverty and educational achievement? What is the relationship between race and crime victimization? What is the relationship between religious affiliation and divorce?

If there is a correlation or relationship between two variables, then a change in one variable is associated with a change in the other variable. When both variables change in the same direction, the correlation is positive. For example, in general, the more sexual partners a person has, the greater the risk of contracting a sexually transmissible disease (STD). As variable A (number of sexual partners) increases, variable B (chance of contracting an STD) also increases. Similarly, as the number of sexual partners decreases, the chance of contracting an STD decreases. Notice that in both cases, the variables change in the same direction, suggesting a positive correlation (see Figure A.1).

When two variables change in opposite directions, the correlation is negative. For example, there is a negative correlation between condom use and contracting STDs. In other words, as condom use increases, the chance of contracting an STD decreases (see Figure A.2).

The relationship between two variables may also be curvilinear, which means that it varies in both the same and opposite directions. For example, suppose a researcher finds that after drinking one alcoholic beverage, research participants are more prone to violent behavior. After two drinks, violent behavior is even more likely, and this trend continues for three and four drinks. So far, the correlation between alcohol consumption and violent behavior is positive. After the research participants have five alcoholic drinks, however, they become less prone to violent behavior. After six and seven drinks, the likelihood of engaging in violent behavior decreases further. Now the correlation between alcohol consumption and violent behavior is negative. Because the correlation

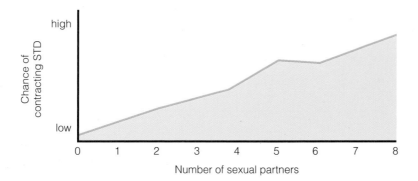

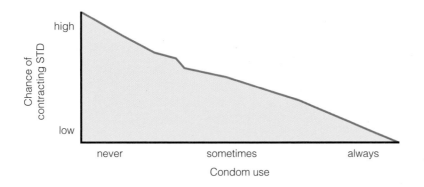

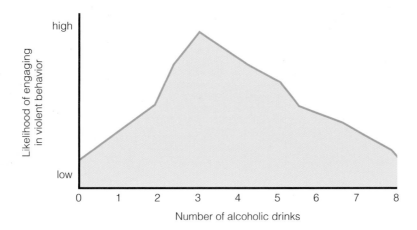

changed from positive to negative, we say that the correlation is curvilinear (the correlation may also change from negative to positive) (see Figure A.3).

A fourth type of correlation is called a spurious correlation. Such a correlation exists when two variables appear to be related, but the apparent relationship occurs only because each variable is related to a third variable. When the third variable is controlled through a statistical method in which the variable is held constant, the apparent relationship between the first two variables disappears. For example, blacks have a lower average life expectancy than whites do. Thus, race and life expectancy appear to be related. This apparent correlation exists, however, because both race and life expectancy are related to socioeconomic status. Because blacks are more likely than whites to be impoverished, they are less likely to have adequate nutrition and medical care.

Causation

If the data analysis reveals that two variables are correlated, we know only that a change in one variable is associated with a change in another variable. We cannot assume, however, that a change in one variable causes a change in the other variable unless our data collection and analysis are specifically designed to assess causation. The research method that best assesses causality is the experimental method (discussed in Chapter 1).

To demonstrate causality, three conditions must be met. First, the data analysis must demonstrate that variable A is correlated with variable B. Second, the data analysis must demonstrate that the observed correlation is not spurious. Third, the analysis must demonstrate that the presumed cause (variable A) occurs or changes before the presumed effect (variable B). In other words, the cause must precede the effect.

It is extremely difficult to establish causality in social science research. Therefore, much social research is descriptive or correlative, rather than causative. Nevertheless, many people make the mistake of interpreting a correlation as a statement of causation. As you read correlative research findings, remember the following adage: "Correlation does not equal causation."

Reliability and Validity

Assessing reliability and validity is an important aspect of data analysis. *Reliability* refers to the consistency of the measuring instrument or technique; that is, the degree to which the way information is obtained produces the same results if repeated. Measures of reliability are made on scales and indexes (such as those in the *Self and Society* features in this text) and on information-gathering techniques, such as the survey methods described in Chapter 1.

Various statistical methods are used to determine reliability. A frequently used method is called the "test-retest method." The researcher gathers data on the same sample of people twice (usually 1 or 2 weeks apart) using a particular instrument or method and then correlates the results. To the degree that the results of the two tests are the same (or highly correlated), the instrument or method is considered reliable.

Measures that are perfectly reliable may be absolutely useless unless they also have a high validity. *Validity* refers to the extent to which an instrument or device measures what it intends to measure. For example, police officers administer "Breathalyzer" tests to determine the level of alcohol in a person's system. The Breathalyzer is a valid test for measuring alcohol consumption.

Validity measures are important in research that uses scales or indexes as measuring instruments. Validity measures are also important in assessing the accuracy of self-report data that are obtained in survey research. For example, survey research on high-risk sexual behaviors associated with the spread of HIV relies heavily on self-report data on topics such as number of sexual partners, types of sexual activities, and condom use. Yet how valid are these data? Do survey respondents underreport the number of their sexual partners? Do people who say they use a condom every time they engage in intercourse really use a condom every time? Because of the difficulties in validating self-reports of number of sexual partners and condom use, we may not be able to answer these questions.

Ethical Guidelines in Social Problems Research

Social scientists are responsible for following ethical standards designed to protect the dignity and welfare of people who participate in research. These ethical guidelines include the following:

1. Freedom from coercion to participate. Research participants have the right to decline to participate in a research study or to discontinue participation at any time during the study. For example, professors who are conducting research using college students should not require their students to participate in their research.
2. Informed consent. Researchers are required to inform potential participants of any aspect of the research that might influence a subject's willingness to participate. After

informing potential participants about the nature of the research, researchers typically ask participants to sign a consent form indicating that the participants are informed about the research and agree to participate in it.

3. Deception and debriefing. Sometimes the researcher must disguise the purpose of the research to obtain valid data. Researchers may deceive participants as to the purpose or nature of a study only if there is no other way to study the problem. When deceit is used, participants should be informed of this deception (debriefed) as soon as possible. Participants should be given a complete and honest description of the study and why deception was necessary.

4. Protection from harm. Researchers must protect participants from any physical and psychological harm that might result from participating in a research study. This is both a moral and a legal obligation. It would not be ethical, for example, for a researcher studying drinking and driving behavior to observe an intoxicated individual leaving a bar, getting into the driver's seat of a car, and driving away.

   Researchers are also obligated to respect the privacy rights of research participants. If anonymity is promised, it should be kept. Anonymity is maintained in mail surveys by identifying questionnaires with a number coding system rather than with the participants' names. When such anonymity is not possible, as is the case with face-to-face interviews, researchers should tell participants that the information they provide will be treated as confidential. Although interviews may be summarized and excerpts quoted in published material, the identity of the individual participants is not revealed. If a research participant experiences either physical or psychological harm as a result of participation in a research study, the researcher is ethically obligated to provide remediation for the harm.

5. Reporting of research. Ethical guidelines also govern the reporting of research results. Researchers must make research reports freely available to the public. In these reports, a researcher should fully describe all evidence obtained in the study, regardless of whether the evidence supports the researcher's hypothesis. The raw data collected by the researcher should be made available to other researchers who might request it for purposes of analysis. Finally, published research reports should include a description of the sponsorship of the research study, its purpose, and all sources of financial support.

Glossary

abortion The intentional termination of a pregnancy.

absolute poverty The lack of resources necessary for material well-being—most importantly, food and water, but also housing, land, and health care.

acculturation The process of adopting the culture of a group different from the one in which a person was originally raised.

achieved status A status that society assigns to an individual on the basis of factors over which the individual has some control.

acid rain The mixture of precipitation with air pollutants, such as sulfur dioxide and nitrogen oxide.

acquaintance rape Rape committed by someone known to the victim.

adaptive discrimination Discrimination that is based on the prejudice of others.

affirmative action A broad range of policies and practices in the workplace and educational institutions to promote equal opportunity as well as diversity.

ageism Negative stereotyping, prejudice, and discrimination based on a person's or group's perceived chronological age.

ageism by invisibility Occurs when older adults are not included in advertising and educational materials.

alienation A sense of powerlessness and meaninglessness in people's lives.

alternative certification programs Programs whereby college graduates with degrees in fields other than education can become certified if they have "life experience" in industry, the military, or other relevant jobs.

American Recovery and Reinvestment Act of 2009 Legislation that provided an economic stimulus package of $787 billion to create and save jobs and to reinvigorate the U.S. economy.

androgyny Having both traditionally defined feminine and masculine characteristics.

anomie A state of normlessness in which norms and values are weak or unclear.

anti-miscegenation laws Laws banning interracial marriage until 1967, when the Supreme Court (in *Loving v. Virginia*) declared these laws unconstitutional.

arbitration Dispute settlement in which a neutral third party listens to evidence and arguments presented by conflicting groups and arrives at a decision that the parties have agreed in advance to accept.

ascribed status A status that society assigns to an individual on the basis of factors over which the individual has no control.

assimilation The process by which formerly distinct and separate groups merge and become integrated as one.

automation Dominant in an industrial society, the replacement of human labor with machinery and equipment that is self-operating.

aversive racism A subtle form of prejudice that involves feelings of discomfort, uneasiness, disgust, fear, and pro-white attitudes.

baby boomers The generation of Americans born between 1946 and 1964, a period of high birthrates.

barrios Slums in the United States that are occupied primarily by Latinos.

Basic Economic Security Tables Index (BEST) A measure of the basic needs and income workers require for economic security.

behavior-based safety programs A strategy used by business management that attributes health and safety problems in the workplace to workers' behavior, rather than to work processes and conditions.

beliefs Definitions and explanations about what is assumed to be true.

bigamy The criminal offense in the United States of marrying one person while still legally married to another.

bilingual education In the United States, teaching children in both English and their non-English native language.

binge drinking As defined by the U.S. Department of Health and Human Services, drinking five or more drinks on the same occasion on at least one day in the past 30 days prior to the National Survey on Drug Use and Health.

biodiversity The diversity of living organisms on earth.

bioinvasion The intentional or accidental introduction of plant, animal, insect and other species in regions where they are not native.

biphobia When prejudice is directed toward bisexual individuals.

bisexuality The emotional, cognitive, and sexual attraction to members of both sexes.

boy code A set of societal expectations that discourages males from expressing emotion, weakness, or vulnerability, or asking for help.

bullying Bullying "entails an imbalance of power that exists over a long period of time in which the more powerful intimidate or belittle others" (Hurst 2005, p. 1).

capital punishment The state (the federal government or a state) takes the life of a person as punishment for a crime.

capitalism An economic system characterized by private ownership of the means of production and distribution of goods and services for profit in a competitive market.

character education Education that emphasizes the moral and ethical aspects of an individual.

charter schools Schools that originate in contracts, or charters, which articulate a plan of instruction that local or state authorities must approve.

chattel slavery A form of slavery in which slaves are considered property that can be bought and sold.

chemical dependency A condition in which drug use is compulsive and users are unable to stop because of physical and/or psychological dependency.

child abuse The physical or mental injury, sexual abuse, negligent treatment, or maltreatment of a child younger than age 18 by a person who is responsible for the child's welfare.

child labor Involves children performing work that is hazardous, that interferes with a child's education, or that harms a child's health or physical, mental, social, or moral development.

civil union A legal status that entitles same-sex couples who apply for and receive a civil union certificate to nearly all of the benefits available to married couples.

clash of civilizations A hypothesis that the primary source of conflict in the 21st century has shifted away from social class and economic issues and toward conflict between religious and cultural groups, especially those between large-scale civilizations such as the peoples of Western Christianity and Muslim and Orthodox peoples.

classic rape Rape committed by a stranger, with the use of a weapon, resulting in serious bodily injury to the victim.

clearance rate The percentage of crimes in which an arrest and official charge have been made and the case has been turned over to the courts.

club drugs A general term for illicit, often synthetic, drugs commonly used at nightclubs or all-night dances called "raves."

Cold War The state of military tension and political rivalry that existed between the United States and the former Soviet Union from the 1950s through the late 1980s.

coming out The ongoing process whereby a lesbian, gay, or bisexual individual becomes aware of his or her sexuality, accepts and incorporates it into his or her overall sense of self, and shares that information with others such as family, friends, and coworkers.

comparable worth The belief that individuals in occupations, even in different occupations, should be paid equally if the job requires "comparable" levels of education, training, and responsibility.

comprehensive primary health care An approach to health care that focuses on the broader social determinants of health, such as poverty and economic inequality, gender inequality, environment, and community development.

compressed workweek A work arrangement that allows employees to condense their work into fewer days (e.g., four 10-hour days each week).

computer crime Any violation of the law in which a computer is the target or means of criminal activity.

constructivist explanations Those explanations that emphasize the role of leaders of ethnic groups in stirring up hatred toward others external to one's group.

contact hypothesis The idea that contact between groups is necessary for the reduction of prejudice.

corporal punishment The intentional infliction of pain for a perceived misbehavior.

corporate violence The production of unsafe products and the failure of corporations to provide a safe working environment for their employees.

corporate welfare Laws and policies that benefit corporations.

corporatocracy A system of government that serves the interests of corporations and that involves ties between government and business.

couch-homeless Individuals who do not have a home of their own and who stay at the home of family or friends.

covenant marriage A type of marriage (offered in a few states) that requires premarital counseling and that permits divorce only under condition of fault or after a marital separation of more than two years.

crack A crystallized illegal drug product produced by boiling a mixture of baking soda, water, and cocaine.

crime An act, or the omission of an act, that is a violation of a federal, state, or local criminal law for which the state can apply sanctions.

crime rate The number of crimes committed per 100,000 population.

cultural imperialism The indoctrination into the dominant culture of a society.

cultural lag A condition in which the material part of culture changes at a faster rate than the nonmaterial part.

cultural sexism The ways in which the culture of society perpetuates the subordination of an individual or group based on the sex classification of that individual or group.

culture The meanings and ways of life that characterize a society, including beliefs, values, norms, sanctions, and symbols.

cyber-bullying The use of electronic devices (e.g., websites, e-mail, instant messaging, text messaging) to send or post negative or hurtful messages or images about an individual or a group.

cybernation Dominant in postindustrial societies, the use of machines to control other machines.

cycle of abuse A pattern of abuse in which a violent or abusive episode is followed by a makeup period when the abuser expresses sorrow and asks for forgiveness and "one more chance," before another instance of abuse occurs.

date-rape drugs Drugs that are used to render victims incapable of resisting sexual assaults.

decriminalization The removal of criminal penalties for a behavior, as in the decriminalization of drug use.

deep ecology The view that maintaining the earth's natural systems should take precedence over human needs, that nature has a value independent of human existence, and that humans have no right to dominate the earth and its living inhabitants.

Defense of Marriage Act (DOMA) Federal legislation that states that marriage is a legal union between one man and one woman and denies federal recognition of same-sex marriage.

deforestation The conversion of forestland to nonforestland.

deinstitutionalization The removal of individuals with psychiatric disorders from mental hospitals and large residential institutions to outpatient community mental health centers.

demand reduction One of two strategies in the U.S. war on drugs (the other is supply reduction), demand reduction focuses on reducing the demand for drugs through treatment, prevention, and research.

demographic transition theory A theory that attributes population growth patterns to changes in birth rates and death rates associated with the process of industrialization.

dependent variable The variable that the researcher wants to explain; the variable of interest.

deregulation The reduction of government control over, for example, certain drugs.

desertification The degradation of semiarid land, which results in the expansion of desert land that is unusable for agriculture.

deterrence The use of harm or the threat of harm to prevent unwanted behaviors.

devaluation hypothesis The hypothesis that women are paid less because the work they perform is socially defined as less valuable than the work men perform.

developed countries Countries that have relatively high gross national income per capita and have diverse economies made up of many different industries.

developing countries Countries that have relatively low gross national income per capita, with simpler economies that often rely on a few agricultural products.

discrimination Actions or practices that result in differential treatment of categories of individuals.

divorce mediation A process in which divorcing couples meet with a neutral third party (mediator) who assists the individuals in resolving issues such as property division, child custody, child support, and spousal support in a way that minimizes conflict and encourages cooperation.

domestic partners See *domestic partnerships.*

domestic partnership A status granted to unmarried couples, including gay and lesbian couples, by some states, counties, cities, and workplaces that conveys various rights and responsibilities.

domestic terrorism Domestic terrorism, sometimes called insurgent terrorism, occurs when the terrorist act involves victims, targets, institutions, governments, or citizens from one country.

doubling time The time required for a population to double in size from a given base year if the current rate of growth continues.

drug Any substance other than food that alters the structure or functioning of a living organism when it enters the bloodstream.

drug abuse The violation of social standards of acceptable drug use, resulting in adverse physiological, psychological, and/or social consequences.

drug courts Special courts that divert drug offenders to treatment programs in lieu of probation or incarceration.

dual-use technologies Defense-funded technological innovations with commercial and civilian use.

earned income tax credit (EITC) A refundable tax credit based on a working family's income and number of children.

e-commerce The buying and selling of goods and services over the Internet.

elderly support ratio The ratio of working-age adults (15 to 64) to adults ages 65 and older in a population.

e-learning Learning in which, by time or place, the learner is separated from the teacher.

economic institution The structure and means by which a society produces, distributes, and consumes goods and services.

ecosystems The complex and dynamic relationships between forms of life and the environments they inhabit.

ecoterrorism Any crime intended to protect wildlife or the environment that is violent, puts human life at risk, or results in damages of $10,000 or more.

education dividend The additional benefit of universal education for women is that it reduces the death rate of children under 5 years of age.

elder abuse The physical or psychological abuse, financial exploitation, or medical abuse or neglect of the elderly.

emotion work Work that involves caring for, negotiating, and empathizing with people.

Employment Non-Discrimination Act (ENDA) A proposed federal bill that would protect LGBTs from workplace discrimination; has been up for congressional debate on a number of occasions since 1994, but has never been signed into law.

environmental footprint The demands that humanity makes on the earth's natural resources, often measured in terms of global hectares (gha) per person.

environmental injustice Also known as **environmental racism**, the tendency for marginalized populations and communities to disproportionately experience adversity due to environmental problems.

environmental racism See *environmental injustice.*

environmental refugees Individuals who have migrated because they can no longer secure a livelihood as a result of deforestation, desertification, soil erosion, and other environmental problems.

Equal Employment Opportunity Commission (EEOC) A U.S. federal agency charged with ending employment discrimination in the United States that is responsible for enforcing laws against discrimination, including Title VII of the 1964 Civil Rights Act that prohibits employment discrimination on the basis of race, color, religion, sex, or national origin.

equal rights amendment (ERA) The proposed 28th amendment to the Constitution, which states that "equality of rights under the law shall not be denied or abridged by the United States, or by any state, on account of sex."

ethnicity A shared cultural heritage or nationality.

Every Child Deserves a Family Act This piece of federal legislation would remove obstacles to non-heterosexual (as well as transgendered) individuals providing loving homes for adoption or foster care.

e-waste Discarded electrical appliances and electronic equipment.

experiments A research method that involves manipulating the independent variable to determine how it affects the dependent variable.

expressive roles Roles into which women are traditionally socialized (i.e., nurturing and emotionally supportive roles).

expulsion Occurs when a dominant group forces a subordinate group to leave the country or to live only in designated areas of the country.

extreme poverty Living on less than $1.25 a day.

Faith-Based and Neighborhood Partnerships A program in which faith-based and other neighborhood organizations receive federal funding for programs that serve the needy, such as homeless services and food aid programs.

familism The view that the family unit is more important than individual interests.

family A kinship system of all relatives living together or recognized as a social unit, including adopted people.

Family and Medical Leave Act (FMLA) A federal law that requires public agencies and companies with 50 or more employees to provide eligible workers with up to 12 weeks of job-protected, unpaid leave so that they can care for an ill child, spouse, or parent; stay home to care for their newborn, newly adopted, or newly placed child; or take time off when they are seriously ill, and up to 26 weeks of unpaid leave to care for a seriously ill or injured family member who is in the armed forces, including the National Guard or Reserves.

Family Support Agreement In China, a voluntary contract between older parents and adult children that specifies the details of how the adult children will provide parental care.

feminism The belief that men and women should have equal rights and responsibilities.

feminist criminology An approach that focuses on how the subordinate position of women in society affects their criminal behavior and victimization.

feminization of poverty The disproportionate distribution of poverty among women.

fetal alcohol syndrome A syndrome characterized by serious physical and mental handicaps as a result of maternal drinking during pregnancy.

field research Research that involves observing and studying social behavior in settings in which it occurs naturally.

flextime A work arrangement that allows employees to begin and end the workday at different times so long as 40 hours per week are maintained.

forced labor Also known as slavery, any work that is performed under the threat of punishment and is undertaken involuntarily.

four assurances The four assurances refer to a state's commitment to improving teacher quality, raising academic standards, intervening in failing schools, and developing assessment databases in return for federal dollars.

fracking Hydraulic fracturing, commonly referred to as "fracking," involves injecting a mixture of water, sand, and chemicals into drilled wells to crack shale rock and release natural gas into the well.

free trade agreements Pacts between two countries or among a group of countries that make it easier to trade goods across national boundaries by reducing or eliminating restrictions on exports and tariffs (or taxes) on imported goods and protecting intellectual property rights.

future shock The state of confusion resulting from rapid scientific and technological changes that unravel our traditional values and beliefs.

gateway drug A drug (e.g., marijuana) that is believed to lead to the use of other drugs (e.g., cocaine).

gay A term that can refer to either women or men who prefer same-sex partners.

gay pride Demonstrative and cultural expressions of gay activism that include celebrations, marches, demonstrations, or other cultural activities promoting gay rights.

gay-straight alliances (GSAs) School-sponsored clubs led by middle or high schools, that strive to address anti-LGBT name-calling and promote respect for all students.

gender The social definitions and expectations associated with being female or male.

gender expression The way in which a person presents her- or himself as a gendered individual (i.e., masculine, feminine, or androgynous) in society. A person could, for example, have a gender identity as male but nonetheless present their gender as female.

gender roles Patterns of socially defined behaviors and expectations associated with being female or male.

gendercide The disproportionate aborting of female fetuses.

gene monopolies Exclusive control over a particular gene as a result of government patents.

gene therapy The transplantation of a healthy gene to replace a defective or missing gene.

genetic engineering The manipulation of an organism's genes in such a way that the natural outcome is altered.

genetic exception laws Laws that require genetic information to be handled separately from other medical information.

Genetic Information Nondiscrimination Act of 2008 A federal law that prohibits discrimination in health coverage or employment based on genetic information.

genetic screening The use of genetic maps to detect predispositions to human traits or disease(s).

Geneva Conventions A set of international treaties that govern the behavior of states during wartime, including the treatment of prisoners of war.

genocide The deliberate, systematic annihilation of an entire nation or people.

ghettos Slums in the United States that are occupied primarily by African Americans.

glass ceiling An invisible barrier that prevents women and other minorities from moving into top corporate positions.

glass escalator effect The tendency for men seeking or working in traditionally female occupations to benefit from their minority status.

GLBT See *LGBT*.

global economy An interconnected network of economic activity that transcends national borders.

global warming The increasing average global air temperature, caused mainly by the accumulation of various gases (greenhouse gases) that collect in the atmosphere.

globalization The growing economic, political, and social interconnectedness among societies throughout the world.

green energy Also known as clean energy, energy that is nonpolluting and/or renewable, such as solar power, wind power, biofuel, and hydrogen.

greenhouse gases Gases (primarily carbon dioxide, methane, and nitrous oxide) that accumulate in the atmosphere and act like the glass in a greenhouse, holding heat from the sun close to the earth.

Green Revolving Funds (GRFs) College and university funds that are dedicated to financing cost-saving energy-efficiency upgrades and other projects that decrease resource use and minimize environmental impacts.

green schools Schools that are "in harmony with the natural environment."

greenwashing The way in which environmentally and socially damaging companies portray their corporate image and products as being "environmentally friendly" or socially responsible.

guerrilla warfare Warfare in which organized groups oppose domestic or foreign governments and their military forces; often involves small groups of individuals who use camouflage and underground tunnels to hide until they are ready to execute a surprise attack.

harm reduction A recent public health position that advocates reducing the harmful consequences of drug use for the user as well as for society as a whole.

hate crime An unlawful act of violence motivated by prejudice or bias.

Head Start Begun in 1965 to help preschool children from the most disadvantaged homes, Head Start provides an integrated program of health care, parental involvement, education, and social services for qualifying children.

health According to the World Health Organization, "a state of complete physical, mental, and social well-being."

heavy drinking As defined by the U.S. Department of Health and Human Services, five or more drinks on the same occasion on each of five or more days in the past 30 days prior to the National Survey on Drug Use and Health.

heterosexism A form of oppression that refers to a belief system that gives power and privilege to heterosexuals, while depriving, oppressing, stigmatizing, and devaluing people who are not heterosexual.

heterosexuality The predominance of emotional, cognitive, and sexual attraction to individuals of the other sex.

homeschooling The education of children at home instead of in a public or private school.

homophobia Negative or hostile attitudes directed toward non-heterosexual sexual behavior, a non-heterosexually identified individual, and communities of non-heterosexuals.

homosexuality The predominance of emotional, cognitive, and sexual attraction to individuals of the same sex.

honor killings Murders, often public, as a result of a female dishonoring, or being perceived to have dishonored, her family or community.

human capital hypothesis The hypothesis that pay differences between females and males are a function of differences in women's and men's levels of education, skills, training, and work experience.

hypothesis A prediction or educated guess about how one variable is related to another variable.

identity theft The use of someone else's identification (e.g., Social Security number, birth date) to obtain credit or other economic rewards.

in vitro fertilization (IVF) The union of an egg and a sperm in an artificial setting such as a laboratory dish.

incapacitation A criminal justice philosophy that argues that recidivism can be reduced by placing offenders in prison so that they are unable to commit further crimes against the general public.

independent variable The variable that is expected to explain change in the dependent variable.

index offenses Crimes identified by the FBI as the most serious including personal or violent crimes (homicide, assault, rape, and robbery) and property crimes (larceny, motor vehicle theft, burglary, and arson).

individual discrimination The unfair or unequal treatment of individuals because of their group membership.

individualism The tendency to focus on one's individual self-interests and personal happiness rather than on the interests of one's family and community.

industrialization The replacement of hand tools, human labor, and animal labor with machines run by steam, gasoline, and electric power.

infant mortality rate The number of deaths of live-born infants under 1 year of age per 1,000 live births (in any given year).

insider trading The use of privileged (i.e., nonpublic) information by an employee of an organization that gives that employee an unfair advantage in buying, selling, and trading stocks or other securities.

institution An established and enduring pattern of social relationships.

institutional discrimination Discrimination in which institutional policies and procedures result in unequal treatment of and opportunities for minorities.

instrumental roles Roles into which men are traditionally socialized (i.e., task-oriented roles).

integration hypothesis A theory that the only way to achieve quality education for all racial and ethnic groups is to desegregate the schools.

intergenerational poverty Poverty that is transmitted from one generation to the next.

internalized homophobia (or internalized heterosexism) The internalization of negative messages about homosexuality by lesbian, gay, and bisexual individuals as a result of direct or indirect social rejection and stigmatization.

Internet An international information infrastructure available through universities, research institutes, government agencies, libraries, and businesses.

Internet piracy Illegally downloading or distributing copyrighted material (e.g., music, games, software).

INTERPOL The largest international police organization in the world.

intimate partner violence (IPV) Actual or threatened violent crimes committed against individuals by their current or former spouses, cohabiting partners, boyfriends, or girlfriends.

job burnout Prolonged job stress that can cause or contribute to high blood pressure, ulcers, headaches, anxiety, depression, and other health problems.

job exportation The relocation of jobs to other countries where products can be produced more cheaply.

Kyoto Protocol The first international agreement to place legally binding limits on greenhouse gas emissions from developed countries.

labor unions Worker advocacy organizations that developed to protect workers and represent them at negotiations between management and labor.

larceny Larceny is simple theft; it does not entail force or the use of force, or breaking and entering.

latent functions Consequences that are unintended and often hidden.

learning health systems The result of electronic records whereby physicians can look across patient populations and identify successful treatments or detect harmful interactions.

least developed countries The poorest countries of the world.

legalization Making prohibited behaviors legal; for example, legalizing drug use or prostitution.

lesbian A woman who prefers same-sex partners.

LGBT, LGBTQ, and LGBTQI Terms used to refer collectively to lesbian, gay, bisexual, transgender, questioning or "queer," and/or intersexed individuals.

life expectancy The average number of years that individuals born in a given year can expect to live.

light pollution Artificial lighting that is annoying, unnecessary, and/or harmful to life forms on earth.

living apart together (LAT) relationships An emerging family form in which couples—married or unmarried—live apart in separate residences.

living wage laws Laws that require state or municipal contractors, recipients of public subsidies or tax breaks, or, in some cases, all businesses to pay employees wages that are significantly above the federal minimum, enabling families to live above the poverty line.

long-term unemployment rate The share of the unemployed who have been out of work for 27 weeks or more.

malware A general term that includes any spyware, viruses, and adware that are installed on an owner's computer without their knowledge.

managed care Any medical insurance plan that controls costs through monitoring and controlling the decisions of health care providers.

manifest functions Consequences that are intended and commonly recognized.

marital decline perspective A pessimistic view of the current state of marriage that includes the beliefs that (1) personal happiness has become more important than marital commitment and family obligations, and (2) the decline in lifelong marriage and the increase in single-parent families have contributed to a variety of social problems.

marital resiliency perspective A view of the current state of marriage that includes the beliefs that (1) poverty, unemployment, poorly funded schools, discrimination, and the lack of basic services (such as health insurance and child care) represent more serious threats to the well-being of children and adults than does the decline in married two-parent families, and (2) divorce provides a second chance for happiness for adults and an escape from dysfunctional and aversive home environments for many children.

master status The status that is considered the most significant in a person's social identity.

maternal mortality rate A measure of deaths that result from complications associated with pregnancy, childbirth, and unsafe abortion.

Matthew Shepard and James Byrd, Jr. Hate Crimes Prevention Act (HCPA) This new law expands the original 1969 federal hate crimes law to cover hate crimes based on actual or perceived sexual orientation, gender, gender identity, and disability.

McDonaldization The process by which principles of the fast-food industry (efficiency, calculability, predictability, and control through technology) are being applied to more sectors of society, particularly the workplace.

means-tested programs Assistance programs that have eligibility requirements based on income.

mechanization Dominant in an agricultural society, the use of tools to accomplish tasks previously done by hand.

Medicaid A public health insurance program, jointly funded by the federal and state governments, that provides health insurance coverage for the poor who meet eligibility requirements.

medicalization Defining or labeling behaviors and conditions as medical problems.

Medicare A federally funded program that provides health insurance benefits to the elderly, disabled, and those with advanced kidney disease.

membership communities Internet sites where participation requires membership and members regularly communicate with one another for personal and/or professional reasons.

mental health The successful performance of mental function, resulting in productive activities, fulfilling relationships with other people, and the ability to adapt to change and to cope with adversity.

mental illness Refers collectively to all mental disorders, which are characterized by sustained patterns of abnormal thinking, mood (emotion), or behaviors that are accompanied by significant distress and/or impairment in daily functioning.

meta-analysis Meta-analysis combines the results of several studies addressing a research question; i.e., it is the analysis of analyses.

microcredit programs The provision of loans to people who are generally excluded from traditional credit services because of their low socio-economic status.

military-industrial complex A term used by Dwight D. Eisenhower to connote the close association between the military and defense industries.

Millennium Development Goals Eight goals that comprise an international agenda for reducing poverty and improving lives.

minority group A category of people who have unequal access to positions of power, prestige, and wealth in a society and who tend to be targets of prejudice and discrimination.

minority stress theory Explains that when an individual experiences the social environment as emotionally or physically threatening due to social stigma, the result is an increased risk for mental health problems.

modern racism A subtle form of racism that involves the belief that serious discrimination in America no longer exists, that any continuing racial inequality is the fault of minority group members, and that the demands for affirmative action for minorities are unfair and unjustified.

monogamy Marriage between two partners; the only legal form of marriage in the United States.

morbidity Illnesses, symptoms, and the impairments they produce.

mortality Death.

motherhood penalty The tendency for women with children, particularly young children, to be disadvantaged in hiring, wages, and the like compared to women without children.

multicultural education Education that includes all racial and ethnic groups in the school curriculum, thereby promoting awareness and appreciation for cultural diversity.

Multidimensional Poverty Index A measure of serious deprivation in the dimensions of health, education, and living standards that combines the number of deprived and the intensity of their deprivation.

multiple chemical sensitivity Also known as "environmental illness," a condition whereby individuals experience adverse reactions when exposed to low levels of chemicals found in everyday substances.

mutually assured destruction (MAD) A Cold War doctrine referring to the capacity of two nuclear states to destroy each other, thus reducing the risk that either state will initiate war.

National Coming Out Day Celebrated on October 11, this day is recognized in many countries as a day to raise awareness of the LGBT population and foster discussion of gay rights issues.

National Day of Silence A day during which students do not speak in recognition of the daily harassment that LBGT students endure.

nativist extremist groups Organizations that not only advocate restrictive immigration policy, but also encourage their members to use vigilante tactics to confront or harass suspected undocumented immigrants.

naturalized citizens Immigrants who apply for and meet the requirements for U.S. citizenship.

needle exchange programs Programs designed to reduce transmission of HIV by providing intravenous drug users with new, sterile syringes in exchange for used, contaminated syringes.

neglect A form of abuse involving the failure to provide adequate attention, supervision, nutrition, hygiene, health care, and a safe and clean living environment for a minor child or a dependent elderly individual.

net neutrality A principle that holds that Internet users should be able to visit any website and access any content without Internet service provider interference.

no-fault divorce A divorce that is granted based on the claim that there are irreconcilable differences within a marriage (as opposed to one spouse being legally at fault for the marital breakup).

norms Socially defined rules of behavior, including folkways, mores, and laws.

nuclear nonproliferation Efforts to prevent the spread of nuclear weapons, or the materials and technology necessary for the production of nuclear weapons.

nuclear winter The predicted result of a thermonuclear war whereby thick clouds of radioactive dust and particles would block out vital sunlight, lower temperature in the northern hemisphere, and lead to the death of most living things on earth.

objective element of a social problem Awareness of social conditions through one's own life experiences and through reports in the media.

occupational sex segregation The concentration of women in certain occupations and men in other occupations.

oppression The use of power to create inequality and limit access to resources, which impedes the physical and/or emotional well-being of individuals or groups of people.

organized crime Criminal activity conducted by members of a hierarchically arranged structure devoted primarily to making money through illegal means.

outsource See *outsourcing*.

outsourcing A practice in which a business subcontracts with a third party to provide business services, often in low-wage countries such as China and India.

overt discrimination Discrimination that occurs because of an individual's own prejudicial attitudes.

parental alienation The intentional efforts of one parent to turn a child against the other parent and essentially destroy any positive relationship a child has with the other parent.

parity In health care, a concept requiring equality between mental health care insurance coverage and other health care coverage.

parole Parole entails release from prison, for a specific time period and subject to certain conditions, before the inmate's sentence is finished.

partial birth abortions Also called an intact dilation and extraction (D & E) abortion, the procedure may entail delivering the limbs and the torso of the fetus before it has expired.

participation gap The tendency for racial and ethnic minorities to participate in information and communication technologies (e.g., using smartphones to access the Internet rather than a computer) that place them in a disadvantaged position (e.g., difficult to research a term paper on a smartphone).

patriarchy A male-dominated family system that is reflected in the tradition of wives taking their husband's last name and children taking their father's name.

penetration rate The percentage of people who have access to and use the internet in a particular area.

pink-collar jobs Jobs that offer few benefits, often have low prestige, and are disproportionately held by women.

pinkwashing The practice of using the color pink and pink ribbons to indicate a company is helping to fight breast cancer, even when the company may be using chemicals linked to cancer.

planned obsolescence The manufacturing of products that are intended to become inoperative or outdated in a fairly short period of time.

pluralism A state in which racial and ethnic groups maintain their distinctness but respect each other and have equal access to social resources.

polyandry The concurrent marriage of one woman to two or more men.

polygamy A form of marriage in which one person may have two or more spouses.

polygyny A form of marriage in which one husband has more than one wife.

population momentum Continued population growth as a result of past high fertility rates that have resulted in a large number of young women who are currently entering their childbearing years.

postindustrialization The shift from an industrial economy dominated by manufacturing jobs to an economy dominated by service-oriented, information-intensive occupations.

postmodernism A worldview that questions the validity of rational thinking and the scientific enterprise.

post-traumatic stress disorder (PTSD) A set of symptoms that may result from any traumatic experience, including crime victimization, war, natural disasters, or abuses.

prejudice Negative attitudes and feelings toward or about an entire category of people.

primary deviance Deviant behavior committed before a person is caught and labeled an offender.

primary groups Usually small numbers of individuals characterized by intimate and informal interaction.

primordial explanations Those explanations that emphasize the existence of "ancient hatreds" rooted in deep psychological or cultural differences between ethnic groups, often involving a history of grievance and victimization, real or imagined, by the enemy group.

privilege When a group has a special advantage or benefits as a result of cultural, economic, societal, legal, and political factors.

probation The conditional release of an offender who, for a specific time period and subject to certain conditions, remains under court supervision in the community.

pronatalism A cultural value that promotes having children.

public housing Federally subsidized housing that is owned and operated by local public housing authorities (PHAs).

race A category of people who are believed to share distinct physical characteristics that are deemed socially significant.

racial profiling The law enforcement practice of targeting suspects on the basis of race.

racism The belief that race accounts for differences in human character and ability and that a particular race is superior to others.

Racism 2.0 A form of racism that allows for and celebrates the achievements of individuals of color who are viewed as having "transcended" their minority status.

radical environmental movement A grassroots movement of individuals and groups that employ unconventional and often illegal means of protecting wildlife or the environment.

recession A significant decline in economic activity spread across the economy and lasting for at least six months.

recidivism A return to criminal behavior by a former inmate, most often measured by re-arrest, reconviction, or re-incarceration.

refined divorce rate The number of divorces per 1,000 married women.

registered partnerships Federally recognized relationships that convey most but not all the rights of marriage.

rehabilitation A criminal justice philosophy that argues that recidivism can be reduced by changing the criminal through such programs as substance abuse counseling, job training, education, and so on.

relative poverty The lack of material and economic resources compared with some other population.

replacement-level fertility The level of fertility at which a population exactly replaces itself from one generation to the next; currently, the number is 2.1 births per woman (slightly more than 2 because not all female children will live long enough to reach their reproductive years).

Respect for Marriage Act (RMA) A bill that, if passed, would overturn DOMA and grant federal recognition to same-sex marriages, regardless of the state laws in which they reside.

restorative justice A philosophy primarily concerned with reconciling conflict between the victim, the offender, and the community.

roles The set of rights, obligations, and expectations associated with a status.

sample A portion of the population, selected to be representative so that the information from the sample can be generalized to a larger population.

sanctions Social consequences for conforming to or violating norms.

sandwich generation A generation of people who care for their aging parents while also taking care of their own children.

school vouchers Tax credits that are transferred to the public or private school that parents select for their child.

science The process of discovering, explaining, and predicting natural or social phenomena.

second shift The household work and child care that employed parents (usually women) do when they return home from their jobs.

secondary deviance Deviant behavior that results from being caught and labeled as an offender.

secondary group Involving small or large numbers of individuals, groups that are task-oriented and are characterized by impersonal and formal interaction.

Section 8 housing A housing assistance program in which federal rent subsidies are provided either to tenants (in the form of certificates and vouchers) or to private landlords.

security dilemma A characteristic of the international state system that gives rise to unstable relations between states; as State A secures its borders and interests, its behavior may decrease the security of other states and cause them to engage in behavior that decreases A's security.

segregation The physical separation of two groups in residence, workplace, and social functions.

selective primary health care An approach to health care that focuses on using specific interventions to target specific health problems.

self-fulfilling prophecy A concept referring to the tendency for people to act in a manner consistent with the expectations of others.

Semantic Web Sometimes called Web 3.0, a version of the Internet in which pages not only contain information but also describe the interrelationship between pages; sometimes called smart media.

serial monogamy A succession of marriages in which a person has more than one spouse over a lifetime but is legally married to only one person at a time.

sex A person's biological classification as male or female.

sexism The belief that innate psychological, behavioral, and/or intellectual differences exist between women and men and that these differences connote the superiority of one group and the inferiority of the other.

sexual harassment In reference to workplace harassment, when an employer requires sexual favors in exchange for a promotion, salary increase, or any other employee benefit and/or the existence of a hostile environment that unreasonably interferes with job performance.

sexual orientation A person's emotional and sexual attractions, relationships, self-identity, and behavior.

sexual orientation change efforts (SOCE) Collectively refers to reparative, conversion, and reorientation therapies, according to the APA.

shaken baby syndrome A form of child abuse whereby the caretaker shakes a baby to the point of causing the child to experience brain or retinal hemorrhage.

single-payer health care A health care system in which a single tax-financed public insurance program replaces private insurance companies.

slums Concentrated areas of poverty and poor housing in urban areas.

social group Two or more people who have a common identity, interact, and form a social relationship.

social movement An organized group of individuals with a common purpose to either promote or resist social change through collective action.

social problem A social condition that a segment of society views as harmful to members of society and in need of remedy.

Social Security Also called "Old Age, Survivors, Disability, and Health Insurance," a federal insurance program established in 1935 that protects against loss of income due to retirement, disability, or death.

socialism An economic system characterized by state ownership of the means of production and distribution of goods and services.

sociological imagination The ability to see the connections between our personal lives and the social world in which we live.

state The organization of the central government and government agencies such as the military, police, and regulatory agencies.

State Children's Health Insurance Program (SCHIP) A public health insurance program, jointly funded by the federal and state governments, that provides health insurance coverage for children whose families meet income eligibility standards.

status A position that a person occupies within a social group.

STEM An acronym for science, technology, engineering, and mathematics.

stem cells Undifferentiated cells that can produce any type of cell in the human body.

stereotype threat The tendency of minorities and women to perform poorly on high-stakes tests because of the anxiety created by the fear that a negative performance will validate societal stereotypes about one's member group.

stereotypes Exaggerations or generalizations about the characteristics and behavior of a particular group.

stigma A discrediting label that affects an individual's self-concept and disqualifies that person from full social acceptance.

structural sexism The ways in which the organization of society, and specifically its institutions, subordinate individuals and groups based on their sex classification.

structure The way society is organized including institutions, social groups, statuses, and roles.

structured choice Choices that are limited by the structure of society.

subjective element of a social problem The belief that a particular social condition is harmful to society, or to a segment of society, and that it should and can be changed.

subprime mortgages High-interest or adjustable-rate mortgages that require little money down and are issued to borrowers with poor credit ratings or limited credit history.

Supplemental Nutrition Assistance Program (SNAP) The largest U.S. food assistance program.

supply reduction One of two strategies in the U.S. war on drugs (the other is demand reduction), supply reduction concentrates on reducing the supply of drugs available on the streets through international efforts, interdiction, and domestic law enforcement.

survey research A research method that involves eliciting information from respondents through questions.

sustainable development Occurs when human populations can have fulfilling lives without degrading the planet.

sweatshops Work environments that are characterized by less-than-minimum wage pay, excessively long hours of work (often without overtime pay), unsafe or inhumane working conditions, abusive treatment of workers by employers, and/or the lack of worker organizations aimed to negotiate better working conditions.

symbol Something that represents something else.

tar sands Large, naturally occurring deposits of sand, clay, water and a dense form of petroleum (that looks like tar).

tar sands oil Oil that results from converting tar sands into liquid fuel. It is known as the world's dirtiest oil because producing it requires energy and generates high levels of greenhouse gases (that cause global warming and climate change), and also leaves behind large amounts of toxic waste.

technological dualism The tendency for technology to have both positive and negative consequences.

technological fix The use of scientific principles and technology to solve social problems.

technology Activities that apply the principles of science and mechanics to the solutions of a specific problem.

technology-induced diseases Diseases that result from the use of technological devices, products, and/or chemicals.

telecommuting A work arrangement involving the use of information technology that allows employees to work part- or full-time at home or at a satellite office.

telepresencing A sophisticated technology which allows life-sized participants in the virtual presence of one another to realistically communicate through broadcast quality sound and images.

Temporary Assistance for Needy Families (TANF) A federal cash welfare program that involves work requirements and a five-year lifetime limit.

terrorism The premeditated use or threatened use of violence by an individual or group to gain a political objective.

therapeutic cloning Use of stem cells to produce body cells that can be used to grow needed organs or tissues; regenerative cloning.

therapeutic communities Organizations in which approximately 35 to 500 individuals reside for up to 15 months to abstain from drugs, develop marketable skills, and receive counseling.

total fertility rates The average lifetime number of births per woman in a population.

transgender(ed) individual A transgender(ed) individual is a person whose sense of gender identity—masculine or feminine—is inconsistent with their birth (sometimes called chromosomal) sex (male or female).

transnational corporations Also known as multinational corporations, corporations that have their home base in one country and branches, or affiliates, in other countries.

transnational crime Criminal activity that occurs across one or more national borders.

transnational terrorism Terrorism that occurs when a terrorist act in one country involves victims, targets, institutions, governments, or citizens of another country.

under-5 mortality rate The rate of deaths of children under age 5.

underemployment Unemployed workers as well as (1) those working part-time but who wish to work full-time, (2) those who want to work but have been discouraged from searching by their lack of success, and (3) others who are neither working nor seeking work but who want and are available to work and have looked for employment in the last year. Also refers to the employment of workers with high skills and/or educational attainment working in low-skill or low-wage jobs.

unemployment To be currently without employment, actively seeking employment, and available for employment, according to U.S. measures of unemployment.

unemployment insurance A federal-state program that temporarily provides laid-off workers with a portion of their paychecks.

union density The percentage of workers who belong to unions.

universal health care A system of health care, typically financed by the government, that ensures health care coverage for all citizens.

value-added measurement (VAM). VAM is the use of student achievement data to assess teacher effectiveness.

values Social agreements about what is considered good and bad, right and wrong, desirable and undesirable.

variable Any measurable event, characteristic, or property that varies or is subject to change.

victimless crimes Illegal activities that have no complaining participant(s) and are often thought of as crimes against morality, such as prostitution.

vivisection The practice of cutting into or otherwise harming living, non-human animals for the purpose of scientific research.

war Organized armed violence aimed at a social group in pursuit of an objective.

wealth The total assets of an individual or household minus liabilities.

wealthfare Laws and policies that benefit the rich.

weapons of mass destruction (WMD) Chemical, biological, and nuclear weapons that have the capacity to kill large numbers of people indiscriminately.

Web 2.0 A platform for millions of users to express themselves online in the common areas of cyberspace.

white-collar crime Includes both *occupational crime*, in which individuals commit crimes in the course of their employment, and *corporate crime*, in which corporations violate the law in the interest of maximizing profit.

workers' compensation Also known as workers' comp, an insurance program that provides medical workers' compensation and living expenses for people with work-related injuries or illnesses.

Workforce Investment Act Legislation passed in 1998 that provides a wide array of programs and services designed to assist individuals to prepare for and find employment.

working poor Individuals who spend at least 27 weeks per year in the labor force (working or looking for work) but whose income falls below the official poverty level.

References

Chapter 1

Anwar, Yasmin. 2009. "Fighting Global Poverty Is Fastest-Growing Minor." *UC Berkeley News*, March 10. Available at http://www.berkeley.edu/news

Associated Press. 2006 (September 7). "Florida Appeals Court Upholds Ban of Veil in Driver's License Photo." *Religious News*. Pew Forum on Religion and Public Life. Available at http://pewforum.org/

BBC (British Broadcasting Company). 1989. "On This Day: Massacre at Tiananmen Square." Available at http://news.bbc.co.uk

Blumer, Herbert. 1971. "Social Problems as Collective Behavior." *Social Problems* 8(3):298–306.

Burns, Laura. 2009. "Students Oppose Contracts with Russell Athletic." *The Phoenix*, March 25. Available at http://loyolaphoenix.com

Caldas, Stephen, and Carl L. Bankston III. 1999. "Black and White TV: Race, Television Viewing, and Academic Achievement." *Sociological Spectrum* 19:39–61.

Canedy, Dana. 2002. "Lifting Veil for Photo ID Goes Too Far, Driver Says." *New York Times*, June 27. Available at http://www.nytimes.com

Catalysts for Change. 2008. "Changing Behavior." July 27. Available at http://www.nytimes.com

Centers for Disease Control and Prevention. 2008 (August 1). Trends in HIV- and STD-Related Risk Behaviors among High School Students—United States, 1991–2007. *Morbidity and Mortality Weekly Report* 57(30):817–822.

Centers for Disease Control and Prevention. 2010 (June 4). Youth Risk Behavior Surveillance, United States, 2009. *Morbidity and Mortality Weekly Report* 59 NO. SS-5.

Dolan, Maura. 2009. "California Supreme Court Looks Unlikely to Kill Proposition 8." *Los Angeles Times*, March 6. Available at http://www.latimes.com

The Eleanor Roosevelt Papers. 2008. Teaching Eleanor Roosevelt: American Youth Congress. Available at http://www.nps.gov

Engel, Robin Shepard, and Robert E. Worden. 2003. "Police Officers' Attitudes, Behavior, and Supervisory Influences: An Analysis of Problem Solving." *Criminology* 41:131–166.

Ferrara, Leigh, Ann Friedman, April Rabkin, Amaya Rivera, Cameron Scott, Marisa Taylor, and Marcus Wohlsen. 2006. "Extra Credit: Campus Activism 2006." *Mother Jones News*, September/October. Available at http://motherjones.com

Fleming, Zachary. 2003. "The Thrill of It All." In *In Their Own Words*, ed. Paul Cromwell, 99–107. Los Angeles, CA: Roxbury.

Fuller-Thomson, Esme, and Angela Dalton. A. 2011. "Suicidal Ideation among Individuals Whose Parents Have Divorced" *Psychiatry Research*, January 5.

Gallup Poll. 2011. "Most Important Problem." February 25. Available at http://www.gallup.com

Garrison, Jessica. 2009. "Gay Couples Hold Vigils Urging Justices to End Prop. 8." *Los Angeles Times*, March 5. Available at http://articles.latimes.com

Gee, Gilbert, Barbara Curbow, Margaret Ensminger, Joan Griffin, et al. 2006. "Are You Positive? The Relationship of Minority Composition to Workplace Drug and Alcohol Testing." *Journal of Drug Issues* 35(4):755–779.

GSAN (Gay-Straight Alliance Network). 2009. Available at http://gsanetwork.org

Hass, Christopher. 2009. "A Call to Serve: President Obama Signs the Edward M. Kennedy Serve America Act." Available at http://my.barackobama.com

Hewlett, Sylvia Ann. 1992. *When the Bough Breaks: The Cost of Neglecting Our Children*. New York: Harper Perennial.

Jacobs, Bruce A. 2003. "Researching Crack Dealers." In *In Their Own Words*, ed. Paul Cromwell, 1–11. Los Angeles, CA: Roxbury.

James, Susan Donaldson. 2007. "Students Use Civil Rights Tactics to Combat Global Warming." ABC News, January 19. Available at http://abcnews.go.com

KillerCoke.org. 2005. "Coke Campaign at Grinell College, Iowa." Available at http://killercoke.org

Kmec, Julie A. 2003. "Minority Job Concentration and Wages." *Social Problems* 50:38–59.

Liebowitz, Debra K. 2009 (May 5). "Student Recycling Activist Raises the Bar in Miami Beach." Available at http://www.miamiherald.com

Merton, Robert K. 1968. *Social Theory and Social Structure*. New York: Free Press.

Mendes, Elizabeth. 2011. "U.S. Satisfaction Remains Near 12-Month Low." January 14. Available at http://www.gallup.com/

Mills, C. Wright. 1959. *The Sociological Imagination*. London: Oxford University Press.

Newman, Jessica Clark, Don Des Jarlais, Charles F. Turner, Jay Gribble, Philip Cooley, and Denice Paone. 2002. "The Differential Effects of Face-to-Face and Computer Interview Modes." *American Journal of Public Health* 92(2):294–297.

Obama, Barack. 2011. "State of the Nation Address." January 26. Available at http://www.guardian.co.uk

Palacios, Wilson R., and Melissa E. Fenwick. 2003. "'E'" Is for Ecstasy." In *In Their Own Words*, ed. Paul Cromwell, 277–283. Los Angeles, CA: Roxbury.

Pope, Carl. 2011. "No, BP Won't Make it Right." *Huffington Post*, February 25. Available at http://www.huffingtonpost.com

Recovery.gov. 2011. "Overview of Funding." Available at http://www.recovery.gov

Reiman, Jeffrey, and Paul Leighton. 2010. *The Rich Get Richer and the Poor Get Prison*, 9th ed. Boston, MA: Allyn and Bacon.

Rifkind, Hugo. 2009. "Student Activism Is Back. *Times Online*, February 16. Available at http://women.timesonline.co.uk

Saad, Lydia. 2007 (January 5). "Local TV Is No. 1 Source of News for Americans." Gallup Poll. The Gallup Organization. Available at http://www.galluppoll.com

Schlosser, Jim. 2000. "Activist Recalls 'Catalyst for Civil Rights.'" *Greensboro News and Record*, February 2. Available at http://www.sitins.com

*The authors and Wadsworth acknowledge that some of the Internet sources may have become unstable; that is, they are no longer hot links to the intended reference. In that case, readers may want to access the article through the search engine or archives of the homepage cited (e.g., fbi.gov, cbsnews.com).

The Scranton Report. 1971. The Report of the President's Commission on Campus Unrest. Washington, DC: U.S. Government Printing Office.

Shaw, Desair. 2009. "Survey Says: Many Newspapers Won't Be Missed." *USA Today*, March 13. Available at http://content.usatoday.com

Simi, Pete, and Robert Futrell. 2009. "Negotiating White Power." *Social Problems* 56(1):98–110.

Snyder, Susan. 2011. "Dickinson College Students Protest School's Handling of Sex Assaults." Philie.com, March 03. Available at http://articles.philly.com

Sykes, Marvin. 1960. "Negro College Students Sit at Woolworth Lunch Counter." *Greensboro Record*, February 2. Available at http://www.sitins.com

Thomas, W. I. 1931/1966. "The Relation of Research to the Social Process." In *W. I. Thomas on Social Organization and Social Personality*, ed. Morris Janowitz, 289–305. Chicago: University of Chicago Press.

Troyer, Ronald J., and Gerald E. Markle. 1984. "Coffee Drinking: An Emerging Social Problem." *Social Problems* 31:403–416.

Ukers, William H. 1935. *All About Tea*, vol. 1. New York: Tea and Coffee Trade Journal Co.

Weir, Sara, and Constance Faulkner. 2004. *Voices of a New Generation: A Feminist Anthology*. Boston, MA: Pearson Education Inc.

Wilson, John. 1983. *Social Theory*. Englewood Cliffs, NJ: Prentice Hall.

Chapter 2

Alan Guttmacher Institute. 2004. New Evidence from Africa, Asia, and Latin America Underscores Impact of Violence against Women. News release, December 10. Available at http://www.guttmacher.org

Allen, P. L. 2000. *The Wages of Sin: Sex and Disease, Past and Present*. Chicago: University of Chicago Press.

Alvarez, Lizette, and Erik Eckholm. 2009. "Purple Heart Is Ruled Out for Traumatic Stress." *New York Times*, January 27, p. A1.

American College Health Association. 2011. American College Health Association National College Health Assessment Fall 2010 reference group data report. Baltimore: American College Health Association.

American Psychiatric Association. 2000. *Diagnostic and Statistical Manual of Mental Disorders*. 4th ed., text revision. Washington, DC: American Psychiatric Association.

Anderson, Gerard F. and David A. Squires. 2010 (June). "Measuring the U.S. Health System: A Cross-National Comparison." *The Commonwealth Fund*. Publication 1412 Vol. 90. Available at http://www.commonwealthfund.org

Angell, Marcia. 2003. "Statement of Dr. Marcia Angell Introducing the U.S. National Health Insurance Act." Physicians for a National Health Program. Available at http://www.pnhp.org

Arehart-Treichel, Joan. 2008. "Psychiatrists and Farmers: Alliance in the Making?" *Psychiatric News* 43(10):15.

Arias, Elizabeth. 2010. "United States Life Tables by Hispanic Origin." National Center for Health Statistics. *Vital Stat* 2(152).

Associated Press. 2007. "Mom: Social Workers Threaten to Take Away 254-Pound Son." *The Daily Reflector*, March 23, p. B2.

AVERT. 2004 (November 30). HIV and AIDS Discrimination and Stigma. Available at http://www.avert.org/aidsstigma.htm

Barker, Kristin. 2002. "Self-Help Literature and the Making of an Illness Identity: The Case of Fibromyalgia Syndrome (FMS)." *Social Problems* 49(3):279–300.

Braine, Theresa. 2011. "Race against Time to Develop New Antibiotics." *Bulletin of the World Health Organization* 89:88–89.

Centers for Disease Control and Prevention. 2011 (January 14). CDC Health Disparities and Inequalities Report—United States, 2011. MMWR 60:all.

Centers for Disease Control and Prevention. 2010a. "HIV Transmission." Available at http://www.cdc.gov

Centers for Disease Control and Prevention. 2010b. *HIV Surveillance Report, 2008;* vol. 20. Available at http://www.cdc.gov

Centers for Disease Control and Prevention. 2010c. (November 19). "Syringe Exchange Programs—United States, 2008." *Morbidity and Mortality Weekly Report* 59(45):1488–1491.

Centers for Medicare and Medicaid Services. 2011 (January 13). NHE Fact Sheet. U.S. Department of Health and Human Services. Available at http://www.cms.hhs.gov

Chandler, C. K. 2005. *Animal Assisted Therapy in Counseling*. New York: Routledge.

Chandra, Anita, and Cynthia S. Minkovitz. 2006. "Stigma Starts Early: Gender Differences in Teen Willingness to Use Mental Health Services." *Journal of Adolescent Health* 38(6):754.e1–754.e8.

Children's Defense Fund. 2000. The State of America's Children Yearbook 2000. Available at http://www.childrensdefense.org

Cockerham, William C. 2007. *Medical Sociology*, 10th ed. Upper Saddle River, NJ: Prentice Hall.

Combes, Marie-Laure. 2007. "French Government Wants Food Warning." *Associated Press*, March 2. Available at http://www.washingtonpost.com

Cotton, Ann, Kenneth R. Stanton, Zoltan J. Acs, and Mary Lovegrove. 2007. 2006 UB Obesity Report Card. Baltimore, MD: University of Baltimore. Available at http://www.ubalt.edu

Davis, Karen, Cathy Schoen, Stephen C. Schoenbaum, Michelle M. Doty, Alyssa L. Holmgren, Jennifer L. Kriss, and Katherine K. Shea. 2007 (May). *Mirror, Mirror on the Wall: An International Update on the Comparative Performance of American Health Care*. Commonwealth Fund. Available at http://commonwealthfund.org

DeNavas, Carmen, Bernadette D. Proctor, and Jessica C. Smith. 2011. Income, Poverty, and Health Insurance Coverage in the United States: 2010. *Current Population Reports* P60–239. U.S. Census Bureau. Available at http://www.census.gov

Dingfelder, Sadie F. 2009a (June). "Stigma: Alive and Well." *Monitor on Psychology* 40(6):56.

Dingfelder, Sadie F. 2009b (June). "The Military's War on Stigma." *Monitor on Psychology* 40(6):52.

Epstein, Samuel S. 2006. *What's in Your Milk?* Victoria, BC, Canada: Trafford Publishing.

Families USA. 2007 (January 9). "No Bargain: Medicare Drug Plans Deliver Higher Prices." Available at http://www.familiesusa.org

Farmer, Paul, Julio Frenk, Felicia M. Knaul, Lawrence N. Shulman, George Alleyne, Lance Armstrong, Rifat Atun, Douglas Blayney, Lincoln Chen, Richard Feachem, Mary Go-spodarowicz, Julie Gralow, Sanjay Gupta, Ana Langer, Julian Lob-Levyt, Claire Neal, Anthony Mbewu, Dina Mired, Peter Piot, K. Srinath Reddy, Jeffrey D. Sachs, Mahmoud Sarhan, and John R. Seffrin. 2010 (October 2). "Expansion of Cancer Care and Control in Countries of Low and Middle Income: A Call to Action." *Lancet* 376(9747):1186–93.

Feachum, Richard G. A. 2000. "Poverty and Inequality: A Proper Focus for the New Century." *International Journal of Public Health* 78:1–2.

Fine, Aubrey H. 2010. "Forward." Pp. xix–xxi in Aubrey H. Fine (Ed.), *Handbook on Animal-Assisted Therapy*, 3rd ed. Burlington MA: Academic Press.

Fischer, Edward, and Amerigo Farina. 1995. "Attitudes toward Seeking Professional Psychological Help: A Shortened Form and Considerations for Research." *Journal of College Student Development* 36(4):368–373.

Fischman, J. 2010 (September 12). "The Pressure of Race." *The Chronicle of Higher Education*. Available at http://chronicle.com

Gallagher, Robert P. 2010. National Survey of Counseling Center Directors. The International Association of Counseling Services. Available at http://www.iacsinc.org

Goldberg, Lisa. 2005. "Educating Seniors on Safe Sex." *Baltimore Sun*, March 1. Available at http://www.baltimoresun.com

Goldstein, Michael S. 1999. "The Origins of the Health Movement." In *Health, Illness, and Healing: Society, Social Context, and Self*, eds. Kathy Charmaz and Debora A. Paterniti, 31–41. Los Angeles, CA: Roxbury.

Grossman, Amy. 2009. "A Birth Pill." *New York Times*, May 9. Available at http://www.nytimes.com

Gruttadaro, Darcy. 2005. "Federal Leaders Call on Schools to Help." *NAMI Advocate*, Winter, pp. 7–9.

Guardian. 2007. "Obese Boy Stays with Mother." *Guardian Unlimited*, February 27. Available at http://www.guardian.co.uk

Halperin, D. T., M. J. Steiner, M. M. Cassell, E. C. Green, N. Hearst, D. Kirby, H. D. Gayle, and W. Cates. 2004. "The Time Has Come for Common Ground on Preventing Sexual Transmission of HIV." *The Lancet* 364:1913–1915.

Harrison, Joel A. 2008. "How Much Is the Sick U.S. Health Care System Costing You?" *Dollars and Sense*, May/June. Available at http://dollarsandsense.com

Harrison, Mary M. 2002. "Silence That Promotes Stigma." *Teaching Tolerance* 21:52–53.

Hawkes, Corinna. 2006. "Uneven Dietary Development: Linking the Policies and Processes of Globalization with the Nutrition Transition, Obesity, and Diet-Related Chronic Diseases." *Globalization and Health* 2:4.

Himmelstein, D. U., D. Thorne, E. Warren, and S. Woolhandler. 2009. "Medical Bankruptcy in the United States, 2007: Results of a National Study." *The American Journal of Medicine* 122(8):741–46.

Hull, Anne, and Dana Priest. 2007. "'It Is Just Not Walter Reed': Soldiers Share Troubling Stories of Military Health Care across U.S." *Washington Post*, March 5. Available at http://washingtonpost.com

Human Rights Watch. 2005 (March 3). U.S. Gag on Needle Exchange Harms U.N. AIDS Efforts. Available at http://www.hrw.org

International Center for Research on Women. 2010 (December). "Child Marriage Facts and Figures." Available at http://www.icrw.org

James et al. 2007

Jenkins, Chris L. 2008. "Law Equalizes Coverage for Mental, Physical Care." *Washington Post*, October 10, p. B01.

Johnson, Cary Alan. 2007. *Off the Map: How HIV/AIDS Programming Is Failing Same-Sex Practicing People in Africa*. New York: International Gay and Lesbian Human Rights Commission.

Kaiser Commission on Medicaid and the Uninsured. 2010 (September). *The Uninsured and the Difference Health Insurance Makes*. Washington, DC: Kaiser Family Foundation.

Kaiser Family Foundation. 2010. The Global HIV/AIDS Epidemic. Fact Sheet. Available at http://www.kff.org

Kaiser Family Foundation. 2009a (June). Health Tracking Poll. Available at http:www.kff.org

Kaiser Family Foundation. 2009b. 2009 Survey of Americans on HIV/AIDS: Summary of Findings on the Domestic Epidemic. Available at http://www.kff.org

Kaiser Family Foundation. 2007. How Changes in Medical Technology Affect Health Care Costs. Available at http://www.kff.org

Kaiser Family Foundation/Harvard School of Public Health. 2011 (January). The Public's Health Care Agenda for the 112th Congress. Available at http://www.kff.org

Kaiser Family Foundation/HRET. 2010. Employer Health Benefits 2010 Annual Survey. Available at http://www.kff.org

Katel, Peter. 2010. "Food Safety." *CQ Researcher* 20(44): all.

Kolata, Gina. 2007. "A Surprising Secret to a Long Life: Stay in School." *New York Times*, January 3, A1, A16.

Komiya, Noboru, Glenn E. Good, and Nancy B. Sherrod. 2000. "Emotional Openness as a Predictor of College Students' Attitudes toward Seeking Psychological Help." *Journal of Counseling Psychology* 47(1):138–143.

Kruger, K. A., and J. A. Serpell. 2010. "Animal-Assisted Interventions in Mental Health: Definitions and Theoretical Foundations." In Aubrey H. Fine (Ed.), *Handbook on Animal-Assisted Therapy*, 3rd ed., pp. 33–48. San Diego: Academic Press.

Lantz, Paula M., James S. House, James M. Lepkowski, David R. Williams, Richard P. Mero, and Jieming Chen. 1998. "Socioeconomic Factors, Health Behaviors, and Mortality: Results from a Nationally Representative Prospective Study of U.S. Adults." *Journal of the American Medical Association* 279:1703–1708.

Lee, Kelley. 2003. "Introduction." In *Health Impacts of Globalization*, ed. Kelley Lee, 1–10. New York: Palgrave MacMillan.

Lerer, Leonard B., Alan D. Lopez, Tord Kjellstrom, and Derek Yach. 1998. "Health for All: Analyzing Health Status and Determinants." *World Health Statistics Quarterly* 51:7–20.

Light, Donald W., and Rebecca Warburton. 2011 (February 7). "Demythologizing the High Costs of Pharmaceutical Research." *Biosocieties*, pp. 1–17.

Link, Bruce G., and Jo Phelan. 2001. "Social Conditions as Fundamental Causes of Disease." In *Readings in Medical Sociology*, 2nd ed., William C. Cockerham, Michael Glasser, and Linda S. Heuser, eds. (pp. 3–17). Upper Saddle River, NJ: Prentice Hall.

Mahar, Maggie. 2006. *Money-Driven Medicine*. New York: Harper-Collins.

Mayer, Lindsay Renick. 2009. "Insurers Fight Public Health Plan." Capitol Eye Blog, June 18. Center for Responsive Politics. Available at http://www.opensecrets.org

MMWR (Morbidity and Mortality Weekly Report). 2011 (June 24). "National HIV Testing Day—June 27, 2011." 60(24):805.

Moisse, Kate. 2011 (June 23). "James Verone: The Medical Motive for His $1 Bank Robbery." *ABC News*. Available at http://abcnews.go.com

Murphy, Elaine M. 2003. "Being Born Female Is Dangerous for Your Health." *American Psychologist* 58(3):205–210.

Nader, Ralph. 2011 (January 25). "Overuse of Antibiotics." *Common Dreams*. Available at http://www.commondreams.org

Nader, Ralph. 2009 (July 25). "Health Care Hypocrisy." *Common Dreams*. Available at http://www.commondreams.org

National Center for Health Statistics. 2011. Health, United States, 2010 with Special Feature on Death and Dying. Hyattsville, MD: U.S. Government Printing Office.

The National Coalition on Health Care. 2009. "Health Insurance Costs." Available at http://www.nchc.org

National Conference of State Legislators. 2010. "Tanning Restrictions for Minors—A State-by-State Comparison." Available at http://www.ncsl.org

National Institute of Mental Health. 2008. The Numbers Count: Mental Disorders in America. Available at http://www.nimh.nih.gov

Ninan, Ann. 2003 (March). "Without My Consent: Women and HIV-Related Stigma in India." Population Reference Bureau. Available at http://prb.org

Olshansky, S. J., D. J. Pasaro, R. C. Hershow, J. Layden, B. A. Carnes, J. Brody, L. Hayflick, R. N. Butler, D. B. Allison, and D. S. Ludwig. 2005. "A Potential Decline in Life Expectancy in the United States in the 21st Century." *New England Journal of Medicine* 352(11):1138–1145.

Orr, Andrea. 2009 (July 14). "Insured . . . and Broke." Economic Policy Institute. Available at http://www.epi.org

Oxfam GB. 2004 (March 19). The Cost of Childbirth. Press release. Available at http://www.oxfam.org.uk/

Park, Madison. 2009 (September 18). "45,000 American Deaths Associated with Lack of Insurance." CNN. Available at http://articles.cnn.com

Paruzzolo, Silvia, Rekha Mehra, Aslihan Kes, and Charles Ashbaugh. 2010. Targeting Poverty and Gender Inequality to Improve Maternal Health. International Center for Research on Women. Available at http://www.icrw.org

Peters, Sharon. 2011. "Animals Can Assist in Psychotherapy." *USA Today*, January 17. Available at http://www.usatoday.com

Pleis, J. R., B. W. Ward, and J. W. Lucas. 2010 (December). "Summary Health Statistics for U.S. Adults: National Interview Survey, 2009." Vital and Health Statistics, Series 10 (249). National Center for Health Statistics. Washington DC: U.S. Government Printing Office.

Potter, Wendell. 2010. *Deadly Spin: An Insurance Company Insider Speaks Out on How Corporate PR is Killing Health Care and Deceiving Americans*. New York: Bloomsbury Press.

Potter, Wendell. 2009 (June 24). "The Health Care Industry vs. Health Reform." Center for Media and Democracy. Available at http://www.prwatch.org

Pryor, J. H., S. Hurtado, L. DeAngelo, L. Paluki Blake, and S. Tran. 2010. *The American Freshman: National Norms Fall 2010*. Los Angeles: Higher Education Research Institute, UCLA.

Quadagno, Jill. 2004. "Why the United States Has No National Health Insurance: Stakeholder Mobilization against the Welfare State 1945–1996." *Journal of Health and Social Behavior* 45:25–44.

Reinberg, Steven. 2009 (July 28). "Tanning Beds Get Highest Carcinogen Rating." *U.S. News & World Report*. Available at http://health.usnews.com

Sachs, Jeffrey D. 2005. "A Practical Plan to End Poverty." *Washington Post*, January 17, p. A17.

Sager, Alan, and Deborah Socolar. 2004 (October 28). 2003 U.S. Prescription Drug Prices 81 Percent Higher Than in Other Wealthy Nations. *Data Brief No. 7*. Boston University School of Public Health.

Sanders, David, and Mickey Chopra. 2003. "Globalization and the Challenge of Health for All: A View from Sub-Saharan Africa." In *Health Impacts of Globalization*, ed. Kelley Lee, 105–19. New York: Palgrave Macmillan.

Sanders, Jim. 2005. "Bill Would OK Condoms in Prisons." *The Sacramento Bee*, February 26. Available at http://www.sacbee.com

Scal, Peter, and Robert Town. 2007. "Losing Insurance and Using the Emergency Department: Critical Effect of Transition to Adulthood for Youth with Chronic Conditions." *Journal of Adolescent Health* 40 (2 Suppl. 1):S4.

Schlosser, Eric. 2002. *Fast Food Nation*. New York: HarperCollins.

Sered, Susan Starr, and Rushika Fernandopulle. 2005. *Uninsured in America: Life and Death in the Land of Opportunity*. Berkeley and Los Angeles, CA: University of California Press.

Sidel, Victor W., and Barry S. Levy. 2002. "The Health and Social Consequences of Diversion of Economic Resources to War and Preparation for War." In *War or Health: A Reader*, Ilkka Taipale, P. Helena Makela, Kati Juva, and Vappu Taipale, eds. (pp. 208–21). New York: Palgrave MacMillan.

Singer, Peter, and Jim Mason. 2006. *The Way We Eat: Why Our Food Choices Matter*. Emmaus, PA: Rodale.

SourceWatch. 2010 (November 8). "Children's Food and Beverage Advertising Initiative." Available at http://www.sourcewatch.org

Springen, Karen. 2006. "Health Hazards: How Mounting Medical Costs Are Plunging More Families into Debilitating Debt and Why Insurance Doesn't Always Keep Them Out of Bankruptcy." *Newsweek*, August 25. Available at www.msnbc.com/newsweek

"Stark Introduces Constitutional Amendment to Establish Right to Health Care." 2005. Press release, March 3. Available at http://www.house.gov

Stein, R., and C. Connolly. 2004. "Medicare Changes Policy on Obesity, Some Treatments May Be Covered." *Washington Post*, July 16. Available at http://www.washingtonpostonline.com

Substance Abuse and Mental Health Services Administration. 2010a. Mental Health, United States, 2008. Rockville, MD: Center for Mental Health Services, Substance Abuse and Mental Health Services Administration.

Substance Abuse and Mental Health Services Administration. 2010b. Results from the 2009 National Survey on Drug Use and Health: Mental Health Findings. Rockville, MD: Department of Health and Human Services.

Szasz, Thomas. 1961/1970. *The Myth of Mental Illness: Foundations of a Theory of Personal Conduct*. New York: Harper & Row.

Thomson, George, and Nick Wilson. 2005. "Policy Lessons from Comparing Mortality from Two Global Forces: International Terrorism and Tobacco." *Globalization and Health* 1:18.

Trust for America's Health. 2008. *F as in Fat: How Obesity Policies Are Failing in America*. Available at http://healthyamericans.org

UNICEF. 2010. *Progress for Children: Achieving the MDGs with Equity*. Available at http://www.unicef.org

Urichuk, L. with D. Anderson. 2003. *Improving Mental Health through Animal-Assisted Therapy*. Edmonton, Alberta: The Chimo Project.

U.S. Department of Health and Human Services. 2001. Mental Health: Culture, Race, and Ethnicity—A Supplement to Mental Health: A Report of the Surgeon General. Rockville, MD: U.S. Government Printing Office.

U.S. Department of Health and Human Services. 1999. Mental Health: A Report of the Surgeon General—Executive Summary. Rockville, MD: U.S. Government Printing Office.

Van de Water. 2011 (January 7). "Debunking False Claims about Health Reform, Jobs, and the Deficit." Center on Budget and Policy Priorities. Available at http://www.cbpp.org

Vestal, Christine. 2011 (January 14). "Healthcare Budgets in Critical Condition." Stateline. Available at http://stateline.org

Weeks, Jennifer. 2007 (January 12). "Factory Farms." *The CQ Researcher* 17(2).

Weitz, Rose. 2010. *The Sociology of Health, Illness, and Health Care: A Critical Approach*. 5th ed. Belmont, CA: Wadsworth/Cengage.

White, Frank. 2003. "Can International Public Health Law Help to Prevent War?" *Bulletin of the World Health Organization* 81(3):228.

Williams, David R. 2003. "The Health of Men: Structured Inequalities and Opportunities." *American Journal of Public Health* 93(5):724–731.

Williams, D. R., M. B. McClellan, and A. M. Rivlin. 2010 (August). "Beyond the Affordable Care Act: Achieving Real Improvements in Americans' Health." *Health Affairs* 29(8): 1481–1488.

World Health Organization. 2011a. World Health Statistics 2011. Available at http://www.who.int

World Health Organization. 2011b (Updated June 2011). "The Top 10 Causes of Death." *Fact Sheet No. 310*. Available at http://www.who.int

World Health Organization. 2010a. *Trends in Maternal Mortality: 1990 to 2008*. Geneva, Switzerland: WHO Press.

World Health Organization. 2010b. Mental Health and Development: Targeting People with Mental Health Conditions as a Vulnerable Group. Available at http://www.who.org

World Health Organization. 2009a. Global Health Risks: Mortality and Burden of Disease Attributable to Selected Major Risks. Available at http://www.who.int

World Health Organization. 2009b. Women and Health. Available at http://www.who.int

World Health Organization. 2002. The World Health Report 2002. Available at http://www.who.int

World Health Organization. 2000. The World Health Report 2000. Available at http://www.who.int

World Health Organization. 1946. Constitution of the World Health Organization. New York: World Health Organization Interim Commission.

World Health Organization and UNICEF. 2010. Progress in Sanitation and Drinking Water, 2010 Update. Available at http://www.wssinfo.org

Chapter 3

Abrams, Jim. 2010. "Congress Narrows Gap in Cocaine Sentences." *Associated Press*, July 28. http://www.yahoo.com

Abadinsky, Howard. 2008. *Drugs: An Introduction*. Belmont, CA: Wadsworth.

Alcoholics Anonymous. 2010. *AA Fact File*. Available at http://www.aa.org

Alfonsi, Sharyn, and Hanna Siegel. 2010. "Heroin Use in Suburbs on the Rise." ABC News, March 29. Available at http://abcnews.go.com

Aloise-Young, Patricia, Michael D. Slater, and Courtney C. Cruickshank. 2006. "Mediators and Moderators of Magazine Advertisement Effects on Adolescent Cigarette Smoking." *Journal of Health Communication* 11:281–300.

American College Health Association. 2011. *National College Health Assessment II*. Reference Group Executive Summary, Fall 2010. Baltimore, Maryland: American College Health Association.

Armstrong, Elizabeth, and Christina McCarroll. 2004. "Girls Lead in Teen Alcohol Use." *Seattle Times*, August 14. Available at http://seattletimes.nwsource.com

Associated Press (AP). 2010. "New Meth Formula Avoids Anti-Drug Laws." MSNBC, August 24. Available at http://www.msnbc.msn.com/

Balsa, A. I, J. F. Homer, and M. T. French. 2009. "The Health Effects of Parental Problem Drinking on Adult Children." *Journal of Mental Health Policy and Economics* 12(2): 55–66.

Becker, H. S. 1966. *Outsiders: Studies in the Sociology of Deviance*. New York: Free Press.

Behrendt, S, H.-U. Wittchen, M. Höfler, R. Lieb, and K. Beesdo. 2009. "Transitions from First Substance Use to Substance Use Disorders in Adolescence: Is Early Onset Associated with a Rapid Escalation?" *Drug and Alcohol Dependence* 99:68–78.

Better Way Foundation. 2009. Available at http://www.abwf-ct.org/

Bonnie, Richard J., Kathleen Stratton, and Robert B. Wallace (eds.). 2007. *Ending the Tobacco Problem: A Blueprint for the Nation*. Committee on Reducing Tobacco Use and the Board on Population Health, and Public Health Practice. Institute of Medicine. Washington, DC.

Brownell, Ginanne. 2009. "'Special K' Goes Global." *Global Post*, May 22. Available at http://huffingtonpost.com

Burke, Jason. 2006. "Europeans Turn to Cocaine and Alcohol as Cannabis Loses Favour." *Guardian Unlimited*, October 15. Available at http://observer.guardian.co.uk

Cattan, Nacha. 2010. "How Mexican Drug Gangs Use YouTube against Rival Groups." *Christian Science Monitor*, November 5. Available at http://www.csmonitor.com

CASA (National Center on Addiction and Substance Abuse). 2010. "Behind Bars II, Substance Abuse and America's Prison Population." New York: Columbia University.

CASA. 2009. "The Impact of Substance Abuse on Federal, State, and Local Budgets." New York: Columbia University.

Caulkins, Jonathan, Peter Reuter, Martin Y. Iguchi, and James Chiesa. 2005. "How Goes the 'War on Drugs'? An Assessment of U.S. Drug Problems and Policy." Rand Drug Policy Research Center. Available at http://www.rand.org/pubs

CDC (Centers for Disease Control). 2011a. "Alcohol in Public Policy." Centers for Disease Control and Prevention. Available at http://www.cdc.gov/alcohol/

CDC. 2011b. "Targeting the Nation's Leading Killer at a Glance 2011." Centers for Disease Control and Prevention. Available at http://www.cdc.gov/chronicdiseases

CDC. 2011c. "Drug-Induced Deaths Overtake Alcohol and Firearms, CDC Reports." Centers for Disease Control and Prevention. January 20. Available at http://www.drugalert.org/news/2011/01/20

CDC. 2006. "Fetal Alcohol Spectrum Disorders." Available at http://www.cdc.gov

CNN. 2009. "Penalties for Drug-Related Crime in Asia." Available at http://edition.cnn.com

Copes, Heigh, Andy Hochstetler, and J. Patrick Williams. 2008. "'We Weren't Like No Regular Dope Fiends': Negotiating Hustler and Crackhead Identities." *Social Problems* 55(2):254–270.

DEA (Drug Enforcement Administration). 2011 (March 1). "Chemicals Used in 'Spice' and 'K2' Type Products Now Under Federal Control and Regulation." Washington, DC, News Release.

DEA. 2010. "Fiction: Drug Production Does Not Damage the Environment." Just Think Twice: Facts and Fiction. Available at http://www.justthinktwice.com

Degenhardt, Louisa, Wai-Tat Chiu, Nancy Sampson, Ronald C. Kessler, James C. Anthony, Matthias Angermeyer, Ronny Bruffaerts, Giovanni de Girolamo, Oye Gureje, Yueqin Huang, Aimee Karam, Stanislav Kostyuchenko, Jean Pierre Lepine, Maria Elena Medina Mora, Yehuda Neumark, J. Hans Ormel, Alejandra Pinto-Meza, José Posada-Villa, Dan J. Stein, Tadashi Takeshima, and J. Elisabeth Wells. 2008. "Toward a Global View of Alcohol, Tobacco, Cannabis, and Cocaine Use: Findings from the WHO World Mental Health Surveys." *PLoS Medicine* 5(1):1053–1077.

Dekel, Rachel, Rami Benbenishty, and Yair Amram. 2004. "Therapeutic Communities for Drug Addicts: Prediction of Long-Term Outcomes." *Addictive Behaviors* 29(9):1833–1837.

Dinan, Stephen, and Ben Conery. 2009. "DEA Pot Raids Go On; Obama Opposes." *The Washington Times*, February 5. Available at http://www.washingtontimes.com

Dinno, Alexis, and Stanton Glantz. 2009. "Tobacco Control Policies are Egalitarian: A Vulnerabilities Perspective on Clean Indoor Air Laws, Cigarette Prices, and Tobacco Use Disparities." *Social Science & Medicine* 68:1439–1447.

Drug Policy Alliance. 2007. *State by State*. Available at http://www.drugpolicy.org/statebystate/

Drug Policy Alliance. 2003. *Drug Policy around the World: The Netherlands*. Available at http://www.drugpolicy.org/global

Duke, Steven, and Albert C. Gross. 1994. *America's Longest War: Rethinking Our Tragic Crusade Against Drugs*. New York: G. P. Putnam.

Elias, Marilyn. 2009. "Secondhand Smoke May Double Likelihood of Depression." *USA Today*, March 4. Available at http://www.usatoday.com/news.health

EMCDDA (European Monitoring Centre for Drugs and Drug Addiction). 2010. The State of the Drugs Problem in Europe. Luxembourg: Office for Official Publications of the European Communities.

EMCDDA–Europol. 2009. "Methamphetamine A European Union Perspective in the Global Context." Available at http://www.emcdda.europa.eu

ESPAD (European School Survey Project on Alcohol and Other Drugs). 2009. *Stockholm: The Swedish Council for Information on Alcohol and Other Drugs*. Available at www.espad.org

Fact Sheet. 2009 (June 22). "Fact Sheet: The Family Smoking Prevention and Tobacco Control Act of 2009." Available at http://www.whitehouse.gov/the_press_office

Fay, Calvina. 2010 (March 2). "K2 Poses Dangers and Should Be Illegal." Available at http://articles.cnn.com/2010-03-02

Feagin, Joe R., and C. B. Feagin. 1994. *Social Problems*. Englewood Cliffs, NJ: Prentice Hall.

Fears, Darryl. 2009. "A Racial Shift in Drug-Crime Prisoners." *The Washington Post*, April 15. Available at http://www.washingtonpost.com

Filkins, Dexter. 2009. "Poppies a Target in Fight against Taliban." *The New York Times*, April 29. Available at http://www.nytimes.com

Firshein, Janet. 2003. "The Role of Biology and Genetics." *Moyers on Addiction*. Public Broadcasting System (PBS). Available at http://www.pbs.org

Flanzer, Jerry P. 2005. "Alcohol and Other Drugs Are Key Causal Agents of Violence." In *Current Controversies on Family Violence*, 2nd ed., Donileen R. Loseke, Richard J. Gelles, and Mary M. Cavanaugh, eds. (pp. 163–174).

FDA (Food and Drug Administration). 2011a. "Menthol Cigarettes and Public Health: Review of the Scientific Evidence and Recommendations." Tobacco Products Scientific Advisory Committee. Available at http://www.fda.gov

FDA 2011b. "Cigarette Health Warnings." Available at http://www.fda.gov

Freeman, Dan, Merrie Brucks, and Melanie Wallendorf. 2005. "Young Children's Understanding of Cigarette Smoking." *Addiction* 100(10):1537–1545.

Friedman-Rudovsky, Jean. 2009. "Red Bull's New Cola: A Kick from Cocaine?" *Time/CNN*, May 25. Available at http://www.time.com

Gilbert, R., C. S. Widom, K. Browne, D. Fergusson, E. Webb, and S. Janson. 2009. "Burden and Consequence of Child Maltreatment in High-Income Countries." *The Lancet* 73(9657):68–81.

Goodman, J. David, 2010. "Popular New Drinking Game Raises Question, 'Who's 'Icing'" Who?" *The New York Times*, June 8. Available at http://nytimes.com

Goodnough, Abby, and Katie Zezima, 2011. "An Alarming New Stimulant, Legal in Many States." *The New York Times*, July 16. Available at http://www.nytimes.com

Goodnough, Abby. 2010. "Caffeine and Alcohol Drink Is Potent Mix for young." *The New York Times*, October 26. Available at http://www.nytimes.com

Greenwald, Glenn. 2009. "Drug Decriminalization in Portugal: Lessons for Creating Fair and Successful Drug Policies." April 2. CATO Institute: Washington, DC. Available at http://www.cato.org

Gusfield, Joseph. 1963. *Symbolic Crusade: Status Politics and the American Temperance Movement*. Urbana, IL: University of Illinois Press.

Heinrich, Henry. 2009. "Obama Drug Policy to Do More to Ease Health Risks." Reuters, March 16. Available at: http://www.reuters.com

Hingson, Ralph W., Timothy Heeren, and Michael R. Winter. 2006. "Age at Drinking Onset and Alcohol Dependence." *Archives of Pediatrics & Adolescent Medicine* 160:739–746.

Human Rights Watch. 2007. "Reforming the Rockefeller Drug Laws." Available at http://www.hrw.org/campaigns/drugs

Jarvik, M. 1990. "The Drug Dilemma: Manipulating the Demand." *Science* 250:387–392.

Jervis, Rick. 2009. "YouTube Riddled with Drug Cartel Videos, Messages." *USA Today*, April 9. Available at http://www.usatoday.com

Johnson, K., and Z. Pan, L. Young, J. Vanderhoff, S. Shamblen, T. Browne, K. Linfield, and G. Suresh. 2008. "Therapeutic Community Drug Treatment Success in Peru: A Follow-Up Outcome Study." *Substance Abuse Treatment and Prevention* (December) 3,3:26.

Kelly, John, Robert Stout, William Zywiak, and Robert Schneider. 2006. "A 3-Year Study of Addiction Mutual-Help Group Participation Following Intensive Outpatient Treatment." *Alcoholism: Clinical and Experimental Research* 30:1381–1392.

Kearns-Bodkin, J. N., and K. E. Leonard. 2008. "Relationship Functioning among Adult Children of Alcoholics." *Journal of Studies on Alcohol and Drugs* 69(6): 941–950.

King, Ryan S., and Jill Pasquarella. 2009. Drug Courts: A Review of the Evidence. The Sentencing Project, April. Available at http://www.sentencingproject.org

Kraft, Scott. 2009. "Pursuing Smugglers, Border Agents Become Trackers." *Los Angeles Times*, May 12. Available at http://www.latimes.com/news

Layton, Lyndsey. 2010. "New FDA Rules Will Greatly Restrict Tobacco Advertising and Sales." *The Washington Post*, March 19. Available at http://www.washingtonpost.com

Lee, Yon M. Abdel-Ghany. 2004. "American Youth Consumption of Licit and Illicit Substances." *International Journal of Consumer Studies* 28(5):454–465.

Leff, Lisa, and Marcus Wohlsen. 2010. "Pot Activists Vow to Push Legalization in 2012." *The Washington Post*, November 3.

MacCoun, Robert J., and Peter Reuter. 2001. "Does Europe Do It Better? Lessons from Holland, Britain and Switzerland." In *Solutions to Social Problems*, eds. D. Stanley Eitzen and Craig S. Leedham, pp. 260–264. Boston: Allyn and Bacon.

Marin Institute. 2006. "Lawsuits against the Industry." Available at http://www.marininstitute.org

Margolis, Robert D., and Joan E. Zweben. 2011. *Treating Patients with Alcohol and Other Drug Problems: An Integrated Approach*. Chapter 3: "Models and Theories of Addiction." Washington DC: American Psychological Association.

Mauer, Marc. 2009. "The Changing Racial Dynamics of the War on Drugs." The Sentencing Project, April 2009. Washington, DC. Available at http://www.sentencingproject.org

Mendoza, Martha. 2010. "U.S. Drug War Has Met None of Its Goals." *Associated Press*, May 13. Available at http://www.msnbc.msn.com

Miron, Jeffrey A., and Elina Tetelbaum. 2009. "The Dangers of the Drinking Age." *Forbes*. Available at http://www.forbes.com

MTF (Monitoring the Future). 2011. *National Results on Adolescent Drug Use*. The University of Michigan, Institute for Social Research. Ann Arbor, Michigan. Available at http://monitoringthefuture.org

Morgan, Patricia A. 1978. "The Legislation of Drug Law: Economic Crisis and Social Control." *Journal of Drug Issues* 8:53–62.

MADD (Mothers Against Drunk Driving). 2011. "Why 21? Addressing Underage Drinking." Available at http://www.madd.org

McDonald. Thomas. 2010. "College Crowd Smokes 'Spice' that Imitates Pot." August. *Charlotte Observer*. Available at http://www.charlotteobserver.com

McMillen, Matt. 2011. "'Bath Salts' Drug Trend: Expert Q&A." WebMD: Mental Health. Available at http://www.webmd.com

Mulvey, Edward P. 2011. "Highlights from Pathways to Desistance: A Longitudinal Study of Serious Adolescent Offenders." U.S. Department of Justice, March 11. Available at http://ncjrs.gov

Myers, Matthew. 2011 (April 25). "FDA Acts to Protect Public Health by Extending Authority over Tobacco Products, Including E-Cigarettes." Available at http://www.tobaccofreekids.org

Myers, Matthew L. 2009a (June 22). "President Obama Delivers Historic Victory for America's Kids and Health over Tobacco." Available at http://www.tobaccofreekids.org

Myers, Matthew L. 2009b. "U.S. Court of Appeals Affirms 2006 Lower Court Ruling That Tobacco Companies Committed Fraud for Five Decades and Lied about the Dangers of Smoking." Press Office Release, May 22. Available at http://www.tobaccofreekids.org

National Drug Intelligence Center. 2004. "Drug Abuse and Mental Illness." Available at http://www.justice.gov

National Institute on Alcohol Abuse and Alcoholism. 2000 (April). *Mechanisms of Addiction*. Alcohol Alert. National Institute on Alcohol Abuse and Alcoholism no. 47. Available at http://www.niaaa.nih.gov

Nebehay, Stephanie. 2011. "Alcohol Kills More than AIDS, TB, or Violence: WHO." Reuters, February 11. Available at http://www.reuters.com

Newport, Frank. 2011. "U.S. Drinking Rate Edges Up Slightly to 25-Year High." Gallup Poll, July 30. Available at http://www.gallup.com/poll/141656

NIDA (National Institute of Drug Abuse). 2010. "Commonly Abused Drugs." Available at http://www.nida.nih.gov

NIDA. 2006. *Principles of Drug Addiction Treatment: A Research Based Guide*. Available at http://drugabuse.gov/PODAT

NIDA. 2005. "Heroin Abuse and Addiction. Research Report Series," NIH Publication 05-4165. Available at http://www.nida.nih/gov

Nordland, Rod. 2010. "U.S. Turns a Blind Eye to Opium in Afghan Town." *The New York Times*, March 20. Available at http://www.nytimes.com

NSDUH. National Survey on Drug Use and Health. 2011. Substance Abuse and Mental Health Services Administration. "Results from the 2010: Volume I. Summary of National Findings." Office of Applied Studies, NSDUH Series H-38A, HHS Publication No. SMA 10-4856. Rockville, MD.

NSDUH. 2008 (September). "Results from the 2007 National Survey on Drug Use and Health: National Findings." Office of Applied Studies, NSDUH Series H-34, DHHS Publication No. SMA 08-4343). Rockville, MD. Available at http://oas.samhsa.gov

NSDUH (National Survey on Drug Use and Health). 2004 (February 13). "Alcohol Dependence or Abuse among Parents with Children Living at Home." Office of Applied Studies, Substance Abuse and Mental Health Services Administration. Available at http://www.oas.samhsa.gov

Obama, Barack. 2011. "Epidemic: Responding To America's Prescription Drug Abuse Crisis." Available at http://www.whitehousedrugpolicy.gov

ONDCP (Office of National Drug Control Policy). 2011. "FY 2010 Funding Highlights." Available at http://www.whitehousedrugpolicy.gov

ONDCP. 2009. *National Youth Anti-Drug Media Campaign*. Available at http://www.theantidrug.com/about.asp

ONDCP. 2008. *National Drug Control Strategy*. Available at http://www.whitehousedrugpolicy.gov

ONDCP. 2006. *Methamphetamine*. Available at http://www.whitehousedrugpolicy.gov/drugfact/methamphetamine

Peters, Jeremy W. 2009. "Albany Takes Step to Repeal '70s-Era Drug Laws." *The New York Times*, March 5. Available at http://www.nytimes.com

Pew Research Center. 2002. *Illegal Drugs*. Pew Research Center for the People and Press Survey. Available at http://www.pollingreport.com

Phoenix House. 2011. "The Phoenix House Way." Available at http://www.phoenixhouse.org

Pignal, Stanley. 2010. "Dutch Look at Weeding out Cannabis Cafes." *Financial Times*, October 8. Available at http://www.ft.com

Primack, Brian A., James E. Bost, Stephanie R. Land, and Michael J. Fine. 2007. "Volume of Tobacco Advertising in African American Markets: Systematic Review and Meta-Analysis." *Public Health Reports* 122(5):607–615.

Reid, Angus. 2010. "Americans Decry War on Drugs, Blame Mexico for Allowing Cartels to Grow." *Public Opinion*, July 21. Available at http://www.angus-reid.com/polls

Rorabaugh, W. J. 1979. *The Alcoholic Republic: An American Tradition*. New York: Oxford University Press.

SAMHSA (Substance Abuse and Mental Health Services Administration). 2010. "Parents on Probation or Parole." *The NSDUH Report*, March 11. Available at http://oas.samhsa.gov

SAMHSA. 2009. "Children Living with Substance-Dependent or Substance-Abusing Parents: 2002–2007." *The NSDUH Report*, April 16. Available at http://oas.samhsa.gov

SAMHSA. 2007. "Parental Substance Abuse Raises Children's Risk." *Practice What You Preach*, February 20. Available at http://www.family.samhsa.gov

Sargant, James, Thomas Wills, Mike Stoolmiller, Jennifer Gibson, and Frederick Gibbons. 2006. "Alcohol Use in Motion Pictures and Its Relation with Early Onset Teen Drinking." *Journal of Studies on Alcohol* 77(1):54–66.

Sheldon, Tony. 2000. "Cannabis Use among Dutch Youth." *British Medical Journal* 321:655.

Siegel, Larry. 2006. *Criminology*. Belmont, CA: Wadsworth.

Snyder, Leslie B., Frances Fleming Miller, Michael Slater, Helen Sun, and Yuliya Strizhakova. 2006. "Effects of Alcohol Advertising Exposure on Drinking among Youth." *Archives of Pediatrics & Adolescent Medicine* 160(1):18–24.

Sohn, Emily. 2010. "Side Effects of Drugs in Water Still Murky." Discovery News, September 28. Available at: http://news.discovery.com

Strunin, Lee, Kirstin Lindeman, Enrico Tempesta, Pierluigi Ascani, Simona Avan, and Luca Parisi. 2010. "Familial Drinking in Italy: Harmful or Protective Factors?" *Addiction Research and Theory* 18(3):344–358.

Szalavitz, Maia. 2009. "Drugs in Portugal: Did Decriminalization Work?" *Time*, April 26. Available at: http://www.time.com

Taifia, Nkechi. 2006 (May). "The 'Crack/Powder Disparity': Can the International Race Convention Provide a Basis for Relief?" American Constitution Society for Law and Policy White Paper. Available at http://acslaw.org

Tarter, Ralph E., Michael Vanyukov, Levent Kirisci, Maureen Reynolds, and Duncan B. Clark. 2006. "Predictors of Marijuana Use in Adolescents Before and After Licit Drug Use: Examination of the Gateway Hypothesis." *American Journal of Psychiatry* 163:2134–2140.

Terry-McElrath, Yvonne M., Melanie A. Wakefield, Sherry Emery, Henry Saffer, Glen Szczypka, Patrick M. O'Malley, Lloyd D. Johnston, Frank J. Chaloupka, and Brian R. Flay. 2007. "State Anti-Tobacco Advertising and Smoking Outcomes by Gender and Race/Ethnicity." *Ethnicity and Health* 12(4):339–362.

The Economist. 2009 (March 7). "A Toker's Guide." Available at http://www.economist.com

Thio, Alex. 2007. *Deviant Behavior*. Boston: Allyn and Bacon.

Timeline. 2001. "Timeline of Tobacco Litigation." Fox News, March 8. Available at http://www.foxnews.com

Tobacco Free Kids. 2009 (February 18). "Deadly in Pink: Big Tobacco Steps Up Its Targeting of Women and Girls." Available at http://www.tobaccofreekids.org

U.S. Department of Justice. 2010. "The Impact of Drugs on Society." *National Drug Threat Assessment 2010*. Available at http://www.justice.gov/ndic/pubs38

Van Dyck, C., and R. Byck. 1982. "Cocaine." *Scientific American* 246:128–141.

Wakefield, Melanie A., Sarah Durkin, Matthew J. Spittal, Mohammad Siahpush, Michelle Scollo, Julie A. Simpson, Simon Chapman, Victoria White, and David Hill. 2008. "Impact of Tobacco Control Policies and Mass Media Campaigns on Monthly Adult Smoking Prevalence." *American Journal of Public Health* 98(8):1443–1450.

WDR (World Drug Report). 2010. *Executive Summary*. United Nations Office on Drugs and Crime (UNODC). New York: United Nations Publication, Sales No. E., 10.XI.13).

WDR (World Drug Report). 2008. *Executive Summary*. Office of Drug Control and Crime Prevention. United Nations. Available at www.unodc.org/documents/wdr/WDR_2008/WDR_2008_eng_web.pdf

Wechsler, William, and Toben F. Nelson. 2008. "What We Have Learned from the Harvard School of Public Health College Alcohol Study: Focusing Attention on College Student Alcohol Consumption and the Environmental Conditions That Promote It." *Journal of Alcohol Studies* (July):1–9.

Weitzman, Elissa, Henry Wechsler, and Toben F. Nelson. 2003 (January 21). *Environment, Not Education, a Stronger Predictor of Binge Drinking Behavior Among College Freshman*. Press Release, Harvard School of Public Health.

Werner, Erica. 2009. "Do Smokers Cost Society Money?" *USA Today*, April 8. Available at http://www.usatoday.com/news/health

West, Steven L., and Keri K. O'Neal. 2004. "Project D.A.R.E. Outcome Effectiveness Revisited." *American Journal of Public Health* 94(6):1027–1029.

Wheeler, Mark. 2008. "UCLA Issues New Report on Prop. 36." *UCLA Newsroom*, October 14. Available at http://newsroom.ucla.edu

Williams, Jenny, Frank J. Chaloupka, and Henry Wechsler. 2005. "Are There Differential Effects of Price and Policy on College Students' Drinking Intensity?" *Contemporary Economic Policy* 23(1):78–90.

Willing, Richard. 2002. "Study Shows Alcohol Is Main Problem for Addicts." *USA Today*, October 3, p. B4.

Willing, Richard. 2004. "Lawsuits Target Alcohol Industry." *USA Today*, May 13. Available at http://www.usatoday.com

Wilson, Joy Johnson. 1999. "Summary of the Attorneys General Master Tobacco Settlement Agreement." National Conference of State Legislators—AFI Health Committee. March. Available at http://academic.udayton.edu/health

Witters, Weldon, Peter Venturelli, and Glen Hanson. 1992. *Drugs and Society*, 3rd ed. Boston: Jones & Bartlett.

WHO (World Health Organization). 2011a. "Data and Statistics." Geneva: World Health Organization.

WHO. 2011b. "Tobacco." February. Geneva: World Health Organization. Available at http://www.who.int

WHO. 2011c. *Systematic Review of the Link Between Tobacco and Poverty*. Geneva: World Health Organization.

WHO. 2011d. *Global Status Report on Alcohol and Health 2011*. Geneva: World Health Organization. Available at http://www.who.int/substance_abuse

WHO. 2011e. *WHO Report on the Global Tobacco Epidemic, 2011*. Geneva: World Health Organization.

WHO. 2008. *WHO Report on the Global Tobacco Epidemic, 2008: The MPOWER Package*. Geneva: World Health Organization, 2008.

Wysong, Earl, Richard Aniskiewicz, and David Wright. 1994. "Truth and Dare: Tracking Drug Education to Graduation and as Symbolic Politics." *Social Problems* 41:448–468.

Zailckas, Koren. 2005. *Smashed: Story of a Drunken Girlhood*. New York: Viking.

Zanis, David A., Donna. M. Coviello, Jacqueline J. Lloyd, and Barry L. Nazar. 2009. "Predictors of Drug Treatment Completion among Parole Violators." *Journal of Psychoactive Drugs* 41–2 (June): 173–80.

Zezima, Matie. 2010. "Beer with Kick is Caught in FDA's Net." 2010. *The New York Times*, November 28. Available at http://www.nytimes.com

Zickler, Patrick. 2003. "Study Demonstrates That Marijuana Smokers Experience Significant Withdrawal." *NIDA Notes* 17:7, 10.

Chapter 4

ABA (American Board of Anesthesiology). 2011. "Anesthesiologists and Capital Punishment." Available at http://www.deathpenaltyinfo.org

Amnesty International. 2011a. "Abolish the Death Penalty, 2011." Amnesty International, London, UK. Available at http://www.amnesty.org

Amnesty International. 2011b. "Figures on the Death Penalty." Available at http://www.amnesty.org

Amnesty International. 2009. "Death Sentences and Executions." Available at http://www.amnesty.org

Austin, Andrew D. 2011. "Discretionary Budget Authority by Subfunction: An Overview." Washington, DC: Congressional Research Service.

Barkan, Steven. 2006. *Criminology*. Upper Saddle River, NJ: Prentice Hall.

Baze v. Rees, 553 U.S. 35, 2008. U.S. Supreme Court. Available at http://www.supremecourtus.gov

Barboza, David. 2009. "Death Sentences Given in Chinese Milk Scandal." *The New York Times*, February 2. Available at http://www.nytimes.com

Becker, Howard S. 1963. *Outsiders: Studies in the Sociology of Deviance*. New York: Free Press.

Beiser, Vince. 2009. "Study in Contrepreneurship." *Miller-McCune Magazine*, April 18. Available at http://www.miller-mccune.com/culture_society/study-in-contrepreneurship-1098

Bell, Kerryn E. 2009. "Gender and Gangs: A Quantitative Comparison." *Crime and Delinquency* 55(3):363–387.

BJS (Bureau of Justice Statistics). 2011a. Criminal Victimization, 2010. Available at http://bjs.ojp.usdoj.gov

BJS. 2010b. "Total Correctional Population." Bureau of Justice Statistics. Available at http://bjs.ojp.usdoj.gov

BJS. 2011c. *Capital Punishment*. Available at http://bjs.ojp.usdoj.gov

BJS. 2010a. *Four Measures of Serious Violent Crime*. Available at http://bjs.ojp.usdoj.gov

BJS. 2010b. V*ictims*. Available at http://bjs.ojp.usdoj.gov

Carlson, Darren K. 2005. "Americans Deal with Crime by Steering Clear." Gallup Poll, November 22. Available at http://www.gallup.com

Carlson, Jr., Joseph R. 2009. "Prison Nurseries: A Pathway to Crime-Free Futures." *Corrections Compendium* 34(1):17–24.

Carroll, Joseph. 2007. "How Americans Protect Themselves from Crime." Gallup Poll, October 26. Available at http://www.gallup.com

CDCP (Centers for Disease Control and Prevention). 2008. "Youth Risk Behavior Surveillance." *Morbidity and Mortality Weekly Report* June 6, pp. 1–131.

Chen, Elsa Y. 2008. "The Liberation Hypothesis and Racial and Ethnic Disparities in the Application of California's Three Strikes Law." *Journal of Ethnicity in Criminal Justice* 6 (2):83–102.

Chan, Sewell. 2010. "Congress Rethinks Its Ban on Internet Gambling." *The New York Times*, July 28. Available at http://www.nytimes.com

Chesney-Lind, Meda, and Randall G. Shelden. 2004. *Girls, Delinquency, and Juvenile Justice*. Belmont, CA: Wadsworth.

Coates, Sam. 2005. "Rader Gets 175 Years for BTK Slayings." *Washington Post*, August 19, p. A3. Available at http://www.washingtonpost.com

Cohen, Mark A., and Alex R. Piquero. 2009. "New Evidence on the Monetary Value of Saving a High Risk Youth." *Journal of Quantitative Criminology* 25:25–49.

Conklin, John E. 2007. *Criminology*, 9th ed. Boston: Allyn and Bacon.

COPS (Community Oriented Policing Services). 2009. *Community Policing Defined*. Available at http://www.cops.usdoj.gov

Correctional Association. 2009. *Education from the Inside Out: The Multiple Benefits of College Programs in Prison*. The Correctional Association of New York. Available at http://www.correctionalassociation.org

D'Alessio, David, and Lisa Stolzenberg. 2002. "A Multilevel Analysis of the Relationship between Labor Surplus and Pretrial Incarceration." *Social Problems* 49:178–193.

Davey, Monica. 2010. "Safety Is Issue as Budget Cuts Free Prisoners." *The New York Times*, March 4. Available at http://www.nytimes.com

District of Columbia v. Heller, 554 U.S. 570, 2008. Available at http://supreme.justia.com

DPIC (Death Penalty Information Center). 2011. "Facts about the Death Penalty." Available at http://www.deathpenaltyinfo.org

Durose, Matthew R., Erica L. Smith, and Patrick A. Langan. 2007. "Contacts Between Police and the Public, 2005." *Bureau of Justice Statistics Special Report*, April. U.S. Department of Justice. NCJ 215243.

The Economist. 2008 (October 30). "The Oldest Conundrum." Available at http://www.economist.com

Egley Jr., Arlen, and James C. Howell. 2011. "Highlights of the 2009. National Youth Gang Survey." June. Washington, DC: U. S. Department of Justice, Office of Juvenile Justice and Delinquency Prevention.

Eith, Christine and Matthew Durose. 2011. *Contact between Police and the Public, 2008.* Special report by the Bureau of Justice Statistics. Available at http://bjs.ojp.usdoj.gov

Erikson, Kai T. 1966. *Wayward Puritans.* New York: Wiley.

Europol. 2011. "Frequently Asked Questions." Available at http://www.europol.europa.eu

FBI (Federal Bureau of Investigation). 2011. "Crime in the United States, 2010." *Annual Uniform Crime Report.* Washington, DC: U.S. Government Printing Office.

FBI. 2009a. *National Incident-Based Reporting System.* April. Available at http://www.fbi.gov

FBI. 2008. "Innocence Lost Sting II." *Headline Archives,* October 27. Available at http://www.fbi.gov

FDA (U.S. Food and Drug Administration). 2008. "Melamine Pet Food Recall of 2007." Available at http://www.fda.gov/Animal Veterinary

Felson, Marcus. 2002. *Crime and Everyday Life,* 3rd ed. Thousand Oaks, CA: Sage.

Fight Crime. 2011. *Fight Crime: Invest in Kids.* Available at http://www.fightcrime.org

Florida Criminal Code. 2009. Available at http://www.leg.state.fl.us/statutes

Ford, Jason. 2005. "Substance Use, Social Bond, and Delinquency." *Sociological Inquiry* 75 (1):109–128

FTC (Federal Trade Commission). 2011 (March). "Consumer Sentinel Network Data Book." Available at http://www.ftc.gov

Frank, Ted. 2011. "Refutation of Toyota Sudden Acceleration Hysteria Doesn't Stop Toyota Sudden Acceleration Litigation." *Forum,* May 2. Available at http://www.pointoflaw.com

Gallup Poll. 2010a. "In U.S., 65% Support Death Penalty in Cases of Murder," November 8. Available at http://www.gallup.com

Gallup Poll. 2010b. "Nearly 4 in 10 Americans Still Fear Walking Alone at Night." *Gallup Poll,* November 5. Available at http://www.gallup.com

Gallup Poll. 2007 "Gallup's Pulse of Democracy: Crime." Available at http:www.galluppoll.com

Greenblatt, Alan. 2008. "Second Chance Programs Quietly Gain Acceptance." *Congressional Quarterly Weekly,* September 15. Available at www.cq.com

Greenburg, Zack O'Malley. 2009. "America's Most Dangerous Cities." *Forbes,* April 23. Available at http://www.forbes.com

Hagerty, Barbara Bradley. 2010. "Can Your Genes Make You Murder?" National Public Radio, July 1. Available at http:// www.npr.org

Hanrahan, Mark. 2011. "Christian Longo, Death Row Inmate, Fights for Right to Donate his Organs after Execution." *The Huffington Post,* April 21. Available at http://www.huffingtonpost.com

Hartney, Christopher, and Linh Vuong. 2009. Created Equal: Racial and Ethnic Disparities in the U.S. Criminal Justice System. Oakland, CA: National Council on Crime and Delinquency.

Heimer, Karen, Stacy Wittrock, and Halime Unal. 2005. "Economic Marginalization and the Gender Gap in Crime." In *Gender and Crime: Patterns of Victimization and Offending,* Karen Heimer and Candace Kruttschnitt, eds. (pp. 115–136). New York University Press.

The Herald Sun. 2008 (January 17). "Jessica's Victim Impact Statement." Available at http://www.news.com.au/heraldsun

Hilzenrath, David. 2011. "Goldman Sachs Subpoenaed." *The Washington Post,* June 2. Available at http://www.washingtonpost.com

Hirschi, Travis. 1969. *Causes of Delinquency.* Berkeley, CA: University of California Press.

ICC (International Chamber of Commerce). 2011. "IMB Piracy Reporting Center." Available at http://www.icc-ccs.org

ICCC (Internet Crime Complaint Center). 2011. "2010 Internet Crime Report." Available at http://www.ic3.gov

ICPC (International Centre for the Prevention of Crime). 2009. *About ICPC.* Available at http://www.crime-prevention-intl.org/

The Innocence Project. 2011. "The Causes of Wrongful Conviction." Available at http://www.innocenceproject.org

INTERPOL. 2011a (April 4). "INTERPOL's Six Priority Crime Areas." Available at http://www.INTERPOL.int

INTERPOL. 2011b. "INTERPOL's Four Core Functions." Available at http://www.INTERPOL.int

INTERPOL. 2011c. "Trafficking in Human Beings." Available at http://www.INTERPOL.int

IRJ (Institute on Race and Justice). 2011. "Racial Profiling Data Collection Resource Center: Legislation and Litigation." Northeastern University. Available at http://www.racialprofilinganalysis.neu.edu

Jacobs, David, Zhenchao Qian, Jason Carmichael, and Stephanie Kent. 2007. "Who Survives on Death Row? An Individual and Contextual Analysis." *American Sociological Review* 72:610–632.

Katel, Peter. 2011. "Crime on Campus." February 4. *CQ Researcher.* CQ Press.

Kennedy, Kelli. 2011. "Massive Nationwide Medicare Bust: 111 Charged for Scams Worth $225 Million." Huffpost Health. *Huffington Post,* February 17.

Kerley, Kent, Michael Benson, Matthew Lee, and Francis Cullen. 2004. "Race, Criminal Justice Contact, and Adult Position in the Social Stratification System." *Social Problems* 51(4):549–568.

Klapper, Bradley. 2005. "UN Told Governments Must Combat Internet Child Pornography." Associated Press, April 14. Available at http://informationweek.com

Krisberg, Barry, Christopher Hartney, Angela Wolf, and Fabiana Silva. 2009 (February 12). "Youth Violence Myths and Realities. A Tale of Three Cities." National Council on Crime and Delinquency. Anne E. Casey Foundation.

Kubrin, Charis E. 2005. "Gangsters, Thugs, and Hustlas: Identity and the Code of the Street in Rap Music." *Social Problems* 52(3): 360–378.

Kubrin, Charis, and Ronald Weitzer. 2003. "Retaliatory Homicide: Concentrated Disadvantage and Neighborhood Culture." *Social Problems* 50:157–180.

Lambert, Lisa. 2011. "Jailbreak: US States Seek to Escape Prison Costs." *Reuters,* May 20. Available at http://in.reuters.com/article

Latimer, Jeff, Craig Dowden, and Danielle Muise. 2005. "The Effectiveness of Restorative Justice Practices: A Meta-Analysis." *The Prison Journal* 85:127–144.

Levy, Steven. 2006. "An Identity Heist the Size of Texas." *Newsweek,* June 12, p. 18.

Lichtblau, Eric, David Johnston, and Ron Nixon. 2008. "F.B.I. Struggles to Handle Financial Fraud Cases." *The New York Times,* October 19. Available at http://www.nytimes.com

Liptak, Adam, and Lisa Faye Petak. 2011. "Juvenile Killers in Jail for Life Seek a Reprieve." *The New York Times,* April 20. Available at http://www.nytimes.com

Lott, John R., Jr. 2003. "Guns Are an Effective Means of Self-Defense." In *Gun Control,* Helen Cothran, ed. (pp. 86–93). Farmington Hills, MI: Greenhaven Press.

MAD DADS. 2011. *Taking It to the Streets.* Available at http://www.maddads.com

Marks, Alexandria. 2006. "Prosecutions Drop for US White Collar Crime." *The Christian Science Monitor,* August 31. Available at http://www.csmonitor.com

Maynard, Micheline. 2010. "Toyota Cited $100 Million Savings after Limiting Recall." *The New York Times,* February 21. Available at http://www.nytimes.com

MacAskill, Ewan, and Xan Rice. 2011. "Somali Pirates Kill Four U.S. Hostages." *The Guardian,* February 22. Available at http://www.guardian.co.uk

Merton, Robert. 1957. *Social Theory and Social Structure.* Glencoe, IL: Free Press.

Moore, Solomon. 2009. "Prison Spending Outpaces All but Medicaid." *The New York Times,* March 3. Available at http://www.nytimes.com

Mundy, Alicia, and Brent Kendall. 2011. "'Insider' Is Charged at FDA." *The Wall Street Journal,* March 30. Available at http://online.wsj.com

National Gang Intelligence Center. 2009. *National Gang Threat Assessment 2009.* Product No. 2009-M0335-001. Washington, DC.

National Night Out. 2011. *What Is National Night Out?* Available at http://www.nationaltownwatch.org/nno

National Research Council. 1994. *Violence in Urban America: Mobilizing a Response.* Washington, DC: National Academy Press.

NCMEC (National Center on Missing and Exploited Children). 2007. *Online Enticement Laws Vary between States*. Available at http://www.missingkids.com

NCPC. 2005. "Preventing Crime Saves Money." Available at http://www.ncpc.org

National Crime Victimization Survey (NCVS). 2010. *National Crime Victimization Survey*. Bureau of Justice Statistics. Available at http://bjs.ojp.usdoj.gov

Newport, Frank. 2010. "Americans Want BP to Pay All Losses, No Matter the Cost." Gallup, June 15. Available at http://www.gallup.com

NWCCC (National White Collar Crime Center). 2010. "National Public Survey on White Collar Crime." Available at http://crimesurvey.nw3c.org

OJJDP (Office of Juvenile Justice and Delinquency Prevention). 2011. "Highlights of a 2009 National Youth Gang Survey." Available at http://www.ncjrs.gov

PBB (Puppies Behind Bars). 2009. *About Us*. Available at http://www.puppiesbehindbars.com/

PEP (Prison Entrepreneurship Program). 2009. *About PEP*. Available at http://www.prisonentrepreneurship.org/

Pertossi, Mayra. 2000 (September 27). *Analysis: Argentine Crime Rate Soars*. Available at http://news.excite.com

Pew. 2011a (April). "State of Recidivism: The Revolving Door of America's Prisons." Pew Charitable Trust. Available at http://pewresearch.org

Pew. 2011b (January 13). "Views of Gun Control—A Detailed Demographic Background." Available at http://pewresearch.org

Pew. 2009. "One in 31: The Long Reach of American Correction." Washington, DC: The Pew Charitable Trust. Available at http://www.pewtrusts.org

Pew. 2007. "Public Safety, Public Spending." Washington, DC: The Pew Charitable Trust. Available at http://www.pewtrusts.org

Pham, Alex. 2011. "Study Says Hacker May Have Stolen Information from 24.6 Million Additional Accounts." *Los Angeles Times*, May 2. Available at http://www.latimes.com

Porter, Nicole D. 2011. "The State of Sentencing, 2010." February. Washington, DC: The Sentencing Project. Available at http://sentencingproject.org

Pridemore, William Alex, and Sang-Weon Kim. 2007. "Socioeconomic Change and Homicide in a Transitional Society" *Sociological Quarterly* 48:229–251.

PUP (Prison University Project). 2011. *About Us*. Available at http://www.prisonuniversityproject.org

Polaris Project. 2011. "International Trafficking." Available at http://www.polarisproject.org

Pollack, Buffy. 2007. "Staying Safe—Home Security Trends." *Mail Tribune*, June 28. Available at http://www.mailtribune.com

Reiman, Jeffrey, and Paul Leighton. 2010. *The Rich Get Richer and the Poor Get Prison*. Boston: Allyn and Bacon.

Romano, Lois. 2005. "More Complete Portrait of BTK Suspect Is Emerging." *Washington Post*, March 5. Available at http://www.washingtonpost.com

Rubin, Paul H. 2002. "The Death Penalty and Deterrence." *Forum*, Winter, pp. 10–12.

Sakahara, Tim. 2011. "Proposal Would Ban Kids from Buying All Toy Guns." Hawaii NEWSNOW, January 9. Available at http://www.hawaiinewsnow.com

Saad, Lydia. 2010. "Nearly 4 in 10 Americans Still Fear Walking Alone at Night." Gallup Poll, November 5. Available at http://www.gallup.com

Salow, Julie. 2009. "AIG's Six Year Saga of Alleged Fraud." *Huffington Post*, April 2. Available at http://www.huffingtonpost.com

Sampson, Robert J., Jeffrey D. Morenoff and Stephen W. Raudenbush. 2005. "Social Anatomy of Racial and Ethnic Disparities in Violence." *American Journal of Public Health* 95(2):224–232.

Schecter, Anna, Brian Ross, and Justin Rood. 2009. "The Executive Who Brought Down AIG." ABC News, March 30. Available at http://www.abcnews.go.com

Schelzig, Erik. 2007 (September 19). "Court Ruling Halts Tennessee Executions." Available at http://www.wral.com

Schweinhart, Lawrence J. 2007. "Crime Prevention by the High/Scope Perry Preschool Program." *Victims and Offenders* 2:141–160.

Shapland, Joanna, and Matthew Hall. 2007. "What Do We Know About the Effects of Crime on Victims?" University of Sheffield: Great Britain. *International Review of Victimology* 14:175–217.

Shaw, Greg M., and Kathryn. 2009. "The Polls—Trends Confidence in Law Enforcement." *Public Opinion Quarterly* 73 (1): 199–220.

Shelley, Louise. 2007. "Terrorism, Transnational Crime and Corruption Center." American University. Available at http://www.american.edu/traccc/

Sherman, Lawrence. 2003. "Reasons for Emotion." *Criminology* 42:1–37.

Serrano, Richard A. 2011. "More Than 100 Charged Nationwide with Medicare Fraud." *Los Angeles Times*, February 11.

Siegel, Larry. 2006. C*riminology*, 9th ed. Belmont, CA: Wadsworth.

Siegel, Larry. 2009. *Criminology*, 10th ed. Belmont, CA: Cengage Learning Wadsworth.

Steinhauer, Jennifer. 2009. "To Cut Costs, States Relax Prison Policies." *The New York Times*, March 25. Available at http://www.nytimes.com

Surgeon General. 2002. "Cost-Effectiveness." In *Youth Violence: A Report of the Surgeon General*. Available at http://www.mentalhealth.org

Sutherland, Edwin H. 1939. *Criminology*. Philadelphia: Lippincott.

Thio, Alex. 2007. *Deviant Behavior*, 9th ed. Boston: Allyn and Bacon.

Thio, Alex. 2004. *Deviant Behavior*, 7th ed. Boston: Allyn and Bacon.

Truman, Jennifer L. 2011. *Criminal Victimization, 2010*. U.S. Department of Justice, Office of Justice Programs. Bureau of Justice Statistics. NCJ 235508.

Turner, Wendy. 2007. "Experiences of Offenders in Prison Canine Programs." *Federal Probation* 71(1):38–43.

UNOS (United Organ Sharing Network). 2011. "Transplant Trends." Available at http://www.unos.org/

U.S. Census Bureau. 2011. *Statistical Abstract of the United States*, 131st edition. Washington, DC: U.S. Government Printing Office.

U.S. Congress. 2011. "Bill Summary and Status." Available at http://thomas.loc.gov

U.S. Department of Justice. 2011. "Programs Summary: Internet Crimes Against Children Task Force Program." Available at http://www.ojjdp.gov

U.S. Department of Justice. 2008. "Serial Murder: Multi-Disciplinary Perspectives for Investigators." Washington, DC: Behavioral Analysis Unit, National Center for the Analysis of Violent Crime.

U.S. Department of Justice. 2003. "Global Crime Issues." In *International Center Global Crimes Issues*, National Institute of Justice. Washington, DC: U.S. Government Printing Office.

U.S. Department of State. 2008 (June 4). "Trafficking in Persons Report." Available at http://www.state.gov

Van de Kamp, John. 2009. "California Can't Afford the Death Penalty." *Los Angeles Times*, June 10. Available at http://www.latimes.com

Victim Statements. 2009. *U.S. v. Bernard L. Madoff*. 2009. U.S. Department of Justice. Available at http://www.pbs.org

VORP (Victim-Offender Reconciliation Program). 2009. *About Victim-Offender Mediation and Reconciliation*. Available at http://www.vorp.com

Vu, Pauline. 2007. "Executions Halted as Doctors Balk." *Stateline*, March 21. Available at http://www.stateline.org

Wagley, John R. 2006. *Transnational Organized Crime: Principal Threats and U.S. Responses*. Congressional Research Service: The Library of Congress.

Weaver, Jay. 2011. "Fans Make Medicare Fraud Sweeps in Miami, Nationwide." *The Miami Herald*, February 17. Available at http://www.miamiherald.com

Webb, Jim. 2011. "The National Criminal Justice Commission Act." Available at http://webb.senate.gov

Weed and Seed. 2011. *Weed and Seed*. Available at http://www.ojp.usdoj.gov

Williams, Linda. 1984. "The Classic Rape: When Do Victims Report?" *Social Problems* 31:459–467.

Williams, Pete. 2007. "Court Overturns D.C. Handgun Ban." MSNBC, March 9. Available at http://www.msnbc.msn.com

Winslow, Robert W., and Sheldon X. Zhang. 2008. *Criminology: A Global Perspective*. Upper Saddle River, NJ: Prentice Hall.

Wilson, Charles. 2010. "Indiana Child Porn Ring: Feds Bust International Online Operation, More Than 50 Arrested." *The Huffington Post,* May 26.

Women's Prison Association (WPA). 2011. "Quick Facts: Women and Criminal Justice 2009." New York: Institute on Women and Criminal Justice.

Wright, Darlene, and Kevin Fitzpatrick. 2006. "Violence and Minority Youth: The Effects of Risk and Asset Factors among African American Children and Adolescents." *Adolescence* 41(162):251–263

Chapter 5

ABC News. 2011 (June 23). "Census 2010: One-Quarter of Gay Couples Raising Children." Available at http://abcnews.go.com

Ahrons, C. 2004. *We're Still Family: What Grown Children Have to Say about Their Parents' Divorce.* New York: HarperCollins.

Amato, Paul. 1999. "The Postdivorce Society: How Divorce Is Shaping the Family and Other Forms of Social Organization." In *The Postdivorce Family: Children, Parenting, and Society,* R. A. Thompson and P. R. Amato, eds. (pp. 161–190). Thousand Oaks, CA: Sage.

Amato, Paul. 2003. "The Consequences of Divorce for Adults and Children." In *Family in Transition,* 12th ed., Arlene S. Skolnick and Jerome H. Skolnick, eds. (pp. 190–213). Boston: Allyn and Bacon.

Amato, Paul. 2004. "Tension between Institutional and Individual Views of Marriage." *Journal of Marriage and Family* 66:959–965.

Amato, P. R. 2010. "Research on Divorce: Continuing Trends and New Developments." *Journal of Marriage and Family* 72:650–666.

Amato, P. R., and J. Cheadle. 2005. "The Long Reach of Divorce: Divorce and Child Well-Being across Three Generations." *Journal of Marriage and the Family* 67:191–206.

Amato, P. R., A. Booth, D. R. Johnson, and S. J. Rogers. 2007. *Alone Together: How Marriage in America Is Changing.* Cambridge MA: Harvard University Press.

American College Health Association. 2011. *American College Health Association National College Health Assessment Fall 2010 Reference Group Data Report.* Baltimore: American College Health Association.

American Humane Association. 2010 (April 20). "Orange County Animal Services and Harbor House Create First Pets and Women's Shelter (PAWS) Program in Central Florida." News Release. Available at http://www .americanhumane.org

Anderson, Kristin L. 1997. "Gender, Status, and Domestic Violence: An Integration of Feminist and Family Violence Approaches." *Journal of Marriage and the Family* 59:655–669.

Applewhite, Ashton. 2003. "Covenant Marriage Would Not Benefit the Family." In *The Family: Opposing Viewpoints,* Auriana Ojeda, ed. (pp. 189–195). Farmington Hill, MI: Greenhaven Press.

Ascione, F. R. 2007. "Emerging Research on Animal Abuse as a Risk Factor for Intimate Partner Violence. In *Intimate Partner Violence,* K. Kendall-Tackett and S. Giacomoni, eds. (pp. 3.1–3.17). Kingston, NJ: Civic Research Institute.

Ascione, Frank R., and Kenneth Shapiro. 2009. "People and Animals, Kindness and Cruelty: Research Directions and Policy Implications." *Journal of Social Issues* 65(3):569–587.

Baker, 2006. "The Power of Stories/Stories about Power: Why Therapists and Clients Should Read Stories about Parental Alienation Syndrome." *The American Journal of Family Therapy* 34:191–203.

Baker, Amy J. L. 2007. *Adult Children of Parental Alienation Syndrome: Breaking the Ties that Bind.* New York: W. W. Norton & Co.

Baker, Amy J. L., and Jaclyn Chambers. 2011. "Adult Recall of Childhood Exposure to Parental Conflict: Unpacking the Black Box of Parental Alienation." *Journal of Divorce & Remarriage* 52(1):55–76.

Bernstein, Nina. 2007. "Polygamy, Practiced in Secrecy, Follows Africans to New York." *New York Times,* March 23, p. A1.

Bonach, Kathryn. 2009. "Empirical Support for the Application of the Forgiveness Intervention Model to Postdivorce Coparenting." *Journal of Divorce & Remarriage* 50(1):38–54.

Bureau of Justice Statistics. 2011. "Intimate Partner Violence in the U.S.: Victim Characteristics." Available at http://bjs.ojp .usdoj.gov/victims.cfm/content/intimate

Bureau of Labor Statistics. 2011. *Employment Characteristics of Families in 2010.* Available at http://www.bls.gov

Carr, D., and K. W. Springer. 2010. "Advances in Families and Health Research in the 21st Century." *Journal of Marriage and Family* 72:743–761.

Carrington, Victoria. 2002. *New Times: New Families.* Dordrecht, the Netherlands: Kluwer Academic.

Carter, Lucy S. 2010. *Batterer Intervention: Doing the Work and Measuring the Progress.* Family Violence and Prevention Fund. Available at http://www.endabuse .org

Catalano, Shannan. 2006. *Intimate Partner Violence in the United States.* Bureau of Justice Statistics. Available at http://www .ojp.usdoj.gov

Centers for Disease Control and Prevention. 2010. *Understanding Child Maltreatment.* Fact Sheet. Available at http//www.cdc .gov

Cherlin, A. J. 2010. "Demographic Trends in the United States: A Review of Research in the 2000s." *Journal of Marriage and Family* 72:403–419.

Cherlin, Andrew J. 2009. *The Marriage-Go-Round: The State of Marriage and Family in America Today.* New York: Alfred A. Knopf.

Coontz, Stephanie. 2005a. *Marriage, a History.* New York: Penguin Books.

Coontz, Stephanie. 2005b. "For Better, for Worse." *Washington Post,* May 1. Available at http:// www.washingtonpost.com

Coontz, Stephanie. 2004. "The World Historical Transformation of Marriage." *Journal of Marriage and Family* 66(4):974–979.

Coontz, Stephanie. 2000. "Marriage: Then and Now." *Phi Kappa Phi Journal* 80:10–15.

Coontz, Stephanie. 1997. *The Way We Really Are.* New York: Perseus.

Coontz, Stephanie. 1992. *The Way We Never Were: American Families and the Nostalgia Trap.* New York: Basic.

Daniel, Elycia. 2005. "Sexual Abuse of Males." In *Sexual Assault: The Victims, the Perpetrators, and the Criminal Justice System,* Frances P. Reddington and Betsy Wright Kreisel, eds. (pp. 133–140). Durham, NC: Carolina Academic Press.

Davis, Lisa Selin. 2009. "Everything but the Ring." *Time* (May 25):57–58.

Decuzzi, A., D. Knox, and M. Zusman. 2004. "The Effect of Parental Divorce on Relationships with Parents and Romantic Partners of College Students." Roundtable Discussion, Southern Sociological Society, Atlanta, April 17.

DeGue, Sarah. 2009 (June). "Is Animal Cruelty a 'Red Flag' for Family Violence? Investigating Co-occurring Violence toward Children, Partners, and Pets." *Journal of Interpersonal Violence* 24(6):1033–1056.

Demo, David H., Mark A. Fine, and Lawrence H. Ganong. 2000. "Divorce as a Family Stressor." In *Families and Change: Coping with Stressful Events and Transitions,* 2nd ed., P. C. McKenry and S. J. Price, eds. (pp. 279–302). Thousand Oaks, CA: Sage.

Dennison, R. P., and S. Koerner. 2008. "A Look at Hopes and Worries about Marriage: The Views of Adolescents Following a Parental Divorce." *Journal of Divorce & Remarriage* 48:91–107.

Doyle, Joseph. 2007. "Child Protection and Child Outcomes: Measuring the Effects of Foster Care." *American Economic Review* 97(5):1583–1610.

Edin, Kathryn. 2000. "What Do Low-Income Single Mothers Say about Marriage?" *Social Problems* 47(1):112–133.

Emery, Robert E. 1999. "Postdivorce Family Life for Children: An Overview of Research and Some Implications for Policy." In *The Postdivorce Family: Children, Parenting, and Society,* R. A. Thompson and P. R. Amato, eds. (pp. 3–27). Thousand Oaks, CA: Sage.

Emery, Robert E., David Sbarra, and Tara Grover. 2005. "Divorce Mediation: Research and Reflections." *Family Court Review* 43(1):22–37.

Federal Bureau of Investigation. 2010. *Crime in the United States, 2009.* Available at http://www .fbi.gov

Fincham, F. D., J. Hall, and S. R. H. Beach. 2006. "Forgiveness in Marriage: Current Status and Future Directions." *Family Relations* 55:415–427.

Fogle, Jean M. 2003. "Domestic Violence Hurts Dogs, Too." *Dog Fancy,* April, p. 12.

Follingstad, D. R., and M. Edmundson. 2010. "Is Psychological Abuse Reciprocal in Intimate Relationships? Data from a National Sample of American Adults." *Journal of Family Violence* 25:495–508.

Foubert, J. D., E. E. Godin, and J. L. Tatum. 2010. "In Their Own Words: Sophomore College Men Describe Attitude and Behavior Changes Resulting from a Rape Prevention Program 2 Years after Their Participation." *Journal of Interpersonal Violence* 25:2237–2257.

Fowler, K. A., and D. Westen. 2011. "Subtyping Male Perpetrators of Intimate Partner Violence." *Journal of Interpersonal Violence* 26(4):607–639.

Gadalla, Tahany M. 2009. "Impact of Marital Dissolution on Men's and Women's Income: A Longitudinal Study." *Journal of Divorce & Remarriage* 50(1):55–65.

Gartrell, Nanette K., Henny M. W. Bos, and Naomi G. Goldberg. 2010. "Adolescents of the U.S. National Longitudinal Lesbian Family Study: Sexual Orientation, Sexual Behavior, and Sexual Risk Exposure." *Archives of Sexual Behavior*, online November 6.

Gelles, Richard J. 2000. "Violence, Abuse, and Neglect in Families." In *Families and Change: Coping with Stressful Events and Transitions*, 2nd ed., P. C. McKenry and S. J. Price, eds. (pp. 183–207). Thousand Oaks, CA: Sage.

Gilbert, Neil. 2003. "Working Families: Hearth to Market." In *All Our Families*, 2nd ed., M. A. Mason, A. Skolnick, and S. D. Sugarman, eds. (pp. 220–243). New York: Oxford University Press.

Global Initiative to End All Corporal Punishment of Children. 2010. *Global Report 2010: Ending Legalised Violence against Children*. Available at http://www.endcorporalpunishment.org

Grogan-Kaylor, Andrew, and Melanie Otis. 2007. "The Predictors of Parental Use of Corporal Punishment." *Family Relations* 56:80–91.

Grych, John H. 2005. "Interparental Conflict as a Risk Factor for Child Maladjustment: Implications for the Development of Prevention Programs." *Family Court Review* 43(1):97–108.

Hackstaff, Karla B. 2003. "Divorce Culture: A Quest for Relational Equality in Marriage." In *Family in Transition*, 12th ed., Arlene S. Skolnick and Jerome H. Skolnick, eds. (pp. 178–190). Boston: Allyn and Bacon.

Hamilton, Brady E., Joyce A. Martin, and Stephanie Ventura. 2010. "Births: Preliminary Data for 2009." *National Vital Statistics Reports* 59(3). Available at http://www.cdc.gov/nchs

Hawkins, Alan J., Jason S. Carroll, William J. Doherty, and Brian Willoughby. 2004. "A Comprehensive Framework for Marriage Education." *Family Relations* 53(5):547–558.

Hewlett, Sylvia Ann, and Cornel West. 1998. *The War against Parents: What We Can Do for Beleaguered Moms and Dads*. Boston: Houghton Mifflin.

Hochschild, Arlie Russell. 1997. *The Time Bind: When Work Becomes Home and Home Becomes Work*. New York: Henry Holt.

Hochschild, Arlie Russell. 1989. *The Second Shift: Working Parents and the Revolution at Home*. New York: Viking.

Jackson, Shelly, Lynette Feder, David R. Forde, Robert C. Davis, Christopher D. Maxwell, and Bruce G. Taylor. 2003 (June). *Batterer Intervention Programs: Where Do We Go from Here?*

U.S. Department of Justice. Available at http://www.usdoj.gov

Jalovaara, M. 2003. "The Joint Effects of Marriage Partners' Socioeconomic Positions on the Risk of Divorce." *Demography* 40:67–81.

Jasinski, J. L., L. M. Williams, and J. Siegel. 2000. "Childhood Physical and Sexual Abuse as Risk Factors for Heavy Drinking among African-American Women: A Prospective Study." *Child Abuse and Neglect* 24:1061–1071.

Jekielek, Susan M. 1998. "Parental Conflict, Marital Disruption, and Children's Emotional Well-Being." *Social Forces* 76:905–935.

Johnson, Michael P. 2001. "Patriarchal Terrorism and Common Couple Violence: Two Forms of Violence against Women." In *Men and Masculinity: A Text Reader*, T. F. Cohen, ed. (pp. 248–260). Belmont, CA: Wadsworth.

Johnson, Michael P., and Kathleen Ferraro. 2003. "Research on Domestic Violence in the 1990s: Making Distinctions." In *Family in Transition*, 12th ed., A. S. Skolnick and J. H. Skolnick, eds. (pp. 493–514). Boston: Allyn and Bacon.

Kalmijn, Matthijs, and Christiaan W. S. Monden. 2006. "Are the Negative Effects of Divorce on Well-Being Dependent on Marital Quality?" *Journal of Marriage and the Family* 68:1197–1213.

Kaufman, Joan, and Edward Zigler. 1992. "The Prevention of Child Maltreatment: Programming, Research, and Policy." In *Prevention of Child Maltreatment: Developmental and Ecological Perspectives*, Diane J. Willis, E. Wayne Holden, and Mindy Rosenberg, eds. (pp. 269–295). New York: John Wiley.

Kitzmann, K. M., N. K. Gaylord, A. R. Holt, and E. D. Kenny. 2003. "Child Witnesses to Domestic Violence: A Meta-Analytic Review." *Journal of Clinical and Consulting Psychology* 71:339–352.

Knox, David (with Kermit Leggett). 1998. *The Divorced Dad's Survival Book: How to Stay Connected with Your Kids*. New York: Insight Books.

Knox, D., and S. Hall. 2010 "Relationship and Sexual Behaviors of a Sample of 2,922 University Students." Unpublished data. Department of Sociology, East Carolina University, and Department of Family and Consumer Sciences, Ball State University.

Koch, Wendy. 2009. "Fees Cut Down Private Adoptions." *USA Today*, April 27, 1A.

Lacey, K.K., D. G. Saunders, and L. Zhang. 2011. "A Comparison of Women of Color and Non-Hispanic White Women on Factors Related to Leaving a Violent Relationship." *Journal of Interpersonal Violence* 26:1036–1055.

LaFraniere, Sharon. 2005. "Entrenched Epidemic: Wife-Beatings in Africa." *New York Times*, August 11, pp. A1 and A8.

Langer, Gary. 2011 (March 18). "Support for Gay Marriage Reaches a Milestone." ABC News. Available at http://abcnews.go.com

Lara, Adair, 2005. "One for the Price of Two: Some Couples Find Their Marriages Thrive When They Share Separate Quarters." *San Francisco Chronicle*, June 29. Available at http://www.sfgate.com

Laungani, P. 2005. "Changing Patterns of Family Life in India." In *Families in Global Perspective*, J. L. Roopnarine and U. P. Gielen, eds. (pp. 85–103). Boston: Pearson, Allyn and Bacon.

Levin, Irene. 2004. "Living Apart Together: A New Family Form." *Current Sociology* 52(2):223–240.

Lewin, Tamar. 2000. "Fears for Children's Well-Being Complicates a Debate over Marriage." *New York Times*, November 4. Available at http://www.nytimes.com

Lindsey, Linda L. 2005. *Gender Roles: A Sociological Perspective*, 4th ed. Upper Saddle River, NJ: Pearson Prentice Hall.

Lloyd, Sally A. 2000. "Intimate Violence: Paradoxes of Romance, Conflict, and Control." *National Forum* 80(4):19–22.

Lloyd, Sally A., and Beth C. Emery. 2000. *The Dark Side of Courtship: Physical and Sexual Aggression*. Thousand Oaks, CA: Sage.

Mason, Mary Ann, Arlene Skolnick, and Stephen D. Sugarman. 2003. "Introduction." In *All Our Families*, 2nd ed., Mary Ann Mason, Arlene Skolnick, and Stephen D. Sugarman, eds. (pp. 1–13). New York: Oxford University Press.

Mental Health America. 2003. *Effective Discipline Techniques for Parents: Alternatives to Spanking*. Strengthening Families Fact Sheet. Available at http://www.nmha.org

Morin, Rich. 2011. (February 16). "The Public Renders a Split Verdict on Changes in Family Structure." Pew Research Center. Available at http://pewresearch.org

National Center for Injury Prevention and Control. 2011. *Understanding Intimate Partner Violence*. Available at http://www.cdc.gov

National Marriage Project. 2009. *The State of Our Unions: Marriage in America 2009*. Rutgers, the State University of New Jersey: The National Marriage Project.

Nelson, B. S., and K. S. Wampler. 2000. "Systemic Effects of Trauma in Clinic Couples: An Exploratory Study of Secondary Trauma Resulting from Childhood Abuse." *Journal of Marriage and Family Counseling* 26:171–184.

Nock, Steven L. 1995. "Commitment and Dependency in Marriage." *Journal of Marriage and the Family* 57:503–514.

OECD. 2010. OECD Family Database. Available at http://www.oecd.org

Parker, K. 2011. "A Portrait of Stepfamilies." Pew Research Center. Available at http://pewsocialtrends.org

Parker, Marcie R., Edward Bergmark, Mark Attridge, and Jude Miller-Burke. 2000. "Domestic Violence and its Effect on Children." *National Council on Family Relations Report* 45(4):F6–F7.

Pasley, Kay, and Carmelle Minton. 2001. "Generative Fathering after Divorce and Remarriage: Beyond the 'Disappearing Dad.'" In *Men and Masculinity: A Text Reader*, T. F. Cohen, ed. (pp. 239–248). Belmont CA: Wadsworth.

Pew Research Center. 2008. "Women Call the Shots at Home: Public Mixed on Gender Roles in Jobs." Available at http://pewresearch.org

Population Reference Bureau. 2011. *The World's Women and Girls 2011 Data Sheet.* Available at http://www.prb.org

Ricci, L., A. Giantris, P. Merriam, S. Hodge, and T. Doyle. 2003. "Abusive Head Trauma in Maine Infants: Medical, Child Protective, and Law Enforcement Analysis." *Child Abuse and Neglect* 27:271–283.

Rubin, D. M., C. W. Christian, L. T. Bilaniuk, K. A. Zaxyczny, and D. R. Durbin. 2003. "Occult Head Injury in High-Risk Abused Children." *Pediatrics* 111:1382–1386.

Russell, D. E. 1990. *Rape in Marriage.* Bloomington: Indiana University Press.

Saad, Lydia. 2011 (May 31). "Doctor-Assisted Suicide is Moral Issue Dividing Americans Most." Available at http://www.gallup.com

Scott, K. L., and D. A. Wolfe. 2000. "Change among Batterers: Examining Men's Success Stories." *Journal of Interpersonal Violence* 15:827–842.

Shepard, Melanie F., and James A. Campbell. 1992. "The Abusive Behavior Inventory: A Measure of Psychological and Physical Abuse." *Journal of Interpersonal Violence* 7(3):291–305.

Simonelli, C. J., T. Mullis, A. N. Elliott, and T. W. Pierce. 2002. "Abuse by Siblings and Subsequent Experiences of Violence within the Dating Relationship." *Journal of Interpersonal Violence* 17:103–121.

Smith, J. 2003. "Shaken Baby Syndrome." *Orthopaedic Nursing* 22:196–205.

Steimle, Brynn M., and Stephen F. Duncan. 2004. "Formative Evaluation of a Family Life Education Web Site." *Family Relations* 53(4):367–376.

Stone, R. D. 2004. *No Secrets, No Lies: How Black Families Can Heal from Sexual Abuse.* New York: Broadway Books.

Straus, Murray. 2010. "Prevalence, Societal Causes, and Trends in Corporal Punishment by Parents in World Perspective." *Law and Contemporary Problems* 73(1):1–30.

Straus, Murray. 2000. "Corporal Punishment and Primary Prevention of Physical Abuse." *Child Abuse and Neglect* 24:1109–1114.

Sullivan, Erin. 2010. "Abused Pasco Dog Taken in by Victim Advocate Now Pays It Forward." *St. Petersburg Times,* January 23. Available at http://www.tampabay.com

Swan, S. C., L. J. Gambone, J. E. Caldwell, T. P. Sullivan, and D. L Snow. 2008. "A Review of Research on Women's Use of Violence with Male Intimate Partners." *Violence and Victims* 23:301–315.

Sweeney, M. M. 2010. "Remarriage and Stepfamilies: Strategic Sites for Family Scholarship in the 21st Century." *Journal of Marriage and the Family* 72:667–684.

Swiss, Liam, and Celine Le Bourdais. 2009. "Father-Child Contact after Separation: The Influence of Living Arrangements." *Journal of Family Issues* 30(5):623–652.

Teaster, Pamela B., Tyler A. Dugar, Marta S. Mendiondo, Erin L. Abner, Kara A. Cecil, and Joanne M. Otto. 2006 (February). *The 2004 Survey of State Adult Protective Services: Abuse of Adults 60 Years of Age and Older.* National Center on Elder Abuse. Washington, DC.

Trinder, L. 2008. "Maternal Gate Closing and Gate Opening in Postdivorce Families." *Journal of Family Issues* 29:1298–1298.

Truman, Jennifer L. 2011. *Criminal Victimization, 2010.* National Crime Victimization Survey. Bureau of Justice Statistics Bulletin.

Ulman, A. 2003. "Violence by Children against Mothers in Relation to Violence between Parents and Corporal Punishment by Parents." *Journal of Comparative Family Studies* 34:41–56.

Umberson, D., K. L. Anderson, K. Williams, and M. D. Chen. 2003. "Relationship Dynamics, Emotion State, and Domestic Violence: A Stress and Masculine Perspective." *Journal of Marriage and the Family* 65:233–247.

UNICEF. 2010. *Child Disciplinary Practices at Home: Evidence from a Range of Low- and Middle-Income Countries.* Available at http://www.childinfo.org

United Nations Development Programme. 2009 (November 23). "Ending Violence Against Women Helps Achieve Development Goals." Available at http://www.beta.undp.org

U.S. Census Bureau. 2010. *America's Families and Living Arrangements: 2010.* Available at http://www.census.gov

U.S. Department of Health and Human Services. Administration on Children, Youth, and Families. 2010. *Child Maltreatment 2009.* Washington, DC: U.S. Government Printing Office.

Walker, Alexis J. 2001. "Refracted Knowledge: Viewing Families through the Prism of Social Science." In *Understanding Families into the New Millennium: A Decade in Review,* Robert M. Milardo, ed. (pp. 52–65). Minneapolis, MN: National Council on Family Relations.

Wallerstein, Judith S. 2003. "Children of Divorce: A Society in Search of Policy." In *All Our Families,* 2nd ed., Mary Ann Mason, Arlene Skolnick, and Stephen D. Sugarman, eds. (pp. 66–95). New York: Oxford University Press.

Wang, Wendy, and Paul Taylor. 2011 (March 19). "For Millennials, Parenthood Trumps Marriage." Pew Research Center. Available at http://pewsocialtrends.org

Whiffen, V. E., J. M. Thompson, and J. A. Aube. 2000. "Mediators of the Link between Childhood Sexual Abuse and Adult Depressive Symptoms." *Journal of Interpersonal Violence* 15:1100–1120.

Whitehurst, Dorothy H., Stephen O'Keefe, and Robert A. Wilson. 2008. "Divorced and Separated Parents in Conflict: Results from a True Experiment Effect of a Court Mandated Parenting Education Program." *Journal of Divorce & Remarriage* 48(3/4):127–144.

Williams, K., and A. Dunne-Bryant. 2006. "Divorce and Adult Psychological Well-Being: Clarifying the Role of Gender and Child Age." *Journal of Marriage and the Family* 68:1178–1196.

Yun, I., D. Ball, and H. Lim. 2011. "Disentangling the Relationship between Child Maltreatment and Violent Delinquency: Using a Nationally Representative Sample." *Journal of Interpersonal Violence* 26(1):88–110.

Zeitzen, Miriam K. 2008. *Polygamy: A Cross-Cultural Analysis.* Oxford: Berg.

Chapter 6

Administration for Children and Families. 2002. *Early Head Start Benefits Children and Families.* U.S. Department of Health and Human Services. Available at http://www.acf.hhs.gov

Albelda, Randy, M. V. Lee Badgett, Alyssa Schneebaum, and Gary J. Gates. 2009. "Poverty in the Lesbian, Gay, and Bisexual Community." The Williams Institute. Available at http://www.law.ucla.edu/williamsinstitute

Albelda, Randy, and Chris Tilly. 1997. *Glass Ceilings and Bottomless Pits: Women's Work, Women's Poverty.* Boston: South End Press.

Alex-Assensoh, Yvette. 1995. "Myths about Race and the Underclass." *Urban Affairs Review* 31:3–19.

Anderson, Sarah, Chuck Collins, Sam Pizzigati, and Kevin Shih. 2010 (September 1). *CEO Pay and the Great Recession: 17th Annual Executive Compensation Survey.* Washington DC: Institute for Policy Studies.

Anderson, Sarah, John Cavanagh, Chuck Collins, Sam Pizzigati, and Mike Lapham. 2008. *Executive Excess: 15th Annual CEO Compensation Survey.* Boston: Institute for Policy Studies and United for a Fair Economy.

Bickel, G., M. Nord, C. Price, W. Hamilton, and J. Cook. 2000. *United States Department of Agriculture Guide to Measuring Household Food Security.* Alexandria, VA: U.S. Department of Agriculture, Food and Nutrition Service.

Boston, Rob. 2005 (February 15). "Faith-Based 'Flim-Flam' Initiative Didn't Have a Prayer Says Former White House Aide." Americans United for a Separation of Church and State. Available at http://blog.au.org

Buchheit, Paul. 2011. "The Mindless Mantra of Wall Street: The Corporate Tax Rate is Too High." *Common Dreams,* April 8. Available at http://www.commondreams.org

Briggs, Vernon M., Jr. 1998. "American-Style Capitalism and Income Disparity: The Challenge of Social Anarchy." *Journal of Economic Issues* 32(2):473–481.

Chandy, Laurence, and Geoffrey Gertz. 2011 (January). *Poverty in Numbers: The Changing State of Global Poverty from 2005 to 2015.* Policy Brief 2011–01. Washington DC: The Brookings Institute.

Children's Defense Fund. 2003. *Children in the United States.* Available at http://www.childrensdefense.org

City Mayors Society. 2011 (January). "Hunger and Homelessness Remain Most Pressing Issues for U.S. Cities." Available at http://www.citymayors.com

Coleman-Jensen, Alisha, Mark Nord, Margaret Andrews, and Steven Carlson. 2011. *Household Food Security in the United States, 2010.* USDA Economic Research Service. Available at http://www.ers.usda.gov

Corak, Miles. 2006. "Do Poor Children Become Poor Adults? Lessons from a Cross-Country Comparison of Generational Earnings Mobility. In *Dynamics of Inequality and Poverty: Research on Economic Inequality*, eds. J. Creedy and G. Kalb:143–188). Elsevier.

Davies, James B., Susanna Sandstrom, Anthony Shorrocks, and Edward N. Wolff. 2006 (December 5). *The World Distribution of Household Wealth*. United Nations University-World Institute for Development Economics Research. Available at http://www.wider.unu.edu

Davis, Kingsley, and Wilbert Moore. 1945. "Some Principles of Stratification." *American Sociological Review* 10:242–249.

DeNavas-Walt, Carmen, Bernadette D. Proctor, and Jessica C. Smith. 2011. *Income, Poverty, and Health Insurance in the United States: 2010*. U.S. Census Bureau, Current Population Reports P60-239. Washington, DC: U.S. Government Printing Office.

DeNavas-Walt, Carmen, Bernadette D. Proctor, and Cheryl Hill Lee. 2006. *Income, Poverty, and Health Insurance Coverage in the United States: 2005*. U.S. Census Bureau, Current Population Reports P60-231. Washington, DC: U.S. Government Printing Office.

Deng, Francis M. 1998. "The Cow and the Thing Called 'What': Dinka Cultural Perspectives on Wealth and Poverty." *Journal of International Affairs* 52(1):101–115.

Dordick, Gwendolyn. 1997. *Something Left to Lose: Personal Relations and Survival Among New York's Homeless*. Philadelphia: Temple University Press.

Dowd, Maureen. 2005. "United States of Shame." *New York Times*, September 3. Available at http://www.nytimes.com

Dvorak, Petula. 2009. "Increase Seen in Attacks on Homeless." *Washington Post*, February 5, p. DZ01.

Economic Policy Institute. 2011. *The State of Working America*. Washington DC: Economic Policy Institute. Available at http://www.stateofworkingamerica.org

Edin, Kathryn, and Laura Lein. 1977. *Making Ends Meet*. New York: Russell Sage Foundation.

Epstein, William M. 2004. "Cleavage in American Attitudes toward Social Welfare." *Journal of Sociology and Social Welfare* 31(4):177–201.

FAO (Food and Agriculture Organization). 2010. *The State of Food Insecurity in the World*. Available at http://www.fao.org

Forster, Michael, and Marco Mira d'Ercole. 2005 (March 10). "Income Distribution and Poverty in OECD Countries in the Second Half of the 1990s." Organization for Economic Cooperation and Development. Available at http://www.oecd.org

Gans, Herbert. 1972. "The Positive Functions of Poverty." *American Journal of Sociology* 78:275–289.

Goldberg, Alison, Chuck Collins, Sam Pizzigati, and Scott Kinger. 2011. *Unnecessary Austerity, Unnecessary Shutdown*. Washington DC: Institute for Policy Studies.

Gonzalez, David. 2005. "From Margins of Society to Center of Tragedy." *New York Times*, September 2. Available at http://www.nytimes.com

Grunwald, Michael. 2006 (August 27). "The Housing Crisis Goes Suburban." *Washington Post*, August 27. Available at http://www.washingtonpost.com

Hartman, Chris. 2011. "Income Inequality." Available at http://www.incomeinequality.org

van den Heuvel, Katrina. 2011. "Putting Poverty on the Agenda." *The Nation*, January 17. Available at http://www.thenation.com

Hoback, Alan, and Scott Anderson. 2007. "Proposed Method for Estimating Local Population of Precariously Housed." National Coalition for the Homeless. Available at http://www.nationalhomeless.org

hooks, bell. 2000. *Where We Stand: Class Matters*. New York: Routledge.

International Labour Organization. 2008a. *Global Wage Report 2008/2009*. Geneva: International Labour Office.

International Labour Organization. 2008b. *World of Work Report 2008: Income Inequalities in the Age of Financial Globalization*. Geneva: International Labour Office.

Jones, Jeffrey M. 2007 (February 9). "Public: Family of 4 Needs to Earn Average of $52,000 to Get By." Gallup News Service. Available at http://www.galluppoll.com

Kennedy, Bruce P., Ichiro Kawachi, Roberta Glass, and Deborah Prothrow-Stith. 1998. "Income Distribution, Socioeconomic Status, and Self-Rated Health in the U.S.: Multilevel Analysis." *British Medical Journal* 317(7163):917–921.

Kocieniewski, David. 2011. "G.E.'s Strategies Let It Avoid Taxes Altogether." *New York Times*, March 24. Available at http://www.nytimes.com

Kraut, Karen, Scott Klinger, and Chuck Collins. 2000. *Choosing the High Road: Businesses That Pay a Living Wage and Prosper*. Boston: United for a Fair Economy.

Leventhal, Tama, and Jeanne Brooks-Gunn. 2003. "Moving to Opportunity: An Experimental Study of Neighborhood Effects on Mental Health." *American Journal of Public Health* 93(9):1576–1585.

Lewan, Todd. 2007. "Unprovoked Beatings of Homeless Soaring." Associated Press, April 8. Available at http://www.breitbart.com

Lichtblau, Eric. 2009. "Attacks of Homeless Bring Push on Hate Crimes Laws." *New York Times*, August 7. Available at http://www.nytimes.com

Llobrera, Joseph, and Bob Zahradnik. 2004. *A HAND UP: How State Earned Income Tax Credits Helped Working Families Escape Poverty in 2004*. Center on Budget and Policy Priorities. Available at http://www.cbpp.org

Luker, Kristin. 1996. *Dubious Conceptions: The Politics of Teenage Pregnancy*. Cambridge, MA: Harvard University Press.

Malatu, Mesfin Samuel, and Carmi Schooler. 2002. "Causal Connections between Socioeconomic Status and Health: Reciprocal Effects and Mediating Mechanisms." *Journal of Health and Social Behavior* 43:22–41.

Mann, Judy. 2000 (May 15). "Demonstrators at the Barricades Aren't Very Subtle, but They Sometimes Win." *Washington Spectator* 26(10):1–3.

Massey, D. S. 1991. "American Apartheid: Segregation and the Making of the American Underclass." *American Journal of Sociology* 96:329–357.

Mayer, Susan E. 1997. *What Money Can't Buy: Family Income and Children's Life Chances*. Cambridge, MA: Harvard University Press.

Mishel, Lawrence, Jared Bernstein, and Heidi Shierholz. 2009. *The State of Working America 2008–2009*. New York: Cornell University Press.

Narayan, Deepa. 2000. *Voices of the Poor: Can Anyone Hear Us?* New York: Oxford University Press.

National Coalition for the Homeless. 2010. *Hate Crimes against the Homeless: America's Growing Tide of Violence*. Available at http://www.nationalhomeless.org

Odede, Kennedy. 2010. "Slumdog Tourism." *The New York Times*, August 10, section A, p. 25. Available at http://www.nytimes.com

OECD. 2011a. *Society at a Glance 2011—OECD Social Indicators*. Available at http:/www/oecd.org

OECD. 2011b (April 6). "Development Aid Reaches an Historic High in 2010." Available at http://www.oecd.org.

Office of Family Assistance. 2009. *Temporary Assistance for Needy Families Program (TANF): Eighth Annual Report to Congress*. Available at http://www.acf.hhs.gov

Oxfam. 2006 (November). *Our Generation's Choice*. Oxfam Briefing Paper. Available at http://www.oxfam.org

Oxfam. 2005. *Paying the Price: Why Rich Countries Must Invest Now in a War on Poverty*. Available at http://www.oxfam.org

Pew Research Center. 2010 (September 17). "Few Say Religion Shapes Immigration, Environment Views." Available at http://people-press.org

Pew Research Center. 2011 (July 26). "Wealth Gaps Rise to Record Highs Between Whites, Blacks, Hispanics." Available at http://pewsocialtrends.org

Popenoe, David. 2008. *The State of Our Unions 2007: The Social Health of Marriage in America*. Rutgers, The State University of New Jersey: The National Marriage Project.

Pugh, Tony. 2007. "U.S. Economy Leaving Record Numbers in Severe Poverty." *McClatchy Newspapers*, February 22. Available at http://www.mcclatchydc.com

Ramos, Alcida Rita, Rafael Guerreiro Osorio, and Jose Pimenta. 2009. "Indigenising Development." *Poverty in Focus* 17(May):pp. 3–5. International Policy Centre for Inclusive Growth.

Roberts, John. 2005 (January 17). *Thai Government Puts Tourism Ahead of the Poor in Tsunami Relief Effort*. World Socialist Web Site. Available at http://www.wsws.org

Robinson, Phyllis. 2009 (June 9). "Urgent Action: Protest Massacre of Indigenous People in Peru!" Small Farmers. Big Change. Available at http://smallfarmersbigchange.coop

Roseland, Mark, and Lena Soots. 2007. "Strengthening Local Economies." In *2007 State of the World*, Linda Starke, ed., (152–169). New York: W. W. Norton & Co.

Rothstein, Richard. 2004. *Class and Schools*. Washington, DC: Economic Policy Institute.

Satterthwaite, David, and Gordon McGranahan. 2007. "Providing Clean Water and Sanitation." In *2007 State of the World: Our Urban Future*, L. Starke, ed. (pp. 26–45). New York: W. W. Norton & Company.

Scalzi, John. 2005 (September 3). "Being Poor." Available at http://whatever.scalzi .com/2005/09/03/being-poor/

Schifferes, Steve. 2004. "Can Globalization Be Tamed?" *BBC News Online*, February 24. Available at http://www.bbc.co.uk

Seccombe, Karen. 2001. "Families in Poverty in the 1990s: Trends, Causes, Consequences, and Lessons Learned." In *Understanding Families into the New Millennium: A Decade in Review*, Robert M. Milardo, ed. (pp. 313–332). Minneapolis, MN: National Council on Family Relations.

Shierholz, Heidi. 2011 (September 7). "The U.S. Doesn't Lack the Right Workers, it Lacks Work." The Economic Policy Institute Blog. Available at http://www.epi.org

Sobolewski, Juliana M., and Paul R. Amato. 2005. "Economic Hardship in the Family of Origin and Children's Psychological Well-Being in Adulthood." *Journal of Marriage and Family* 67(1):141–156.

Stocking, Barbara. 2005 (January 5). *The Tsunami and the Bigger Picture*. Oxfam. Available at http://www.oxfam.org

Susskind, Yifat. 2005 (May). *Ending Poverty, Promoting Development: MADRE Criticizes the United Nations Millennium Development Goals*. Available at http://www.madre.org

Turner, Margery Austin, Susan J. Popkin, G. Thomas Kingsley, and Deborah Kaye. 2005 (April). *Distressed Public Housing: What It Costs to Do Nothing*. The Urban Institute. Available at http://www.urban.org

UNDP (United Nations Development Programme). 2010. *Human Development Report 2010*. Available at http://hdr.undp.org

UNDP. 2006. *Human Development Report 2006*. New York: Palgrave Macmillan.

UNDP. 1997. *Human Development Report 1997*. New York: Oxford University Press.

UNICEF. 2006 (September). *Progress for Children: A Report Card on Water and Sanitation*. Number 5. Available at http://www.unicef.org

UN-Habitat. 2010. *State of the World's Cities 2010/2011*. Available at http://www.unhabitat .org

United Nations. 2005. *Report on the World Social Situation 2005*. New York: United Nations.

United Nations Population Fund. 2002. *State of World Population 2002: People, Poverty, and Possibilities*. New York: United Nations.

U.S. Census Bureau. 2011a. *Poverty Thresholds for 2010*. Available at http://www.census.gov

U.S. Census Bureau. 2011b. *Current Population Survey, Annual Social and Economic Supplement*. Table POV01. Available at http:// www.census.gov

U.S. Census Bureau. 2011c. *2010 American Community Survey*. Available at http://www .census.gov U.S. Census Bureau. 2006. Annual Social and Economic Supplement. *Current Population Survey*. Available at http://www .census.gov

USDA Food and Nutrition Service. 2011. *Program Data, Supplemental Nutrition Assistance Program*. Available at http://www.fns.usda.gov

WHO/UNICEF. *Progress on Sanitation and Drinking-Water 2010 Update*. 2010. Geneva: WHO Press.

Wider Opportunities for Women. 2010. *The Basic Economic Security Tables for the United States*. Washington DC: Wider Opportunities for Women.

Wilson, William J. 1996. *When Work Disappears: The World of the New Urban Poor*. New York: Knopf.

Wilson, William J. 1987. *The Truly Disadvantaged: The Inner City, the Underclass, and Public Policy*. Chicago: University of Chicago Press.

World Bank. 2005. *Global Monitoring Report 2005*. Available at http://www.worldbank.org

World Bank. 2001. *World Development Report: Attacking Poverty, 2000/2001*. Herndon, VA: World Bank and Oxford University Press.

World Health Organization. 2002. *The World Health Report 2002*. Available at http://www .who.int/pub/en

World Population News Service. 2003. "Reducing Poverty Is Key to Global Stability." *Popline*, May–June, p. 4.

Wolff, Edward N. 2010 (March). "Recent Trends in Household Wealth in the United States: Rising Debt and the Middle-Class Squeeze—an Update to 2007." Levy Economics Institute of Bard College, Working Paper No. 589.

Wright, Erik Olin, and Joel Rogers. 2011. *American Society: How It Really Works*. New York: W. W. Norton & Co.

Zedlewski, Sheila R. 2003. *Work and Barriers to Work among Welfare Recipients in 2002*. Urban Institute. Available at http://www .urban.org

Zedlewski, Sheila R., and Kelly Rader. 2005 (March 31). *Feeding America's Low-Income Children*. New Federalism: National Survey of America's Families, No. B-65. Urban Institute. Available at http://www.urban.org

Chapter 7

AFL-CIO. 2011. *Death on the Job: The Toll of Neglect*, 20th ed. Available at http://www.aflcio.org

Austin, Colin. 2002. "The Struggle for Health in Times of Plenty." In *The Human Cost of Food: Farmworkers' Lives, Labor, and Advocacy*, C. D. Thompson Jr. and M. F. Wiggins, eds. (pp. 198–217). Austin: University of Texas Press.

Baily, Martin Neil, and Douglas J. Elliott. 2009 (June 15). "The U.S. Financial and Economic Crisis: Where Does It Stand and Where Do We Go from Here?" The Brookings Institution. Available at http://www.brookings.edu

Bales, Kevin. 1999. *Disposable People: New Slavery in the Global Economy*. Berkeley: University of California Press.

Barstow, David, and Lowell Bergman. 2003. "Deaths on the Job, Slaps on the Wrist." *New York Times Online*, January 10. Available at http://www.nytimes.com

Bassi, Laurie J., and Jens Ludwig. 2000. "School-to-Work Programs in the United States: A Multi-Firm Case Study of Training, Benefits, and Costs." *Industrial and Labor Relations Review* 53(2):219–239.

Benjamin, Medea. 1998. *What's Fair About Fair Labor Association (FLA)?* Sweatshop Watch. Available at http://www.sweatshopwatch.org

Bonior, David. 2006. "Undermining Democracy: Worker Repression in the United States." *Multinational Monitor* 27(4). Available at http:// www.essential.org/monitor

Brand, Jennie E., and Sarah A. Burgard. 2008. "Job Displacement and Social Participation over the Lifecourse: Findings for a Cohort of Joiners." *Social Forces* 87(1):211–242.

Bureau of Labor Statistics. 2011a. "The Employment Situation—April 2011." *Employment Situation Summary*. Available at http://www .bls.gov

Bureau of Labor Statistics. 2011b. Census of Fatal Occupational Injuries Chart, 1992–2009 (revised data). Available at http://www.bls.gov

Bureau of Labor Statistics. 2011c. Union Members 2010. Available at http://www.bls.gov

Bureau of Labor Statistics. 2010 (November 9). "Nonfatal Occupational Injuries and Illnesses Requiring Days Away from Work, 2009." Available at http://www.bls.gov

Butterworth, P., L. S. Leach, L. Strazdins, S. C. Olesen, B. Rodgers, and D. H. Broom. 2011. "The Psychosocial Quality of Work Determines Whether Employment has Benefits for Mental Health: Results from a Longitudinal National Household Survey." *Occupational and Environmental Medicine*. Advance online publication. Doi:10.1136/oem.2010.059030.

Cantor, David, Jane Waldfogel, Jeffrey Kerwin, Mareena McKinley Wright, Kerry Levin, John Rauch, Tracey Hagerty, and Martha Stapelton Kudela. 2001. *Balancing the Needs of Families and Employers: The Family and Medical Leave Surveys, 2000 Update*. U.S. Department of Labor. Available at http://www.dol.gov

Caston, Richard J. 1998. *Life in a Business-Oriented Society: A Sociological Perspective*. Boston: Allyn and Bacon.

Center for Responsive Politics. 2009 (February 4). "TARP Recipients Paid Out $114 Million for Politicking Last Year." Capital Eye Blog. Available at http://www.opensecrets.org

Cernasky, Rachel. 2003 (December). "Slavery: Alive and Thriving in the World Today—The Satya Interview with Kevin Bales." In *Law & Ethics in the Business Environment*, 6th ed., Terry Halbert and Elaine Ingulli, eds. (pp. 172–173). Mason, Ohio: South-Western Cengage Learning.

Cockburn, Andrew. 2003. "21st Century Slaves." *National Geographic*, September, pp. 2–11, 18–24.

Council of Economic Advisors. 2010 (March). *Work-Life Balance and the Economics of Workplace Flexibility*, Christina Romer, ed. Executive Office of the President. Available at http://www.whitehouse.gov

Dorell, Oren. 2011. "Report Blames Massey for W. Va. Mine Explosion." *USA Today*, May 19. Available at http://www.usatoday.com

Ebeling, Richard M. 2009 (February 12). "Capitalism the Solution, Not Cause of the Current Economic Crisis." American Institute for Economic Research. Available at http://www.aier.org

Faux, Jeff. 2008 (February 29). "Overhauling NAFTA." Economic Policy Institute. Available at http://www.epi.org

FLA Watch. 2007a. *FLA Watch: Monitoring the Fair Labor Association*. Available at http://www.flawatch.org

FLA Watch. 2007b. *About FLA Watch*. Available at http://www.flawatch.org

Fram, Alan. 2007. Poll: *A Fifth Vacation with Laptop*. Associated Press, June 3. Yahoo News. Available at http://news.yahoo.com

Frederick, James, and Nancy Lessin. 2000. "Blame the Worker: The Rise of Behavior-Based Safety Programs." *Multinational Monitor* 21(11). Available at http://www.essential.org/monitor

Galinsky, Ellen, James T. Bond, Kelly Sakai, Stacy S. Kim, and Nicole Giuntoli. 2008. *2008 National Study of Employers*. New York: Families and Work Institute.

Galinsky, Ellen, Kerstin Aumann, James T. Bond. 2009. *Times Are Changing: Gender and Generation at Work and Home*. New York: Families and Work Institute.

"The Garment Industry." 2001. Sweatshop Watch. Available at http://www.change.org

George, Kathy. 2003 (December 1). *Myanmar: Unocal Faces Landmark Trial over Slavery*. CorpWatch. Available at http://www.corpwatch.org

Gordon, David M. 1996. *Fat and Mean: The Corporate Squeeze of Working Americans and the Myth of Managerial "Downsizing."* New York: Free Press.

Greenhouse, Steven. 2008. *The Big Squeeze*. New York: Alfred A. Knopf.

Hall, Charles A. S., and John W. Day, Jr. 2009. "Revising the Limits to Growth After Peak Oil." *American Scientist* (May–June):230–237.

Harris Poll. 2010 (November 11). "Americans Still Cutting Back on the Small Things to Save Money." Available at http://www.harrisinteractive.com

Harris Poll. 2009 (June 30). "Americans Are Purchasing More Generic Brands and Brown Bagging It to Save Money." Available at http://www.harrisinteractive.com

Heymann, Jody. Alison Earle, and Jeffrey Hayes. 2007. *The Work, Family and Equity Index*. Montreal, QC: The Project on Global Working Families and The Institute for Health and Social Policy.

Huffstutter, P. J. 2009. "Struggling Cities Cancel Fourth of July Fireworks." *Los Angeles Times*, June 29. Available at http://www.latimes.com

Human Rights Watch. 2010. *Fields of Peril: Child Labor in US Agriculture*. Available at http://www.hrw.org

Human Rights Watch. 2009 (January). *The Employee Free Choice Act: A Human Rights Imperative*. Available at http://www.hrw.org

Human Rights Watch. 2007 (May). *Discounting Rights: Wal-Mart's Violation of US Workers' Right to Freedom of Association*. Volume 19, No. 2 (G). Available at http://hrw.org

Ilg, Randy. 2011 (May). "How Long before the Unemployed Find Jobs or Quit Looking?" Bureau of Labor Statistics. Available at http://www.bls.gov

International Confederation of Free Trade Unions. 2011. *Annual Survey of Violations of Trade Union Rights 2011*. Available at http://www.icftu.org

ILO (International Labour Organization. 2011. *Global Employment Trends 2011: The Challenge of a Jobs Recovery*. Geneva: International Labour Office.

ILO. 2010. *Accelerating Action against Child Labour*. Geneva: International Labour Office.

ILO. 2009. Recovering from the Crisis: A Global Jobs Pact. Available at http://www.ilo.org

Institute for Global Labour & Human Rights. 2011 (March 23). "Triangle Returns: Young Women Continue to Die in Locked Sweatshops." Available at http://www.globallabourrights.org

Isidore, Chris. 2011. "The Great Recession's Lost Generation." *CNNMoney*, May 17. Available at http://money.cnn.com

Jensen, Derrick. 2002. "The Disenchanted Kingdom: George Ritzer on the Disappearance of Authentic Culture." *The Sun*, June, pp. 38–53.

Kelly, Erin L., Phyllis Moen, and Eric Tranby. 2011. "Changing Workplaces to Reduce Work-Family Conflict: Schedule Control in a White Collar Organization." *American Sociological Review* 76(2): 265–290.

Lenski, Gerard, and J. Lenski. 1987. *Human Societies: An Introduction to Macrosociology*, 5th ed. New York: McGraw-Hill.

Leonard, Bill. 1996 (July). "From School to Work: Partnerships Smooth the Transition." *HR Magazine* (Society for Human Resource Management). Available at http://www.shrm.org

MacEnulty, Pat. 2005 (September). "An Offer They Can't Refuse: John Perkins on His Former Life as an Economic Hit Man." *The Sun* 357:4–13.

Maher, Kris. 2011. "Mine Probe Faults Massey." *Wall Street Journal*, May 20. Available at http://online.wsj.com

Martinson, Karin, and Pamela Holcomb. 2007. *Innovative Employment Approaches and Programs for Low-Income Families*. Washington, DC: The Urban Institute.

Mehta, Chirag, and Nik Theodore. 2005 (December). *Undermining the Right to Organize: Employer Behavior during Union Representation Campaigns*. American Rights at Work. Available at http://www.americanrightsatwork.org

Mendenhall, Ruby, Ariel Kalil, Laurel J. Spindel, and Cassandra M. D. Hart. 2008. "Job Loss at Mid-Life: Managers and Executives Face the 'New-Risk Economy.'" *Social Forces* 87(1):185–209.

Miers, Suzanne. 2003. *Slavery in the Twentieth Century: The Evolution of a Global Problem*. Walnut Creek, CA: AltaMira Press.

Moloney, Anastasia. 2005. "Terror as Anti-Union Strategy: The Violent Suppression of Labor Rights in Colombia." *Multinational Monitor* 26(3–4).

Multinational Monitor. 2000. "Big Business for Reform." *Multinational Monitor* 21(11). Available at http://www.essential.org

National Labor Committee. 2007. "Senate Minority Leader Senator Harry Reid, Congressman Bernie Sanders, AFL-CIO and Others Endorse Anti-Sweatshop Bill." Available at http://www.nlcnet.org

National Partnership for Women and Families. 2008. "Family and Medical Leave Act." Available at http://www.nationalpartnership.org

Newport, Frank. 2011 (March 31). "Americans' Top Job-Creation Idea: Stop Sending Work Overseas." Gallup. Available at http://www.gallup.com

"New OSHA Policy Relieves Employees." 1998. *Labor Relations Bulletin* no. 687, p. 8.

Norris, Floyd. 2011 (August 5). "As Corporate Profits Rise, Workers' Income Declines." *New York Times*. Available at http://www.nytimes.com

Ordonez, Franco. 2010 (June 25). "A Worker's Grueling Day." *Charlotte Observer*. Available at http://www.charlotteobserver.com

O'Rourke, Dara. 2003. "Outsourcing Regulation: Analyzing Nongovernmental Systems of Labor Standards and Monitoring." *Policy Studies Journal* 31(1):1–29.

Parenti, Michael. 2007 (February 16). *Mystery: How Wealth Creates Poverty in the World*. Common Dreams NewsCenter. Available at http://www.commondreams.org

Perkins, John. 2004. *Confessions of an Economic Hit Man*. San Francisco: Berrett-Koehler Publishers, Inc.

Pew Research Center. 2011 (March 3). "Fewer are Angry at Government, But Discontent Remains High." Available at http://people-press.org

Pew Research Center. 2010. "'Social' Not So Negative, 'Capitalism' Not So Positive." Available at http://people-press.org

Pew Research Center. 2009. *Trends in Political Values and Core Attitudes: 1987–2009*. Washington DC: The Pew Research Center for People & the Press.

Pugh, Tony. 2009 (June 25). "Recession's Toll: Most Recent College Grads Working Low-Skill Jobs." McClatchy News. Available at http://www.mcclatchydc.com

Ritzer, George. 1995. *The McDonaldization of Society: An Investigation into the Changing Character of Contemporary Social Life*. Thousand Oaks, CA: Pine Forge Press.

Roach, Brian. 2007. *Corporate Power in a Global Economy*. Medford, MA: Global Development and Environment Institute, Tufts University.

Runyan, Carol W., Michael Schulman, Janet Dal Santo, Michael Bowling, Robert Agans, and Ta Myduc. 2007. "Work-Related Hazards and Workplace Safety of U.S. Adolescents Employed in the Retail and Service Sectors." *Pediatrics* 119(3):526–534.

Saad, Lydia. 2010 (August 30). "On the Job Stress is U.S. Workers' Biggest Complaint." Gallup Poll. Available at http://www.gallup.com

Schaeffer, Robert K. 2003. *Understanding Globalization: The Social Consequences of Political, Economic, and Environmental Change*, 2nd ed. Lanham, MD: Rowman & Littlefield.

Schieman, Scott, Melissa Milkie, and Paul Glavin. 2009. "When Work Interferes with Life: the Social Distribution of Work-Nonwork Interference and the Influence of Work-Related Demands and Resources." *American Sociological Review* 74:966–987.

Scott, Robert E., and David Ratner. 2005 (July 20). *NAFTA'S Cautionary Tale*. Economic Policy Institute Briefing Paper 214. Available at http://www.epi.org

"Sex Trade Enslaves Millions of Women, Youth." 2003. *Popline* 25:6.

Shierholz, Heidi. 2011 (May 11). "Continuing Dearth of Job Opportunities Leaves Many Workers Still Sidelined." Economic Policy Institute. Available at http://www.epi.org

Shipler, David K. 2005. *The Working Poor*. New York: Vintage Books.

Skinner, E. Benjamin. 2008. *A Crime So Monstrous: Face-to-Face with Modern-Day Slavery*. New York: Free Press.

Strully, Kate W. 2009. "Job Loss and Health in the U.S. Labor Market." *Demography* 46(2):221–247.

Students and Scholars Against Corporate Misbehavior. 2005 (August 12). *Looking for Mickey Mouse's Conscience: A Survey of the Working Conditions of Disney Factories in China*. Available at http://www.nlcnet.org

SweatFree Communities. n.d. "Adopted Policies." Available at http://www.sweatfree.org

Tate, Deborah. 2007 (February 14). *U.S. Lawmakers Seek to Crack Down on Foreign Sweatshops*. The National Labor Committee. Available at http://www.nlcnet.org

Thompson, Charles D., Jr. 2002. "Introduction." In *The Human Cost of Food: Farmworkers' Lives, Labor, and Advocacy*, C. D. Thompson, Jr., and M. F. Wiggins, eds. (pp. 2–19). Austin: University of Texas Press.

Turner, Anna, and John Irons. 2009 (July). "Mass Layoffs at Highest Level since at Least 1995." Economic Policy Institute. Available at http://epi.org

Uchitelle, Louis. 2006. *The Disposable American: Layoffs and Their Consequences*. New York: Knopf.

UNICEF. 2009. *Children and Conflict in a Changing World*. New York: UNICEF.

United Nations. 2005. *The Millennium Development Goals Report*. New York: United Nations.

Watkins, Marilyn. 2002. *Building Winnable Strategies for Paid Family Leave in the United States*. Economic Opportunity Institute. Available at http://www.econop.org

Weiss, Tara. 2009. "Some New Grads Are Glad There Are No Jobs." *Forbes*, May 27. Available at http://www.Forbes.com

Chapter 8

AASA (American Association of School Administrators). 2009. *Bullying at School and Online*. Education.com. Available at http://www.education.com

Adams, Caralee. 2011. "Obama Calls Community Colleges 'Key to the Future.'" *Education Week*, October 5. Available at http://blogs.edweek.org

AFT (American Federation of Teachers). 2009. *Building Minds, Minding Buildings: A Union's Roadmap to Green and Sustainable Schools*. Available at http://www.aft.org

AFT. 2006. *Building Minds, Minding Buildings: Turning Crumbling Schools into Environments for Learning*. Available at http://www.aft.org

ACSFA. (Advisory Committee on Student Financial Assistance). 2010. *The Rising Price of Inequality*. June. Washington, DC Available at http://www.ed.gov/acsfa

Arum, Richard, and Josipa Roksa. 2011. *Academically Adrift: Limited Learning on College Campuses*. Chicago: University of Chicago Press.

Baker, Peter, and David M. Herszenhorn. 2010. "Obama Signs Overhaul of Student Loan Program." *The New York Times,* March 30. Available at http://www.nytimes.com

Barton, Paul E. 2005. *One Third of a Nation: Rising Drop-Out Rates and Declining Opportunities*. Princeton, NJ: Educational Testing Service.

Barton, Paul E. 2004. "Why Does the Gap Persist?" *Educational Leadership* 62(3):9–13.

Bausell, Carole Viongrad, and Elizabeth Klemick. 2007. "Tracking U.S. Trends." *Education Week* 26(30):42–44.

Benenson Strategy Group. Benenson, Joel. 2009. "Hi-Tech Cheating: Cell Phones and Cheating in Schools: A National Poll." Benenson Strategy Group Commissioned by Common Sense Media. Available at http://www.commonsensemedia.org

Berger, Joseph. 2007. "Fighting over When Public Pay Private Tuition for Disabled." *New York Times*, March 21. Available at http://www.nytimes.com

BLS (Bureau of Labor Statistics). 2011. "Education Pays." Available at http://www.bls.gov

Blankinship, Donna Gordan. 2009. "New CEO: Gates Foundation Learns from Experiments." *Forbes*, May 28. Available at http://www.forbes.com

Booker, Kimberly, and Angela Mitchell. 2011. "Patterns in Recidivism and Discretionary Placement in Disciplinary Alternative Education: The Impact of Gender, Ethnicity, Age, and Special Education Status." *Education and Treatment of Children* 34(2):193–208.

Bowens, Dan. 2009. "North Carolina High Court Upholds Mandatory Year-Around Schools." Available at http://www.wral.com

Boyd, Donald, Pamela Grossman, Hamilton Lankford, Susanna Loeb, and James Wyckoff. 2009. "Who Leaves? Teacher Attrition and Student Achievement." Working Paper No. 23. National Center for the Analysis of Longitudinal Data in Education Research.

Braun, Henry, Lauren **Chapman, and** Sailesh Vezzu. 2010. "The Black-White Achievement Gap Revisited." *Education Policy Analysis Archives* 18(21):1–95.

Bridgeland, John M., John DiIulio, Jr., and Karen Burke Morison. 2006. *The Silent Epidemic*. A Report by Civic Enterprises in association with Peter D. Hart Research Association. The Bill & Melinda Gates Foundation.

Broughman, S. P., N. L. Swaim, and C. A. Hryczaniuk. (2011). *Characteristics of Private Schools in the United States: Results From the 2009–10 Private School Universe Survey* (NCES 2011-339). National Center for Education Statistics, Institute of Education Sciences, U.S. Department of Education. Washington, DC.

Bushaw, William J., and Shane Lopez. 2010. "A Time for Change: The 42nd Annual Phi Delta Kappa/Gallup Poll of the Public's Attitudes toward the Public Schools." *Kappan Magazine*, September. Available at http://www.kappanmagazine.org

CCSSI (Common Core State Standards Initiative). 2011. Available at http://www.corestandards.org/

CDF (Children's Defense Fund). 2010 (May 28). State of America's Children 2010 Report. Available at http://www.childrensdefense.org

Coleman, James S., J. E. Campbell, L. Hobson, J. McPartland, A. Mood, F. Weinfield, and R. York. 1966. *Equality of Educational Opportunity*. Washington, DC: U.S. Government Printing Office.

Corbett, Christianne, Catherine Hill, and Andresse St. Rose. 2008. *Where the Girls Are*. Washington, DC: American Association of University Women.

CNCS (Corporation for National and Community Service). 2010 (April). "Edward M. Kennedy Serve America Act One Year Later." Available at http://www.nationalservice.gov

CNCS (Corporation for National Community Service). 2008 (January). "The Impact of Service-Learning: A Review of Current Research." Issue Brief. Available at http://www.learnandserve.gov

Darling-Hammond, Linda. 2011. "Teacher Preparation Is Essential to TFA's Future." *Education Week* 30(24):25–26.

Demarest, Elizabeth. 2011. "The Urgent Need to Reframe Education Policy." *Education Week* 30(28):1

DeNeui, Daniel, and Tiffany Dodge. 2006. "Asynchronous Learning Networks and Student Outcomes." *Journal of Instructional Psychology* 33(4):256–259.

Dobbs, Michael. 2005. "Youngest Students Most Likely to Be Expelled." *Washington Post*, May 16. Available at http://www.washingtonpost.com

EFA Global Monitoring Report. 2009. *Overcoming Inequality: Why Governance Matters*. UNESCO Publishing. Available at http://unesdoc.unesco.org

Eliot, Lise. 2010. *The Myth of PINK & BLUE Brains*. Educational Leadership 68(3):32–36.

ELT (Expanded Learning Time). 2009. "Guiding Principles." Available at http://www.mass2020.org

ERC (Educational Research Center). 2011. *Quality Counts*. Editorial Products in Educational Research Center. Available at http://www.edweek.org/rc

Fletcher, Robert S. 1943. *History of Oberlin College to the Civil War*. Oberlin, OH: Oberlin College Press.

Flexner, Eleanor. 1972. *Century of Struggle: The Women's Rights Movement in the United States*. New York: Atheneum.

Forster, Greg. 2009. *A Win-Win Solution: The Empirical Evidence on How Vouchers Affect Public Schools*. Friedman Foundation for Educational Choice. Available at http://www.friedmanfoundation.org

Gabriel, Rachael and Jessica Lester. 2010 (December 15). "Public Displays of Teacher Effectiveness." *Education Week* 30 (15):1

Goldenberg, Claude. 2008. "Teaching English Language Learners." *American Educator* (Summer): 8–11, 14–19, 22–23, 42–44.

Greenhouse, Linda. 2007. "Supreme Court Votes to Limit the Use of Race in Integration Plans." *New York Times*, June 29. Available at http://www.nytimes.com

Hopkinson, Natalie. 2011. "The McEducation of the Negro." *The Root*, January 3. Available at http://www.theroot.com

Hall, Daria, and Shana Kennedy. 2006. *Primary Progress, Secondary Challenge: A State-by-State Look at Student Achievement Patterns*. Washington, DC: The Education Trust.

Hanushek, Eric A., Steven G. Rivkin, and John J. Kain. 2005. "Teachers, Schools and Academic Achievement." *Econometrics* 73(2):417–458.

Haberman, Martin. 1991. "The Pedagogy of Poverty versus Good Teaching." *Phi Delta Kappan*. Available at https://www.det.nsw.edu.au

Human Development Report). 2010. *Human Development Report 2010*. Available at http://hdr.undp.org

Hurst, Marianne. 2005. "When It Comes to Bullying, There Are No Boundaries." *Education Week*, February 8. Available at http://www.edweek.org

Independent Sector. 2002. *Engaging Youth in Lifelong Service*. Newsroom. Available at http://www.independentsector.org

IES (Institute of Education Sciences). 2010. *Efficacy of Schoolwide Programs to Promote Social and Character Development and Reduce Problem Behavior in Elementary School Children*. Report from the Social and Character Development Research Program. Washington, DC: U.S. Department of Education.

Josephson Institute of Ethics. 2011. "The Ethics of American Youth: 2010." Josephson Institute Center for Youth Ethics. Available at available at http://charactercounts.org

Kahlenberg, Richard D. 2006. "A New Way on School Integration." *Issue Brief*. The Century Foundation. Available at http://www.equaleducation.org

Kanter, Rosabeth Moss. 1972. "The Organization Child: Experience Management in a Nursery School." *Sociology of Education* 45:186–211.

Karrer, Paul, 2011. "A Letter to My President—the One I Voted for..." *Education Week* (30):19, 23.

Kohn, Alfie. 2011. "How Education Reform Traps Poor Children." *Education Week* 30(29):32–33.

Kozol, Jonathan. 2005. *The Shame of the Nation*. New York: Crown.

Kozol, Jonathan. 1991. *Savage Inequalities: Children in America's Schools*. New York: Crown.

Lewin, Tamar. 2009. "Community Colleges Challenge Hierarchy with 4-Year Degrees." *The New York Times*, May 3. Available at http://www.nytimes.com

Lickona, Thomas, and Matthew Davidson. 2005. *A Report to the Nation: Smart and Good High Schools*. Available at http://www.cortland.edu

Lubienski, Sarah Theule, and Christopher Lubienski. 2006. "School Sector and Academic Achievement: A Multi-Level Analysis of NAEP Mathematics Data." *American Educational Research Journal* 43(4):651–698.

Manzo, Kathleen K. 2005. "College-Based High Schools Fill Growing Need." *Education Week*, May 25. Available at http://www.edweek.org

Maps.com. 2010. "Literacy Rates Worldwide Map." Available at http://www.maps.com

Mau, Rosalind Y. 1992. "The Validity and Devolution of a Concept: Student Alienation." *Adolescence* 27(107):739–740.

McCrummen, Stephanie. 2011. "In N.C., a New Battle on School Integration." *The Washington Post*. January 12: A1.

McDonnell, Sanford. 2009 (October 3). "America's Crisis of Character—And What to Do About It." Available at http://www.edweek.org

McKinsey and Company 2009. *The Economic Impact of the Achievement Gap in America Schools*. Social Sector Office: McKinsey & Company. Available at http://www.mckinsey.com

McNeil, Michele. 2011a. "Proportion of Schools Falling Short on AYP Rises, Report Says." *Education Week* 30(30):22.

McNeil, Michele. 2011b. "States: Stimulus Aid Sparked Progress on Goals." *Education Week* 30 (32):16–17.

McNichol, Elizabeth, Phil Oliff, and Nicholas Johnson. 2011. "States Continue to Feel Recession's Impact." Center on Budget and Policy Priorities. Available at http://www.cbpp.org

Mead, Sara. 2011 (March 30). "Let's Chill Out about D.C. Vouchers." Available at http://www.educweek.org

Mead, Sara. 2006. "The Truth about Boys and Girls." *Education Sector*, June. Available at http://www.educationsector.org

Merton, Robert K. 1968. *Social Theory and Social Structure*. New York: Free Press.

MetLife. 2008. *MetLife Survey of the American Teacher: Past, Present and Future*. Available at http://www.metlife.com

Milkie, Melissa A., and Catherine H. Warner. 2011. "Classroom Learning Environments and Mental Health of First Grade Children." *Journal of Health and Social Behavior* 42:4–22.

Minder, Raphael. 2011. "In Troubled Spain, Boom Times for Foreign Languages." *The New York Times*, March 29. Available at http://www.nytimes.com

Mitchell, Kenneth. 2010. "Taking Teacher Evaluation to Extremes." *Education Week* 30(15):1.

Molnar, A., Boninger, F., Wilkinson, G., Fogarty, J., & Geary, S. (2010). *Effectively Embedded: Schools and the Machinery of Modern Marketing*. The Thirteenth Annual Report on Schoolhouse Commercializing Trends: 2009–2010. Boulder, CO: National Education Policy Center.

Morse, Jodie. 2002. "Learning While Black." *Time*, May 27, pp. 50–52.

MPR (Mathematica Policy Research, Inc.). 2007. "Early Head Start Research and Evaluation." Available at http://mathematica-mpr.com

Muller, Chandra, and Katherine Schiller. 2000. "Leveling the Playing Field?" *Sociology of Education* 73:196–218.

NAEP (National Assessment of Educational Progress). 2010. *The Nation's Report Card*. Available at http://nces.ed.gov/nationsreportcard

NBPTS (National Board for Professional Teaching Standards). 2007. *Impact of National Board Certification*. Available at http://www.nbpts.org

NCD (National Council on Disability). 2000. *Executive Summary*. Available at http://www.ncd.gov

NCES (National Center for Educational Statistics). 2011a. *Digest of Education Statistics, 2010*. U.S. Department of Education. Available at http://nces.ed.gov

NCES. 2011b. *The Condition of Education, 2011*. U.S. Department of Education. Available at http://nces.ed.gov

NCES. 2011c. *Indicators of School Crime and Safety 2010*. U.S. Department of Education. Available at http://nces.ed.gov

NCES. 2011d. Projections of Education Statistics to 2019. Available at http://nces.ed.gov

NCES. 2009. *The Condition of Education, 2008*. U.S. Department of Education. Available at http://nces.ed.gov

NCES. 2007. *Effectiveness of Reading and Mathematics Software Products*. Report to Congress. Washington, DC: U.S. Department of Education. NCES 2007–4005.

NCPPHE (The National Center for Public Policy and Higher Education). 2010. *Squeeze Play 2010: Continued Public Anxiety on Cost, Harsher Judgments on How Colleges Are Run*. Available at http://highereducation.org

NCPPHE. 2009. *Measuring Up 2008: The National Report Card on Higher Education*. Available at http://highereducation.org

NSBA (National School Board Association). 2007. *Cleveland Voucher Program*. Available at http://www.nsba.org

OECD (Organization for Economic Cooperation and Development). 2010. *Education at a Glance 2010*. Available at http://www.oecd.org

OECD (Organization for Economic Cooperation and Development). 2009. *Education at a Glance 2008*. Available at http://www.oecd.org

Office of Head Start. 2011. *Head Start Program Fact Sheet, 2010*. Available at http://www.acf.hhs.gov/programs

Olson, Lynn. 2007. "Paying Attention Early On." *Education Week* 26(17):29–31.

Orfield, Gary, and Chungmei Lee. 2006. *Racial Transformation and the Changing Nature of Segregation*. Cambridge, MA: Harvard University: The Civil Rights Project.

Orthner, Dennis K., Hinckley Jones-Sanpei, Elizabeth C. Hair, Kristin Moore, Randal D. Day, and Kelleen Kaye. 2009. "Marital and Parental Relationship Quality and Educational Outcomes for Youth." *Marriage and Family Review* 45 (2/3):249–269.

Peskin, Melissa Fleschler, Susan R. Tortolero, and Christine M. Markham. 2006. "Bullying and Victimization among Latino and Hispanic Adolescents." *Adolescence* 41(163):467–484.

Pew, Social Trends Staff. 2011. "Is College Worth It?" *Pew Social and Demographic Trends*, May 15. Available at http://pewsocialtrends.org

Picciano, Anthony G., and Jeff Seaman. 2009. *K-12 Online Learning: A 2008 Follow-up of the Survey of U.S. School District Administrators*. Available at http://www.sloanconsortium.org

Poliakoff, Anne Rogers. 2006. "Closing the Gap: An Overview." *ASCD InfoBrief* 44 (January):1–10. The Association for Supervision and Curricular Development.

Ramierz-Valles, Jesus, and Amanda Brown. 2003. "Latinos' Community Involvement in HIV/AIDS: Organizational and Individual Perspectives on Volunteering." *AIDS Education and Prevention* 15(Suppl. 1):90–104.

Ravitch, Diane. 2010. *The Death and Life of the Great American School System: How Testing and Choice Are Undermining Education*. New York, NY: Basic Books.

Reardon, Sean F., Allison Atteberry, Nicole Arshan, and Michal Kurlaend. 2009. *Effects of the California High School Exit Exam on Student Persistent, Achievement and Graduation*. Stanford: Institute for Research on Education Policy & Practice. Available at http://www.stanford.edu

Riehl, Carolyn. 2004. "Bridges to the Future: Contributions of Qualitative Research to the Sociology of Education." In *Schools and Society*, Jeanne Ballantine and Joan Spade, eds. (pp. 56–72). Belmont, CA: Thomson Wadsworth.

Rodriguez, Richard. 1990. "Searching for Roots in a Changing World." In *Social Problems Today*, James M. Henslin, ed. (pp. 202–213). Englewood Cliffs, NJ: Prentice-Hall.

Rosenthal, Robert, and Lenore Jacobson. 1968. *Pygmalion in the Classroom: Teacher Expectations and Pupils' Intellectual Development*. New York: Holt, Rinehart & Winston.

Sadovnik, Alan. 2004. "Theories in the Sociology of Education." In *Schools and Society*, Jeanne Ballantine and Joan Spade, eds. (pp. 7–26). Belmont, CA: Thomson Wadsworth.

Save the Children. 2011. "An Uneducated Girl Is a Girl in Darkness." Available at http://www.savethechildren.org

Sawchuk, Stephen. 2011a. "New Teacher Distribution Methods Hold Promise." *Education Week* 29(35):16–17.

Sawchuk, Stephen. 2011b. "Districts Scrutinizing Teaching Applicants' Potential." *Education Week* 30(33):1,11.

SBAC (Smarter Balanced Assessment Consortium). 2011. "Smarter Balanced Assessment Consortium." Available at http://www.k12.wa.us/smarter

Schott Report. 2010. Yes We Can: *The 2010 Schott 50 State Report on Black Males in Public Education*. Schott Foundation for Public Education. Available at http://www.blackboysreport.org

SFSF (State Fiscal Stabilization Fund). 2011. "Initial Annual Report." U.S. Department of Education. Available at http://www2.ed.gov/

Stancill, Jane. 2011. "School Cuts Go Before Judge." *News and Observer*, June 22. Available at http://www.newsandobserver.com

Stotsky, Sandra. 2009. "The Academic Quality of Teachers: A Civil Rights Issue." *Education Week*, June 26. Available at http://www.edweek.org

Strauss, Valerie. 2009. "Colleges Consider 3-Year Degrees to Save Undergrads Time, Money." *The Washington Post*, May 23. Available at http://www.washingtonpost.com

Talbert, Kent. 2010. "No Child Left Behind Act: Not An 'Unfunded Mandate.'" *Education Law Review*, June 7. Available at http://www.educationlawreview.com

Tanner, C. K. 2008. "Explaining Relationships among Student Outcomes and the School's Physical Environment." *Journal of Advanced Academics* 19:444–471.

Tyler, John H., and Magnus Lofstrom. 2009. "Finishing High School: Alternative Pathways and Dropout Recovery." *The Future of Children* 19(1):78–102.

Tyre, Peg. 2008. *The Trouble with Boys: A Surprising Report Card on Our Sons, Their Problems at School, and What Parents and Educators Must Do*. New York: Crown Publishing Group.

UFT (United Federation of Teachers). 2010 (January). "Separate and Unequal: The Failure of New York City Charter Schools to Serve the City's Neediest Students." Available at http://www.uft.org

UNESCO (United Nations Educational Scientific and Cultural Organization). 2011. "Literacy." Available at http://www.unesco.org/

U.S. Census Bureau. 2011. *Statistical Abstracts of the United States*. Washington, DC: U.S. Government Printing Office.

U. S. Department of Education. 2011a (February 14). "Fiscal Year 2012 Budget Summary." Available at http://www2.ed.gov

U.S. Department of Education. 2011b. *Blueprint for Reform: The Reauthorization of the Elementary and Secondary Education Act*. Available at http://www2.ed.gov

U.S. Department of Education. 2010. Enhancing Education through Technology (ED-TECH) State Program. Available at http://www2.ed.gov

Viadero, Debra. 2006. "Rags to Riches in U.S. Largely a Myth, Scholars Write." *Education Week*, October 25. Available at http://www.edweek.org

Viadero, Debra. 2005. "Study Sees Positive Effects of Teacher Certification." *Education Week*, April 27. Available at http://www.edweek.org

Vigdor, Jacob L., and Helen F. Ladd. 2010 (June). "Scaling the Digital Divide: Home Computer Technology and Student Achievement." The National Bureau of Economic Research,

Working Paper No. 16078. Available at http://papers.nber.org/

Waggoner, Martha. 2009. "Judge: 'Academic Genocide' in Halifax Schools." *News & Observer*, March 19. Available at http://www.newsobserver.com/

Walsh, Mark, and Erik W. Robelen. 2009. "Supreme Court Backs Reimbursement for Private Tuition." *Education Week*, June 22. Available at http://www.edweek.org

Wei, C. C., L. Berkner, S. He, S. Lew, M. Cominole, and P. Siegel. 2009. *National Postsecondary Student Aid Study: Student Financial Aid Estimates for 2007–08: First Look* (NCES 2009–166). Washington DC: National Center for Education Statistics, Institute of Education Sciences, U.S. Department of Education.

Winters, Rebecca. 2001. "From Home to Harvard." *Time*, September 11. Available at http://www.time.com

Wong, Jennifer S. 2009. *No Bullies Allowed: Understanding Peer Victimization, the Impacts on Delinquency, and the Effectiveness of Prevention Programs*. Rand Corporation. Available at http://www.rand.org

Xu, Zeyu, Jane Hannaway, and Colin Taylor. 2007. *Making a Difference? The Effects of Teach for America in High School*. Working Paper No. 17. National Center for the Analysis of Longitudinal Data in Education Research.

Zehr, Mary Ann. 2009. "Community-Service Opportunities Expanded." *Education Week*, April 29. Available at http://www.edweek.org

Zigler, Edward, Sally Styfco, and Elizabeth Gilman. 2004. "The National Head Start Program for Disadvantaged Preschoolers." In *Schools and Society*, Jeanne Ballantine and Joan Spade, eds. (pp. 341–346). Belmont, CA: Thomas Wadsworth.

Chapter 9

Allport, G. W. 1954. *The Nature of Prejudice*. Cambridge, MA: Addison-Wesley.

American Council on Education and American Association of University Professors. 2000. *Does Diversity Make a Difference? Three Research Studies on Diversity in College Classrooms*. Washington, DC: American Council on Education and American Association of University Professors.

Armario, Christine. 2011. "Feds: All Kids, Legal or Not, Deserve K-12 Education." *Chron*, May 7. Available at http://chron.com

Associated Press. 2011. "FBI Investigate Fatal Rundown of Black Miss. Man." *NPR*, August 17. Available at http://www.npr.org

Associated Press. 2001. "In-State Tuition for Illegal Immigrants in Conn. Gives Some University Option for First Time." *Washington Post*, July 11. Available at http://www.washingtonpost.com

Balko, Radley. 2009 (July 6). "The El Paso Miracle." Reasononline. Available at http://www.reason.com

Bauer, Mary, and Sarah Reynolds. 2009. *Under Siege: Life for Low-Income Latinos in the South*. Montgomery, AL: The Southern Poverty Law Center.

Beirich, Heidi. 2011 (Spring). "The Year in Nativism." *Intelligence Report* (151):35–39.

Beirich, Heidi. 2007 (Summer). "Getting Immigration Facts Straight." *Intelligence Report*. Available at http://www.splcenter.org

Brace, C. Loring. 2005. *"Race" Is a Four-Letter Word*. New York: Oxford University Press.

Brooks, Roy. 2004. *Atonement and Forgiveness: A New Model for Black Reparations*. Berkeley: University of California Press.

Brown University Steering Committee on Slavery and Justice. 2007. *Slavery and Justice*. Providence, RI: Brown University.

Campus Tolerance Foundation. 2009. "Intolerance on America's College Campuses? A 10-School Comparison." Available at http://campustolerance.org

CBCNews.ca. 2009 (May 25). "Girl Watched Skinhead Videos and Talked of How to Kill, Hearing Told." Available at http://www.cbc.ca/canada

CNN. 2009 (June 18). "Senate Approves Resolution Apology for Slavery." Available at http://www.cnn.com

Conley, Dalton. 2002. "The Importance of Being White." *Newsday*, October 13. Available at http://www.newsday.com

Conley, Dalton. 1999. *Being Black, Living in the Red: Race, Wealth, and Social Policy in America*. Berkeley: University of California Press.

Dawkins, Marcia. 2010 (June 6). "'Loving,' Hating and Interracial Relationships." *Huffington Post*. Available at http://www.huffingtonpost.com

Dees, Morris. 2000 (December 28). Personal correspondence. Morris Dees, co-founder of the Southern Poverty Law Center, 400 Washington Avenue, Montgomery, AL 36104.

Deslatte, Melinda. 2009. "Keith Bardwell Quits: Justice of the Peace Who Refused to Give Interracial Couple Marriage License Resigns." *Huffington Post*, November 3. Available at http://www.huffingtonpost.com

Dwyer, Devin. 2011. "Opponents of Illegal Immigration Target Birthright Citizenship." *ABC News*, January 5. Available at http://www.abcnews.go.com

EEOC (Equal Employment Opportunity Commission). 2011 (June 22). "A. C. Widenhouse Sued by EEOC for Racial Harassment." Press Release. Available at http://www.eeoc.gov

EEOC. 2007. "EEOC and Walgreens Resolve Lawsuit." Press release. Available at http://www.eeoc.gov

Ennis, Sharon R., Merarys Rios-Vargas, and Nora G. Albert. 2011 (May). "The Hispanic Population: 2010." *2010 Census Briefs*. U.S. Census Bureau. Available at http://www.census.gov

Esposito, John L. 2011. "Getting It Right about Islam and American Muslims." *Huffington Post*, May 25. Available at http://www.huffingtonpost.com

FBI (Federal Bureau of Investigation). 2010. *Hate Crime Statistics 2009*. Available at http://www.fbi.gov

FBI. 2008 (Fall). "Intelligence Briefs. FBI Report Confirms Extremist Activity in U.S. Military." *Intelligence Reports* (131):5–6.

Fry, Richard. 2009. "Sharp Growth in Suburban Minority Enrollment Yields Modest Gains in School Diversity." Pew Hispanic Center. Available at http://www.pewhispaniccenter.org

Gaertner, Samuel L., and John F. Dovidio. 2000. *Reducing Intergroup Bias: The Common Ingroup Identity Model*. Philadelphia: Taylor & Francis.

Gallup Organization. 2011a. *Race Relations*. Available at http://www.gallup.com

Gallup Organization. 2011b. *Immigration*. Available at http://www.gallup.com

Glaser, Jack, Jay Dixit, and Donald P. Green. 2002. "Studying Hate Crime with the Internet: What Makes Racists Advocate Racial Violence?" *Journal of Social Issues* 58(1):177–193.

Glaubke, Christina Roman, and Katharine Heintz-Krowles. 2004. *Fall Colors: Prime Time Diversity Report 2003–04*. Oakland, CA: Children Now & the Media Program.

Goldstein, Joseph. 1999. "Sunbeams." *The Sun* 277, January, p. 48.

Greenhouse, Steven. 2005. "Wal-Mart to Pay U.S. $11 Million in Lawsuit on Illegal Workers." *New York Times*, March 19. Available at http://www.nytimes.com

Grieco, Elizabeth M., and Edward Trevelyan. 2010 (October). "Place of Birth of the Foreign-Born Population: 2009." *American Community Survey Briefs*. Available at http://www.census.gov

Gryn, Thomas A., and Luke J. Larsen. 2010 (October). "Nativity Status and Citizenship in the United States: 2009." *American Community Survey Briefs*. Available at http://www.census.gov

Gurin, Patricia. 1999. "New Research on the Benefits of Diversity in College and Beyond: An Empirical Analysis." *Diversity Digest*, Spring, pp. 5–15.

Healey, Joseph F. 1997. *Race, Ethnicity, and Gender in the United States: Inequality, Group Conflict, and Power*. Thousand Oaks, CA: Pine Forge Press.

"Hear and Now." 2000 (Fall). *Teaching Tolerance*, p. 5.

Higginbotham, Elizabeth, and Margaret L. Anderson. 2012. "The Social Construction of Race and Ethnicity." In *Race and Ethnicity in Society*, 3rd ed., E. Higginbotham and M. L. Anderson, eds. (pp. 3–6). Belmont, CA: Wadsworth, Cengage Learning.

Hodgkinson, Harold L. 1995. "What Should We Call People? Race, Class, and the Census for 2000." *Phi Delta Kappa*, October, pp. 173–179.

Hoefer, Michael, Nancy Rytina, and Bryan C. Baker. 2011 (February). "Estimates of the Unauthorized Immigrant Population Residing in the United States: January 2010." U.S. Department of Homeland Security. Available at http://www.dhs.gov

Holthouse, David. 2006. "A Few Bad Men." *Intelligence Report*, July 7. Available at http://www.splcenter.org

Holzer, Harry, and David Neumark. 2000. "Assessing Affirmative Action." *Journal of Economic Literature* 38(3):483–568.

hooks, bell. 2000. *Where We Stand: Class Matters*. New York: Routledge.

Humes, Karen R., Nicholas A. Jones, and Roberto R. Ramirez. 2011 (March). "Overview of Race and Hispanic Origin: 2010. *2010 Census Briefs*. Available at http://www.census.gov

Humphreys, Debra. 1999. "Diversity and the College Curriculum: How Colleges and Universities Are Preparing Students for a Changing World." *Diversity-Web*. Available at http://www.inform.umd.edu

Hwang, Jenny. 2010 (June 17). "Arizona's Anti-Immigration Law Is Also Anti-Faith." Immigration Policy Center. Available at http://www.immigrationpolicycenter.org.

Jaschik, Scott. 2011. "Court: Michigan Voters Can't Ban Affirmative Action." *USA Today*, July 5. Available at http://www.usatoday.com

Jay, Gregory. 2005 (March 17). "Introduction to Whiteness Studies." Available at http://www.uwm.edu

Jensen, Derrick. 2001. "Saving the Indigenous Soul: An Interview with Martin Prechtel." *The Sun* 304(April):4–15.

Kandel, William A. 2011 (January 18). "The U.S. Foreign-Born Population: Trends and Selected Characteristics." Congressional Research Service. Available at http://www.crs.gov

King, Joyce E. 2000 (Fall). "A Moral Choice." *Teaching Tolerance* 18:14–15.

Kovac, Amy. 2009. "Transcript of Rev. Lowery's Inaugural Benediction." *Washington Post*, January 20. Available at http://voices.washingtonpost.com

Kozol, Jonathan. 1991. *Savage Inequalities: Children in America's Schools*. New York: Crown.

Kremer, James D., Kathleen A. Moccio, and Joseph W. Hammell. 2009. *Severing a Lifeline: The Neglect of Citizen Children in America's Immigration Enforcement Policy*. Minneapolis: Dorsey & Whitney LLP.

Leadership Council on Civil Rights Education Fund. 2009. *Confronting the New Faces of Hate: Hate Crimes in America*. Available at http://www.civilrights.org

Levin, Jack, and Jack McDevitt. 1995. "Landmark Study Reveals Hate Crimes Vary Significantly by Offender Motivation." *Klanwatch Intelligence Report*, August, pp. 7–9.

Lollock, Lisa. 2001 (March). *The Foreign-Born Population in the United States: March 2000*. Current Population Reports P20–534. Washington DC: U.S. Bureau of the Census.

Mahony, Edmund H., and Josh Kovner. 2009 (June 30). "U.S. Supreme Court Rules in Favor of New Haven Firefighters." *The Hartford Courant*. Available at http://www.courant.com

Marger, Martin N. 2012. *Race & Ethnic Relations: American and Global Perspectives*, 9th ed. Belmont CA: Wadsworth, Cengage Learning.

Maril, Robert Lee. 2011. *The Fence: National Security, Public Safety, and Illegal Immigration along the U.S.-Mexico Border*. Lubbock Texas: Texas Tech University Press.

Maril, Robert Lee. 2004. *Patrolling Chaos: The U.S. Border Patrol in Deep South Texas*. Lubbock, TX: Texas Tech University Press.

Massey, Douglas S., and Garvey Lundy. 2001. "Use of Black English and Racial Discrimination in

Urban Housing Markets: New Methods and Findings." *Urban Affairs Review* 36(4):452–469.

McIntosh, Peggy. 1990 (Winter). "White Privilege: Unpacking the Invisible Knapsack." *Independent School* 49(2):31–35.

Morrison, Pat. 2002. "September 11: A Year Later—American Muslims Are Determined Not to Let Hostility Win." *National Catholic Reporter* 38(38):9–10.

Moser, Bob. 2004. "The Battle of 'Georgiafornia.'" *Intelligence Report* 116:40–50.

Mukhopadhyay, Carol C., Rosemary Henze, and Yolanda T. Moses. 2007. *How Real Is Race?* Lanham, MD: Rowman & Littlefield Education.

NBC News/Wall Street Journal. 2010. "NBC News/Wall Street Journal Survey." Study # 9500. Available at http://online.wsj.com

Olson, James S. 2003. *Equality Deferred: Race, Ethnicity, and Immigration in America Since 1945*. Belmont, CA: Wadsworth/Thomson.

Orfield, Gary. 2001 (July). *Schools More Separate: Consequences of a Decade of Resegregation.* Cambridge, MA: Harvard University, Civil Rights Project.

Ossorio, Pilar, and Troy Duster. 2005. "Race and Genetics." *American Psychologist* 60(1):115–128.

Pager, Devah. 2003. "The Mark of a Criminal Record." *American Journal of Sociology* 108(5):937–975.

Passel, Jeffrey S., and D'Vera Cohn. 2011. "Unauthorized Immigrant Population: National and State Trends." Pew Hispanic Research Center. Available at http://www.pewhispanic.org

Passel, Jeffrey S., Wendy Wang, and Paul Taylor. 2010 (June 4). "Marrying Out: One-in-Seven New U.S. Marriages is Interracial or Interethnic." Pew Research Center. Available at http://pewresearchcenter.org

Passel, Jeffrey S., and D'Vera Cohn. 2009. "A Portrait of Unauthorized Immigrants in the United States." Pew Hispanic Research Center. Available at http://www.pewhispanic.org

Passel, Jeffrey S., and Paul Taylor. 2009. "Who's Hispanic?" Pew Hispanic Research Center. Available at http://www.pewhispanic.org

Paul, Pamela. 2003. "Attitudes toward Affirmative Action." *American Demographics* 25(4):18–19.

Pew Research Center. 2011a (February 24). "Public Favors Tougher Border Controls and Path to Citizenship." Available at http://people-press.org

Pew Research Center. 2011b (May 4). *Beyond Red vs. Blue: The Political Typology.* Available at http://people-press.org

Pew Research Center. 2010 (January 12). "Blacks Upbeat about Black Progress, Prospects." Available at http://pewsocialtrends.org

Pew Research Center. 2005 (September 8). *Huge Racial Divide over Katrina and Its Consequences.* Available at http://www.people-press.org

Picca, Leslie, and Joe R. Feagin. 2007. *Two-Faced Racism.* New York: Routledge.

Plous, S. 2003. "Ten Myths About Affirmative Action." In *Understanding Prejudice and Discrimination*, S. Plous, ed. (pp. 206–212). New York: McGraw-Hill.

Pollin, Robert. 2011. "Economic Prospects: Can We Stop Blaming Immigrants?" *New Labor Forum* 20(1):86–89.

Potok, Mark. 2011 (Spring). "The Year in Hate & Extremism." *Intelligence Report* (141):41–67.

Pryor, John H., Sylvia Hurtado, Linda DeAngelo, Laura Paluki Blake, and Serge Tran. 2010. *The American Freshman: National Norms Fall 2010.* Los Angeles: Higher Education Research Institute, UCLA.

Rosenfeld, Michael J., and Byung-Soo Kim. 2005 (August). "The Independence of Young Adults and the Rise of Interracial and Same-Sex Unions." *American Sociological Review* 70:541–562.

Saulny, Susan. 2011. "Census Data presents Rise in Multiracial Population of Youths." *New York Times*, March 24. Available at http://www.nytimes.com

Schiller, Bradley R. 2004. *The Economics of Poverty and Discrimination*, 9th ed. Upper Saddle River, NJ: Pearson Education.

Schmidt, Peter. 2004 (January 30). "New Pressure Put on Colleges to End Legacies in Admissions." *Chronicle of Higher Education* 50(21):A1.

Schuman, Howard, and Maria Krysan. 1999. "A Historical Note on Whites' Beliefs About Racial Inequality." *American Sociological Review* 64:847–855.

Shierholz, Heidi. 2010 (February 4). "Immigration and Wages—Methodological Advancements Confirm Modest Gains for Native Workers." EPI Briefing Paper #255. Available at http://www.epi.org

Shipler, David K. 1998. "Subtle vs. Overt Racism." *Washington Spectator* 24(6):1–3.

Sidanius, Jim, Shana Levin, Colette Van Laar, and David O. Sears. 2010. *The Diversity Challenge: Social Identity and Intergroup Relations on the College Campus.* New York: Russell Sage Foundation.

SPLC (Southern Poverty Law Center). 2009. "SPLC Files Complaint after Worker Sexually Assaulted, Brutalized by Manager." *SPLC Report* 39 (2):5.

SPLC. 2007. *Close to Slavery: Guestworker Programs in the United States.* Montgomery, AL: Southern Poverty Law Center.

SPLC. 2005 (June). "Center Wins $1.35 Million Judgment against Violent Border Vigilantes." *SPLC Report* 35(2):3.

SPLC. 2004 (Winter). "Neo-Nazi Label Woos Teens with Hate-Music Sampler." *Intelligence Report* 116:5.

SPLC. 2000 (Spring). "Hate on Campus." *Intelligence Report* 98:6–15.

Tanneeru, Manav. 2007 (May 11). "Asian-Americans' Diverse Voices Share Similar Stories." CNN.com. Available at http://www.cnn.com

Tavernise, Sabrina. 2011. "In Census, Young Americans Increasingly Diverse." *New York Times*, February 4. Available at http://www.nytimes.com

Teaching Tolerance. 2011 (Spring). "10 Myths About Immigration." Available at http://www.tolerance.org

Thompson, Ginger. 2009. "After Losing Freedom, Some Immigrants Face Loss of Custody of Their Children." *New York Times*, April 23. Available at http://nytimes.com

Tolbert, Caroline J., and John A. Grummel. 2003. "Revisiting the Racial Threat Hypothesis: White Voter Support for California's Proposition 209." *State Politics and Policy Quarterly* 3(2):183–202, 215–216.

Tramonte, Lynn. 2011 (April). *Debunking the Myth of "Sanctuary Cities."* Immigration Policy Studies. Available at http://www.immigrationpolicy.org

Turner, Margery Austin, Stephen L. Ross, George Galster, and John Yinger. 2002. *Discrimination in Metropolitan Housing Markets.* Washington, DC: Urban Institute.

Turner, Margery Austin, and Karina Fortuny. 2009. *Residential Segregation and Low-Income Working Families.* Washington, DC: Urban Institute.

Turn It Down. 2009 (May 3). "Social Networking: A Place for Hate?" Available at http://turnitdown.newcomm.org

U.S. Census Bureau (2011a; generated by C. Schacht, July 3). *2010 Census Data.* Available at http://2010.census.gov/2010census/data/index.php

U.S. Census Bureau. 2011b. Current Population Survey. Table A-3. "Mean Earnings of Workers 18 Years and Over, by Educational Attainment, Race, and Hispanic Origin: 1975–2009." Available at http://www.census.gov

U.S. Citizenship and Immigration Services. 2011. *A Guide to Naturalization.* Available at http://www.uscis.gov

U.S. Customs and Border Protection. 2011 (January 5). "Border Patrol Overview." Available at http://www.cbp.gov

U.S. Department of Labor. 2002. *Facts on Executive Order 11246 Affirmative Action.* Available at http://www.dol.gov

Van Ausdale, Debra, and Joe R. Feagin. 2001. *The First R: How Children Learn Race and Racism.* Lanham, MD: Rowman & Littlefield.

Washington Post. 2009. "Washington Post-ABC News Poll." Available at http://www.washingtonpost.com

Williams, Eddie N., and Milton D. Morris. 1993. "Racism and Our Future." In *Race in America: The Struggle for Equality*, Herbert Hill and James E. Jones Jr., eds. (pp. 417–424). Madison: University of Wisconsin Press.

Williams, Richard, Reynold Nesiba, and Eileen Diaz McConnell. 2005. "The Changing Face of Inequality in Home Mortgage Lending." *Social Problems* 52(2):181–208.

Wilson, William J. 1987. *The Truly Disadvantaged: The Inner City, the Underclass and Public Policy.* Chicago: University of Chicago Press.

Winter, Greg. 2003a. "Schools Resegregate, Study Finds." *New York Times*, January 21. Available at http://www.nytimes.com

Winter, Greg. 2003b. "U. of Michigan Alters Admissions Use of Race." *New York Times*, August 29. Available at http://www.nytimes.com

Wise, Tim. 2009. *Between Barack and a Hard Place: Racism and White Denial in the Age of Obama.* San Francisco: City Light Books.

Zinn, Howard. 1993. "Columbus and the Doctrine of Discovery." In *Systemic Crisis: Problems in Society, Politics, and World Order*, William D. Perdue, ed. (pp. 351–357). Fort Worth, TX: Harcourt Brace Jovanovich.

Chapter 10

AAUW (American Association of University Women). 2011a. *Why So Few?* Available at http://www.aauw.org

AAUW. 2011b. *Gender Wage Gap Map.* Available at http://www.aauw.org

AAUW. 2011c. "Title IX Case Goes to Court Tuesday with AAUW Legal Advocacy Fund Support." Press Release. May 31. Available at http://www.aauw.org

AAUW. 2007 (April 23). *Pay Gap Exists as Early as One Year out of College, Research Says.* Available at http://www.aauw.org

Abernathy, Michael. 2003. *Male Bashing on TV.* Tolerance in the News. Available at http://www.tolerance.org

Abraham, Tamara, and Jennifer Madison. 2011. "Nailing a Trend: From Gwen Stefani to Jennifer Lopez, How J Crew Boss Took Her Lead from the Stars in Giving Son a Manicure." *The Daily Mail*, April 15. Available at http://www.dailymail.co.uk

ACLU (American Civil Liberties Union). 2011a (April 12). "On Equal Pay Day, ACLU Welcomes Reintroduction of Paycheck Fairness Act." Available at http://www.aclu.org

ACLU. 2011b (June 20). "Jessica Gonzales v. U.S.A." Available at http://www.aclu.org

Adams, Jimi. 2007. "Stained Glass Makes the Ceiling Visible." *Gender and Society* 21(1):80–105.

Allam, Hannah. 2010. "Iraq Election Increases Women in Parliament—Regardless of Vote Count." *Christian Science Monitor,* March 24. Available at http://www.csmonitor.com

Anderson, Margaret L. 1997. *Thinking about Women,* 4th ed. New York: Macmillan.

Askari, Sabrina F., Mirian Liss, Mindy J. Erchull, Samantha E. Staebell, and Sarah J. Axelson. 2010. "Men Want Equality, but Women Don't Expect It: Young Adults' Expectations for Participation in Household and Child Care Chores." *Psychology* of *Women Quarterly* 34(2): 243–252.

Basenhausen, Kurt. 2011. "The World's Highest Paid Athletes." *Forbes,* May 31. Available at http://blogs.forbes.com

Begley, Sharon. 2000. "The Stereotype Trap." *Newsweek,* November 6, pp. 66–68.

Bennett, Rosemary. 2009 (April 24). "Compulsory Audits on Equal Pay Will Force Firms to Give Women More." Available at http://business.timesonline.co.uk

Bertrand, Marianne, Claudia Goldin, and Lawrence F. Katz. 2009 (January). "Dynamics of the Gender Gap for Young Professionals in the Financial and Corporate Sectors." Working Paper. Available at http://www.economics.harvard.edu

Bianchi, Suzanne M., Melissa A. Milkie, Liana C. Sayer, and John Robinson. 2000. "Is Anyone Doing the Housework? Trends in the Gender Division of Household Labor." *Social Forces* 79:191–228.

Billitteri, Thomas J. 2008. "Gender Pay Gap." *CQ Researcher* 8(11):241–264. Available at http://www.cqresearcher.com

Bittman, Michael, and Judy Wajcman. 2000. "The Rush Hour: The Character of Leisure Time and Gender Equity." *Social Forces* 79:165–189.

Bly, Robert. 1990. *Iron John: A Book about Men.* Boston: Addison-Wesley.

Bourin, Lenny, and Bill Blakemore. 2008. "More Men Take Traditionally Female Jobs." ABC World News, September 1. Available at http://abcnews.go.com

Boylan, Jenny Finney. 2007. "I Want to Wake up." Speech given to the National Press Club. May. Available at http://jenniferboylan.blogspot.com

Brady, David, and Denise Kall. 2008. "Nearly Universal, but Somewhat Distinct: The Feminization of Poverty in Affluent Western Democracies, 1969–2000." *Social Science Research* 37:976–1,007.

Burnett, Victoria. 2008. "Defense Minister's New Baby Confirms Symbolism of Parity in Spain." *New York Times*, May 20. Available at http://www.nytimes.com

CAWP (Center for American Women and Politics). 2011a. *Women in Elected Office, 2011.* Available at http://www.cawp.rutgers.edu

CAWP. 2011b. *Women in the U.S. Congress 2011.* Available at http://www.cawp.rutgers.edu

Ceci, Stephen J., Wendy M. Williams, and Susan M. Barnett. 2009. "Women's Underrepresentation in Science: Socio-Cultural and Biological Considerations." *Psychological Bulletin* 135 (2):218–261.

Cockburn, Patrick. 2010. "The Death Sentence that Drags Dua Back into a Bloody Feud." *The Independent*, April 28. Available at http://www.independent.co.uk

Cohen, Theodore. 2001. *Men and Masculinity.* Belmont, CA: Wadsworth.

Cohn, D'vera. 2011 (February 7). "India Offers Three Gender Options." Pew Research Center. Available at http://pewsocialtrends.org

Coltrane, Scott, and Michelle Adams. 2008. *Gender and Families.* Lanham, MD: Rowman and Littlefield.

Connolly, Ceci. 2005. "Access to Abortion Pared at State Level." *Washington Post,* August 29, p. A1.

Cook, Carolyn. 2009. "ERA Would End Women's Second-Class Citizenship: Only Three More States Are Needed to Declare Gender Bias Unconstitutional." *The Philadelphia Inquirer*, April 12. Available at http://www.philly.com

Correll, Shelly J., Stephen Benard, and In Paik. 2007. "Getting a Job: Is There a Motherhood Penalty?" *American Journal of Sociology* 112(5):1,297–1,338.

Cosgrove, Maureen. 2011. "Ohio Lawmakers Approve Bill Prohibiting Abortion after Fetal Heartbeat Detection." *Jurist,* June 29. Available at http://jurist.org

Daily Mail. 2011. "Are These the Most PC Parents in the World? The Couple Raising a 'Genderless Baby'. . . to Protect His (or Her) Right to Choice." *The Daily Mail Online*, May 25. Available at http://www.dailymail.co.uk

Davis, Linsey. 2011 (May 29). "Genderless Baby: A 'Tribute to Liberty' or Unwilling Conscript into War on Sex Roles." Available at http://abcnews.go.com

Davis, James A., Tom W. Smith, and Peter V. Marsden. 2002. *General Social Surveys, 1972–2002: 2nd ICPSR Version.* Chicago: National Opinion Research Center.

Dee, Thomas S. 2006 (Fall). "How a Teacher's Gender Affects Boys and Girls." *Education Next.* Available at http://www.educationnext.org

EEOC (Equal Employment Opportunity Commission). 2011a (July 27). "Area Temps Agrees to Pay $650,000 for Profiling Applicants by Race, Sex, National Origin and Age." Available at http://www.eeoc.gov

EEOC (Equal Employment Opportunity Commission). 2011b. *Charge Statistics: FY 1997 through FY2010.* Available at http://www.eeoc.gov

EEOC. 2009. "Sexual Harassment." Available at http://www.eeoc.gov

ERA (Equal Rights Amendment). 2011. *The Equal Rights Amendment.* Available at http://www.equalrightsamendment.org

European Commission. 2011. *Report on Progress on Equality between Men and Women in 2010.* Luxembourg: Publications Office of the European Union. Available at http://ec.europa.eu

Faludi, Susan. 2008. "Think the Gender War Is Over? Think Again." *The New York Times*, June 15. Available at http://www.nytimes.com

Faludi, Susan. 1991. *Backlash: The Undeclared War against American Women.* New York: Crown.

Filkins, Dexter. 2009. "Afghan Women Protest New Law on Home Life." *The New York Times*, April 16. Available at http://www.nytimes.com

Fitzpatrick, Maureen, and Barbara McPherson. 2010. "Coloring within the lines: Gender Stereotypes in Contemporary Coloring Books." *Sex Roles* 62 (1/2):127–137.

Foerstel, Karen. 2008. "Women's Rights: Are Violence and Discrimination against Women Declining?" *Global Researcher* 2(5):115–147.

Foser, Jamison. 2010. "Sunday Shows Offer Excuses, Not Reasons, for Male-Dominated Guest Lists." *Media Matters*, June 15. Available at http://mediamatters.org

Gallagher, Sally K. 2004. "The Marginalization of Evangelical Feminism." *Sociology of Religion* 65:215–237.

Gault, Barbara. 2010. "The Wage Gap and Occupational Segregation." Institute for Women's Policy Research. Available at http://www.iwpr.org

Gender Spectrum. 2011. Available at http://www.genderspectrum.org

GMMP (Global Media Monitoring Project). 2010. *Who Makes the News?* September. Available at http://www.whomakesthenews.org

Goffman, Erving. 1963. *Stigma*. Englewood Cliffs, NJ: Prentice Hall.

Gonzales v. The United States of America. 2009. Inter-American Commission on Human Rights. Case No. 12.626. Available at http://www.aclu.org

Goodstein, Laurie. 2009."U.S. Nuns Facing Vatican Scrutiny." *The New York Times*, July 2. Available at http://www.nytimes.com

Grogan, Sarah. 2008. *Body Image: Understanding Body Dissatisfaction in Men, Women and Children*. New York: Routledge.

Gupta, Sanjay. 2003. "Why Men Die Young." *Time*, May 12, p. 84.

Guy, Mary Ellen, and Meredith A. Newman. 2004. "Women's Jobs, Men's Jobs: Sex Segregation and Emotional Labor." *Public Administration Review* 64:289–299.

Haines, Errin. 2006 (November 1). "Father Convicted in Genital Mutilation." Available at http://www.breitbart.com/news

Hamilton, Mykol C., David Anderson, Michelle Broaddus, and Kate Young. 2006. "Gender Stereotyping and Under-Representation of Female Characters in 200 Popular Children's Picture Books: A Twenty-First Century Update." *Sex Roles* 55:757–765.

Haub, Carl. 2011. "A First for Census Taking: The Third Sex." January 21. Population Reference Bureau: *Behind the Numbers*. Available at: http://prbblog.org.

Hausmann, Ricardo, Laura D. Tyson, and Saadia Zahidi. 2010. *The Global Gender Gap Report 2010*. Geneva: World Economic Forum.

Hegewisch, Ariane, Claudia Williams, and Amber Henderson. 2011. The Gender Wage Gap: 2010. Institute for Women's Policy Research. Available at http://www.iwpr.org/

Heyzer, Noeleen. 2003. "Enlisting African Women to Fight AIDS." *Washington Post*, July 8. Available at http://www.globalpolicy.org

Hochschild, Arlie. 1989. *The Second Shift*. London: Penguin.

HRC (Human Rights Campaign). 2005. *Transgender Americans: A Handbook for Understanding*. Available at http://www.hrc.org

Hvistendahl, Mara. 2011. *Unnatural Selection: Choosing Boys over Girls, and the Consequences of a World Full of Men*. Philadelphia: Public Affairs.

IASC (Inter–Agency Standing Committee). 2009. "IASC Policy Statement Gender Equality in Humanitarian Action." Available at http://www.humanitarianinfo.org

IDEA (International Institute for Democracy and Electoral Assistance). 2009. "Quotas for Women." Available at http://www.idea.int

ILO (International Labour Organization). 2011a. *Equality at Work: The Continuing Challenge*. Available at http://www.ilo.org

ILO. 2011b. *Global Unemployment Trends 2011*. Available at http://www.ilo.org

ILO (International Labour Organization). 2009 (March). *Global Employment Trends for Women*. Available at http://www.ilo.org

ILO News. 2009. "ILO Warns Economic Crisis Could Generate up to 22 Million More Unemployed Women in 2009, Jeopardize Equality Gains at Work and at Home." Available at http://www.ilocarib.org.tt

IPC (International Poverty Centre). 2008. *Poverty in Focus: Gender Equality*. No. 13. Available at http://www.undp-povertycentre.org

ITUC (International Trade Union Commission). 2011 (March 8). *Decisions for Work: An Examination of the Factors Influencing Women's Decisions for Work*. Available http://www.ituc-csi.org

ITUC (International Trade Union Commission). 2009. *Gender (In)Equality in the Labour Market: An Overview of Global Trends and Development*. Available http://www.ituc-csi.org

IWPR (Institute for Women's Policy Research). 2011a. (April). *Fact Sheet: The Gender Wage Gap, 2010*. Available at http://www.iwpr.org

IWPR. 2011b. (April). *Fact Sheet: The Gender Gap by Occupation*. Available at http://www.iwpr

Jackson, Janna. 2010. "'Dangerous Presumptions': How Single-Sex Schooling Reifies False Notions of Sex, Gender, and Sexuality." *Gender and Education* 22(2):227–238.

Jackson, Robert, and Meredith A. Newman. 2004. "Sexual Harassment in the Federal Workplace Revisited: Influences on Sexual Harassment." *Public Administration Review* 64(6):705–717.

Jaschik, Scott. 2010 (November 3). "Arizona Bans Affirmative Action." *Inside Higher Education*. Available at http://www.insidehighered.com

Jose, Paul, and Isobel Brown. 2009. "When Does the Gender Difference in Rumination Begin? Gender and Age Differences in the Use of Rumination by Adolescents." *Journal of Youth and Adolescence* 37:180–192.

Judd, Terri. 2008. "Barbaric 'Honour Killings' Become the Weapon to Subjugate Women in Iraq." *The Independent*, April 28. Available at http://www.independent.co.uk

Kaufman, David. 2009. "Introducing America's First Black, Female Rabbi." *Time*, June 6. Available at http://www.time.com

Kimmel, Michael. 2011 (Winter). "Gay Bashing Is about Masculinity." *Voice Male*. Available at http://www.voicemalemagazine.org

Kimmel, Michael. 2008. *Guyland*. New York: Harper Collins.

Kravets, David. 2007 (February 6). *Court Says Wal-Mart Must Face Bias Trial*. Available at http://www.breitbart.com

Gorney, Cynthia. 2011. "Too Young to Wed." *National Geographic*, June. Available at http://ngm.nationalgeographic.com

Grant, Jaime M., Lisa A. Mottet, Justine Tants, Jack Harrison, Jody L. Herman, and Mara Ketsling. 2011. *Injustice at Every Turn: A Report of the National TransGender Discrimination Survey*. Washington, DC: National Center for Transgender Equality and the National Gay and Lesbian Task Force.

Lacey, Marc. 2008. "A Lifestyle Distinct: The Muxe of Mexico." *The New York Times*, December 7. Available at http://www.nytimes.com

LCCREF (Leadership Conference on Civil Rights Education Fund). 2009. "Confronting the New Faces of Hate: Hate Crimes in America." Available at http://www.civilrights.org

Lepkowska, Dorothy. 2008. "Playing Fair?" *The Guardian*, December 16. Available at http://www.guardian.co.uk/

Leslie, David W. 2007 (March). "The Reshaping of America's Academic Workforce." *Research Dialogue* 87. New York: TIAA-CREF Institute. Available at http://www.tiaa-crefinstitute.org

Levin, Diane E., and Jean Kilbourne. 2009. *So Sexy So Soon: The New Sexualized Childhood and What Parents Can Do to Protect Their Kids*. New York: Random House.

Livingston, Gretchen, and Kim Parker. 2011 (June 15). "A Tale of Two Fathers: More Are Active, but More Are Absent." Pew Research Center. Available at http://pewsocialtrends.org

Lopez-Claros, Augusto, and Saadia Zahidi. 2005. *Women's Empowerment: Measuring the Global Gender Gap*. World Economic Forum.

Lutheran World Federation. 2009 (January 29). "Lutheran Woman Bishop Jeruma-Grinberga Succeeds Jagucki in Great Britain." Available at http://www.lutheranworld.org

Madera, Juan M., Michelle M. Hebl, and Randi C. Martin. 2009. "Gender and Letters of Recommendation for Academia: Agentic and Communal Differences." *Journal of Applied Psychology* 94(6): 1,591–1,599.

Mahapatra, Dhananjay. 2008. "Customary Payments, Gifts Not Dowry: SC." *The Times of India*, February. Available at http://timesofindia.indiatimes.com

Mannino, Celia Anna, and Francine M. Deutsch. 2007. "Changing the Division of Household Labor: A Negotiated Process between Partners." *Sex Roles* 56:309–324.

McGill. 2011 (July 12). "Two-Spirited People." Available at http://www.mcgill.ca

Media Matters. 2010. "Limbaugh: With Kagan Nomination, 'Obama Has Chosen Himself a Different Gender'—'She Is a Pure Academic Elitist Radical.'" *Media Matters TV*, May 10. Available at http://mediamatters.org

Media Matters. 2009. "Limbaugh Asserts 'Chicks . . . Have Chickified the News'; Again Refers to Female Reporter as Infobabe." Available at http://mediamatters.org

Mehmood, Isha. 2009 (January 29). "Lilly Ledbetter Fair Pay Act Becomes Law." Available at http://www.civilrights.org

Misra, Joya, Jennifer Hickes Lundquist, Elissa Holmes, and Stephanie Agiomavritis. 2011. "The Ivory Ceiling of Service Work." *Academe* 97:22–26.

Messner, Michael A., and Jeffrey Montez de Oca. 2005. "The Male Consumer as Loser: Beer and Liquor Ads in Mega Sports Media Events." *Signs* 30:1,879–1,909.

Mianecki, Julie. 2011. "Federal Guidelines Issued to Fight Campus Sexual Violence." *Los Angeles Times*, April 5. Available at http://articles.latimes.com/

MKP (ManKind Project). 2010a. "What We Stand For." Available at http://www.mkp.org

MKP (ManKind Project). 2010b. "The New Warrior Training Adventure." Available at http://www.mkp.org

Mnet (Media Awareness Network). 2009. "Beauty and Body Image in the Media." Available http://www.media-awareness.ca

Models. 2011. "Andrej Pejic, Biography." Available at http://models.com/models/andrej-pejic

Moen, Phyllis, and Yan Yu. 2000. "Effective Work/Life Strategies: Working Couples, Working Conditions, Gender, and Life Quality." *Social Problems* 47:291–326.

Moreno, Pedro C. 2009 (July 3). "Report from Rio: Men and Boys in Gender Equality." Available at http://www.huffingtonpost.com

Morin, Rich, and Paul Taylor. 2008 (September 15). *Revisiting the Mommy Wars: Politics, Gender and Parenthood.* Available at http://pewsocialtrends.org

Moskowitz, Clara. 2010. "When Teachers Highlight Gender, Kids Pickup Stereotypes." November 16. *Live Science.* Available at http://www.livescience.com

Nanda, Serena. 2000. *Gender Diversity: Cross-Cultural Variations.* Long Grove, IL: Wavelane Press.

NCFM (National Coalition of Free Men). 2011. "About Us." Available at http://www.ncfm.org

NCFM. 2009. "Philosophy." Available at http://www.ncfm.org

NCES (National Center for Educational Statistics). 2011. *The Condition of Education, 2011.* U.S. Department of Education. Available at http://nces.ed.gov

NCES. 2009. "Fast Facts: What Is Title IX?" Available at http://nces.ed.gov

NCWGE (National Coalition for Women and Girls in Education). 2009. National Coalition for Women and Girls in Education Recommendations for the Obama Administration and the 111th Congress." Available at http://www.ncwge.org/PDF

NOMAS (National Organization for Men Against Sexism). 2009. "Statement of Principles." Available at http://www.nomas.org

NSBA (National School Boards Association) 2011. "Female Student Whose Picture Was Excluded from Yearbook because She Wore a Tux Sues Mississippi School District." *Recent Developments in School Law.* Available at http://legal clips.nsba.org

Nordberg, Jenny. 2010. "Afghan Boys Are Prized, So Girls Live the Part." *The New York Times,* September 20. Available at http://www.nytimes.com

Nosek, B. A., F. L. Smyth, N. Sriram, N. M. Lindner, T. Devos, A. Ayala, Y. Bar-Anan, R. Bergh, H. Cai, K. Gonsalkorale, S. Kesebir, N. Maliszewski, F. Neto, E. Olli, J. Park, K. Schnabel, K. Shiomura, B. Tulbure, R. W. Wiers, M. Somogyi, N. Akrami, B. Ekehammar, M. Vianello, M. R. Banaji, and A. G. Greenwald. 2009. "National Differences in Gender-Science Stereotypes Predict National Sex Differences in Science and Math Achievement." *Proceedings of the National Academy of Sciences* 106:10,593–10,597.

Padavic, Irene, and Barbara Reskin. 2002. *Men and Women at Work,* 2nd ed. Thousand Oaks, CA: Pine Forge Press.

Parks, Janet B., and Mary Ann Roberton. 2000. "Development and Validation of an Instrument to Measure Attitudes toward Sexist/Nonsexist Language." *Sex Roles* 42(5/6):415–438.

Parks, Janet B., and Mary Ann Roberton. 2001. "Inventory of Attitudes toward Sexist/Nonsexist Language—General (IASNL-G): A Correction in Scoring Procedures." Erratum. *Sex Roles* 44(3/4):253.

Pew Research Center. 2007 (July 12). "From 1997 to 2007 Fewer Mothers Prefer Full-time Work." Available at http://www.pewresearch.org

Polachek, Soloman W. 2006. "How the Life-Cycle Human Capital Model Explains Why the Gender Gap Narrowed." In *The Declining Significance of Race,* Francine D. Blau, Mary C. Brinton, and David B. Grusky, eds. (pp. 102–124). New York: Russell Sage.

Pollack, William. 2000a. *Real Boys' Voices.* New York: Random House.

Pollack, William. 2000b. "The Columbine Syndrome." *National Forum* 80:39–42.

Pozner, Jennifer L. 2009 (July 8). "Hot and Bothering: Media Treatment of Sarah Palin." National Public Radio. Available at http://www.npr.org

Quist-Areton, Ofeibea. 2003. "Fighting Prejudice and Sexual Harassment of Girls in Schools." *All Africa,* June 12. Available at http://www.globalpolicy.org

Rauhala. Emily. 2011. "A Powerhouse Province Wants to Relax China's One-Child Policy—But Don't Bet on a Baby Boom." *Time Online,* July 12. Available at http://globalspin.blogs.time.com/

Renzetti, Claire, and Daniel Curran. 2003. *Women, Men and Society.* Boston: Allyn and Bacon.

Reskin, Barbara, and Debra McBrier. 2000. "Why Not Ascription? Organizations' Employment of Male and Female Managers." *American Sociological Review* 65:210–233.

Rose, Stephen J., and Heidi Hartman. 2008 (February). "Still a Man's Labor Market: The Long-Term Earnings Gap." IWPR# C366. New York: Institute for Women's Policy Research.

Rudnanski, Ryan. 2011 (April 10). "Augusta National Golf Club Won't Alter Tradition and Allow Women in Locker Room." Available at http://bleacherreport.com/

Sadker, David, and Karen Zittleman. 2009. *Still Failing at Fairness: How Gender Bias Cheats Boys and Girls in Schools.* New York: Simon and Schuster.

Sanchez, Diana T., and Jennifer Crocker. 2005. "How Investment in Gender Ideals Affects Well-Being: The Role of External Contingencies of Self-Worth." *Psychology of Women Quarterly* 29:63–77.

Sayman, Donna M. 2007. "The Elimination of Sexism and Stereotyping in Occupational Education." *Journal of Men's Studies* 15(1):19–30.

Sherrill, Andrew. 2009 (April 28). "Women's Pay." Testimony before the Joint Economic Committee. U.S. Congress. Washington, DC.

United States Government Accountability Office.

Simpson, Ruth. 2005. "Men in Non-Traditional Occupations: Career Entry, Career Orientation, and Experience of Role Strain." *Gender Work and Organization* 12(4):363–380.

Smith, Melanie. 2006. "Is Church Too Feminine for Men?" *The Decatur Daily,* July 1. Available at http://www.decaturdaily.com

Snyder, Karrie Ann, and Adam Isaiah Green. 2008. "Revisiting the Glass Escalator: The Case of Gender Segregation in a Female Dominated Occupation." *Social Problems* 55(2):271–299.

SOC (Security on Campus). 2011. "Campus Sexual Violence Elimination Act (Campus SaVE Act) Summary." Available at http://www.securityoncampus.org

Social Watch. 2011. *Social Watch Report 2010: After the Fall.* Available at http://www.socialwatch.org

Social Watch. 2009. "Gender: 20th Century Debts, 21st Century Shame." *Social Watch Report 2008.* Available at http://www.socialwatch.org

Sonnert, Gerhard, Mary Frank Fox, and Kristen Adkins. 2007. "Undergraduate Women in Science and Engineering: Effects of Faculty, Fields, and Institutions Over Time." *Social Science Quarterly* 88 (5):1,333–1,356.

Soroptimist International. 2010 (March). *White Paper: Women and Poverty.* Available at http://www.soroptimist.org

Stohr, Greg. 2011. "Wal-Mart Million-Worker Bias Suit Thrown Out by High Court." June 20. Available at http://www.bloomberg.com

Tenenbaum, Harriet R. 2009. "You'd Be Good at That: Gender Patterns in Parent-Child Talk about Courses." *Social Development* 18(2):447–463.

Uggen, C., and A. Blackstone. 2004. "Sexual Harassment as a Gendered Expression of Power." *American Sociological Review* 69:64–92.

UNESCO (United Nations Educational, Scientific and Cultural Organization). 2011. "EFA Global Monitoring Report." Available at http://www.unesco.org

UNICEF. 2011. *The State of the World's Children: 2010.* United Nations: United Nations Children's Fund. Available at http://www.unicef.org

UNICEF. 2007. *The State of the World's Children: 2007.* United Nations: United Nations Children's Fund. Available at http://www.unicef.org

UNIFEM (United Nations Development Fund for Women). 2007. "Harmful Traditional Practices." Available at http://www.unifem.org

United Nations. 2011. *The Millennium Development Goals Report 2011.* Available at http://www.un.org

UN Women. 2011. "About UN Women." *United Nations Entity for Gender Equality and the Empowerment of Women.* Available at http://www.unwomen.org

U.S. Census Bureau. 2011. *Statistical Abstract of the United States: 2010.* 128th ed. Washington, DC: U.S. Government Printing Office.

U.S. Census Bureau. 2009. *Statistical Abstract of the United States: 2008.* 128th ed. Washington, DC: U.S. Government Printing Office.

Vandello, Joseph A., Jennifer K. Bosson, Dov Cohen, Rochelle M. Burnaford, and Jonathan R. Weaver. 2008. "Precarious Manhood." *Journal of Personality and Social Psychology* 95 (6):1,325–1,339.

Venkatesan, J. 2009. "Supreme Court Anguish over Dowry Deaths." *The Hindu*, June 2. Available at http://www.hindu.com

Vincent, Norah. 2006. *Self-Made Man*. New York: Viking Penguin.

Vogel, Chris. 2008. "A New Retreat for the ManKind Project Houston." *Houston Press*, June 26. Available at http://www.houstonpress.com

Warbelow, Sarah. 2011. "Connecticut Governor Signs Transgender Nondiscrimination Bill." July 6. *HRC Back Story*. Available at http://www.hrcbackstory.org

Weeks, Linton. 2011. "The End of Gender?" National Public Radio, June 23. Available at http://www.npr.org

White, Alan, and Karl Witty. 2009. "Men's Under Use of Health Services—Finding Alternative Approaches." *Journal of Men's Health* 6(2):95–97.

Witterick, Kathy. 2011. "Baby Storm's Mom on Gender, Parenting and the Media." *The Vancouver Sun*, May 28. Available at http://www.vancouversun.com/news

WHO (World Health Organization). 2011. Global Sector Strategy on HIV/AIDS 2011–2015. Available at http://whqlibdoc.who.ints

WHO. 2010. "Gender, Women and Primary Heal Care Renewal." Discussion Paper. Available at http://whqlibdoc.who.int

WHO. 2009. "Ten Facts About Women's Health." Available at http://www.who.int

WHO. 2008. *Eliminating Female Genital Mutilation: An Interagency Statement*. Available at http://whqlibdoc.who.int

WHO. 2006. *Gender and Women's Mental Health*. Available at http://www.who.int/mental_health

Williams, Christine L. 2007. "The Glass Escalator: Hidden Advantages for Men in the 'Female' Occupations." In *Men's Lives*, 7th ed., Michael S. Kimmel and Michael Messner, eds. (pp. 242–255). Boston: Allyn and Bacon.

Williams, Joan. 2000. *Unbending Gender: Why Family and Work Conflict and What to Do About It*. Oxford: Oxford University Press.

Williams, Kristi, and Debra Umberson. 2004. "Marital Status, Marital Transitions, and Health: A Gendered Life Course Perspective." *Journal of Health and Social Behavior* 45:81–98.

Wood, Wendy and Alice H. Eagly. 2002. "A Cross-Cultural Analysis of the Behavior of Women and Men: Implications for the Origins of Sex Differences." *Psychological Bulletin* 128(5):699–727.

Yoder, P. Stanley, N. Abderrahim, and A. Zhuzhuni. 2004. *Female Genital Cutting in the Demographic and Health Surveys: A Critical and Comparative Analysis*. DHS Comparative Reports 7. Calverton, MD: ORC Macro.

York, E. Anne. 2008. "Gender Differences in the College and Career Aspiration of High School Valedictorians." *Journal of Advanced Academics* 19(4):578–580.

Zakrzewski, Paul. 2005. "Daddy, What Did You Do in the Men's Movement?" *Boston Globe*, June 19. Available at http://www.bostonglobe.com

Zuber, Helene. 2010. "Spain's Defense Minister Makes Her Mark in a Macho World." *Spiegel International*, April 15. Available at http://www.spiegel.de/international/

Chapter 11

AAA (American Anthropological Association). 2004. *Statement on Marriage and the Family*, February 26. Available at http://www.aaanet.org

ACLU (American Civil Liberties Union). 2011 (July 22). "'Don't Ask, Don't Tell' Repeal Certified by President, Defense Secretary and Joint Chiefs Chairman." Available at http://www.aclu.org

ACLU. 2010. (June 16) "Coalition Letter to the House Ways and Means Committee in Support of HR 1681 The Every Child Deserves a Family Act." http://www.aclu.org

ACLU. 2003. (June 26). "Getting Rid of Sodomy Laws: History and Strategy that Led to the Lawrence Decision." Available at http://www.aclu.org

Allport, G. W. 1954. *The Nature of Prejudice*. Cambridge, MA: Addison-Wesley.

AMA (American Medical Association). 2011. "AMA Policy Regarding Sexual Orientation." *GLBT Advisory Committee*. Available at http://www.ama-assn.org

Amato, Paul R. 2004. "Tension between Institutional and Individual Views of Marriage." *Journal of Marriage and Family* 66:959–965.

APA Task Force on Appropriate Therapeutic Responses to Sexual Orientation. (2009). *Report of the Task Force on Appropriate Therapeutic Responses to Sexual Orientation*. Washington, DC: American Psychological Association.

APA (American Psychological Association). 2008. "Answers to Your Questions: For a Better Understanding of Sexual Orientation and Homosexuality." Washington, DC: APA. Available at http://www.apa.org

American College Health Association. 2011. "National College Health Assessment Reference Group Executive Summary." Available at http://www.achancha.org

Association of Welcoming and Affirming Baptists (AWAB). 2006. "Who We Are . . . Gay-Affirming Baptists? No, This Is Not an Oxymoron!" Available at http://www.wabaptists.org

Badgett, M.V. Lee, Ilan H. Meyer, Gary J. Gates, Nan D. Hunter, Jennifer C. Pizer, and Brad Sears. 2011. (July 20). "Written Testimony of the Williams Institute, UCLA School of Law." Hearings on s. 598, *The Respect for Marriage Act: Assessing the Impact of DOMA on American Families*. Available at http://www3.law.ucla.edu/williamsinstitute

Badgett, Lee. 2009. "Domestic Partnership Benefits and Obligations Act." Williams Institute. Available at http://www.law.ucla.edu/williamsinstitute

Bayer, Ronald. 1987. *Homosexuality and American Psychiatry: The Politics of Diagnosis*. 2nd ed. Princeton, NJ: Princeton University Press.

Bergman, K., R. J. Rubio, R. J. Green, and E. Padron. 2010. "Gay Men Who Become Fathers via Surrogacy: The Transition to Parenthood." *Journal of GLBT Family Studies* 6: 111–141.

Besen, Wayne. 2010. "Ex-Gay Group Should Repent, Not Revel." *Huffington Post*, June 17. Available at http://www.huffingtonpost.com

Bettencourt, Rebecca G. 2009. (Jan 9). *The Transgender Umbrella*. San Luis Obispo, CA: California Polytechnic University.

Bobbe, Judith. 2002. "Treatment with Lesbian Alcoholics: Healing Shame and Internalized Homophobia for Ongoing Sobriety." *Health and Social Work* 27(3):218–223.

Bockting, Walter, Autumn Benner, and Eli Coleman. 2009. "Gay and Bisexual Identity Development among Female-to-Male Transsexuals in North America: Emergence of a Transgender Sexuality." *Archives of Sexual Behavior* 38:688–701.

Bonds-Raacke, Jennifer M., Elizabeth T. Cady, Rebecca Schlegel, Richard J. Harris, and Lindsey Firebaugh. 2007. "Can Ellen and Will Improve Attitudes toward Homosexuals?" *Journal of Homosexuality* 53:19–34.

Bosker, Bianca. 2011. "Apple Pulls Controversial 'Gay Cure' App." *Huffington Post*, March 23. Available at http://www.huffingtonpost.com

Brown, Devin. 2011. "Tracy Morgan's Gay Rant Riles Up Twitter and LGBT Community." *Huffington Post*, June 10. Available at http://www.huffingtonpost.com

Brown, Michael J., and Ernesto Henriquez. 2008. "Socio-Demographic Predictors of Attitudes towards Gays and Lesbians." *Individual Differences Research* 6:193–202.

Broverman, Neal. 2010. "McMillen: I Was Sent to Fake Prom." *Advocate*, April 5. Available at http://www.advocate.com

Bruce-Jones, Eddie, and Lucas P. Itaborahy. 2011. "State-Sponsored Homophobia: A World Survey of Laws Criminalizing Same-Sex Sexual Acts between Consenting Adults." International Lesbian, Gay, Bisexual, Trans and Intersex Association (ILGA). Available at http://old.ilga.org

Burn, Shawn M. 2000. "Heterosexuals' Use of 'Fag' and 'Queer' to Deride One Another: A Contributor to Heterosexism and Stigma." *Journal of Homosexuality* 40:1–11.

Burn, Shawn M., Kelly Kadlec, and Ryan Rexer. 2005. "Effects of Subtle Heterosexism on Gays, Lesbians, and Bisexuals." *Journal of Homosexuality* 49:23–38.

Cahill, Sean. 2007. "The Coming GLBT Senior Boom." *The Gay & Lesbian Review*, January–February, pp. 19–21.

Cahill, Sean, Ellen Mitra, and Sarah Tobias. 2002. *Family Policy: Issues Affecting Gay, Lesbian, Bisexual, and Transgender Families*. National Gay and Lesbian Task Force Policy Institute. Available at http://www.thetaskforce.org

Caldwell, Alicia A. 2011. "Same-Sex Couples Denied Immigration Benefits by U.S." *Huffington Post*, March 30. Available at http://www.huffingtonpost.com

California Supreme Court. 2011. "In re Marriage Cases." Case No. 2147999. The Supreme Court of the State of California. Judicial Council Coordination Processing No. 4365. Available at http://outandequal.org/

Carney, Michael P. 2007 (September 5). "The Employment Non-Discrimination Act: Testimony by Officer Michael P. Carney." Springfield Massachusetts. Available at http://edlabor.house.gov

Carpenter, Christopher, and Gary J. Gates. 2008. "Gay and Lesbian Partnership: Evidence from California." *Demography* 45:573–590.

Casey, Bob. 2011. "Focus on the Family Stands Up for Bullying." *Huffington Post,* September 8. Available at http://www.huffingtonpost.com

Chamberlain, Jared, Monica K. Miller, and Brian H. Bornstein. 2008. "The Rights and Responsibilities of Gay and Lesbian Parents: Legal Developments, Psychological Research, and Policy Implications." *Social Issues and Policy Review* 2(1):103–126.

Cianciotto, Jason, and Sean Cahill. 2007. "Anatomy of a Pseudo-Science." *The Gay & Lesbian Review,* July–August, pp. 22–24.

Cianciotto, Jason, and Sean Cahill. 2006. *Youth in the Crosshairs: The Third Wave of Ex-Gay Activism.* National Gay and Lesbian Task Force Policy Institute. Available at http://www.thetaskforce.org

Ciarlante, Mitru, and Kim Fountain. 2010. "Why It Matters: Rethinking Victim Assistance for Lesbian, Gay, Bisexual, Transgender, and Queer Victims of Hate Violence & Intimate Partner Violence." National Center for Victims of Crime and the New York City Anti-Violence Project. Available at http://www.avp.org

CNN. 2009. (October 28) "Obama Signs Hate Crimes Bill into Law." Available at http://articles.cnn.com

Confessore, Nicholas, and Michael Barbaro. 2011. "New York Allows Same-Sex Marriage, Becoming Largest State to Pass Law." *The New York Times,* June 24. Available at http://www.nytimes.com/

Coogan, Steve. 2011. "Michael Strahan, Fiancé Film Same-Sex Marriage PSA." *USA Today,* June 12. Available at http://content.usatoday.com

D'Augelli, Anthony R., Arnold H. Grossman, and Michael T. Starks. 2006. "Childhood Gender Atypicality, Victimization, and PTSD among Lesbian, Gay, and Bisexual Youth." *Journal of Interpersonal Violence* 21:1462–1482.

DHMH (Department of Health and Mental Hygiene). 2011. (February 10). "Announcement Letter to State Birth Registrars." State of Maryland. Available at http://data.lambdalegal.org

Diamond, Lisa. M. 2008. "Female Bisexuality from Adolescence to Adulthood: Results from a 10-Year Longitudinal Study. *Developmental Psychology* 44: 5–14.

Dixon, Robyn. 2011. "Uganda Lawmakers Remove Death Penalty Clausefrom Anti-Gay Bill." *The Los Angeles Times,* May 12. Available at http://articles.latimes.com

Dougherty, Jill. 2011. "U.N. Council Passes Gay Rights Resolution." CNN, June 17. Available at http://articles.cnn.com

Duke University Student Affairs. 2011. "Lavender Graduation." Available at http://www.studentaffairs.duke.edu

Durkheim, Emile. 1993 [1938]. "The Normal and the Pathological." In *Social Deviance,* Henry N. Pontell, ed. (pp. 33–63). Englewood Cliffs, NJ: Prentice-Hall. (Originally published in *The Rules of Sociological Method.*)

Eliason, Mickey. 2001. "Bi-Negativity: The Stigma Facing Bisexual Men." *Journal of Bisexuality* 1:136–154.

Ellison, Jesse. 2008. "Be Like Others: Director Tanaz Eshaghian on Sundance, Sex Changes, and the Ayatollah." *New York,* January 24. Available at http://nymag.com

Eskridge, William N. 2008. *Dishonorable Passions: Sodomy Laws in American 1861–2003.* NY: Viking Penguin Group.

Equality Forum. 2011. "FORTUNE 500 Project." Available at http://www.equalityforum.com

Esterberg, K. 1997. *Lesbian and Bisexual Identities: Constructing Communities, Constructing Selves.* Philadelphia: Temple University Press.

Exodus International. 2011. "Fact Sheet." Available at http://exodusinternational.org

Falomir-Pichastor, Juan Manuel, and Gabriel Mugny. 2009. "I'm Not Gay . . . I'm a Real Man!": Heterosexual Men's Gender Self-Esteem and Sexual Prejudice." *Personality and Social Psychology Bulletin* 35:1233–1243.

FBI (Federal Bureau of Investigation). 2011. "Federal Officials Discuss Hate Crimes Protections with LGBT Community." U.S. Attorney's Office. Available at http://www.fbi.gov

FBI. 2010 (November). *Hate Crime Statistics, 2009.* US Department of Justice. Available at http://www.fbi.gov

Fidas, Deena. 2009. "At the Water Cooler." *Equality* (Spring):23, 29.

Fone, Byrne. 2000. *Homophobia: A History.* New York: Henry Holt.

Frank, Barney. 1997. "Foreword." In *Private Lives, Public Conflicts: Battles over Gay Rights in American Communities,* by J. W. Button, B. A. Rienzo, and K. D. Wald, eds. (pp. xi). Washington, DC: CQ Press.

Frank, David John, and Elizabeth H. McEneaney. 1999. "The Individualization of Society and the Liberalization of State Policies on Same-Sex Relations, 1984–1995." *Social Forces* 77(3):911–944.

Frank, Nathaniel. 2010. (August). "Don't Ask, Don't Tell: Detailing the Damage." The Palm Center. Available at http://www.palmcenter.org

Friedman, Mark S., Gary F. Koeske, Anthony J. Silvestre, Wynne S. Korr, and Edward W. Sites. 2006. "The Impact of Gender-Role Nonconforming Behavior, Bullying, and Social Support on Suicidality among Gay Male Youth." *Journal of Adolescent Health* 38:621– 623.

Gallagher, John. 1995. (November 14). "Counterattack: U.S. Protest Keeps Homophobia at Bay in South Africa." *The Advocate* 694:30.

Gallup Organization. 2009. "Pulse on Democracy: Homosexual Relations." Available at http://www.gallup.com

Gardner, Lisa A., and Ryan M. Roemerman. 2011. "Iowa College Climate Survey: The Life Experiences of Lesbian, Gay, Bisexual, Transgender & Straight Allied (LGBTA) Students at Iowa's Colleges and Universities." Iowa Pride Network. Available at http://www.iowapridenetwork.org

Gates, Gary J. 2011. "How Many People are Lesbian, Gay, Bisexual, and Transgender?" Williams Institute. Available at http://www.law.ucla.edu

Gilman, Stephen E., Susan D. Cochran, Vickie M. Mays, Michael Hughes, David Ostrow, and Ronald C. Kessler. 2001. "Risk of Psychiatric Disorders among Individuals Reporting Same-Sex Sexual Partners in the National Comorbidity Survey." *American Journal of Public Health* 91(6):933–939.

GLSEN. (Gay, Lesbian, and Straight Education Network). 2011a. "Damaging Language." Available at http://thinkb4youspeak.com/

GLSEN. 2011b. "What We Do." Available at http://www.glsen.org

GLSEN. 2010. The 2009 National School Climate Survey. Available at http://www.glsen.org

Goodstein, Laurie. 2010. "Lutherans Offer Warm Welcome to Gay Pastors." *The New York Times,* July 25. Available at http://www.nytimes.com

Greenberg, Bradley S. (1988). "Some Uncommon Television Images and the Drench Hypothesis: Television as a Social Issue." Oskamp, Stuart (Ed), (1988). *Applied Social Psychology Annual,* Vol. 8 (pp. 88–102). Thousand Oaks, CA, US: Sage Publications.

Greenberg, Quinlan, & Rosner. 2011. (March 15). "Voters Find Republican Defense of DOMA Unfair and Unnecessary: Findings from a National Survey." Available at http://www.hrc.org

Guadalupe, Krishna L., and Doman Lum. 2005. *Multidimensional Contextual Practice: Diversity and Transcendence.* Belmont, CA: Thomson Brooks/Cole, p. 11.

Hagerty, Barbara B. 2011. "A Peek Inside the Westboro Baptist Church." National Public Radio, August 1. Available at http://www.npr.org

Haider-Markel, Donald P., and Mark R. Joslyn. 2008. "Beliefs about the Origins of Homosexuality and Support for Gay Rights: An Empirical Test of Attribution Theory." *Public Opinion Quarterly* 72:291–310.

Hamilton, Julie. 2011. "Anti-Gay?! NARTH President Addresses Misperceptions about NARTH." Available at http://narth.com

Harper, Gary W., Nadine Jernewall, and Maria C. Zea. 2004. "Giving Voice to Emerging Science and Theory for Lesbian, Gay, and Bisexual People of Color." *Cultural Diversity and Ethnic Minority Psychology* 10:187–199.

Harris Interactive and GLSEN. 2005. *From Teasing to Torment: School Climate in America, A Survey of Students and Teachers.* New York: GLSEN.

Herek, Gregory M. 2009. "Facts about Homosexuality and Child Molestation." Available at http://psychology.ucdavis.edu

Herek, Gregory M. 2004. "Beyond 'Homophobia': Thinking about Sexual Prejudice and Stigma in the Twenty-First Century." *Sexuality Research & Social Policy: A Journal of the NSRC* 1:6–24.

Herek, Gregory M. 2002. "Gender Gaps in Public Opinion about Lesbians and Gay Men." *Public Opinion Quarterly* 66:40–66.

Herek, Gregory M. 2000a. "The Psychology of Sexual Prejudice." *Current Directions in Psychological Science* 9:19–22.

Herek, Gregory M. 2000b. "Sexual Prejudice and Gender: Do Heterosexuals' Attitudes toward Lesbians and Gay Men Differ?" *Journal of Social Issues* 56:251–266.

Herek, Gregory M., and Garnets, Linda D. 2007. "Sexual Orientation and Mental Health." *Annual Review of Clinical Psychology* 3:353–375.

Herman, Joanne. 2009. *Transgender Explained for Those Who Are Not*. Bloomington: Author House.

Hewlett, Sylvia Ann, and Karen Sumberg. 2011. "For LGBT Workers, Being 'Out' Brings Advantages." *Harvard Business Review* 89(7/8): n/a.

Hicks, Stephen. 2006. "Maternal Men—Perverts and Deviants? Making Sense of Gay Men as Foster Carers and Adopters." *Family Studies* 2:93–114.

Hostetler, Andrew J. 2009. "Single by Choice? Assessing and Understanding Voluntary Singlehood among Mature Gay Men." *Journal of Homosexuality* 56:499–531.

HRC (Human Rights Campaign). 2011a. "Faith Positions." Available at http://www.hrc.org

HRC. 2011b. Statewide Employment Laws & Policies. Available at http://www.hrc.org

HRC. 2011c. "Marriage Equality & Other Relationship Recognition Laws." Available at http://www.hrc.org

HRC. 2011d. "Professional Organizations on LGBT Parenting." Available at http://www.hrc.org

HRC. 2011e. "Parenting Laws: Joint and Second Parent Adoption." Available at http://www.hrc.org

HRC. 2011f. "Statewide School Laws and Policies." Available at http://www.hrc.org

HRC. 2010. "Statewide Marriage Prohibitions." Available at http://www.hrc.org

HRC. 2009a. "The State of the Workplace for Lesbian, Gay, Bisexual, and Transgender Americans, 2007–2008." Available at http://www.hrc.org

HRC. 2009b. "State Hate Crime Laws." Available at http://www.hrc.org

Human Rights Watch. 2001 (March). *World Report 2001*. Available at http://www.hrw.org

Humphrey, Tom. 2011. (April 21). "'Don't Say Gay' Bill Clears Senate Panel." Available at http://www.knoxnews.com

It Gets Better Project. 2011. "What Is the It Gets Better Project?" Available at http://www.itgetsbetter.org

James, Susan D. 2011. "Census 2010: One-Quarter of Gay Couples Raising Children." ABC News. Available at http://abcnews.go.com

Joannides, Paul. (2011). *The Guide to Getting It On*. Waldport: Goofy Foot Press.

Johnson, Chris. 2011. "ENDA Makes Debut in 112 Congress." *Washington Blade*, April 6. Available at http://www.washingtonblade.com

Johnson, Renee M., Jeremy D. Kidd, Erin C. Dunn, Jennifer G. Green, Heather L. Corliss, and Deborah Bowen. 2011. "Associations between Caregiver Support, Bullying, and Depressive Symptomatology among Sexual Minority and Heterosexual Girls: Results from the 2008 Boston Youth Survey." *Journal of School Violence* 10:185–200.

Jones, Jeffrey M. 2011. "Support for Legal Gay Relations Hits New High." Gallup Organization. Available at http://www.gallup.com/

Jones, Jeffrey M. 2010. "Americans' Opposition to Gay Marriage Eases Slightly." Gallup Organization. Available at http://www.gallup.com

Karimi, Faith. 2011. "Alleged Rape, Killing of Gay Rights Campaigner Sparks Call for Action." CNN. Available at http://www.cnn.com

Kimmel, Michael. 2011. "Gay Bashing Is about Masculinity." Winter. *Voice Male*. Available at http://www.voicemalemagazine.org

Kinsey, A. C., W. B. Pomeroy, and C. E. Martin. 1948. *Sexual Behavior in the Human Male*. Philadelphia: W. B. Saunders.

Kinsey, A. C., W. B. Pomeroy, C. E. Martin, and P. H. Gebhard. 1953. *Sexual Behavior in the Human Female*. Philadelphia: W. B. Saunders.

Kirkpatrick, R. C. 2000. "The Evolution of Human Sexual Behavior." *Current Anthropology* 41(3):385–414.

Kosciw, Joseph G., and Elizabeth M. Diaz. 2005. *The 2005 National School Climate Survey*. New York: GLSEN.

Lambda Legal. 2011a. "Health and Medical Organization Statements on Sexual Orientation, Gender Identity/Expression and 'Reparative Therapy.'" Available at http://data.lambdalegal.org

Lambda Legal. 2011b. "Preventing Censorship of LGBT Information in Public School Libraries." Available at http://data.lambdalegal.org

Life. 2009. "Out and Proud: Gay Celebrities." Available at http://www.life.com

Louderback, L. A., and B. E. Whitley. 1997. "Perceived Erotic Value of Homosexuality and Sex-Role Attitudes as Mediators of Sex Differences in Heterosexual College Students' Attitudes toward Lesbians and Gay Men." *Journal of Sex Research* 34:175–182.

Matthews, Joe. 2010. "Prop 8: The Most Wasteful Campaign in History." NBC Los Angeles. Available at http://www.nbclosangeles.com

McGreevy, Patrick. 2011. "New State Law Requires Textbooks to Include Gays' Achievements." *The Los Angeles Times*, July 15. Available at http://articles.latimes.com

McKinley, Jesse. 2010. "Suicides Put Light on Pressures of Gay Teenagers." *The New York Times*, October 3. Available at http://www.nytimes.com/

Mears, Bill. 2011. "Court Upholds Gay Judge's Ruling on Proposition 8." CNN, June 14. Available at http://articles.cnn.com

Meyer, Ilan. 2003. "Prejudice, Social Stress, and Mental Health in Lesbian, Gay, and Bisexual Populations: Conceptual Issues and Research Evidence." *Psychological Bulletin* 129: 674–697.

Miller, Patti, McCrae A. Parker, Eileen Espejo, and Sarah Grossman-Swenson. 2002. *Fall Colors: Prime Time Diversity Report 2001–02*. Oakland CA: Children Now and The Media Program.

Mohipp, C., and M. M. Morry. 2004. "Relationship of Symbolic Beliefs and Prior Contact to Heterosexuals' Attitudes toward Gay Men and Lesbian Women." *Canadian Journal of Behavioral Science* 36(1):36–44.

Morales, Lymari. 2011. (May 27). "Adults Estimate that 25% of Americans Are Gay or Lesbian." Gallup Organization. Available at http://www.gallup.com

Morales, Lymari. 2010. (December 9). "In U.S., 67% Support Repealing 'Don't Ask, Don't Tell.'" Gallup Organization. Available at http://www.gallup.com

Morales, Lymari. 2009. (May 29) "Knowing Someone Gay/Lesbian Affects Views of Gay Issues." Gallup Organization. Available at http://www.gallup.com

Moulton, B. (2011). "HRC Supports 'Every Child Deserves a Family Act'." Human Rights Campaign. Available at http://www.hrcbackstory.org

Mungin, Lateef. 2011. (May 27). "Expert: Use Gay Slurs Controversy to Tackle Homophobia in Sports." Available at http://www.cnn.com

Murphy, Tim. 2011. "Has Your State Banned Sodomy?" *Mother Jones,* April 19. Available at http://motherjones.com/

Mustanski, Brian S., Robert Garofalo, and Erin M. Emerson. 2010. "Mental Health Disorders, Psychological Distress, and Suicidality in a Diverse Sample of Lesbian, Gay, Bisexual, and Transgender Youths." *American Journal of Public Health* 100:2426–2432.

NARTH (National Association for Research & Therapy of Homosexuality). 2011. "NARTH Mission Statement." Available at http://narth.com

National Coalition of Anti-Violence Programs. 2011. *Hate Violence against Lesbian, Gay, Bisexual, Transgender, Queer, and HIV-Affected Communities in the United States in 2010*. Available at http://www.avp.org

National Coalition of Anti-Violence Programs. 2008. *Anti-Lesbian, Gay, Bisexual and Transgender Violence in 2007*. New York: National Coalition of Anti-Violence Programs.

National Gay and Lesbian Task Force. 2009 (July 1). "State Nondiscrimination Laws in the U.S." Available at http://www.thetaskforce.org

Newcomb, Michael E., and Brian Mustanski. 2010. "Internalized Homophobia and Internalizing Mental Health Problems: A Meta-analytic Review." *Clinical Psychology Review*, 30:1,019–1,029.

Newport, Frank. 2011. (May 20). "For the First Time, Majority of Americans Favor Legal Gay Marriage." Gallup Organization. Available at http://www.gallup.com

O'Keefe, Ed. 2010. "'Don't Ask, Don't Tell' is Repealed by Senate; Bill Awaits Obama's Signing." *The Washington Post*, December 19. Available at http://www.washingtonpost.com

Olson, Laura R., Cadge, Wendy, and Harrison, James T. 2006. "Religion and Public Opinion about Same-Sex Marriage." *Social Science Quarterly* 87: 340–360.

Ottosson, Daniel. 2008. *A World Survey of Laws Prohibiting Same-Sex Activity between Consenting Adults*. International Lesbian and Gay Association. Available at http://www.ilga.org

Padilla, Yolanda C., Catherine Crisp, and Donna Lynn Rew. 2010. "Parental Acceptance and Illegal Drug Use among Gay, Lesbian, and Bisexual Adolescents: Results from a National Survey." *Social Work* 55:265–275.

Page, S. 2011. "Gay Candidates Gain Acceptance." *USA Today*, July 20. Available at http://www.usatoday.com

Palm Center. 2009. "Countries that Allow Military Service by Openly Gay People." Available at http://www.palmcenter.org

Patterson, Charlotte J. 2009. "Children of Lesbian and Gay Parents: Psychology, Law, and Policy." *American Psychologist* 64:727–736.

Pew Research Center. 2011. (May 13). "Most Say Homosexuality Should Be Accepted by Society." Available at http://pewresearch.org

Pollack, William. 2000a. *Real Boys' Voices*. New York: Random House.

Pollack, William. 2000b. "The Columbine Syndrome." *National Forum* 80:39–42.

Poteat,V. Paul, and Craig D. DiGiovanni. 2010. "When Biased Language Use is Associated with Bullying and Dominance Behavior: The Moderating Effect of Prejudice." *Journal of Youth and Adolescence* 39:1123–1133.

Price, Jammie, and Michael G. Dalecki. 1998. "The Social Basis of Homophobia: An Empirical Illustration." *Sociological Spectrum* 18:143–159.

Rankin, Sue, Genevieve Weber, Warren Blumenfeld, and Somjen Frazer. 2010. *State of Higher Education for Lesbian, Gay, Bisexual & Transgender People*. Campus Pride. Available at http://www.campuspride.org

Rapado, Donna, and Janie Campbell. 2011 (July 23). "New Anti-Bullying Rule in Effect for Miami-Dade Schools. NBC Miami. Available at http://www.nbcmiami.com

Röndahl, Gerd, and Sune Innala. 2008. "To Hide or Not to Hide, That Is the Question!" *Journal of Homosexuality*, 52: 211–233.

Rosky, Clifford, Christy Mallory, Jenni Smith, and M. V. Lee Badgett. 2011 (January). "Employment Discrimination against LGBT Utahns: Executive Summary." The Williams Institute. Available at http://www3.law.ucla.edu

Rothblum, Esther D. 2000. "Sexual Orientation and Sex in Women's Lives: Conceptual and Methodological Issues." *Journal of Social Issues*, 56: 193–204.

Rust, P. C. R. 2002. "Bisexuality: The State of the Union." *Annual Review of Sex Research* 13:180–240.

Ryan, Caitlin, David Huebner, Rafael M. Diaz, and Jorge Sanchez. 2009. "Family Rejection as a Predictor of Negative Health Outcomes in White and Latino Lesbian, Gay, and Bisexual Young Adults." *Pediatrics* 123(1):346–352.

Saad, Lydia. 2010. "Americans' Acceptance of Gay Relations Crosses 50% Threshold." Gallup Organization. Available at http://www.gallup.com

Saad, Lydia. 2007 (May 29). "Tolerance for Gay Rights at High Water Mark." Gallup News Service. Available at http://www.gallup.com

Salvage, Charlie, and Sheryl Gay Stolberg. 2011. "In Shift, U.S. Says Marriage Act Blocks Gay Rights." *The New York Times*, February 23, 2011, Available at http:///www.nytimes.com

Savage, Dan. 2010. "Savage Love." Chicago: *The Onion*, October 13.

Savage, Charlie and Sheryl Gay Stolberg. 2011. "In Shift, U.S. Says Marriage Act Blocks Gay Rights." *The New York Times*, February 23, 2011, Available at http://www.nytimes.com

Savin-Williams, R. C. 2006. "Who's Gay? Does It Matter?" *Current Directions in Psychological Science* 15:40–44.

Schiappa, E., P. B. Gregg, and D. E. Hewes. 2005. "The Parasocial Contact Hypothesis." *Communication Monographs* 72(1):92–115.

Schleifer, David. (2006). "Make Me Feel Mighty Real: Gay Female-to-Male Transgenderists Negotiating Sex, Gender, and Sexuality." *Sexualities* 9:57–75.

Schemo, Diana J. 2007. "Lessons on Homosexuality Move into the Classroom." *The New York Times*, August 15. Available at http://www.nytimes.com

Schwartz, John. 2011. "After New York, New Look at Defense of Marriage Act." *The New York Times*, June 27. Available at http://www.nytimes.com

Sears, Tim, and Christy Mallory. 2011. "Evidence of Employment Discrimination on the Basis of Sexual Orientation in State and Local Government: Complaints Filed with State Enforcement Agencies 2003–2007." July. Williams Institute. Available at http://www3.law.ucla.edu/

Shackelford, Todd K., and Avi Besser. 2007. "Predicting Attitudes toward Homosexuality: Insights from Personality Psychology." *Individual Differences Research* 5:106–114.

Shilts, Randy. 1982. *The Mayor of Castro Street: The Life and Times of Harvey Milk*. New York: St. Martin's Press.

Smith, Sara J., Amber M. Axelton, and Donald A. Saucier. 2009. "The Effects of Contact on Sexual Prejudice: A Meta-Analysis." *Sex Roles* 61:178–191.

Smith, Tom W. 2011. "Cross–National Differences in Attitudes towards Homosexuality." April, GSS Cross-National Report No. 31. University of Chicago: National Opinion Research Center. Available at http://www3.law.ucla.edu/williamsinstitute/

Stanley, Kim. 2009. *Resilience, Minority Stress, and Same Sex Populations: Toward a Fuller Picture*. Available at http://www.proquest.com

Stone, Andrea. 2011. "Pentagon Discharged Hundreds of Service Members Under 'Don't Ask, Don't Tell' in Fiscal 2010: Report." *Huffington Post*, March 24. Available at http://www.huffingtonpost.com/

Sue, Derald W. 2010 *Microaggressions in Everyday Life: Race, Gender, and Sexual Orientation*. Hoboken, NJ: John Wiley & Sons, Inc.: Hoboken, NJ.

Sullivan, A. 1997. "The Conservative Case." In *Same-Sex Marriage: Pro and Con*, A. Sullivan, ed. (pp. 146–154). New York: Vintage Books.

Summers, Bryce B. 2010. "Factor Structure and Validity of the Lesbian, Gay, and Bisexual Knowledge and Attitude Scale for Heterosexuals (LGB-KASH). Proquest Dissertations and Theses. (UMI No. 3425038)

Swearer, Susan M., Rhonda K. Turner, Jami E. Givens, and William S. Pollack. 2008. "'You're So Gay!': Do Different Forms of Bullying Matter for Adolescent Males?" *School Psychology Review* 37:160–173.

Szymanski, Dawn M., Susan Kashubeck-West, and Jill Meyer. 2008. "Internalized Heterosexism: Measurement, Psychosocial Correlates, and Research Directions." *The Counseling Psychologist*, 36:525–574.

Tasker, Fiona, Helen Barrett, and Frederica De Simone. 2010. "'Coming Out Tales': Adults Sons and Daughters' Feeling about Their Gay Father's Identity." *Australian and New Zealand Journal of Family Therapy* 31(4):326–337.

Times Research. 2011. "Interactive: Gay Marriage Chronology." *Los Angeles Times*. Available at http://www.latimes.com

Thompson, Cooper. 1995. "A New Vision of Masculinity." In *Race, Class, and Gender in the United States*, 3rd ed., P. S. Rothenberg, ed. (pp. 475–481). New York: St. Martin's Press.

Tobias, Sarah, and Sean Cahill. 2003. *School Lunches, the Wright Brothers, and Gay Families*. National Gay and Lesbian Task Force. Available at http://www.thetaskforce.org

Toomey, Russell B., Caitlin Ryan, Rafael M. Diaz, Noel A. Card, and Stephen T. Russell. 2010. "Gender Non-Conforming Lesbian, Gay, Bisexual, and Transgender Youth: School Victimization and Young Adult Psychosocial Adjustment." *Developmental Psychology* 46:1580–1589.

U.S. Court of Appeals, District 3. 2010. *In re: Matter of Adoption of X.X.G. AND N.R.G. The State of Florida*. September 22. No. 3D08-3044. Available at http://www.3dca.flcourts.org/

United States Census Bureau. 2010. "The Census: A Snapshot." Available at http://2010.census.gov

University of Alabama. 2011. "Safe Zone." Available at http://main.uab.edu/

Varjas, Kris, Brian Dew, Megan Marshall, Emily Graybill, Anneliese Singh, Joel Meyers, and Lamar BircKbichler. 2008. "Bullying in Schools towards Sexual Minority Youth." *Journal of School Violence*, 7: 59–86.

Vrangalova, Zhana, and Ritch C. Savin-Williams. 2010. "Correlates of Same-Sex Sexuality in Heterosexually Identified Young Adults." *Journal of Sex Research*, 47: 92–102.

Wilcox, Clyde, and Robin Wolpert. 2000. "Gay Rights in the Public Sphere: Public Opinion on Gay and Lesbian Equality." In *The Politics of Gay Rights*, Craig A. Rimmerman, Kenneth D. Wald, and Clyde Wilcox, eds. (pp. 409–432). Chicago: University of Chicago Press.

WWMT News. 2011 (May 4). "Saugatuck Considers Including Discussion of Homosexuality in Sex-Ed." Available at http://www.wwmt.com

Whitehead, Andrew L. 2010. "Sacred Rites and Civil Rights: Religion's Effect on Attitudes toward Same-Sex Unions and the Perceived Cause of Homosexuality." *Social Science Quarterly* 91:63–79.

Worthington, Roger L., Frank R. Dillon, and Ann M. Becker-Shutte. 2005. "Development, Reliability, and Validity of the Lesbian, Gay, and Bisexual Knowledge and Attitudes Scale for Heterosexuals (LGB-KASH)." *Journal of Counseling Psychology* 52:104–118.

Chapter 12

American Humane Association. 2011. "Pet Over-population." Available at http://www .americanhumane.org

ASPCA. 2011. "Position Statement on Mandatory Spay/Neuter Laws." Available at http://www .aspca.org

Barot, Sneha. 2011 (Spring). "Unsafe Abortion: The Missing Link in Global Efforts to Improve Maternal Health." *Guttmacher Policy Review* 14(2):24–28.

Bongaarts, John, and Susan Cotts Watkins. 1996. "Social Interactions and Contemporary Fertility Transitions." *Population and Development Review* 22(4):639–682.

Branigan, Tania. 2011. "China Considers Relaxing One-Child Policy." *The Guardian,* March 8. Available at http://www.guardian.co.uk

Brenoff, Ann. 2011 (May 26). "Foreclosures' Other Victims: Abandoned Pets." AOL Real Estate. Available at http://realestate.aol.com

Chou, Rita Jing-Ann. 2011. "Filial Piety by Contract? The Emergence, Implementation, and Implications of the 'Family Support Agreement' in China." *The Gerontologist* 51(1):3–16.

Cincotta, Richard, Robert Engelman, and Daniele Anastasion. 2003. *The Security Demographic: Population and Civil Conflict After the Cold War.* Washington, DC: Population Action International.

Cloutier-Fisher, Denise, Karen Kobayashi, and Andre Smith. 2011. "The Subjective Dimension of Social Isolation: A Qualitative Investigation of Older Adults' Experiences in Small Social Support Networks." *Journal of Aging Studies,* doi:10.1016/j.jaging.2011.03.012

Coate, Stephen and Brian Knight. 2010. "Pet Overpopulation: An Economic Analysis." *The B.E. Journal of Economic Analysis & Policy* 10(1) (Advances), Article 106. Available at http://www.bepress.com

Dunn, Mark. 2008. "Darlington Will Have Australia's First Vertical Cemetery." *Herald Sun,* November 21. Available at http://www .heraldsun.com

Edwards, Kathryn A., Anna Turner, and Alexander Hertel-Fernandez. 2011. *A Young Person's Guide to Social Security.* Economic Policy Institute. Available at http://www.epi.org

Engelman, Robert. 2011 (July 18). "The World at 7 Billion: Can We Stop Growing Now?" *Yale Environment 360.* Available at http://e360. yale.edu

Farrell, Paul. 2009 (January 26). "Peak Oil? Global Warming? No, It's 'Boomsday'!" *MarketWatch.* Available at http://www.marketwatch.com

File, Thom, and Sarah Crissey. 2010 (May). "Voter Registration in the Election of November 2008." *Current Population Reports* P20-562. U.S. Census Bureau. Available at http://www .census.gov

Franko, Kantelle. 2011. "Colorado to Allow Alkaline Hydrolysis, An Alternative to Corpse Cremation or Burial." *Huffington Post,* June 2. Available at http://huffingtonpost.com

Frost, Ashley E., and F. Nii-Amoo Dodoo. 2009. "Men Are Missing from African Family Planning." *Contexts: Understanding People in Their Social Worlds* 8(1):44–49.

Frynes-Clinton, Jane. 2011. "Problem of Grave Concern." *Courier Mail,* June 16. Available at http://www.couriermail.com.au

Greenblatt, Alan. 2011. "Aging Population." *CQ Researcher* 21(25):577–600. Los Angeles: Sage.

Greenfield, Beth. 2011. "The True Costs of Owning a Pet." *Forbes.com,* May 24. Available at http://www.forbes.com

Gullette, Margaret M. 2011. *Agewise: Fighting the New Ageism in America.* Chicago: University of Chicago Press.

Here and Now. 2009 (October 18). "Japan's High-Tech Graveyard in the Sky." Public Radio International. Available at http://www .pri.org

Hollander, Dore. 2007. "Women Who Are Fecund but Do Not Wish to Have Children Outnumber the Involuntarily Childless." *Perspectives on Sexual and Reproductive Health* 39(2):120.

Humane Society. 2010 (December 16). "Austin City Council Prohibits Retail Sales of Dogs and Cats." Available at http://www.humansociety .org

Humane Society. 2009 (November 23). "HSUS Pet Overpopulation Estimates." Available at http:// www.humanesociety.org

Jacobsen, Linda A., Mary Kent, Marlene Lee, and Mark Mather. 2011. "America's Aging Population." *Population Bulletin* 66(1):all. Population Reference Bureau. Available at http://www .prb.org

Jones, Jeffrey M. 2011 (June 1). "Most Workers Expect to Keep Working after Retirement Age." Gallup Organization. Available at http://www .gallup.com

Kidd, Andrew. 2009. "Shelters See Rise in Abandoned Pets as College Students' Year Ends." *Fox News,* May 9. Available at http://www .Foxnews.com

Kornadt, Anna E., and Klaus Rothermund. 2010. "Constructs of Aging: Assessing Evaluative Age Stereotypes in Different Life Domains." *Educational Gerontology* 36(6).

Kornblau, Melissa. 2009. "Social Security Systems around the World." *Today's Research on Aging* No. 15: all. Population Reference Bureau. Available at http://www.prb.org

Lahey, Johanna. 2008. "Age, Women, and Hiring: An Experimental Study." *Journal of Human Resources* 43:30–56.

Lalasz, Robert. 2005 (July). *Baby Bonus Credited with Boosting Australia's Fertility Rate.* Population Reference Bureau. Available at http:// www.prb.org

Leonard, Matt. 2009. "The Kindest Cut." *Earth Island Journal.* Available at http://www .earthislandjournal.org

Livernash, Robert, and Eric Rodenburg. 1998. "Population Change, Resources, and the Environment." *Population Bulletin* 53(1):1–36.

McFalls, Joseph A. 2007 (March). "Population: A Lively Introduction, 5th ed." *Population Bulletin* 62(1).

Mesce, Deborah, and Donna Clifton. 2011. *Abortion: Facts and Figures 2011.* Washington DC: Population Reference Bureau.

Morrissey, Monique. 2011 (January 26). "Beyond 'Normal': Raising the Retirement Age is the Wrong Approach for Social Security." *EPI Briefing Paper #287.* Available at http://www.epi.org

National Council on Pet Population Study and Policy. 2009. "The Top Ten Reasons for Pet Relinquishment to Shelters in the United States." Available at http://petpopulation.org

Nelson, Todd D. 2011. "Ageism: The Strange Case of Prejudice Against the Older You." In *Disability and Aging Discrimination,* R. L. Wiener and S. L. Willborn, eds. (p. 37). New York: Springer Science + Business Media.

Newport, Frank. 2011 (April 25). "In U.S., 53% Worry about Having Enough Money in Retirement." Gallup Organization. Available at http://www.gallup.com

Palmore, Erdman B. 2004. "Research Note: Ageism in Canada and the United States." *Journal of Cross-Cultural Gerontology* 19(1):41–46.

Parks, Kristin. 2005. "Choosing Childlessness: Weber's Typology of Action and Motives of the Voluntarily Childless." *Sociological Inquiry* 75(3):372–402.

Population Reference Bureau. 2010. *World Population Data Sheet.* Washington, DC: Population Reference Bureau. Available at http://prb.org

Population Reference Bureau. 2007. *World Population Data Sheet.* Washington, DC: Population Reference Bureau. Available at http:// www.prb.org

Population Reference Bureau. 2004. "Transitions in World Population." *Population Bulletin* 59(1).

Ryerson, William N. 2011 (March 11). "Family Planning: Looking Beyond Access." *Science* 331: 1265.

Santora, Marc. 2010. "City Cemeteries Face Gridlock." *New York Times,* August 13. Available at http://www.nytimes.com

Schueller, Jane. 2005 (August). "Boys and Changing Gender Roles." YouthNet. *YouthLens* 16. Available at http://www.fhi.org

Scott, Laura. 2009. *Two Is Enough: A Couple's Guide to Living Childless by Choice.* Berkeley, CA: Seal Press.

Shikina, Rob. 2011. "Bill Mandates 'Fixing' of Cats, Dogs Before Sale." *Star Advertiser,* April 17. Available at http://www .staradvertiser.com

Smith, Gar. 2009. "Planet Girth." *Earth Island Journal* 24(2):15.

Social Security Administration. 2011 (July). "Monthly Statistical Snapshot." Available at http://www.ssa.gov

Social Security Trustees. 2011. *The 2011 Annual Report of the Board of Trustees of the Federal Old Age and Survivors Insurance and Federal Disability Insurance Trust Funds.* Washington DC: U.S. Government Printing Office.

Sonfield, Adam. 2011 (Winter). "The Case for Insurance Coverage of Contraceptive Services and Supplies without Cost-Sharing." *Guttmacher Policy Review* 14(11):7–15.

SUNY College of Environmental Science and Forestry. 2009. "Worst Environmental Problem? Overpopulation, Experts Say." *ScienceDaily*, April 20. Available at http://www.sciencedaily.com

Szinovacz, Maximiliane E. 2011. "Introduction: The Aging Workforce: Challenges for Societies, Employers, and Older Workers." *Journal of Aging & Social Policy* 23(2):95–100.

United Nations. 2011. *World Population Prospects: The 2010 Revision*. Available at http://www.un.org/esa/population/publications/publications.htm

United Nations. 2009. *World Population Prospects: The 2008 Revision*. New York: United Nations.

United Nations Population Division. 2009. "What Would It Take to Accelerate Fertility Decline in the Least Developed Countries?" United Nations Population Division Policy Brief No. 2009/1. New York: United Nations.

Weeks, John R. 2012. *Population: An Introduction to Concepts and Issues*, 11th ed. Belmont, CA: Wadsworth, Cengage Learning.

Weiland, Katherine. 2005. *Breeding Insecurity: Global Security Implications of Rapid Population Growth*. Washington, DC: Population Institute.

Wiseman, Paul. 2005. "Towns Hope Cash-for-Babies Incentives Boost Populations." *USA Today*, July 28. Available at http://www.usatoday.com

Women's Studies Project. 2003. *Women's Voices, Women's Lives: The Impact of Family Planning*. Family Health International. Available at http://www.fhi.org

Chapter 13

American Lung Association. 2011. *State of the Air: 2011*. Available at http://lungaction.org

American Lung Association. 2005. *Lung Disease Data in Culturally Diverse Communities: 2005*. Available at http://www.lungusa.org

Aukema, J. E., B. Leung, K. Kovacs, C. Chivers, and K. O. Britton, et al. 2011. "Economic Impacts of Non-Native Forest Insects in the Continental United States." *PLoS ONE* 6(9):e24587. Available at http://www.plooneorg

Barbosa, Luiz C. 2009. "Theories in Environmental Sociology." In *Twenty Lessons in Environmental Sociology*, Kenneth A. Gould and Tammy L. Lewis, eds. (pp. 25–44). New York: Oxford University Press.

Beckel, Michael. 2010. "Congressmen Maintain Massive Portfolio of Oil and Gas Investments." *Open Secrets Blog*, August 27. Available at http://www.opensecrets.org

Beinecke, Frances. 2009 (Spring). "Debunking the Myth of Clean Coal." *OnEarth*. Available at http://www.onearth.org

Brecher, Jeremy. 2011 (January 4). "Climate Protection Strategy: Beyond Business as Usual." Labor Network for Sustainability." Available at http://www.labor4sustainability.org

Betts, Kellyn S. 2011. "Plastics and Food Sources: Dietary Intervention to Reduce BPA and DEHP." *Environmental Health Perspectives* 119(7): A306.

Blatt, Harvey. 2005. *America's Environmental Report Card: Are We Making the Grade?* Cambridge, MA: MIT Press.

BP. 2011. *BP Statistical Review of World Energy*. Available at http://www.bp.com

Bradshaw, Nancy. 2010 (March). *Fragrance-Free Policy Management Presentation*. Women's College Hospital, University of Toronto.

Brody, Julia Gree, Kirsten B. Moysich, Olivier Humblet, Kathleen R. Attfield, Gregory P. Beehler, and Ruthann A. Rudel. 2007. "Environmental Pollutants and Breast Cancer." *Cancer* 109(S12):2667–2711.

Brown, Lester R. 2007. "Distillery Demand for Grain to Fuel Cars Vastly Understated: World May Be Facing Highest Grain Prices in History." *Earth Policy News,* January 4. Available at http://www.earthpolicy.org

Brown, Lester R., and Jennifer Mitchell. 1998. "Building a New Economy." In *State of the World 1998*, Lester R. Brown, Christopher Flavin, and Hilary French, eds. (pp. 168–187). New York: W. W. Norton.

Brulle, Robert J. 2009. "U.S. Environmental Movements." In *Twenty Lessons in Environmental Sociology*, Kenneth A. Gould and Tammy L. Lewis, eds. (pp. 211–227). New York: Oxford University Press.

Bruno, Kenny, and Joshua Karliner. 2002. *Earthsummit.biz: The Corporate Takeover of Sustainable Development*. CorpWatch and Food First Books. Available at http://www.corpwatch.org

Bullard, Robert D. 2000. *Dumping in Dixie: Race, Class, and Environmental Quality*, 3rd ed. Boulder, CO: Westview Press.

Bullard, Robert D., Paul Mohai, Robin Saha, and Beverly Wright. 2007 (March). *Toxic Wastes and Race at Twenty 1987–2007*. Cleveland, Ohio: United Church of Christ.

Caress, Stanley M., and Anne C. Steinemann. 2004. "A National Population Study of the Prevalence of Multiple Chemical Sensitivity." *Archives of Environmental Health* 59(6):300–305.

Carlson, Scott. 2006. "In Search of the Sustainable Campus." *The Chronicle of Higher Education* LIII(9):A10–A12, A14.

Chafe, Zoe. 2006. "Weather-Related Disasters Affect Millions." In *Vital Signs*, L. Starke ed. (pp. 44–45). New York: W. W. Norton & Co.

Chafe, Zoe. 2005. "Bioinvasions." In *State of the World 2005*, L. Starke, ed. (pp. 60–61). New York: W. W. Norton.

Cheeseman, Gina-Marie. 2007. "Plastic Shopping Bags Being Banned." *The Online Journal*, June 27. Available at http://www.onlinejournal.com

Chepesiuk, Ron. 2009. "Missing the Dark: Health Effects of Light Pollution." *Environmental Health Perspectives* 117(1):20, A20–A27.

Cincotta, Richard P., and Robert Engelman. 2000. *Human Population and the Future of Biological Diversity*. Washington, DC: Population Action International.

Clarke, Tony. 2002. "Twilight of the Corporation." In *Social Problems, Annual Editions 02/03*, 30th ed., Kurt Finster-Busch, ed. (pp. 41–45). Guilford, CT: McGraw-Hill/Dushkin.

Clemmitt, Marcia. 2011. Nuclear Power. *CQ Researcher* 21(22):all.

Cooper, Arnie. 2004. "Twenty-Eight Words That Could Change the World: Robert Hinkley's Plan to Tame Corporate Power." *The Sun* 345(September):4–11.

Coyle, Kevin. 2005. *Environmental Literacy in America*. Washington, DC: The National Environmental Education and Training Foundation.

Denson, Bryan. 2000. "Shadowy Saboteurs." *IRE Journal* 23(May-June):12–14.

Edwards, Bob, and Adam Driscoll. 2009. "From Farms to Factories: The Environmental Consequences of Swine Industrialization in North Carolina." In *Twenty Lessons in Environmental Sociology*, Kenneth A. Gould and Tammy L. Lewis, eds. (pp. 153–175). New York: Oxford University Press.

Elk, Mike. 2011 (September 5). "Which is More Likely to Rebuild the Labor Market: Environmental Allies or 6,000 Temp Jobs?" *Working in These Times*. Available at http://www.inthesetimes.com

Energy Information Administration. 2010. *Annual Energy Review*. Available at http://www.eia.gov

Environmental Defense Fund. 2011 (August 9). "Washington Gets it Right with First-Ever National Fuel Standards for Trucks and Buses." Press Release. Available at http://www.environmentaldefense.net

Environmental Defense Fund. 2001. "A Prescription for Reducing the Damage Caused by Dams." *Environmental Defense* 32(2).

Environmental Working Group. 2005 (July 14). *Body Burden: The Pollution in Newborns*. Available at http://www.ewg.org

EPA (U.S. Environmental Protection Agency). 2011a. *National Priorities List (NPL)*. Available at http://www.epa.gov/superfund/sites

EPA. 2011b. *Electronics Waste Management in the United States through 2009*. Available at http://www.epa.gov

EPA. 2010. *Municipal Solid Waste Generation, Recycling, and Disposal in the United States: Facts and Figures for 2009*. Available at http://www.epa.gov

EPA. 2004. *What You Need to Know about Mercury in Fish and Shellfish*. Available at http://www.epa.gov

Ewing, B., D. Moore, S. Goldfinger, A. Oursler, A. Reed, and M. Wackernagel. 2010. *The Ecological Footprint Atlas 2010*. Oakland: Global Footprint Network.

Fisher, Brandy E. 1999. "Focus: Most Unwanted." *Environmental Health Perspectives* 107(1). Available at http://ehpnet1.niehs.nih.gov/docs/1999/107-1/focus-abs.html

Flavin, Christopher, and Robert Engelman. 2009. "The Perfect Storm." In *State of the World*. Worldwatch Institute.

Flavin, Chris, and Molly Hull Aeck. 2005 (September 15). "Cleaner, Greener, and Richer." Tom.Paine.com. Available at http://www.tompaine.com/articles/2005/09/15/cleaner_greener_and_richer.php

Food and Drug Administration. 2010. *Pesticide Residue Monitoring Program Results and Discussion FY 2008*. Available at http://www.fda.gov

Food & Water Watch and Network for New Energy Choices. 2007. *The Rush to Ethanol: Not All Biofuels Are Created Equal.* Available at http://www.newenergychoices.org

Foster, Joanna. 2011. "Impact of Gulf Spill's Underwater Dispersants Examined." *New York Times,* August 26. Available at http://nytimes.com

French, Hilary. 2000. *Vanishing Borders: Protecting the Planet in the Age of Globalization.* New York: W. W. Norton.

Gallup Organization. 2011. *Environment.* Available at http://www.gallup.com/poll

Gardner, Gary. 2005. "Forest Loss Continues." In *Vital Signs 2005,* Linda Starke, ed. (pp. 92–93). New York: W. W. Norton.

Gearhart, Jeff. 2009 (September 16). "Toxic Chemicals in Everyday Products: Find Out What's in the Stuff You Use." *Alternet.* Available at http://www.alternet.org

Global Footprint Network. 2010. *2010 Annual Report.* Oakland, CA.: Global Footprint Network. Available at http://www.footprintnetwork.org

Global Humanitarian Forum. 2009. *Human Impact Report: Climate Change—The Anatomy of a Silent Crisis.* Geneva: Global Humanitarian Forum.

Global Invasive Species Database. 2006. "Felis Catus." Available at http://www.issg.org

Goodstein, Laurie. 2005. "Evangelical Leaders Swing Influence Behind Effort to Combat Global Warming." *New York Times,* March 10. Available at http://www.nytimes.com

Gottlieb, Roger S. 2003a. "Saving the World: Religion and Politics in the Environmental Movement." In *Liberating Faith,* Roger S. Gottlieb, ed. (pp. 491–512). Lanham, MD: Rowman & Littlefield.

Gottlieb, Roger S. 2003b. "This Sacred Earth: Religion and Environmentalism." In *Liberating Faith,* Roger S. Gottlieb, ed. (pp. 489–490). Lanham, MD: Rowman & Littlefield.

Hager, Nicky, and Bob Burton. 2000. *Secrets and Lies: The Anatomy of an Anti-Environmental PR Campaign.* Monroe, ME: Common Courage Press.

Hefling, Kimberly. 2007. "Hearing Planned Today in Lejeune Water Case." *Marine Corps Times,* June 12. Available at http://www.marinecorpstimes.com

Hilgenkamp, Kathryn. 2005. *Environmental Health: Ecological Perspectives.* Sudbury, MA: Jones and Bartlett Publishers.

Holzman, David C. 2011. "Methane Found in Well Water Near Fracking Sites." *Environmental Health Perspectives* 119(7):A289.

Hunter, Lori M. 2001. *The Environmental Implications of Population Dynamics.* Santa Monica, CA: Rand Corporation.

Intergovernmental Panel on Climate Change. 2007a. *Climate Change 2007: The Physical Science Basis.* United Nations Environmental Programme and the World Meteorological Organization. Available at http://www.ipcc.ch/

Intergovernmental Panel on Climate Change. 2007b. *Climate Change 2007: Impacts, Adaptation and Vulnerability.* United Nations

Environmental Programme and the World Meteorological Organization. Available at http://www.ipcc.ch/

IUCN. 2011. *2011 IUCN Red List of Threatened Species.* Available at http://www.iucnredlist.org

Jamail, Dahr. 2011. "Fukushima Radiation Alarms Doctors." *Common Dreams,* August 20. Available at http://www.commondreams.org

Janofsky, Michael. 2005. "Pentagon Is Asking Congress to Loosen Environmental Laws." *New York Times,* May 11. Available at http://www.nytimes.com

Jensen, Derrick. 2001. "A Weakened World Cannot Forgive Us: An Interview with Kathleen Dean Moore." *The Sun* 303 (March):4–13.

Kahn, Chris. 2011. "Hurricane Irene Power Outages: Electricity Blackouts Affect 4 Million Homes and Businesses." *Huffington Post,* August 28. Available at http://www.huffingtonpost.com

Kaplan, Sheila, and Jim Morris. 2000. "Kids at Risk." *U.S. News and World Report,* June 19, pp. 47–53.

Kitasei, Saya. 2011. "Wind Power Growth Continues to Break Records Despite Recession." In *Vital Signs,* Linda Starke, ed. (pp. 26–28). Washington DC: Worldwatch Institute.

Klinkenborg, Verlyn. 2008. "Our Vanishing Night." *National Geographic* (November): 102–123.

Knoell, Carly. 2007 (August 9). "Malaria: Climbing in Elevation as Temperature Rises." *Population Connection.* Available at www.populationconnection.org

Lamm, Richard. 2006. "The Culture of Growth and the Culture of Limits." *Conservation Biology* 20(2):269–271.

Leahy, Stephen. 2009 (May 21). "Alien Species Eroding Ecosystems and Livelihoods." Interpress Service News Agency. Available at http://www.ipsnews.net

Leitzell, Katherine. 2011 (May 3). "When Will the Arctic Lose its Sea Ice?" National Snow and Ice Data Center. Available at http://nsidc.org

Little, Amanda Griscom. 2005. "Maathai on the Prize: An Interview with Nobel Peace Prize Winner Wangari Maathai." *Grist Magazine,* February 15. Available at http://www.grist.org

Lubitow, Amy, and Mia Davis. 2011. "Pastel Injustice: The Corporate Use of Pinkwashing for Profit." *Environmental Justice* 4(2):139–144.

Mackinder, Evan. 2010. "Pro-Environment Groups Outmatched, Outspent in Battle Over Climate Change Legislation." *Open Secrets Blog,* August 23. Available at http://www.opensecrets.org

Makower, Joel, Ron Pernick, and Clint Wilder. 2009. *Clean Energy Trends 2009.* Clean Edge. Available at http://www.cleanedge.com

Malcolm, Jay R., Canran Liu, Ronald P. Neilson, Lara Hansen, and Lee Hannah. 2006. "Global Warming and Extinctions of Endemic Species from Biodiversity Hotspots." *Conservation Biology* 20(2): 538–548.

Martinez, Edico. 2011. "Actress Daryl Hannah Arrested for a Greater Freedom." CBS News, August 31. Available at http://www.cbsnews.com

McCarthy, Michael. 2008. "Cleared: Jury Decides That Threat of Global Warming Justifies Breaking the Law." *The Independent,* September 11. Available at http://www.independent.co.uk

McCormick, James. 2011. "Nuclear Illinois Helped Shape Obama View of Energy in Dealings with Exelon." *Bloomberg,* March 23. Available at http://www.bloomberg.com

McDaniel, Carl N. 2005. *Wisdom for a Livable Planet.* San Antonio, TX: Trinity University Press.

McGinn, Anne Platt. 2000. "Endocrine Disrupters Raise Concern." In *Vital Signs 2000,* Lester R. Brown, Michael Renner, and Brian Halweil, eds. (pp. 130–131). New York: W. W. Norton.

McKibben, Bill. 2008 (October 24). "Meltdown: A Global Warming Travelogue." CNN.com. Available at http://www.cnn.com

McMichael, Anthony J., Kirk R. Smith, and Carlos F. Corvalan. 2000. "The Sustainability Transition: A New Challenge." *Bulletin of the World Health Organization* 78(9):1067.

Mead, Leila. 1998. "Radioactive Wastelands." *The Green Guide* 53(April 14):1–3.

Millennium Ecosystem Assessment. 2005. *Ecosystems and Human Well-Being: Synthesis.* Washington, DC: Island Press.

Miller, G. Tyler, Jr., and Scott E. Spoolman. 2009. *Living in the Environment,* 16th ed. Belmont, CA: Brooks/Cole, Cengage Learning.

Miranda, M. L., D. A. Hastings, J. E. Aldy, and W. H. Schlesinger. 2011. "The Environmental Justice Dimensions of Climate Change." *Environmental Justice* 4(1):17–25.

Mooney, Chris. 2005. "Some Like It Hot." *Mother Jones,* May-June. Available at http://www.MotherJones.com

Mulrow, John, and Alexander Ochs (with Shakuntala Makhijani). 2011. "Glacial Melt and Ocean Warming Drive Sea Level Upward." In *Vital Signs,* Linda Starke, ed. (pp. 43–46). Washington DC: Worldwatch Institute.

Murtaugh, Paul, and Michael Schlax. 2009. "Reproduction and the Carbon Legacies of Individuals." *Global Environmental Change* 19:14–20.

National Assessment Synthesis Team. 2000. *Climate Change Impacts on the United States: The Potential Consequences of Climate Variability and Change.* Washington, DC: U.S. Global Change Research Program.

National Commission on the BP Deepwater Horizon Oil Spill and Offshore Drilling. 2011. *Deepwater: The Gulf Oil Disaster and the Future of Offshore Drilling.* Available at http://www.oilspillcommission.gov

National Geographic. 2007 (June 5). "Top Ten Tips to Fight Global Warming." *Green Guide.* Available at http://www.thegreenguide.com

NRDC (Natural Resources Defense Council). 2011 (April). *The BP Oil Disaster at One Year: A Straightforward Assessment of What We Know, What We Don't, and What Questions Need to Be Answered.* Available at http://www.nrdc.org

Newport, Frank. 2010. "Americans' Global Warming Concerns Continue to Drop." Gallup Poll, March 11. Available at http://www.gallup.com

"Nukes Rebuked." 2000. *Washington Spectator* 26(13):4.

Pearce, Fred. 2005. "Climate Warming as Siberia Melts." *New Scientist,* August 11. Available at http://www.NewScientist.com

Pew Center on the States. 2009. *State of the States 2009.* Available at http://www.stateline.org

Pimentel, D., S. Cooperstein, H. Randell, D. Filiberto, S. Sorrentino, B. Kaye, C. Nicklin, J. Yagi, J. Brian, J. O'Hern, A. Habas, and C. Weinstein. 2007. "Ecology of Increasing Diseases: Population Growth and Environmental Degradation." *Human Ecology* 35(6):653–668.

Prah, Pamela M. 2007 (July 30). "States Forge Ahead on Immigration, Global Warming." *Stateline.* Available at http://www.stateline.org

Price, Tom. 2006. "The New Environmentalism." *CQ Researcher* 16(42):987–1007.

Pryor, John H., S. Hurtado, L. DeAngelo, L. Paluki Blake, and S. Tran. 2010. *The American Freshman: National Norms Fall 2010.* Los Angeles: Higher Education Research Institute, UCLA.

Public Citizen. 2005 (February). *NAFTA Chapter 11 Investor-State Cases: Lessons for the Central America Free Trade Agreement. Public Citizens Global Trade Watch Publication E9014.* Available at http://www.citizen.org

Reese, April. 2001 (February). *Africa's Struggle with Desertification.* Population Reference Bureau. Available at http://www.prb.org

Renner, Michael. 2004. "Moving toward a Less Consumptive Economy." In *State of the World 2004,* Linda Starke, ed. (pp. 96–119). New York: W. W. Norton.

Rifkin, Jeremy. 2004. *The European Dream: How Europe's Vision of the Future Is Quietly Eclipsing the American Dream.* New York: Tarcher/Penguin.

Rogers, Sherry A. 2002. *Detoxify or Die.* Sarasota, FL: Sand Key.

Russell, James. 2011. "Record Growth in Photovoltaic Capacity and Momentum Builds for Concentrating Solar Power." In *Vital Signs,* Linda Starke, ed. (pp. 29–31). Washington DC: Worldwatch Institute.

Sawin, Janet. 2004. "Making Better Energy Choices." In *State of the World 2004,* Linda Starke, ed. (pp. 24–43). New York: W. W. Norton.

Scavia, Donald. 2011 (September 2). "Dead Zones in Gulf of Mexico and Other Waters Require a Tougher Approach: Donald Scavia." *Nola.com.* Available at http://www.nola.com

Schneider, Mycle, Antony Froggatt, and Steve Thomas. 2011. *World Nuclear Industry Status Report: Nuclear Power in a Post-Fukushima World.* Washington D.C.: Worldwatch Institute.

Shapiro, Isaac, and John Irons. 2011. *Regulation, Employment, and the Economy.* EPI Briefing Paper #305. Washington, DC. Economic Policy Institute.

Schapiro, Mark. 2007. *Exposed: The Toxic Chemistry of Everyday Products and What's at Stake for American Power.* White River Junction, Vermont: Chelsea Green Publishing.

Schapiro, Mark. 2004. "New Power for 'Old Europe.'" *The Nation,* December 27, pp. 11–16.

Schneider, Mycle, Antony Frogatt, and Steve Thomas. 2011. *The World Nuclear Industry*

Status Report 2010–2011: Nuclear Power in a Post Fukushima World. Washington DC: Worldwatch Institute.

Schwartz-Nobel, Loretta. 2007. *Poisoned Nation.* New York: St. Martin's Press.

Sheppard, Kate. 2011. "Did ExxonMobil Break Its Promise to Stop Funding Climate Change Deniers?" *Mother Jones,* June 28. Available at http://motherjones.com

Shrank, Samuel. 2011. "Growth of Biofuel Production Slows." In *Vital Signs,* Linda Starke, ed. (pp. 16–18). Washington DC: Worldwatch Institute.

Sinks, Thomas. 2007 (June 12). *Statement by Thomas Sinks, PhD, Deputy Director, Agency for Toxic Substances and Disease Registry on ATSDR's Activities at U.S. Marine Corps Base Camp Lejeune before Committee on Energy and Commerce Subcommittee on Oversight and Investigations United States House of Representatives.* Available at http://www.hhs.gov

Soraghan, Mike. 2011. "Baffled about Fracking? You're Not Alone." *New York Times,* May 13. Available at http://www.nytimes.com

Stoll, Michael. 2005 (September-October). "A Green Agenda for Cities." *E Magazine* 16(5). Available at http://www.emagazine.com

Sustainable Endowments Institute. 2011. *Greening the Bottom Line: The Trend toward Green Revolving Funds on Campus.* Cambridge MA: Sustainable Endowments Institute.

Swift, Anthony, Susan Casey-Lefkowitz, & Elizabeth Shope. 2011. *Tar Sands Pipelines Safety Risks.* Natural Resources Defense Council. Available at http://www.nrdc.org

TerraChoice Group Inc. 2009. *The Seven Sins of Greenwashing: Environmental Claims in Consumer Markets.* Available at http://sinsofgreen-washing.org

UNDP (United Nations Development Programme). 2010. *Human Development Report 2010.* Available at http://hdr.undp.org

United Nations Development Programme. 2007. *Human Development Report 2007/2008: Fighting Climate Change: Human Solidarity in a Divided World.* New York: Palgrave Macmillan.

UNEP (United Nations Environment Programme). 2008. *Vital Forest Graphics.* UNEP/GRID-Arendal. Available at http://www.fao.org

UNEP. 2007. *GEO Yearbook 2007: An Overview of Our Changing Environment.* Available at http://www.unep.org

U.S. Department of Health and Human Services. 2011. *12th Report on Carcinogens.* Washington, DC: Public Health Service.

U.S. Energy Administration. 2010. "State Nuclear Profiles." Available at http://www.eia.gov

U.S. Geological Survey. 2007 (September 7). "Future Retreat of Arctic Ice Will Lower Polar Bear Populations and Limit Their Distribution." *USGS Newsroom.* Available at http://www.usgs.gov

Vajpeyi, Dhirendra K. 1995. "External Factors Influencing Environmental Policymaking: Role of Multilateral Development Aid Agencies." In *Environmental Policies in the Third World: A Comparative Analysis,* O. P. Dwivedi and Dhirendra K. Vajpeyi, eds. (pp. 24–45). Westport, CT: Greenwood Press.

Vedantam, Shankar. 2005a. "Nuclear Plants Not Keeping Track of Waste." *Washington Post,* April 19. Available at http://www.washingtonpost.com

Vedantam, Shankar. 2005b. "Storage Plan Approved for Nuclear Waste." *Washington Post,* September 10. Available at http://www.washingtonpost.com

Westerling, A. L., H. G. Hidalgo, D. R. Cayan, and T. W. Swetnam. 2006. "Warming and Earlier Spring Increase Western U.S. Forest Wildfire Activity." *Science* 313 (5789):940–943.

Wire, Thomas. 2009 (August.) "Fewer Emitters, Lower Emissions, Less Cost." Optimum Population Trust. Available at http://www.optimumpopulation.org

Woodward, Colin. 2007. "Curbing Climate Change." *CQ Global Researcher* 1(2):27–50. Available at http://www.globalresearcher.com

World Health Organization. 2010. *WHO Guidelines for Indoor Air Quality: Selected Pollutants.* Available at http://www.euro.who.int

World Health Organization. 2005. *Indoor Air Pollution and Health.* Fact Sheet. Available at http://www.who.org

World Resources Institute. 2000. *World Resources 2000–2001: People and Ecosystems—The Fraying Web of Life.* Washington, DC: World Resources Institute.

World Water Assessment Program. 2009. *World Water Development Report 3: Water in a Changing World.* Available at http://www.unesco.org

WWF (World Wildlife Federation). 2008. *Living Planet Report.* World Wildlife Fund, Zoological Society of London, and Global Footprint Network. Available at http://www.panda.org

Zelman, Joanna. 2011. "50 Million Environmental Refugees by 2020, Experts Predict." *Huffington Post,* February 22. Available at http://www.huffingtonpost.com

Zimmer, Carl. 2011. "How Many Species? A Study Says 8.7 Million, but It's Tricky." *New York Times,* August 23. Available at http://www.nytimes.com

Chapter 14

AAEM (American Academy of Environmental Medicine). 2009 (May 8). "Genetically Modified Foods." Available at http://www.aaemonline.org

Abcarian, Robin. 2009. "A New Push to Define 'Person,' and to Outlaw Abortion in the Process." *Los Angeles Times,* September 28. Available at http://www.latimes.com

ACLU (American Civil Liberties Union). 2005. *Utah Businesses, Free Speech Groups, and Individuals Challenge Restrictions on Internet Speech. Privacy and Technology.* Available at http://www.aclu.org/Privacy

Archibold, Randal C. 2009. "Octuplets, 6 Siblings, and Many Questions." *New York Times,* February 3. Available at http://www.nytimes.com

Ashton, Jennifer. 2010. "Man Goes Home with 'Total Artificial Heart.'" CBS News, May 22. Available at http:www//cbsnews.com

Associated Press. 2010. "China's Internet Censorship." CBS News, January 11. Available at http://www.cbsnews.com

Associated Press. 2009. "U.N. Gives Up Global Cloning Ban." CBS News, February 11. Available at http://www.cbsnews.com

Atkinson, Robert D., and Daniel D. Castro. 2008 (October). "Digital Quality of Life: Understanding the Personal and Social Benefits of the Information Technology Revolution." The Information and Technology Foundation. Available at http://www.itif.org

AT&T (American Telephone and Telegraph). 2011. "AT&T Small Business Technology Poll." Available at http://www.att.com

Attewell, Paul, Belkis Suazo-Garcia, and Juan Battle. 2003. "Computers and Young Children: Social Benefit or Social Problem?" Social Forces 82:277–296.

Barbash, Fred. 2005. "High Court Takes up Abortion Consent Case." Washington Post, May 25. Available at http://www.washingtonppost.com

Barker, Colin. 2007. "The Top 10 IT Disasters of All Time." ZDNetUK, November 27. Available at http://www.zdnet.com

BBC. 2009. "On This Day: July 21, 1969." Available at http://news.bbc.co.uk/onthisday

BBC. 2006. "Google Censors Itself for China." BBC News, January 25. Available at http://newsvote.bbc.co.uk

Behm-Morawitz, Elizabeth, and Dana Mastro. 2009. "The Effects of the Sexualizaton of Female Video Game Characters on Gender Stereotyping and Female Self-Concept." Sex Roles 61:808–823.

Bell, Daniel. 1973. The Coming of Post-Industrial Society: A Venture in Social Forecasting. New York: Basic Books.

Beniger, James R. 1993. "The Control Revolution." In Technology and the Future, Albert H. Teich, ed. (pp. 40–65). New York: St. Martin's Press.

BLS (Bureau of Labor Statistics). 2010 (May). "More than 75 Percent of American Households Own Computers." Consumer Expenditures 1(4):1. Available at http://www/bls.org

Bollier, David. 2009 (July 22). "Deadly Medical Monopolies." On the Commons. Available at http://onthecommons.org

Bowman, Lee. 2011. "Animal Testing: Biomedical Researchers Using Millions of Animals Yearly." Scripps Howard News Service, May 7.

Brown, David. 2006. "For 1st Woman with Bionic Arm, a New Life Is Within Reach." Washington Post, September 14, p. A01.

Budde, Paul. 2009. "2008 Global Broadband—M-Commerce, E-Commerce & E-Payments." Available at http://www.researchandmarkets.com

Buchanan, Allen, Dan Brock, Norman Daniels, and Daniel Wikler. 2000. From Chance to Choice: Genetics and Justice. New York: Cambridge University Press.

Busari, Stephanie. 2009 (July 28). "Nicaragua Abortion Ban 'Cruel and Inhuman Disgrace.'" CNN Health. Available at http://www.cnn.com/2009/HEALTH

Bush, Corlann G. 1993. "Women and the Assessment of Technology." In Technology and the Future, Albert H. Teich, ed. (pp. 192–214). New York: St. Martin's Press.

Cauley, Leslie. 2009. "What's Broadband? Billions in Stimulus Funds Are at Stake." USA Today, April 7. Available at http://www.usatoday.com/money

Ceruzzi, Paul. 1993. "An Unforeseen Revolution." In Technology and the Future, Albert H. Teich, ed. (pp. 160–174). New York: St. Martin's Press.

Chan, Sewell. 2007. "New Scanners for Tracking City Workers." New York Times, January 23. Available at http://www.nytimes.com

Chen, Shirong. 2011. "China Tightens Internet Censorship Controls." British Broadcasting Corporation. Available at http://www.bbc.co.uk

Children's Hospital Boston staff. 2009. Available at http://childrenshospitalblog.org/stem-cell-research-a-fathers-story

Clarke, Adele E. 1990. "Controversy and the Development of Reproductive Sciences." Social Problems 37(1):18–37.

Clifford, Stephanie. 2009. "Ads Follow Web Users, and Get More Personal." The New York Times, July 31. Available at http://www.nytimes.com

Coalition for Genetic Fairness. 2004. "Genetic Discrimination: How Genetic Discrimination Affects Real People." National Partnership for Women and Families on behalf of the Coalition for Genetic Fairness. Available at http://www.nationalpartnership.org

Cohen, S. P. 2002. "Can Pets Function as Family Members?" Western Journal of Nursing Research 24:621–638.

Crichton, Michael. 2007. "Patenting Life." New York Times, February 13. Available at http://www.nytimes.com

David-Ferdon, Corrine, and Marci Feldman Hertz. 2009. Electronic Media and Youth Violence: A CDC Issue Brief for Researchers. Atlanta, GA: Centers for Disease Control.

Davis, Devra. 2010. Disconnect: The Truth about Cell Phone Radiation, What the Industry Has Done to Hide It, and How to Protect Your Family. Boston: Dutton Books.

Della Cava, Marco R. 2010. "Some Ditch Social Networks to Reclaim Time, Privacy." USA Today, February 10. Available at http://www.usatoday.com

DeNoon, Daniel. 2005 (March 8). Study: Computer Design Flaws May Create Dangerous Hospital Errors. Available at http://my.webmd.com

Department of Health and Human Services. 2009 (April 6). "The Genetic Information Nondiscrimination Act of 2008: Information for Researchers and Health Care Professionals." Available at http://www.genome.gov

Dewan, Shaila. 2010. "To Court Blacks, Foes of Abortion Make Racial Case." The New York Times, February 26. Available at http://www.nytimes.com

Dignan, Larry. 2010. "Offshoring's Toll: IT Departments to Endure Jobless Recovery through 2014." ZDNet, November 18. Available at http://www.zdnet.com/

Durkheim, Emile. 1973/1925. Moral Education. New York: Free Press.

Dutta, Soumitra, and Irene Mia. 2011. The Global Information Technology Report 2010–2011. World Economic Forum. Available at http://reports.weforum.org

Eckholm, Erik. 2011. "Across Country, Law Makers Push Abortion Curbs." The New York Times, January 21. Available at http://www.nytimes.com

Eibert, Mark D. 1998. "Clone Wars." Reason 30(2):52–54.

Eilperin, Juliet, and Rick Weiss. 2003. "House Votes to Prohibit All Human Cloning." Washington Post, February 28. Available at http://www.washingtonpost.com

Eisenberg, Anne. 2009. "Better Vision, with a Telescope Inside the Eye." The New York Times, July 19. Available at http://www.nytimes.com

ESA (Entertainment Software Association). 2011. "Industry Facts." Available at http://www.theesa.com/facts

ETC Group. 2003. Contamination by Genetically Modified Maize in Mexico Much Worse Than Feared. Available at http://www.etcgroup.org

Evans, M. D. R., and Jonathan Kelley. 2011. "U.S. Attitudes toward Human Embryonic Stem Cell Research." Nature Biotechnology 29:484–488.

FBI (Federal Bureau of Investigation). 2011. "Statement before the House Judiciary Subcommittee on Crime, Terrorism, and Homeland Security." Gordon Snow, Assistant Director, Federal Bureau of Investigation. Washington, DC.

Ferdowsian, Hope. 2010 (November 7). "Animal Research: Why We Need Alternatives." The Chronicle of Higher Education. Available at http://chronicle.com

Flecknell, Paul A. 2010. "Do Mice Have a Pain Face?" Nature Methods 7:437–438.

Fox, Susannah. 2009 (October 26). "The Social Life of Health Information." Pew Internet and the American Life Project. Available at http://www.pewinternet.org

Freudenheim. 2010. "The Doctor Will See You Now. Please Log On." The New York Times, May 29. Available at http://www.nytimes.com

Friedman, Thomas L. 2005. The World Is Flat. New York: Farrar, Strauss, and Giroux.

FTC (Federal Trade Commission). 2010. "Facts for Consumers." Washington, DC. Available at http://www.ftc.gov

Galama, Titus, and James Hosek. 2008. U.S. Competitiveness in Science and Technology. Rand Corporation. Available at http://www.rand.org

Gallup. 2000. Abortion Issues. Available at http://www.gallup.com/poll/indicators

Ganapati, Priya. 2010. "Anybots Robot Will Go to the Office for You." Wired Magazine, May 18. Available at http:///www.wired.com

Genetics and Public Policy Center. 2010. "Frequently Asked Questions." John Hopkins University, Berman Institute of Bioethics. Washington, DC.

Genomics Law Report. 2011. "Myriad Gene Patent Litigation." A Publication of Robinson Bradshaw and Hinson. Available at http://www.genomicslawreport.com

Gibbs, Nancy. 2003. "The Secret of Life." Time, February 17, pp. 42–45.

Goodman, Paul. 1993. "Can Technology Be Humane?" In Technology and the Future, Albert H. Teich, ed. (pp. 239–255). New York: St. Martin's Press.

Gottlieb, Scott. 2000. "Abortion Pill Is Approved for Sale in United States." *British Medical Journal* 321:851.

Grady, Denise. 2008. "Parents Torn over Fate of Frozen Embryos." *New York Times*, December 4. Available at http://www.nytimes.com

Greenhouse, Linda. 2007. "Justices Back Ban on Method of Abortion." *New York Times*, April 19. Available at http://www.nytimes.com

Greenhouse, Linda. 2006. "Justices Reaffirm Emergency Access to Abortion." *New York Times*, January 19. Available at http://www.nytimes.com

Grossman, Lev. 2006. "Time's Person of the Year: You." *Time*, December 13. Available at http://www.time.com

Guttmacher Institute. 2011 (July 1). "State Policies in Brief: Bans on 'Patiala Birth'' Abortion." Available at http://www.guttmacher.org

Hanson, Lawrence A. 2010 (November 7). "Animal Research: Groupthink in Both Camps." *The Chronicle of Higher Education.* Available at http://chronicle.com

Harmon, Amy. 2006. "That Wild Streak? Maybe It Runs in the Family." *New York Times*, June 15. Available at http://www.nytimes.com

Harris Interactive. 2011. "Trouble for Traditional Media—Both Print and Television." October 28. Available at http://www.harrisinteractive.com

Harris Poll. 2010 (June 8). "What We Love and Hate about America." Available at http://www.harrisinteractive.com/

Harris, Gardiner. 2009. "Some Stem Cell Research Limits Lifted." *The New York Times*, April 18. Available at http://www.nytimes.com

Heitmann, John. 2011 (July 21). "Big Three Weigh in on Online Privacy: FTC, FCC and NTIA Testify at Privacy Hearing." Available at http://www.telecomlawmonitor.com

Helft, Miguel. 2009. "At First, Funny Videos. Now, a Reference Tool." *The New York Times*, January 18. Available at http://www.nytimes.com

Henry, Bill, and Roarke Pulcino. 2009. "Individual Differences and Study-Specific Characteristics Influencing Attitudes about the Use of Animals in Medical Research." *Society and Animals* 17:305–324.

Holland, Earle. 2010 (November 7). "Animal Research: Activists' Wishful Thinking, Primitive Reasoning." *The Chronicle of Higher Education* Available at http://chronicle.com

Horrigan, John. 2009 (May 7). "Wireless Connectivity Has Drawn Many Users More Deeply into Digital Life." Washington, DC: Pew Internet and American Life Project.

Horrigan, John. (May 18). 2008. "The Internet and Consumer Choice." Washington, DC: Pew Internet and American Life Project.

Houlahan, Brent. 2006. (September 18) "Telepresence Defined by Brent Houlahan with HSL's Thoughts and Analysis." Human Productivity Lab. Available at http://www.humanproductivitylab.com/

Human Cloning Prohibition Act of 2007. U.S. House of Representatives, Washington, DC. Available at http://www.govtrack.us

Human Genome Project. 2011. "Genetic Disease Information." Available at http://www.ornl.gov

ITIF (Information Technology and Innovation Foundation). 2011 (July). "The Atlantic Century II." European-American Business Council (February). Available at http://www.itif.org

ITIF (Information Technology and Innovation Foundation). 2009. "Benchmarking EU & U.S. Innovation and Competitiveness." European-American Business Council (February). Available at http://www.itif.org

ICCVAM (Interagency Coordinating Committee on the Validation of Alternative Methods). 2011. "Since You Asked: Alternatives to Animal Testing." Available at http://www.niehs.nih.gov

IFR (International Federation of Robotics). 2011. "Continuing Growth of Robot Sales Worldwide." *Press Releases.* Available at http://www.ifr.org

Internet Freedom Preservation Act. 2009. Available at http://www.congress.org

Internet Statistics. 2011a. *The Internet Big Picture.* Available at http://www.internetworldstats.com

Internet Statistics. 2011b. "Top 20 Countries with the Highest Number of Internet Users." Available at http://www.internetworldstats.com

Jaschik, Scott. 2008. "If You Text in Class, This Prof Will Leave." *Inside Higher Ed,* April 2. Available at http://www.insidehighered.com

Jones, Jeffrey M. 2009 (July 17). "Majority of Americans Say Space Program Costs Justified." Gallup Poll. Available at http://www.gallup.com

Jordan, Tim, and Paul Taylor. 1998. "A Sociology of Hackers." *Sociological Review* 46(4):757–778.

Joseph, Binoy. 2010 (December). "Global Economic Crisis and Labour: Need for Broadening the Boundaries and Frontiers of Labour Laws." *Rajagiri Journal of Social Development* 2(1):25–42.

Kahn, A. 1997. "Clone Mammals . . . Clone Man?" *Nature* 386:119.

Kaiser Family Foundation, 2010. *Generation M²: Media Lives of 8- to 18-Year Olds*

Kaiser, Jocelyn. 2011. "More Briefs Coming in Stem Cell Lawsuit." *Science Magazine*, May 9. Available at http://news.sciencemag.org

Kalil, Thomas. 2009. "Letter to President-Elect Obama." Available at http://www.nsf.gov/nsb/publications/2009/01_10_stem_rec_obama.pdf

Kaplan, Karen. 2009. "Corn Fortified with Vitamins Devised by Scientists." *Los Angeles Times*, April 29. Available at http://www.latimes.com

Kaplan, Carl S. 2000. "The Year I Technology Law." *The New York Times,* December 22. Available at http://www.nytimes.com

Kharfen, Michael. 2006. "1 of 3 and 1 in 6 Pre-Teens Are Victims of Cyber-Bulling." Available at http://www.fightcrime.org

Kirkpatrick, David D. 2009. "Abortion Fight Complicates Debate on Health Care." *The New York Times*, September 29. Available at http://www.nytimes.com

Klotz, Joseph. 2004. *The Politics of Internet Communication.* Lanham, MD: Rowman and Littlefield.

Kopytoff, Verne G. 2011. "Dell's Earnings More Than Double on Sales of Computing Gear to Big Corporations." *The New York Times,* February 15. Available at http://www.nytimes.com

Kuhn, Thomas. 1973. *The Structure of Scientific Revolutions.* Chicago: University of Chicago Press.

Legay, F. 2001 (January). "Genetics: Should Genetic Information Be Treated Separately?" *Virtual Mentor*, p. E5. Available at http://virtualmentor.ama-assn.org

Lemonick, Michael. 2006. "Are We Losing Our Edge?" *Time,* February 13, pp. 22–33.

Lemonick, Michael, and Dick Thompson. 1999. "Racing to Map Our DNA." *Time Daily* 153: 1–6. Available at http://www.time.com

Lenhart, Amanda. 2008 (December 1). "Adults and Video Games." Available at http://www.pewinternet.org

Library of Congress. 2011. "Bill Summary and Status." Available at http://thomas.loc.gov

Lohr, Steve. 2011. "Carrots, Sticks and Digital Health Records." *The New York Times,* February 26. Available at http://www.nytimes.com

Lohr, Steve. 2009. "Wal-Mart Plans to Market Digital Health Records System." *The New York Times,* March 11. Available at http://www.nytimes.com

Lohr, Steve. 2008. "Health Care That Puts a Computer on the Team." *The New York Times,* December 27. Available at http://www.nytimes.com

Lohr, Steve. 2006. "Outsourcing Is Climbing Skills Ladder." *New York Times*, February 16. Available at http://www.nytimes.com

Lynch, Colum. 2005. "U.N. Backs Human Cloning Ban." *Washington Post,* March 8. Available at http://www.washingtonpost.com

Lynch, John, Jennifer Bevan, Paul Achter, Tim Harris, and Celeste M. Condit. 2008. "A Preliminary Study of How Multiple Exposures to Messages about Genetics Impact Lay Attitudes Towards Racial and Genetic Discrimination." *New Genetics and Society* 27(1):43–56.

MacMillan, Robert. 2005. "My Cubicle, My Cell." *Washington Post,* May 19. Available at http://www.washingtonpost.com

Magee, Mike. 2009 (July 17). "Apple, Microsoft, Others Sued over Touchpad Products." Available at http://www.tgdaily.com

Malamud, Ofer, and Cristian Pop-Eleches. 2010. "Home Computer Use and the Development of Human Capital." National Bureau of Economic Research Working Paper 15814. Available at http://www.nber.org

Markoff, John. 2008. "Thieves Winning Online War, Maybe Even in Your Computer." *The New York Times*, December 5. Available at http://www.nytimes.com

Marin, Rick. 2011. "A Man, His Son, and the Fight for Angry Birds." *The New York Times,* April 28. Available at http://www.nytimes.com

Mateescu, Oana. 2010. "Introduction: Life in the Web." *Journal of Comparative Research in Anthropology and Sociology* 1(2):1–21.

Mayer, Sue. 2002. "Are Gene Patents in the Public Interest?" *BIO-IT World*, November 12. Available at http://www.bio-itworld.com

McCarthy, Tom, and Scott Michels. 2009. "Lori Drew MySpace Suicide Hoax Conviction Thrown Out." ABC News, July 2. Available at http://abcnews.go.com

McCollum, Sean. 2011. "Getting Past the 'Digital Divide.'" Teaching Tolerance 39. Available at http://www.tolerance.org

McCormick, S. J., and A. Richard. 1994. "Blastomere Separation." Hastings Center Report, March-April, pp. 14–16.

McDermott, John. 1993. "Technology: The Opiate of the Intellectuals." In Technology and the Future, Albert H. Teich, ed. (pp. 89–107). New York: St. Martin's Press.

McCullagh, Declan. 2010. "FBI Wants Records Kept of Web Sites Visited." CNET News, February 5. Available at http://news.cnet.com

McFarling, Usha L. 1998. "Bioethicists Warn Human Cloning Will Be Difficult to Stop." Raleigh News and Observer, November 18, p. A5.

McGann, Rob. 2005. (January 6). "Most Active Web Users Are Young, Affluent." Jupiter Research. Available at http://www.clickz.com

McKinley, Jr., James C. 2010. "Strict Abortion Measures Enacted in Oklahoma." The New York Times, April 27. Available at http://www.nytimes.com

Meinert, Dori. 2011 (June 1). "Make Telecommuting Payoff." Human Resource Management 56(6):1. Available at http://www.shrm.org

Merton, Robert K. 1973. "The Normative Structure of Science." In The Sociology of Science, Robert K. Merton, ed. Chicago: University of Chicago Press.

Mesthene, Emmanuel G. 1993. "The Role of Technology in Society." In Technology and the Future, Albert H. Teich, ed. (pp. 73–88). New York: St. Martin's Press.

Michaels, Philip, and Jason Snell. 2010. "Apple Tallies Record Revenue on Mac, iPad, and iPhone Sales." MacWorld, July 20. Available at http://www.macworld.com

Mooney, Chris, and Sheril Kirshenbaum. 2009. Unscientific America: How Scientific Literacy Threatens Our Future. Philadelphia PA: Basic Books.

Murphie, Andrew, and John Potts. 2003. Culture and Technology. New York: Palgrave Macmillan.

National Sleep Foundation. 2011 (March 7). "Annual Sleep in America Poll Exploring Connections with Communications Technology Use and Sleep." Available at http://www.sleepfoundation.org

Nature News. 2011 (February 23). "Animal Research: Battle Scars." Nature 470 (7335):452–453.

NCSL (National Conference of State Legislatures). 2008. "Human Cloning Laws." Available at http://ncsl.org

Neuman, William, and Andrew Pollack. "Farmers Cope with Roundup Resistant Weeds." The New York Times, May 3. Available at http://www.nytimes.com

NIC (National Intelligence Council). 2003. "The Global Technology Revolution, Preface and Summary." Rand Corporation. Available at http://www.rand.org

Nie, Norman, Alberto Simpser, Irena Stepanikova, and Lu Zheng. 2004 (December). Ten Years after the Birth of the Internet: How Do Americans Use the Internet in Their Daily Lives? Stanford Institute for the Quantitative Study of Society.

NIEHS (National Institute of Environmental Health Sciences) 2009 (April 27). "Countries Unite to Reduce Animal Use in Product Toxicity Testing Worldwide" Press Release. Available at http://www.niehs.nih.gov

Nielsen. 2009a (March). "Social Networks and Blogs Now 4th Most Popular Online Activity, Ahead of Personal Email, Nielsen Reports." Available at http://www.nielsen.com

Nielsen. 2009b (February). "Twitter Posts Meteoric 1,382% YoY Growth." Available at http://www.marketingcharts.com

NIH (National Institute of Health). 2010. "Scientists Model Autism and Test Potential Treatments Using Human Induced Pluripotent Stem Cells." NIH Director's New Innovator Award, A.R. Muotri. Available at http://stem-cells.nih.gov

NSF (National Science Foundation). 2010. "Science and Engineering Indicators, 2010." Available at http://www.nsf.gov

NNI (National Nanotechnology Initiative). 2009. Available at http://www.nano.gov

Ogburn, William F. 1957. "Cultural Lag as Theory." Sociology and Social Research 41:167–174.

OpenNet. 2011. OpenNet Initiative: Country Profiles. Available at http://optenet.net

Ophir, Eyal, Charles Nass, and Anthony D. Wagner. 2009. "Cognitive Control in Media Multitaskers." Proceeds of the National Academy of Science 106(37):15583–15587.

Optenet. 2008. "2008 International Internet Trends Study." Available at http://www.optenet.com

ORNL (Oak Ridge National Laboratory). 2011. "Medicine and the New Genetics." Office of Biological and Environmental Research. Available at http://www.ornl.gov

Park, Alice. 2009. "Stem-Cell Research: The Quest Resumes." Time, January 29. Available at http://www.time.com

Pethokoukis, James M. 2004. "Meet Your New Co-Worker." U.S. News and World Report, March 7. http://www.usnews.com

Pew. 2011a (May 12). The Social Life of Health Information. Pew Internet and American Life Project. Available at http://pewinternet.org

Pew. 2011b (February 28). Peer-to-Peer Healthcare. Pew Internet and American Life Project. Available at http://pewinternet.org

Pew. 2011c (March 3). "Fewer Are Angry at Government, but Discontent Remains High." Section 3: Attitudes toward Social Issues. Available at http://people-press.org

Pew. 2010 (December 16). Generations 2010. Pew Internet and American Life Project. Available at http://pewinternet.org

Pew. 2009a (July 9). "Public Praises Science; Scientists Fault Public, Media Scientific Achievements Less Prominent Than a Decade Ago." Available at http://people-press.org/reports/pdf

Pew. 2009b (July 29). "56% of All Americans Have Accessed the Internet by Wireless Means." Available at http://www.pewinternet.org

Pew. 2009c (November 4). "Social Isolation and the New Technology." Available at http://www.pewresearch.org

Pew. 2007. Trends in Political Values and Core Attitudes: 1987–2007. Washington, DC: The Pew Research Center.

Pew. 2006. Awareness of Genetically Modified Food Has Declined over the Last Five Years. Available at http://pewagbiotech.org

Pollack, Andrew. 2011. "U.S. Approves Genetically Modified Alfalfa." The New York Times, January 27. Available at http://www.nytimes.com

Postman, Neil. 1992. Technopoly: The Surrender of Culture to Technology. New York: Alfred A. Knopf.

Power, Emma. 2008. "Furry Families: Making a Human-Dog Family through Home." Social and Cultural Geography 9(5):535–555.

Preston, Jennifer. 2011. "Facebook Officials Keep Quiet on Its Role in Revolts." The New York Times, February 14. Available at http://www.nytimes.com

Price, Tom 2008. "Science in America: Are We Falling Behind in Science and Technology?" CQ Researcher 18(2):24–48.

Rabino, Isaac. 1998. The Biotech Future. American Scientist 86(2):110–112.

Randewich, Noel. 2011 (July 20). "Intel Backs Off PC Market Outlook, Shares Slide." Available at http://old.news.yahoo.com

Research Animal Resources. 2003. "Ethics and Alternatives." University of Minnesota. Available at http://www.ahc.umn.edu

Richmond, Riva. 2010. "Malware Hits Computerized Industrial Equipment." The New York Times, September 24. Available at http://www.nytimes.com

Richtel, Matt. 2009. "Drivers and Legislators Dismiss Cellphone Risks." The New York Times, July 19. Available at http://www.nytimes.com

Rigby, Bill, and Richard Chang. 2011. "Novell Wins Appeal in Microsoft Antitrust Lawsuit." Reuters, May 3. Available at http://www.reuters.com

Rizzo, Alessandra, 2009. "RU-486 Abortion Drug to Be Allowed in Italy Despite Vatican's Protests." The Huffington Post, July 31. Available at http://www.huffingtonpost.com

Roe v. Wade. 1973. 410 U.S. 113.

Rovner, Julie, and Jenny Gold. 2009 (July 25). "Obama Lifts Limit on Funding Stem Cell Research." NPR. Available at http://www.npr.org

Saad, Lydia. 2010. "Four moral issues sharply divide Americans." May 26. Gallup Poll. Available at http://www.gallup.com

Sang-Hun, Choe, and John Markoff. 2009. "Cyberattacks Jam Government and Commercial Web Sites in U.S. and South Korea." The New York Times, July 9. Available at http://www.nytimes.com

Sarno, David. 2010. "Google Sues U.S. over Bidding for E-Mail Contract." Los Angeles Times, November 2. Available at http://articles.latimes.com

Schaefer, Naomi. 2001. "The Coming Internet Privacy Scrum." *American Enterprise* 12:50–51.

Semantic Media. 2011. "Semantic Media—Smart Media for the Semantic Web." Available at http://semanticmedia.org

Sharkey, Joe. 2009. "A Meeting in New York? Can't We Videoconference?" *The New York Times*, May 12. Available at http://www.nytimes.com

Singh, Susheela, Deirdre Wulf, Rubina Hussain, Akinrinola Bankole, and Gilda Sedgh. 2009. *Abortion Worldwide: A Decade of Uneven Progress*. New York, NY: Guttmacher Institute Available at http://www.guttmacher.org

Smeal, Eleanor. 2007. "Supreme Court Ruling in Abortion Ban Case Disastrous for Women's Health and Safety." Press Release, April 18. Available at http://www.feminist.org

Smith, Aaron. 2011 (March 17). "The Internet and Campaign 2010." Available at http://www.pewinternet.org

Smith, Aaron. 2010 (October 14). "Americans and Their Gadgets." Available at http://www.pewinternet.org

Smith, David. 2008. "Addiction to Internet 'Is An Illness.'" *The Observer*, March 23. Available at http://www.guardian.co.uk

Spaeth, Matt. 2011. "What You Need to Know about GMOs." *Food Integrity Now*, March 8. Available at http://www.foodintegritynow.org

Steinhauer, Jennifer. 2011. "Under Banner of Fiscal Restraint, Republicans Plan New Abortion Bills." *The New York Times*, February 8. Available at http://www.nytimes.com

Stelter, Brian, and Miguel Helft. 2009. "Deal Brings TV Shows and Movies to YouTube." *The New York Times*, April 17. Available at http://www.nytimes.com

Stem Cell Research Advancement Act. 2011. Available at http://www.gpo.gov

Sterns, Olivia. 2009. "Paralyzed from the Chest Down, Man Bikes 500 Miles for Stem Cell Research." *Huffington Post*, July 29. Available at http://www.huffingtonpost.com

Supreme Court of Wisconsin. 2010. *Schill et al. v. Wisconsin Rapids School District*. July 16. Case No. 2008AP967-AC. Available at http://www.wicourts.gov

Syncardia. 2011 (February 15). "1st U.S. Total Artificial Heart Patient Discharged Home Using Freedom Driver Receives Dual Transplant." Available at http://www.syncardia.com/

Thas, Angela, Chat Garcia Ramilo, and Cheekay Garcia Cinco. 2007. *Gender and ICT*. United Nations Development Programme. Bangkok, Thailand: Pacific Development Information Programme. Available at http://www.apdip.net

The Economist. 2009 (September 3). "Keeping Pirates at Bay." Available at http://www.economist.com

The Economist. 2008 (December 18). "Pocket World in Figures." Available at http://www.economist.com

Toffler, Alvin. 1970. *Future Shock*. New York: Random House.

UCL (University College London). 2008 (July 11). "Information Behaviour of the Researcher of the Future." Available at http://www.bl.uk/news/pdf/googlegen.pdf

United Nations Environmental Programme (UNEP). 2011 (May 12). "The International Treaty on Damage Resulting from Living Modified Organisms Receives Sixteen Signatures." Available at http://www.cbd.int

United Nations. 2010. "United Nations E-Government Survey 2010." Available at http://unpan1.un.org

U.S. Attorney's Office. 2011. "Tucson Man Sentenced to More than 17 Years in Prison for Distribution and Possession of Child Pornography." Federal Bureau of Investigation. *Press Release*, March 30. Available at http://www.fbi.gov

U.S. Bureau of the Census. 2011a. *Statistical Abstract of the United States, 2010*, 126th ed. Washington, DC: U.S. Government Printing Office.

U.S. Bureau of the Census. 2011b (January 14). "U.S. Census Bureau News." U.S. Department of Commerce. Available at http://www2.census.gov

U.S. Bureau of the Census. 2011c (May 26). "E-Stats." Available at http://www.census.gov

U.S. Bureau of the Census. 2009. *Statistical Abstract of the United States, 2008*, 126th ed. Washington, DC: U.S. Government Printing Office.

U.S. Citizenship and Immigration Services. 2011 (March 24). "H-1B Specialty Occupations, DOD Cooperative Research and Development Project Workers, and Fashion Models." Available at http://www.uscis.gov

U.S. Department of Commerce. 2010. *Exploring the Digital Nation: Home Broadband Internet Adoption in the United States*. November. Available at http://www.esa.doc.gov

U.S. Department of Justice. 2011 (April 22). "Joint Status Report on Microsoft's Compliance with the Final Judgment." *Document 927*. Available at http://www.justice.gov

U.S. Human Genome Project. 2007. *Medicine and the New Genetics*. Available at http://www.ornl.gov

U.S. Newswire. 2003 (October 2). *Feminists Condemn House Passage of Deceptive Abortion Ban, Urge Activists to March on Washington*. Available at http://www.usnewswire.com

Vance, Ashlee. 2009a. "Microsoft Profit Falls for First Time in 23 Years." *The New York Times*, April 24. Available at http://www.nytimes.com

Vance, Ashlee. 2009b. "Microsoft Mapping Course to a Jetsons-Style Future." *The New York Times*, March 2. Available at http://www.nytimes.com

Wakefield, Jane. 2010. "World Wakes Up to Digital Divide." *BBC New*, March 19. Available at http://news.bbc.co.uk

Wayne, Teddy. 2010. "Age Gap Narrows on Social Networks." *The New York Times*, December 26. Available at http://www.nytimes.com

Weinberg, Alvin. 1966. "Can Technology Replace Social Engineering?" *University of Chicago Magazine* 59(October):6–10.

Welsh, Jonathan. 2009. "Late on a Car Loan? Meet the Disabler." *The Wall Street Journal*, March 25. Available at http://www.online.wsj.com

Welter, Cole H. 1997. "Technological Segregation: A Peek through the Looking Glass at the Rich and Poor in an Information Age." *Arts Education Policy Review* 99(2):1–6.

White, D. Steven, Angappa Gunasekaran, Timothy P. Shea, and Godwin C. Ariguzo. 2011. "Mapping the Global Digital Divide." *International Journal of Business Information Systems* 7:207–219.

White House. 2003 (November 5). *President Bush Signs Partial Birth Abortion Ban Act of 2003*. Available at http://www.whitehouse.gov

Wikileaks. 2011. "Keep Us Strong." Available at http://www.wikileaks.org

Widman, Jake. 2011. "10 Massive Security Breaches." *Information Week*, March 12. Available at http://www.informationweek.com

Willing, Richard. 2005. "Federal Judge Blocks 'Partial Birth Abortion' Ban." *USA Today*, June 1. Available at http://www.usatoday.com

Wilson, Robin. 2011 (February 23). "90% of Biomedical Scientists in Poll Say Animals Are 'Essential' in Research." *The Chronicle of Higher Education*. Available at http://chronicle.com

Winner, Langdon. 1993. "Artifact/Ideas as Political Culture." In *Technology and the Future*, Albert H. Teich, ed. (pp. 283–294). New York: St. Martin's Press.

Wisconsin Supreme Court. 2010. *Schill et al. v. Wisconsin Rapids School District*. July 16. Case No. 2008AP967-AC. Available at http://www.wicourts.gov

Wolf, Gary. 2010. "The Data-Driven Life." *The New York Times*, April 28. Available at http://www.nytimes.com

World at Work. 2009. "Telework Trendlines 2009." Available at www.worldatwork.org

World Bank. 2009. *Global Economic Prospects: Technology Diffusion in the Developing World*. Washington DC: The World Bank.

World Hunger Education Service. 2011. "2011 World Hunger and Poverty Facts and Statistics." Available at http://www.worldhunger.org

Wortham, Jenna. 2010. "Facebook Tops 500 Million Users." *The New York Times*. July 21. Available at http://www.nytimes.com

Young, Kimberly, and Cristiano N. de Abreu. (2011). *Internet Addiction: A Handbook and Guide to Evaluation and Treatment*. Hoboken, NJ: John Wiley and Sons.

Chapter 15

Alexander, Karen, and Mary E. Hawkesworth, eds. 2008. *War and Terror: Feminist Perspectives*. Chicago: University of Chicago Press.

Alvarez, Lizette. 2009a. "G.I. Jane Breaks the Combat Barrier." *The New York Times*, August 15. Available at http://www.nytimes.com

Alvarez, Lizette. 2009b. "Suicides of Soldiers Reach High of Nearly 3 Decades." *New York Times*, January 29. Available at http://www.nytimes.com

American Jewish Committee. 2010. *2009 Annual Survey of American Jewish Opinion*. Available at http://www.ajc.org

Animals at Arms. 2010. "Animals at Arms." *Army-Technology*, December 22. Available at http://www.army-technology.com

Annan, Jeannie, Christopher Blattman, Khristopher Carlson, and Dyan Mazurana. 2008. "Findings from the Survey of War-Affected Youth (SWAY) Phase II." Available at https://wikis.uit.tufts.edu/confluence/display

Arms Control Association. 2003 (May). *The Nuclear Proliferation Treaty at a Glance.* Available at http://www.armscontrol.org/factsheets

Associated Press. 2005 (March 15). *Pentagon Tests Negative for Anthrax.* MSNBC. Available at http://www.msnbc.msn.com

Associated Press. 2008 (August 6). "U.S. Officials: Scientist Was Anthrax Killer." MSNBC. Available at http://www.msnbc.msn.com

Associated Press. 2009 (December 1). "Obama Details Afghan War Plan, Troop Increases." MSNBC. Available at http://www.msnbc.msn.com

Atomic Archive. 2011. *Arms Control Treaties.* Available at http://www.atomicarchive.com

Baker, Peter, Helene Cooper, and Mark Mazzetti. 2011. "Bin Laden Is Dead, Obama Says." *The New York Times,* May 1. Available at http://www.nytimes.com

Baliunas, Sallie. 2004. "Anthrax Is a Serious Threat." In *Biological Warfare,* William Dudley, ed. (pp. 53–58). Farmington Hills, MA: Greenhaven Press.

Barkan, Steven, and Lynne Snowden. 2001. *Collective Violence.* Boston: Allyn and Bacon.

Barnes, Steve. 2004. "No Cameras in Bombing Trial." *New York Times,* January 29, p. 24.

Barstow, David. 2008. "One Man's Military-Industrial-Media Complex." *New York Times,* November 29. Available at http://www.nytimes.com

BBC. 2000. "UN Admits Rwanda Genocide Failure." BBC News, April 15. Available at http://news.bbc.co.uk

BBC. 2007a. "Hamas Takes Full Control of Gaza." BBC News, June 15. Available at http://news.bbc.co.uk

BBC. 2007b. "Fresh Clashes Engulf Lebanon Camp." BBC News, June 1. Available at http://news.bbc.co.uk

BBC. 2009. "Gaza Crisis: Key Maps & Timelines." BBC News, January 18. Available at http://news.bbc.co.uk

BBC. 2011 (September 23). "Israeli-Palestinian Talks Must Resume—Mideast Quartet." Available at http://www.bbc.co.uk

Belasco, Amy. 2011. The Cost of Iraq, Afghanistan, and Other Global War on Terror Operations Since 9/11. Congressional Research Service, March 29. Available at http://www.fas.org

Bercovitch, Jacob, ed. 2003. *Studies in International Mediation: Advances in Foreign Policy Analysis.* New York: Palgrave Macmillan.

Bergen, Peter. 2002. *Holy War, Inc.: Inside the Secret World of Osama bin Laden.* New York: Free Press.

Berrigan, Frida. 2009. "We Arm the World." *In These Times,* January 2. Available at http://www.inthesetimes.com

Berrigan, Frida, and William Hartung. 2005. "U.S. Weapons at War: Promoting Freedom or Fueling Conflict?" *World Policy Institute Report.* Available at http://www.worldpolicy.org/projects

Billitteri, Thomas J. 2010. "Drone Warfare: Are Strikes by Unmanned Aircraft Ethical?" *CQ Researcher* 20 (028):653–676.

Bjorgo, Tore. 2003. *Root Causes of Terrorism.* Paper presented at the International Expert Meeting, June 9–11. Oslo, Norway: Norwegian Institute of International Affairs.

Blum, Gabriella, and Philip Heymann. 2010 (June 27). "Law and the Policy of Targeted Killing." *Harvard National Security Journal* 1:145–170.

Borum, Randy. 2003. "Understanding the Terrorist Mind-Set." *FBI Law Enforcement Bulletin,* July, pp. 7–10.

Boustany, Nora. 2007. "Janjaweed Using Rape as 'Integral' Weapon in Darfur, Aid Group Says." *Washington Post,* July 3. Available at http://www.washingtonpost.com

Brauer, Jurgen. 2003. "On the Economics of Terrorism." *Phi Kappa Phi Forum,* Spring, pp. 38–41.

Braun, Stephen. 2011. "U.S. Defense Sales to Bahrain Rose Before Crackdown." ABC News, June 11. Available at http://abcnews.go.com

Brinkley, Joel, and David E. Sanger. 2005. "North Koreans Agree to Resume Nuclear Talks." *New York Times,* July 10. Available at http://www.nytimes.com

Broad, William J., and David E. Sanger. 2009 (October 3). "Report Says Iran Has Data to Make a Nuclear Bomb." Available at http://www.nytimes.com

Brown, Michael E., Sean M. Lynn-Jones, and Steven E. Miller, eds. 1996. *Debating the Democratic Peace.* Cambridge, MA: MIT Press.

Bumiller, Elisabeth. 2011. "The Dogs of War: Beloved Comrades in Afghanistan." *The New York Times,* May 11. Available at http://www.nytimes.com

Buncombe, Andrew. 2009. "End of Sri Lanka's Civil War Brings Back Tourists." *The Independent,* August 16. Available at http://www.independent.co.uk

Burns, Robert. 2011. "Panetta: U.S. within Reach of Defeating al Qaeda." *The Washington Times,* July 9. Available at http://www.washingtontimes.com

Carnegie Endowment for International Peace. 2009. *World Nuclear Arsenals 2009.* Available at http://www.carengieendowment.org

Carneiro, Robert L. 1994. "War and Peace: Alternating Realities in Human History." In *Studying War: Anthropological Perspectives,* S. P. Reyna and R. E. Downs, eds. (pp. 3–27). Langhorne, PA: Gordon & Breach.

Cary, Peter. 2010 (October 19). *The Pentagon, Informational Operations and Media Development.* Center for International Media Assistance. Available for http://ics.leeds.ac.uk

Center for Defense Information. 2003. *Military Almanac.* Available at http://www.cdi.org

CNN. 2001 (January 31). *Libyan Bomber Sentenced to Life.* Available at http://www.europe.cnn.com

CNN. 2003a (August 10). *Army Begins Chemical Weapons Burn.* Available at http://www.cnn.com

CNN. 2003b (February 26). *Poll: Muslims Call U.S. Ruthless, Arrogant.* Available at http://www.cnn.com

CNN. 2009 (December 31). *Charges Dismissed against Iraq Contractors.* Available at http://www.cnn.com

CNN/Opinion Research Corporation Poll. 2007 (June 22–24). *Iraq.* Available at http://www.pollingreport.com/iraq3.htm

Cohen, Ronald. 1986. "War and Peace Proneness in Pre- and Post-industrial States." In *Peace and War: Cross-Cultural Perspectives,* M. L. Foster and R. A. Rubinstein, eds. (pp. 253–267). New Brunswick, NJ: Transaction Books.

Cohn, Marjorie. 2011. "The Targeted Assassination of Osama Bin Laden." *Common Dreams,* May 10. Available at http://www.commondreams.org

Conflict Research Consortium. 2003. *Mediation.* Available at http://www.colorado.edu/conflict/peace

Cowell, Alan, and A. G. Sulzberger. 2009. "Lockerbie Convict Returns to Jubilant Welcome." *The New York Times,* August 20. Available at http://www.nytimes.com

Dao, James. 2009. "Veterans Affairs Faces Surge of Disability Claims." *New York Times,* July 12. Available at http://www.nytimes.com

Dareini, Ali Akbar. 2009. "Iran Missile Test: Ahmadinejad Says It's within Israel's Range." *Huffington Post,* May 20. Available at http://www.huffingtonpost.com

Davenport, Christian. 2005. "Guard's New Pitch: Fighting Words." *Washington Post,* April 28. Available at http://www.washingtonpost.com

Davies, Nick, and David Leigh. 2010. "Afghanistan War Logs: Massive Leak of Secret Files Exposes Truth of Occupation." *The Guardian,* July 25. Available at http://www.guardian.co.uk

Dedman, Bill. 2006 (October 24). *In Limbo: Cases Are Few against Gitmo Detainees.* Available at http://www.msnbc.msn.com

Deen, Thalif. 2000 (September 9). *Inequality Primary Cause of Wars, Says Annan.* Available at http://www.hartford-hwp.com/archives

Defense News. 2011. "Major Cyber Attack is Act of War: Pentagon Repot." *Defense News,* May 31. Available at http://www.defensenews.com

DeYoung, Karen, and Walter Pincus. 2008. (October 3). *Iraq: U.S. to Fund Pro-American Publicity in Iraqi Media.* Available at http://www.corpwatch.org

Dixon, William J. 1994. "Democracy and the Peaceful Settlement of International Conflict." *American Political Science Review* 88(1):14–32.

Donnelly, John. 2011 (January 24). "More Troops Lost to Suicide." Available at http://www.congress.org

Dwyer, Devin. 2011. "Afghanistan War Costs Loom Large Over Obama Troops Announcement." ABC News, June 22. Available at http://abcnews.go.com

Environmental Media Services. 2002 (October 7). *Environmental Impacts of War.* Available at http://www.ems.org

Erlanger, Steven. 2011. "Talks on Iran's Nuclear Program Close With No Progress." *New York Times*, January 22. Available at http://www.nytimes.com

Feder, Don. 2003. "Islamic Beliefs Led to the Attack on America." In *The Terrorist Attack on America*, Mary E. Williams, ed. (pp. 20–23). Farmington Hills, MA: Greenhaven Press.

Frankel, Rebecca. 2011a. "War Dog: There's a Reason Why They Brought One to Get Osama Bin Laden." *Foreign Policy*, May 4. Available at http://www.foreignpolicy.com/articles

Frankel, Rebecca. 2011b. "War Dogs: The Legend of the Bin Laden Hunter Continues." *Foreign Policy*, May 12. Available at http://www.foreignpolicy.com

Frey, Josh. 2003 (August 12). "Anti-Swimmer Dolphins Ready to Defend Gulf." Available at http://www.navy.mil

Friedman, Emily, and Richard Esposito. 2009. "Fort Hood Gunman Who Killed 12, Wounded 30 Survived Gun Battle." ABC News, November 5. Available at http://abcnews.go.com

Funke, Odelia. 1994. "National Security and the Environment." In *Environmental Policy in the 1990s: Toward a New Agenda*, 2nd ed., Norman J. Vig and Michael E. Kraft, eds. (pp. 323–345). Washington, DC: Congressional Quarterly.

Gamel, Kim, and Lee Keath. 2011. "Muammar Gaddafi Dead: Libya Dictator Maddened West, Captured, Killed in Sirte." *The Huffington Post*, October 20. Available at http://www.huffingtonpost.com

Gardner, Simon. 2007 (February 22). *Sri Lanka Says Sinks Rebel Boats on Truce Anniversary.* Available at http://www.reuters.com

Garrett, Laurie. 2001. "The Nightmare of Bioterrorism." *Foreign Affairs* 80:76.

German, Erik. 2011. "Flipper Goes to War." *The Daily*, July 18. Available at http://www.thedaily.com

Gettleman, Jeffrey. 2008. "Rape Victims' Words Help Jolt Congo into Change." *New York Times*, October 18. Available at http://www.nytimes.com

Gettleman, Jeffrey. 2009. "Symbol of Unhealed Congo—Male Rape Victims." *New York Times*, August 4. Available at http://www.nytimes.com

Geneva Graduate Institute for International Studies. 2006. *Small Arms Survey, 2006.* New York: Oxford University Press.

Gioseffi, Daniela. 1993. "Introduction." In *On Prejudice: A Global Perspective*, Daniela Gioseffi, ed. (pp. xi–l). New York: Anchor Books, Doubleday.

Goodwin, Jeff. 2006. "A Theory of Categorical Terrorism." *Social Forces* 84(4): 2027–2046.

Goure, Don. 2003. "First Casualties? NATO, the U.N." MSNBC News, March 20. Available at http://www.msnbc.com/news

Greenburg, Jan Crawford, and Ariane de Vogue. 2008. "Supreme Court: Guantanamo Detainees Have Rights in Court." ABC News, June 12. Available at http://abcnews.go.com

GreenKarat. 2007. *Mining for Gems.* Available at http://www.greenkarat.com

Hefling, Kimberly. 2011. "Former Marine, Advocate Kills Self after War Tour." 2011. *The Jacksonville Daily News*, April 15. Available at http://www.jdnews.com

Herszenhorn, David. 2008. "Estimates of Iraq War Cost Were Not Close to Ballpark." *New York Times*, March 19. Available at http://www.nytimes.com

Hewitt, J. Joseph, Jonathan Wilkenfeld, and Ted Robert Gurr. 2010. *Peace and Conflict 2010: Executive Summary.* College Park, MD: Center for International Development and Conflict Management.

Hooks, Gregory, and Leonard E. Bloomquist. 1992. "The Legacy of World War II for Regional Growth and Decline: The Effects of Wartime Investments on U.S. Manufacturing, 1947–72." *Social Forces* 71(2):303–337.

Horsley, Scott, and Kelly McEvers. 2011. "President Obama: All Troops Out of Iraq by Dec. 31." National Public Radio, October 21. Available at http://www.npr.org

Howard, Michael. 2007. "Children of War: The Generation Traumatised by Violence in Iraq." *The Guardian*, February 6. Available at http://www.guardian.co.uk

Huntington, Samuel. 1996. *The Clash of Civilizations and the Remaking of World Order.* New York: Simon and Schuster.

Inglesby, Thomas, Donald Henderson, John Bartlett, Michael Archer, et al. 1999. "Anthrax as a Biological Weapon: Medical and Public Health Management." *Journal of the American Medical Association* 281:1735–1745.

ICBL (International Campaign to Ban Landmines). 2011. *Making the Treaties Universal.* Available at http://www.icbl.org

Iraq Coalition Casualty Count. 2011. *Iraq Coalition Military Fatalities By Year.* Available at http://icasualties.org

Jeon, Arthur. 2011. "German Shepherd? Belgian Malinois? Navy SEAL Hero Dog Is Top Secret." *Global Animal*, May 5. Available at http://www.globalanimal.org

Johnston, David, and John Broder. 2007. "F.B.I. Says Guards Killed 14 Iraqis without Cause." *The New York Times*, November 14. Available at http://www.nytimes.com

Johnson, David, and Scott Shane. 2009. "U.S. Monitored Fort Hood Suspect Before Shooting." *The New York Times*, November 10. Available at http://www.nytimes.com

Kemper, Bob. 2003. "Agency Wages Media Battle." *Chicago Tribune*, April 7. Available at http://www.chicagotribune.com

Kerr, Paul K., and Mary Beth Nikitin. 2009. *Pakistan's Nuclear Weapons: Proliferation and Security Issues.* Congressional Research Service, June 12. Available at http://www.opencrs.com

Klare, Michael. 2001. *Resource Wars: The New Landscape of Global Conflict.* New York: Metropolitan Books.

Knickerbocker, Brad. 2002. "Return of the Military-Industrial Complex?" *Christian Science Monitor*, February 13. Available at http://www.csmonitor.com

Lamont, Beth. 2001. "The New Mandate for UN Peacekeeping." *The Humanist* 61:39–41.

Langley, Robert. n.d. "Chemical Sniffing Dogs Deployed along Borders." Available at http://usgovinfo.about.com

Laqueur, Walter. 2006. "The Terrorism to Come." In *Annual Editions 05–06*, Kurt Finsterbusch, ed. (pp. 169–176). Dubuque, IA: McGraw-Hill/Dushkin.

Large, Tim. 2006. *Crisis Profile: What's Going on in Northern Uganda?* Thomas Reuters Foundation. Available at http://www.alertnet.org

Larsen, Kaj. 2011. "Harnessing the Military Power of Animal Intelligence." CNN, July 31. Available at http://articles.cnn.com

Lederer, Edith. 2005. "Annan Lays out Sweeping Changes to U.N." Associated Press, May 20. Available at http://www.apnews.com

Leigh, David. 2010. "Iraq War Logs Reveal 15,000 Previously Unlisted Civilian Deaths." *The Guardian*, October 22. Available at http://www.guardian.co.uk

Leigh, David, James Ball, Ian Cobain, and James Burke. 2011. "Guantanamo Leaks Lift Lid on World's most Controversial Prison. April 24. *The Guardian*. Available at http://www.guardian.co.uk

Leinwand, Donna. 2003. "Sea Lions Called to Duty in Persian Gulf." *USA Today*, February 16. Available at http://www.usatoday.com

Levy, Jack S. 2001. "Theories of Interstate and Intrastate War: A Levels of Analysis Approach." In *Turbulent Peace: The Challenges of Managing International Conflict*, Chester A. Crocker, Fen Osler Hampson, and Pamela Aall, eds. (pp. 3–27). Washington, DC: U.S. Institute of Peace.

Leland, Anne, and Mari-Jana Oboroceanu. 2010 (February 26). *American War and Military Operations Casualties: Lists and Statistics.* Washington, DC: Congressional Research Service. Available at http://www.fas.org

Levy, Clifford J., and Peter Baker. 2009. "U.S.–Russian Nuclear Agreement Is First Step in Broad Effort." *The Washington Post*, July 6. Available at http://www.washingtonpost.com

Li, Xigen, and Ralph Izard. 2003. "Media in a Crisis Situation Involving National Interest: A Content Analysis of Major U.S. Newspapers' and TV Networks' Coverage of the 9/11 Tragedy." *Newspaper Research Journal* 24:1–16.

Lindner, Andrew M. 2009. "Among the Troops: Seeing the Iraq War through Three Journalistic Vantage Points." *Social Problems* 56(1):21–48.

MacAskill, Ewen, Ed Pilkington, and Jon Watts. 2006. "Despair at UN over Selection of 'Faceless' Ban Ki-moon as General Secretary." *The Guardian*, October 7. Available at http://www.guardian.co.uk

Macfarquhar, Neil. 2010. "Change Will Not Come Easily to the Security Council." *The New York Times*, November 8. Available at http://www.nytimes.com

Marshall, Leon. 2007 (July 16). *Elephants "Learn" to Avoid Land Mines in War-Torn Angola.* National Geographic News. Available at http://www.nationalgeographic.com

Mazzetti, Mark, and Scott Shane. 2009a. "Interrogation Memos Detail Harsh Tactics by the C.I.A." *New York Times*, April 16. Available at http://www.nytimes.com

Mazzetti, Mark, and Scott Shane. 2009b. "C.I.A. Abuse Cases Detailed in Report on Detainees." *The New York Times*, August 24. Available at http://www.nytimes.com

McEvers, Kelly. 2011. "Iraqi Leader Reconsiders U.S. Troop Withdrawal." National Public Radio, May 29. Available at http://www.npr.org

McGirk, Tim. 2006. "Collateral Damage or Civilian Massacre in Haditha?" *Time*, March 19. Available at http://www.time.com

Military. 2007. *Tuition Assistance (TA) Program Overview*. Available at http://education.military.com

Military Leadership Diversity Commission. 2011 (March). *From Representation to Inclusion. Diversity Leadership for the 21st Century Military*. Final Report. Available at http://mldc.whs.mil/

Military Professional Resources Inc. 2011. *Info-Center: Brochures*. Available at http://www.mpri.com

Miller, Susan. 1993. "A Human Horror Story." *Newsweek*, December 27, p. 17.

Miller, T. Christian. 2007. "Private Contractors Outnumber U.S. Troops in Iraq." *Los Angeles Times*, July 4. Available at http://www.latimes.com

Millman, Jason. 2008. "Industry Applauds New Dual-Use Rule." *Hartford Business Journal*, September 29. Available at http://www.hartfordbusiness.com

Mogelson, Luke. 2011. "A Beast in the Heart of Every Fighting Man." April 27. *New York Times*. Available at http://www.nytimes.com

Montalván, Luis Carlos. 2011. *Until Tuesday: A Wounded Warrior and the Golden Retriever Who Saved Him*. New York: Hyperion.

Morales, Lymari. 2009. (July 2). *Americans' Worry about Terrorism Nears 5-Year Low*. Available at http://www.gallup.com

Morgan, Jenny, and Alic Behrendt. 2009. *Silent Suffering: The Psychological Impact of War, HIV, and Other High-Risk Situations on Girls and Boys in West and Central Africa*. Woking: Plan. Available at http://plan-international.org

Moroney, Jennifer D. P., Joe Hogler, Benjamin Bahney, Kim Cragin, David R. Howell, Charlotte Lynch, and Rebecca Zimmerman. 2009. "Building Partner Capacity to Combat Weapons of Mass Destruction." Rand National Defense Research Institute. Available at http://www.rand.org

Mueller, John. 2006. *Overblown: How Politicians and the Terrorism Industry Inflate National Security Threats, and Why We Believe Them*. New York: Free Press.

Myers, Steven Lee. 2009. "Women at Arms: Living and Fighting alongside Men, and Fitting In." *The New York Times*, August 16. Available at http://www.nytimes.com

Myers-Brown, Karen, Kathleen Walker, and Judith A. Myers-Walls. 2000. "Children's Reactions to International Conflict: A Cross-Cultural Analysis." Paper presented at the National Council of Family Relations, Minneapolis, November 20.

NASP (National Association of School Psychologists). 2003. *Children and Fear of War and Terrorism*. Available at http://www.nasponline.org

NCPSD (National Center for Posttraumatic Stress Disorder). 2007. *What Is Post Traumatic Stress Disorder?* Available at http://www.ncptsd.va.gov

NCTC (National Counterterrorism Center). 2011 (April 30). *2010 Report on Terrorism*. Available at http://www.nctc.gov

National Priorities Project. 2011. *Cost of War: Trade Offs*. Available at http://costofwar.com

NPR (National Public Radio). 2009. "The Navy's Other Seals . . . Dolphins." National Public Radio, December 5. Available http://www.npr.org

Newport, Frank. 2011a (February 15). *Americans Remain Divided on Defense Spending*. Available at http://www.gallup.com

Newport, Frank. 2011b (March 9). *Republicans and Democrats Disagree on Muslim Hearings*. Available at http://www.gallup.com

NTI (Nuclear Threat Initiative). 2011. *North Korea Profile*. Available at http://www.nti.org

Ochmanek, David, and Lowell H. Schwartz. 2008. "The Challenge of Nuclear-Armed Regional Adversaries." Rand Project Air Force. Available at http://www.rand.org

Office of the Coordinator for Counterterrorism. 2011 (May 19). "Foreign Terrorist Organizations." Available at http://www.state.gov/s/ct

Office of Management and Budget. 2011. *Table 32-1: Policy Budget Authority and Outlays by Function, Category, and Program*. Available at http://www.gpoaccess.gov/usbudget/fy12

Office of Weapons Removal and Abatement. 2010. *To Walk the Earth in Safety: The United States' Commitment to Humanitarian Mine Action and Conventional Weapons Destruction*. Available at http://www.state.gov

Oliver, Amy. 2011. "Mammals with a Porpoise . . . Meet the Dolphins and Sea Lions Who Go to War with the U.S. Navy." *The Daily Mail*, July 6. Available at http://www.dailymail.co.uk

Paul, Annie Murphy. 1998. "Psychology's Own Peace Corps." *Psychology Today* 31:56–60.

Perlo-Freeman, Sam, Julian Cooper, Olawale Ismail, Elisabeth Skons, and Carina Solmirano. 2011. *Chapter 4: Military Expenditure*. Available at http://www.sipri.org

Permanent Court of Arbitration. 2011. *About Us* and *Cases*. Available at http://www.pca-cpa.org

Pew Research Center. 2011a (April 3). *A Nation of Flag Wavers*. Available at http://www.pewresearch.org

Pew Research Center. 2011b (May 17). *Obama's Challenge in the Muslim World: Arab Spring Fails To Improve U.S. Image*. Pet Global Attitudes Project. Available at http://www.pewresearch.org

Pew Research Center for the People & the Press. 2010 (August 24). "Public Remains Conflicted over Islam." Available at http://www.people-press-org

Pew Research Center for the People & the Press. 2009 (April 24). *Public Remains Divided over Use of Torture*. Available at http://www.people-press.org

Porter, Bruce D. 1994. *War and the Rise of the State: The Military Foundations of Modern Politics*. New York: Free Press.

Powell, Bill, and Tim McGirk. 2005. "The Man Who Sold the Bomb." *Time*, February 14, pp. 22–31.

Price, Eluned. 2004. "They Served and Suffered for Us." *The Telegraph*, November 1. Available at http://www.telegraph.co.uk

Priest, Dana, and William M. Arkin. 2010. "Top Secret America: A Hidden World, Growing Beyond Control." *The Washington Post*, July 19. Available at http://www.washingtonpost.com

Puckett, Neal, and Haytham Faraj. 2009 (June 25). *Navy-Marine Corps Court Hears Appeal in Wuterich v. U.S.* Available at http://www.puckettfaraj.com

Rashid, Ahmed. 2000. *Taliban: Militant Islam, Oil, and Fundamentalism in Central Asia*. New Haven, CT: Yale University Press.

Rasler, Karen, and William R. Thompson. 2005. *Puzzles of the Democratic Peace: Theory, Geopolitics, and the Transformation of World Politics*. New York: Palgrave Macmillan.

Renner, Michael. 2000. "Number of Wars on Upswing." In *Vital Signs: The Environmental Trends That Are Shaping Our Future*, Linda Starke, ed. (pp. 110–111). New York: W. W. Norton.

Renner, Michael. 1993. "Environmental Dimensions of Disarmament and Conversion." In *Real Security: Converting the Defense Economy and Building Peace*, Karl Cassady and Gregory A. Bischak, eds. (pp. 88–132). Albany: State University of New York Press.

Reuters. 2008. "Case Dropped against Officer Accused in Iraq Killings." *New York Times*, June 18. Available at http://www.nytimes.com

RIA Novosti. 2010 (November 28). "Russia Opens New Chemical Weapons Destruction Plant." Available at http://en.rian.ru

Rizzo, Jennifer. 2010. "Veteran's Affairs Faces Daunting Job of Reducing Medical Claims Backlog." CNN, December 20. Available at http://www.cnn.com

Romero, Anthony. 2003. "Civil Liberties Should Not Be Restricted During Wartime." In *The Terrorist Attack on America*, Mary Williams, ed. (pp. 27–34). Farmington Hills, MA: Greenhaven Press.

Rosenberg, Tina. 2000. "The Unbearable Memories of a U.N. Peacekeeper." *New York Times*, October 8, pp. 4, 14.

Roughton, Randy. 2011. "Fallen Marine's Family Adopts His Best Friend." February 3. U.S. Air Force Official Website, Available at http://www.af.mil/news/story

Ryan, Jason, and Huma Khan. 2011. "In Reversal, Obama Orders Guantanamo Military Trial for 9/11 Mastermind Khalid Sheikh Mohammed." ABC News, April 4. Available at http://www.abcnews.go.com

Saad, Lydia. 2010 (January 13). *U.S. Fear of Terrorism Steady after Christmas Attack*. Available at http://poll.gallup.com

Saad, Lydia. 2011 (May 4). *Majority in U.S. Say Bin Laden's Death Makes America Safer*. Available at http://www.gallup.com

Salem, Paul. 2007. "Dealing with Iran's Rapid Rise in Regional Influence." *The Japan Times*, February 22. Available at http://www.carnegieendowment.org

Salladay, Robert. 2003 (April 7). *Anti-War Patriots Find They Need to Reclaim Words, Symbols, Even U.S. Flag from Conservatives*. Available at http://www.commondreams.org

Sang-Hun, Choe, and John Markoff. 2009. "Cyberattacks Jam Government and Commercial Web Sites in U.S. and South Korea." *The New York Times*, July 8. Available at http://www.nytimes.com

Sanok, Stephanie, and Nathan Freier. 2010. "The End of Operation Iraqi Freedom and DoD's Future in Iraq." Center for Strategic and International Studies, September 1. Available at http://csis.org/publication

Savage, Charlie. 2009. "Accused 9/11 Mastermind to Face Civilian Trial in N.Y." *New York Times*, November 13. Available at http://www.nytimes.com

Savage, Charlie, William Glaberson, and Andrew W. Lehren. 2011. "Classified Files Offer New Insights Into Detainees." *The New York Times*, April 24. Available at http://www.nytimes.com

Save the Children. 2009. *Last in Line, Last in School 2009*. Available at http://www.savethechildren.org

Save the Children. 2007. *State of the World's Mothers: Saving the Lives of Children Under Five*. Available at http://www.savethechildren.org

Save the Children. 2005. *One World, One Wish: The Campaign to Help Children and Women Affected by War*. Available at http://www.savethechildren.org

Scheff, Thomas. 1994. *Bloody Revenge*. Boulder, CO: Westview Press.

Schmitt, Eric. 2005. "Pentagon Seeks to Shut Down Bases Across Nation." *New York Times*, May 14. Available at http://www.nytimes.com

Schneier, Bruce. 2008. "America's Dilemma: Close Security Holes, or Exploit Them Ourselves." *Wired*, May 1. Available at http://www.wired.com

Schroeder, Matt. 2007. "The Illicit Arms Trade." Washington, DC: Federation of American Scientists. Available at http://www.fas.org

Schwartz, Moshe, and Joyprada Swain. 2011. *Department of Defense Contractors in Afghanistan and Iraq: Background and Analysis*. Washington, DC: Congressional Research Service. Available at http://www.fas.org

Schultz, George P., William J. Perry, Henry A. Kissinger, and Sam Nunn. 2007. "A World Free of Nuclear Weapons." *Wall Street Journal*, January 4. Available at http://www.wsj.com

Sciutto, Jim. 2011. "Obama Orders Start to U.S. Troop Withdrawal from Afghanistan." ABC News, June 22. Available at http://abcnews.go.com

Sengupta, Somini. 2009. "Dossier Gives Details of Mumbai Attacks." *New York Times*, January 6. Available at http://www.nytimes.com

Shanahan, John J. 1995. "Director's Letter." *Defense Monitor* 24(6):8.

Shanker, Thom, and David E. Sanger. 2009. "Pakistan Is Rapidly Adding Nuclear Arms, U.S. Says." *New York Times*, May 17. Available at http://www.nytimes.com

Sheridan, Mary Beth, and William Branigin. 2010. "Senate Ratifies New U.S.–Russia Nuclear Weapons Treaty." *Washington Post*, December 22. Available at http://washingtonpost.com

Shrader, Katherine. 2005. "WMD Commission Releases Scathing Report." *Washington Post*, March 31. Available at http://www.washingtonpost.com

Shuster, Mike. 2009. *Experts Disagree over Iran's Nuclear Intentions*. NPR, March 2. Available at http://www.npr.org

Silverstein, Ken. 2007. "Six Questions for Joost Hiltermann on Blowback from the Iraq-Iran War." *Harper's Magazine*, July 5. Available at http://www.harpers.org

Simon, Scott. 2003 (March 29). "Marine Mammals on Active Duty Navy Uses Dolphins, Sea Lions to Patrol Waters in Persian Gulf." Available at http://www.npr.org

SIPRI (Stockholm International Peace Research Institute). 2011a. *SIPRI Yearbook 2011: Armaments, Disarmament, and International Security*. Oxford, UK: Oxford University Press.

SIPRI (Stockholm International Peace Research Institute). 2011b. *Recent Trends in Military Expenditure*. Available at http://www.sipri.org

Skocpol, Theda. 1994. *Social Revolutions in the Modern World*. Cambridge, UK: Cambridge University Press.

Slevin, Peter. 2009. "U.S. To Announce Transfer of Detainees to Ill. Prison." *Washington Post*, December 15. Available at http://www.washingtonpost.com

Small Arms Survey. 2011. *States of Security*. Cambridge: Cambridge University Press. Available at http://www.smallarmssurvey.org

Smith, Craig. 2004. "Libya to Pay More to French in '89 Bombing." *New York Times*, January 9, p. 6.

Smith, Lamar. 2003. "Restricting Civil Liberties During Wartime Is Justifiable." In *The Terrorist Attack on America*, Mary Williams, ed. (pp. 23–26). Farmington Hills, MA: Greenhaven Press.

Starr, J. R., and D. C. Stoll. 1989. *U.S. Foreign Policy on Water Resources in the Middle East*. Washington, DC: Center for Strategic and International Studies.

Stiglitz, Joseph E., and Linda J. Bilmes. 2010. "The True Cost of the Iraq War: $3 Trillion and Beyond." *The Washington Post*, September 5. Available at http://www.washingtonpost.com

Stimson Center. 2007. *Reducing Nuclear Dangers in South Asia*. Available at http://www.stimson.org

Stolberg, Sheryl. 2010. "Obama Signs Away 'Don't Ask, Don't Tell.'" *New York Times*, December 22. Available at http://www.nytimes.com

Strobel, Warren, David Kaplan, Richard Newman, Kevin Whitelaw, and Thomas Grose. 2001. "A War in the Shadows." *U.S. News and World Report* 130:22.

Tanielian, Terri, Lisa H. Jaycox, Terry L. Schell, Grant N. Marshall, Audrey Burnam, Christine Eibner, Benjamin R. Karney, Lisa S. Meredith, Jeanne S. Ringe, Mary E. Vaiana, and the Invisible Wounds Study Team. 2008. "Invisible Wounds of War: Summary and Recommendations for Addressing Psychological and Cognitive Injuries." Available at http://www.rand.org

Tavernise, Sabrina. 2007. "U.S. Contractor Banned by Iraq over Shootings." *New York Times*, September 18. Available at http://www.nytimes.com

Top Secret America. 2010. "Top Secret America: A Washington Post Investigation." *The Washington Post*. Available at http://www.washingtonpost.com

Tremlett, Giles. 2011. "Eta Declares Permanent Ceasefire." *The Guardian*, January 20. Available at http://www.guardian.co.uk

Treo. 2010. "Treo the Dog Awarded Animal VC." The Telegraph, February 6. Available at http://www.telegraph.co.uk

Tyson, Ann Scott. 2009. "Media Ban Lifted for Bodies' Return." *Washington Post*, February 27. Available at http://www.washingtonpost.com

UNAMO (United National Mine Action Office). 2011 (June). "UNMAO Regional Fact Sheet." *Southern Sudan*. Available at http://www.sudan-map.org

UNICEF (United Nations Children's Fund). 2011. *The State of the World's Children 2011: Adolescence—An Age of Opportunity*. Available at http://www.unicef.org

United Nations. 2011. *United Nations Peacekeeping Operations*. Available at http://www.un.org/Depts/dpko/dpko/bnote.htm

United Nations. 2003. *Some Questions and Answers*. Available at http://www.unicef.org

United Nations. 1948. *Convention on the Prevention and Punishment of the Crime of Genocide*. Available at http://www.un.org

United Nations Security Council. 2011 (March 17). *Resolution 1973.S/RES/1973*. Available at http://www.un.org

UNUSA (United Nations Association of the United States of America). 2004. *Landmines Overview*. Available at http://www.unausa.org

U.S. Army Medical Command. 2007 (May). *Mental Health Advisory Team IV Findings*. Available at http://www.armymedicine.army.mil

U.S. Department of Homeland Security. 2011. *FY 2011 Budget-in Brief*. Available at http://www.dhs.gov

U.S. Department of State. 2011. The United States' Leadership in Conventional Weapons Destruction." *Fact Sheet: Bureau of Political-Military Affairs*. February 14, Available at http://www.state.gov

U.S. Department of State. 2007. *Small Arms/Light Weapons Destruction*. Available at http://www.state.gov

U.S. Navy. 2010. "Navy Marine Mammal Program Excels during Frontier Sentinel." Available at http://www.navy.mil/search/display.asp?story_id=53979

Vesely, Milan. 2001. "UN Peacekeepers: Warriors or Victims?" *African Business* 261:8–10.

Viner, Katharine. 2002. "Feminism as Imperialism." *The Guardian*, September 21. Available at http://www/guardian.co.uk

Wali, Sarah O., and Deena A. Sami. 2011. "Egyptian Police Using U.S.-Made Tear Gas against Demonstrators." ABC News, January 28. Available at http://abcnews.go.com

Walker, Mark. 2011. "Military: Wuterich Trial Postponed Indefinitely." *North County Times*, June 14. Available at http://www.nctimes.com

Walker, Paul F., and Jonathan B. Tucker. 2006. "The Real Chemical Threat." *The Los Angeles Times*, April 1. Available at http://cns.miis.edu

Waters, Rob. 2005. "The Psychic Costs of War." *Psychotherapy Networker*, March–April, pp. 1–3.

Watkins, Thomas. 2007. "Haditha Hearings Enter Fourth Day." *Time*, May 11. Available at http://www.time.com

Wax, Emily. 2003. "War Horror: Rape Ruining Women's Health." *Miami Herald*, November 3. Available at http://www.miami.com

Weimann, Gabriel. 2006. *Terror on the Internet: The New Arena, the New Challenges*. Washington, DC: U.S. Institute of Peace Press.

Wein, Lawrence. 2009. "Counting the Walking Wounded." *New York Times*, January 25. Available at http://www.nytimes.com

Williams, Mary E., ed. 2003. *The Terrorist Attack on America*. Farmington Hills, MA: Greenhaven Press.

Wilson, Michael. 2010. "Shahzad Gets Life Term for Times Square Bombing Attempt." *New York Times*, October 5. Available at http://www.nytimes.com

Wise, Lindsey. 2011. "Marine Who Pushed Suicide Prevention Took Own Life." *Houston Chronicle*, April 9. Available at http://www.chron.com/news/

Woehrle, Lynne M., Patrick G. Coy, and Gregory M. Maney. 2008. *Contesting Patriotism: Culture, Power, and Strategy in the Peace Movement*. Lanham, MD: Rowman & Littlefield Publishers, Inc.

Worsnip, Patrick. 2011. "South Sudan Admitted to U.N. as 193rd Member." *Reuters UK*, July 14. Available at http://uk.reuters.com

Zakaria, Fareed. 2000. "The New Twilight Struggle." *Newsweek*, October 12. Available at http://www.msnbc.com/news

Zoroya, Gregg. 2006. "Lifesaving Knowledge, Innovation Emerge in War Clinic." *USA Today*, March 27. Available at http://www.usatoday.com

Name Index

Note: Page numbers in *italics* refer to information in tables, figures, illustrations, and captions.

Subject Index

Note: Page numbers in **boldface** type denote definitions. Page numbers in *italics* denote tables, figures, illustrations and captions.